Forest Canopies

These are indeed exciting times.
Canopy science has made biology grandly
three-dimensional; no other discipline, save possibly
the study of abyssal benthos, offers more unexplored
spaces, species diversity, and combined physical and
intellectual adventure.

Edward O. Wilson
University Research Professor Emeritus
Harvard University, 7 October 2002

Illustration by Marie Stile.

Forest Canopies

Second Edition

Edited by

Margaret D. Lowman
Professor of Environmental Studies
New College of Florida;
TREE Foundation
Sarasota, Florida

H. Bruce Rinker
Center for Canopy Ecology
Marie Selby Botanical Gardens;
TREE Foundation
Sarasota, Florida

ELSEVIER
ACADEMIC
PRESS

Amsterdam • Boston • Heidelberg • London
New York • Oxford • Paris • San Diego
San Francisco • Singapore • Sydney • Tokyo

Elsevier Academic Press
200 Wheeler Road, 6th Floor,
Burlington, MA 01803, USA
525 B Street, Suite 1900, San Diego, California 92101-4495, USA
84 Theobald's Road, London WC1X 8RR, UK

This book is printed on acid-free paper.

Library of Congress Cataloging-in-Publication Data
Application submitted

British Library Cataloguing in Publication Data
A catalogue record for this book is available from the British Library

ISBN: 0-12-457553-6

For all information on all Academic Press publications
visit our website at www.academicpress.com

Printed and bound by CPI Group (UK) Ltd, Croydon, CR0 4YY

Transferred to Digital Print 2011

To our parents,
Alice and John Lowman,
Ruth D. and Harry B. Rinker, Jr.;
and to all our colleagues "in the treetops" around the world
who risk life and limb in their advocacy of
the High Frontier

Contents

I Structures of Forest Canopies

Introduction   1
Margaret D. Lowman and H. Bruce Rinker

1. The Nature of Forest Canopies   3
Nalini M. Nadkarni, Geoffrey G. Parker, H. Bruce Rinker, and David M. Jarzen

> Empty Space: Another View of Forest Canopy Structure: Roman Dial, Nalini M. Nadkarni, and Judith B. Cushing
> Verticality and Habitat Analysis: MacArthur and Wilson's Biogeography Theory Revisited: H. Bruce Rinker

2. Tropical Microclimate Considerations   24
Stephen R. Madigosky

3. Quantifying and Visualizing Canopy Structure in Tall Forests: Methods and a Case Study   49
Robert Van Pelt, Stephen C. Sillett, and Nalini M. Nadkarni

> "Canopy Trekking": A Ground-Independent, Rope-Based Method for Horizontal Movement through Forest Canopies: Roman Dial, Stephen C. Sillett, and Jim C. Spickler

4. Vertical Organization of Canopy Biota   73
David C. Shaw

> Macaws: Dispersers in a Tropical Habitat: Sharon Matola
> Vertical Stratification among Neotropical Migrants: H. Bruce Rinker

5. Age-Related Development of Canopy Structure and Its Ecological Functions   102
Hiroaki T. Ishii, Robert Van Pelt, Geoffrey G. Parker, and Nalini M. Nadkarni

> Measuring Canopy Structure: The Forest Canopy Database Project: Nalini M. Nadkarni and Judith B. Cushing

IV Conservation and Forest Canopies

Contributors

Marie E. Antoine, Department of Biological Sciences, Humboldt State University, Arcata, CA 95521

Connie Barlow, c/o Koenigsberg, 15206 263rd Avenue, SE, Issaquah, WA 98027

Paula J. Benshoff, Myakka River State Park, 13207 S.R. 72, Sarasota, FL 34241

David H. Benzing, Ph.D., Department of Biology, Oberlin College, Oberlin, OH 44074

Edward A. Burgess, Princeton University, Princeton, NJ 08544; TREE Foundation, P.O. Box 48839, Sarasota, FL 34230

James Burgess, Burgundy Center for Wildlife Studies, Capon Bridge, WV 26711; TREE Foundation, P.O. Box 48839, Sarasota, FL 34230

Judith B. Cushing, Ph.D., The Evergreen State College, Olympia, WA 98505

David L. Dilcher, Ph.D., NAS, Paleobotany and Palynology Laboratory, Florida Museum of Natural History, P.O. Box 117800, Gainesville, FL 32611-7800

Roman Dial, Ph.D., Department of Environmental Science, Alaska Pacific University, 4101 University Drive, Anchorage, AK 99504

Kristina A. Ernest, Ph.D., Department of Biological Sciences, Central Washington University, Ellensburg, WA 98926

Nathan Erwin, O. Orkin Insect Zoo, MRC 158, Smithsonian Institution, P.O. Box 37012, Washington, DC 20013-7012

Terry L. Erwin, Ph.D., Department of Entomology, MRC 169, National Museum of Natural History, Smithsonian Institution, Washington, DC 20560

Steven J. Fonte, Department of Agronomy and Range Science, University of California, One Shields Avenue, Davis, CA 95616

Jennifer L. Funk, Department of Biological Sciences, Stanford University, Stanford, CA 94305-5020

Wesley E. Higgins, Ph.D., Systematics Division, Marie Selby Botanical Gardens, 811 South Palm Avenue, Sarasota, FL 34236

Bruce K. Holst, Herbarium, Marie Selby Botanical Gardens, 811 South Palm Avenue, Sarasota, FL 34236

Peter J. Horchler, German Federal Institute of Hydrology, Ecological Interactions Section, Am Mainzer Tor 1, D-56068 Koblenz, Germany

Mark D. Hunter, D.Phil., Institute of Ecology, University of Georgia, Athens, GA 30602-2202

Hiroaki T. Ishii, Division of Forest Resources, Graduate School of Science and Technology, Kobe University, Kobe 657-8501 Japan

David M. Jarzen, Ph.D., Research Associate, Florida Museum of Natural History, Dickinson Hall, P.O. Box 117800, Gainesville, FL 32611-7800

Beth A. Kaplin, Ph.D., Center for Tropical Ecology and Conservation, Department of Environmental Studies, Antioch New England Graduate School, 40 Avon Street, Keene, NH 03431

Cheryl D. Knott, Ph.D., Department of Anthropology, Harvard University, Peabody Museum, 53D, 11 Divinity Avenue, Cambridge, MA 02138

Donna J. Krabill, Department of Education, Marie Selby Botanical Gardens, 811 South Palm Avenue, Sarasota, FL 34236

Timothy G. Laman, Ph.D., Arnold Arboretum, Harvard University, 22 Divinity Avenue, Cambridge, MA 02138

Manuel T. Lerdau, Ph.D., Department of Ecology and Evolution, State University of New York, Stony Brook, NY 11794-5245

T.A. Lott, Paleobotany and Palynology Laboratory, Florida Museum of Natural History, P.O. Box 117800, Gainesville, FL 32611-7800

Thomas E. Lovejoy, Ph.D., AAAS, Am. Phil. Soc., The H. John Heinz III Center for Science, Economics, and the Environment, 1001 Pennsylvania Avenue, NW, Suite 735 South, Washington, DC 20004

Margaret D. Lowman, Ph.D., Department of Environmental Studies, New College of Florida, Sarasota, FL 34232; TREE Foundation, P.O. Box 48839, Sarasota, FL 34230

Stephen R. Madigosky, Ed.D., Department of Biology and Environmental Science, Widener University, Chester, PA 19013

Jay R. Malcolm, Ph.D., Faculty of Forestry, University of Toronto, 33 Willcocks Street, Toronto, ON M5S 3B3 Canada

Lynn Margulis, Ph.D., NAS, Department of Geosciences, University of Massachusetts, Amherst, MA 01003-9297

Sharon Matola, Belize Zoo and Tropical Education Centre, Mile 29 Western Highway, P.O. Box 1787, Belize City, Belize

William R. Miller, Ph.D., Department of Biology, Chestnut Hill College, 9601 Germantown Avenue, Philadelphia, PA 19118

Andrew Mitchell, FRGS, Global Canopy Programme, John Krebs Field Station, University of Oxford, Wytham, Oxford OX2 8QJ UK

Wilfried Morawetz, Universitat Leipzig, Institut fur Botanik, Spezielle Botanik, Johannisallee 21-23, D-04103 Leipzig, Germany

Nalini M. Nadkarni, Ph.D., The Evergreen State College, Olympia, WA 98505

Geoffrey G. Parker, Ph.D., Smithsonian Environmental Research Center, 647 Contee's Wharf Road, P.O. Box 28, Edgewater, MD 21037

David W. Pearce, OBE, D. Sc., University College London, Gower Street, London WC1E 6BT UK

Romina Rader, CRC Rainforest, James Cook University, P.O. Box 6811, Cairns QLD 4870 Australia

Barbara C. Reynolds, Ph.D., Department of Environmental Studies, CPO #2330, University of North Carolina, One University Heights, Asheville, NC 28804-8511

Barbara A. Richardson, Research Associate Luquillo LTER, 165 Braid Road, Edinburgh EH10 6JE Scotland

H. Bruce Rinker, Ph.D., Center for Canopy Ecology, Marie Selby Botanical Gardens, 811 South Palm Avenue, Sarasota, FL 34236; TREE Foundation, P.O. Box 48839, Sarasota, FL 34230

Joan Roughgarden, Department of Biological Sciences, Stanford University, Stanford, CA 94305-5020

Dale A. Russell, Ph.D., Department of Marine, Earth and Atmospheric Sciences, North Carolina State University, Raleigh, NC 27695-8208

Timothy D. Schowalter, Ph.D., Department of Entomology, Louisiana State University, Baton Rouge, LA 70803

Katherine Secoy, FRGS, Global Canopy Programme, John Krebs Field Station, University of Oxford, Wytham, Oxford OX2 8QJ UK

David C. Shaw, Ph.D., Wind River Canopy Crane Research Facility, University of Washington, 1262 Hemlock Road, Carson, WA 98610

Stephen C. Sillett, Ph.D., Department of Biological Sciences, Humboldt State University, Arcata, CA 95521-8299

Jim C. Spickler, Ph.D., Eco-Ascension Research and Consulting, P.O. Box 202, Arcata, CA 95518

Marie Stile, Pratt Institute, 277 Washington Avenue, Apt. #5E, Brooklyn, NY 11205

Howard L. Taylor, 1812 Wood Hollow Court, Sarasota, FL 34235-9146

Robert Van Pelt, Ph.D., College of Forest Resources, Box 352100, University of Washington, Seattle, WA 98195

David Evans Walter, Department of Biological Sciences, University of Alberta, Edmonton, Alberta T6G 2E9 Canada

Xin Wang, Paleobotany and Palynology Laboratory, Florida Museum of Natural History, P.O. Box 117800, Gainesville, FL 32611-7800

Qi Wang, Paleobotany and Palynology Laboratory, Florida Museum of Natural History, P.O. Box 117800, Gainesville, FL 32611-7800

David M. Watson, Ph.D., Ecology and Ornithology, The Johnstone Centre and School of Environmental and Information Sciences, Charles Sturt University, P.O. Box 789, Thurgoona NSW 2640 Australia

Jessica Hope Whiteside, Department of Earth and Environmental Sciences, Lamont-Doherty Earth Observatory of Columbia University, 61 Route 9W, Palisades, NY 10964-8000

Edward O. Wilson, Ph.D., Museum of Comparative Zoology, Harvard University, 26 Oxford Street, Cambridge, MA 02138-2902

Michelle L. Zjhra, Ph.D., Department of Biology, Georgia Southern University, P.O. Box 08042, Statesboro, GA 30460

Foreword

When I started working in tropical forests in the mid-1960s, the canopy rendered me the biologist's equivalent of Tantalus from the very outset. Frank Chapman's *My Tropical Air Castle* (1929, D. Appleton & Company, New York) and Marston Bates' *The Forest and the Sea* (1965, Random House, New York) had already stirred the imagination. So had Colin Pittendrigh's hypothesis that bromeliads arose in xeric environments and then invaded the canopy. Early on I met the colorful Jorge Boshell, Bates' successor as director at Villavicencio, who had solved the riddle of jungle yellow fever's ability to vault from its canopy cycle with howler monkeys to people far below on the forest floor. Boshell had noted a cloud of blue *Haemagogus* mosquitoes, normally canopy dwellers, swarming around woodsmen who had felled a canopy tree. Today there are ominous echoes of this critical observation in a tale of deforestation with disquieting portent, namely occasional outbreaks of the Zaire *Ebola* virus. For me, however, the canopy will always be symbolized by the exquisite but deadly metallic blue mosquitoes and the less ominous *Sabethes* mosquitoes that float on flanges rather like an entomological equivalent of a South Pacific outrigger canoe.

It is not surprising, then, that even after Boshell solved the riddle of the Sylvan yellow fever, tropical epidemiologists have had a compulsion about canopy studies. At the mouth of the Amazon, the Belem Virus Laboratory erected a modest 35-meter tower where one could glimpse canopy birds close at hand aided by the same fearlessness for which Galápagos wildlife is renowned. There was also an ingenious counterweighted rope "elevator" so someone could easily zip up into the canopy to put out or collect back sentinel animals, such as chickens, to detect the presence of any arthropod-borne viruses. Philip Humphrey and I set out a couple of mistnets rigged like a sail to catch birds in the upper reaches of the forest. Imagine our delight on the very first day when we showed this new technique to Helmut Sick, dean of South American ornithology, and caught two swifts, *Brachyura spinicauda*, the very species Sick had come north to collect for Brazil's Museu Nacional.

These experiences merely highlight how very little is known about the forest canopy and how challenging it is to do canopy research. While I expect no one would disagree that canopy biology is still in its infancy, the field has undergone an intellectual radiation in recent years, as evident in this second edition of *Forest Canopies*. Physically, it may be represented by the use of cranes that the Smithsonian Tropical Research Institute has installed in Panama and by studies of the very physiology of the forest canopy as part of the Large-Scale Biosphere Atmosphere (LBA) project in the Brazilian Amazon. Published nearly 10 years after the first edition, this second edition of *Forest Canopies* provides up-to-date literature reviews on structure and function, organisms, ecological processes, and conservation. Exciting topic additions include chapters and sidebars on microbial canopies; tardigrades, rotifers, and other canopy microfauna; linkages between canopies and soils; paleocanopies and evolutionary relics; and a sense of wonder from multi-disciplinary studies in the treetops of the world. Intellectually, a wide field of canopy studies is represented by this volume, and a rich smorgasbord it is indeed!

There is no better evidence than canopy biology that the age of exploration is not over. We can anticipate a diverse panoply of discoveries emanating from this field. Some will be of serious practical benefit to ourselves as living organisms. Others will illuminate aspects of biology never before dreamed of. Yet others will astound with their beauty or be intrinsically fascinating. And one cannot but wonder what may never be learned because of continued high rates of tropical deforestation. In all cases, it will be clear that canopy biology, as a recognized field of intellectual endeavor, began with the first edition of this book and now expands splendidly with its second edition.

THOMAS E. LOVEJOY
The H. John Heinz III Center for Science, Economics, and the Environment
1001 Pennsylvania Avenue, NW, Suite 735 South
Washington, DC 20004

Preface

The first edition of *Forest Canopies* was published in 1995. Nalini Nadkarni and I are indebted to Chuck Crumly, formerly with Academic Press, who doggedly encouraged us to persist as editors of that first edition. In that volume, 31 canopy scientists were tracked down around the world to complete the chapters that offered an overview of recent research in the treetops. Scientists were literally hauled from their ropes or traced into their canopy towers to complete that first-ever volume on the exciting biology of the uppermost section of forests. As editors, Nalini Nadkarni and I felt like detectives in some cases, attempting to locate authors for specific chapters to round out the volume. One of my best "lecture" stories involves my first meeting with Charley Munn that occurred by chance at the top of the canopy walkway at ACEER (Amazon Center for Education and Environmental Research), now called ACTS (Amazon Conservatory for Tropical Studies), in Peru. What a great place to seal the deal on our bird chapter! In the early 1990s, relatively few accomplished canopy biologists existed, and the field was youthful in terms of any extensive observational or experimental studies.

The pioneers from the first volume, however, achieved a wonderful goal: they provided an excellent introductory text that inspired subsequent students, researchers, and educators to expand their conventional perspectives of the forest floor into the canopy. When Chuck Crumly called in 2001 to remind me that the first volume was out of print, I initially considered deleting his e-mail. It had been a daunting task to find and harness that first group of canopy biologists to complete the first edition. Most of them were rightfully consumed with both the challenges of getting into the canopy as well as their passion for finding answers to important questions. In spite of their busy schedules, that initial group produced a lasting volume of information, inspiration, and thought-provoking discussion that launched a second generation of canopy biologists.

In pulling together this second edition, Bruce Rinker brought the enthusiasm and creativity of a new generation of canopy biologists into our editorial team. Nalini was busy hanging from trees in the Pacific Northwest, so Bruce took on the challenge as co-editor. He cajoled, reminded, e-mailed, and persuaded a larger, more international group of canopy scientists to write sections of this volume. For this second edition, we had the technological advantage of e-mail that was not available for the first volume when so many researchers were working in remote forests without field stations or permanent canopy access sites. For the present volume, we also had the resources of the International Canopy Network, the Global Canopy Programme, and other networks only recently developed to integrate our international community.

The final product of our combined efforts is a volume significantly different from the first edition. Thanks to the advances in canopy access during the past decade, the field of available authors almost doubled from what existed 10 years ago. And perhaps even more exciting, it is possible to trace the students and collaborators of those first pioneering canopy studies, and observe the evolution and healthy growth of this professional group of canopy scientists.

We hope you will agree with us that this volume is exciting and timely. It advocates the conservation of forest habitats yet it also explains some of the basic ecology and physiology of the organisms that exist in forest canopies. It hints at the unknown. It suggests new areas of study. It goes into depth on important subjects such as invertebrates, climates, structure, nutrient cycling, and decay. It touches on new subjects such as microbes, tardigrades, ecotourism, and stromatolites. Innovative methods and ideas are summarized in sidebars throughout most chapters to make the text more varied and user-friendly. Our hope is that this second edition will serve to inspire a third generation of canopy scientists, as some of us advance toward retirement age.

Is it really possible for canopy biologists to retire? We think not. We all look forward to future canopy conferences, integrated canopy research stations, and other volumes that will promote and illuminate this important region of the planet, absolutely essential to the quality of life for us and for the generations to follow.

Thank you for sharing with us this exciting field of exploration of the forest canopy.

Yours from the treetops,
"Canopymeg"
MARGARET D. LOWMAN
Sarasota, Florida

Introduction

Alfred Russel Wallace in *Tropical Nature* (1878, Macmillan and Company, London) wrote compellingly about the architecture of an equatorial forest:

> The observer new to the scene would perhaps be first struck by the varied yet symmetrical trunks, which rise up with perfect straightness to a great height without a branch, and which, being placed at a considerable average distance apart, gives an impression similar to that produced by the columns of some enormous building. Overhead, at a height, perhaps, of a hundred feet, is an almost broken canopy of foliage formed by the meeting together of these great trees and their interlacing branches; and this canopy is usually so dense that but an indistinct glimmer of the sky is to be seen, and even the intense tropical sunlight only penetrates to the ground subdued and broken up into scattered fragments. There is a weird gloom and a solemn silence, which combine to produce a sense of the vast—the primeval—almost of the infinite. It is a world in which man seems an intruder.

Only during the past 25 years has our understanding of canopy ecology expanded beyond this 19th-century, ground-based perspective. Yet, even this perspective hints at some of the opportunities anticipated overhead by explorers in the 19th and 20th centuries.

Our views of forests have changed dramatically during the past century. After the term *ecosystem* was proposed and defined by the British ecologist A.G. Tansley in 1935, scientists moved dramatically beyond the descriptive ecologies of Charles Darwin's era into a wide spectrum of system analysis. Elements of this analysis include spatial/temporal issues, experimental treatments, phylogenies based on molecular genetics, conservation strategies, ecological linkages, energetics, and much more. Many of the views purported in this text would have been impossible to imagine prior to our ecosystems approach to large natural stands of trees. Though many discoveries about temperate and tropical forests have been made in the past century, we still have not established a firm understanding of the complex ecological circuitry between processes in the canopy and on the forest floor.

For scientists and educators, entering the living laboratory of the world's forests has always been a relatively easy affair; however, accessing the treetops proved fairly difficult until the 1990s, the defining decade for the emerging science of canopy ecology. In addition to single-rope technique (SRT), other methods to get into the treetops were proffered, including airships, canopy rafts, sleds, cranes, towers, tram-lines, and walkways, that allowed easy access for scientific research and instruction in this lofty biological frontier. The result of our combined efforts since then is an awe-inspiring picture of temperate and tropical forests as integrated systems top to bottom. Much of the productivity and biological richness of forests is housed in the canopy. The treetops are also a place of wonder and challenge as we continue to decipher their aerial mysteries.

Though far from comprehensive, the second edition of *Forest Canopies* provides literature reviews divided into four major sections of study: structures of forest canopies (Part I); organisms in forest canopies (Part II); ecological processes in forest canopies (Part III); and conservation and forest canopies (Part IV). Ironically, we hope that this dichotomy, with respective chapters and sidebars representing the work of nearly 60 researchers around the world, will accentuate some of that integration. The summaries presented here may help managers and conservationists, politicians, lawyers, naturalists, citizens' action groups, educators, ethicists, and others to collaborate on governing policies, cutting regimens, wilderness designations, and other issues related to the sensitivities of the ecological circuits operating in forest systems. Since the publication of the first edition in 1995, we have made Herculean strides in our knowledge base on the ecology and evolution of forest canopies. We hope that the second edition does some justice to this extraordinary advancement.

Though it is a book about discovery and opportunity, this edition of *Forest Canopies* also highlights some of the great unknowns in forest canopies around the globe. Despite our numerous insights and successes in the past 25 years, much work remains for our concerted efforts to define the vast and near-infinite aspects of the High Frontier. We hope that this volume will underscore directions and needs for the next generation of devoted canopy ecologists.

We are indebted to many colleagues who assisted with countless aspects of this challenging project: the exceptional contributors to this book; the staff of New College (Sarasota, FL) and the Marie Selby Botanical Gardens (Sarasota, FL); participants at the Third International Forest Canopy Conference (Cairns, Australia) in June 2002; Chuck Crumly (previously senior publishing editor for Life Sciences at Elsevier/Academic Press); Susan A. Jarzen (administrative assistant); Beth Kaplin (professor at Antioch New England Graduate School and director of the Center for Tropical Ecology and Conservation); Saul Lowitt (biostatistician); Nalini Nadkarni (ICAN president and professor at Evergreen State College); Peter Ochs (Edgar Bronfman Professor of Modern Judaic Studies at the University of Virginia); Roy H. Park, Jr. (president and chairman of Triad Foundation, Inc.); Kelly Sonnack (editorial assistant for Life Sciences at Elsevier/Academic Press); E.O. Wilson (professor emeritus at Harvard University); and TREE Foundation. Joseph Connell (professor at the University of California, Santa Barbara), Francis Hallé (professor emeritus at the University of Montpellier, France), and Thomas E. Lovejoy (president of the Heinz Center for Science, Economics, and the Environment) afforded lasting inspiration throughout our respective careers. Mark D. Hunter (professor at the University of Georgia's Institute of Ecology) and Timothy D. Schowalter (professor and entomologist at Louisiana State University) provided mentorship through our canopy collaborations under the auspices of a National Science Foundation ecosystems grant. We are especially grateful to our artists: Marie Stile for the beautiful frontispiece and Ringling School of Art and Design students, Bronwyn Coffeen, Jace Gostisha, Ji Sun Hyun, and Anna-Dawn Maynard, for their lovely black-and-white illustrations found in the section introductions and throughout the text.

Finally, heartfelt thanks are given to our multi-talented, witty, and dedicated project editor Susan Fernandez without whose insights, queries, and challenges we would have floundered long ago. She represented the best that the publishing world can offer: friend, mentor, and confidante. Many of the successes of this book were due entirely to Susan's devoted efforts.

In characteristic clear-minded fashion, Alfred Russel Wallace wrote that tropical nature produces in the observer a sense of the vast and infinite. Forests around the world constitute an ecological type that requires both our intellectual undertakings and emotional vision for their long-term conservation. It is a critical time for collaboration among all ecological disciplines—indeed, between research and education—to halt declining biodiversity and to address accelerating scientific illiteracy, two faces of the same coin of societal ignorance.

Perhaps that is the ultimate goal of canopy research—all scientific research for that matter—to produce a sense of the vast and the infinite and to promote our sense of wonder, a curiosity that needs to be fed by experience to be long-lived. We are confident that the second edition of

Forest Canopies will help to guide our ascent into the trees, climbing branches heretofore inaccessible but filled with challenge for the explorer in us all.

MARGARET D. LOWMAN AND H. BRUCE RINKER
Sarasota, Florida

I

Structures of Forest Canopies

Architectural organization helps to determine the dynamics of a forest ecosystem. Biological diversity, microclimate, ecological processes, succession, and even its evolutionary history can be determined, at least in part, by assessing forest structure. In the first section of *Forest Canopies*, we set the stage for the entire volume with an overview of the nature and composition of forests from a canopy perspective. All of the authors in Section I have grappled with the notion of forest structure for many years. Until a decade ago, forests were often viewed as homogeneous compositions of tall trees broken into delineated strata of vegetation called the canopy, the understory, the ground or herbaceous layer, and the forest floor. Characteristics, such as aboveground and belowground diversity, and even processes, such as herbivory, were considered relatively uniform in forests throughout the world. The major reason for this sweeping generalization was perhaps due to the fact that early foresters based their assumptions on observations of the first two meters of forest structure, those lower regions that they could touch, collect, and measure with ease.

Since these early days, however, canopy biologists now recognize that 95 percent of complex forest structure exists overhead and out of reach of ground-based bipeds. Such a relatively simple conclusion has now led us to entirely different assumptions about forest dynamics, all based on our increased knowledge of forest structure in the upper regions. Armed with the understanding that forest structure is extremely multifarious, scientists have begun to quantify the immense variability among forest types, among layers ranging from understory to emergent levels within one forest type, and with respect to factors such as age, light levels, evolutionary status, genetics, and other variables.

In Chapter 1, Nalini Nadkarni, Geoffrey Parker, Bruce Rinker, and David Jarzen detail the dynamics and services of forest canopies; they also discuss forest management from a treetops perspective and create a "map" for future directions in canopy studies. In Chapter 2, Steve Madigosky focuses on key aspects of forest microclimate, advocating an integrative, holistic approach to this vital aspect of the forest ecosystem. Bob Van Pelt, Steve Sillett, and Nalini Nadkarni propose a simple sampling protocol in Chapter 3 to obtain three-dimensional information about forests, using *Eucalyptus* and *Pseudotsuga* stands as case studies.

In Chapter 4, Dave Shaw summarizes the principles of vertical stratification of biota within forest canopies, employing extensive use of the international canopy crane network and other access techniques. Hiroaki Ishii, Bob Van Pelt, Geoffrey Parker, and Nalini Nadkarni review in Chapter 5 some of the canopy processes that drive the development of structural complexity as a forest stand matures; they also present a three-dimensional perspective of the structural development of temperate forest ecosystems. In Chapter 6, the final chapter of Section I, Dave Dilcher, T.A. Lott, XinWang, and Qi Wang discuss how trees have evolved independently and repeatedly over hundreds of millions of years; thus, the species composition of forest landscapes and their contribution to the nature of the forest canopy are clearly dynamic through time as well as space.

These six chapters, along with their various sidebars of information, provide insight into the structure of forest canopies in four dimensions: a complex, three-dimensional ecology, changing over time. With the evolutionary/ecological perspectives presented in Section I, we can then examine in detail some of the organisms and processes operating in forest canopies around the world, and even touch on their management and conservation, offered in subsequent sections of this second edition of *Forest Canopies*.

CHAPTER 1

The Nature of Forest Canopies

Nalini M. Nadkarni, Geoffrey G. Parker, H. Bruce Rinker, and David M. Jarzen

To know the forest, we must study it in all aspects, as birds soaring above its roof, as earth-bound bipeds creeping slowly over its roots.
—*Alexander F. Skutch,* A Naturalist in Costa Rica, *1992*

Definition and Scope of Canopy Studies

The term "canopy" connotes an uppermost layer, a covering for an important person or sacred object, the enclosure over an airplane cockpit, the surface of a parachute, the topmost ornamentation in architecture, or any tent-like covering. In the early literature concerning forest tree-tops, the canopy was defined simply as the topmost layer of vegetation (Richards 1954). The forest canopy, however, is now considered a structurally complex and ecologically critical sub-system of the forest, and is defined as "the combination of all foliage, twigs, fine branches, their attending flora and fauna, the interstices (air), and their environment" (Parker 1995; Moffett 2000). For many critical canopy functions (e.g., interception of rainfall, absorption of light, uptake of gases, and provision of wildlife habitat), all plants contribute. Similarly, the forest environment changes continuously from top to bottom, so gradients cannot be subdivided objectively. In mixed-species, multi-aged stands, the upper layers alone do not represent canopy structure and all its microhabitats, microclimates, and exchange processes. Thus, researchers now recognize that the forest canopy is part of the forest ecosystem as a whole (Parker and Brown 2000).

Non-forest vegetation also supports canopies, such as kelp "forests," algal mats on a river bed, beds of sea grasses, orchards, lawns, wheat fields, and stromatolite aggregates (Dayton 1985; Margulis 2001; Moffett 2001). In this chapter, however, we will generally restrict our discussion to canopies associated with tall, woody plants; that is, terrestrial forests. In this context, the term *canopy* denotes forest community architecture as well as species composition, nutrient cycling, energy transfer, and plant-animal interactions from the ground to the forest-atmosphere interface.

In this chapter we also discuss the development of canopy studies and summarize the general features of canopy composition, structure, and distribution. We identify ways that humans and forest canopies have been linked through history and in the present, and then discuss the dependence of our species on forest canopies for fiber, fuel, food, medicines, and spirituality. We describe the involvement of forest canopies in pressing issues for anthropogenic change, such as biodiversity declines, carbon storage, and landscape management. Our summaries are based on the state of knowledge of canopies within the broad geographical range of forest vegetation, from equatorial vegetation to boreal and tree-line forests. These dynamic canopies change with seasons, disturbances, and long-term global patterns.

3

History and Development of Canopy Studies

Historically, forest canopies have been among the most poorly understood regions of our planet. Over the past three decades, however, they have been explored by increasing numbers of researchers. Initially, people who sought the thrill of climbing and followed the lure of discovering new species dominated canopy studies. Early European explorers hired climbers and trained monkeys to collect specimens of exotic "air plants" that grew out of their reach (Trichon 2002). Pioneering ecologists who worked in old-growth forests of the Pacific Northwest applied mountain-climbing techniques for safe access to the canopies of tall trees (Denison 1973). These techniques were modified for use in the tropics, including the creation of a "canopy web" in the rainforests of Costa Rica. Strong cables stretched among three emergent trees provided access to observe and document canopy tree pollination in unparalleled detail (Perry 1978).

A wide variety of other access tools have since been developed, making canopy study a more attractive option for scientific research. The development of climbing methods (such as the canopy raft, walkways, and cranes) and of ground-based methods such as insecticidal fogging has permitted researchers to maximize efforts of recording, analyzing, and communicating canopy data (Moffett and Lowman 1995). In 2002, 11 canopy cranes were in operation in temperate and tropical forests (Ozanne et al. 2003). Researchers step into a gondola and are pulled aloft to the height of the supporting tower, swung in any direction, move in and out along the jib, and are deposited back on the ground. This enables unprecedented access to the "outer envelope" of the canopy, where considerable material and energy exchange occurs (Parker et al. 1992; Basset et al. 2003a). Operation Canopée, a French-sponsored team that uses lighter-than-air apparati, takes teams of researchers to the rainforest treetops by use of a dirigible-driven raft and sled (Rinker et al. 1995). Canopy walkways in temperate and tropical regions are used both to raise awareness among ecotourists of the importance of forests and provide access for scientific study. In 2003, at least eight arboreal walkways were in use in the eastern United States, several of

Table 1-1   Canopy Walkways in the Eastern United States

Year/Name	Location	Forest Type	Primary Use
1991: Hopkins Forest	Williams College, Amherst, MA http://www.williams.edu	Temperate Deciduous Forest	Student Research/Education
1992: Hampshire	Hampshire College, Amherst, MA http://www.hampshire.edu	Temperate Deciduous Forest	Student Research/Education
1993: Coweeta	Coweeta Hydrological Laboratory, Otto, NC http://sparc.ecology.uga.edu	Temperate Deciduous Forest	Research
1994: Selby	Marie Selby Botanical Gardens, Sarasota, FL http://www.selby.org	Subtropical Forest	Education
1995: Millbrook	Millbrook School, Millbrook, NY http://www.millbrook.org	Temperate Deciduous Forest	Student Research/Education
1999: EcoTarium	EcoTarium, Worcester, MA http://www.ecotarium.org	Temperate Deciduous Forest	Education
2000: Myakka	Myakka River State Park, Sarasota, FL http://www.myakkariver.org	Subtropical Oak/ Palm Hammock	Education
2001: Burgundy	Burgundy Center for Wildlife Studies, Capon Bridge, WV http://www.camppage.com/bcws	Temperate Deciduous Forest	Student Research/ Education

which were designated exclusively for research (see Table 1–1); numerous others exist in tropical locations that are used mainly for ecotourism and education.

Ground-based inventory techniques include insecticidal fogging with pyrethrum-based knockdown chemicals to capture arboreal invertebrates (Erwin 1982). Detailed canopy structure has been mapped with a backpack-held laser range finder that takes thousands of measurements every second (Parker, Harding, and Berger 2004). New remote sensing technologies have made possible large-scale and detailed understanding of canopy structure and function. Particularly useful for probing the interior structure of forest canopies are the new LIDAR (Light Detection and Ranging) systems (Lefsky et al. 2002) that can be deployed from airplanes (Blair et al. 1999), the space shuttle (Garvin et al. 1998), and satellites (Zwally et al. 2002). The availability of new and rapidly developing technologies that record data on the physical environment continually (e.g., high-speed anemometers and gas sensors), coupled with advanced nanotechnology, is making it possible to study canopy functions at scales well beyond the leaf and branch levels of earlier researchers. Thus, in many instances, canopy access no longer limits our understanding of canopy biology.

As these techniques were being developed in the 1980s, canopy research was at an early stage of development (see Table 1–2). It was characterized by descriptive studies, researchers who worked in isolation, lack of harmonized field protocols and data collection methods, and the perception that questions addressed were mainly curiosity driven (Nadkarni and Parker 1994). As access methods improved, however, researchers from many scientific disciplines began contributing to the field, including ecosystem and landscape ecology, meteorology, zoology, botany, landscape ecology, and conservation biology (Lowman and Nadkarni 1995). At the turn of the 21st century, as a more mature field, canopy investigations frequently involve multiple researchers who address process-oriented questions. Interdisciplinary research groups coalesced to approach canopy questions from new and different spatial scales. Canopy researchers have begun to validate predictive models and relate findings to those of other disciplines, which allows them to respond to the "what if" questions about pressing societal concerns posed by policy-makers (see Table 1-3).

The number of scientific publications on canopy structure and function has grown at a disproportionately rapid pace relative to the general field of biology (Nadkarni and Parker 1994). In parallel, interest in the forest canopy from policy-makers, resource managers, and the general public has generated books, popular articles, and films related to increasing concerns about loss of

Table 1-2 The Development of the Field of Forest Canopy Studies (based on the types of studies carried out, the characteristics of data, and the logistics to carry out research based on the stage of maturity)

	Stage	Types of Studies	Data Characteristics	Logistics
Relevance to Compelling Social Questions ↓	Young	Descriptive	Few datasets Individually designed Technological tools	Personal, individual communication Informal networks
	Maturing	Process-oriented	Use of database models and technology	Formal communication Courses Graduate programs

	Mature	Predictive	Linked, harmonized,
			   knowledge base with
			   connectivity to many
			   datasets
				Jointly funded programs

Table 1-3 International Issues that Pertain to Forest Canopy Research

Issue	International Forum	URL
CO_2 uptake and sequestration	Kyoto Protocols of the UN	http://unfccc.int/resource/docs/convkp/kpeng.pdf
Conservation of biodiversity	Rio de Janeiro Convention on Biodiversity	http://www.biodiv.org/biosafety/ratification.asp
Ozone and UV-B radiation	Montreal Protocol	http://www.unep.org/ozone/montreal.shtml
Global timber supply	UNECE Timber Committee	http://www.unece.org/trade/timber/Welcome.html
Biodiversity hotspots	Conservation International	http://www.conservation.org
Trade of biological specimens	CITES Convention	http://www.cites.org/eng/disc/text.shtml
Climatic change	World Climatic Change Conference	http://www.wccc2003.org/index_e.htm
Forest protection	World Wildlife Fund	http://www.panda.org/about_wwf/what_we_do/forests/
Ecotourism	International Ecotourism Standard	http://www.ecotourism.org.au/ies/ies.cfm

biodiversity, global atmospheric change, endangered species habitat, and sustainable forest management (Lugo and Scatena 1992; Benzing 1998).

In the last two decades, canopy studies have developed rapidly as a result of seven activities:

1. Development of safe, non-destructive access techniques;
2. Establishment of formal and informal networks within and outside canopy research;
3. Active collaborations among canopy scientists to visualize, process, and archive canopy datasets;
4. Increased use of experimental approaches to research;
5. A growing information base from which to draw theory and synthetic works;
6. Recognition that canopy studies yield insights into global environmental issues such as climate change, maintenance of biodiversity, and sustainable forest management;
7. Heightened awareness of the general public to the excitement and importance of canopy studies.

The study of the forest canopy has reached a critical stage in its development from a young "frontier" area of study to become a vibrant and coalescing field of investigation and communication (Nadkarni 2001) that has fundamentally altered the way we think about forest ecosystems. Global efforts to investigate the impacts of atmospheric change on biodiversity and ecosystem function are being staged in forest canopies. Work on the effects of elevated CO_2 in mature, complex forest canopies is enhancing the predictive value of climate change models and will help answer questions about carbon sources and sinks (Pepin and Körner 2002). Canopy studies are a driving force in the debate on global species richness, biodiversity loss, and the synergistic effects of global change and habitat disturbance (Ozanne et al. 2003).

Canopy research has also challenged certain long-held concepts of forest ecosystem process and function, such as increased specialization in the tropics; specificity of pollinator-plant relationships; and general applicability of the intermediate disturbance hypothesis. Interactions at the biosphere-atmosphere interface have been revised, such as links between leaf-shedding mechanisms and fire vulnerability and connections between isoprene production and atmospheric chemistry. Recent developments in canopy studies led to the realization that forest canopies can play a central role in the development and testing of predictive models that address major environmental policy issues.

The Character and the Environment of Forest Canopies

Forests are unique among other vegetation types in being "dense, extensive, tall, and perennial" (Shuttleworth 1989). Beyond these similarities, forest canopies occur in a tremendous variety of shapes, sizes, complexity, species, and structures (Moffett 2001). Canopies found in different habitats have strikingly different appearances and functions (Figure 1-1). Canopies encompass an enormous range of sizes, from the giant coastal redwoods to the pygmy forests of the northern California coast. They also differ in phenology, pollinator and seed dispersal mechanisms, and patterns of nutrient cycling.

Despite these differences, forest canopies share several attributes in terms of both structure and function. Compared to the outside of a forest, the interior is dark, moist, quiet, and still. The structural complexity of canopies creates a variety of micro-environments for the biota that are directly

Figure 1-1 A collage of canopies at varying scales: (A) Stromatolites at low tide, Hamelin Pool, Shark Bay, Western Australia (photo credit: D.M. Jarzen); (B) Lichens near Shaw's Woods, Renfrew County, Ontario, Canada (photo credit: D.M. Jarzen); (C) *Equisetum sylvaticum* L., sterile stems, Mer Bleu, Ontario, Canada (photo credit: D.M. Jarzen); (D) Wheat fields, Saskatchewan, Canada (photo credit: D.M. Jarzen); (E) Heath, sclerophyll vegetation at Cape Naturaliste, Western Australia (photo credit: D.M. Jarzen); (F) Coniferous forest, Wind River Canopy Crane Facility, Carson, WA (photo credit: H. Bruce Rinker);

Continued

Figure 1-1—*Continued*   (G) Temperate deciduous forest in early spring, Millbrook School Forest Canopy Walkway, Millbrook, NY (photo credit: H. Bruce Rinker); (H) Subtropical oak/palm hammock, Myakka River Canopy Walkway, Sarasota, FL (photo credit: Ray Villares); (I) Tropical rainforest, Australian Canopy Crane Facility, Cairns, Australia (photo credit: H. Bruce Rinker); (J) Tropical rainforest, Amazon Conservatory for Tropical Studies, Iquitos, Peru (photo credit: H. Bruce Rinker); (K) Composite showing katydid, epiphylly, and herbivory from the Upper Amazon of Peru; assemblages, known as biofilms, of bacteria and other microorganisms on leaf surfaces constitute a new ecological science called biofilm ecology or even phyllosphere microbiology (photo credit: H. Bruce Rinker).

and indirectly associated with them (Parker 1995). The temperature and relative humidity of the branch and twig environment can change greatly over very short distances (1 to 100 cm) and can affect the distribution and phenology of canopy-dwelling species (Freiberg 1996). Canopies modify average values of temperature, relative humidity, wind speed, light, sound, and turbulence. However, the variance of these factors is also altered by the canopy, which is an important metric for the ecological and evolutionary environment of organisms. Canopies modify the character of their surroundings from interactions of both active and passive surfaces. In general, the canopy modifies the environment of the forest interior by making it more buffered from incoming sunlight, wind, and temperature (Parker 1995).

Composition and Diversity of Forest Canopies

The canopy of many forest ecosystems fosters extremely diverse plant communities, including vascular and non-vascular epiphytes, hemi-epiphytes, and parasites (Madison 1977; Benzing 1990; Putz and Mooney 1990). Canopy-dwelling plants contribute substantially to overall forest biodiversity by providing resources for a diverse assemblage of arboreal vertebrates, invertebrates, and microbes (Nadkarni and Matelson 1989; Stork et al. 1997). This diversity exists at different taxonomic levels. For example, vascular epiphytes belong to over 28,500 species, equivalent to about 10 percent of all named vascular plants. They occur in only 84 plant families, however, and over 60 percent of these fall into the three most common epiphytic plant families: Orchidaceae, Bromeliaceae, and Araceae (Kress 1986).

Global estimates of 30 million invertebrate species by Erwin (1982) were largely based on early canopy fogging studies in lowland tropical forests. These estimates were a key driver in the formulation of species coexistence and habitat specialization models. Subsequent studies of herbivorous forest insects, suggesting much lower levels of host specificity, have resulted in revised estimates of 2 to 6 million (Novotny et al. 2002; Basset et al. 2003b). A relatively high proportion of invertebrates (ca. 20 to 25 percent) is unique to the canopy (Sørensen 2003), although this varies with forest type and stand history. Invertebrates appear to stratify vertically at the species level, though there may be overlap between forest floor and canopy at the genus and family level (Longino and Nadkarni 1990).

The driving forces behind such diversity are not well understood, but much is due to the great amount of structural and microclimatic complexity that can exist within tree crowns. For example, the twigs of a single tree present an enormously different microclimate and substrate characteristics (in terms of shape, size, turnover rate, and chemistry) from the inner branches or trunk of the same tree. This could lead to the evolution of specialized holdfast mechanisms, nutrient and water acquisition techniques, and modes of dispersal, even within a single tree. For example, a speciose epiphytic plant group such as Orchidaceae includes specialists that grow on twigs rather than trunks (Chase 1987).

Structure of Forest Canopies

Measuring forest canopy structure is key to understanding many aspects of forest ecology (Spies 1998). Ecological data are inherently spatial, and most research involves making observations about structural elements. The complex three-dimensional dynamic structure of forest landscapes, however, requires that researchers deal selectively with a subset of components of the whole forest. In effect, each research question invokes its own "structural filter." This brings to mind the childhood poem "The Blind Men and the Elephant." Each blind man chose a different part of the elephant to examine—the tusk, the tail, the leg, or the trunk—and each came to wildly different conclusions based on his particular investigation, experience, and spatial model of the pachyderm.

Similarly, canopy structure has been categorized in a number of ways. Much of the early canopy classification was based on species or life forms in which constituents are measured based on their life habit (e.g., shrub, epiphytic, or vine) (Webb 1959). Profile diagrams were generated by destructive sampling of portions of the forest, providing a "two-dimensional slice" of habitat (Richards 1954). Those who documented tree architecture supplied a means of classifying the "idealized" form of trees, based on the anatomy and growth forms of the species (Hallé 1995). Concerned with the distribution of crown associates such as epiphytes or bird nests, many researchers have categorized individual tree crowns by zones (e.g., Catling et al. 1986). Current research in canopy structure generalizes the components and arrangements of forest canopy structural elements to ease the exchange of data between researchers and to provide ways to compare canopy structure across sites (Cushing et al. 2003).

Of all structural components, the form of the outer canopy is probably the most critical in terms of effects on whole-ecosystem processes. This outer layer is often a heavily folded collection of surfaces, much like a cerebral cortex, and therefore presents an aerodynamically rough plane upon which the atmosphere interacts. One of the measurements recently developed to quantify this outer canopy is the "rumple factor," or the ratio of the surface area of the outer canopy to the surface area of the ground, which effectively integrates the relative complexity of a forest canopy. Thus, a tall and complex temperate old-growth coniferous forest canopy that supports chasms and canyons reaching all the way to the forest floor would have a much larger rumple factor than a tropical montane forest canopy that supports a contiguous and tightly knit forest canopy with few breaks (Brokaw et al. in press).

Development and Dynamics of Forest Canopies

Though they often appear constant and immutable, canopies are dynamic and change in structure and function at a variety of time scales. At very short time scales, canopy structure may be drastically altered and some functions immediately changed depending on the character and intensity of disturbance. Over an annual cycle, canopies respond to seasonality of temperature and water availability with changes in leaf presence, from complete long-term leaf exchange (many evergreen forests) to complete deciduousness. On an interannual scale, canopies change through the ingrowth, increase, and death of constituent crowns. These changes are quantified by measurements in the number, size, connectivity, and character of the parts.

Over a much longer time, canopies may undergo a series of developmental changes with some common patterns (Figure 1-2). These stages are parallel to the successional patterns recognized for stems (Oliver and Larson 1990). Following a stand-replacement event, open ground is colonized with leaves. The rapidity of this stage depends on local environment; it ends when individual crowns interfere with coverage growth (crown closure). The next stage is of rapid and coordinated vertical growth of crowns, where the plane of competition for light moves at first rapidly and then more slowly away from the ground. This stage ends as the crowns approach their maximum height, which is often controlled by site factors such as nutrition, water availability, and disturbances. At this stage, the assemblage of crowns may appear as a single elevated, nearly uniform layer, with very little foliage in the bottom. The vertical distribution of canopy material may be termed "top-heavy" or "stratified"—the canopy surface is very smooth. Small differences in initial position and competition for light then begin to cause crowns to differentiate, such that some increase in height and/or volume relative to neighbors while others decline. The canopy surface becomes more rumpled. The loss of the crowns from competitive mortality creates numerous small openings, enhancing light transmittance to the forest floor and stimulating the growth of shade-tolerant understory and overstory species. Together, the crowns of these species are a second, albeit slower-growing, vertical wave of canopy material. At the beginning of this stage, the vertical canopy profile may have distinct peaks in both the understory and overstory. As time passes and canopy material accumulates, the peaks become less distinct.

In the last stage, overstory trees die, leaving a large opening in the canopy, which stimulates regeneration. Unlike the wave of vertical growth stimulated by the competitive death of young crowns, here the regeneration is local. Because of the large gaps and continued crown differentiation, the canopy surface at this point is increasingly complex. Continued mortality of the overstory results in a distribution of localities in different stages of regeneration with a concomitant spectrum of local vertical regeneration. The structure of the canopy is then dominated by features with a large spatial scale—that is, big crowns and big gaps. The vertical canopy profile can

EMPTY SPACE: ANOTHER VIEW OF FOREST CANOPY STRUCTURE

Roman Dial, Nalini M. Nadkarni, and Judith B. Cushing

Between the solid canopy elements that make up a forest is space—empty space surrounding the wood and foliage. Indeed, the trees *in addition to* the negative space around them essentially define the forest canopy. Empty space is where rain falls, sunlight passes, winds blow, and gasses circulate. It is this airy matrix that provides travel routes to flying and gliding animals, as well as to wind-dispersed pollen and seeds. Emmons (Emmons and Gentry 1983; Emmons 1995) suggested the evolution of that quintessential canopy creature, the vertebrate glider, depends on sufficient canopy space, while Freiberg (1997) experimentally compared the effects of empty space and canopy elements on microclimate. However, until recently, the free space of forests has been virtually ignored quantitatively. In fact, Moffett (2001) implored canopy scientists to put the "canopy into canopy biology" by measuring space.

Canopy space has been an implicit or explicit component of canopy studies for half a century or more. One of the earliest controversies in canopy science, the presence or absence of "strata" (Richards 1952; Parker and Brown 2000), depends on the relationship between canopy elements and the empty space around them. The classic measure of leaf area index advanced by Watson (1947) actually calculates the proportion of foliage samples to empty space samples, while Terborgh (1985) used ideas about light passing through empty space to construct a model of canopy strata. Even the cornerstone of thinking about forest dynamics, the canopy "gap," required rethinking in terms of various types of spaces that might puncture forest canopies (Liberman et al. 1989). Later Parker (1995) and Connell et al. (1997) erected a discrete classification system for canopy space. Perhaps the highest technology used in studies of forest canopy structure has been airborne scanning laser, or LIDAR (Lefsky et al. 1999), which can measure canopy space with a continuous variable.

Recently, another application of laser technology has measured empty space from within forest canopies (Cushing et al. 2003; Dial et al. in press). By locating 50- to 150-meter horizontal traverses as near the canopy surface as possible, then suspending vertical transects from these, we have been able to sample the empty space in a variety of forests. Palm Pilots receive downloads of distances and directions from handheld digital laser range finders and flux compasses, as well as text entries of canopy elements using Graffiti software. In this way, we can map the open space crudely but systematically, bounding objects at 1- to 2-meter vertical intervals from the forest floor to near the outer canopy surface. These data are used to quantify the amount and vertical distribution of forest canopy space.

Figure 1 shows the results of some of these studies for six forests around the Pacific at different latitudes: tropical (Borneo and Costa Rica), temperate (Australia and Washington), and boreal (Hemlock and Spruce, Alaska). These figures represent multiple transects from individual sites but suggest questions about canopy structure: Do sites of similar climate and age have similar structure, despite biogeographical differences? Does canopy structure map onto canopy element composition? What is the relationship between empty space and canopy functions and processes? Clearly more studies, such as ground-based LIDAR and further in-canopy measures, are needed to answer these important questions about canopy structure from the viewpoint of negative space.

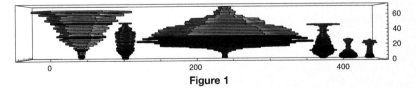

Figure 1

Continued

EMPTY SPACE: ANOTHER VIEW OF FOREST CANOPY STRUCTURE–*cont'd*

References

Connell, J.H., Lowman, M.D., and Noble, I.R. (1997). Subcanopy gaps in temperate and tropical forests. *Australian J of Eco* **22**, 163–168.

Cushing, J, Nadkarni, N, Bond, B., and Dial, R. (2003). How trees and forests inform biodiversity and ecosystem informatics. *Comp. in Sci. and Eng.* **5**, 32–43.

Dial, R., Bloodworth, B., Lee, A., Boyne, P., and Heys, J. (In press). The distribution of free space and its relation to canopy composition at six forest sites. *Forest Sci.*

Emmons, L.H. (1995). Mammals of rainforest canopies. *In* "Forest Canopies" (M.D. Lowman and N.M. Nadkarni, eds.), pp. 199–223. Academic Press, New York, NY.

Emmons, L.H. and Gentry, A.H. (1983). Tropical forest structure and the distribution of gliding and prehensile-tailed vertebrates. *Amer. Nat.* **121**, 513–524.

Freiberg, M. (1997). Spatial and temporal pattern of temperature and humidity of a tropical premontane rain forest tree in Costa Rica. *Selbyana* **18**, 77–84.

Lefsky, M.A., Cohen, W.B., Acker, S.A., Parker, G.G., Spies, T.A., and Harding, D. (1999). LIDAR remote sensing of the canopy structure and biophysical properties of Douglas-fir Western Hemlock Forests. *Rem. Sens. of the Environ.* **70**, 339–361.

Lieberman, M., Lieberman, D., and Peralta, R. (1989). Forests are not just Swiss cheese: canopy stereogeometry of non-gaps in tropical forests. *Ecology* **70**, 550–552.

Moffett, M.W. (2001). The nature and limits of canopy biology. *Selbyana* **22**, 155–179.

Parker, G.G. (1995). Structure and microclimate of forest canopies. *In* "Forest Canopies" (M. D. Lowman and N. M. Nadkarni, Eds.), pp. 73–106. Academic Press, New York, NY.

Parker, G.G. and Brown, M.J. (2000). Forest canopy stratification: is it useful? *Amer. Nat.* **155**, 473–484.

Richards, P.W. (1952). "The Tropical Rainforest: An Ecological Study." Cambridge University Press, Cambridge.

Terborgh, J. (1985). The vertical component of plant species diversity in temperate and tropical forests. *Amer. Nat.* **126**, 760–776.

Watson, D. J. (1947). Comparative physiological studies on the growth of field crops. I. Variation in net assimilation rate and leaf area between species and varieties, and within and between years. *Ann. Bot.* **11**, 41–76.

be more or less uniform, or, more commonly, concentrated near the forest floor ("bottom-heavy"). The canopy complexity is highest at this stage.

Human Services Provided by Forest Canopies

Forest canopies satisfy or enhance many human needs, most of which were overlooked or under-valued until the characteristics of canopy biota and processes were quantified by canopy researchers who focused on ecological values. Economic valuation is a continuing challenge for conservationists and canopy researchers, as it is almost impossible to place a dollar value on these often subtle and long-term benefits. We divide these provisions or enhancements into four broad categories, following Blair and Ballard (1996):

1. **Ecosystem Services:** Canopy biota enhance watershed integrity, nutrient cycling, erosion control, carbon sequestering, and climate stability.
2. **Economic Benefits:** Many plants, animals, and microorganisms provide us with food, shelter, fuel, clothing, and medicines. Approximately 50 percent of our medicines are derived from plants, and 25 percent of all prescription drugs have their origins in tropical forests (Bennett 1992). For example, the rosy periwinkle (*Catharanthus roseus*) from Madagascar produces scores of alkaloids, two of which led to major breakthroughs in cancer treatment.

general developmental stages of canopies

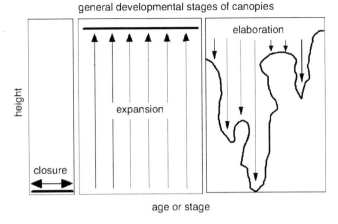

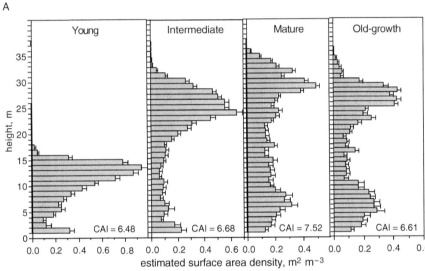

Figure 1-2 Development of Forest Canopies. (A) This is a schematic view of the major processes involved in the general developmental of forest canopy structure. (B) This is the vertical profile of surface area density in four developmental stages of closed-canopy mixed-species temperate deciduous forests on the mid-Atlantic coastal plain in the United States.

3. **Aesthetic Importance:** The canopy provides inspiration and a sense of wonder about the natural world. Gardeners gain pleasure from rearing canopy-dwelling plants such as orchids and bromeliads on their kitchen windowsills; birdwatchers travel to remote ecotourism sites to spot treetop species; recreational tree-climbers gain a sense of adventure using access techniques devised by canopy researchers.
4. **Ethical Values:** The preservation of species is a mandate felt by many humans. The large number of rare and recently discovered species in the canopy makes it an especially important locale to deal with the issues pertaining to ethical values of biota.

The influence of canopies in terms of providing ecosystem services is far out of proportion to their absolute size. For example, canopies have less than one-thousandth the weight of the overlying atmosphere (2×10^4 g/m^2 for a medium-sized forest compared to 10^7 g/m^2 for its atmosphere). Due to their reflective and absorptive properties, however, they control much of the world's energy balance. Most forests reflect less than 10 to 20 percent of solar radiation, absorbing the remainder to heat air, drive photosynthesis, and evaporate water. Because the large amount of energy required to evaporate water is released when the vapor condenses, the production, consumption, and movement of water vapor are critical energy transport pathways. The planet's forests are carbon collectors, scrubbing carbon dioxide from the air during photosynthesis and converting it into sugars, starch, and other plant products. Thus, cutting and burning an old-growth forest rapidly pulses carbon back into the atmosphere as a greenhouse gas that took decades or centuries to store (Ozanne et al. 2003).

Forest canopies also support organisms that add available nutrients from sources that originate outside the ecosystem. Early work in nitrogen-limited forests of the Pacific Northwest demonstrated that arboreal epiphytes are capable of fixing 1–7 kilograms per hectare per year of gaseous nitrogen (N) and passing it to other ecosystem members via leaching and epiphyte litterfall (Denison 1973). In tropical montane forests, up to 60 percent of nitrate-N and up to 25 percent of ammonium-N can be captured and retained from atmospheric sources by epiphytic bryophytes (Clark et al. 1998).

Structural elements of the forest canopy provide important habitats for wildlife species, many of which are specific to old-growth forests. Elements such as large horizontal branches, thick mats of epiphytic moss and accompanying canopy humus, and foliose lichens tend to occur almost exclusively in primary forests (Ishii and Ford 2001). Bird use of epiphytes in tropical montane forests, for example, includes the gathering of fruits, pollen, and nectar for nutrients and energy; collecting foliage and lichens for nest structural material and nest camouflage; and foraging for invertebrates (Remsen and Parker 1984; Nadkarni and Matelson 1989). In temperate coniferous forests, epiphyte mats are microhabitats that are extremely rich in diversity and abundance of invertebrates (Winchester and Ring 1999). Predators dominate many of these canopy assemblages, keeping herbivorous arthropod populations at a low level, thus inadvertently serving to benefit human forestry practices (Schowalter 1989). Certain species valued by humans because of their rarity appear to depend on old-growth forest structures, such as the "bolsters" of moss on old-growth forest branches. The seagoing marbled murrelet (*Brachyramphus marmoratus*), for example, travels up to 100 km inland from the Pacific Ocean to create nests on these moss mats (Whitworth et al. 2000).

Spirituality and Life-Giving Forces

Humans have long recognized the irreplaceable functions of trees, often assigning them a sacred value in their cultures and linking them to spirituality and religion. In sacred writings, trees are presented as symbols or manifestations of divine knowledge, connecting heaven to earth. They have power to bestow eternal life or renew the life force (Thich Nat Hanh 1995). In fact, in the folklore of many peoples of the world, a great tree—evergreen, ever-blooming, and ever-bearing—exists whose fruit assures its eater of immortality or enlightenment, often guarded by a dragon or serpent. To different groups, different trees were identified with this tree of life: e.g., the apple tree of the Celts, the peach tree of the Chinese, the centerpiece Tree of Life of the Old Testament, and the date palm of the Semites (Leach 1972). The Buddha achieved enlightenment under the Bodhi tree (*Ficus religiosa*) thousands of years ago. In Egypt, the ancient gods sat in a tall sycamore known as the "Tree of Life." Most Guiana tribes have a myth about a wonderful tree bearing all the food plants. Versions of this same tale have been recorded among the Cuna and Choco Indians of Bolivia and the Witoto of the Putumayo River (Leach 1972). Many events significant to humans, such as proclamations, funerals, and treaty signings, are staged in the presence of grand trees.

Many faiths emphasize a sense of stewardship for trees and forests. Druids valued their ancient groves and viewed some plant associations as holy (e.g., oaks and mistletoe). The Jewish Talmud instructs ". . . and the Lord God planted a Garden in Eden, and so you, too, when you come to Israel, shall do nothing before you have planted. . . ."(Bernstein 1998). Within Christianity, the Association of African Earthkeeping Churches, a "green force" of 150 African independent churches (with a combined membership of two million followers), has tree-planting embedded in its liturgical practices. This group, which operates on the basis of religious motivation and mobilization, has been responsible for planting millions of trees in the central and southeastern regions of Zimbabwe since the early 1990s (Hessel and Ruether 2000).

Trees are linked closely to a number of spiritual concepts, such as enlightenment, silence and stillness, and the hidden aspects within each person. Their very forms, with their roots in the soil and their foliage reaching toward the sun, remind us of the connection between the earth and non-earth. Their dendritic forms are ubiquitous in nature—streams and rivers, cave systems, blood vessels, lungs, family trees, and temple hierarchies. As with all living beings, trees breathe. Through photosynthesis, they help supply the most basic of human needs—clean air to breathe and stored carbohydrates for energy. Also, the spiritual life force of many religions comes by way of air and respiration. The word "spirituality" is derived from the Latin word *spirare*, meaning "to breathe." The Hebrew word for breath, *nesheema*, shares the same root as the word for soul, or *neshama*. Zen Buddhists use the action of paying attention to air entering and exiting the nostrils as the initiation and maintenance of a meditative state. Silence is a powerful medium for reflection. Buddhist silence, or *samantha*, is critical to spiritual development because it allows the practitioner to stop, calm, and concentrate. This stillness is evident when we gaze at a tree on a windless summer day. Trees are rooted in the ground and make no sounds; they epitomize *samantha*.

Trees also help humans tell time; they spell the seasons. Nothing tells us about the passage of time more clearly than autumn colors, the tender green of emerging spring buds, or the delicate filigree of snow on tiny twigs. Forests manifest the dynamism of nature, the need to accept change even if change seems to be destructive for the short term. When a tree falls, we remind ourselves that this is the nature of the forest and of life. Later, seedlings grow in the light created by its fallen parent, manifesting the interweaving of death with life.

Healing and Health

Another powerful way that trees benefit humans is their use as agents and inspiration for healing. Quinine from the cinchona tree (*Cinchona officianalis*) is used to treat malaria, and taxol from the Pacific Yew (*Taxus brevifolius*) helps cure ovarian cancer. Cancer patients who face rounds of chemotherapy might be encouraged to learn that trees can sustain tumors ("burls") for centuries. Amputee victims might gain heart by knowing that trees lose limbs and adapt to the loss by growing epicormic branches (those arising from dormant buds that emerge in response to light). Further, since the death of individual trees creates light gaps for young saplings in the rainforests, trees might inspire the dying because of the continuity of life that they symbolize. In the shamanistic cultures of Amazonia, healing ceremonies take place in the shadow of great kapok trees (*Ceiba pentandra*).

Quantitative studies by behavioral psychologists on patients with views of trees from their hospital room windows reinforce the healing properties of tree images (Ulrich 1984). Patients who had a window view of a tree had significantly fewer days of recovery, a significantly smaller number of calls for narcotic medication, and a significantly fewer number of complications following surgery. These results suggested that the visual presence of trees might reduce stress and enhance the recovery process. The results have led to some changes in hospital design (Ulrich et al. 1991), as well as larger-scale shifts in designing living space for urban senior adults (Takano et al. 2002) to include green spaces within residential areas.

Forests have also inspired countless authors, artists, musicians, poets, and engineers to express the aesthetics invoked by tree forms or to make analogies to human emotions and the human condition. The symbolic power of trees is manifested by their portrayal on hundreds of postage stamps, currency and coinage, and the national flags of eight countries. Hundreds of musical compositions have taken their themes from emotions evoked from trees. The life force of trees is presented in such writings as J.R.R. Tolkien's *The Lord of the Rings* and C.S. Lewis's *The Chronicles of Narnia*. Many children's books, such as Lynne Cherry's *The Great Kapok Tree*, provide tales about global biodiversity with strong conservation messages for readers young and old.

Human Management and Conservation of Forest Canopies

The history of humans is inextricably linked with forests. Most places of human habitation are or were at some time in the past forested. Repeatedly, humanity has cleared space for itself in the midst of forests, and forests have had continuous involvement with our species throughout human history. Modern humans retain a partly genetic predisposition for natural settings with savanna-like properties such as spatial openness, copses of trees and shrubs, and relatively uniform grassy ground surfaces (Kellert and Wilson 1993). From the beginnings of plant husbandry to our present destruction of vast areas of tropical forests for fiber and fuel, humans have sought to maximize the amounts of materials they can extract in the shortest amounts of time (Harrison 1992). The long generation time of trees relative to that of humans has made the conservation of canopy habitat more pressing for forestry than for other types of agriculture and domestic animal production.

Canopy research and forest management are closely linked but are still in early stages of collaboration (Rook et al. 1985; Berg et al.1996; Franklin et al. 2001). The development of forest canopy structure as stands mature is of interest both to forest canopy researchers and to foresters interested in stand management. How tree crowns develop the astonishing complexity they display in old-growth forests of the Pacific Northwest, for example, has been of interest to both ground-bound foresters and forest canopy researchers. Recent canopy research on the development of epicormic shoots shows how these shoots allow a long-lived tree to regenerate itself over time. This renewal of ephemeral branch structures during the long life of the tree canopy is initiated after crown closure (Ishii and Ford 2001).

Many current timber-harvesting practices aggressively disrupt the complex vertical stratification of microhabitats and resource use in the canopy (Davis and Sutton 1998) and can substantially alter nutrient, carbon, and invertebrate dynamics (Schowalter et al. 1981). Selective logging in tropical forests may not always reduce tree diversity, but it can modify pollinator activities, which then reduces the reproductive potential of both logged and adjacent unlogged lands (Ghazoul and McLeish 2001). Canopy specialists are more affected by these activities than forest generalists. For example, insectivorous birds suffer greater effects than do other guilds as a consequence of impacts on their prey (Thiollay 1992). Thus, canopy organisms are highly susceptible to human disturbances linked to forest management (Ozanne et al. 2003).

Alternatives to traditional forestry practices such as clear-cutting in the Pacific Northwest and other regions of the United States are being developed. These include variable density thinning, the restoration of old-growth structures to younger stands, and preserving riparian vegetation. These practices allow the mitigating effects of the canopy to carry out the functions that the primary intact forest performed, including modifying microclimate at canopy and ground levels, which then provides habitat for wildlife and creates higher variability of biotic and abiotic aspects of the environment (Spies and Franklin 1991).

VERTICALITY AND HABITAT ANALYSIS: MACARTHUR AND WILSON'S BIOGEOGRAPHY THEORY REVISITED

H. Bruce Rinker

The most striking and important fact for us in regard to the inhabitants of islands, is their affinity to those of the nearest mainland, without being actually the same species.

—**Charles Darwin, 1860,** On the Origin of Species by Means of Natural Selection

MacArthur and Wilson (1967) proposed "the possibility of a theory of biogeography at the species level" that expressed the relationship between insularity and biodiversity in quantitative terms. They noted that islands constantly gain (immigration) and lose (extinction) species based on their areas and distances to the mainland or other biodiversity sources. Area alone, however, is an insufficient measure of species richness as oceanic islands are very different from "habitat islands," fragments of vegetation in a mosaic of terrestrial environments. Among others, Wu and Vankat (1995) critiqued the theory and debated its validity and its application to oceanic and continental islands as well as to a variety of insular habitats such as city parks and wildlife corridors. Weaknesses in the MacArthur and Wilson theory soon came to our attention. All species cannot be treated the same. Immigration and extinction probabilities are not the same for all species. The theory assumes random distribution of biodiversity in undisturbed habitat while the sustained occurrence of a particular species may necessitate the presence (or absence) of other types of organisms. All areas of island habitats are not equally accessible or preferable for species' colonization.

Generally speaking, the MacArthur-Wilson equation is inadequate because it is two-dimensional and does not clearly address the widely recognized multidimensional aspects of landscape structure and what Janzen calls "the eternal external threat" (1986): the external nonhuman and unintentional human threats to island preserves. Today, a more accurate approach to the tough complexities of living (as opposed to theoretical) islands is that of Forman's *Land Mosaics* (1995). Studies on dendrobatid frogs and their reliance on canopy bromeliad tanks, for instance, or on euglossine bees and their complex interdependencies on orchids and Brazil nut trees illustrate the formula's inadequacies. Like fractals, *all* the co-variables governing species richness may approach infinity, depending on spatial-temporal scales.

Canopy ecology presented two related shifts in biogeography perspective apropos to this discussion:

- The term *canopy* is now defined as the combination of all leaves, twigs, and small branches in a stand of vegetation (Parker 1995). Indeed, to be even more expansive, a forest canopy is all aboveground vegetation including the crowns and trunks of its trees, its biota, and its ecological processes. The meaning of canopy may be expanded eventually to include the entire column of life sandwiched between atmosphere and bedrock including soil biodiversity!
- The term *canopy* emphasizes the verticality in a stand of vegetation. No matter which spatial-temporal scales are applied to forest habitat analysis, a complex stratification of biota exists that is ignored by a strict interpretation of the MacArthur-Wilson formula. Area is a simple two-dimensional measure, but organisms live in a sometimes baffling complexity of four dimensions: space plus change through time.

Continued

VERTICALITY AND HABITAT ANALYSIS: MACARTHUR AND WILSON'S BIOGEOGRAPHY THEORY REVISITED–*cont'd*

Organisms are found across a four-dimensional landscape. In the past decade, canopy work with mammals (e.g., Malcolm 1995), neotropical migrants (e.g., Rinker 2001), and arthropods (e.g., Stork et al. 1997) have confirmed that reducing biological complexities to a species-area curve is an inadequate reflection of the natural world. With the emergence of canopy ecology, the dynamic vertical aspects of habitats (e.g., architecture and light penetration in aging forests) provide a richer detail about the distribution of biota.

Ecosystems may no longer be treated as predictable machines with scientific theories as crisp, explanatory models for organisms. They cannot be approached as small-number assemblages (i.e., individuals in a collection) or as large-number means (i.e., averages in a collection). Neither extreme adequately describes the *system*. Living systems are stochastic, uncertain, and somewhat chaotic; their coupled generalizations cannot be ossified into species-area regressions. Systems are too difficult to grasp by word or formula so their theories are really signposts to what really is. They point the way to more research, more study, and more experimentation; but they are never enough in themselves.

When MacArthur and Wilson wrote *The Theory of Island Biogeography* in 1967, canopy ecology had not yet emerged as a formal discipline with established protocols and firm definitions. Canopy ecologists have since learned that forests are messy systems profiled like a cerebral cortex and that verticality is one of a number of features important for the distribution of ecosystem biota. Canopy ecology has disclosed the mythological aspects of the MacArthur-Wilson formula: it is only a stepping-stone toward a better understanding of ecosystem complexity.

References

Forman, R.T.T. (1995). "Land Mosaics: The Ecology of Landscapes and Regions." Cambridge University Press, Cambridge.

Janzen, D.H. (1986). The eternal external threat. *In* "Conservation Biology: The Science of Scarcity and Diversity" (M. E. Soulé, Ed.), pp. 286–303. Sinauer Associates, Sunderland, MA.

MacArthur, R.H. and Wilson, E.O. (1967). "The Theory of Island Biogeography." Princeton University Press, Princeton.

Malcolm, J.R. (1995). Forest structure and the abundance and diversity of neotropical small mammals. In "Forest Canopies" (M.D. Lowman and N.M. Nadkarni, Eds.), pp. 179–197. Academic Press, San Diego.

Parker, G.G. (1995). Structure and microclimate of forest canopies. *In* "Forest Canopies" (M.D. Lowman and N.M. Nadkarni, Eds.), pp. 73–106. Academic Press, San Diego.

Rinker, H.B. (2001). The use of a forest canopy walkway for studying habitat selection by neotropical migrants. *Selbyana* **22**, 89–96.

Stork, N.E., Adis, J., and Didham, R.K. (1997.) "Canopy Anthropods." Chapman and Hall, London.

Wu, J. and Vankat, J.L. (1995). Island biogeography: Theory and applications. *In* "Encyclopedia of Environmental Biology" (W.A. Nierenberg, Ed.), pp. 371–379. Academic Press, San Diego.

Another issue that relates forest canopy research to forest management concerns the effects of forest fragmentation on forests and their associated canopy biota. Some canopy researchers have shifted their research locales away from primary forests in order to examine questions posed by trees growing in secondary forests or those isolated in pastures or other agricultural settings. They have begun to address the question of whether a few trees or even a single tree crown can function in a pasture or clear-cuts in terms of the biota it might support. Can pollinator and disperser animals cross open spaces? Can epiphytes found in old-growth trees disperse and maintain themselves, or will the epiphytes ultimately be only those that can survive under edge conditions?

Recent experimental work involving the transplantation of lichen thalli to tree canopies that have been isolated in clear-cuts has shown that the microclimate is not inimical to the survival and reproduction of these "old-growth" dependents. This suggests that dispersal to the isolated tree crowns is the limiting factor rather than some element of microclimate change (McCune et al. 1996; Sillett et al. 2000).

An additional link between canopy research and forest management is the dispersion of pathogens and parasites in primary and managed forests. Canopy researchers, studying the ecology and evolution of tree parasites, have presented forest pathologists with insights into how they might be controlled. For example, canopy-level research at the Wind River Canopy Crane Research Facility in Carson, Washington, has documented the natural history of a major parasitic plant, the western dwarf mistletoe (*Arceuthobium campylopodum*) (Shaw et al. 2000). Researchers have identified the life history stages when infection occurs and the distance through which these explosive dispersal plants are able to project their seeds. This information, coupled with stand-level infection rates and species distribution information collected from the ground and from the crane, provides forest managers with powerful tools to control this parasite.

The harvest of secondary forest products, some of which are removed from the canopy, is of interest to horticulturists, foresters, and ecologists. The sustainable removal of non-timber resources can provide human resources from forests without threatening the integrity of forests (Bennett 1992). One pressing conservation issue in the Pacific Northwest is the over-exploitation of mosses from wild forests for the commercial floral/horticulture industry. The increasing market for secondary forest products such as epiphytic mosses (as well as mushrooms, terrestrial shrubs, and ferns) places unsustainable pressure on these plants, especially in old-growth forests (Peck 1997). Due to the long-term nature of epiphyte regeneration (Nadkarni 2000; Cobb et al. 2001), management efforts to "farm" rather than "mine" epiphytic mosses must be explored. Those who use canopy resources are beginning to recognize that their tools and approaches must change, both in carrying out research and in communicating with forest managers. Canopy researchers can meet the challenge to make this transition in an environmentally sound way.

The natural and seemingly innate affinity that humans have for trees can result in a sense of curiosity, conservation, and stewardship about forests. For example, in the 1990s, forest activists recognized that "tree-sitting" can be an effective part of their conservation campaigns, as such tree-sittings received media attention (Durbin 1996). More specifically, the forest canopy can serve as a focal point to raise awareness and care of the forest as a whole. In the last decade, many popular articles, television programs, and films (such as the 1998 National Geographic Society film *Heroes of the High Frontier*) have been produced about forest canopy researchers and their work. Public interest comes from a wide range of age groups, particularly children, who have a natural love for climbing trees and exploring little-known places. Many people who are involved in environmental education and outreach have found that the forest canopy can serve as a focal point to raise awareness and mindfulness of the forest as a whole because of the youth of the science, the beauty of the habitat, and the instinctive tug that it has on humans.

Future Directions for Forest Canopy Studies

Prioritizing certain activities will help the growing field of forest canopy studies progress efficiently. Enabling canopy researchers to communicate with each other and with those outside the field is an important element in addressing these issues, so strengthening existing networks such as the International Canopy Network and the Global Canopy Programme is critical. In addition, the establishment of graduate-level training programs that pertain to canopy study will generate and maintain a healthy discipline. Moreover, a database for scientific references, protocols, and experimental results should be maintained with assistance from computer scientists and systems modelers. Formal procedures to identify particular forest sites of critical concern should be generated, and these should include communication of prioritized sites to conservation groups and policy-makers. Instilling a sense of wonder and appreciation for organisms and interactions in non-scientists via educators and the media is another important avenue for forest canopy conservation.

As an ethical obligation to society at large, canopy researchers must take time from their projects to provide information in accessible forms so that the general public can understand forest canopies and the importance of the diversity and ecology of the organisms that live there. This could take the form of popular articles for children's magazines or publications for lay adults, informal talks at a local naturalist center or bird-watching club, or expertise given to politicians and land managers on local environmental issues related to forest canopies. Such indirect activities, though rarely rewarded in traditional academic systems, are fundamental to improving societal attitudes about research and conservation.

Conclusions

Until recently, our perspective on forest ecology was founded on ground-based information, which left us largely ignorant about canopy biodiversity and processes. Only in the past decade has our understanding of treetop ecology expanded substantially, in large part because of the access techniques that canopy researchers developed. We have come to understand that canopies are diverse, spatially organized in complex ways, and provide a variety of functions, many of which are of concern to humans. Forests across the face of the planet are disappearing at rapid rates. However, researchers, conservationists, science educators, grassroots organizers, ethicists, politicians, religious leaders, and others are all calling for responsible stewardship of our natural resources. Canopy ecology is an emerging science that has helped us clarify needs and directions.

References

Basset, Y., Horlyck, V., and Wright, S.J. (2003a.) "Studying Forest Canopies from Above: The International Canopy Crane Network." Smithsonian Tropical Research Institute and United Nations Environment Programme, Panama City.

Basset, Y., Kitching, R., Miller, S., and Novotny, V. (Eds.). (2003b). "Arthropods of Tropical Forests: Spatio-Temporal Dynamics and Resource Use in the Canopy." Cambridge University Press, Cambridge.

Bennett, B. (1992). Uses of epiphytes, lianas and parasites by the Shuar people of Amazonian Ecuador. *Selbyana* **13**, 99–114.

Benzing, D.H. (1990). "Vascular Epiphytes: General Biology and Related Biota." Cambridge University Press, Cambridge.

Benzing, D.H. (1998). Vulnerabilities of tropical forests to climate change: the significance of resident epiphytes. *Climatic Change* **39**, 519–540.

Berg, D.R., Brown, T.K., and Blessing, B. (1996). Silvicultural system design with emphasis on the forest canopy. *Northwest Sci.* **70**, 31–36.

Bernstein, E., Ed. (1998). "Ecology and the Jewish Spirit: Where Nature and the Sacred Meet." Jewish Lights Publishing, Woodstock.

Blair, R.B. and Ballard, H.L. (1996). "Conservation Biology: A Hands-On Introduction to Biodiversity." Stanford University Press, Stanford.

Blair, J.B., Rabine, D.L., and Hofton, M.A. (1999). The Laser Vegetation Imaging Sensor: a medium-altitude, digitization-only, airborne laser altimeter for mapping vegetation and topography. *ISPRS J. Photogram. Rem. Sens.* **54**, 115–122.

Brokaw, N., Fraver, S., Grear, J.S., Thompson, J., Zimmerman, J.K., Waide, R.B., Everham, E.M. III, Hubbell, S.P., and Foster, R.B. Disturbance and canopy structure in two tropical forests. (In press). *In* "Tropical

Forest Diversity and Dynamism: Results from a Long-Term Tropical Forest Network" (E. Losos and E.G. Leigh Jr., Eds.). University of Chicago Press, Chicago.

Catling, P.M., Brownell, V.R., and Lefkovitch, L.P. (1986). Epiphytic orchids in a Belizean grapefruit orchard: Distribution, colonization, and association. *Lindleyana* **1**, 194–202.

Chase, M.W. (1987). Obligate twig epiphytes in the Oncidiinae and other neotropical orchids. *Selbyana* **10**, 24–30.

Clark, K.L., Nadkarni, N.M., Schaefer, D.A., and Gholz, H.L. (1998). Atmospheric deposition and net retention of ions by the canopy in a tropical montane forest, Monteverde, Costa Rica. *J. Trop. Ecol.* **14**, 27–45.

Cobb, A.R., Nadkarni, N.M., Ramsey, G.A., and Svoboda, A.J. (2001). Recolonization of bigleaf maple branches by epiphytic bryophytes following experimental disturbance. *Can. J. Bot.* **79**, 1–8.

Cushing, J.B., Bond, B., Dial, R., and Nadkarni, N.M. (2003). How trees and forests inform biodiversity and ecosystem informatics. *Comp. in Sci. and Eng.* **5**, 32–43.

Davis, A.J. and Sutton, S.E. (1998). The effects of rain forest canopy loss on arboreal dung beetles in Borneo: implications for the measurement of biodiversity in derived tropical ecosystems. *Diversity and Distributions* **4**, 167–173.

Davis, M.B. (1996). "Eastern Old Growth." Island Press. Washington, D.C.

Dayton, P.K. (1985). Ecology of kelp communities. *Ann. Rev. Ecol. Syst.* **16**, 215–245.

Denison, W.C. (1973). Life in tall trees. *Sci. Amer.* **228**, 74–80.

Durbin, K. (1996). "Tree Huggers: Victory, Defeat, and Renewal." The Mountaineers, Seattle.

Erwin, T. (1982). Tropical forests: their richness in Coleoptera and other arthropod species. *Coleopterists Bulletin* **36**, 74–75.

Franklin, J.F., Spies, T.A., Van Pelt, R., Carey, A., Thornburgh, D., Berg, D.R., Lindenmayer, D., Harmon, M., Keeton, W., and Shaw, D.C. (2001). Disturbances and the structural development of natural forest ecosystems with some implications for silviculture. *For. Ecol. Manage.* **155**, 399–423.

Freiberg, M. (1996). Phenotypic expression of epiphytic Gesneriaceae under different microclimatic conditions in Costa Rica. *Ecotropica* **2**, 49–57.

Garvin, J., Bufton, J., Blair, B., Harding, D., Luthcke, S., Frawley, J., and Rowlands, D. (1998). Observations of the Earth's topography from the Shuttle Laser Altimeter (SLA): laser-pulse echo-recovery measurements of terrestrial surfaces. *Phys. Chem.Earth Sci.* **23**, 1053–1068.

Ghazoul, J. and McLeish, M. (2001). Reproductive ecology of tropical forest trees in logged and fragmented habitats in Thailand and Costa Rica. *Plant Ecology* **153**, 335–345.

Hallé, F. (1995). Canopy architecture in tropical trees: a pictorial approach. *In* "Forest Canopies" (M.D. Lowman and N.M. Nadkarni, Eds.), pp. 27–44. Academic Press, San Diego.

Harrison, R.P. (1992). "Forests: The Shadow of Civilization." University of Chicago Press, Chicago.

Hayden, B.P. (1998). Ecosystem feedbacks on climate at the landscape scale. *Phil. Trans. R. Soc. Lond. B* **353**, 5–18.

Hessel, D.T. and Reuther, R.R., Eds. (2000). "Christianity and Ecology." Harvard University Press, Cambridge, MA.

Ishii, H. and Ford, E.D. (2001). The role of epicormic shoot production in maintaining foliage in old *Pseudotsuga menziesii* (Douglas-fir) trees. *Can. J. Bot.* **79**, 251–264.

Kellert, S.R. and Wilson, E.O. (1993). "The Biophilia Hypothesis." Island Press, Washington, D.C.

Kress, W.J. (1986). The systematic distribution of vascular epiphytes: an update. *Selbyana* **9**, 2–22.

Leach, M. (1972). "Standard Dictionary of Folklore, Mythology, and Legend." Harper and Row, San Francisco.

Lefsky, M.A., Cohen, W.B., Harding, D.J. and Parker, G.G. (2002). Lidar remote sensing for ecosystem studies. *BioScience* **52**, 19–30.

Longino, J. and Nadkarni, N.M. (1990). A comparison of ground and canopy leaf litter ants (*Hymenoptera Formicidae*) in a neotropical montane forest. *Psyche* **97**, 81–94.

Lowman, M.D. and Nadkarni, N.M. (1995). "Forest Canopies." Academic Press, San Diego.

Lugo, A.E. and Scatena, F. (1992). Epiphytes and climate change research in the Caribbean: a proposal. *Selbyana* **13**, 123–130.

Margulis, L. (2001). What is canopy biology? A microbial perspective. *Selbyana* **22**, 232–235.

McCune, B., Derr, C.C., Muir, P.S., Shirazi, A.M., Sillett, S.C., and Daly, W.J. (1996). Lichen pendants for transplant and growth experiments. *Lichenologist* **28**, 161–169.

Moffett, M.W. (2000). What's "up?" A critical look at the basic terms of canopy biology. *Biotropica* **32**, 569–596.

Moffett, M.W. (2001). The nature and limits of canopy biology. *Selbyana* **22**, 155–179.

Moffett, M.W. and Lowman, M.D. (1995). Canopy access techniques. *In* "Forest Canopies" (M.D. Lowman and N.M. Nadkarni, Eds.), pp. 3–26. Academic Press, San Diego.

Nadkarni, N.M. (2000). Colonization of stripped branch surfaces by epiphytes in a lower montane cloud forest, Monteverde, Costa Rica. *Biotropica* **32**, 358–363.

Nadkarni, N.M. (2001). Enhancement of forest canopy research, education, and conservation in the new millennium. *Plant Ecology* **153**, 361–367.

Nadkarni, N.M. (2002). Trees and spirituality: an exploration. *Northwest Dharma News* **15**, 10–13.

Nadkarni, N.M. and Matelson, T.J. (1989). Bird use of epiphyte resources in neotropical trees. *Condor* **69**, 891–907.

Nadkarni, N.M. and Parker, G.G. (1994). A profile of forest canopy science and scientists—who we are, what we want to know, and obstacles we face: results of an international survey. *Selbyana* **15**, 38–50.

Novotny, V., Basset, Y., Miller, S.E., Weiblen, G.D., Bremer, B., Cizek, L., and Drozd, P. (2002). Low host specificity of herbivorous insects in a tropical forest. *Nature* **416**, 841–844.

Oliver, C.D. and Larson, B.C. (1990). "Forest Stand Dynamics." McGraw-Hill, New York.

Orians, G.H. (1980). Habitat selection: general theory and applications to human behavior. *In* "The Evolution of Human Social Behavior" (J.S. Lockard, Ed.) pp. 32–45. Elsevier North-Holland.

Ozanne, C.M.P., Anhuf, D., Boulter, S.L., Keller, M. Kitching, R.L., Korner, C., Meinzer, F.C., Mitchell, A.W., Nakashizuka, T., Silva Dias, P.L., Stork, N.E., Wright, S.J., and Yoshimura, M. (2003). Biodiversity meets the atmosphere: a global view of forest canopies. *Science* **301**, 183–186.

Parker, G.G. (1995). Structure and microclimate of forest canopies. *In* "Forest Canopies" (M.D. Lowman and N.M. Nadkarni, Eds.), pp. 73–106. Academic Press, San Diego.

Parker, G.G. (1997). Canopy structure and light environment of an old-growth Douglas-fir/western hemlock forest. *Northwest Sci.* **71**, 261–270.

Parker, G.G. and Brown, M.J. (2000). Forest canopy stratification: is it useful? *American Naturalist* **155**, 473–484

Parker, G.G., Harding, D.J., Berger M. (2004). A portable laser altimeter for rapid determination of forest canopy structure. *J. Appl. Ecol.* (in press.)

Parker, G.G., Smith, A.P., and Hogan. K.P. (1992). Access to the upper forest canopy with a large tower crane. *BioScience* **42**, 664–670.

Peck, J.E. (1997). Commercial moss harvest in northwestern Oregon: describing the epiphyte communities. *Northwest Sci.* **71**, 186–195.

Pepin, S. and Körner, C. (2002). Web-FACE: a new canopy free-air CO_2 enrichment system for tall trees in mature forests. *Oecologia* **133**, 1–9

Perry, D.R. (1978). A method of access into the crowns of emergent and canopy trees. *Biotropica* **10**, 155–157.

Remsen, J.V. Jr. and Parker, T.A. III. (1984). Arboreal dead-leaf-searching birds of the neotropics. *Condor* **86**, 36–41.

Richards, P.W. (1954). "The Tropical Rain Forest." Cambridge University Press, Cambridge.

Rinker, H.B., Lowman, M.D., and Moffett, M.W. (1995). Africa from the treetops. *Amer. Biol. Teacher* **57**, 393–401.

Rook, D.A., Grace, J.C., Beets, P.N., Whitehead, D., Sanantonio, D., and Madgwick, H.A.I. (1985). Forest canopy design: biological models and management implications. *In* "Attributes of Trees as Crops" (M.G.R. Cannell and J.E. Jackson, Eds.), pp. 507–524. Abbots Ripton, Huntingdon.

Schowalter, T.D. (1989). Canopy arthropod community structure and herbivory in old-growth and regenerating forests in western Oregon. *Can. J. For. Res.* **19**, 318–322.

Schowalter, T.D., Webb, J.W., and Crossley, D.A. Jr. (1981). Community structure and nutrient content of canopy arthropods in clearcut and uncut forest ecosystems. *Ecology* **62**, 1010–1019.

Shaw, D.C., Freeman, E.A., and Mathiasen, R.L. (2000). Evaluating the accuracy of ground-based hemlock dwarf mistletoe rating: a case study using the Wind River Canopy Crane. *West. J. For Appl.* **15**, 8–14.

Shuttleworth, W.J. (1989). Micrometeorology of temperate and tropical forests. *Philos. Trans. R. Soc. London, Ser. B.* **324**, 299–334.

Sillett, S.C., McCune, B., Peck, J.E., Rambo, T.R., and Ruchty, A. (2000). Dispersal limitations of epiphytic lichens result in species dependent on old-growth forests. *Ecol. Appl.* **10**, 789–799.

Sillett, S.C., and Van Pelt, R. (2000). A redwood tree whose crown is a forest canopy. *Northwest Sci.* **74**, 34–44.

Skutch, A.F. (1992). "A naturalist in Costa Rica." University Press of Florida, Gainesville.

Sørensen. L.L. (2003). Stratification of the spider fauna in a Tanzanian forest. *In:* "Arthropods of Tropical Forests: Spatio-Temporal Dynamics and Resource Use in the Canopy" (Y. Basset, V. Novotny, S.E. Miller, and R.L. Kitching, Eds.), pp. 92–101. Cambridge University Press, Cambridge.

Spies, T. (1998). Forest structure: a key to the ecosystem. *Northwest Sci.* (Special Issue): **72**, 34–39.

Spies, T.A., and Franklin, J.F. (1991). The structure of natural young, mature, and old-growth Douglas-fir forests in Oregon and Washington. *In* "Wildlife and Vegetation of Unmanaged Douglas-Fir Forests (L.F. Ruggiero, K.B. Aubry, and M.H. Brooks, Eds.), pp. 533–544. Pacific Northwest Research Station, Portland.

Stork, N.E., Adis, J.A., and Didham, R.K. (Eds.) (1997). "Canopy Arthropods." Chapman and Hall, London.

Takano, T., Nakamura, K., and Watanabe, M. (2002). Urban residential environments and senior citizens' longevity in megacity areas: the importance of walkable green spaces. *J. Epidemial Comm. Health* **56**, 913–918.

Thich Nhat Hanh. (1995). "Living Buddha, Living Christ." Riverhead Books, New York.

Thiollay, J.M. (1989) Area requirements of the conservation of rain forest raptors and game birds in French Guiana. *Conserv. Biol.* **3**:128-137.

Trichon, V. (2002). Monkeys as canopy collectors. *In* "Forest Canopy Handbook" (A.W. Mitchell, K. Secoy, and T. Jackson, Eds.), pp. 11–12. Global Canopy Programme, Oxford.

Ulrich, R.S. (1984). View through a window may influence recovery from surgery. *Science* **224**, 420–421.

Ulrich, R.S., Simons, R.F., Losito, B.D., Fiorito, E., Miles, M.A., and Zelson, M. (1991). Stress recovery during exposure to natural and urban environments. *J. Environ. Psych.* **11**, 201–230.

Webb, L.J. (1959). A physiognomic classification of Australian rain forests. *J. Ecol.* **47**, 551–570.

Whitworth, D.L., Nelson, S.K., Newman, S.H., Van Vliet, G.B., and Smith, W.P. (2000). Foraging distances of radio-marked Marbled murrelets from inland areas in southeast Alaska. *Condor* **102**, 452–456.

Winchester, N., and Ring, R.A. (1999). The biodiversity of arthropods from northern temperate ancient coastal rainforests: conservation lessons from the high canopy. *Selbyana* **20**, 268–275.

Zwally, H.J., Schutz, R., Abdalati, W., Abshire, J., Bentley, C., Bufton, J., Harding, D., Herring, T., Minster, B., Spinhirne, J., and Thomas, R. (2002). ICESat's laser measurements of polar ice, atmosphere, ocean, and land. *J. Geodynamics* **34**, 405–445.

CHAPTER 2

Tropical Microclimatic Considerations

Stephen R. Madigosky

The morning wind forever blows, the poem of creation is uninterrupted;
but few are the ears that hear it.
—*Henry David Thoreau*, Walden, Where I Lived, and What I Lived For, *1854*

Introduction

The significant role of rainforests in regulating local, regional, and global climate has become a focus of contemporary environmental research. Initially, interests were in defining rainforests in the context of a global system, essentially a physical ecosystem approach. Much of the early work concentrated on gross descriptive accounts, structural characteristics, and taxonomic aspects of the rainforest environment. Richards' (1952) classic account of this environment helped place perspective on what had been accomplished to that point and paved the way for those addressing aspects of rainforest climate. However, establishing the link between rainforest vegetation and climate would prove to be difficult. The architectural complexity of rainforests, the lack of reliable instrumentation that can perform under high temperatures and humidity, and the operational constraints imposed by trees measuring 30 to 80 meters or more are considerable obstacles to overcome. Yet progress has been made on all of these fronts and information retrieved within the last several decades by canopy scientists has helped greatly increase our understanding of this environment.

As technological advances began to provide the methods and tools needed to assess tropical forest climate, an approach to treat the entire canopy as a bulk unit resulted (Rosenberg et al. 1983). Oke's (1987) text on boundary layer climates helped provide a theoretical framework for which all aspects of climate could be assessed. Since then, a wealth of information has materialized addressing a conglomerate of topics related to rainforests as they impact climate at all levels and scales (Nenderson-Sellers 1987; Molion 1987a, 1987b; Paegle 1987; Shukla 1987; Lauer 1989; Richey et al.1989; Maslin and Burns 2000). Currently, a push toward understanding rainforest microclimate is underway, as seen in the eclectic nature of articles published on this topic over the past two decades (Shuttleworth et al. 1988; Kira and Yoda 1989; Shuttleworth 1989; Wright et al. 1992; Parker 1995; Cabral et al. 1996; Culf et al. 1996; Freiberg 1998; Williams-Linera et al. 1998; Szarzynski and Anhuf 2001).

A major interest that has materialized concerns how the modification of rainforest environments may be linked to local, regional, and global climate change. In this respect, a number of attempts utilizing models have been employed to assess such disturbance (Dickinson and Henderson-Sellers 1988; Lean and Warrilow 1989; Shukla et al. 1990; Chu et al. 1994; Grace et al. 1996; Kruijt et al. 1996; Lean et al. 1996; Nobre et al. 1996; Wright et al. 1996). Given the current state of tropical deforestation and the wholesale alteration of land throughout the tropics, an urgent need exists to assess the consequences of such activity (Madigosky and Grant 1996; Whitmore 1997). This has served as the impetus for critically examining many aspects of the rainforest environment.

The information presented in this chapter is an attempt to review key aspects of forest micro-climate as it relates specifically to tropical forests, including intracanopy structure and dynamics, temperature and humidity gradients, radiation budgets, canopy/wind interactions, along with reference to canopy light gaps. Temperate canopy architecture and aspects of microclimate have been treated in other works (e.g., Runkle 1985; Shuttleworth 1989).

Architectural Consideration of Rainforest Canopies

The size, shape, and density of canopy elements help to determine the degree of rainfall interception, vertical radiation regime, and corresponding humidity and wind field profiles. In this respect, elements of structure (the vertical and horizontal architecture) and natural forces that act upon such (wind, rainfall, humidity, and carbon flux) are used to characterize and define forests. Although rainforest canopies are one of the most biologically diverse and complex systems on the planet, they cover only some 6 percent of all terrestrial lands (Oldfield 2002).

It is little wonder that attention is now being focused on this environment. In fact, the interest in defining canopy structure may be pieced together by exploring the abundance of literature compiled just within the past several decades (Tomlinson and Gill 1973; Hallé 1974; Dransfield 1978; Hallé 1978; Hallé et al. 1978; Jenik 1978; Tomlinson 1978; Bourgeron 1983; Oke 1987; Tomlinson 1983). Aligned to this is the movement to reconcile structural elements to function and forest processes (Hubbell and Foster 1986; Shuttleworth 1989; Roberts et al. 1993; Parker 1995) as well as their relevance to global climate change (Richey et al. 1989; Shukla et al. 1990; Walther et al. 2002).

An understanding of basic forest architecture helps to assess internal aspects of canopies. Hallé and Oldeman's (1970) classic description of tree growth characteristics helped establish the basis for classifying diverse models as a point of reference. Since then, the idea of an architectural continuum has emerged where species garner greater significance when placed in ecological context. This seems reasonable since growth throughout the canopy is adaptive and, therefore, may result from forces other than genetic predisposition. This interplay has been viewed as reiteration or a growth response to an environmental cue. When viewed collectively, architecture and reiteration forces provide a more complete picture of canopy dynamics.

Collectively, canopies are multi-dimensional structures that can be defined according to space and time. They possess enormous vertical and horizontal complexity. They are a culmination of evolution, the result of environmental and biological interactions working together in ways that are difficult to interpret. Moreover, in tropical forest canopies, the degree of diversity adds yet another dimension of complexity. Here, some of the highest numbers of species have been recorded, each contributing, often subtly, to aspects of forest structure.

An intriguing characteristic of tropical forests is the degree of vertical dimension: trees towering to 30 to 84 meters. Because of their immense vertical size, rainforest trees create zones or niches that may be recognized by the conditions they support. The height and the nature of the architecture exert a considerable influence on heat exchange, mass, and momentum within the canopy. Since most exchange occurs between leaves at various heights, the density of leaves at these levels is an important factor to consider. Because large trees have a multi-layered structure, this sets up different meteorological conditions at different heights.

Spatial heterogeneity, as typically encountered within tropical forests, is due in part to the different ages of trees and succession events that occur as a result of tree fall and other forest disturbance. In forests where disturbance is routinely encountered, a mosaic of patches (the presence of different species) is apparent, often a result of gaps formed at different times (Bourgeron 1983). This can be visually noted from a vantage-point overlooking the upper canopy as when flying over primary rainforest. A patchwork of greens of different shades with an occasional flowering tree gives a two-dimensional clue of the extent of heterogeneity (Richards 1952; Whitmore 1975; Hubbell 1979; Clark 1994; Denslow and Hartshorn 1994; Lieberman and Lieberman 1994). Hubbell (1979) noted low adult tree density with uniform spatial distribution among tropical tree species in the dry forests of Costa Rica. An explanation given for such dispersion pointed to the possibility of seeds being removed from the mother plant by predators, thus lowering the over-all number of individuals of a particular species (Janzen 1970). In such instances, a resultant void occurs in which other species have the potential to invade, thereby increasing diversity. However, in this instance, the demographic examination of Hubbell's plot did not substantiate this claim (Hubbell 1979). Nonetheless, Janzen's suggestion has merit as an explanation of how seed dispersal promotes heterogeneity of species in tropical and temperate forests.

The spatial distribution of species throughout tropical canopies can also be explained by recent and historic land use practices (Jordon 1987; Saldarriaga 1987; Scott 1987; Unruh 1991). The discovery of charcoal and artifacts at the La Selva Biological Station (Horn and Sanford 1993) suggests that this forest was probably modified sometime within the past 1100 to 2430 years (Clark 1994). Consequently, any assessment of plant demography should reflect such activity. The composition and structure of the forests may also be a result of ecological responses to recent climate change. There appears to be supporting evidence that a broad range of organisms with diverse geographical distributions may be changing in some biomes (Walther et al. 2002; Post 2003). A number of informative works have been published on the Area de Conservacion Guanacaste, and Janzen's (2002) most recent work provides an informative chronology of forest succession on tropical dry forests of northern Costa Rica.

Forest structure can also be analyzed through careful examination of such physical properties as edaphic variation. Clark (1994) lists four soil types that differ in nutrient composition from the La Selva Reserve and discusses the presence/absence of particular plant species as a result of these differences. Additionally, evidence suggests that the distribution or clumping of plants within tropical forests may be a function of topographic heterogeneity (Lieberman and Lieberman 1994). Variation in seasonal parameters can also impart considerable influence on the structural development of forests (Garcia-Martino et al. 1996). Each of these factors contribute to the diverse spatial heterogeneity—hence, structure—of the rainforest canopy in unique and unusual ways.

Temperature within the Tropical Canopy

Considerable information has been published on aspects of temperature and humidity profiles in tropical forests (Richards 1952; Kira and Yoda 1989; Windsor 1990; Sanford et al. 1994; Wright et al. 1996). Although tropical canopies display great heterogeneity, little spatial variation exists in the amount of radiation received over a defined horizontal surface. However, this can change, especially concerning short wavelengths, since the amount of radiation produced by the sun is not constant. Excluding this fact, there are usually only slight differences in temperature above and within the immediate reaches of canopy. Exceptions arise from the creation of large-scale gaps occurring naturally or more often anthropogenically. With the loss of dominant vegetation, wide disparity in temperatures may be encountered as compared to those from undisturbed tropical forests (Shukla et al. 1990; Unruh 1991; Culf et al. 1996). Typically, the average intracanopy temperature reported for undisturbed equatorial rainforests falls within 24°C to 27°C (Lauer 1989). Variation in weekly, monthly, and yearly temperature profiles in this zone is negligible (see Figure 2-1). Even at the June solstice, the noon sun is never less than 63.5° above the horizon at latitude 3°S. This assures that equatorial zones receive a surplus of radiation as compared to zones at higher latitudes. As one progresses north or south from the equator, the change in

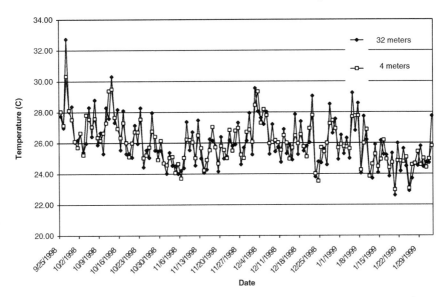

Figure 2-1 Typical profiles depicting mean daily temperature at 4 and 32 meters at the Amazon Conservatory for Tropical Studies (ACTS), Iquitos, Peru (S3° 14′ W72° 55′). (From S.R. Madigosky and I. Vatnick, unpublished data.)

temperature amounts to approximately 1°C per 1000 km as compared to 5°C per 1000 km in middle latitudes (Lauer 1989). The static temperatures are undoubtedly a result of the consistent day length and solar radiation received year round in these regions.

Changes in vertical temperature and net radiation regimes within tropical canopies have been scrutinized from a variety of locations (Molion 1987a; Shuttleworth 1989; Lyra et al. 1992; Barker 1996; Madigosky and Vatnick 1999; Szarynski and Anhuf 2001). Typically, ground-level temperatures remain about 0.5°C to 2.5°C cooler than those encountered at the top of the canopy. The slightly lower temperatures near the forest floor correspond to reduced light (short wave radiation) entering these areas. Hence, average daily temperatures tend to rise with an increase in height above the forest floor. This relates directly to the density of vegetation restricting passage of short wave radiation and also corresponds to the thermal mass associated with the entire canopy system. Together with the enhanced emission of long wave radiation, air temperatures are quite stable.

The greatest variation between ground and upper canopy temperatures is experienced between 4 PM and 6 PM after incoming radiation has had ample time to exert its maximum effect on the upper canopy where incident radiation is greatest. An inspection of hourly temperatures obtained from a primary rainforest in northeast Peru reveals the dynamic nature that occurs in a typical day (see Figure 2-2). Distinct diurnal shifts in temperature are encountered throughout the entire canopy and often conform to a polynomial expression of third order or higher. The temperature profiles, most notably in the lower reaches of the canopy, are routinely inverted. In such instances, minimum air currents pass throughout the lower third of the canopy, thus partially isolating the understory, limiting air exchange between horizontal and vertical surfaces. Episodes such as those brought about by sudden downpours or forced convection cells will sometimes break the stagnation and help circulate air. The presence of cloud cover, especially over prolonged periods, may also lessen conditions that promote inversions. Clouds reduce incoming short wave radiation entering the canopy and thus may dramatically distort the distinct thermal profile maintained within tropical forests. In instances where frontal systems envelop the forest over a period of several days, the difference in temperature throughout the intracanopy is negligible or nonexistent.

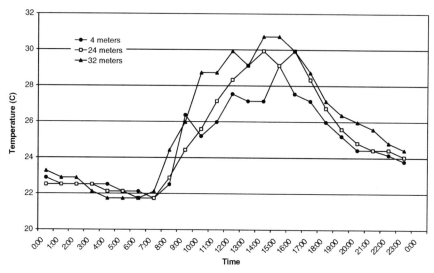

Figure 2-2 Daily temperature profiles at elevations 4, 24, and 32 meters recorded at the Amazon Conservatory for Tropical Studies (ACTS), Iquitos, Peru (S3° 14′ W72° 55′) 21 June 2002. (From S.R. Madigosky and I. Vatnick, unpublished data.)

The movement of air circulating in the upper canopy during the day will vary according to the general atmospheric conditions, crown architecture, leaf area index, and surface topography. The greater heating of the upper crown surface allows this region to display higher temperatures than in the cooler denser air within the lower third of the canopy (Szarzynski and Anhuf 2001). At night, the absence of solar radiation exacerbates radiative cooling throughout the canopy that advances thermal stability (a decoupling) from the overlying atmosphere. Processes that suppress such cooling (e.g., cloud cover, storm convection, and turbulence) can restrict and even prevent thermal exchange from occurring altogether.

Canopy Humidity

The absolute humidity within rainforests is maintained at high levels nearly all the time. However, relative humidity (the amount of water held by air at a given temperature relative to the saturation point) varies inversely with temperature; as a result, diurnal and annual variation is noted when temperature change occurs (see Figure 2-3). Yet, most rainforests maintain high moisture conditions throughout the year, even when rainfall is scarce or absent. Dense rainforest vegetation plays an important role in maintaining the moisture in the soil as well as throughout the intracanopy. Because of this, it is a rare occasion when tropical forests fall below 75 percent relative humidity (Lauer 1989).

Generally, higher relative humidity levels are encountered near the forest floor where temperatures tend to be lowest (see Figure 2-4). In combination with restricted airflow, this sets up conditions that support and maintain high moisture and minimal evaporation at these levels. For full appreciation of the degree of change that occurs in tropical forests, it is necessary to monitor changes in humidity on an hourly basis over extended periods. Such observations, albeit rare, provide a precise picture of the diurnal nature within the canopy. In one such instance, relative humidity monitored over a nine-year period in a lowland tropical forest in Peru indicated tremendous uniformity with levels rarely dropping below 70 percent in the lower

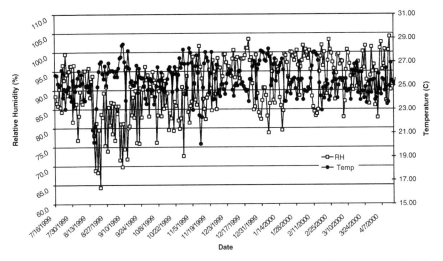

Figure 2-3 Temperature and relative humidity profiles recorded at the Amazon Conservatory for Tropical Studies (ACTS), Iquitos, Peru (S3° 14′ W72° 55′) 21 June 2002. (From S.R. Madigosky and I. Vatnick, unpublished data.)

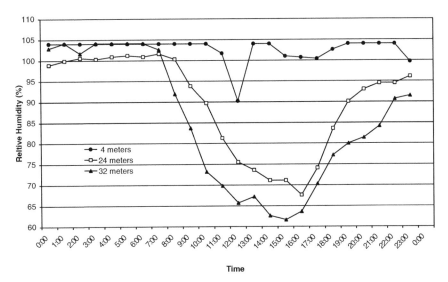

Figure 2-4 Relative humidity (RH) profiles at elevations 4, 24, and 34 meters recorded at the Amazon Conservatory for Tropical Studies (ACTS), Iquitos, Peru (S3° 14′ W72° 55′) 21 June 2002. (From S.R. Madigosky and I. Vatnick, unpublished data.)

canopy (personal observation). While changes do occur on a daily basis, they are negligible and are at or near saturation much of the time.

The spatial and temporal distribution of relative humidity near the upper boundary layer of the canopy is a great deal more variable than on the forest floor. Nighttime temperatures result in the highest levels of relative humidity, and they are routinely at or near saturation. At daybreak, however, conditions reverse as the sun exerts its influence at the top of the irregular canopy boundary. This accentuates the different relative humidity values vertically throughout the forest. The difference in relative humidity at the two extremes (upper boundary and forests floor) often varies by 5 to 20 percent (Windsor 1990), and is at its maximum sometimes around 11:00 AM to 3:00 PM (Dolman et al. 1991). This, of course, will vary for different locations depending on the individual site characteristics, such as canopy architecture, leaf density, and other related features. The maximum difference in relative humidity recorded at other lowland rainforest sites can be as high as 50 to 60 percent (Kira and Yoda 1989). Nonetheless, only a careful examination at all levels of the canopy will reveal the true nature of diurnal variation for a particular location.

At times, when seasonal rainfall diminishes within the Amazon Basin (in July to September at S3°W72°), relative humidity is still maintained within 50 percent of saturation and commonly much higher than that (Madigosky and Vatnick 1999). This undoubtedly is site-specific and contingent upon the fluctuation of seasonal rainfall. For instance, monthly precipitation patterns in La Selva, Costa Rica (N10°26' W83°59'), Banco National Park, Ivory Coast (N5°15' W3°56'), San Carlos, Venezuela (N1°54' W 67°03'), and Barro Colorado Island, Panama (N9°09' W79°51') (as cited in Sanford et al. 1994) show distinct seasonal patterns in rainfall. It is reasonable to infer high forest humidity gradients when periods of high rainfall are encountered. Windsor's (1990) report on Barro Colorado Island supports this premise.

Canopy/Wind Interactions

Ascertaining just how a canopy system will react to turbulence is difficult. Many factors help direct or distort the movement of air currents above and within the intracanopy of the tropical forest environment. Some of them include the vertical nature of the canopy, the three-dimensional geometry of the stand, species composition, ground surface topography, the irregular or rough nature of the upper boundary layer, foliage bend in response to wind generated forces, and upper surface canopy albedo. Collectively, these factors exert considerable influence on the upper surface boundary of the canopy and act to distort the natural flow of air. They often help break the streamline or parallel flow of air riding over the forest into a jumble of eddies and vortices characteristic of turbulent flow.

The wind field in the upper boundary layer of the tropical forest canopy is controlled by frictional drag imposed on the freely moving air by the myriad of densely packed leaves and the rigid architecture of the stand. This helps considerably to direct the movement of air. The roughness or irregularity of the upper boundary layer is important in determining the extent of flow into the lower reaches of the canopy. Greater frictional drag is encountered with increasing depth of penetration into the canopy until little to no effect is evident at or near ground level. The extent of the vertical profile and density of stems per unit area help determine how and where wind is diminished (Dolman 1986). This, of course, assumes that other forces such as thermal convection are negligible. In addition, there is considerable variation in the structure of different rainforest zones, some displaying minimal cover near the lower reaches of the canopy. In this instance, typical wind patterns tend to break down (McNaughton 1989). Forests possessing very thin crowns also offer little resistance to turbulence extending into the middle reaches of the canopy (Leuthart et al. 1994).

Nonetheless, tropical forests are very difficult to characterize because they possess such a complex multifaceted vertical and horizontal structure. What applies to one canopy may have little relation to that of others, even with similar vegetation in the same basic area. A thorough understanding of the vertical and horizontal nature of structural dynamics of the forest is essential. Although imperfect, methods exist to approximate vertical wind flow over a forested area, although conditions of neutral atmospheric stability are assumed. One of the most noted and accurate is the logarithmic wind profile equation that depicts vertical logarithmic decay. This expression conforms to the following:

$$\bar{u}_z = \frac{u^*}{k} \ln \frac{(z - d)}{z_0} \qquad (1)$$

where u_z represents the mean wind speed (m s^{-1}) at some height above the ground z, u^* is the friction velocity (m s^{-1}), k denotes Von Karman's constant (0.41), z_0 the roughness of the surface (m), and d the zero plane displacement, which is the level of bulk drag exerted by the vegetation. d is also linked to wind speed since vegetation has the tendency to respond proportionately to wind speed. Since shearing stress is proportional to the square of the wind velocity at a given height, the term u_* can be brought in as a means of expressing the domain under which the square law holds true such that:

$$u_*^2 = \tau/\rho \qquad (2)$$

where (τ) is the air forces riding over the surface that exert a shearing stress that can be expressed as a pressure (ρ) (force per unit surface area). The term u_* may then be evaluated from wind profile measurements to determine τ to further explore other flux measurements. Another way to characterize wind movement is to consider aerodynamic conductance g_a. Here:

$$u_*^2 = \frac{u_*^2}{u} \qquad (3)$$

where u denotes wind speed above the canopy (m s^{-1}) and u_* the friction velocity (m s^{-1}) that may be generated from the wind profile equation by taking several height measurements above the canopy. A more complete discussion of the effect of three-dimensional aerodynamic resistance on turbulent flow along with bulk boundary layer resistance within canopies is presented in Oke (1987) and Sellers (1987).

For the most part, severe turbulence is encountered infrequently within tropical rainforest canopies, especially within the lower reaches of the canopy (Molion 1987b; Kira and Yoda 1989). Yet when a strong wind field envelops a tropical forest, the effects can be devastating. In one instance, over 1 percent of emergent trees were felled as a result of an ephemeral squall recorded at the Amazon Conservatory for Tropical Studies (ACTS), Iquitos, Peru (S3°14′44″ W72°55′28″) in November of 2002 (personal observation). This event was quite significant since the degree of disturbance reported for tropical canopies amounts to 1 to 2 percent annually (Runkle 1985; Denslow 1987; Whitmore 1991). Although this type of event is unusual at this latitude, it can dramatically alter the physical nature of the forest and, consequently, the microclimate for many years. In fact, Hartshorn (1980) suggested that these events are essential for emergent species to reach maturity. Typically, the dense rainforest canopy acts as a substantial barrier against such events. Yet strong winds coupled with saturated soil conditions—as those seasonally found in a rainforest environment—leave shallow rooted trees vulnerable to collapse.

More subtle turbulence can result from the pronounced diurnal cycle in many rainforests. Winds tend to be weaker during evening hours owing to nocturnal cooling of the atmosphere throughout the canopy. As thermal gradients that drive horizontal turbulence weaken, corresponding pressure gradients diminish, thus reducing the wind field.

By day, the Earth's surface is heated by the sun; consequently, there is an upward movement or transfer of heat into the cooler atmosphere above the canopy. As a result, substantial mixing occurs vertically, and this influence may extend a considerable distance above the upper reaches of the canopy. Within the context of the intracanopy, mixing of thermal and moisture profiles occurs within the upper two-thirds of the forest with little change in the lower reaches of the forest (Szarzynski and Anhuf 2001). The vertical movement of air throughout the intracanopy environment serves to link the deeper recesses of the forest with the upper atmosphere. Although movement is often restricted, there are instances such as precipitation events that perpetuate vertical air movement from the lower to the upper levels of the canopy. In such cases, it appears that the thermal mass of the lower region allows convective movement upward. The transfer of latent heat and saturated air then pass into the upper reaches of the canopy and eventually into the freely moving atmosphere (Leuthart et al. 1994).

Conversely, at night, when the Earth's surface begins to cool more rapidly than the surrounding atmosphere, a downward transfer of heat occurs. Radiative cooling of the Earth's surface causes a decoupling from the overlying atmosphere, resulting in a stable density profile above the canopy (Szarzynski and Anhuf 2001). Several hours after sunset, moisture-laden air pushes downward and begins to condense, appearing as clouds descending within the forest. This movement, although slight, may account for vertical mixing between the upper boundary layers of the canopy. At sunrise, the reverse takes place, lifting the moisture back into the upper reaches of the canopy and beyond.

A few reports suggest that wind speed profiles near the upper boundary surface of the rainforest canopy reveal very little resistance to wind (Kira and Yoda 1989; Leuthart et al. 1994), but this can be a result of the thinning nature of the crown in some emergent trees. Turbulence within the upper reaches of the canopy can also be affected by slight variations in vegetative cover. The surface albedo of the upper boundary often varies from location to location owing to the collage of colors or contrasting surface types displayed by different species within the forest. Light gaps created by tree fall or topographic erosion can exacerbate changes in canopy temperature profiles. This variation has the potential to create small-scale temperature anomalies that can cause differences in horizontal pressure resulting in air movement or wind. This is one way in which thermal energy is converted into the kinetic energy of a wind system. The kinetic energy is available to participate in a cascade of events that eventually transfers energy at microscale levels into air movement or turbulence (eddies and vortices) throughout the canopy.

Neotropical Fronts and Cold Episodes

The daily temperature profile in undisturbed equatorial lowland rainforests is typically regarded as stable and, at times, even monotonous. This is because there is little seasonal variation in sun-earth geometry throughout the year (Turton et al. 1999). Consequently, daily, seasonal, and yearly temperature regimes are much less variable compared to biomes at higher altitudes and latitudes. Temperature profiles within undisturbed Amazonia lowland rainforest, although relatively stable, follow a daily cycle that routinely changes by only 5°C to 10°C per 24 hours (Madigosky and Vatnick 1999; Szarzynski and Anhuf 2001). Such change is achieved as a result of unreflected solar radiation, the extent of rainfall, presence of cloud cover, surface albedo of the canopy, and the extent that fronts envelop extended areas of forest. Distinct seasonal changes throughout much of lowland Amazonia, although slight, are brought about by extensive cloud cover that drives processes to influence the temperature and corresponding wind movement within the forest. In an undisturbed forested area in northeast Peru (S3° W72°), the last eight years of temperature data indicate that certain months are notably cooler than in others (personal observation). It is in these periods that the abundance of rainfall tends to be greatest and the incidence of solar radiation (cal cm^{-2} day^{-1}) reaching the canopy somewhat diminished.

The prolonged and continuous presence of cloud cover and rainfall enables temperatures within the forest to drop by several degrees over extended periods. Temperature profiles reported by others working near Manaus, Brazil (S3° W60°), indicate similar patterns. The average global solar radiation in this region has been reported to be almost 500 cal cm^{-2} day^{-1} in

September and October (mean monthly average) but substantially lower in February (267 cal cm^{-2} day^{-1}) (Ribeiro et al. 1982). Research conducted by Leopoldo et al. (1993) at the Ducke Forest Reserve 26 kilometers north of Manaus also showed a relation between temperature and insolation at different months over a three-year period. The subtle differences outlined above are important in ascertaining complex biological cycles that occur within tropical forests, yet there are few extended datasets available in such regions for researchers to consult.

There are also dramatic short-term cooling episodes where large expanses of equatorial forests are blanketed by frontal systems that cause significant temperature depressions. Typically, the Amazon Basin maintains surface temperatures of near 25°C to 27°C (Parmenter 1976). When regional cold fronts originating from the south push through the forest, temperatures may drop by as much as 10°C to 15°C. These episodes usually are encountered in July and August near the neotropical equatorial zones and last for two to seven days (see Figure 2-5). During these periods, animals, especially poikilotherms, are scarcely observed as they take cover in the underbrush in an attempt to conserve energy resources. In addition, well-established vertical temperature isotherms normally observed within the forest at various levels are temporarily eliminated during these episodes only to return when the frontal system has passed. Strong downpours and low stratiform cloudiness often accompany this phenomenon over extensive areas of forest.

On much rarer occasions, strong frontal systems originating from the southernmost regions of South America bring freezing temperatures into the inner tropical rainforest regions. Such events are unusual, causing temperatures to plummet below freezing in areas that harbor tropical rainforest vegetation. Such episodes can wreak havoc both in lowland tropical areas and in elevated Andean agricultural zones. One such event occurred on July 12, 1975, first entering Chile and Argentina and eventually crossing the equator nine days later (Parmenter 1976). Freezing temperatures were recorded in southern Brazil on two successive nights with much of the mainland experiencing 15°C temperatures and winds varying from 3 to 10 m/s. In rainforests located in marginal tropical zones influenced by coastal trade winds (southern Brazil, northeastern Australia, and the Philippines), the coldest monthly average temperatures may drop to 18°C and at times 10°C (Lauer 1989).

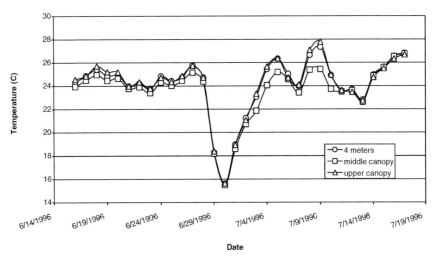

Figure 2-5 Temperature profiles during a frontal system at the Amazon Conservatory for Tropical Studies (ACTS) Iquitos, Peru, 1996. Similar periods of depressed temperatures have been observed each year from 1994 to the present. (From S.R. Madigosky and I. Vatnick, 1999, *Selbyana*, reprinted with permission.)

Forest Gap Dynamics

Researchers have historically recognized the significance of natural disturbances as part of an overall process that helps contribute to and maintain aspects of forest structure and community dynamics. The notion that forests age in a particular manner is now well accepted, but the degree to which internal mechanisms such as tree fall direct such activities has just recently gained attention. In the tropics, early studies tended toward how forests respond subsequent to anthropogenic disturbance (Ewel 1970, 1976). The importance of Ewel's work centers on how remnants of natural vegetation may be utilized and managed to maximize an altered landscape for future generations. Given the status of tropical deforestation, this continues to be a valuable topic for theoretical and practical pursuit (Williams-Linera et al.1998).

Although valuable, these studies do not address natural forest regeneration processes that arise from tree fall openings. The necessity to unravel this puzzle has been pursued by individuals in a variety of locations throughout neotropical America (Grubb 1977; Hartshorn 1978, 1980; Denslow 1980; Orians 1982; Augspurger 1984; Vitousek and Denslow 1986; Brokaw 1982, 1985; Augspurger and Franson 1988; Lawton 1990). The extensive research conducted on this topic has cumulatively helped establish important features of tree fall, light gaps, and patch dynamics to natural succession and maturation of forested communities (Runkle 1985; Canham 1988; Canham et al. 1990; Lawton 1990; Belsky and Canham 1994; Emborg 1998). The creation of gaps within forests is one of the most important events in the natural aging process of a forest. Gaps play an important role in establishing resource allocation, in altering microclimate dynamics, and ultimately in determining how effectively resources are recycled and distributed. Gaps also play a profound role in promoting seed germination. In one instance, Lawton and Putz (1988) showed that seeds buried prior to gap formation have a much better chance of actually producing seedlings than seeds dispersed subsequent to the creation of a gap. They also noted that many seedlings originate as epiphytes in tree crowns prior to becoming established as a result of a disturbance such as tree fall (gap formation).

Microclimate and Gaps

Many factors help determine the nature of the microclimate in a forest gap. The most obvious is the size of the gap. In general, the larger the gap. the greater the degree of influence on forest microclimate (Lee 1978; Chazdon and Fetcher 1984; Brokaw 1985; Lawton 1990; Denslow and Hartshorn 1994). For instance, the amount and intensity of light tends to be greater in a large gap than in smaller forest openings. As a result, the temperature of large gaps is considerably greater, most notably toward the central portion of the opening. Because the incidence of solar radiation is more intense in this area, humidity can also be influenced. Typically, relative humidity in undisturbed closed canopy primary rainforest can be at or near 100 percent for extended periods (Madigosky and Vatnick 1999). Usually, this is not the case in large well-developed gaps. Here, intense solar radiation elevates soil temperatures, thereby increasing soil evaporation and lowering relative humidity. The influence of wind within the canopy is also increased as a function of gap size. Clearly, gaps provide ecological benefits for species that favor open conditions, such as shade-intolerant species.

Many tropical tree gaps are formed as a result of disturbance driven by strong winds (Uhl et al. 1988), waterlogged soils (Brokaw 1982; Denslow and Hartshorn 1994), landslides (Garwood et al. 1979) and as a result of the natural aging process (Runkle 1985; Lawton and Putz 1988). These factors contribute dramatically to the vegetative mosaic encountered within tropical primary forests. The distribution of uneven-aged gaps is often characterized by the extent of the area they cover. Specifically, the creation of these disturbances is a function of size and numbers of trees felled, surface topography, canopy geometry, gap orientation, azimuth, and estimated turnover rate (average number of years between the creation of successive gaps). In most instances, the underlying factors for creating gaps of similar size and dimension are random. Consequently, the standardization of such disturbance is problematic (Brokaw 1985). Yet, there are instances where comparisons between gaps from different forest types have been made; gaps measured from tropical elfin forests tend to be smaller than those found in lowland tropical forests (Lawton and Putz 1988).

Considering structural complexity alone, Hallé et al. (1978) outlined 23 distinct architectural tree models, capitalizing on key diagnostic features. How trees within a particular architectural class contribute to tree gaps is largely unknown, but a diameter/height expression may offer a simple way to characterize forest openings related to tree fall. The expression D/H, where D is the diameter of the gap created by tree fall and H denotes the height of the surrounding forest, can be used to quantify the opening size of the gap (Geiger 1965).

Light Gap Parameters

One reason for characterizing gap openings is to assess the degree of light within various layers of the lower canopy. This is a difficult task because light entering these levels is ephemeral and has varied spectral quality. As a general rule, light received at lower levels via gaps is of high temporal variability, owing to the random nature of sunflecks (small patches of solar radiation) through which light enters. The amount of light entering gaps tends to be more heterogeneous than that received by well-covered understory vegetation (Fetcher et al. 1994). Consequently, the amount of useful radiation (400 to 700 nm) available to plants within gaps is best assessed on a case-by-case basis.

To some extent, functionally similar gaps may be compared in order to gain some semblance of consistent measurement in different forest types. Recent studies have indicated that the central portion of the gap has the highest percentage of light. Light analysis of large gaps in the forest of La Selva, Costa Rica, has shown 8.6 to 23.3 percent of full sunlight received near the center of 275 to 335 m^2 clearings on sunny days (Denslow and Hartshorn 1994). Peripheral zones were on the order of 2.8 to 11.1 percent of full sunlight, and surrounding understory conditions 0.4 to 2.4 percent full sunlight from the same area (Denslow 1980). Interestingly, these figures are in line with those reported for locations in the lowland tropics (Chazdon and Fetcher 1984) montane rainforests in Costa Rica (Lawton 1990), Hawaiian Island forests (Pearcy 1983), and even temperate forests in Denmark (Emborg 1998). The amount of light received within forests and the quality of that light helps determine the overall productivity at all levels and scales. Within the lower recesses of the canopy, wavelengths that are most useful for the photosynthetic process may often be lacking (Bjvrkman and Ludlow 1972). This absence not only reduces the incidence of functional light to germinating seeds, but it can also decrease gross primary production of plants already established. In a sense, gaps provide unexpected opportunity in the form of useful wavelengths for seed germination and a potential release mechanism for vegetation subdued by neighboring competitors.

On a much larger scale, clouds also serve to lower the transmission of direct sunlight to the canopy. The amount of blockage is proportional to cloud thickness and duration of coverage over an area. One way of conceptualizing the relationship between radiation and cloudiness is to determine the percent possible sunshine. This may be accomplished using the following expression:

$$\frac{R_s}{R_{so}} = a+b\,\frac{n}{N} \qquad (4)$$

where R_s denotes the amount of solar radiation received over a prescribed duration, R_{so} the theoretical solar radiation reaching Earth without the influence of an atmosphere, n the precise amount of time sunshine is received, N the theoretical duration (time) of sunshine, with a and b representing empirical constants (Rosenberg et al. 1983). Typical values for a and b will vary dramatically depending on location and altitude. Rosenberg et al. (1983) presented contours of percent sunshine over the United States between 20 and 90 percent depending on the latitude and season. The lack of this type of information for particular locations throughout the neotropics makes it difficult to relate the expression to tropical forests. It is possible, however, for well-monitored sites such as Barro Colorado Island in Panama to utilize such expressions. The recent availability of satellite data has potentially made information of this nature available for those seeking it.

Radiation Budget

The Earth receives 99 percent of its total radiation budget from the sun. Yet only a fraction of this—some 47 percent—is actually received on the surface of the Earth, and most of this within the equatorial regions between the Tropic of Cancer and the Tropic of Capricorn (N23° S23°). The rest of this energy is reflected back into space or is absorbed by the atmosphere. The amount of global radiation received above the atmosphere prior to absorption or scattering is 1367 watts/m^2 at a 90° angle on the side facing the sun (Graedel and Crutzen 1993). Since Earth radiates as a sphere but absorbs as a circle, 1367 watts/m^2 must be divided by 4 to calculate the energy budget. Hence, the amount of energy received is 341.75 watts/m^2, of which 102.5 is reflected back, on average. This represents a 30 percent reduction, which essentially equates to Earth's albedo. A graphic representation of this is depicted in Figure 2-6. To achieve radiation balance, approximately 239 watts/m^2 is absorbed and reemitted (Hoyt and Schatten 1997).

This is fairly consistent, but variation may occur as a result of a change in the ratio of direct and indirect light received in any one area. Thus, the amount of photosynthetically available radiation (PAR) reaching the canopy can be altered within short time periods contingent upon atmospheric influence. Not all radiation received on Earth is of the same nature. In addition, the absorption of solar radiation by the atmosphere is selective, as is that absorbed by canopy vegetation. Visible light (that used by the photosynthetic mechanisms of plants) is received nearly undiminished but can be altered by water droplets, ice droplets, particulates, and general atmospheric pollutants. Within the visible spectrum, blue light (short wavelengths) tends to be altered more easily than red light (longer wavelengths) because short wavelengths are preferentially scattered. Other forms of radiation such as ultraviolet are significantly reduced upon entering the upper atmosphere (stratosphere), and infrared is largely absorbed in the lower troposphere by carbon dioxide and water.

By far, clouds and general atmospheric conditions have the greatest impact upon the transmission of short wave radiation (wavelengths up to 3 µm—that is, a large part of infrared) and all of the wavelengths where insolation exceeds the Earth's thermal radiation, which peaks at about 10 µm. As much as 70 percent or more of incoming short wave radiation may be depleted in its passage through the atmosphere before actually reaching the top of the rainforest canopy (Leuthart et al. 1994). Haze, particulate matter, and a host of general pollutants all serve to further diminish the degree of short wave radiation reaching the canopy. Upon reaching the top of the canopy, the vast majority of short wave radiation is effectively absorbed by the vegetation.

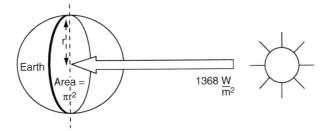

Incoming solar radiation

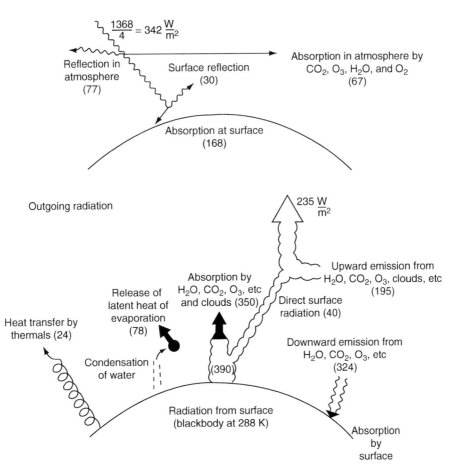

Figure 2-6 Global average mean radiation and energy balance per unit of Earth's surface. The numbers in parentheses are the energy in units of W^{m-2} typically involved in each path. (From Finlayson-Pits, B.J., and Pits, J.N., Jr. (2000). "Chemistry of the Upper and Lower Atmosphere," Academic Press, reprinted with permission.)

In fact, the transmission of short wave radiation into the canopy displays an almost logarithmic decay with increasing depth of penetration (Oke 1987). This may be in part a result of the irregular nature of the upper boundary of the canopy surface (Shuttleworth 1989) and the efficiency of vegetation in capturing radiation (Ghuman and Lal 1987). Nonetheless, once short wave radiation enters the intracanopy, there is minimal loss due to reflection and scattering within the vertical profile. Reports on how efficiently canopy vegetation is likely to absorb different forms of radiation show that short wave radiation and photosynthetically active radiation are absorbed within the top 20 meters of a 40-meter-high canopy system (Parker 1995). Leuthart et al. (1994) found that as much as 80 percent of the short wave radiation is eliminated in this upper zone. This figure remained consistent throughout their study period, regardless of the variability of incident short wave radiation at the top of the canopy.

In addition, Leuthart et al. (1994) indicate a reduction of more than 96 percent of the total short wave radiation just above the forest floor. These results are nearly identical to short wave extinction coefficients observed in the central Amazon region in Brazil (Leopoldo et al. 1993), rainforests in Sri Lanka (Ashton 1992) and pine forests in North Carolina (Gay et al. 1971). Additionally, they are very similar to vertical profiles of the transmittance of photosythetically available radiation found above 26 different forest locations (Parker 1995). Other studies have reported that as much as 99 percent of the total light flux entering a rainforest is eliminated before reaching the forest floor (Leigh 1975). Considering this information, it is not difficult to see why plants have evolved strategies to take advantage of gaps created by tree fall. The extreme depression of photosynthetically available radiation (380 to 750 nm) at the base of the forest floor makes it nearly impossible for plants to become established without the aid of some sort of opening or disturbance.

Little doubt remains as to the importance of light quantity extending through the forest. Of equal importance is the spectral quality of light that passes downward through the canopy, for it is these wavelengths that control the photosynthetic environment. The segment of electromagnetic radiation most useful for photosynthesis is a narrow band that ranges from 380 to 750 nm. This is the range where plants most effectively absorb electromagnetic radiation. Outside of this range (e.g., short wave radiation in the near infrared region), absorption may occur, but not as strongly. Correspondingly, leaves minimally transmit and reflect solar radiation between 400 and 1100 nm (Lee 1987). Therefore, the amount of electromagnetic radiation absorbed by any canopy is a function of the total leaf area and its ability to absorb, reflect, and transmit radiation. This process may be visualized in the following manner:

$$A = 1 - r - t \qquad (5)$$

where A is the absorptance, r the reflectance, and t the transmittance. In particular, dense forest canopies have a great capacity to assimilate solar radiation entering the upper boundaries, especially within certain ranges of the electromagnetic spectrum. This has profound implications for understanding growth and development processes within tropical forests and has been reviewed by several scientists throughout varied habitats (Ghuman and Lal 1987; Kira and Yoda 1989; Lawton 1990; Parker 2000).

The radiation budget of a forest canopy is difficult to determine owing to internal radiative absorption, transmission, reflection, and emission beyond the initial exchange of energy that occurs at the upper layer boundary surface. Since the transmission of short wave radiation decreases as a function of depth of penetration into the canopy, a modification of Beer's Law can be used to assess the attenuation of this radiation through the canopy using the following expression:

$$K \downarrow z = K \downarrow_0 e^{-aA_l(z)} \qquad (6)$$

where $K \downarrow z$ is the average short wave radiation at a particular canopy depth (z), $K \downarrow_0$ is the short wave radiation at the upper boundary of the canopy, $A_{l\ (z)}$ represents the leaf area from the top to a particular depth z, and a represents the coefficient of atmospheric extinction. The applicability of this equation in the context of a rainforest canopy may present some concerns owing to the heterogeneous nature of the canopy. Hence, the exponent $A_{I(z)}$ is not defined but can reasonably be regarded as a bulk transmission quantity such that $\beta = aA_{I(z)}$. To solve for bulk extinction and assess the logarithmic decay with depth of penetration of solar (short wave) radiation, as defined by Beer's Law, the following expression may be applied.

$$\beta = \ln \frac{[K \downarrow_z]}{K \downarrow_o} \tag{7}$$

This provides an expression for the exponent shown in the first equation, thus unifying leaf area accumulation and the basic atmospheric coefficient to a depth within the canopy.

As noted, a goodly amount of short wave radiation is absorbed upon reaching the canopy; the interception of light and its consequential scattering by vegetation is a reality that can be interpreted via a series of mathematical expressions. A comprehensive treatment of the expressions used to determine radiation balance is dealt with in reviews presented by Coakle and Chylek (1975), Meador and Weaver (1980), and Arya (1988).

A bit more difficult to discern is the fate of long wave radiation (3 to 100 μm) that enters the canopy, primarily due to the collage of vegetative types encountered in a primary rainforest canopy. Each surface is different in its ability to emit, absorb, reflect, and transmit radiation and therefore falls short as that of a perfect blackbody. Natural surfaces such as bulk canopy leaves are "gray" radiative surfaces, each with its own characteristic pattern of output. This, too, has the potential of changing with height. The emission of long wave radiation flux (R) from such surfaces as leaves can be determined through the use of the Stefan-Boltzmann Law:

$$R_L = -\varepsilon\sigma T^4 \tag{8}$$

where ε is the emissivity for different surface types, σ the Stefan-Boltzmann constant, and T the temperature of the emitting body (in Kelvin). For clear-sky conditions, the expected emissivity usually ranges between 0.6 and 0.73 (Oke 1987). If, however, cloudy conditions prevail, as they often do in tropical forests, the canopy dynamics may be altered dramatically. Clouds essentially emit at blackbody capacity; consequently, atmospheric emissivity is pronounced during these periods and may remain so for extended periods (Leuthart et al. 1994). At the top of the canopy the net long wave radiation is most frequently at a loss, but this tends to lessen vertically with penetration into the canopy (Arya 1988).

Precipitation

Forests trap, use, transport, and distribute considerable quantities of water on both temporal and spatial scales. In particular, rainforests are one of the most efficient systems for retaining and recirculating fresh water on the planet. Still, there is substantial disparity in the amount of rainfall received each year. For instance, the tropical forests on Barro Colorado Island, Panama, receive on average 2612 mm of rainfall annually; however, records as low as 1679 mm and as high as 4133 mm have been recorded (Windsor 1990). In general, tropical rainforests usually receive considerably more rainfall compared to other biomes. The average amount of annual precipitation received globally is approximately 1040 mm (Oke 1987), whereas rainfall recorded throughout the Amazon Basin varies from about 2000 to nearly 4000 mm annually (Salati 1987). A comparison of rainfall recorded within select tropical forests is presented in Table 2-1. On average, rainforests receive 2000 to 3000 mm of precipitation annually although amounts as high as of 10000 to 13000 mm have been recorded from Africa along the coast near Guinea, between

Table 2-1   Recorded Rainfall in Select Tropical Forests

Location	Forest classification	Latitude	Longitude	mm/year	Elevation	Reference
Cozumel, Mexico	Tropical Rainforest	20° 31' N	86° 55' W	1575		Lauer 1989
Los Tuxtlas Tropical Rain Forest Reserve, Veraacruz, Mexico	Tropical Rainforest	18° 35-37' N	95° 03-07' W	~4900		Williams-Linera et al. 1998
El Verde in the Luquillo Mountains, Puerto Rico	Tropical Rainforest	18° 19' N	65° 49' W	3524	400 m	Garcia-Martino et al. 1996
Rio Blanco in the Luquillo Mountains, Puerto Rico	Tropical Rainforest	18° 17' N	65° 47' W	3752	439 m	Garcia-Martino et al. 1996
Pico del Este in the Luquillo Mountains, Puerto Rico	Tropical Rainforest	18° 15' N	65° 45' W	4436	1051 m	Garcia-Martino et al. 1996
Legazpi, Philippines	Tropical Rainforest	13° 10' N	123° 45' E	3370		Lauer 1989
Mangalore, India	Tropical Rainforest	12° 55' N	74° 47' E	3293		Lauer 1989
Iloilo, Philippines	Tropical Rainforest	10° 45' N	122° 33' E	2248		Lauer 1989
Finca La Selva, Costa Rica	Lowland Tropical Rainforest	10° 26' N	83° 26' W	~4000	37-100 m	cited in Chazdon and Fetcher 1984
Monteverde Cloud Forest Reserve, Cordillera de Tilaran, Costa Rica	Leeward Cloud Forest	10° 18' N	84° 48' W	2519	~1500 m	Clark et al. 1998
Monteverde Cloud Forest Reserve, Costa Rica	Montane Forest	10° 12' N	84° 42' W	~3000		Lawton 1990
Barro Colorado Island, Panama	Tropical Rainforest Island	9° 10' N	79° 51' W	~2612		Windsor 1990
La Mucuy, Merida, Venezuela	Lower Montane Cloud Forest	8° 30' N	71° 45' W	2959	2300 m	Ataroff 1998
Central Cordillera, Hornitos Weather Station, Panama	Lower Montane Forest	app 8° N	app 82° W	3327	1340 m	Cavellier et al. 1996
Parque Natural Metropolitano Republic of Panama	Lowland Tropical Rainforest	app 8° N	app 79° W	1740		Wright and Colley 1994
Okomu Oil Palm Co. Ltd., 75 km SW of Benin City, Nigeria	Tropical Rainforest	6° 24' N	5° 12' E	2409		Ghuman and Lal 1987
Sandakan, Malaysia	Tropical Rainforest	5° 53' N	118° 4' E	2995		Lauer 1989

Location	Forest type	Latitude	Longitude		Altitude	Reference
Chinchina, Colombia	Tropical Rainforest	4° 59′ N	75° 37′ W	2675		Lauer 1989
Bati Apoi Forest Reserve, Brunei	Dipterocarp Forest	4° 31′ N	115° 8′ E	~4000		Barker 1996
Douala, Cameroon	Tropical Rainforest	4° 00′ N	9° 45′ E	3902		Lauer 1989
Rio Surumoni, Venezuela	Neotropical Rainforest	3° 10′ N	65° 40′ W	~2700	2358 m	Szarzynski and Anhuf 2001
Amazon, French Guiana	Tropical Rainforest	3°-5° N	52-54°W	~3500		Roche 1981
Northern Congo-Ombrophilous Evergreen Forest	Temperate Rainforest	1° 37′ N	18° 04′ W	1860		Lyra et al. 1992
Yangambi, Zaire	Tropical Rainforest	0° 47′ N	24° 20′ E	1654		Lauer 1989
Eala, Zaire	Tropical Rainforest	0° 02′ N	18° 22′ E	1755		Lauer 1989
Pontianak, Indonesia	Tropical Rainforest	0° 05′ S	109° 15′ E	3167		Lauer 1989
Kericho, Sambret, Kenya	Montane Forest	0° 22′ S	35° 15′ E	2139	2073 m	Edwards and Blackie 1981
Kimakia, Kenya	Bamboo Forest with some Evergreens	0° 48′ S	36° 45′E	2235	2438 m	Edwards and Blackie 1981
Amazonian Lowlands, Yasuni National Park, Ecuador	Tropical Evergreen Forest	app 1° S	app 76° W	~3200	~250 m	cited in Burnham et al. 2001
Fonte Boa, Brazil	Tropical Rainforest	2° 33′ S	65° 59′ W	2801		Lauer 1989
Ducke Forest Reserve, Manous, Brazil	Tropical Evergreen Forest	3° 08′ S	60° 02′ W	2178		Leopoldo et al. 1993
Amazon Conservatory for Tropical Studies, Iquitos Peru	Tropical Rainforest	3° 14′ S	72° 55′ W	3492	2428 m	Madigosky and Vastnick 1999
Mbeya, Tanzania	Pretected Forest	8° 50′ S	33° 20′E	1733		Edwards and Blackie 1981
Australian Canopy Crane Facility, North Queensland, Australia	Lowland Tropical Rainforest	16° 07′ S	145° 27′ E	~3500		cited in Turton et al. 1999
Cairns, Queensland	Tropical Rainforest	16° 57′ S	145° 45′ E	2205		Lauer 1989
Danbulla State Forest, Atherton Tableland, Queensland, Australia	Tropical Rainforest (rainforest boundary)	17° 12′ S	145° 36′ E	1634	720 m	cited in Turton and Sexton 1996
Brisbane, Queensland, Australia	Lower Montane Cloud Forest	28° 17′ S	152° 24′ E	~1350	1000 m	Hutley et al. 1997
Atherton Tableland, Northeastern Australia	Tropical Rainforest Remnant	17° 22′ S	145° 35′ E	~2000	760 m	cited in Turton and Friburger 1997

Panama and Ecuador, the Khasia Mountains of Burma, and at Mt. Waialeale on Hawaii (Lauer 1989). What these areas have in common is a westerly flowing air mass within the equatorial trough.

Much of the total rainfall received by forests never reaches the forest floor. Instead it is intercepted, retained on vegetation for a short duration, absorbed by plants, or evaporated from the intracanopy. A deciduous forest canopy typically intercepts as much as 10 to 25 percent of the total annual precipitation (Oke 1987). Attempts to quantify interception within a rainforest environment have proven to be somewhat problematic. Measurements and modeling of interception within Amazonian forests indicate losses of approximately 10 to 12 percent (Lloyd et al. 1988). Yet results derived from tropical forests in Indonesia and Brazil suggest a 20 percent loss due to interception (Calder et al. 1986) or even higher (Salati 1987). The significant discrepancy between these findings warrants further attention.

Mean annual precipitation of forested areas is an important measure that helps to characterize a given forest community, but this reveals little information on how or in what manner rainfall is received and distributed within a forest. Such information is important because it provides a more precise picture of how plants may regulate reproductive cycles according to seasonal cycles (phenology). This also helps describe the seasonal variation that may initially not be so apparent. It is a rare instance where this type of information exists; an exception is the Barro Colorado Island facility on Panama, where a nearly complete record of hourly, daily, and monthly rainfall has been recorded since January 1929 to the present (Windsor 1990). Few areas within tropical forests report such complete information.

Acquisition of this nature requires long-term meteorological monitoring often in remote sites under harsh environmental conditions. The recent development of compact weatherproof computerized instrumentation is beginning to yield valuable datasets for micrometeorological assessment. Exact pulses and spikes of rainfall and corresponding parameters can now be monitored with little effort for extended periods with minimal retrieval requirements. Information such as the behavior of localized atmospheric conditions resulting from torrential downpours indicates a strong influence on the vertical wind field above and within the canopy and promotes isotrophic conditions throughout the entire canopy (personal observation). This is just one example of the type of information that may be gathered. Assessment of monthly and yearly rainfall records is valuable in determining forest phenology as well as how vegetation responds to large-scale climatic events such as El Niño/La Niña cycles.

Conclusions

The tropical regions of Central and South America, Africa, Southeast Asia, and the Pacific Islands support some of the most complex and biologically diverse forests on the planet. Although they differ in species composition and general architecture, they share the common attributes of ample rainfall, high humidity, uniform temperatures, and tremendous biological diversity. The dominant feature of many tropical forests is their imposing canopy architecture. Massive crowns of emergent trees tower over younger and less dominant species that create an irregular heterogeneous boundary on which most other interactions depend. In this regard, no two forests or parcels are exactly alike, which helps explain why at this point in our history we know so little about tropical habitat.

Microclimatic gradients from varied locations behave in a somewhat predictable and consistent manner. In particular, temperature profiles are very stable throughout tropical equatorial lowland forests; the intracanopy temperatures remain amazingly stable over prolonged periods (for weeks, months, and even years) with slight variation. Undoubtedly, this stability has contributed much to the tremendous diversity and growth characteristics common to tropical forests.

Precipitation and humidity profiles are also fairly stable, although the degree to which they can change over prescribed periods is somewhat more pronounced, especially when considering differences in the lower and upper canopy. These parameters follow diurnal patterns that relate to radiation input, crown characteristics, and the extent of the vertical dimension. Then, too, the amount of time it takes for regeneration to occur owing to the creation of gaps and patch dynamics is not consistent spatially or temporally between forests. Understanding the significance of how gaps contribute to the physical nature of different forests certainly adds additional evolutionary complexity to consider when assessing all dynamics of the rainforest environment.

In an attempt to understand the many facets of rainforest canopies, we have sometimes resorted to confronting topics in a reductionist fashion, essentially seeing the trees without noticing the forest. This is certainly a rational way of approaching such complexity, especially given the state of affairs concerning the loss of tropical forests. Indeed, in doing so, we run the risk of missing the true value of the information we gather. This is not to downplay the importance of research that has taken place to this point, but rather to call attention to the current state of our condition. What is now needed, however, is an integrative and holistic approach so that we may place rainforests in their proper context and importance. Only then will we be able to make the important connections between all aspects of climate and biological events as they relate to our own global existence.

Acknowledgments

This work is dedicated to René, Meredith, and Paul. The author would like to express great thanks to Widener University for continued support through grants and departmental funds for well over a decade. Specifically, the insights of Lawrence Panek, Lawrence Buck, and Marc Brodkin helped see that funding was provided when most needed. Great appreciation is extended to Itzick Vatnick for continued assistance over the years. Much thanks is extended to CONOPAC (Conservation de la Naturaleza del Peru, A.C.) along with Peter Jensen and Pam Bucar de Arevalo (Explorama Tours) for their in-country support and to Harry J. Augensen and Wülff D. Heintz, who provided critical review of this manuscript. To my undergraduate and graduate students who accompanied me in the field, a great deal of appreciation is extended, especially to Brandon Driscoll, who assisted with information retrieval.

References

Arya, S.P.S. (1988). "Introduction to Micrometeorology." Academic Press, San Diego.

Ashton, P.M. (1992). Some measurements of the microclimate within a Sri Lankan tropical rainforest. *Agricultural and Forest Meteorology* **59**, 217–235.

Ataroff, S. (1998). Importance of cloud-water in Venezuelan Andean cloud forest water dynamics. *In* "Proceedings of International Conference on Fog and Fog Collection, Vancouver, Canada." (R.S. Schemenauer and H. Bridgman, Eds.), pp. 25–28. International Development Research Center (IDRC), Ottawa.

Augspurger, C.K. (1984). Seedling survival among tropical tree species: interactions of dispersal distance, light gaps, and pathogens. *Ecology* **65**, 1705–1712.

Augspurger, C.K. and Franson, S. E. (1988). Input of wind dispersed seeds into light gaps and forest sites in a Neotropical forest. *J. Trop. Ecol.* **4**, 239–252.

Barker, M.G. (1996). Vertical profiles in a Brunei rain forest: I. Microclimate associated with a canopy tree. *J Trop. For. Sci.* **8(4)**, 505–519.

Belsky, A.J. and Canham, C.D. (1994). Forest gaps and isolated savanna trees. *Bioscience* **44**, 77–84.

Bjvrkman, O. and Ludlow, M.M. (1972). Characterization of the light climate on the floor of a Queensland rainforest. *Carnegie Institute Washington Yearbook* **71**, 94–102.

Bourgeron, P.S. (1983). Spatial aspects of vegetation structure. *In* "Ecosystems of the World 14A, Tropical Rain Forest Ecosystems: Structure and Function" (F.B. Golley, Ed.), pp. 29–47. Elsevier Scientific Publishing Company, Amsterdam, Oxford, and New York.

Brokaw, N.V.L. (1982). The definition of treefall gap and its effect on measures of forest dynamics. *Biotropica* **14**, 158–160.

Brokaw, N.V.L. (1985). Treefalls, regrowth, and community structure in tropical forests. *In* "The Ecology of Natural Disturbance and Patch Dynamics" (S.T.A. Picket and S.P. White, Eds.), pp. 53–69. Academic Press, Orlando and London.

Burnham, R.J., Pitman, C.A., Johnson, K.R., and Wilf, P. (2001). Habitat-related error in estimating temperatures form leaf margins in a humid tropical forest. *American Journal of Botany* **88(6)**, 1096–1102.

Cabral, O.M.R., McWilliam, A.L.C., and Roberts, J.M. (1996). In-canopy microclimate of Amazonian forest and estimates of transpiration. *In* "Amazonian Deforestation and Climate" (J.H.C. Gash, C.A. Nobre, J.M. Roberts, and R.L. Victoria, Eds.), pp. 207–219. Institute of Hydrology, Wallingford.

Calder, I.R., Wright, I.R., and Murdiyarso, D. (1986). A study of evapotranspiration from tropical forests: West Java. *Journal of Hydrology* **89**, 13–31.

Canham, C.D. (1988). Growth and architecture of shade-tolerant trees: response to canopy gaps. *Ecology* **69**, 786–795.

Canham, C.D., Denslow, J.S., Platt, W.J., Runkle, J.R., Spies, T.A., and White, P.S. (1990). Light regimes beneath closed canopies and treefall gaps in temperate and tropical forests. *Can. J. For. Res.* **20**, 620–631.

Cavelier, J., Solis, D., and Jaramillo, M.A. (1996). Fog interception in montane forest across the Central Cordillera of Panama. *Journal of Tropical Ecology* **12**, 357–369.

Chazdon, R.L. and Fetcher, N. (1984). Light environments of tropical forests. *In* "Physiological Ecology of Plants of the Wet Tropics" (E. Medina, H.A. Mooney, and C. Vasquez-Yanes, Eds.), pp. 27–36. W. Junk, The Hague.

Chu, P-S., Yu, Z-P., and Hastenrath, S. (1994). Detecting climate change concurrent with deforestation in the Amazon Basin: which way has it gone? *Bulletin of the American Meteorological Society* **75(4)**, 579–583.

Clark, D.A. (1994). Plant demography. *In* "La Selva: Ecology and Natural History of a Neotropical Rain Forest" (L.A. McDade, K.S. Bawa, H.A. Hespenheide, and G.S. Hartshorn, Eds.), pp. 90–105. University of Chicago Press, Chicago.

Clark, K.L., Nadkarni, N.M., Schaeffer, D., and Gholz, H.L. (1998). Atmospheric deposition and net retention of ions by the canopy in a tropical montane rain forest, Monteverde, Costa Rica. *J Trop. Ecol.* **14**, 27–45.

Coakle, J.A. and Chylek, P. Jr. (1975). The two stream approximation in radiative transfer: including the angle of incident radiation. *Journal of Atmospheric Science* **32**, 409–418.

Culf, A.D., Esteves, J.L., Marques-Filho, A. de O., and Da Rocha, H.R. (1996). Radiation, temperature and humidity over forest and pasture in Amazonia. *In* "Amazonian Deforestation and Climate" (J.H.C. Gash, C.A. Nobre, J.M. Roberts, and R.L. Victoria, Eds.), pp. 175–191. Institute of Hydrology, Wallingford.

Denslow, J.S. (1980). Gap partitioning among tropical rainforest trees. *Biotropia* **12**, 47–55.

Denslow, J.S. (1987). Tropical rain forest gaps and tree species diversity. *Annual Review of Ecology and Systematics* **18**, 431–451.

Denslow, J.S. and Hartshorn, G.S. (1994). Tree-fall gap environments and forest dynamic processes. *In* "La Selva: Ecology and Natural History of a Neotropical Rain Forest" (L.A. McDade, K.S. Bawa, H.A. Hespenheide, and G.S. Hartshorn, Eds.), pp. 120–127. University of Chicago Press, Chicago.

Dickinson, R.E. and Henderson-Sellers, A. (1988). Modeling tropical deforestation: a study of GCM land-surface parametrizations. *Q.J.R. Meteorol. Soc.* **114**, 439–462.

Dolman, A.J. (1986). Estimates of surface roughness length and zero plane displacement for a foliated and non-foliated oak canopy. *Agric. For. Meteorol.* **36**, 241–248.

Dolman, A.J., Gash, J.H.C., Roberts, J., and Shuttleworth, J.W. (1991). Stomatal and surface conductance of tropical rainforest. *Agricultural and Forest Meteorology* **54**, 303–318.

Dransfield, J. (1978). Growth forms of rain forest palms. *In* "Tropical Trees as Living Systems" (P.B. Tomlinson and M.H. Zimmerman, Eds.), pp. 247–268. Cambridge University Press, New York.

Edwards, K.A. and Blackie, J.R. (1981). Results of the east African catchment experiments 1958–1974. *In* "Tropical Agricultural Hydrology" (R. Lal and E.W. Russell, Eds.), pp. 163–188. John Wiley & Sons, New York.

Emborg, J. (1998). Understory light conditions and regeneration with respect to the structural dynamics of a near-natural temperate forest in Denmark. *Forest Ecology and Management* **106**, 83–95.

Ewel, J.J. (1970). Biomass changes in early tropical succession. *Turrialba* **21**, 110–112.

Ewel, J.J. (1976). Litter fall and leaf decomposition in a tropical forest succession in eastern Guatemala. *J. Ecology* **64**, 293–408.

Fetcher, N., Oberbauer, S.F., and Chazdon, R.L. (1994). Physiological ecology of plants. *In* "La Selva: Ecology and Natural History of a Neotropical Rain Forest" (L.A. McDade, K.S. Bawa, H.A. Hespenheide, and G.S. Hartshorn, Eds.), pp. 128–141. University of Chicago Press, Chicago.

Finlayson-Pitts, B.J., Pitts, J.N. Jr. (2000). "Chemistry of the Upper and Lower Atmosphere" Academic Press, San Diego.

Freiberg, E. (1998). Microclimatic parameters influencing nitrogen fixation in the phyllosphere in a Costa Rican premontane rain forest. *Oecologia* **17**, 9–18

Garcia-Martino, A.R., Warner, G.S., Scatena, F.N., Civco, D.L. (1996). Rainfall, runoff and elevation relationships in the Luquillo Mountains of Puerto Rico. *Caribbean Journal of Science* **32(4)**, 413–424.

Garwood, N.C., Janos, D.P., and Brokaw, N. (1979). Earthquake caused landslides: a major disturbance to tropical forests. *Science* **205**, 997–999.

Gay, L.W., Knoerr, K.R., and Braaten, M.O. (1971). Solar radiation variability on the floor of a pine plantation. *Agric. Meteorol.* **8**, 39–50.

Geiger, R. (1965). "The Climate Near the Ground." Harvard University Press, Cambridge, MA.

Ghuman, B.S. and Lal, R. (1987). Effects of deforestation on soil properties and microclimate of a high rain forest in southern Nigeria. *In* "The Geophysiology of Amazonia: Vegetation and Climate Interactions" (R.E. Dickinson, Ed.), pp. 225–244. John Wiley and Sons, New York.

Grace, J., Lloyd, J., McIntyre, J., Miranda, A.C., Meir, P., and Miranda, H.S. (1996). Carbon dioxide flux over Amazonian rain forest in Rondonia. *In* "Amazonian Deforestation and Climate" (J.H.C. Gash, C.A. Nobre, J.M. Roberts, and R.L. Roberts, Eds.), pp. 307–318. Institute of Hydrology, Wallingford.

Graedel, T.E. and Crutzen, P.J. (1993). "Atmosphere Change: An Earth System Perspective." W.H. Freeman and Company, New York.

Grubb, P.J. (1977). The maintenance of species richness in plant communities: the importance of the regeneration niche. *Biol. Rev.* **52**, 107–145.

Hallé, F. (1974). Architecture of trees in the rain forest of Morobe District, New Guinea. *Biotropia* **6**, 43–50.

Hallé, F. (1978). Architectural variation at the specific level in tropical trees. *In* "Tropical Trees as Living Systems" (P.B. Tomlinson and M.H. Zimmerman, Eds.), pp. 209–231. Cambridge University Press, New York.

Hallé, F. and Oldeman, R.A.A. (1970). Essai sur l'architecture et la dynamique de croissance des arbres tropicaux. *Monographies de Botanique et de Biologie Vegetale* No. 6. Paris, Masson.

Hallé, F., Oldeman, R.A.A., and Tomlinson, P.B. (1978). "Tropical Trees and Forests: An Architectural Analysis." Springer-Verlag, Berlin.

Hartshorn, G.S. (1978). Tree falls and tropical forest dynamics. *In* "Tropical Trees as Living Systems" (P.B. Tomlinson and M.H. Zimmerman, Eds.), pp. 617–638. Cambridge University Press, London.

Hartshorn, G.S. (1980). Neotropical forest dynamics. *Biotropica* **12**, 23–30.

Henderson-Sellers, A. (1987). Effects of change in land use on climate in the humid tropics. *In* "The Geophysiology of Amazonia: Vegetation and Climate Interactions" (R.E. Dickinson, Ed.), pp. 463–496. John Wiley & Sons, New York.

Horn, S.P. and Sanford, R.L. Jr. (1993). Holocene fires in Costa Rica. *Biotropia* **24**, 354–361.

Hoyt, D.V. and Schatten, K.H. (1997). "The Role of the Sun in Climate Change." Oxford University Press, New York.

Hubbell, S.P. (1979). A general hypothesis of species diversity. *American Naturalist* **113**, 81–101.

Hubbell, S.P. and Foster, R.B. (1986). Canopy gaps and the dynamics of a neotropical forest. In "Plant Ecology" (M.J. Crawley, Ed.), pp. 77–96. Blackwell Scientific, Oxford.

Hutley, L.B., Doley, D., Yates, D.J., and Boonsaner, A. (1997). Water balance of an Australian subtropical rainforest at altitude: the ecological and physiological significance of intercepted cloud and fog. *Aust. J. Bot.* **45**, 311–329.

Janzen, D.H. (1970). Herbivores and the number of tree species in tropical forests. *American Naturalist* **109**, 501–578.

Janzen, D.H. (2002). Tropical dry forest: Area de Conservacion Guanacaste, northwestern Costa Rica. *In* "Handbook of Ecological Restoration, Restoration in Practice" (M.R. Perrow and A.J. Davy, Eds.), Vol. 2, pp. 559–583. Cambridge University Press, Cambridge.

Jenik, J. (1978). Roots and root systems in tropical trees: morphologic and ecologic aspects. *In* "Tropical Trees as Living Systems" (P.B. Tomlinson and M.H. Zimmerman, Eds.), pp. 232–349. Cambridge University Press, Cambridge and New York.

Jordon, C.F. (1987). Shifting cultivation: slash and burn agriculture near San Carlos de Rio Negro, Venezuela. *In* "Amazonian Rain Forests: Ecosystem Disturbance and Recovery" (C.F. Jordon, Ed.), pp. 9–23. Springer-Verlag, New York.

Kira, T. and Yoda, K. (1989). Vertical stratification in microclimate. *In* "Ecosystems of the world 14B, Tropical Rain Forest Ecosystems: Biogeographical and Ecological Studies" (H. Leith and M.J. Werger, Eds.), pp. 55–71. Elsevier Science Publishers, Amsterdam, Oxford, New York, Tokyo.

Kruijt, B., Lloyd, J., Grace, G., McIntyre, J.A., Farquhar, G.D., Miranda, A.C., and McCraken, P. (1996). Sources and sinks of CO_2 in Rondonia tropical rainforest. *In* "Amazonian Deforestation and Climate" (J.H.C. Gash, C.A. Nobre, J.M. Roberts, and R.L. Victoria, Eds.), pp. 331–351. Institute of Hydrology, Wallingford.

Lauer, W. (1989). Climate and weather. *In* "Ecosystems of the World 14B, Tropical Rain Forest Ecosystems: Biogeographical and Ecological Studies" (H. Lieth and M.J.A. Werger, Eds.), pp. 7–53. Elsevier Science Publishers, Amsterdam, Oxford, New York, Tokyo.

Lawton, R.O. (1990). Canopy gaps and light penetration into a wind-exposed tropical lower montane rain forest. *Can. J. Forest Research* **20**, 659–667.

Lawton, R.O. and Putz, F.E. (1988). Natural disturbance and gap-phase regeneration in a wind-exposed tropical cloud forest. *Ecology* **69(3)**, 764–777.

Lean, J., Bunton, C.A., Nobre, C.A., and Rowntree, P.R. (1996). The simulated impact of Amazonian deforestation on climate using measured ABRACOS vegetation characteristics. *In* "Amazonian Deforestation and Climate" (J.H.C. Gash, C.A. Nobre, J.M. Roberts, and R.L. Victoria, Eds.), pp. 549–576. Institute of Hydrology, Wallingford.

Lean, J. and Warrilow, D.A., (1989). Simulation of the regional climatic impact of Amazon deforestation. *Nature* **342**, 411–413.

Lee, D.W. (1987). The spectral distribution of radiation in two neotropical rainforests. *Biotropia* **19(2)**, 161–166.

Lee, R. (1978). "Forest Microclimatology." Columbia University Press, New York.

Leigh, E.G. Jr. (1975). Structure and climate in tropical rain forest. *Annu. Rev. Ecol. Syst.* **6**, 67–86.

Leopoldo, P.R., Chaves, J.G., and Franken, W.K. (1993). Solar energy budgets in central Amazonian ecosystems: a comparison between natural forest and bare soil areas. *Forest Ecol. and Manage.* **59**, 313–328.

Leuthart, C.A., Spencer, H.T., Mountain K.R., and Kosnik, K.E. (1994). Hydrologica and energy balance characteristics of the Amazonian Tropical Rainforest, Iquitos, Peru. Pt 1. Micrometerology. (Unpublished manuscript).

Lieberman, M. and Lieberman, D. (1994). Patterns of density and dispersion of forest trees. *In* "La Selva: Ecology and Natural History of a Neotropical Rain Forest" (L.A. McDade, K.S. Bawa, H.A. Hespenheide, and G.S. Hartshorn, Eds.), pp. 106–119. University of Chicago Press, Chicago.

Lloyd, C.R., Gash, J.H.C., Shuttleworth, W.J., and Marques, F.A.O. (1988). The measurement and modeling of rainfall interception by Amazonian rain forest. *Agric. For. Meteorol.* **43**, 277–294.

Lyra, R., Druilhet, A., Benech, B., and Bouka-Biona, C. (1992). Dynamics above a dense equatorial rain forest from the surface boundary layer to the free atmosphere. *Journal of Geophysical Research* **97(12)**, 12953–12965.

Madigosky, S.R. and Grant, B.W. (1996). Biodiversity loss: a problem or symptom of our time? *In* "Forests: A Global Perspective" (S.K. Majumdar, E.W. Miller, and F.J. Brenner, Eds.), pp. 226–241. Pennsylvania Academy of Sciences, Easton, PA.

Madigosky, S.R. and Vatnick, I. (1999). Microclimatic characteristics of a primary tropical Amazonian rain forest, ACEER, Iquitos, Peru. *Selbyana* **21(1–2)**, 165–172.

Maslin, M.A. and Burns, S.J. (2000). Reconstruction of the Amazon Basin effective moisture availability over the past 14,000 years. *Science* **290**, 2285–2287.

McNaughton, K.G. (1989). Micrometeorology of shelter belts and forest edges. *Philos. Trans. R. Soc. London, Ser. B* **324**, 351–368.

Meador, W.E. and Weaver, W.R. (1980). Two-stream approximations to radiative transfer in planetary atmospheres: a unified description of existing methods and a new improvement. *Journal of Atmospheric Science* **37**, 630–643.

Molion, L.C.B. (1987a). Micrometeorology of an Amazonian Rain Forest. *In* "The Geophysiology of Amazonia: Vegetation and Climate Interactions" (R.E. Dickinson, Ed.), pp. 255–272. John Wiley & Sons, New York.

Molion, L.C.B. (1987b). On the dynamic climatology of the Amazon Basin and associated rain-producing mechanisms. *In* "The Geophysiology of Amazonia: Vegetation and Climate Interactions" (R.E. Dickinson, Ed.), pp. 391–407. John Wiley & Sons, New York.

Nobre, C.A., Fisch, G., DaRocha, H.R., Lyra, R F. da F., Da Rocha, E.P., Da Costa, A.C.L., and Ubarana, V.N. (1996). Observations of the atmospheric boundary layer in Rondonia. *In* "Amazonian Deforestation and Climate" (J.H.C. Gash, C.A. Nobre, J.M. Roberts, and R. L. Vicotria, Eds.), pp. 413–424. Institute of Hydrology, Wallingford.

Oldfield, S. (2002). "Rainforest." Massachusetts Institute of Technology Press, Cambridge

Oke, T.R. (1987). "Boundary Layer Climates." Methuen, London and New York.

Orians, G.H. (1982). The influence of tree-falls in tropical forests in tree species richness. *J. Trop. Ecol.* **23**, 255–279.

Paegle. J. (1987). Interactions between convective and large-scale motions over Amazonia. *In* "The geophysiology of Amazonia: Vegetation and Climate Interactions" (R.E. Dickinson, Ed.), pp. 347–390. John Wiley & Sons, New York.

Parker, G.G. (1995). Structure and microclimate of forest canopies. *In* "Forest Canopies" (M.D. Lowman and N.M. Nadkarni, Eds.), pp. 73–86. Academic Press, San Diego.

Parker, G.G. (2000). Forest canopy stratification: is it useful? *The American Naturalist* **155(4)**, 473–484.

Parmenter, F.C. (1976). A southern hemisphere cold front passage at the equator. *Bulletin American Meteorological Society* **57(12)**, 1435–1440.

Pearcy, R.W. (1983). The light environment and growth of C3 and C4 tree species in the understory of a Hawaiian forest. *Oecologia* **58**, 19–25.

Post, E. (2003). Large-scale climate synchronizes the timing of flowering by multiple species. *Ecology* **84(2)**, 277–281.

Richards, P.W. (1952). "The Tropical Rain Forest: An Ecological Study." Cambridge University Press, Cambridge.

Richey, J.E., Nobre, C., Deser, C. (1989). Amazon River discharge and climate variability: 1903–1985. *Science* **246**, 101–103.

Ribeiro, M.N.G., Salati, E., and Villa Nova, N.A. (1982). Radiacao solar disponvel em Manaus e sua relacao com a duracao do brilho solar. *Acta Amazon* **12**, 338–345.

Roberts, S.D., Long, J.N., and Smith, F.W. (1993). Canopy stratification and leaf area efficiency: a conceptualization. *Forest Ecology and Management* **60**, 143–156.

Roche, M.A. (1981). Watershed investigations for development of forest resources of the Amazon region in French Guiana. *In* "Tropical Agricultural Hydrology" (R. Lal and E.W. Russell, Eds.), pp. 75–82. John Wiley & Sons, New York.

Rosenberg, N.J., Blad, B.B., and Verma, S.B. (1983). "Microclimate: The Biological Environment," Second Ed. John Wiley & Sons, New York.

Runkle, J.R. (1985). Disturbance regimes in temperate forests. *In* "The Ecology of Natural Disturbance and Patch Dynamics" (S.T.A. Picket and P.S. White, Eds), pp. 17–33. Academic Press, Orlando and London.

Salati, E. (1987). The forest and the hydrological cycle. *In* "The Geophysiology of Amazonia: Vegetation and Climate Interactions" (R.E. Dickinson, Ed.), pp. 273–296. John Wiley & Sons, New York.

Saldarriaga, J.G. (1987). Recovery following shifting cultivation: a century of succession in the upper Rio Negro. *In* "Amazonian Rain Forests: Ecosystem Disturbance and Recovery" (C.F. Jordon, Ed.), pp. 24–33. Springer-Verlag, New York.

Sanford, R.L. Jr., Paaby, P., Luvall, J.C., and Phillips, E. (1994). Climate, geomorphology, and aquatic systems. *In* "LaSelva: Ecology and Natural History of a Neotropical Rain Forest" (L.A. McDade, K.S. Bawa, H.A. Hespenheide, and G.S. Hartshorn, Eds.), pp. 19–33. University of Chicago Press, Chicago.

Scott, G.A.J. (1987). Shifting cultivation where land is limited: Campa Indian agriculture in the Gran Pajonal of Peru. *In* "Amazonian Rain Forest: Ecosystem Disturbance and Recovery" (C.F. Jordon, Ed.), pp. 34–45. Springer-Verlag, New York.

Sellers, P.J. (1987). Modeling effects of vegetation on climate. *In* "The Geophysiology of Amazonia: Vegetation and Climate Interactions" (R.E. Dickinson, Ed.), pp. 297–344. John Wiley & Sons, New York.

Shukla, J. (1987). General circulation modeling and the tropics. *In* "The Geophysiology of Amazonia: Vegetation and Climate Interactions" (R.E. Dickinson, Ed.), pp. 409–461. John Wiley & Sons, New York.

Shukla, J., Nobre, C., and Sellers, P. (1990). Amazon deforestation and climate change. *Science* **247**, 1322–1325.

Shuttleworth, W.J. (1989). Micrometeorology of temperate and tropical forests. *Philos. Trans. R. Soc. London, Ser. B.* **324**, 299–334.

Shuttleworth, W.J., Gash, J.H.C., Lloyd, C.R., McNeill, D.D., Moore, C.J., and Wallace, J.S. (1988). An integrated micrometeorological system for evaporation measurement. *Agric. For. Meteorol.* **43**, 295–317.

Szarzynski, J. and Anhuf, D. (2001). Micrometeorological conditions and canopy energy exchanges of a neotropical rain forest (Surumoni-Crane Project, Venezuela). *Plant Ecology* **153**, 231–239.

Tomlinson, P.B. (1978). Branching and axis differentiation in tropical trees. *In* "Tropical Trees as Living Systems" (P.B. Tomlinson and M.H. Zimmerman, Eds.), pp. 187–207. Cambridge Univeristy Press, Cambridge and New York.

Tomlinson, P.B. (1983). Structural elements of the rain forest. *In* "Ecosystems of the World 14A, Tropical Rain Forest Ecosystems: Structure and Function" (F.B. Golley, Ed.), pp. 9–28. Elsevier Scientific Publishing Company, Amsterdam.

Tomlinson, P.B., and Gill, A.M. (1973). Growth habits of tropical trees: some guiding principles. *In* "Tropical Forest Ecosystems in Africa and South America: A Comparative Review" (B.J. Meggers, E.S. Ayensu, and W.D. Duckworth, Eds.), pp. 129–143. Smithsonian Institution Press, Washington, DC.

Turton, S.M., Freiburger, H.J. (1997). Edge aspect effects on the microclimate of a small tropical forest remnant on the Atherton Tableland, Northeastern Australia. *In* "Tropical Forest Remnants; Ecology, Management and Conservation of Fragmented Communities" (W.F. Laurance and R.O. Bierregaard Jr., Eds.), pp. 45–54. University of Chicago Press, Chicago.

Turton, S.M. and Sexton, G.J. (1996). Environmental gradients across four rainforest: open forest boundaries in northeastern Queensland. *Australian Journal of Ecology* **21**, 245–254.

Turton, S.M., Tapper, N.J., and Soddell, J. (1999). Proposal to measure diurnal and seasonal energetics above and within a lowland tropical rain forest canopy in northeastern Australia. *Selbyana* **20(2)**, 345–349.

Uhl, C., Clark, K., Dezzeo, N., and Maquirino, P. (1988). Vegetation dynamics in Amazonian treefall gaps. *Ecology* **69**, 751–763.

Unruh, J.D. (1991). Canopy structure in natural and agroforest successions in Amazonia. *Tropical Ecology* **32(2)**, 168–181.

Vitousek, P.M. and Denslow, J.S. (1986). Nitrogen and phosphorus availability in treefall gaps of a lowland tropical rainforest. *J. Ecol.* **74**, 1167–1178.

Walther, G-R., Post, E., Convey, P., Menzel, A., Parmesan, C., Beebee, T.J.C., Fromentin, J-M., Hoegh-Guldberg, O., and Bairlein, F. (2002). Ecological responses to recent climate change. *Nature* **416**, 389–395.

Whitmore, T.C. (1975). "Tropical Rain Forests of the Far East." Clarendon Press, Oxford.

Whitmore, T.C. (1991). Tropical rain forest dynamics and its implications for management. *In* "Rain Forest Regeneration and Management" (A. Gomez-Pompa, T.C. Whitmore, and M. Hadley, Eds.), pp. 67–89. UNESCO, Paris.

Whitmore, T.C. (1997). Tropical forest disturbance, disappearance, and species loss. *In* "Tropical Forest Remnants: Ecology, Management, and Conservation of Fragmented Communities" (W.F. Laurance and R.O. Bierregaard Jr., Eds.), pp. 3–11. University of Chicago Press, Chicago.

Williams-Linera, G., Dominguez-Gastelu, V., and Garcia-Zurita, M.E. (1998). Microenvironment and Floristics of different edges in a fragmented tropical rainforest. *Conservation Biology* **12(5)**, 1091–1102.

Windsor, D.M. (1990). Climate and moisture variability in a tropical forest: long-term records from Barro Colorado Island, Panama. Smithsonian Contributions to the Earth Sciences, No. 29. Smithsonian Institution Press, Washington, DC.

Wright, S.D., Colley, M. (1994). "Assessment of Biological Diversity and Microclimate of the Tropical Forest Canopy: Phase I." United Nations Environment Programme/Smithsonian Tropical Research Institute Project.

Wright, I.R., Gash, J.H.C., DaRocha, H.R., Roberts, J.M. (1996). Modeling surface conductance for Amazonian pasture and forest. *In* "Amazonian Deforestation and Climate" (J.H.C. Gash, C.A. Nobre, J.M. Roberts, and R.L. Victoria, Eds.), pp. 437–458. Institute of Hydrology, Wallingford.

Wright, I.R., Gash, J.H.C., DaRocha, H.R., Shuttleworth, W.J., Nobre, C.A., Maitelli, G.T., Zamparoni, C.A.G.P., and Carvalho, P.R.A. (1992). Dry season micrometeorology of central Amazonia ranchland. *Q.J.R. Meteorol. Soc.* **118**, 1083–1099.

CHAPTER 3

Quantifying and Visualizing Canopy Structure in Tall Forests: Methods and a Case Study

Robert Van Pelt, Stephen C. Sillett, and Nalini M. Nadkarni

The mighty trees getting their food are seen to be wide awake,
every needle thrilling in the welcome nourishing storms, chanting
and bowing low in glorious harmony, while every raindrop and snowflake is seen as a
beneficent messenger from the sky.
—*John Muir*, Our National Parks, *1901*

Introduction

In 1970, one of the first canopy research projects began in an old-growth *Pseudotsuga menziesii* forest. The goal was to assess the biomass and surface area of its entire canopy as part of an ecosystem-level study of forest dynamics. Supported by the International Biological Program, ten tall *Pseudotsuga* trees were climbed, and every branch was mapped and estimated for foliar biomass and leaf area. This pioneering research used rock-climbing techniques (e.g., lag screws and webbing ladders) to gain access to the tree crowns (Denison et al. 1972). The study generated an unprecedented amount of within-canopy structure data. At that time, however, such large datasets were difficult to handle. Most of the original data are now available only on obsolete punch cards; only a small portion have been analyzed and published (Pike et al. 1977; Massman 1982). Although canopy access techniques and data management methods have greatly evolved, the data collected for that study remain the most detailed ever collected on the crown structure of *Pseudotsuga menziesii*.

Since then, other researchers have pursued similar goals in other forests. Modern techniques for gaining access to forest canopies are diverse and varied (Moffett and Lowman 1995, Mitchell et al. 2002). With nearly a dozen canopy cranes and scores of canopy walkways and towers maintained around the world, access is no longer limited to the adventurous (Nadkarni 1995). However, rope-based techniques remain the best way to gain the particular type of canopy access to trunks and branches of tall trees required by researchers who study canopy structure.

Mapping the three-dimensional (3-D) structure of trees has been achieved from the ground using lasers with some success (Sumida et al. 2001; Nychka and Nadkarni 2003). These techniques involve mapping the 3-D coordinates (x,y,z, or d,α,θ) of each bifurcation, or node, within the tree down to some minimum size. The primary limitation is that mapping each bifurcation within a tree is possible only with small trees and trees whose structure is not hidden by foliage or epiphytes. Larger trees can be mapped with canopy cranes, but access to the near-bole space within tree crowns is often restricted—the bulk of the measurements must still be taken remotely. A further limitation to mapping 3-D structure as nodes and segments is that the resultant tree is a stick-figure shape—there is no thickness to the segments.

Many aspects of tree physiology, epiphyte ecology, and stand-level forest dynamics can greatly benefit from whole-tree estimates of surface area, wood volume, and biomass. Surface area estimates are needed to determine carbon production for trees with photosynthetic bark and to estimate epiphyte habitat with tree crowns (Pike et al. 1977; Lyons et al. 2000; Ellyson and Sillett 2003). Although most conifers carry much of their volume within a single stem, others—along with most angiosperm trees—contain much of their wood volume in branches within the crown. To assess these accurately requires knowing branch diameters within the tree, an extremely difficult task from the ground or even from the gondola of a crane.

Biomass estimates in forests require quantification of components such as trunks, branches, and foliage. In tall-stature forests, this has been accomplished by felling one to many trees for samples. While the crowns of large trees are destroyed in the felling process, a small proportion of branches may be relatively undamaged. These can often be sufficient to develop regression equations for various components with the crown, but not for whole-tree estimates.

Forest ecologists need a comprehensive, nondestructive sampling regime to obtain accurate whole-tree estimates of 3-D structure for any forest type. In this chapter, we propose a three-tiered approach to obtain 3-D structural information in forests:

1. **Stand Mapping:** To gain information on the forest as a whole and to put individual trees in a stand-scale perspective;
2. **Crown Mapping:** To gain detailed 3-D information on individual trees; and
3. **Analyses:** To scale up from subsamples to whole-tree and stand levels.

We follow this protocol with a case study comparing two very tall but structurally dissimilar forests: a *Eucalyptus regnans* forest in southern Australia and a *Pseudotsuga menziesii* forest in the Pacific Northwest of North America. We present this comparison to illustrate the power and diversity of these techniques, many of which are being presented here for the first time.

3-D Structural Mapping Protocol for Forests

Stand Mapping

Site selection varies greatly with study objectives. In our work, sample areas at each site were located in large, relatively flat blocks of forest with plot boundaries located well away from clear-cut edges to reduce confounding factors such as edge influences, slope, and aspect. Ideally, each potential study site has geographic information system (GIS) and/or aerial photos available. Stands are discrete, homogeneous blocks of forests that are defined locally. In coniferous forests, stands are often identifiable from aerial photos or even detailed satellite imagery. In other forests, this may have to be done through ground-based reconnaissance. Stand boundaries are obtained or generated using GIS, and the resulting GIS-generated polygons are used as the study site.

At each site, transects are the sample units and are used as replicates. Maximum tree height measured during reconnaissance is used to determine transect size. Transects must be large enough to capture the stand-level variability but small enough to maximize efficiency. The transect dimensions are 3×0.3 times the dominant tree height, following Kuiper (1994) and Van Pelt (1995). For example, a stand in which the dominant tree height is 50 m is sampled by 150 m $\times$ 15 m transects. The number of transects used at a given site will vary depending on resources and time, but three or four usually suffice. Transects are located inside stand polygons with random starting points and directions. These are then located on the ground using GPS and surveyed in with rebar at the recorded endpoints.

A metric tape is stretched down the center of the plot and staked at either end with rebar. Any piece of dead wood (greater than 5 cm) intercepted by the tape is measured. Measurements may include species, diameter, decay class, piece length, and beginning and end points where each log intersects the tape. Diameter is measured perpendicular to the central axis of the log at the interception point. These data are used to calculate log volume in the entire plot (Harmon et al. 1986).

Shrubs are sampled in a subplot centered on the whole plot. Width of the subplot is one-tenth that of the whole plot (i.e., 0.03 times the mean height of canopy dominants). Species, X-value, basal diameter, and height can be recorded for shrubs and trees with diameter at breast height (DBH) less than 5 cm and taller than 50 cm. DBH (rather than basal diameter) is recorded for large shrubs.

All trees within the plot greater than 5 cm in diameter at breast height are mapped and measured for DBH. The DBH is measured at 1.37 m above the high point of ground (i.e., ground level on the uphill side of the tree). The species, status (live or dead), and a unique identification number are recorded for each stem. For each stem, the perpendicular distance between the center of the stem and the tape is recorded as the Y value. One side of the tape is chosen to have positive Y values, the other to have negative. The value on the tape perpendicular to the center of the stem is recorded as the X value. Compasses are set perpendicular to the direction of the transect to maintain accuracy in the Y value measurements.

Data sheets and a stem map are prepared from these initial measurements for double-checking in the field. This two-step process allows one to correct errors of location, identification, and measurement. A laser range finder (e.g., Impulse 200LR, Laser Technologies, Inc.) is used to measure total tree heights and heights of crown bases. The crown base is defined as the point above which living foliage surrounds more than one-third of the trunk. Crown radii are measured from the center of the trunk to the tip of the longest living branch. For large trees, or trees with highly asymmetrical crowns, eight or more radii must be measured; one radius in each cardinal direction suffices for most trees.

In each stand, a subset of trees of each species is chosen for stem volume measurements. The random sample must include the full range of tree sizes. A survey laser is used to measure trunk diameters every 3 m along the entire stem. These data are used to supplement those collected during crown mapping (see below) to develop regression equations that predict trunk volume from height and DBH. In the case of trees with large buttresses, basal maps must be prepared for each stem to be converted to a functional diameter (see the Analyses section below).

Crown Mapping

Several trees of each species are randomly chosen for detailed study. Again, these should span the size range of the species. Single rope techniques are used to gain access to tree crowns and permit climbers to measure directly locations and dimensions of all reiterations and branches. Very tall trees in old-growth forests are accessed via a bow that is used to shoot an arrow over sturdy branches in the crown. A vertical range up to 85 m is possible with this technique. The arrows are tied to fishing filament (14- to 20-lb test), which is attached to a spinning reel on the front of the bow. If the shot misses, or the arrow goes over undesired or unsafe branches, the arrow is lowered to the ground, untied, and the filament is retrieved.

After a successful shot, the filament is used to pull a nylon line followed by a climbing rope (9 mm) over the branches. When both ends of the rope are on the ground, one end of the rope is anchored with a knot and the other end is climbed via single rope technique (Perry 1978). Once the branch supporting the climbing rope is reached, rope techniques developed by arborists are used to gain access to progressively higher branches (Jepson 2000). A pulley is secured near the treetop with a webbing sling through which the climbing rope is passed and lowered to the ground along two clear paths on opposite sides of the crown. By double-tying the midpoint of the rope above the pulley, a team of two climbers can gain access to all parts of the crown for recording data. Between climbing sessions, the rope is replaced by the nylon cord.

A metric tape is stretched from average ground level to the treetop. If the tape cannot be anchored on the ground, it is tied to a nearby tree or shrub at the proper height. Metal tags are attached to the main trunk at 5-m intervals for use as future benchmarks in height measurements of trunks, limbs, and branches once the metric tape has been removed. The tree's total height is recorded to the nearest decimeter, and main trunk diameters are measured at each tag. Near the base of large trees, additional diameters are often needed for accurate estimation of wood volume.

Reiterated trunks are defined as accessory trunks arising from the main trunk, other reiterated trunks, or limbs (Sillett 1999; Sillett and Van Pelt 2000). They are vertically oriented stems with their own branches and are architecturally indistinguishable from freestanding trees except for their locations within the crown of the larger, supporting tree. This follows the terminology of Hallé et al. (1978), but in the narrow sense of using this term only for complete reiterations. The following data are recorded for each reiterated trunk: basal diameter, base height, base distance, base azimuth, top height, top distance, and top azimuth. Distances and azimuths are referenced to the center of the main trunk. If the reiteration is more than 5 m tall, or if there are structural anomalies, additional diameters, distances, and azimuths are measured.

Limbs are defined as large branches that arise from the main trunk and give rise to reiterated trunks. They consist of limb segments, and they terminate in other limb segments or reiterated trunks. Branch segments are defined as accessible sections of large branches that arise from the main trunk and give rise to other branch segments or branches. Limbs behave differently from branches physiologically and are thus kept separate on data sheets. Both limb segments and branch segments receive the same measurements: the diameter at each end is measured as well as the height, distance, and azimuth of each end. Smaller branches can also arise along the length of a branch segment, but larger branches and trunks will cause the branch segment to be divided into additional segments.

Limb segments, branch segments, and branches are named according to their origins. For example, a branch consists of two segments and four branches (see Figure 3-1). The first branch segment is named M-1 because it arises from the main trunk (M). Branch segment M-1 gives rise to branches 101 and 104 as well as branch segment 1-2. The origin of branch 101 is listed as 1 on the data sheet because it arises from the distal end of branch segment M-1. Branch segment 1-2 gives rise to two branches, 102 and 103. Their origins are listed as 2 on the data sheet because they arise from the distal end of branch segment 1-2. Branch 104 is too small to deflect the branch segment significantly, so its origin is listed as M-1 on the data sheet. To know exactly where branch 104 originates, the distance along the branch segment M-1 is also recorded on the data sheet.

Each branch receives a unique number within a given tree. The following measurements are taken for each branch greater than 2 cm in diameter: basal height, basal diameter, extension (i.e., slope-corrected horizontal distance from trunk to branch tip), percent foliated, azimuth from trunk, overall slope, and curvature. Overall slope is measured in degrees, positive (up) or negative (down), in five-degree increments. A two-letter code is used to further describe curvature. The first letter refers to the shape of the portion of the branch near the trunk, and the second letter refers to the shape of the rest of the branch. There are three different designations (O = orthotropic, upward tending, G = geotropic, downward tending, and N = neutral), resulting in nine possible codes (see Figure 3-2).

Each branch is classed as either original or epicormic. Original branches are those formed when the treetop was at the height of the branch (i.e., piths of original branches are continuous with the pith of the trunk). Epicormic branches are those formed below the top of the tree (i.e., they originated from the cambium). However, the process of epicormic branching often

Figure 3-1 Photo of terminal end of limb segment being measured and a diagrammatic illustration of same system with limb segments and branches delineated.

results in many branches originating from the same location (Ishii and Wilson 2001). On old trees, these branch systems can become very well developed, sometimes containing a dozen or more epicormic branches emerging from a single locus on the trunk. In these instances, the entire epicormic system is given one number, but all of the branch parameters are separately recorded.

On young trees or older trees of many species, branches are simple and conform to a genetically programmed architectural model (Hallé et al. 1978). On such simple branches, biomass parameters of the branch can be predicted from measurements of basal diameter, extension, and percent of foliage cover. As trees age, branches sustain damage from disturbances and reiterate to become asymmetric and individualistic (Ishii et al. 2002). Biomass estimates of these complex branches require three additional measurements:

1. Number of bifurcations in which both forks exceed 4 cm in diameter;
2. Total axis length greater than 4 cm in diameter; and
3. Number of foliar units (see below).

Live axes are distinguished from dead axes, and their lengths are estimated to the nearest 0.5 m.

A *foliar unit* is a species-specific, naturally occurring unit that consists of repeating clusters of stems and foliage that can be counted to quantify foliage on individual branches. This modifies previous methods used to quantify foliage in *Pseudotsuga menziesii* (Pike et al. 1977; Massman 1982). In some tree species, leaf density within a foliar unit varies from tree to tree and along the height gradient, so calibration is needed to maintain accuracy and repeatability. At the beginning of the study, the crew assembles in a tree to discuss the foliar unit for that particular species. Also during sampling, a subset of randomly chosen foliar units are removed from each tree and brought to the ground for further calibration. These destructive samples are then brought to the laboratory

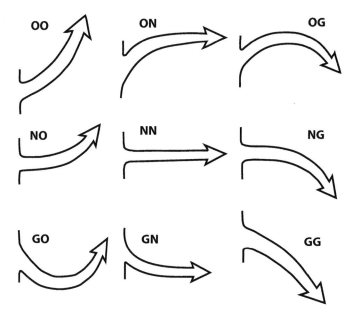

Figure 3-2 Diagrams of the nine possible branch-shape designations. The first letter refers to the basal half of the branch and the second letter refers to the outer half of the branch. O = orthotropic or upward tending, G = geotropic or downward tending, and N= neutral.

for dissection (see the Analyses section below). Trees in which the foliage on branches is sampled via foliar units do not need to have the percent foliated number recorded.

The spread, depth, and centroid of foliage on each branch are also estimated to improve visualizations and modeling of the crown mapping data since the foliage may not be centered on the branch. Spread is the greatest width of foliage associated with a branch in a horizontal plane perpendicular to the branch azimuth. Depth is the greatest length of foliage in a vertical plane. The centroid is the center of mass of all foliage associated with a single branch. Its location is defined in three dimensions by recording its height, distance, and azimuth relative to the main trunk.

To account for all of the foliage on a tree, including leaves on branches less than 2 cm in diameter, live branches in 0 to 1 and 1 to 2 cm diameter classes are counted in 5 m height intervals throughout the crown. In the case of epicormic branch systems, these are given the branch number and height of that system. These data can be further refined by estimating the number of foliar units of these small branches aggregated by height intervals.

As mentioned above, trees with complex branches have a subsample of foliar units removed and brought to the laboratory for dissection. This will result in enough information to complete within-tree and between-tree comparisons. For trees with simple branches, this is not practical. Whole branches are very large in comparison to foliar units, so that the same amount of removal would be destructive to the forest and present a nearly insurmountable task in the laboratory. In these instances, a subset of branches is removed from the set of trees of the same species in the stand, so that the full range of branch sizes is sampled.

Analyses

Trees measured in the field with a laser as well as trees climbed and directly measured are used to develop species-specific equations predicting stem volume from height and DBH. On large trees, DBH can vary greatly between trees due to the presence of buttresses. The addition of height in the regression is essential to help stabilize this variation, especially with old trees. On many tropical trees and *Eucalyptus*, this is not sufficient because the buttresses are too large and variable. One method is to measure above the buttresses with ladders (Condit et al. 1996). Another method is to develop a conversion from measured DBH to a functional DBH, which is calculated by taking the actual basal area and converting back to DBH. This requires using a survey laser and mapping the bases of a subset of trees. The trunk perimeter is mapped, and the area occupied by the fluted base is calculated with a digitizer. The set of diameter measurements is then used with the formula for a conic frustum to estimate the volume of the trunk. A frustum of a cone is simply a piece of a cone that has two parallel ends. The volume of a frustum is calculated as the average cross-sectional area of the two ends multiplied by the length of the frustum. For a trunk of a tree with multiple measured diameters, each section between two measurement points becomes a frustum, all of which are then summed for the total trunk volume. The volumes for branches, limb segments, and reiterations are calculated in the same way.

The size of each tree's crown must be calculated from the ground-based parameters. Simple conic shapes (e.g., cones and paraboloids) are useful for comparing relative crown sizes (Van Pelt and North 1996, Van Pelt and Franklin 2000). While an obvious oversimplification of eccentricities of individual trees, the measured parameters are surprisingly robust in estimating crown volume at the stand-level (Van Pelt and North 1999). The crown radii can be used to calculate a crown-projected area either through mapping individual radii and digitizing or through averaging. This radii is then combined with crown height and the chosen crown form to calculate crown volume.

Sample branches and foliar units are dissected in the laboratory. Components of each branch or foliar unit (leaves, twigs, branchlets, reproductive material, and epiphytes) are separated, oven-dried, and weighed. Component masses are then used as dependent variables, and branch dimensions (basal diameter, extension, and percent foliated) are used as independent variables in stepwise multiple regression analysis. The resulting species-specific equations are used to estimate component masses of all simple branches on each tree. For trees with complex branches, average values for the foliar units are used to estimate foliage biomass of all complex branches on each tree.

To estimate leaf areas, green, undried foliage of each species must be used. A subset of non-overlapping green leaves is counted and digitally scanned to obtain a one-sided estimate of leaf area. These leaves are then dried and weighed, and the specific leaf area (m^2 g) is calculated. This allows the conversion of foliar biomass into leaf area for entire branches and trees.

In the case of thin-barked trees, the cambium may be an important part of the photosynthesis of the whole tree (Pfanz et al. 2000). The calculation of surface area proceeds in the same manner as for estimating volume. The surface area of a frustum is simply the average circumference of the two ends of the frustum multiplied by the length. In order to estimate the surface area on the smaller branches and twigs within the foliar units, these are separated into five categories: 3 to 4 cm, 2 to 3 cm, 1 to 2 cm, 0.5 to 1 cm, and < 0.5 cm, as well as leaves. The smaller branches within the foliar units are separated into five categories to aid in surface area estimations: 3 to 4 cm, 2 to 3 cm, 1 to 2 cm, 0.5 to 1 cm, and < 0.5 cm, as well as leaves. The total mass and linear length of each twig category is recorded for each foliar unit. These are converted to surface area per foliar unit and then combined with the trunk and branch measurements of surface area for a whole-tree surface area estimate.

Destructive samples of foliage biomass, leaf area, and leaf count, whether calculated from foliar units or whole branch estimates, must be applied to the entire trees to come up with whole-tree estimates. In the case of foliar units, because the entire tree was inventoried during the mapping process, values are simply multiplied for a whole-tree estimate. For trees in which entire branches were measured in the field, regression must be used. Independent variables can include branch diameter, branch extension, branch height, and/or percent foliated. The predictive equation is applied to all of the branches within a tree that were not destructively sampled to derive a whole-tree estimate.

Once the tree-based estimates of branch and reiteration volume, foliage biomass, and leaf area are known for all of the mapped trees, we can estimate them for the remaining trees that were not climbed to derive stand-level estimates using multiple, stepwise regression. For each species, potential variables include DBH, height, crown depth, crown diameter, maximum crown radius, height to crown base, crown projection area, and crown volume.

Case Studies in Two Tall Forests

We selected two contrasting old-growth forest stands for detailed study, one in Victoria, Australia (Wallaby Creek), and one in Washington State, United States (Cedar Flats). These stands are similar in stature, but they differ greatly in composition and structure. We established 270 × 27 m plots at each site. Stem maps were made and double-checked prior to climbing any trees.

Site Descriptions

Wallaby Creek   Located in Kinglake National Park, the Wallaby Creek forest straddles the Hume Plateau and helps supply the city of Melbourne with high-quality fresh water (see Table 3-1). There is strong interest in maintaining old-growth forest on the Plateau. Most of the old-growth *Eucalyptus* forests in Victoria have long been felled, but the unique situation of this site has allowed it to persist. It now stands as the tallest angiosperm-dominated forest in the world. The forest is 292 years old, having originated after a catastrophic fire (Ashton 2000), and is dominated by one tree species, *E. regnans* (see Figure 3-3). At least 26 trees over 85 m and at least two over 90 m tall occur here. A *Nothofagus/Atherosperma* rainforest mid-story, which is present in many of the Tasmanian *E. regnans* forests, is not present here. Instead, *Pomaderris aspera*, *Olearia agrophylla*, and tree-ferns (e.g., *Cyathea australis* and *Dicksonia antarctica*) form a dense understory along with occasional *Acacia* trees (*A. dealbata* and *A. melanoxylon*).

We climbed and mapped nine trees at Wallaby Creek. The selected trees represented the full array of sizes and crown structures, which allowed us to extrapolate to the entire stand. The extremely dense understory forced us to alter our standard protocol. Rather than uniformly sample the entire plot, a 20 percent sub-plot (270 × 5.4 m) was used for the *Pomaderris* and *Olearia* component. A complete inventory was made of tree ferns over the entire plot. *Eucalyptus regnans* has an easy-to-recognize foliar unit due to its fast growth and short leaf life (18 months; D. Ashton, personal comm.). Similar foliar units were used for the *Pomaderris* and *Olearia* shrub-trees;

Table 3-1   Summary of Site Characteristics*

Site	Latitude Longitude	Elevation (m)	Age	Slope (%-aspect)	Precipitation (cm)			Temp (°C) Annual (Ave)
					Ann.	Summer	Snowfall	
Wallaby Creek	37° 26' S 145° 11' E	680	292	0-5% S	145	27 (18.6%)	~38	5.0–17.5 (8.9)
Cedar Flats	46° 07' N 122° 01' W	411	~650	0%	317	20 (6.3%)	77	5.0–14.4 (9.7)

*Both sites are on the margins of a Mediterranean climate–with warm, dry summers and mild winters. The Cedar Flats stand is on swampy ground, thus reducing a need for summer precipitation. Precipitation data does not Include condensation from fog, which can add 10-30% at both Sites.

Figure 3-3 Photo of the Wallaby Creek site. Photos are difficult to take once one leaves the roads due to the extremely dense understory. Note that the height of bark exfoliation roughly corresponds to the height of the understory.

these were counted from the ground. For tree ferns, we counted all live fronds on each of the 133 individuals within the plot.

We also quantified whole-tree surface area. The annual exfoliation of bark of *Eucalyptus regnans* allows light to penetrate through the very thin bark to the photosynthetically active bark, which presumably supplements foliar photosynthesis. Near its base, the main trunk retains thick bark, which protects it from low-level fires. This thick bark extends only to the level of the shrub-tree understory before giving way to the thin, smooth bark of the upper tree. We measured the height of this transition on each tree.

Cedar Flats   Located in the Gifford Pinchot National Forest, the Cedar Flats Research Natural Area contains one of the finest *Pseudotsuga menziesii*–dominated forests remaining in the Cascade Mountains (see Table 3-1). The site was originally protected for its representation of *Thuja plicata* swamps, but magnificent 600- to 650-year-old *P. menziesii/Tsuga heterophylla* forests surround the swamps. These are among the largest and tallest known forests in the Pacific Northwest of North America, with individual *Pseudotsuga* up to 90 m tall and 360 cm in diameter (see Figure 3-4). *Thuja plicata* is a common associate, and there is a rich shrub/fern/herb/moss understory topped by *Acer circinatum*. Crowns of the large conifers support lush epiphyte communities dominated by lichens and bryophytes.

We climbed and mapped 17 trees of all three dominant tree species (10 *Pseudotsuga*, 4 *Tsuga*, and 3 *Thuja*). The foliar unit in *P. menziesii* was a 3- to 4-cm diameter branchlet, whose shape varied along the vertical gradient. Foliar units in the upper crown were more compact and dense than those in the lower crown. Because nearly all *Thuja plicata* and *Tsuga heterophylla* branches had simple architecture, basic branch measurements were adequate for foliage estimation in these trees.

Results

Basic Stand Comparisons   Basic stand-level information was summarized for all of the tree species present on the plots, followed by summaries of the dominant species, *E. regnans* and *P. menziesii* (see Table 3-2). Although the density and basal area of *P. menziesii* were only about half as high as *E. regnans*, the mean diameters, heights, and their respective standard deviations were similar. The main structural difference between the two forests was attributable to the absence of tree crowns in the mid-story canopy at Wallaby Creek. The abundance of the small shrub-trees at Wallaby Creek reduces the mean DBH to a value 15 cm smaller than the mean DBH at Cedar Flats, while increasing its standard deviation (see Table 3-2). A similar phenomenon occurred with tree height, although the presence of a tall mid-story provided a more even distribution of tree heights at Cedar Flats, where the standard deviation of height was less than the mean.

These differences in stem number, basal area, and the presence or absence of a mid-story canopy were clearly evident in the trunk diameter distributions of the two forests (see Figure 3-5). The presence of a dominant *Tsuga heterophylla* component at the Cedar Flats site fills the mid-story whereas the large diameter *Thuja plicata* somewhat compensated for the difference in density of the large trees.

Mapped Tree Comparisons   While on ropes, a total of 51,371 measurements were taken on 4,541 branches, 194 limb segments, and 125 reiterated trunks on the 26 mapped trees in this study (see Table 3-3). All tree heights were measured from the ground, but the final number reported is that obtained by direct climbing and measuring. Due to the techniques used to measure tree height and the additional advantage of having lasers, the difference between the ground-based estimates and actual tree heights was small (1.2 percent error).

Clipped foliar units were used to derive the whole-tree foliage biomass estimates for *E. regnans* and *P. menziesii*. Foliage biomass on the foliar units of *E. regnans* averaged 874 g (sd = 263 g, n = 9) while the foliar units of *P. menziesii* averaged 297 g (sd = 66 g, n = 38). The crowns on some of the *P. menziesii* trees at Cedar Flats are among the largest known for that species. A value of over 500 kg of foliage biomass on a single tree is nearly twice that previously reported (Pike et al. 1977). Six of the ten trees climbed had values over 300 kg (see Table 3-3).

The foliar biomass of *Tsuga heterophylla* and *Thuja plicata* were estimated from whole branch samples. The number of collected branches on these two species was too small to develop branch-based estimates of foliage biomass, so data from branches we collected were supplemented with previously published data and equations from felled trees (Brown 1976, 1978; Snell and Anholt 1981; Snell and Max 1985; Means et al. 1994).

Figure 3-4 Photo of the Cedar Flats site. The two rough-barked trees are *Pseudotsuga menziesii*, the others *Tsuga heterophylla*. Note the abundant bryophytes on trunks, branches, and the forest floor.

Table 3-2 Stand Characteristics for Each Site*

Research Site	Stem Count		Mean DBH		σ DBH		Mean Height		σ Height		Basal Area	
	Stand	*Dom*	Stand	*Dom*	Stand	*Dom*	Stand	*Dom*	Stand	*Dom*	Stand	*Dom*
Wallaby Creek	515	51	43.0	205.1	71.6	45.8	18.7	76.6	25.1	6.1	186.6	176.0
Cedar Flats	232	29	58.4	210.8	67.8	50.3	27.4	77.3	23.2	8.2	133.2	86.4

*The first column in each of the six categories listed are shown for all trees, followed by that for the dominant (Dom) tree alone–*Eucalyptus or Psuedotsuga*. Stem count and basal area are per hectare values, and includes all trees >5 cm dbh. Mean diameter and the standard deviation of diameter are in centimeters, mean height and standard deviation of height are in meters and basal area is in m^2.

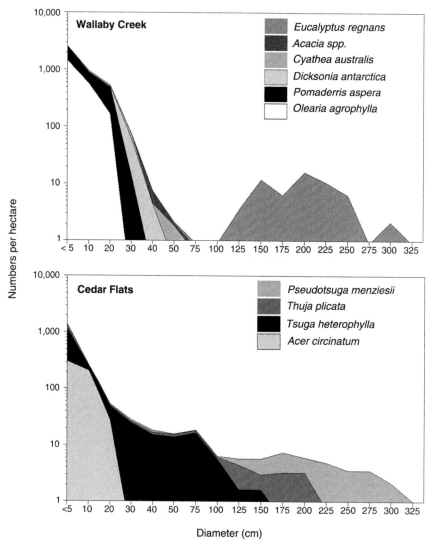

Figure 3-5 Diameter distribution of trees at the two sites. The Y-axis is logarithmic. Note the lack of medium-sized trees at Wallaby Creek.

To determine whole-tree surface area on *E. regnans*, component areas were calculated on all of the trunks, limbs, and branches. The lower portion of the main trunk covered with thick bark was excluded from these calculations. The surface area per foliar unit was multiplied by the total foliar unit count and then added to the remainder of the tree to come up with whole-tree esti-mates of surface area (see Table 3-3).

Visualizations Every measurement taken within the crowns was done in a spatial context so 3-D visualizations were possible. Two perpendicular views of the largest *Eucalyptus* and *Pseudotsuga* within this study were prepared using all of the height, diameter, and extension data for trunks, reiterated trunks, limbs, branch segments, and branches to show these capabilities (see Figure 3-6).

Table 3-3 Summary Statistics of the Mapped Trees from the Two Old-growth Sites*

	DBH	Height	Crown Depth	Crown Spread	# of Branches	# of Limb Segments	Branch Vol	# of Reits	Reit Vol	Main trunk Vol	Total Vol	Foliage Biomass	Bark Surface Area
Wallaby Creek													
Eucalyptus regnans	316	85.5	30.2	21.4	112	33	14.8	2	0.5	134.5	149.7	353 (45)	1,274
	247	91.6	23.6	13.3	49	13	7.1	1	1.2	98.0	106.3	155 (20)	664
	240	79.2	21.2	14.3	63	11	7.2	0	0.0	86.1	93.3	159 (20)	649
	227	79.3	20.7	18.5	54	6	7.1	2	0.2	74.7	82.0	162 (20)	680
	226	85.2	24.2	14.0	54	16	6.7	1	0.1	74.7	81.5	240 (30)	831
	208	82.1	21.1	14.8	73	17	7.2	3	2.7	64.6	74.5	170 (21)	693
	203	75.6	19.3	13.2	36	9	4.5	0	0.0	59.1	63.6	119 (15)	482
	196	78.8	23.7	12.2	24	3	2.1	0	0.0	46.5	48.6	87 (11)	385
	153	73.9	24.4	9.7	23	0	1.9	0	0.0	37.7	39.6	59 (7)	277
Cedar Flats													
Pseudotsuga menziesii	308	86.5	52.0	15.2	426	16	12.2	22	2.1	140.5	154.8	381 (57)	
	281	84.5	30.5	17.8	133	6	9.5	6	0.1	138.0	147.5	206 (30)	
	249	78.1	63.6	14.8	317	3	13.0	6	0.4	130.2	143.7	514 (80)	
	263	79.4	46.9	15.1	225	14	10.3	19	0.6	124.4	135.3	364 (54)	
	266	83.8	60.8	15.4	343	6	11.6	12	1.1	113.2	125.9	397 (60)	
	226	84.5	64.0	16.3	248	6	9.1	8	0.3	108.5	117.9	366 (57)	
	219	83.4	55.6	14.6	330	11	7.5	13	0.3	101.0	108.8	319 (49)	
	231	80.7	44.1	12.7	107	6	3.1	6	0.1	94.8	97.9	245 (75)	
	209	77.7	48.2	10.1	171	2	1.2	2	0.0	57.8	59.0	215 (32)	
	176	70.4	40.0	11.4	72	2	0.7	3	0.0	44.5	45.3	136 (21)	
Thuja plicata	222	51.1	35.7	11.0	356	10	1.7	13	0.5	32.6	34.9	309 (58)	
	204	60.7	46.3	11.6	369	0	1.3	0	0.0	32.0	33.2	288 (57)	
	177	56.0	45.4	12.8	231	4	2.4	6	0.2	25.1	27.6	239 (48)	
Tsuga heterophylla	119	52.8	37.4	11.6	197	0	2.8	0	0.0	17.6	20.4	200 (35)	
	77	48.4	42.0	11.0	139	0	0.7	0	0.0	7.5	8.2	130 (26)	
	82	46.2	40.1	12.0	205	0	0.7	0	0.0	6.8	7.5	149 (24)	
	61	43.0	31.1	10.4	184	0	0.5	0	0.0	3.9	4.4	89 (17)	

*Trees are listed by species in descending order of total volume. DBH is in cm and represents the *functional* diameter (actual basal area reconverted into a diameter). Height and crown dimensions are in m. Reit refers only to reiterated trunks. Volumes are in m³, foliage biomass (σ in parentheses) in kg, and surface area in m². See text for discussion of specific variables.

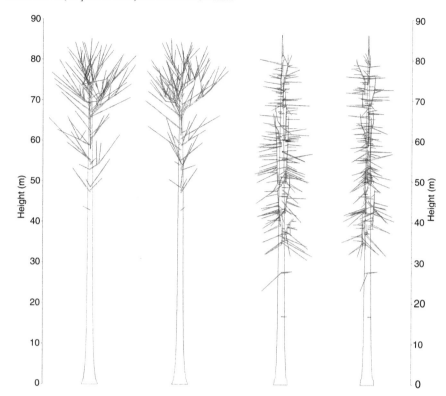

Figure 3-6   Two perpendicular views of the largest tree at each site. The main trunk, all reiterated trunks, limbs, branch segments, and branches are shown.

The 3-D tree maps were used to help prepare a stand profile drawing for each site. These were combined with location data from the stem map, all of the ground-based tree measurements, and detailed sketches and photos made on site (see Figure 3-7). The width of the transect was partly chosen for its usefulness in stand profile visualizations, allowing reasonable illustrations of stem density, tree height, and form. Using the stem map data, we constructed comparative maps including crown projections for the two sites (see Figure 3-7).

Scaling Up to Stand Level   Once whole-tree estimates of branch and reiteration volume, foliage biomass, and leaf area were calculated, we were then able to estimate (via regression) these quantities for the remaining trees that were not climbed to derive stand totals (see Table 3-4). Mapped tree bases for *Eucalyptus* provided the data needed to calculate functional DBH from measured DBH ($r^2 = 0.953$, n = 14). The volume of the main trunk, branches, and limbs were then estimated from the results ($r^2 = 0.978$, n = 14; $r^2 = 0.822$, n = 9). Bark surface area (that portion above the thick-barked base) was also estimated at the stand-level. The thin-bark portion of *Eucalyptus* in the forest was nearly one-quarter the surface area of the leaves. The values for total wood volume and foliage biomass for the *E. regnans* forest are the highest recorded for Australia and are only known to be exceeded by some *P. menziesii* and *Abies procera* forests in the Pacific Northwest of North America, as well as many *Sequoia sempervirens* forests in coastal California (Van Pelt and Franklin 2000; Sawyer et al. 2000).

At Cedar Flats, the wood volume for unmapped trees was predicted from regression equations based on height and diameter (*P. meziesii* $r^2 = 0.949$, n = 42; *T. plicata* $r^2 = 0.956$, n = 23; *T. heterophylla* $r^2 = 0.987$, n = 58). Foliage biomass estimates were regressed on crown volume,

"CANOPY TREKKING": A GROUND-INDEPENDENT, ROPE-BASED METHOD FOR HORIZONTAL MOVEMENT THROUGH FOREST CANOPIES

Roman Dial, Stephen C. Sillett, and Jim C. Spickler

Who among us, having climbed into the canopy, has not wanted to move "just over there" maybe a few meters, maybe into another tree, or maybe even across the entire forest canopy? The canopy raft (Hallé 1990), "web" (Perry and Williams 1981), cranes (Shaw, this volume), booms (Ashton 1995), walkways (see Chapter 23), and zip-lines (Chapter 25) are all realizations of this wish for horizontal freedom within forest canopies. Unfortunately, all of these techniques are expensive with regard to both time and money, are sometimes cumbersome and elaborate, and are potentially harmful to the canopy we wish to study. What many of us really want is the individual freedom to go where we want and to leave little trace of our passage.

The first hint of rope-based horizontal movement came with exposure to arborist techniques for moving within crowns and between trees (Dial and Tobin 1994; Jepson 2000).

Figure 1

Continued

"CANOPY TREKKING"–*cont'd*

Because these techniques use a moving rope rather than single rope technique (reviewed by Moffett and Lowman 1995; Lowman and Wittman 1996), climbers can move through the canopy pulling their ropes along with them, moving for hours or even days in any direction they can situate their ropes.

The principal objective in rope-based horizontal movement through a canopy is to place a climbing rope from one limb across the top of a second limb and to retrieve the rope from under the second limb and back to the climber at the first limb. Then the rope can be anchored, allowing the climber to move horizontally from the first to the second limb. The process of actually moving between limbs is straightforward, and several techniques can be used (Dial and Tobin 1994; Smith and Padgett 1996). What is more problematic is how to (1) get the climbing line over the second limb and (2) retrieve it from under the second limb.

Sillett and his students at Humboldt State University working with inventor Tom Ness came up with a fist-sized mini-grapnel attached to fly-fishing line for retrieving throw bags tossed over distant limbs. This technique described by Ellyson and Sillett (2003) is suitable for distances less than 10 m. For distances greater than 10 m, we employ something called the "magic missile" (see Figure 1). The magic missile is a fiberglass arrow tipped with re-curved tines and a backward pointing barb fired from a 150-pound pull crossbow. The missile i tied to fishing line fed from a crossbow-mounted spinning reel. This setup can be used to retrieve a throw bag line up to 40 m away. We have applied the technique for greater than 500 m horizontal movement through tall canopies in California conifers and hardwood forests in Borneo and Australia (Weintraub 2003).

While orangutans, gibbons, and spider monkeys still mock our technique, it does bring canopy access and movement closer to the ideal of total freedom of movement that is low in weight, cost, and impact to the forest environment.

References

Ashton, P. (1995). Using booms in canopy research. *In* "Forest Canopies" (M.D. Lowman and N.M. Nadkarni, Eds.), pp. 11–12. Academic Press, New York.

Dial, R. and Tobin, S.T. (1994). Description of arborist methods for forest canopy access and movement. *Selbyana* **15**, 24–37.

Ellyson, W.J.T. and Sillett, S.C. (2003). Epiphyte communities on Sitka spruce in an old-growth redwood forest. *The Bryologist* **106**, 197–211.

Hallé, F. (1990). A raft atop the rain forest. *National Geographic* **178**, 129–138.

Jepson, J. (2000). "The Tree Climber's Companion." Beaver Tree Publishing, Longville, Minnesota.

Lowman, M.D. and Wittman, P.K. (1996). Forest canopies: methods, hypotheses, and future directions. *Annual Review of Ecology and Systematics.* **27**, 55–81.

Moffett, M. and Lowman, M.D. (1995). Canopy access techniques. *In* "Forest Canopies" (M.D. Lowman and N.M. Nadkarni, Eds.), pp. 3–26. Academic Press, New York.

Perry, D.R. and Williams, J. (1981). The tropical rain forest canopy: a method providing total access. *Biotropica* **13**, 283–285.

Smith, B. and Padgett, A. (1996). "On Rope: North American Vertical Rope Techniques." National Speleological Society, Huntsville, AL.

Weintraub, B. (2003). Tree snorkeling? A new perspective on forest research. *National Geographic* **203**, 7–8.

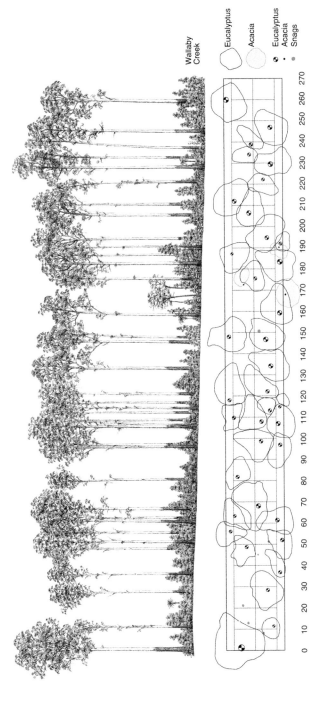

Figure 3-7 Crown profiles and stem maps with crown projections for the two sites. While similar in stature, the two forests are very different architecturally. Wallaby Creek has a higher wood volume but only two-thirds as much foliage. Scale is in meters.

Continued

65

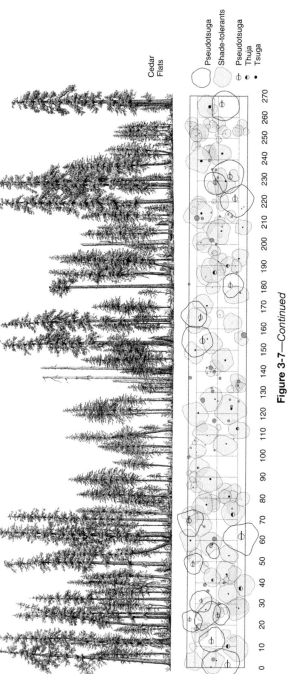

Cedar
Flats

Pseudotsuga
Shade-tolerants

Pseudotsuga
Thuja
Tsuga

Figure 3-7—*Continued*

66

Table 3-4 Summary of Total Stem Numbers, Basal Area, Wood Volume, and Foliage Components at the Two Sites*

Species	Stem Count	Basal Area	Main trunk Volume	Branch Volume	Reiteration Volume	Total Volume	Crown Volume	Foliage Biomass	LAI	SSA
Wallaby Creek										
Eucalyptus regnans	51	176.0	2,902.8	234.3	9.7	3,146.8	101,310	5,931	4.59	24,018
Pomaderris aspera	133	2.6	50.6	34.0	–	84.6	19,510	2,832	4.10	
Olearia agrophylla	145	1.9	16.3	17.9	–	34.2	8,720	610	0.75	
Acacia spp.	4	0.4	2.9	1.1	–	4.0	1,378	124	0.03	
Dicksonia antarctica	155	4.1	–	–	–	–	3,550	991	0.48	
Cyathea australis	27	1.6	–	–	–	–	620	172	0.08	
Total	**515**	**186.6**	**2,972.6**	**287.3**	**9.7**	**3,269.6**	**135,088**	**10,660**	**10.03**	
Cedar Flats										
Pseudotsuga menziesii	29	86.4	1,904.7	116.9	8.7	2,030.3	101,500	6,082	4.26	
Tsuga heterophylla	155	23.4	206.0	18.9	0.7	225.5	76,960	6,949	4.20	
Thuja plicata	23	20.9	273.7	17.3	7.3	298.3	19,100	2,772	2.07	
Acer circinatum	25	2.5	0.9	0.3	–	1.2	5,790	202	0.96	
Total	**232**	**133.2**	**2,385.3**	**153.4**	**16.6**	**2,555.3**	**203,350**	**16,005**	**11.49**	

*Listed are the total stem numbers of each species per hectare, basal area (in m^2 per hectare) by each species, the various wood components (in m^3 per hectare) by species, the main trunk volume, branch volume, crown volume (in m^3 per hectare) by species, foliage biomass (kg per hectare) by species, and leaf area index (LAI – Unitless). The total stem surface area (SSA) per hectare is calculated for only the eucalyptus and excludes the portions of the lower trunks covered by thick bark.

crown height, DBH, height, wood volume, branch volume, and reiteration volume to find the best-fit predictors. For *P. menziesii,* the best predictor was crown volume and the ratio of branch plus reiteration volume to trunk volume. For *T. plicata,* it was DBH, height, and crown height. For *T. heterophylla,* it was DBH and the ratio of crown height to tree height.

Using the ground-based measurements of the crown location of individual trees, the known foliage distribution of the climbed trees, and the foliage biomass estimates for all of the trees, we estimated the vertical distribution of foliage biomass by tree species for the two sites on a per hectare basis (see Figure 3-8). This was accomplished by calculating how much of the crown of each tree will be present at a given height for each 2 m height interval. Based on the crown shape chosen for a given tree (e.g., paraboloid and cone), the proportion of its crown volume was separated into 2-m segments, based on the total crown height, then summed and converted into per-hectare totals.

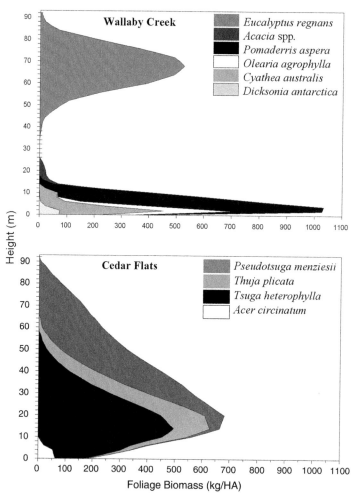

Figure 3-8 Vertical canopy distribution at the two sites. While both forests have foliage distributed for 90 m, it is astounding that the two distributions are nearly the inverse of each other.

Discussion

Accuracy is one obvious strength of direct within-tree mapping over ground-based methods. Trees have been mapped from the ground (Sumida et al. 2001; Nychka and Nadkarni 2003), but this is limited to relatively small trees. Also, only the branching nodes can be mapped; branch diameters cannot be accurately measured from the ground. The resulting stick-figure representation may be useful for some applications but not for others. In the two stands used in this study, for example, it was a challenge to get accurate estimates of tree height, let alone the 3-D coordinates for each node within the crown. Evergreen trees further complicate the problem for ground-based mapping techniques by hiding much of the needed information with foliage.

Biomass estimates have often been obtained by felling trees and then sorting through the debris to estimate such aspects as branch size, wood volume, and foliage biomass (Grier and Logan 1977; Snell and Anholt 1981; D. Ashton, personal communication). The amount of debris and damage to the crowns of large trees after they have fallen reduces the usefulness of the data. Within-tree mapping techniques permit the acquisition of these data with minimal disturbance. Apart from some clipped branches, there is virtually no impact on the forest. This makes the technique useful in many situations where little or no felling is allowed (e.g., national parks and nature reserves).

The use of transects to establish stand maps prior to canopy access enabled us to extrapolate our results from individual trees to the whole stand. Previous studies using within-tree measurements (e.g., Pike et al. 1977; Clement and Shaw 1999; Ishii et al. 2000; Ishii and Wilson 2001) focused on individual trees and were unable to extrapolate to entire stands.

Quantifying the surface area of tree bark, which is easily accomplished using the methods presented here, facilitates studies of photosynthesis and epiphytes. The bark of many trees contains chloroplasts, and bark photosynthesis can substantially contribute to carbon gain at the whole-tree level (Pfanz et al. 2000). Knowing the linear distances between branch nodes is not adequate for such estimations. Only by mapping the diameters of the trunks, limbs, and branches can these estimates be made with a high degree of accuracy. Bark surface area is also useful when attempting to estimate epiphyte biomass on a whole-tree basis. Such estimations require random sampling a known proportion of a tree's surface area and then extrapolating to the whole tree by dividing epiphyte masses by the sampling intensity (Pike et al. 1977; Nadkarni 1984; Rhoades 1981; Hofstede et al. 1993; Ellyson and Sillett 2003). Accurate quantification of bark surface areas on whole trees would greatly improve these estimations.

Quantifying the biomass of foliage in forests has become a high priority for models of ecosystem productivity, stand respiration, and carbon budgets (Chen et al. 2002). Leaf Area Index (LAI), a concept developed in agricultural research for estimating productivity, has often been used as a surrogate for foliar biomass. Its use in forested ecosystems has always been uncertain (Parker 1995; Parker and Brown 2000; Turner et al. 2000; Van Pelt and Franklin 2000). In some simple, plantation forests, LAI is not difficult to obtain. For tall-stature forests, however, these data are nearly impossible to collect from the ground. Optical methods, which are based on the amount of light passing through the canopy, are useless since in complex canopies there is a low correlation between the amount of foliage and the amount of light transmitted by a canopy (Van Pelt and Franklin 2000). Vertical foliar arrangement in forests determines how much light will penetrate much more than the total amount. Direct methods such as line intercept sampling can be are accurate but are extremely time-consuming in tall forests (Thomas and Winner 2000). Remote techniques have promise but are most useful for landscape-level assessments (Lefsky et al. 1999).

The techniques presented here provide a simple, universal protocol for mapping the 3-D structure of forest trees. Since the methods are rope-based, they are not limited to areas that already have a canopy crane, walkway, or other expensive, permanent structure. The nondestructive nature of the protocol allows trees at virtually any location to be considered. The resulting datasets yield information on the 3-D structure of a forest, including foliage biomass, leaf area, bark surface area, and wood volume. They allow quantification of vertical and horizontal distributions of these variables at whole-tree and stand levels.

There are, however, a few disadvantages to our protocol. Chief among them is the need for tree climbers experienced in the use of nondestructive, arborist-style rope techniques. The initial rigging of study trees is particularly difficult for novices. Thus, research budgets must include funds for experienced climbers to rig the trees and train others in safe rope technique. Without proper care, foot traffic and moving ropes involved in tree climbing also have a negative impact on tree surfaces, especially delicate epiphytes. This damage can be minimized by the use of soft-soled shoes and cambium savers (see Jepson 2000).

Another disadvantage of our protocol is a bias toward trees that are safe to climb. Dying and excessively leaning trees are too dangerous and can only be accessed from traverse ropes suspended between sturdy surrounding trees. Thus, our protocol should not be used to study declining forests containing many hazardous trees.

Not only do the methods presented here give an accurate assessment of the amount of foliage of individual trees, but they also yield stand totals. These can be converted into total leaf number or leaf surface area, which are useful for analyzing canopy-atmosphere interactions (Baldocchi et al. 2000), as well as nutrient and gas exchange estimations (Martin et al. 2001).

The strength of our mapping protocol for large trees lies in its simplicity. In this case study, we effectively mapped two tall forests with widely differing architectures. The key is quantification of each major structural component of tree crowns—including the main trunk, reiterated trunks, limbs, branches, and foliar units—and cataloguing these components in a 3-D framework. From this perspective, the differences between the architecture of conifers and angiosperms become trivial; even a banyan (e.g., *Ficus bengalensis*) is mappable using these techniques.

Acknowledgments

This research was supported by the National Science Foundation (BIR 96-30316 and BIR 99-975510) and the Global Forest Society (GF-18-1999-51, GF-2001-18-201, and GF-2001-18-202). Trees were rigged by lead climbers Steve Sillett, Jim Spickler, and Billy Ellyson. Assistant climbers were Marie Antoine, Yoav Bar-ness, Dan Barshis, Matt Dunlap, Kelly Gleason, Tom Greenwood, Mariah Gregori, Karl Gillick, Rich Gvozdz, Brenden Kelley, Steve Rentmeester, Abraham Svoboda, Gary Tryon, Greg Urata, and Robert Van Pelt. Thanks to Peter Frenzen for helping set up accommodations near Cedar Flats in Washington. Thanks also to Ion Maher for providing accommodations and access to facilities at Kinglake National Park in Australia. We thank David Ashton for valuable ecological insights about *Eucalyptus regnans* forests. Judith Cushing and Erik Ordway gave valuable database input to the project. We thank the Computer Applications Laboratory at the Evergreen State College for technical support. We also thank Roman Dial and National Geographic for their assistance.

References

Ashton, D.H. (2000). The environment and plant ecology of the Hume Range, central Victoria. *Proc. Royal Soc. Victoria* **112**, 185–278.
Ashton, D.H. (2002). Personal Communication on February 12th, David Aston visited our site and spent a half-day discussing this forest, his research, and experiences.

Baldocchi, D.J., Finnigan, K., Wilson, U, K.T.P., and Falge, E. (2000). On measuring net ecosystem carbon exchange over tall vegetation in complex terrain. *Boundary-Layer Meteorology* **96**, 257–291.

Brown, J.K. (1976). Predicting crown weights for 11 Rocky Mountain conifers. *In* "Proceedings, Oslo Biomass Studies, IUFRO Congress, June 1976, Oslo, Norway," pp 103–111. IUFRO, Vienna.

Brown, J.K. (1978). Weight and density of crowns of Rocky Mountain conifers. USDA For. Serv. INT Res. Pap. INT-197. 56 pages.

Chen, J., Falk, M., Euskirchen, E., U, K.T.P., Suchanek, T.H., Ustin, S.L., Bond, B.J., Brosofske, K.D., Phillips, N., and Bi, R. (2002). Biophysical controls of carbon flows in three successional Douglas-fir stands based on eddy-covariance measurements. *Tree Phys.* **22**, 169–177.

Clement, J.P., and Shaw, D.C. (1999). Crown structure and the distribution of epiphyte functional group biomass in old-growth *Pseudotsuga menziesii* trees. *Ecoscience* **6**, 243–254.

Condit, R., Hubbell, S.P., and Foster, R.B. (1996). Changes in a tropical forest with a shifting climate: results from a 50 ha permanent census plot in Panama. *J. Trop. Ecol.* **12**, 231–256.

Denison, W.C., Tracy, D.M., Rhoades, F.M., and Sherwood, M. (1972). Direct, nondestructive measurement of biomass and structure in living old-growth Douglas-fir. *In* "Proceedings: Research on Coniferous Forest Ecosystems: A Symposium" (J.F. Franklin, L.J. Dempster, and R.H. Waring, Eds.), pp. 147–158. USDA For. Serv. PNW For. Ran. Exp. Stn. Portland OR.

Ellyson, W.J.T. and Sillett, S.C. (2003). Epiphyte communities on Sitka spruce in an old-growth redwood forest. *Bryologist* 106:197–211.

Grier, C.C. and Logan, R.S. (1977). Old-growth *Pseudotsuga menziesii* communities of a western Oregon watershed: biomass distribution and production budgets. *Ecol. Mono.* **47**, 373–400.

Hallé, F., Oldeman, R.A.A., and Tomlinson, P.B. (1978). "Tropical Trees and Forests: An Architectural Analysis." Springer Verlag, Berlin.

Harmon, M.E., Franklin, J.F., Swanson, F.J., Sollins, P., Gregory, S.V., Lattin, J.D., Anderson, N.H., Cline, S.P., Aumen, N.G., Sedell, J.R., Lienkaemper, G.W., Cromack, K. Jr., and Cummins, K.W. (1986). Ecology of coarse woody debris in temperate ecosystems. *In* "Advances in Ecological Research" (A. MacFadyen and E.D. Ford, Eds.), pp. 133–302. Academic Press, Orlando.

Hofstede, R.G.M., Wolf, J.H.D., and Benzing, D. H. (1993). Epiphytic biomass and nutrient status of a Colombian upper montane rain forest. *Selbyana* **14**, 37–45.

Ishii, H., Clement, J. P., and Shaw, D. C. (2000). Branch growth and crown form in old coastal Douglas-fir. *For. Ecol. Manag.* **131**, 81–91.

Ishii, H., and Wilson, M.E. (2001). Crown structure of old-growth Douglas-fir in the western Cascade Range, Washington. *Can. J. For. Res.* **31**, 1–12.

Ishii, H., Ford, E. D., and Dinnie, C. E. (2002). The role of epicormic shoot production in maintaining foliage in old Pseudotsuga menziesii trees II. Basal reiteration from older branch axes. *Canadian Journal of Botany* **80**, 916–926.

Jepson, J. (2000). "The Tree Climber's Companion." Beaver Tree Publishing, Longville, MN.

Kuiper, L.C. (1994). Architectural analysis of Douglas-fir forests. Ph.D. dissertation. Wageningen Agricultural University, Netherlands. 186 pp.

Lefsky, M.A., Cohen, W.B., Acker, S.A., Parker, G.G., Spies, T.A., and Harding, D. (1999). LIDAR remote sensing of the canopy structure and biophysical properties of Douglas-fir western hemlock forests: concepts and management. *Remote Sens. Env.* **70**, 339–361.

Lyons, B., Nadkarni, N.M., and North, M.P. (2000). Spatial distribution and succession of epiphytes on *Tsuga heterophylla* (western hemlock) in an old-growth Douglas-fir forest. *Can. J. Bot.* **78**, 957–968.

Martin, T.A., Brown, K.J., Kucera, J., Meinzer, F.C., Sprugel, D.G., and Hinckley, T.M. (2001). Control of transpiration in a 220-year-old *Abies amabilis* forest. *For. Ecol. Manag.* **152**, 211–224.

Massman, W.J. (1982). Foliage distribution in old-growth coniferous tree canopies. *Can. J. For. Res.*, **12**, 10–17

Means, J.E., Hansen, H.A., Koerper, G.J., Alaback, P.B., and Klopsch, M.W. (1994). Software for computing plant biomass: BIOPAK users guide. Gen. Tech. Rep. PNW-GTR-340. Portland, OR: U.S. Department of Agriculture, Forest Service, Pacific Northwest Research Station.

Mitchell, A.W., Secoy, K., and Jackson, T. (2002). "Global Canopy Handbook." Global Canopy Programme, Oxford, UK.

Moffett, M.W., and Lowman, M.D. (1995). Canopy access techniques. In "Forest Canopies" (M.D. Lowman and N.M. Nadkarni, Eds.), pp 3–26. Academic Press, NY.

Muir, J. (1901). Our National Parks, p 26. Houghton, Mifflin and Co. New York.

Nadkarni, N.M. (1984). Biomass and mineral capital of epiphytes in an Acer macrophyllum community of a temperate moist coniferous forest, Olympic Peninsula, Washington State. Canadian Journal of Botany **62**, 2223–2228.

Nadkarni, N.M. (1995). Good-bye, Tarzan: the science of life in the treetops gets down to business. The Sciences 35, 28–33.

Nychka, D. and Nadkarni, N.M. (2003). Three-dimensional analysis of the distribution of epiphytes in tropical tree crowns. *Biometrics*. In press.

Parker, G.G. (1995). Structure and microclimate of forest canopies. *In* "Forest Canopies" (M.D. Lowman and N.M. Nadkarni, Eds.), pp 73–106 Academic Press, NY.

Parker, G.G. and Brown, M.J. (2000). Forest canopy stratification: is it useful? *Amer. Nat.* **155**, 473–484.

Perry, D.A. (1978). A method of access into the crowns of emergent and canopy trees. *Biotropica* **10**, 155–157.

Pfanz, H., Aschan, G., and Wittmann, C. (2000). Bark photosynthesis: fact—not fiction. *In* "L'Arbre 2000" (M. Labrecque, Ed.), pp. 117–121. Isabelle Quentin, Montréal, Québec.

Pike, L.H., Rydell, R.A., and Denison, W.C. (1977). A 400-year-old Douglas-fir tree and its epiphytes: biomass, surface area, and their distributions. *Can J. For. Res.* **7**, 680–699.

Rhoades, F.M. (1981). Biomass of epiphytic lichens and bryphytes on *Abies lasiocarpa* on Mt. Baker lava flow, Washington. *Bryologist* **84**, 39–47.

Sawyer, J.O., Sillett, S.C., Libby, W.J., Dawson, T.E., Popenoe, J.H., Largent, D.L., Van Pelt, R., Veirs, S.D. Jr., Noss, R.F., Thornburgh, D.A., and Del Tredici, P. (2000). Redwood trees, communities, and ecosystems: a closer look. *In* "The redwood Forest: History, Ecology, and Conservation of the Coast Redwoods" (R.F. Noss, Ed.), Chapter 4. Island Press, Washington, DC.

Sillett, S.C. (1999). Tree crown structure and vascular epiphyte distribution in *Sequoia sempervirens* rain forest canopies. *Selbyana* **20**, 76–97

Sillett, S.C, and Van Pelt, R. (2000). A redwood tree whose crown is a forest canopy. *Northwest Sci.* **74**, 34–43.

Snell, J.A.K. and Anholt, B.F. (1981). Predicting crown weight of coast Douglas-fir and western hemlock. USDA For. Serv. PNW Res. Pap. PNW-281. 13 pp.

Snell, J.A.K. and Max, T.A. (1985). Estimating the weight of crown segments for old-growth Douglas-fir and western hemlock. USDA For. Serv. PNW Res. Pap. PNW-329. 22 pp.

Sumida, A., Terazawa, I., Togashi, A. and Komiyama, A. (2001). Three-dimensional structure of branches for community-grown trees of a deciduous species, *Castanea crenata*, as related to their crown expansion patterns. *In* "L'Arbre 2000" (M. Labrecque, Ed.), pp. 46–52. Isabelle Quentin, Montréal, Québec.

Thomas, S.C., and Winner, W.E. (2000). Leaf area index in an old-growth Douglas-fir forest: an estimate based on direct structural measurements in the canopy. *Can. J. For. Res.* **30**, 1922–1930.

Turner, D.P., Acker, S.A., Means, J.E., and Garman, S.L. (2000). Assessing alternative allometric algorithms for estimating leaf area of Douglas-fir trees and stands. *For. Ecol. Manag.* **126**, 61–76.

Van Pelt, R. (1995). Understory tree response to canopy gaps in old-growth Douglas-fir forests of the Pacific Northwest. Ph.D. dissertation, University of Washington, Seattle, WA.

Van Pelt, R. and North, M.P. (1996). Analyzing canopy structure in Pacific Northwest old-growth forests using a stand-scale crown model. *Northwest Sci.* **70** (Special Issue), 15–30.

Van Pelt, R., and North, M.P. (1999). Testing a ground-based canopy model using the Wind River Canopy Crane. *Selbyana* **20**, 357–362.

Van Pelt, R., and Franklin, J.F. (2000). Influence of canopy structure on the understory environment in tall, old-growth, conifer forests. *Can. J. For. Res.* **30**, 1231–1245.

CHAPTER 4

Vertical Organization
of Canopy Biota

David C. Shaw

It has not been until recently that ecologists have recognized
a vertical habitat gradation in the vegetation.
—*Joseph C. Dunlavy,*
Studies on the phyto-vertical distribution of birds, Auk 52, *1935*

Introduction

The vertical organization of microclimate, structures, and biota in forests is a recurring theme in scientific investigations and forest description (Allee et al. 1949; Richards 1952; Geiger 1965; Smith 1973; Hallé et al. 1978; Baker and Wilson 2000; Bongers 2001). One aspect of this is the concept of vertical stratification (Richards et al. 1996). Stratification proposes predictable vertical separation of canopy components such as forest leaves and other structures, species, or individual organisms into distinct horizons, layers, or gradients (Smith 1973). Moffett (2000) defined stratification as any non-uniform vertical distribution within vegetation, which can be continuous or discontinuous. When discontinuous, the various layers may be called *strata*. Smith (1973) investigated stratification and concluded that it is generally beneficial in forest trees. He concluded that stratification optimizes light utilization, CO_2 concentrations, pollination, and dispersal; reduces predation on flowers, fruits, and leaves; and increases structural integrity of the forest.

Strict adherence to the proposition of stratification has been criticized as too limiting when in fact there are ecological gradients in three dimensions from forest floor through the canopy that are complex mosaics of biota, microclimate, gaps, and the growth, mortality, and development of trees (Hallé et al. 1978; Parker and Brown 2000). The definitions of stratification often vary with the user, and therefore lose some utility (Parker and Brown 2000; Baker and Wilson 2000). Early naturalists, like Allee et al. (1949), saw stratification in the big picture, as a general phenomenon of animal and plant communities in aquatic (freshwater lakes, oceans, inter-tidal zones) and terrestrial (grasslands, deserts, forests) environments. For example, forests can be divided into soil, forest floor, understory, and overstory.

This chapter includes a short review of the importance of forest composition, age, and structure, as well as microclimate on vertical organization of biota, and then describes general features of functional biotic groups and the primary factors that influence their vertical organization. I use many examples from the Pacific Northwest of North America (PNW) and the Wind River Canopy Crane Research Facility (WRCCRF), where intensive research on canopy ecology has been occurring in a tall stature, old-growth coniferous forest (see Figure 4-1) (Shaw et al. in press).

Forest Composition, Age, and Structure

Forest composition is the fundamental parameter around which all forest biota integrate. However, structure, age, and history are also fundamental in influencing the vertical organization of canopy biota. Forest composition determines substrate for bacteria, fungi, and mechanically dependent plants as well as food and shelter resources for invertebrates and vertebrates. The composition of the forest also influences the microclimate profiles, lower canopy environments, and light quality and quantity. Evergreen conifers allow less light through the canopy than deciduous hardwoods, and this influences forest floor biota and lower canopy environments (Kato and Yamamoto 2002). Differences in seed dispersal mechanisms influence vertebrates that are attracted to trees; for example, conifers have wind-dispersed seeds that attract nuthatches, finches, and crossbills, whereas hardwoods tend to have fruits and nuts that are favored by thrushes, tanagers, and waxwings (Sharpe 1996). Foliage type is important because it influences arthropod and fungi biodiversity, gas and water exchange with the atmosphere, herbivory, and suitability of structure for vertebrates.

Age of individual trees as well as age, or ecological continuity, of the forest have profound influences on forest cohabitants, especially mechanically dependent plants like epiphytes (Barkman 1958; McCune 1993), and wood decay–dependent organisms like cavity-nesting birds (Mannan et al. 1980; Sharpe 1996). Increasing branch and/or tree age is often associated with increasing size and more complex structure, and is linked with higher diversity and abundance of organisms (Clement and Shaw 1999; Nieder et al. 2001). McCune (1993) has demonstrated that epiphyte diversity in Douglas-fir forests increases over time and that some functional groups (N fixers) do not become abundant for several hundred years. The epiphytes sort out along vertical gradients, and as the stand ages, the lower canopy species migrate upward in height. Time is also correlated with increases in tree species diversity, canopy height, and diversity and size of individual tree structures in these Douglas-fir forests (Franklin et al. 2002).

Forest structure, the spatial arrangement of both live and dead tree and plant material, strongly influences the spatial patterns of forest cohabitants due to the fundamental organization of food, shelter, and space. Structure is closely linked with composition, age, and history (including disturbance). Typical tree structures that are vertically organized are leaves, reproductive parts, branches, trunks, butts, roots, and dead wood. Vertical stratification of tree species in forest stands increases structural complexity, accentuates vertical patterns exhibited by other biota, and allows longevity for some species (see Figure 4-2) (Richards 1952; Hallé et al. 1978; Terborgh 1985; Parker et al. 1989; Ishii et al. 2000).

Forest gaps and edges are particularly important in shaping the general vertical organization of canopies, allowing light deeper into the canopy, increasing wind exposure at depth, and facilitating reproduction of certain tree species, lianas, and plants. Edges associated with human-mediated fragmentation of forest ecosystems may have negative consequences on biological diversity and ecosystem function (Franklin and Forman 1987; Laurance and Bierregaard 1997; Foggo et al. 2001), while natural gaps and edges are associated with long-term forest dynamics (Hartshorn 1978; Pickett and White 1985; Lertzman et al. 1996). Epiphyte species restricted to the upper canopy in closed forests can occur almost at ground level along highly exposed edges or in very open forests and woodlands (Benzing 1990; Richards et al. 1996). Lianas can form dense tangles in gaps and along edges (Putz 1995), while gaps and edges influence forest birds (Sharpe 1996).

Leaf characteristics such as size, shape, mass, inclination, chlorophyll content, N content, and photosynthetic capacity vary within a tree crown along gradients of light and other microenvironmental conditions so that the individual plant maximizes carbon gain (Hollinger 1989; Holbrook and Lund 1995; Cermak 1998). Two end-points in this gradient are sun and shade leaves (Kozlowski and Pallardy 1997). Although expressed differently in various tree species, sun

Figure 4-1 The Wind River Canopy Crane Research Facility is a forest ecosystem observatory located in a 60 m tall old-growth temperate coniferous forest in southwest Washington State, USA. The tallest cohort of trees, with complex crowns, are Douglas-fir (*Pseudotusga menziesii*). Western hemlock (*Tsuga heterophylla*) dominates the mid and lower canopy with western redcedar (Thuja plicata) and true firs (*Abies* spp.). Pacific yew (*Taxus brevifolia*) is an abundant small stature conifer that rarely exceeds 20 m in height. Although rainfall exceeds 2,000 mm/yr there are no vascular plant epiphytes in the canopy. Hemlock dwarf mistletoe (*Arceuthobium tsugense*) is the only Angiosperm that occurs above 10 m. Photograph by Jerry Franklin.

leaves are generally denser, arranged more compactly, have higher light-saturated photosynthetic rates, and feature thicker cuticles for resistance to wind desiccation. Shade leaves have less nitrogen and mass and fewer rows of chlorophyll-bearing mesophyll cells. They are thin and large, with lower photosynthetic capacity, few stomates, and are more horizontally oriented to maximize light reception (Holbrook and Lund 1995; Kozlowski and Pallardy 1997).

Forest Micro-Environmental Gradients

The vertical pattern of microenvironmental conditions within forests is important because it influences the distribution of forest biota, behavior of vertebrates, development and growth of tree

A

Figure 4-2 A, The upper canopy of the old-growth forest at Wind River. Note the long deep crowns of these 60 m tall conifers, as well as the low light levels in the understory. Photograph courtesy WRCCRF image archive.

Continued

structures, amount of gas exchange and water release by forest leaves, infection potential for numerous tree parasites, invasiveness of lianas into tree crowns, growth and productivity of epiphytes, biological activity levels of microbes, and numerous other aspects of ecosystem function. The forest influences microclimate, and microclimate influences how and where the forest will grow, so that the interaction of the two results in measured patterns in a dynamic state.

The three-dimensional microenvironment (light, humidity, temperature, and wind) of forests is spatially heterogeneous based on composition and structure of the forest (Geiger 1965; Parker 1995). The density of trees and the number and distribution of gaps in the forest are particularly important. In general, however, as a basic stratification of the forest microclimate, the climate of the upper canopy has more extreme variation in daily environmental characteristics compared to

B

Figure 4-2—*Continued*   B, The lower canopy and forest floor of the old-growth forest at Wind River. The tall shrub, *Acer circinatum*, forms a canopy above shorter shrubs, herbs and ferns. Photograph by Jerry Franklin.

the lower canopy, which is moderated by the overstory structures. Wind, for example, has higher peaks in the upper canopy than down near the forest floor (Figure 4-3). On sunny days, temperature peaks in the upper canopy, just below the tops of the tallest trees, and on cool clear nights, the lowest daily temperature occurs in the upper canopy also. Rainfall may be captured by the canopy and never reach the forest floor. However, Geiger (1965) showed that the lower canopy has more humid conditions than the upper canopy, which may experience light rains and dew, but dries out quickly with exposure and wind.

Light is perhaps the most influential and complex of all canopy microclimatic variables (Canham 1993; Endler 1993; Théry 2001). Most microenvironmental factors correlate with light intensity. Therefore, the effect of microclimate on plant physiology and growth as well as on animal behavior and occurrence is difficult to separate from that of light alone. This makes

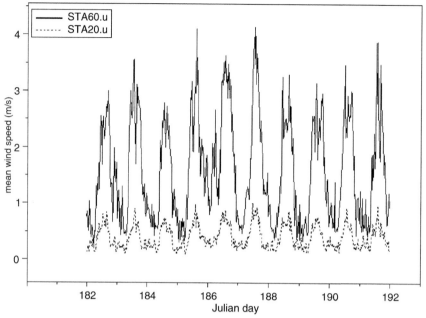

Figure 4-3 Synoptic mean wind speeds at the top of the canopy (STA60.u at 57.0 m) and in the lower canopy (STA20.u at 22.8 m) during early July at the Wind River Canopy Crane Research Facility. Stations are on the tower of the canopy crane. Note the much higher mean wind speeds in the upper canopy. Figure provided by Ken Bible.

light a suitable universal indicator for just about every aspect of forest canopy ecology, but it is very difficult to measure in three-dimensional space. Parker et al. (2002) have proposed a vertical subdivision of canopies into three zones based on the patterns of the mean and variance of vertical light transmittance: bright, transition, and dim. The bright zone (upper canopy) is characterized by high transmittance and low variability, the transition zone (mid-canopy) is where transmittance is most variable and the mean changes rapidly with height, whereas the dim zone (lower canopy) is characterized by low transmittance and variability. Using an age class sequence of Douglas-fir forests near the WRCCRF, Parker et al. (2002) have shown that the bright zone is wide and the dim zone is narrow in older stands due to the vertical differentiation of the canopy with age. Younger forests have a narrow bright zone and a wide dim zone.

Within individual tree crowns are complex microclimatic gradients, often more aligned along the horizontal rather than vertical axis, especially in exposed trees (Sillett et al. 1995; Freiberg 1997). The outer crown creates shade and a barrier to wind, allowing a diversity of organisms to sort out along the interior vs. exterior axis, such as non-vascular and vascular epiphytes (Johansson 1974; Benzing 1990; Sillett 1995; Clement and Shaw 1999) and warblers foraging within tree crowns (MacArthur 1958).

The Vertical Gradient as a Defining Feature of Forests

The vertical gradient is a defining feature of forests because increasing height causes structure and microclimate to be more obviously vertically organized. Height of the canopy and tree size are defining characteristics of forests, yet smaller-stature vegetation types all exhibit vertical patterns (Moffett 2001). The vertical plane is a simple and convenient way to visualize and organize concepts, and it is a first step in understanding the biotic community in a forest.

Fungi and Bacteria

Fungi and bacteria are strongly dependent on substrate; i.e., they consume that which they are on or in, and many are very specialized. Therefore, the vertical distribution of substrate types, in relation to microclimate, will influence the distribution of fungi and bacteria. However, fungi and bacteria are poorly studied in forest canopies, especially their vertical organization. Canopy fungi and bacteria are highly diverse and cause tree diseases and wood decay; live symbiotically in wood, bark and leaves; live epiphytically on all surfaces; decompose perched litter; are important food for mites and other arthropods; absorb nutrients passing through the canopy; and are critical in many other ecological functions. The opportunity for new understanding in forest ecosystems rests on seeking knowledge of microbial biodiversity and ecological roles.

Wood decay fungi (Basidiomycetes; Gilbertson 1980) of live trees may be some of the best-known canopy fungi (particularly in temperate zone forests) because of their economic impacts, ease of study, and macroscopic reproductive structures (conks and mushrooms). Wood decay of live trees is associated with a complex sequence which involves infection ecology of the primary decay fungi, response of the host tree to invasion, and microbial succession within the decay column as well as interactions with invertebrates and vertebrates (Merrill and Shigo 1979; Shigo 1979). Wood decay fungi generally stratify a tree based on position of tree structures and potential entrance courts into the tree; branches, major limbs and trunk(s), top of tree at breaks, lower trunk, tree butt, roots, and major wounds. For example, in an old Douglas-fir tree, I have observed laminated root rot (*Phellinus weiri*) and red brown butt rot (*Phaeolus schweinitzii*)in the roots and butt of the tree, red ring rot (*Phellinus pini*) in the major branches and trunk, and brown top rot (*Fomitopsis cajanderi*) at the broken top. The incidence and abundance of wood decay fungi is a function of tree age (i.e., the older the tree, the more wood decay pockets can be expected). As a forest ages, the total volume of wood decay/hectare in live trees increases also (Foster and Foster 1951).

Surface-living fungi and bacteria (microepiphytes) must be tolerant to light exposure, desiccation, and low nutrients. The vertical distribution within the canopy has been documented in a few studies, which all tend to show a decrease in abundance with height (Reynolds 1972; Carroll 1979a; Stone et al. 1996; de Jager et al. 2001). This is thought to be related to lower available moisture and higher light levels in the upper canopy. However, Carroll et al. (1980) proposed several other factors that might be important in Douglas-fir forests (PNW). Mites that graze on leaf surface fungi are more abundant in the upper canopy and may reduce the standing crop of fungi, while canopy water throughfall is enriched as it passes through the canopy. Therefore, the higher nutrient load in water within the lower canopy may encourage the growth of surface fungi.

Endophytic fungi occur within host tissue, cause no symptoms, are nonpathogenic, may be symbiotic, and inhabit apparently healthy tissues, including leaves, bark, and wood (Stone and Petrini 1997). They are poorly studied, yet they may aid in the defense of the tree because they are antagonistic to herbivorous insects (Carroll 1991). There is little work published on vertical occurrence of these organisms. Gamboa and Bayman (2001) found strong stratification in fungal endophyte communities according to height within *Guarea guidonia* trees in Puerto Rico, while Bernstein and Carroll (1977) found no stratification with height in *Pseudotsuga mensiezii* trees in the Pacific Northwest of North America. Johnson and Whitney (1989) found a negative correlation between density of fungi and height in *Abies balsamea* in Northeast North America. Within an intact forest, Gamboa and Bayman (2001) found higher diversity in all crown layers when compared to similar trees species in open grown conditions, suggesting that high radiation and high water stress may lower endophyte loads within leaves.

Many forest fungi and bacteria disperse via airborne sexual and asexual propagules, or are splashed by raindrops and otherwise water transported. The composition, structure, and resulting microclimatic gradients of the canopy have a profound influence on the transport and survival of these propagules. Airborne fungal spores can be passively or actively launched (Carlile and Watkinson 1994). Once launched, they are dispersed by convection currents that carry spores vertically upward. The vertical location of spore release influences the role of wind and convection in dispersal. In highly stratified canopies, wind may not penetrate into the lower canopy (see Figure 4-4). In general, the effect of eddy diffusion on spores is to increase spore cloud size; however, spore concentration is diluted with increasing distance from the source (Carlile and Watkinson 1994). Humidity also plays a role in spore dispersal. Ascomycetes and Basidiomycetes

Figure 4-4   Smoke released from a Hollywood smoke machine indicates the flow of air in the understory of the forest at WRCCRF and the relative diffusion of fine particles. Note the path is not straight, the smoke hangs together for quite some distance, and seems to accumulate, while dispersing at the end. Imagine the spores of a fungus following this same non-linear path. Photograph courtesy of WRCCRF image archives.

actively discharge spores during damp weather, while *Cladosporium* and *Alternaria* (Deuteromycetes), which have hydrophobic, passively launched spores, release them during dry weather.

Edmonds and Driver (1974) released fluorescent particles and spores of *Fomes annosus* to study dispersion and deposition patterns in relation to meteorological conditions. Atmospheric stability, release time, and presence and density of vegetation influenced dispersion and deposition. The forest canopy had a significant influence on vertical dispersion. Heating of the canopy on sunny days produced an inversion and spores were not able to escape the canopy except in large openings where thermal chimneys carried them up. On clear nights, spores could not escape the canopy due to the temperature inversion that forms above the canopy. Cloudy conditions obscured any inversions, and dispersion through the canopy was limited only by vegetation density. At night or early in the morning is when spore dispersion was greatest. Under these conditions, spores did not escape the canopy but were channeled underneath it. Spore deposition under the forest canopy was greatest during the daytime, however (Edmonds and Driver 1974).

Bacteria are particularly poorly studied in aerial portions of forest trees, but much is known concerning the ecology of plant pathogenic bacteria in crop canopies (Beattie and Lindow 1995). The major environmental stresses for bacteria colonizing leaves are UV radiation and low water availability, which bacteria avoid or tolerate depending on life history and colonization strategy (Beattie and Lindow 1995, 1999). The distinctions among endophyte, epiphyte, and pathogen become blurred due to the variety of colonization strategies, where bacteria may colonize a leaf as an endophyte, and then, after some time, become a pathogen of leaf surfaces. Vertical stratification of bacteria within forest canopies is expected due to the variation in microclimate, leaf type, and other substrates, but is not well studied.

Invertebrates

Invertebrates can be highly mobile, especially adult winged insects. Their vertical occurrence is strongly dependent on food source and microclimate, although predation and behavior are also important. Primary habitats in the canopy include foliage, lichens and bryophytes, vascular epiphytes, bark, wood and decay columns, perched litter and soils, suspended water, flowers, fruits and seeds, and air space. Invertebrate distribution closely follows the stratification of these habitats/structures, although the mobility of many species means they may occur in several habitats during any given day or life cycle. For example, the caterpillar of Johnson's Hairstreak (*Mitoura johnsoni*) in the PNW feeds on the aerial shoots of dwarf mistletoe (*Arceuthobium*) high in tree crowns, whereas the adult butterfly descends to various shrubs for flower rewards (Pyle 2002). On a smaller temporal and spatial scale, Prinzing (2001) found that shifting microclimate mosaics on exposed tree trunks is correlated with occurrence of arthropods at specific sites on the trunk. Diurnal and nocturnal patterns of activity vary with height and microclimate in the biting cycle of mosquitoes (Haddow et al. 1947; Scholl et al. 1979)

Arthropods are reasonably well studied from the perspective of vertical stratification, and most studies, although not all (e.g., see Walter and O'Dowd 1995; Schowalter and Ganio 1998), show stratification (see Figure 4-5). This is particularly strong when contrasting between the understory versus the canopy (Fichter 1939; Adams 1941; De Vries 1988; Sutton 1989; Basset et al. 1992; Schulze et al. 2001). Sutton (1989) summarized our understanding of vertical stratification of insects in tropical rainforests: "In general, it is accepted that the upper canopy is considerably richer in species and individuals than the levels below, although exceptions occur and the effects of topography, diversity of tree architecture, season, and weather may cause local or temporary departures from the norm." Stratification also occurs within the crown of tropical trees. For example, Simon and Linsenmair (2001) sampled arthropods in upper and lower crowns of

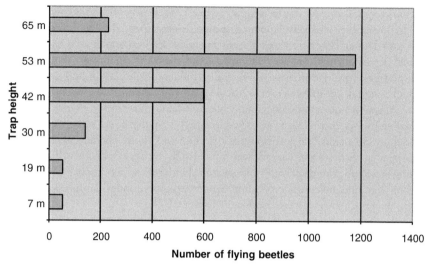

Figure 4-5 Number of flying beetles (Coleoptera) captured by omni-directional plexiglas barrier traps hung on the WRCCRF crane tower at 7 m, 19 m, 30 m, 42 m, 53 m, 65 m, over a single year period. This does not differentiate species patterns but shows the overall stratification of the path flying beetles choose through the forest. The implication of this n = 1 vertical transect is that the upper canopy of this old-growth forest is an optimum travel route. Dominant families included Oedemeridae, Staphylinidae, Elateridae, and Buprestidae. Data provided by David Braun and Melinda Davis.

tropical oaks in Sabah, Borneo, and found distinct communities with Homoptera and ants primarily in the lower canopy, Diptera and Hymenoptera (ants excluded) in the upper canopy, and beetles and several other groups with no differences.

The vertical distribution of flying butterflies in tropical rainforests is a classic example of invertebrate stratification. In the Peruvian rainforests, Papageorgius (1975) studied flight heights of butterflies that use Mullerian mimicry. Wing pattern and color correlated with flight height: the transparent species occurred near the ground; tiger-patterned species were in the understory; red and yellow patterned species flew in the low-mid canopy; blue butterflies were in the upper canopy; and orange species were above the canopy. In Costa Rican rainforests, De Vries (1988), found fruit-feeding nymphalids with short wing lengths and uniform underside patterns occur in the canopy (subfamilies Charaxinae and Nymphalinae), while long wing lengths and underside patterns with eyespots occur in the understory (subfamilies Morphinae and Satyrinae). This reflects the amount of open space (long wing lengths need more open space), variable predation pressures, and was correlated with taxonomy.

Schulze et al. (2001) described the vertical stratification of Lepidoptera in a Bornean rainforest. Fruit-feeding nymphalids showed a decrease in abundance with height in the canopy because rotting fruit is more predictable in the lower canopy levels. Flower-visiting butterflies increased toward higher canopy levels because flowers are more abundant in the upper canopy. In addition, predation by insectivorous aerial-hawking birds may influence butterfly composition and distribution. In this case, Schulze et al. (2001) theorize that butterflies must be stout bodied, strong fliers to occur higher up because aerial-hawking birds were most common in the higher canopy levels. Strong flight is important to escape these birds. In the understory, the nymphalids are slender bodied and not strong fliers.

Insect herbivory is not homogeneous within forest canopies and in individual tree crowns due to insect behavior, leaf nutrient content and toughness, predation factors, and microclimate (Lowman and Heatwole 1992; Basset et al. 1992; Rowe and Pottter 1996; Coley and Barone

1996; Reynolds and Crossley 1997; Williams-Linera and Baltazar 2001). Basset et al. (1992) demonstrated that the rate of herbivory is greater in the understory than in the canopy, even though insect abundance and diversity is higher in the canopy of lowland rainforest in Cameroon. Lowman and Heatwole (1992) reported that eucalypt trees have the highest rate of herbivory in the upper crown compared to other lower crown levels, while Reynolds and Crossley (1997) reported that the lower crowns of red maples had the highest rates of herbivory compared to the upper crowns. Leaf age is also important, as herbivory rates are highest in young and newly developing leaves (Coley 1983; Ernest 1989; Waterman and McKey 1989).

The growth of epiphytes, lianas, and other mechanically dependent plants in the canopy contributes to invertebrate biodiversity, the capture of organic materials and nutrients, and the development of canopy soils (Nadkarni 1994). A complex food web that is an independent subsystem of the forest community can develop (Carroll 1979b). Invertebrates play a critical role in many ecosystem functions of this subsystem, including decomposition, predation, herbivory, pollination, seed dispersal, and plant defense. Plant-invertebrate mutualisms such as epiphyte-ant-gardens and aquatic communities as found in some tank bromeliads illuminate the complexities of canopy life and the interdependencies of invertebrates with other organisms (Dejean et al. 1995; Wallace 1989; Benzing 1990). The vertical distribution of these plants and subsequent influence on structure provide strong controls on the vertical organization of invertebrates.

Canopy soils are a particularly interesting invertebrate habitat due to their patchy spatial distribution and vertical organization (Nadkarni 1994). Soils develop where there is a perch that allows the accumulation of litter, such as under and around epiphytes, on large branches, and in branch crotches. Comparisons of invertebrate communities in forest floor litter and soil to canopy litter and perched soils show similar functional groups of arthropods in both environments but different species of organisms (Longino and Nadkarni 1990; Nadkarni and Longino 1990; Paoletti et al. 1991; Winchester 1997). Forest canopy biodiversity is unique, and invertebrates are serving specialized ecosystem functions—in this case decomposition and nutrient cycling of organic material before it reaches the forest floor. Poorly studied canopy invertebrates such as protozoans, tardigrades, mollusks, and nematodes can be highly specialized and are often associated with mechanically dependent plant communities and perched litter/soil accumulations, reflecting the substrate-driven vertical organization of many invertebrates.

Mechanically Dependent Plants

Epiphytes, mistletoes, climbers, and hemi-epiphytes, including strangler figs, all depend on other plants, usually woody trees or shrubs, for support, and are considered mechanically dependent plants (Richards 1952; Kelly 1985). As plants, they are also united in their dependence on light, water, CO_2, and inorganic nutrients. Their occurrence throughout the vertical structure of a forest reflects the adaptations and ecology of the individual species, with an increasing need for drought tolerance/avoidance with increasing height in the canopy. These organisms can be stationary and restricted to one location, such as an epiphyte, or can dynamically exploit changing microenvironments, such as a vine can do. Most mechanically dependent plants show strong vertical stratification along environmental and substrate gradients. The ecology of these organisms cannot be separated from forest composition, structure, age, disturbance history, and general openness of the canopy, however. For example, vascular epiphytes show little stratification with height in short, open, dry woodlands because there are no strong vertical microclimatic gradients, yet their stratification is well documented in closed rainforest canopies (Benzing 1990).

Epiphytes

Nonvascular epiphytes, lichens, algae, and bryophytes (mosses and liverworts) are often treated separately by ecologists and botanists and show strong vertical organization within canopies due to microenvironment and substrate characteristics (Hale 1952; Richards 1984; Rose 1988; Gradstein and Pocs 1989; Mazimpaka and Lara 1995; Rhoades 1995; McCune et al. 2000; Holz et al. 2002). They do not have roots, they are capable of drying out and rehydrating (poikilohydric), and they depend on atmospheric sources of water and inorganic nutrients (Rhoades 1995). Substrate, inner and outer crown environments, rainfall drainage patterns along the tree branches and trunks, as well as successional patterns and changing microenvironment at individual locations on long-lived branches all influence epiphyte occurrence. The vertical axis is often clearly defined, however, especially in closed canopy forests.

Nonvascular epiphytes are the only epiphytes that occur in an old-growth Douglas-fir/western hemlock forest canopy at the WRCCRF, where a rich community of 111 species of lichens, mosses, and liverworts has developed (McCune et al. 2000). They are clearly stratified along three broad vertical zones:

1. Bryophytes dominate the lower canopy (dim light zone);
2. Cyanolichens (see Figure 4-6) (lichens with a cyanobacteria partner) dominate the mid-canopy (transition light zone); and
3. Alectorioid (*Alectoria, Usnea, Bryoria*) and non-cyanolichens (green algal partner only) dominate the upper canopy (bright light zone) (McCune et al. 1997).

Bryophytes do well in low light, require higher humidity and water, and are less tolerant of desiccation (Burgaz et al. 1994; Mazimpaka and Lara 1995). Most cyanolichens cannot tolerate high levels of UV exposure (Demmig-Adams et al. 1990, Gauslaa and Solhaug 1996), require relatively humid sites, and need liquid water to become physiologically active (Lange et al. 1986). Alectorioid and non-cyanolichens can tolerate higher UV, respond quickly to water availability

Figure 4-6   The cyanolichen *Lobaria oregana* growing in the mid canopy of the forest at the WRCCRF. This lichen is highly stratified within the canopy. It does not occur in the bright light zone of the canopy presumably because it cannot tolerate high UV exposure and excessive desiccation, while requiring a more humid site. Photograph by David Shaw.

by absorbing water from atmospheric humidity, and are more tolerant to wind desiccation and higher temperatures.

Large branches, which occur throughout the vertical crown of old Douglas-fir trees, play a very important role in the distribution of epiphytes, having greater biomass and species diversity than smaller branches as well as allowing bryophytes to occur higher into the canopy (Clement and Shaw 1999). Species diversity varies with height, the top two-thirds of the canopy each having about twice the species of the lower third, while the top two meters of the tallest trees host a unique assemblage of lichens presumably associated with bird roosting, defecations, and propagule transport (McCune et al. 2000). Substrate and other factors play a prominent role in species richness, but height is the factor most likely to explain richness (McCune et al. 2000).

Vascular epiphytes such as orchids, bromeliads, and ferns are also strongly tied to vertical and horizontal microclimate, especially light, which is often hard to uncouple from droughtiness and wind exposure (Benzing 1990). The vertical stratification of vascular epiphyte communities is well documented (Pittendrigh 1948; Richards 1952; Kelly 1985; Benzing 1990; Nieder et al. 2000, 2001). In neotropical rainforests, there is generally a species-poor upper canopy stratum, a species-rich stratum below this, and a fern-rich zone in the lower canopy (Nieder et al. 2001). The ecophysiological requirements and tolerances of vascular epiphytes are the primary control on their vertical positioning within the canopy. Upper canopy epiphytes tend to be CAM (crassulacean acid metabolism) photosynthetic pathway plants that can tolerate drought, desiccation, and high UV (Wallace 1989; Benzing 1990). The branch size, age, and moisture characteristics of the substrate coupled with successional patterns and development of canopy soils are also strong influences (Johansson 1974).

Nieder et al. (2000) surveyed all the vascular epiphytes and hemi-epiphytes, in a tropical rainforest under the Suromoni Canopy Crane on the upper Orinoco River in Venezuela. Their analysis revealed a clear vertical stratification of epiphytes that could be attributed to deterministic factors such as individual plant physiological adaptations and requirements. However, the horizontal distribution (dominant spatial control) was governed by availability of substrate. Pittendrigh (1948) classified the bromeliad epiphytes of forests in Trinidad into three vertically stratified communities: the exposure, sun, and shade-tolerant groups. The exposure plants tended to be drought-adapted, tankless bromeliads or species with large impoundments relative to shoot size, and they occurred in the full sun of the upper canopy in everwet forests while occurring throughout canopies of drier, more open forests. The sun plants occurred in the mid/upper canopy and consisted of tank bromeliads, while the shade-tolerant plants were shallow catchment tank bromeliads that were well watered and had deeper soils available.

Forest age and site environmental factors such as slope position and aspect, proximity to streams or valley bottoms, and overall moisture availability have an influence on the vertical position of epiphytes. Lower canopy organisms gain height with increasing moisture regime and stand age, (defined as ecological continuity as in Barkman 1958) while typical upper canopy plants can occur lower in the canopy in drier or more open forests (Pittendrigh 1948; Hale 1965; Pike et al. 1975; Kelly 1985; Benzing 1990; McCune 1993; Sillett and Nietlich 1996). Cyanolichens are more abundant and occur higher in the canopy in the riparian zone compared to ridgetop forests in mountainous terrain of Oregon (Sillett and Nietlich 1996). The lower tree bases in tropical montane forests of Jamaica have higher species richness and more luxuriant epiphyte community in hollows and bottoms than on slopes (Kelly 1985). Cyclanthaceae seedling establishment heights in a neotropical rainforest were highest in valley bottoms and sites with high forest floor humidity (Cockle 2001).

McCune (1993) developed the similar gradient hypothesis to explain these types of variations on the landscape. Although he focused on Douglas-fir forests of the PNW (Franklin et al. 2002), the hypothesis is applicable to all epiphytes. Three distinct spatial and temporal gradients control epiphyte communities:

1. There are vertical differences in species composition within any given stand due to vertical microclimate and substrate gradients within the forest canopy.
2. There are species-compositional (and relative vertical height) differences among stands differing in moisture regime (wetter is generally more diverse) but of the same composition and age, indicating the importance of geographical setting on epiphyte communities.
3. There are changes (increasing diversity) in species composition (and relative vertical height) through time in a given stand that reflect the changes in stand microclimate and structure as the forest ages, increases in size, and gains height.

This hypothesis is useful in explaining most variation in vertical stratification gradients of epiphyte communities, with perhaps a fourth additional gradient along the axis of "openness" of the canopy (Pittendrigh 1948). A prediction is that there will be a relative displacement downward of the upper (or sun/exposure) canopy epiphytes as the canopy becomes more open.

Mistletoes

Mistletoes are diverse, hemi-parasitic shrubs that utilize the host tree for water, mineral nutrition, and carbohydrates in some cases (Calder and Bernhardt 1983; Benzing 1990; Reid et al. 1995). This ability to tap into the water system of the host may release them from vertical limitations imposed by water availability. Therefore, light/microclimate, physiological adaptations, and infection ecology may provide the strongest controls on the vertical distribution of mistletoes. The mistletoe plant is long-lived; most are stationary and therefore experience a gradient of microclimates as they age. Young plants might be in the exposed outer crown, while older plants may occur in the interior crown as the tree continues to grow. Some mistletoes can grow down branches into the tree trunk, however. Most seeds have a viscous coating and adhere to any substrate. Appropriate infection locations are few as many seeds can only penetrate the host on younger branches with thin bark. There are three groups of seed dispersal types in the more than 1,200 mistletoe species worldwide: (1) avian-dispersed seeds (most mistletoes), (2) explosively-discharged seeds (*Arceuthobium*), and (3) wind-dispersed seeds (*Misodendron*). These dispersal systems have a large influence on the ability of a mistletoe seed to find suitable infection sites within the vertical axis of the canopy.

Most tropical rainforest mistletoes have a marked vertical distribution within closed forest conditions where they occur in the sunniest portions of the taller tree crowns (Richards et al. 1996). In forest openings, or on open-grown trees, the same mistletoes that are limited to upper canopies may occur to ground level. Room (1973) studied *Tapinanthus bangwensis* in cocoa plantations of Ghana. An overstory of native trees is often present in these plantations, and the mistletoe was restricted to unshaded tree canopies in this patchy environment. Room demonstrated experimentally that *T. bangwensis* seeds germinate and establish best in unshaded, rather than shaded, conditions.

Light is correlated with aerial shoot production in North American dwarf mistletoes (e.g., *Arceuthobium tsugense*), where the aerial shoots occur only in the upper sunlit portions of closed-canopy forest, while the aerial shoots occur near ground level at forest edges (Hawksworth and Wiens 1996; Shaw and Weiss 2000). Norton et al. (1997) found that *Alepis flavida* and *Peraxilla tetrapetala* in New Zealand *Nothofagus* forests partitioned the crown of individual hosts. *A. flavida* was restricted to the outer crown, while *P. tetrapetala* was most abundant on inner branches and trunk, and both species were most common in the lower and mid-crown of the hosts. In another study site, *P. colensoi* was most common in the upper portion of the tree crowns (Norton et al. 1997). The authors suggested that forest structure may play a role in this pattern. The forest where *P. colensoi* occurs is relatively dense and even, and bird activity may be focused in the upper canopy, while the canopy where *A. flavida* and *P. tetratpetala* occurred is composed of large scattered *Nothofagus* emergents above the canopy or forming a multi-tiered canopy.

Climbers

Climbers (including lianas, vines, and palms) are a diverse group of plants that are rooted in the ground but use other plants for support to gain height in the canopy (Hegarty 1989; Putz and Mooney 1991; Putz 1995). Their vertical ecology is quite diverse but linked to microclimate profiles, tree species composition (bark texture), tree branching structure, and forest canopy structure. Some vines cover the upper canopy of a tree, suffocating the crown. Others prefer to expand just below the outer crown of a tree or stay in the understory. Four general types of climbing occur (Hegarty 1989):

1. Scramblers often have hooks.
2. Tendril climbers use tendrils and tend to stay in small stature vegetation.
3. Root climbers adhere to trees by penetrating bark, decayed wood, and canopy soils with roots.
4. Twiners are most common, twisting or climbing around boles and branches of trees or even between different trees.

A key factor in vine climbing is the size and distribution of trellises and the ability of the plants to cross between struts (Putz 1995). Vines are strongly responsive to disturbance and light availability, responding to forest gaps, logging, hurricanes, and road building. In some forest gaps or along edges and roadsides, they may form vine carpets over the ground.

On the south island of New Zealand, Baars et al. (1998) found that the height of canopy accessible to lianas (woody climbing plants) was a function of climbing mechanism and liana stem longevity. Twiners, with longer-lived woody stems, reached the highest host canopies. A twiner with herbaceous stems, a tendril climber, and a hook climber were restricted to lower canopy levels. Light environment of the potential host tree, tree height, and support availability were the major features determining susceptibility of trees and shrubs to crown invasion by lianas. In a tropical evergreen forest of the Eastern Ghats in India, Chittibabu and Parthasarathy (2001) studied liana diversity and abundance. Large diameter trees and those with low first trellis height had the most lianas within their crowns.

Hemi-Epiphytes

Hemi-epiphytes are plants rooted in soil at least for some portion of their life, but also have a stage that is completely epiphytic (Kelly 1985; Williams-Linera and Lawton 1995). Some establish in a tree crown and grow down to the ground level (primary hemi-epiphytes). Others establish on the ground, climb into the canopy, and eventually cut connections with the soil (secondary hemi-epiphytes). Stranglers are a unique group of primary hemi-epiphytes that eventually grow to suppress a tree and become free-standing trees themselves. The hemi-epiphytic habit of establishing off the forest floor may be an adaptation to avoid shade and predation in the juvenile stage of the plant, while connecting to the soil as plant size increases may avoid water stress (Williams-Linera and Lawton 1995).

Kelly (1985) found that hemi-epiphytes were most abundant in the mid-canopy of a tropical montane forest in Jamaica. The diversity of hemi-epiphytes precludes any single reason for vertical organization, however. Most hemi-epiphytes have animal-dispersed seed, but some also have wind-dispersed seed. Primary hemi-epiphytes typically begin life associated with branch crotches, large branches with epiphyte mats, and other places where litter accumulates (Williams-Linera and Lawton 1995). Therefore, the distribution of these types of structures may act as a controlling influence on the occurrence of hemi-epiphytes as long as suitable light and water are present. Failure of establishment, and subsequent low competition for space, is considered the primary reason for coexistence of ecologically similar hemi-epiphytic fig species in the dipterocarp forests of Southeast Asia (Laman 1993).

MACAWS: DISPERSERS IN A TROPICAL HABITAT

Sharon Matola

The scarlet macaw, *Ara macao*, is the third largest of the remaining 16 species of macaw parrots (see Figure 1). It is found in tropical forest habitat from southern Mexico into northern South America. In the northernmost part of its geographical range, ornithologists have designated it as a subspecies, *Ara macao cyanoptera*, having more blue in the wings and being larger than its scarlet macaw relatives found south of Nicaragua (Wiedenfeld 1994). Fewer than 1,000 remain in the Selva Maya, the continuous block of tropical forest shared between Belize, Guatemala, and Mexico (Carreon et al. 1998).

The scarlet macaw is principally a seed-eating bird. Since tropical species of trees do not fruit every year, these brilliantly colored macaws forage as part of their feeding ecology. They are known to travel tens of kilometers a day in search of food to sustain their

Figure 1 The scarlet macaw is the third largest macaw parrot in the world and a seed predator/disperser in neotropical systems. Photograph by M.D. Lowman.

populations (Munn 1988). Scarlet macaws have been referred to as "seed predators." However, recent field research has shown their value as seed dispersers in a tropical forest system.

Feeding upon the fruits of the Cohune Palm, *Attalea cohune*, they are known to carry these large, hard seeds away from the parent tree. The seeds then germinate, grow, and thrive (Matola and Sho 2000). Many tropical mammals eat the fruits of the Cohune Palm, and this dispersal mechanism plays an important role in maintaining a healthy forest landscape.

Scarlet macaws frequently nest in the cavities of trees found in riverine forests. They lay between one to three eggs; after 13 weeks, the featherless macaw chicks hatch. Both Macaw parents feed and care for the chicks.

The forest canopy not only provides the scarlet macaw with food, but water is obtained from bromeliads high up in the treetops or from the trunk crotches. Clearly, the tropical forest canopy plays an imperative role in sustaining the remaining populations of scarlet macaws in tropical America. Preserving this important ecosystem is key in seeing that the scarlet macaw will live on for future generations.

References

Carreon, A.G. and Inigo-Elias, E.E. (1998). Final report from the Trinational Workshop for the Conservation of the Scarlet Macaw, *Ara macao*, in the Maya Forest. Chiapas, Mexico. 28–30 September 1998.

Matola, S. and Sho, E. (2000). The Natural History of the Scarlet Macaw, *Ara macao cyanoptera*, and Cultural History of the Central Maya Mountains, Belize. A report to The Wildlife Trust, Palisades, New York, and the Ministry of Natural Resources, Government of Belize.

Munn, C.A. (1988). Macaw Biology in Manu National Park, Peru. *Parrot Letter* **1**, 18–21.

Wiedenfeld, D.A. (1994). A New Subspecies of Scarlet Macaw and its Status and Conservation. *Ornithologia Neotropical* **5**, 99–104.

Vertebrates

Amphibians, reptiles, birds, and mammals are generally at the top tiers of the food chain and are highly mobile. Primary controls on vertical occurrence of vertebrate species include a diversity of factors such as locomotion abilities, forest composition and structure, food location, reproduction requirements, behavior, competition, and predation. Naturalists in both temperate and tropical forests over the past 150+ years have described vertical stratification and habitat partitioning by canopy vertebrates. This focus on vertebrates has reached new levels of sophistication by integrating observation, natural history, species and community ecology, behavior, evolution, climate change and paleo-environments, quantitative analysis and modeling, rare species biology, and conservation in recent decades. In this short review, pains are taken to avoid such complexity.

Arboreality is particularly well developed among vertebrates in tropical rainforests where three out of four species (excluding fish) are fully or partly arboreal (Kays and Allison 2001). Emmons (1995) noted that tropical rainforests have such rich arboreal mammal assemblages because a year-round food base allows for canopy specialization. This probably relates to all the vertebrates since many of them are highly specialized canopy dwellers that use the interconnections of lianas and branches and habitat created by the mechanically dependent plant community (Bourliére 1989; Malcolm 1995; Emmons 1995; Bongers et al. 2001).

In temperate forests and tropical dry forests, canopy specialists are fewer and use of the canopy is more opportunistic. Carey (1996) rated 13 common tree-using mammals in the PNW on a scale of 1 (low use) to 12 (exclusive use). Only one animal received a 12, the red tree vole (*Phenacomys longicaudus*), which consumes Douglas-fir leaves, drinks dew, and is rarely found on the ground. The northern flying squirrel (*Glaucomys sabrinus*), which nests in tree cavities and is a glider, received a 7 because it searches the forest floor for mushrooms and truffles and requires this habitat to survive. The porcupine (*Erethizon dorsatum*) received a 3 because it prefers rock dens and herbaceous plants and climbs trees to supplement its diet with dwarf mistletoe and tree cambium (Carey 1996).

Amphibians and Reptiles

Arboreal amphibians (salamanders and frogs) and reptiles (lizards and snakes) are the least studied of the vertebrates in tropical rainforests (Kays and Allison 2001) and are reflective of two extremes in canopy ecology. Amphibians are generally intolerant of desiccation, whereas reptiles are more tolerant of desiccation. There are no diurnal arboreal frogs present in six neotropical forests studied by Duellman (1989), while most arboreal lizards are diurnal. Amphibians occur higher in the canopy only in the wetter rainforests, such as in the tropics or in the coastal rainforests of temperate regions. Canopy frogs and salamanders are often associated with epiphytes and other mechanically dependent plants, canopy soil and litter accumulations, hollows in trees, and anywhere water can accumulate, such as in tank bromeliads or tree holes (Laurent 1989). Seventeen percent of the anurans at Nouragues, French Guiana, used a constraint water body (water filled tree hole or bromeliad axil) for breeding, while some other species were completely independent of water and a few others laid eggs on foliage overhanging water (Born and Gaucher 2001).

Duellman (1989) divided arboreal lizard microhabitat at Santa Cecilia, Ecuador, first into sun versus shade habitat and then into the following structural features: limbs, trunks, and buttresses of trees, and bushes. The vertical distribution of these microenvironments and forest structural features control vertical occurrence of lizards. Anole lizards in Puerto Rico are stratified within forest canopies, which was related to shade/sun microhabitats and the structural habitat features: animal perch height and perch diameter (Rand 1964; Reagan 1995). *Anolis stratulus* was observed most often in the upper canopy during vertical surveys from towers, yet Reagan (1992) demonstrated that perch size was a more important factor than perch height for determining the vertical position of individuals.

Arboreal snakes are particularly dependent on forest structural features. Climbing performance of rat snakes (*Elaphe obsolete*) was influenced by bark type and vines, the result being a decrease in nest predation of the Acadian Flycatcher (*Empidonax virescens*) if the birds use smooth-barked trees without vines (Mullin and Cooper 2002). The arboreal habit is well represented in tropical snakes (Laurent 1989). The arboreal snakes at Duellman's (1989) six tropical forests generally had equal numbers of diurnal versus nocturnal species. He divided microhabitat for all snakes into categories of aquatic, terrestrial, bushes, and trees, implying these animals sort out along a vertical axis associated with major structural features of the stand. Prey items such as mammals, birds, lizards, frogs, and snails were consumed by arboreal snakes.

Birds

Forest birds have been of considerable interest to scientists, and stratification of avian assemblages in forest canopies is commonly reported, especially with reference to where birds forage (e.g., Dunlavy 1935; Pearson 1971; Karr 1971; Terborgh 1980; Bell 1982; Rinker 2001; Winkler and Preleuthner 2001; Walther 2002a; Shaw et al. 2002). The main factors associated with vertical use are related to life history of the particular species (morphological adaptations, nesting requirements, behavior), vegetation structure, distribution of food resources, vertical microclimate, competition, and predation pressures. Bird anatomy/morphology predisposes them for stratification within a canopy. For example, woodpeckers with stiff tails and specialized feet are adapted for foraging on tree trunks; turkeys for walking and foraging on the ground; red crossbills with a unique bill used for extracting seeds from conifer cones; while kinglets with delicate claws adapted for grasping small branchlets and gleaning small arthropods. Winkler and Preleuthner (2001) have shown that morphological features, such as bill width and depth, tarsus length, and claw characteristics of the third toe, can generally separate many birds into their given stratum in a neotropical rainforest. They did not conclude that morphology alone can explain all factors associated with vertical niche partitioning. Bird behavior, especially foraging behavior, is also very important in determining the vertical preference of a species.

Forest structure emerges as a key feature in forest avian ecology and habitat selection because it is linked to all elements of life history (Cody 1985; Karr 1989). Structural features known to influence bird assemblages include forest height, tree species diversity, bark textures, snags and dead wood, fruit types, leaf characteristics, epiphytes, other dependent plants, gaps, and edges (Karr 1971; Cody 1974, 1985; Robinson and Holmes 1984; Nadkarni and Matelson 1989; Sharpe 1996). Foliage height diversity has been used to explain increasing diversity of birds in forests of increasing height and vertical structural diversity because plant communities of increasing size, diversity, and structure support more available niches (MacArthur and MacArthur 1961).

Tropical rainforest vertical stratification by avian species appears more prominent than in temperate forests (Winkler and Preleuthner 2001). Winkler and Preleuthner (2001) and Walther (2002a) showed that the vertical forest structure of a neotropical rainforest in Venezuela is divided into three strong zones by the avian assemblage: the ground, interior, and canopy. The high contrast in vertical biotic and abiotic environments, corresponding to three common epiphyte zones of exposed, sun, and shade (Pittendrigh 1948), reflects the well-stratified canopy created by tropical forest trees and their associated mechanically dependent plants. The combination of high contrast in vertical microenvironments and year-round food source allows specialization to occur (as in the argument for mammal diversity; see Emmons 1995).

In temperate forests, most avian activity is at the ground level and in the lower canopy (Anderson et al. 1979). However, Shaw et al. (2002), at a single site, divided the 60-m tall canopy at WRCCRF into three vertical zones of equal height and used fixed-area point counts to quantify differences in avian occurrence (see Figure 4-7). During March through October, equal numbers of birds were detected in each zone, while the distribution shifted to the upper canopy during the winter. The upper canopy of this old-growth conifer forest has well-developed tree structures and epiphytic plant (lichen) communities. The environment is open and brightly lit from about 40 m and above, which may create a preferred habitat of sun-warmed foliage and branches in winter. Old-growth temperate conifer forests may provide more habitat features in the upper canopy than young forests.

In general, species near the ground have narrower niche breadth than species living in the higher canopy levels (Orians 1969; Cody 1974; Terborgh 1980). This may be because birds of the canopy must deal with a wider range of microenvironmental conditions and changing food resources. At the Venezuelan Surimoni Canopy Crane site, Walther (2002a, b) and Winkler and Preleuthner (2001) have shown that birds divide the canopy into three distinct strata, rather than the traditional two strata (ground vs. canopy) defined by earlier ornithologists. Using this differentiation, Walther (2002b) demonstrated that mid-story species forage in a broader stratum than the understory or canopy species. He concludes that most avian studies in tropical rainforests do not have the type of canopy access necessary to make accurate analysis of the vertical distribution of the whole avian assemblage.

VERTICAL STRATIFICATION AMONG NEOTROPICAL MIGRANTS

H. Bruce Rinker

Birds often stratify their foraging habits in forest communities, yet few attempts have been made to quantify this vertical distribution due to access difficulties, expense, and the false belief that ground-based observations of canopy birds are adequate. The dense foliage of northern temperate forests during the spring and summer is also a confounding factor for accurate sampling in this ecosystem type, thus necessitating direct canopy access.

Neotropical migrants reflect the regional health of the environment because they may interconnect environmental conditions in one area with another thousands of kilometers away (Marra et al. 1998; Sillett et al. 2000). Consequently, migratory songbirds can be used as a kind of litmus test for the overall soundness of ecological regions (see Figure 1). The heterogeneous vertical elements regulating neotropical bird foraging behavior, distribution, species fragmentation effects, and invasion of exotics are as important and must be studied together with the horizontal components.

Niches used by forest birds occupy a three-dimensional matrix in the ecosystem. Analysis of the entire avian community is indispensable for forest conservation. Recent developments of reliable canopy-access systems make the treetops safely accessible to nearly everyone interested in bird conservation (Rinker 2000, 2001a).

Rinker (2001a, 2001b) documented habitat selection among neotropical migrants in a mature oak-maple forest near Millbrook, New York, using a forest canopy walkway. After four consecutive migration seasons, volunteers banded 139 birds that represented 36 species. Of the 36 species recorded at Millbrook, 35 percent of the total capture to date were exclusively canopy species, 35 percent exclusively ground species, and 31 percent were found in both strata. An earlier study at the Hampshire College Walkway in Massachusetts found 45 percent, 41 percent, and 14 percent, respectively (Stokes 1997).

Rinker's study addresses some of the inherent biases in ground-based netting by sampling canopy and ground migrants simultaneously. The Millbrook project also allowed comparison of forest data with other ecological types of vegetation. Both the Hampshire and the Millbrook

Figure 1   The Kentucky warbler (*Oporornis formosus*) is a common neotropical migrant that nests in moist deciduous woods of southeastern North America but winters in Central America. Photograph by L. John Trott.

pilot studies clearly call for more intensive banding in all layers of forest vegetation. A coordinated effort among existing North American walkways and other canopy access systems is needed for a comprehensive view of the avifaunal forest mosaic.

Long-term banding programs in the treetops of North America and beyond can provide a clearer picture of strata utilization than that acquired through ground-based netting. They can also address issues of impact. For example, what are the long-term effects of aerial walkways on surrounding vegetation and wildlife populations? Understanding the three-dimensional matrix of bird activity in our forests will undoubtedly provide clues about the biological health and integrity of this complex natural resource.

References

Marra, P.P., Hobson, K.A., and Holmes, R.T. (1998). Linking winter and summer events in a migratory bird by using stable-carbon isotopes. *Science* **282**, 1884–1886.

Rinker, H.B. (2000). Conservation in the treetops: environmental action in the emerging science of canopy ecology. *www.actionbioscience.org/environmental/rinker.html*

Rinker, H.B. (2001a). Halfway between heaven and earth: bird conservation in the treetops. *Bird Watchers Digest* **23**(5), 60–64.

Rinker, H.B. (2001b). The use of a forest canopy walkway for studying habitat selection by neotropical migrants. *Selbyana* **22**(1), 89–96.

Sillett, T.S., Holmes, R.T., and Sherry, T.W. (2000). Impacts of a global climate cycle on population dynamics of a migratory songbird. *Science* **288**, 2040–2042.

Stokes, A.E. (1997). Comparison of Mist Net Captures of Migrating Songbirds from Two Strata in a Southern New England Forest. Master's Thesis, University of Massachusetts, Amherst, Massachusetts.

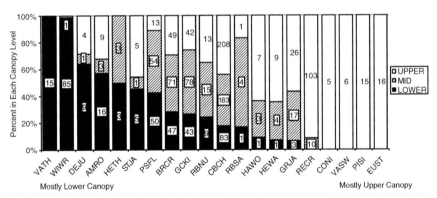

Figure 4-7 The proportion of observations in the lower, mid, and upper canopy of the twenty most common small birds observed at the Wind River Canopy Crane Research Facility during a 3 year period. There is a general stratification of the canopy, but most birds may occur in all zones. Two birds appear to be understory specialists, while 5 birds are generally upper canopy specialists in this forest. Fifteen of these birds were significantly more abundant in certain zones, and therefore considered to stratify the canopy. Species codes are: VATH=Varied Thrush, WIWR=Winter Wren, DEJU=Dark-eyed Junco, AMRO=American Robin, HETH=Hermit Thrush, STJA=Steller's Jay, PSFL=Pacific-slope Flycatcher, BRCR=Brown Creeper, GCKI=Golden-crowned Kinglet, RBNU=Red-breasted Nuthatch, CBCH=Chestnut-back Chickadee, RBSA=Red-breasted Sapsucker, HAWO=Hairy Woodpecker, HEWA=Hermit Warbler, GRJA=Gray Jay, RECR=Red Crossbill, CONI=Common Nighthawk, VASW=Vaux's Swift, PISI=Pine Siskin, EUST=European Starling. (From Shaw et al. 2002).

Mammals

Vertical canopy occurrence in mammals is related to individual life history, animal morphology, behavioral adaptations, season of the year, forest structure, food, predation, shelter, animal psychology, and social structure. Canopy mammals are very difficult to study because many are nocturnal (Malcolm 1995). When vertical components are measured, stratification is often evident (Malcolm 1991; Packer and Layne 1991; Taylor and Lowman 1996; Youlatos and Gasc 2001; Guillemin et al. 2001).

Emmons (1995) and Emmons and Gentry (1983) summarized the importance of forest structure on mammals of the canopy by comparing tropical forest mammals of Africa, Asia, and the neotropics. Tropical rainforests of Africa and the neotropics have closed upper forest canopies in which lianas climb into and interweave within the canopy, thereby forming connectivity. In the Asian tropics, the closed liana influenced portion of the canopy is below more widely scattered taller trees and towering emergents. Gliders are particularly abundant in Asia because the ability to get from isolated crown to isolated crown is actively selected while this ability does not gain much in the closed tropical rainforests of Africa and the neotropics.

At the rainforest research site in Nouragues, French Guiana, two examples of forest canopy vertical use by mammals include arboreal monkeys (Youlatos and Gasc 2001) and marsupials (Guillemin et al. 2001). Large primates tend to use the higher forest layers, eating fruit and leaves. They are less susceptible to aerial predators because of their size. The smaller monkeys stay in the lower canopy and understory where they eat insects and utilize lianas and other vertical supports. They are protected from both ground-based and aerial predators in this habitat. The omnivorous capuchin monkeys exploit all forest layers. This same pattern has been described for other neotropical forests (Terborgh 1983). The marsupial species was divided into three size classes: large (1200 g), medium (300 to 500 g), and small (20 to 100 g). Guillemin et al. (2001) sampled the forest canopy and ground. In the canopy, they captured one species each of small, medium, and large marsupials, whereas they captured two species each of large, medium, and small marsupials on the forest ground.

Bats have access to the open spaces within and above the canopy. Stratification of the forest canopy by bat species is related to density of vegetation (structure), food type and location, foraging behavior, time of day (night), and roost locations (Wilson 1989; Wunder and Carey 1996; Kalcounis et al. 1999; Hayes and Gruver 2000; Kalko and Handley 2001). Although stratification is generally acknowledged, Kalko and Handley (2001) believed that the vertical distribution of bats in neotropical rainforests is less pronounced than that of birds. Most authors agree, however, that forest bats are poorly studied in mid, upper, and above the canopy areas. When comparisons are made between forest floor and canopy habitats, vertical stratification is observed for many, but not all, species (Kalcounis et al. 1999; Hayes and Gruver 2000; Kalko and Handley 2001).

Because of their high wing loading, large bats generally find highly cluttered environments difficult to maneuver in, so are more common in open parts of the forest and above the canopy. Smaller bats with shorter, broader wings can negotiate the dense structure and are often found in the forest interior although not limited to it (Kalko and Handley 2001). Using echolocation detectors in an old-growth forest at the WRCCRF above the canopy and in the upper, lower, and ground levels of the canopy, Hayes and Gruver (2000) showed that *Myotis* bats are most abundant in the lower canopy and absent from the above canopy area. Non-*Myotis* bats are most abundant in the lower and upper canopy but also occur in significant numbers in the ground-level and above-canopy zones. They noted a bimodal activity pattern with increasing use of the higher canopy levels by *Myotis* bats as the night progressed.

Conclusions

A dynamic interaction between plant development and the vertical microclimate generally results in stratification of botanical elements in the forest canopy. Animals and microbes key into different aspects of forest structure (including food distribution) and microclimate. Stratification varies in intensity with species composition, geographical setting, density, stand openness and other forest structural factors. In general, however, forest stratification appears to reach its zenith in closed-canopy, primary (old-growth) tropical rainforests, particularly in the neotropics where the interaction of mechanically dependent plants and forest canopy trees creates a complex of structures and year-round resources below the exposed upper canopy. In temperate zone forests, stratification is best developed in closed canopy stands (see Figure 4-8) and is particularly well developed in old-growth forests.

The vertical positioning of canopy biota and tree structures varies on the landscape along environmental, time, and forest structural gradients. The Similar Gradient Hypothesis (McCune 1993) captures the dynamic gradients (vertical, site moisture regime, canopy openness, age) in the canopy that influence vertical positioning of canopy biota in space and time. Other gradients such as temperature, elevation, and wind exposure are also important if one seeks to explain observed differences in canopy organization across the landscape. Although the Similar Gradient Hypothesis was created to explain epiphyte distribution, it is an excellent unifying model for vertical occurrence of canopy biota and development of tree structures. Given similar compositional gradients:

1. Within any stand of vegetation there exists a vertical gradient of microclimate and a subsequent vertical sorting of composition, structures, and associated biotic species (bacteria and fungi, mechanically dependent plants, invertebrates, and vertebrates).
2. In forests of the same composition and age, there are gradients along landscape moisture regimes (wet to dry axis) that influence composition and relative vertical height of the associated biotic species.

Figure 4-8   A 70-year-old western hemlock forest on the Pacific coast of Washington State, USA. The stand regenerated after a major wind storm blew down the previous forest. Note the highly stratified canopy, most foliage is in the upper 1/3 of the forest and there are no plants in the understory. Photograph by David Shaw.

3. In forests of the same composition and age, there are gradients of canopy openness that influence tree structure and the vertical position of associated biotic species.
4. In a given stand, there is a time gradient that results in changes in the composition and relative vertical height of associated biotic species with forest age (or ecological continuity).

References

Adams, R.H. (1941). Stratification, diurnal and seasonal migration of the animals in a deciduous forest. *Ecological Monographs* **11**, 189–227.

Allee, W.C., Park, O., Emerson, A.E., Park, T., and Schmidt, K.P. (1949). "Principles of Animal Ecology." W.B. Saunders Company, Philadelphia and London.

Anderson, S.H., Shugart, Jr., H.H., and Smith, T.M. (1979). Vertical and temporal habitat utilization within a breeding bird community. *In* "The Role of Insectivorous Birds in Forested Ecosystems" (J.G. Dickson, R.N. Conner, R.R. Fleet, J.A. Jackson, and J.C. Kroll, Eds.), pp. 203-216. Academic Press, New York.

Baars, R., Kelly, D., and Sparrow, A.D. (1998). Liane distribution within native forest remnants in two regions of the South Island, New Zealand. *New Zealand Journal of Ecology* **22**, 71–85.

Baker, P.J. and Wilson, J.S. (2000). A quantitative technique for the identification of canopy stratification in tropical and temperate forests. *Forest Ecology and Management* **127**, 77–86.

Barkman, J.J. (1958). Phytosociology and Ecology of Cryptogamic Epiphytes. Van Gorcum, Assen, The Netherlands.

Basset, Y., Aberlenc, H., and Delvare, G. (1992). Abundance and stratification of foliage arthropods in a lowland rain forest of Cameroon. *Ecological Entomology* **17**, 310–318.

Beattie, G.A. and Lindow, S.E. (1995). The secret life of foliar bacterial pathogens on leaves. *Annual Review of Phytopathology* **33**, 145–172.

Beattie, G.A. and Lindow, S.E. (1999). Bacterial colonization of leaves: a spectrum of strategies. *Phytopathology* **89**, 353–359.

Bell, H.L. (1982). A bird community of New Guinean lowland rainforest. 3. Vertical distribution of the avifauna. *Emu* **82**, 143–162.

Benzing, D.H. (1990). "Vascular Epiphytes." Cambridge University Press, Cambridge, UK.

Bernstein, M.E. and Carroll, G.C. (1977). Internal fungi in old-growth Douglas-fir foliage. *Canadian Journal of Botany* **55**, 644–653.

Bongers, F. (2001). Methods to assess tropical rain forest canopy structure: an overview. *Plant Ecology* **153**, 263–277.

Bongers, F., Charles-Dominique, P., Forget, P., and Thery, M. (2001). "Nouragues, Dynamics and Plant-Animal Interactions in a Neotropical Rainforest." Kluwer Academic Publishers, Dordrecht, Netherlands.

Born, M. and Gaucher, P. (2001). Distribution and life histories of amphibians and reptiles. *In* "Nouragnues, Dynamics and Plant-Animal Interactions in a Neotropical Rainforest" (F. Bongers, P. Charles-Dominique, P. Forget, and M. Thery, Eds.), Chapter 15, pp. 167–184. Kluwer Academic Publishers, Dordrecht, Netherlands.

Bourliére, F. (1989). Mammalian species richness in tropical rainforests. *In* "Vertebrates in Complex Tropical Systems" (M.L. Harmelin-Vivien and F. Bourliere, Eds.), Chapter 6, Springer-Verlag, New York.

Burgaz, A.R., Fuentes, E., and Escudero, A. (1994). Ecology of cryptogamic epiphytes and their communities in deciduous forests in Mediterranean Spain. *Vegetatio* **112**, 73–86.

Calder, M. and Bernhardt, P. (1983). "The Biology of Mistletoes." Academic Press, Sydney.

Canham, J.A. (1993). The color of light in forests and its implications. *Ecological Monographs* **63**, 1–27.

Carey, A.B. (1996). Interactions of Northwest forest canopies and arboreal mammals. *Northwest Science*, **70** (Special Issue), 72–78.

Carlile, M.J. and Watkinson, S.C. (1994). "The Fungi." Academic Press, London.

Carroll, G.C. (1979a). Needle microepiphytes in a Douglas-fir canopy: biomass and distribution patterns. *Canadian Journal of Botany* **57**, 1000–1007.

Carroll, G.C. (1979b). Forest canopies: complex and independent subsystems. *In* "Forests: Fresh Perspectives from Ecosystem Analysis. Proceedings of the 40th Annual Biology Colloquium" (R.H. Waring, Ed.), pp 87–108. Oregon State University, Corvallis.

Carroll, G.C. (1991). Fungal associates of woody plants as insect antagonists in leaves and stems. *In* "Microbial Mediation of Plant-Herbivore Interactions" P. Barbosa, V.A. Krischik, and C.G. Jones, Eds.), pp. 253–271. Wiley, New York.

Carroll, G.C., Carroll, F.E., Pike, L.H., Perkins, J.R., and Sherwood, M. (1980). Biomass and distribution patterns of conifer twig microepiphytes in a Douglas-fir forest. *Canadian Journal of Botany* **58**, 624–630.

Cermak, J. (1998). Leaf distribution in large trees and stands of the floodplain forest in southern *Moravia. Tree Physiology* **18**, 727–737.

Chittibabu, C.V. and Parthasarathy, N. (2001). Liana diversity and host relationships in a tropical evergreen forest in the Indian Eastern Ghats. *Ecological Research* **16**, 519–529.

Clement, J.P. and Shaw, D.C. (1999). Crown structure and the distribution of epiphyte functional group biomass in old-growth *Pseudotsuga menziesii* trees. *EcoScience* **6**, 243–254.

Cockle, A. (2001). The dispersal and recruitment of Cyclanthaceae and Philodendron (Araceae) understorey root-climbing vines. *In* "Nouragnues, Dynamics and Plant-Animal Interactions in a Neoptropical Rainforest" (F. Bongers, P. Charles-Dominique, P. Forget, and M. Thery, Eds.), pp. 251-263. Kluwer Academic Publishers, Dordrecht, Netherlands.

Cody, M.L. (1974). "Competition and the Structure of Bird Communities." Princeton University Press, Princeton, New Jersey.

Cody, M.L. (1985). "Habitat Selection in Birds." Academic Press, Orlando.

Coley, P.D. (1983). Herbivory and defensive characteristics of tree species in a lowland tropical forest. *Ecological Monographs* **53**, 209–233.

Coley, P.D. and Barone, J.A. (1996). Herbivory and plant defenses in tropical forests. *Annual Review Ecology and Systematics* **27**, 305–335.

de Jager, E.S., Wehner, F.C., and Korsten, L. (2001). Microbial ecology of the mango phylloplane. *Microbial Ecology* **42**, 201–207.

De Vries, P.J. (1988). Stratification of fruit-feeding numphalid butterflies in a Costa Rican rainforest. *Journal of Research on the Lepidoptera* **26**, 98–108.

Dejean, A., Olmsted, I., and Snelling, R.R. (1995). Tree-epiphyte-ant relationships in the low inundated forest of Sian Ka'an Biosphere Reserve, Quintana Roo, Mexico. *Biotropica* **27**, 57–70.

Demmig-Adams, B., Maguas, C., Adams III, W.W., Meyer, A., Kilian, E., and Lange, O.L. (1990). Effect of high light on the efficiency of photochemical energy conversion in a variety of lichen species with green and blue-green phycobionts. *Planta* **180**, 400–409.

Duellman, W.E. (1989). Tropical herpetofaunal communities: patterns of community structure in Neotropical rainforests. *In* "Vertebrates in Complex Tropical Systems" (M.L. Harmelin-Vivien and F. Bourliere, Eds.), Chapter 3. Springer-Verlag, New York.

Dunlavy, J.C. (1935). Studies on the phyto-vertical distribution of birds. *Auk* **52**, 425–431.

Edmonds, R.L. and Driver, C.H. (1974). Dispersion and deposition of spores of *Fomes annosus* and fluorescent particles. *Phytopathology* **64**, 1313–1321.

Emmons, L.H. (1995). Mammals of rain forest canopies. In "Forest Canopies" (M.D. Lowman and N.M. Nadkarni, Eds.), pp. 199-223. Academic Press, San Diego.

Emmons, L.H. and Gentry, A.H. (1983). Tropical forest structure and the distribution of gliding and prehensile-tailed vertebrates. *American Naturalist* **121**, 513–524.

Ernest, K.A. (1989). Insect herbivory on a tropical understory tree: effects of leaf age and habitat. *Biotropica* **21**, 194–199.

Endler, J.A. (1993). The color of light in forests and its implications. *Ecological Monographs* **63**, 1–27.

Fichter, E. (1939). An ecological study of Wyoming spruce-fir forest arthropods with special reference to stratification. *Ecological Monographs* **9**, 183–213.

Foggo, A., Ozanne, C.M.P., Speight, M.R., and Hambler, C. (2001). Edge effects and tropical forest canopy invertebrates. *Plant Ecology* **153**, 347–359.

Foster, R.E. and Foster, A.T. (1951). Studies in forest pathology VIII. Decay of western hemlock on the Queen Charlotte Islands, British Columbia. *Canadian Journal of Botany* **29**, 479–521.

Franklin, J.F. and Forman, R.T. (1987). Creating landscape patterns by forest cutting: ecological consequences and principles. *Landscape Ecology* **1**, 5–18.

Franklin, J.F., Spies, T.A., Van Pelt, R., Carey, A.B., Thornburgh, D.A., Berg, D.R., Lindenmayer, D.B., Harmon, M.E., Keeton, W.S., Shaw, D.C., Bible, K. and Chen, J. (2002). Disturbances and structural development of natural forest ecosystems with silvicultural implications, using Douglas-fir forests as an example. *Forest Ecology and Management* **155**, 399–423.

Freiberg, M. (1997). Spatial and temporal pattern of temperature and humidity of a tropical permontane rain forest tree in Costa Rica. *Selbyana* **18**, 77–84.

Gamboa, M.A. and Bayman, P. (2001). Communities of endophytic fungi in leaves of a tropical timber tree (*Guarea guidonia*: Meliaceae). *Biotropica* **33**, 352–360.

Gauslaa, Y. and Solhaug, K.A. (1996). Differences in the susceptibility to light stress between epiphytic lichens of ancient and young boreal forest stands. *Functional Ecology* **10**, 344–354.

Geiger, R. (1965). "The Climate Near the Ground." Harvard University Press, Cambridge, MA.

Gilbertson, R.L. (1980). Wood-rotting fungi of North America. *Mycologia* **72**, 1–49.

Gradstein, S.R. and Pocs, T. (1989). Bryophytes. In "Tropical Rain Forest Ecosystems" (H. Lieth and M.J.A. Werger, Eds.), pp. 311-325. Elsevier Science Publishers, Amsterdam.

Guillemin, M, Atramentowicz, M., and Julien-Laferriére, D. (2001). The marsupial community. *In* "Nouragnues, Dynamics and Plant-Animal Interactions in a Neotropical Rainforest" (F. Bongers, P. Charles-Dominique, P. Forget, and M. Thery, Eds.), pp. 121-128. Kluwer Academic Publishers, Dordrecht, Netherlands.

Haddow, J.J., Gillett, J.D., and Highton, R.B. (1947). The mosquitoes of Bwamba County, Uganda. V. The vertical distribution and biting-cycle of mosquitoes in rain-forest, with further observations on microclimate. *Bulletin of Entomological Research* **37**, 307–330.

Hale, M.E. Jr. (1952). Vertical distribution of cryptogams in a virgin forest in Wisconsin. *Ecology* **33**, 398–406.

Hale, M.E., Jr. (1965). Vertical distribution of cryptogams in a red maple swamp in Connecticut. *The Bryologist* **68**, 193–197.

Hallé, F., Oldeman, R.A.A., and Tomlinson, P.B. (1978). "Tropical Trees and Forests." Springer-Verlag, Berlin.

Hartshorn, G.S. (1978). Tree falls and tropical forest dynamics. *In* "Tropical Trees as Living Systems" (P.B. Tomlinson and M.H. Zimmerman, Eds.) pp. 617-638. Cambridge University Press, Cambridge, UK.

Hawksworth, F.G. and Wiens, D. (1996). Dwarf mistletoes: biology, pathology, and systematics. *U.S. Forest Service, Agriculture Handbook 709*. Washington DC.

Hayes, J.P. and Gruver, J.C. (2000). Vertical stratification of bat activity in an old-growth forest in western Washington. *Northwest Science* **74**, 102–108.

Hegarty, E. (1989). The climbers: lianes and vines. *In* "Tropical Rain Forests Ecosystems. Ecosystems of the World 14B" (H. Lieth and M.J.A. Werger, Eds.), Chapter 18. Elsevier Science, Amsterdam.

Holbrook, N.M. and Lund, C.P. (1995). Photosynthesis in forest canopies. *In* "Forest Canopies" (M.D. Lowman and N.M. Nadkarni, Eds.) pp. 411-430. Academic Press, San Diego.

Hollinger, D.Y. (1989). Canopy organization and foliage photosynthetic capacity in a broad-leaved evergreen montane forest. *Functional Ecology* **3**, 53–62.

Holz, I., Gradstein, S.R., Heinrichs, J., and Kappelle, M. (2002). Bryophyte diversity, microhabitat differentiation, and distribution of life forms in Costa Rican upper montane *Quercus* forest. *Bryologist* **105**, 334–348.

Ishii, H., Reynolds, J.H., Ford, E.D., and Shaw, D.C. (2000). Height growth and vertical development of an old-growth *Pseudotsuga-Tsuga* forest in southwestern Washington State, USA. *Canadian Journal of Forest Research* **30**, 17–24.

Johansson, D. (1974). Ecology of vascular epiphytes in West African rain forest. *Acta Phytogeogr. Suec.* **59**, 1–129.

Johnson, J.A. and Whitney, N.J. (1989). An investigation of needle endophyte colonization patterns with respect to height and compass direction in a single crown of balsam fir (*Abies balsamea*). *Canadian Journal of Botany* **67**, 723–725.

Kalcounis, M.C., Hobson, K.A., Brigham, R.M., and Hecker, K.R. (1999). Bat activity in the boreal forest: importance of stand type and vertical strata. *Journal of Mammology* **80**, 673–682.

Kalko, E.K.V. and Handley, Jr., C.O. (2001). Neotropical bats in the canopy: diversity, community structure, and implications for conservation. *Plant Ecology* **153**, 319–333.

Karr, J.R. (1971). Structure of avian communities in selected Panama and Illinois habitats. *Ecological Monographs* **41**, 207–233.

Karr, J.R. (1989). Birds. *In* "Tropical Rain Forests Ecosystems. Ecosystems of the World 14B" (H. Lieth and M.J.A. Werger, Eds.), Chapter 22. Elsevier, Amsterdam.

Kato, K. and Yamamoto, S. (2002). Branch growth and allocation patterns of saplings of two *Abies* species under different canopy conditions in a subalpine old-growth forest in central Japan. *EcoScience* **9**, 98–105.

Kays, R. and Allison, A. (2001). Arboreal tropical forest vertebrates: current knowledge and research trends. *Plant Ecology* **153**, 109–120.

Kelly, D.L. (1985). Epiphytes and climbers of a Jamaican rain forest: vertical distribution, life forms and life histories. *Journal of Biogeography* **12**, 223–241.

Kozlowski, T.T. and Pallardy, S.G. (1997). "Physiology of Woody Plants." Academic Press, San Diego.

Laman, T.G. (1993). Seedling establishment of the hemiepiphyte *Ficus stupenda* in the Bornean rain forest canopy. *Bull. Ecol. Soc. Amer.* **74** (Supplement), 321.

Lange, O.L., Kilian, E., and Ziegler, H. (1986). Water vapor uptake and photosynthesis of lichens: performance differences in species with green and blue-green algae as phycobionts. *Oecologia* **71**, 104–110.

Laurance, W.F. and Bierregaard, R.O. (1997). "Tropical Forest Remnants: Ecology, Management and Conservation of Fragmented Communities." University of Chicago Press, Chicago.

Laurent, R.F. (1989). Herpetofauna of tropical America and Africa. *In* "Tropical Rain Forest Ecosystems" (H. Lieth and M.J.A. Werger Eds.), Chapter 23. Elsevier Science Publishers, Amsterdam.

Lertzman, K.P., Sutherland, G., Inselberg, A., and Saunders, S. (1996). Canopy gaps and the landscape mosaic in a temperate rainforest. *Ecology* **77**, 1254–1270.

Longino, J.T. and Nadkarni, N.M. (1990). A comparison of ground and canopy leaf litter ants (Hymenoptera: Formicidae) in a neotropical montane forest. *PSYCHE* **97**, 81–93.

Lowman, M.D. and Heatwole, H. (1992). Spatial and temporal variability in defoliation of Australian Eucalypts. *Ecology* **73**, 129–142.

MacArthur, R.H. (1958). Population ecology of some warblers of northeastern coniferous forests. *Ecology* **39**, 599–619.

MacArthur, R.H. and MacArthur, J.W. (1961). On bird species diversity. *Ecology* **42**, 594–598.

Malcolm, J.R. 1991. Comparative abundances of Neotropical small mammals by trap height. Journal of Mammalogy 72, 188–192.

Malcolm, J.R. (1995). Forest structure and the abundance and diversity of neotropical small mammals. *In* "Forest Canopies" (M.D. Lowman and N.M. Nadkarni, Eds.), pp. 179-197. Academic Press, San Diego.

Mannan, R.W., Meslow, E.C., and Wight, H.M. (1980). Use of snags by birds in Douglas-fir forests, Western Oregon. *Journal of Wildlife Management* **44**, 787–797.

Mazimpaka, V. and Lara, F. (1995). Corticolous bryophytes of *Quercus pyrenaica* forests from Gredos Mountains (Spain): vertical distribution and affinity for epiphytic habitats. *Nova Hedwigia* **61**, 431–446.

McCune, B. (1993). Gradients in epiphyte biomass in three *Pseudotsuga: Tsuga* forests of different ages in western Oregon and Washington. *The Bryologist* **96**, 405–411.

McCune, B., Amsberry, K.A., Camacho, F.J., Clery, S., Cole, C., Emerson, C., Felder, G. French, P., Greene, D., Harris, R., Hutten, M., Larson, B., Lesko, M., Majors, S., Markwell, T., Parker, G.G., Pendergrass, K., Peterson, E.B., Peterson, E.T., Platt, J., Proctor, J., Rambo, T., Rosso, A., Shaw, D., Turner, R., and Widmer, M. (1997). Vertical profile of epiphytes in a Pacific Northwest old-growth forest. *Northwest Science* **71**, 145–152.

McCune, B., Rosentreter, R., Ponzetti, J. M., and Shaw, D. C. (2000). Epiphyte habitats in an old conifer forest in western Washington, USA. *The Bryologist* **103**, 417–427.

Merrill, W. and Shigo, A.L. (1979). An expanded concept of tree decay. *Phytopathology* **69**, 1158–1161.

Moffett, M.W. (2000). What's "up"? A critical look at the basic terms of canopy biology. *Biotropica* **32**, 569–596.

Moffett, M.W. (2001). The nature and limits of canopy biology. *Selbyana* **22**, 155–179.

Mullin, S.J. and Cooper, R.J. (2002). Barking up the wrong tree: climbing performance of rat snakes and its implications for depredation of avian nests. *Canadian Journal of Zoology* **80**, 591–595.

Nadkarni, N.M. (1994). Diversity of species and interactions in the upper tree canopy of forest ecosystems. *American Zoologist* **34**, 70–78.

Nadkarni, N.M. and Longino, J.T. (1990). Invertebrates in canopy and ground organic matter in a neotropical montane forest, Costa Rica. *Biotropica* **22**, 286–289.

Nadkarni, N.M. and Matelson, T.J. (1989). Bird use of epiphyte resources in Neotropical trees. *The Condor* **91**, 891–907.

Nieder, J., Engwald, S., Klawun, M., and Barthlott, W. (2000). Spatial distribution of vascular epiphytes (including Hemiepiphytes) in a lowland Amazonian rain forest (Surumoni Crane Plot) of southern Venezuela. *Biotropica* **32**, 385–396.

Nieder, J., Prosperi, J., and Michaloud, G. (2001). Epiphytes and their contribution to canopy diversity. *Plant Ecology* **153**, 51–63.

Norton, D.A., Ladley, J.J., and Owen, H.J. (1997). Distribution and population structure of the loranthaceous mistletoes *Alepis flavida*, *Peraxilla colensoi*, and *Peraxilla tetrapetala* within two New Zealand *Nothofagus* forests. *New Zealand Journal of Botany* **35**, 323–336.

Orians, G.H. (1969). The number of bird species in some tropical forests. *Ecology* **50**, 783–801.

Packer, W.C. and Layne, J.N. (1991). Foraging site preferences and relative arboreality of small rodents in Florida. *American Midland Naturalist* **125**, 187–194.

Paoletti, M., Taylor, R., Stinner, B., Stinner, D., and Benzing, D. (1991). Diversity of soil fauna in the canopy and forest floor of a Venezuelan cloud forest. *Journal of Tropical Ecology* **7**, 373–383.

Papageorgius, C. (1975). Mimicry in neo-tropical butterflies. *American Scientist* **63**, 522–532.

Parker, G.G. (1995). Structure and microclimate of forest canopies. *In* "Forest Canopies" (M.D. Lowman and N.M. Nadkarni, Eds.), pp. 73-106. Academic Press, San Diego.

Parker, G.G., O'Neill, J.P., and Higman, D. (1989). Vertical profile and canopy organization in a mixed deciduous forest. *Vegetatio* **85**, 1–11.

Parker, G.G. and Brown, M.J. (2000). Forest canopy stratification: is it useful? *The American Naturalist* **155**, 473–484.

Parker, G.G., Davis, M.M., and Chapotin, S.M. (2002). Canopy light transmittance in Douglas-fir-western hemlock stands. *Tree Physiology* **22**, 147–158.

Pearson, D.L. (1971). Vertical stratification of birds in a tropical dry forest. *Condor* **73**, 46–55.

Pickett, S.T.A. and White, P.S. (Eds.). (1985). "The Ecology of Natural Disturbance and Patch Dynamics." Academic Press, San Diego.

Pike, L.H., Denison, W.C., Tracy, D.M., Sherwood, M.A. and Rhoades, F.M. (1975). Floristic survey of epiphytic lichens and bryophytes growing on old-growth conifers in western Oregon. *The Bryologist* **7**, 389–402.

Pittendrigh, C.S. (1948). The bromeliad-Anophele-malaria complex in Trinidad. I. The bromeliad flora. *Evolution* **2**, 58–89.

Prinzing, A.J. (2001). Use of shifting microclimatic mosaics by arthropods on exposed tree trunks. *Annals of the Entomological Society of America* **94**, 210–218.

Putz, F.E. and Mooney, H.A. (Eds.). (1991). "The Biology of Vines." Cambridge University Press, Cambridge, UK.

Putz, F.E. (1995). Vines in treetops: consequences of mechanical dependence. *In* "Forest Canopies" (M.D. Lowman and N.M. Nadkarni, Eds.), pp. 311–323. Academic Press, San Diego.

Pyle, R.M. (2002). "The Butterflies of Cascadia." Seattle Audubon Society, Seattle, Washington.

Rand, A.S. (1964). Ecological distribution in anoline lizards of Puerto Rico. *Ecology* **45**, 745–752.

Reagan, D.P. (1992). Congeneric species distribution and abundance in a three-dimensional habitat: the rain forest anoles of Puerto Rico. *Copeia* 392–403.

Reagan, D.P. (1995). Lizard ecology in the canopy of an island rain forest. *In* "Forest Canopies" (M.D. Lowman and N.M. Nadkarni, Eds.), pp. 149–164. Academic Press, San Diego.

Reid, N., Smith, M.S., and Yan, Z. (1995). Ecology and population biology of mistletoes. *In* "Forest Canopies" (M.D. Lowman and N.M. Nadkarni, Eds.), pp. 285–310. Academic Press, San Diego.

Reynolds, B.C. and Crossley, Jr., D.A. (1997). Spatial variation in herbivory by forest canopy arthropods along an elevation gradient. *Environmental Entomology* **26**, 1232–1239.

Reynolds, D.P. (1972). Stratification of tropical epiphylls. Kalikasan, *Philippine Journal of Biology* **1**, 7–10.

Rhoades, F.M. (1995). Nonvascular epiphytes in forest canopies: worldwide distribution, abundance, and ecological roles. *In* "Forest Canopies" (M.D. Lowman and N.M. Nadkarni, Eds.), pp. 353–408. Academic Press, San Diego, USA.

Richards, P.W. (1952). "The Tropical Rain Forest." Cambridge University Press, Cambridge, UK.

Richards, P.W. (1984). The ecology of tropical forest bryophytes. *In* "New Manual of Bryology, Volume 2," (R.M. Schuster, Ed.), pp. 1233–1270. The Hattori Botanical Laboratory, Nichinan.

Richards, P.W., Walsh, R.P.D., Baillie, I.C., and Greig-Smith, P. (1996). "The Tropical Rain Forest: An Ecological Study." Cambridge University Press, Cambridge, UK.

Rinker, H.B. (2001). The use of a forest canopy walkway for studying habitat selection by neotropical migrants. *Selbyana* **22**, 89–96.

Robinson, S.K. and Holmes, R.T. (1984). Effects of plant species and foliage structure on the foraging behavior of forest birds. *The Auk* **101**, 672–684.

Room, P.M. (1973). Ecology of the mistletoe *Tapinanthus bangwensis* growing on cocoa in Ghana. *Journal of Ecology* **61**, 729–742.

Rose, F. (1988). Phytogeographical and ecological aspects of Lobarion communities in Europe. *Botanical Journal of the Linnean Society* **96**, 69–79.

Rowe, W.J. and Potter, D.A. (1996). Vertical stratification of feeding by Japanese beetles within linden tree canopies: selective foraging or height per se? *Oecologia* **108**, 459–466.

Scholl, P.J., DeFoliart, G.R., and Nemenyi, P.B. (1979). Vertical distribution of biting activity by *Aedes triseriatus*. *Annals of the Entomological Society of America* **72**, 537–539.

Schowalter, T.D. and Ganio, L.M. (1998). Vertical and seasonal variation in canopy arthropod communities in an old-growth conifer forest in southwestern Washington, USA. *Bulletin of Entomological Research* **88**, 633–640.

Schulze, C.H., Linsenmair, K.E., and Fiedler, K. (2001). Understory versus canopy: patterns of vertical stratification and diversity among Lepidoptera in a Bornean rain forest. *Plant Ecology* **153**, 133–152.

Sharpe, F. (1996). The biologically significant attributes of forest canopies to small birds. *Northwest Science* **70** (Special Issue), 86–93.

Shaw, D.C. and Weiss, S.B. (2000). Canopy light and the distribution of hemlock dwarf mistletoe (*Arceuthobium tsugense* [Rosendahl] G.N. Jones subsp. *tsugense*) aerial shoots in an old-growth Douglas-fir/western hemlock forest. *Northwest Science* **74**, 306–315.

Shaw, D.C., Freeman, E.A., and Flick, C. (2002). The vertical occurrence of small birds in an old-growth Douglas-fir-western hemlock forest stand. *Northwest Science* **76**, 322–334.

Shaw, D.C., Franklin, J.F., Bible, K., Klopatek, J., Freeman, E., Greene, S., and Parker, G.G. (In press). Ecological setting of the Wind River old-growth forest. *Ecosystems*.

Shigo, A.L. (1979). Tree decay: an expanded concept. US Dept. of Agriculture, *Agricultural Bulletin* 419.

Sillett, S.C. (1995). Branch epiphyte assemblages in the forest interior and on the clearcut edge of a 700-year-old Douglas-fir canopy in western Oregon. *The Bryologist* **98**, 301–312.

Sillett, S.C., Gradstein, S.R., and Griffin, D. (1995). Bryophyte diversity of *Ficus* tree crowns from cloud forest and pasture in Costa Rica. *The Bryologist* **98**, 251–260.

Sillett, S.C. and Neitlich, P.N. (1996). Emerging themes in epiphyte research in westside forests with special reference to cyanolichens. *Northwest Science* **70** (Special Issue), 54–60.

Simon, U. and Linsenmair, K.E. (2001). Arthropods in tropical oaks: differences in their spatial distributions within tree crowns. *Plant Ecology* **153**, 179–191.

Smith, A.P. (1973). Stratification of temperate and tropical forests. *The American Naturalist* **107**, 671–683.

Stone, J.K., Sherwood, M.A., and Carroll, G.C. (1996). Canopy microfungi: function and diversity. *Northwest Science* **70** (Special Issue), 37–45.

Stone, J. and Petrini, O. (1997). Endophytes of forest trees: a model for fungus-plant interactions. *In* "The Mycota V Part B, Plant Relationships" (G.C. Carroll and P. Tudzynski, Eds.), Chapter 8, pp. 129–140. Springer-Verlag, Berlin.

Sutton, S.L. (1989). The spatial distribution of flying insects. *In* "Ecosystems of the World 14B: Tropical Rain Forest Ecosystems" (H. Lieth. and M.J.A. Werger, Eds.), pp. 427-436. Elsevier Science, Amsterdam.

Taylor, P.H. and Lowman, M.D. (1996). Vertical stratification of the small mammal community in a northern hardwood forest. *Selbyana* **17**, 15–21.

Terborgh, J. (1980). Vertical stratification of a Neotropical forest bird community. *In* "Acta XVII Congressus Internationalis Ornithologici" (R. Nöhring, Ed.), pp. 1005–1012. Deutsche Ornithologen-Gesellschaft, Berlin.

Terborgh, J. (1983). "Five New World Primates: A Study in Comparative Ecology." Princeton University Press, Princeton, NJ.

Terborgh, J. (1985). The vertical component of plant species diversity in temperate and tropical forests. *The American Naturalist* **126**, 760–776.

Théry, M. (2001). Forest light and its influence on habitat selection. *Plant Ecology* **153**, 251–261.

Wallace, B.J. (1989). Vascular epiphytism in Australo-Asia. *In* "Ecosystems of the World 14B: Tropical Rain Forest Ecosystems" (H. Lieth. and M.J.A. Werger, Eds.), pp. 261-282. Elsevier, Amsterdam.

Walter, D.E. and O'Dowd, D.J. (1995). Life on the forest phylloplane: hairs, little houses, and myriad mites. *In* "Forest Canopies" (M.D. Lowman and N.M. Nadkarni, Eds.), pp. 325-351. Academic Press, San Diego.

Walther, B.A. (2002a). Vertical stratification and use of vegetation and light habitats by Neotopical forest birds. *Journal für Ornithologie* **143**, 64–81.

Walther, B.A. (2002b). Grounded ground birds and surfing canopy birds: variation of foraging stratum breadth observed in neotropical forest birds and tested with simulation models using boundary constraints. *The Auk* **119**, 658–675.

Waterman, P.G. and McKey, D. (1989). Herbivory and secondary compounds in rain-forest plants. *In* "Ecosystems of the World 14B: Tropical Rain Forest Ecosystems" (H. Lieth. and M.J.A. Werger, Eds.), pp. 513-536. Elesevier Science, Amsterdam.

Williams-Linera, G. and Lawton, R.O. (1995). The ecology of hemiepiphytes in forest canopies. *In* "Forest Canopies" (M.D. Lowman and N.M. Nadkarni, Eds.), pp. 255-283. Forest Canopies. Academic Press, San Diego.

Williams-Linera, G. and Baltazar, A. (2001). Herbivory on young and mature leaves of one temperate deciduous and two tropical evergreen trees in the understory and canopy of a Mexican cloud forest. *Selbyana* **22**, 213–218.

Wilson, D.E. (1989). Bats. *In* "Ecosystems of the World 14B: Tropical Rain Forest Ecosystems" (H. Lieth. and M.J.A. Werger, Eds.), Chapter 20. Elsevier Science, Amsterdam.

Winchester, N.N. (1997). Canopy arthropods of coastal Sitka spruce trees on Vancouver Island, British Columbia, Canada. In "Canopy Arthropods" (N.E. Stork, J.A. Adis, and R.K. Didham, Eds.), Chapman and Hall, New York.

Winkler, H. and Preleuthner, M. (2001). Behaviour and ecology of birds in tropical rain forest canopies. *Plant Ecology* **153**, 193–202.

Wunder, L. and Carey, A.B. (1996). Use of the forest canopy by bats. *Northwest Science* **70** (Special Issue), 79–85.

Youlatos, D. and Gasc, J. (2001). Comparative positional behaviour of five primates. *In* "Nouragnues, Dynamics and Plant-Animal Interactions in a Neoptropical Rainforest" (F. Bongers, P. Charles-Dominique, P. Forget, and M. Thery, Eds.), Chapter 9, pp. 103–114. Kluwer Academic Publishers, Dordrecht, Netherlands.

CHAPTER 5

Age-Related Development of Canopy Structure and Its Ecological Functions

Hiroaki T. Ishii, Robert Van Pelt, Geoffrey G. Parker, and Nalini M. Nadkarni

Structural complexity embodies not only particular types of stand attributes, but also the way they are spatially arranged within stands.
—*Lindenmayer and Franklin*, Conserving Forest Biodiversity, *2003*

Perhaps the word most used in ecology with different meanings is "structure"... Integrative concepts cannot be measured directly, and both their definition and detailed description must be synthesized from studies of a number of systems.
—*Ford*, Scientific Method for Ecological Research, *2000*

Introduction

For many years our view of forests has been two-dimensional and the structure and dynamics of forest ecosystems have been studied and analyzed using data expressed as stem maps that indicate the X and Y coordinates of individual trees along with species and size information (e.g., diameter at breast height). In research that moved toward documenting the three-dimensionality of forests, the vertical variable (Z-coordinate) most often considered was tree height. "Stand structure" usually referred to the species composition and tree size distribution of forest stands and sometimes included information on the two-dimensional spatial distribution of trees and tree height. Because humans are confined to the ground, we have been unable to study in detail the three-dimensional (3-D) structure of forests, including the forest canopy. Instead, we inferred tree growth from diameter increment at breast height, and competitive interactions among trees from analysis of the relationship between horizontal spatial distribution and diameter growth or mortality patterns (e.g., Kenkel 1988; Duncan 1991; Stohlgren 1993; Biondi et al. 1994). However, we have always known that the dynamics of forest ecosystems, including growth and competitive interactions among individual trees, occur in the canopy and not at breast height (Krajicek et al. 1961; Hix and Lorimer 1990; Bravo et al. 2001).

Currently, there are techniques that enable safe and repeated access to the forest canopy to study the 3-D structure of forest ecosystems and ecological processes occurring in the canopy (Moffett and Lowman 1995). The 3-D distribution of tree crowns reflects how trees occupy space in the canopy to capture light resources and drives critical ecological processes such as stand productivity and forest community dynamics. However, measurement of forest canopy structure presents a great challenge (see Box on p.104–108). Visual observation suggests that the canopy of young stands and plantations are structurally homogeneous and simple compared with that of old-growth stands (see Figure 5-1). This difference has been shown through comparative studies of two-dimensional stand structure. Species composition and tree size distributions become more diverse with increasing stand age, and specific structural elements such as large, old trees and snags characterize older stands (reviewed by Franklin et al. 2002). The term "structural complexity" was coined as an integrative concept to represent the complex 3-D structure of old-growth forests. However, many of the measures and indices used to characterize structural complexity such as species composition, tree-size distribution, and abundance of snags and woody debris are derived from two-dimensional, ground-based measurements (e.g., Franklin et al. 1981; Spies and Franklin 1991; Wells et al. 1998; Solomon and Gove 1999; Staudhammer and LeMay 2001). One of the objectives for developing indices of structural complexity is to distinguish quantitatively old-growth forests from younger stands and to establish criteria for enhancing old-growth forest structure in managed stands for conservation purposes, such as creating wildlife habitat (Kohm and Franklin 1997; Smith et al. 1997; Carey et al. 1999b; Hunter 1999; Lindenmayer and Franklin 2002). In order to meet such management objectives, we must integrate various aspects of forest structure, including canopy and below-ground processes, into future ecosystem management strategies.

In this chapter, we review the canopy processes that drive the development of structural complexity with increasing stand age and present a 3-D canopy perspective of structural development of temperate forest ecosystems. Because trees form the basic framework of canopy structure, dynamics of tree crowns define structural development of the forest canopy (see Figure 5-2):

1. Height-growth rate and maximum attainable height of species determine vertical structure.
2. Expansion growth and interactions among individual tree crowns determine horizontal spatial distribution.
3. With increasing stand age, aging and maintenance of individual tree crowns, as well as the effects of small-scale disturbances such as individual tree mortality, crown damage and die-back enhance structural complexity of the canopy.

Developmental processes of canopy structure drive and enhance various ecological functions such as community dynamics and stand productivity. Development of structural complexity enhances biodiversity of forest ecosystems by creating specific structural elements that provide food and habitat for other organisms.

Figure 5-1 Contrasting simple versus complex canopy structure of a naturally regenerated 20-year-old stand of Douglas-fir (a) and a 450-year-old stand (b) in southwestern Washington State.

MEASURING CANOPY STRUCTURE: THE FOREST CANOPY DATABASE PROJECT

Nalini M. Nadkarni and Judy B. Cushing

The increasing interest and realization of the importance of canopy organisms and interactions in whole-forest ecology has generated an unprecedented amount of canopy data (Lowman and Wittman 1996). Both the types and amounts of canopy structure data are changing rapidly. In the past, scientists working alone with simple rope-climbing techniques generated studies that produced fairly small datasets. However, recent access innovations permit multiple teams of scientists to work within the same volume of the canopy. Canopy scientists have to deal with more data, new kinds of data, and the need to share data (Parker 1995). Data collected by canopy research teams will be useful to other scientists (e.g., geographers and land use managers), just as data emanating from allied fields could aid forest canopy researchers (Parker et al. 1992).

Historically, canopy scientists have taken, stored, and analyzed data in independent and idiosyncratic ways. In 1993, a team of forest canopy ecologists and computer scientists received a planning grant from the National Science Foundation's (NSF) Database Activities Program. The project brought together forest canopy researchers, quantitative scientists, and computer scientists to work toward establishing methods to collect, store, display, analyze, and interpret three-dimensional spatial data relating to tree crowns and forest canopies.

A survey we conducted on the perceived obstacles to the advancement of forest canopy studies (Nadkarni et al. 1996) revealed that canopy researchers believe that our understanding of forest canopy biota and processes is not limited by canopy access (as we had anticipated), but rather by two characteristics of canopy data:

1. Lack of quantitative tools that allow canopy researchers to analyze the complex three-dimensional spatial data associated with forest canopy studies, and
2. Lack of harmonized datasets—forest canopy researchers have tended to collect data in non-comparable formats.

Few projects have common methodologies or data formats, so their resulting observations are not easily shared and compared. Thus, the study of the forest canopy is perceived as being held back by the lack of data management tools. However, the relative youth of the field—with its lack of entrenched methods, legacy datasets, and conflicting camps of competing groups—provides a unique opportunity for integrating data management and analysis tools into the research process (Lowman and Nadkarni 1995). The sociology of the discipline is conducive to sharing data; researchers appear openly communicative and supportive of each other's work (Nadkarni and Parker 1995). In particular, documenting tree and forest structure is fundamental to describing and understanding forest ecology, physiology, and forest/atmosphere interactions (Parker 1995). However the task of describing, visualizing, and analyzing structure is extremely difficult because trees are large, irregular, dynamic, and operate on time scales that are different from humans. Statistics and visualization have not been worked out. Trees may not be "individuals" from the perspective of their associated organisms (e.g., epiphytes, birds, and microbes). Rather, branches or individual twigs might be the appropriate sampling unit. Historically, researchers who study tree and forest structure have used a large number of different and non-complementary approaches to describe, visualize, and analyze structure. They have used a variety of different tools and approaches that result in a diversity of ways of seeing the structure (Parker 1995). Thus, the state of the art is that although there are many ways of quantifying forest structure, few are compatible with each other. The time is ripe to bring these together, not to impose a single standard or protocol, but rather to present a way in which forest structural elements from different studies can be integrated to foster comparative work.

Here, we describe the development and some of the applications of the Forest Canopy Database Project. In 1998, we were awarded another NSF grant to develop a database and database tools to enhance the ability of researchers in forest canopy studies to collect, analyze, link, and archive data. The computer database has taken three forms. The first piece is our web-based "Big Canopy Database." This database holds information, field data, and images of use to canopy researchers, educators, and conservationists (http://canopy.evergreen.edu/bcd/)

The second piece is a web-based program called "DataBank," which allows canopy researchers to search for and download field data submitted by other researchers, design field databases and download them for their own use, and to document and archive their own databases. The system thus builds new databases from database components ("templates") that are designed specifically for canopy data. To submit data to the database, a

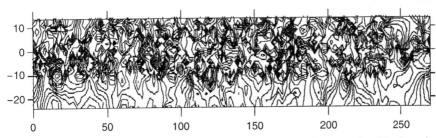

Figure 1 Two-dimensional contour map of the upper canopy from an aerial perspective. The forest is a 950-year-old temperate coniferous forest on Mt. Rainer, Washington State. Graphics produced in Surfer by R. Van Pelt. Data from Van Pelt et al. (in press).

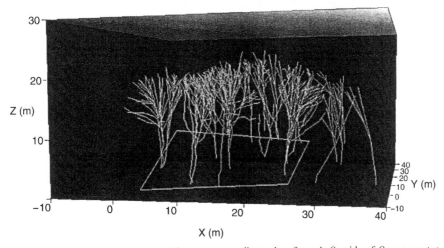

Figure 2 "Stick tree" architecture derived from x,y,z coordinate data from A. Sumida of *Quercus serrata* trees.

Continued

MEASURING CANOPY STRUCTURE: THE FOREST CANOPY DATABASE
PROJECT–*cont'd*

researcher from each study works with a programmer to provide metadata and to structure his or her data to fit one or more existing field data templates, or to generate a new template for novel data types. The current format is implemented in SQLServer, Java, and HTML.

The third component, called "CanopyView," takes data from the database and allows a researcher to select a means of visualizing the data in two or three dimensions. We provide an array of visualization templates and examples from which to choose. For example, we can use simple height and x,y data for branches to create "contour maps" of the forest canopy (i.e., the outer envelope of the trees) (Figure 1). We can also display individual trees as "stick figures" (Figure 2) using the data of A. Sumida (Hokkaido University), in which the x,y,z, coordinates of each node, measured following methods described in Sumida et al. (2002), are connected to provide a schematic of the branching structure of individual trees.

Another approach is to display the distribution of foliage in a forest canopy without regard to which individual tree is attached to a given piece of foliage (Figure 3). Foliar distribution data were provided by G.G. Parker (Smithsonian Environmental Research Center)

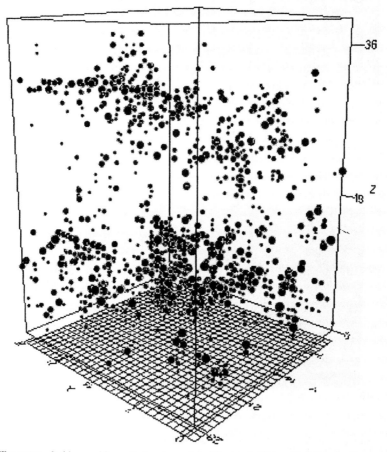

Figure 3 Temperate deciduous old-growth forest stand in Maryland, USA, with foliar density depicted using ArcView. Data from G. Parker.

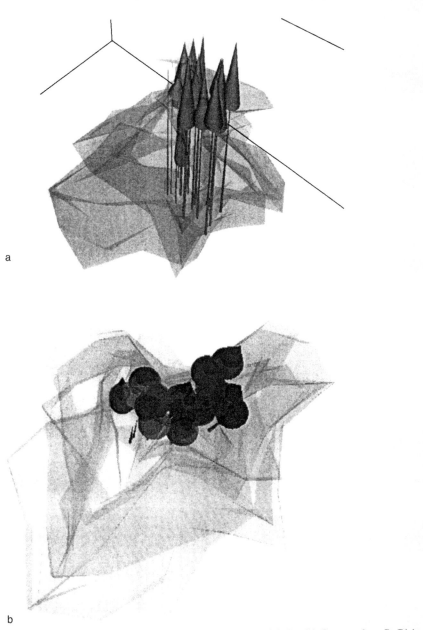

Figure 4 Depiction of trees and airspace around them, side view (a) and aerial view (b). Data are from R. Dial. Graphics produced in *Mathematica* by M. Ficker (Cushing et al. 2003).

using LIDAR technology. Others may be interested in the "airspace" of a forest (i.e., the shape and size of the interstices of air between the solid masses of the forest canopy) (Figure 4), an example of which was created by R. Dial (Alaska Pacific University) for forests of Alaska.

Continued

MEASURING CANOPY STRUCTURE: THE FOREST CANOPY DATABASE
PROJECT–*cont'd*

Our efforts to create a database for the canopy research community will help push forward this emerging field of science. We also believe that our efforts could be viewed as a model for other emerging areas of ecology where data-linking and data-sharing can be effective in integrating results from different studies. This capability will speed the development of the field to more efficiently address both intellectually stimulating and environmentally pressing questions of interest to academics, policy-makers, and the general public. We anticipate that the database and tools can serve as an exemplar for other interdisciplinary and emerging fields of science.

References

Cushing, J.B., Bond, B., Dial, R., and Nadkarni, N.M. (2003). How trees and forests inform biodiversity and ecosystem informatics. *Comput. Sci. Eng.* **5**, 26–31.

Lowman, M.D. and Nadkarni, N.M. (1995). "Forest Canopies." Academic Press, San Diego.

Lowman, M.D. and Wittman, P.K. (1996). Forest canopies: methods, hypotheses, and future directions. *Ann. Rev. Ecol. Syst.* **27**, 55–81.

Nadkarni, N.M. and Parker, G.G. (1994). A profile of forest canopy science and scientists—who we are, what we want to know, and obstacles we face: results of an international survey. *Selbyana* **15**, 38–50.

Nadkarni, N.M., Parker, G.G., Ford, E.D., Cushing, J.B., and Stallman, C. (1996). The International Canopy Network: A pathway for interdisciplinary exchange of scientific information on forest canopies. *Northwest Sci.* **70**, 104–108.

Parker, G.G. (1995). Structure and microclimate of forest canopies. *In* "Forest Canopies" (M. Lowman and N.M. Nadkarni, Eds.) pp. 73–106. Academic Press, San Diego.

Parker, G.G., Smith, A.P., and Hogan, K.P. (1992). Access to the upper forest canopy with a large tower crane. *BioScience* **42**, 664–670.

Sumida, A., Terazawa, I., Togashi, A., and Komiyama, A. (2002). Spatial arrangement of branches in relation to slope and neighbourhood competition. *Ann. Bot.* **89**, 301–310.

Van Pelt, R. and Nadkarni, N.M. (In press). Changes in the vertical and horizontal distribution of canopy structural elements in an age sequence of *Pseudotsuga menziesii* forests in the Pacific Northwest. *For. Sci.*

Deterministic Processes that Drive Development of Canopy Structure

In early stages of stand development, deterministic processes such as timing of establishment following disturbance, height-growth rate, and crown interactions determine canopy structure (Johnson et al. 1994; Dubrasich et al. 1997). In mixed-species natural forests, differences among species in timing of establishment and initial height-growth rates result in vertical stratification of species within the canopy (e.g., Bicknell 1982; Palik and Pregitzer 1991). Early-successional, fast-growing species can establish soon after disturbance, reach the upper canopy, and dominate during early stages of succession. For example, in a 42-year-old forest in Michigan dominated by bigtooth aspen (*Populus grandidentata*) and trembling aspen (*Populus tremuloides*), reconstruction of vertical canopy development revealed that initial height-growth rates were similar among species (Palik and Pregitzer 1991). In this forest, the observed pattern of vertical stratification among species reflected differences in the timing of establishment of each species after disturbance. In a second-

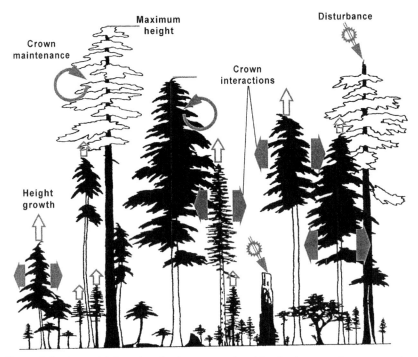

Figure 5-2 Deterministic and stochastic processes that drive development of canopy structure.

ary northern hardwood forest in New Hampshire, early-successional, shade-intolerant species such as pin cherry (*Prunus pensylvania* L.) had faster height-growth rates and longer growing seasons than the mid- and late-successional species and became the tallest trees within the first six years after the stand was cut (Bicknell 1982). Mid-successional species such as trembling aspen, striped maple (*Acer pensylvanicum*), and yellow birch (*Betula alleghaniensis*) occupied intermediate positions in the canopy.

In mixed-species natural forests, each species has an inherent and site-specific maximum height that determines its canopy position (e.g., Thomas 1999; Ishii et al. 2000). Whether individuals reach their potential maximum height or not may also be determined by their demography. For example, in a warm-temperate forest in southern Japan, most canopy tree species reached their potential maximum heights; however, for some species, the estimated maximum height was much greater than the tallest trees observed in the field (Aiba and Kohyama 1996). This was attributed to early mortality in these species (i.e., many individuals died before reaching their potential maximum height). Thus, in this forest, the relative canopy position of each species is determined by both potential maximum height and mortality pattern.

The vertical development of canopy structure with increasing stand age can be inferred by comparing stands of different ages, i.e., a chronosequence approach (see Figure 5-3). In Douglas-fir (*Pseudotsuga menziesii*) and western hemlock (*Tsuga heterophylla*) forests of the Pacific Northwest Coast of North America, development of vertical canopy structure during early-successional stages begins with the dominance of the fast-growing, pioneer species, Douglas-fir, in the upper canopy of mixed-species stands (Wierman and Oliver 1979; Larson 1986). In late-successional stands, canopy height reaches over 70 m. Douglas-fir continues to dominate in the upper canopy as more shade-tolerant species such as western hemlock and western red cedar (*Thuja plicata*) invade the mid- to lower-canopy (Gholz

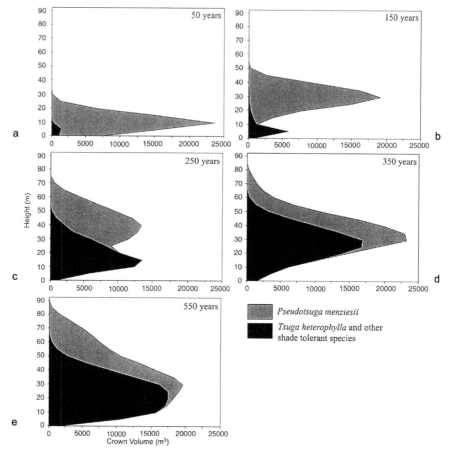

Figure 5-3 Comparison of the vertical distribution of foliage among Douglas-fir–western hemlock forests of different ages in western Washington State. (a) Fifty-year-old stands are nearly pure Douglas-fir. Depending on stand origin and proximity to older stands, shade-tolerant species (largely western hemlock, with vine maple [*Acer circinatum*] present in the understory) may or may not be present at this stage. (b) In 150-year-old stands, the Douglas-fir canopy has a homogeneous and simple structure; most of the canopy trees are nearly the same height and diameter. The understory is more diverse and comprises many shade-tolerant tree species, including western hemlock, western red cedar, and Pacific silver fir. (c) In 250-year-old stands, canopy depth increases and shade-tolerant species are found throughout all canopy heights. (d) In 350-year-old stands, the Douglas-fir canopy is very deep but is only represented by 30 to 60 trees per hectare. Shade-tolerant species represent most of the foliage within all but the uppermost canopy. (e) At 550 years, few large Douglas-fir trees (less than 25 trees per hectare) continue to survive and hold large amounts of foliage in their crowns. Large trees of western hemlock and western red cedar also exist, contributing to increasing stand foliage amount.

et al. 1976; Stewart 1986; Easter and Spies 1994). This results in the development of a deep, continuous canopy comprising both early- and late-successional species.

Crown interactions play an important role in determining the spatial distribution of trees within the canopy (Ford and Sorrensen 1990). "Crown shyness" is a phenomenon where crowns of neighboring trees are separated by small gaps. This has been attributed to mechanical abrasion (Putz et al. 1984; Rudnicki et al. 2003) and to changes in the rate and direction of branch elongation in response to neighbors (Koike 1989; Sumida et al. 2002). "Crown asymmetry," where the crown center is offset from the stem base, results from plasticity in the directional growth of tree crowns. Broad-leaved trees show great plasticity in crown growth, which allows trees to grow toward open areas of the canopy and avoid competition from neighboring crowns. Crown asymmetry can buffer negative effects of crown competition, reducing tree mortality and increasing the mean and variation in tree size within stands (Umeki 1997).

Coniferous trees have less plasticity in crown form and directional growth of the crown. The spatial distribution of coniferous tree crowns is more strongly determined by the outcome of crown competition. In a western hemlock forest in northern Idaho, spatial distribution of trees became increasingly regular with increasing diameter, indicating that crown competition results in mortality of spatially aggregated young trees, leading to increasingly regular spatial distribution as trees increase in size (Moeur 1993). Similarly, in the 450-year-old Douglas-fir–western hemlock forest at the Wind River Canopy Crane Research Facility in southwestern Washington State, tree abundance decreases markedly with increasing canopy height, and horizontal spatial distribution of trees changes from aggregated distributions in the lower canopy to increasingly regular, dispersed distributions with increasing canopy height (Van Pelt and Franklin 2000; Ishii et al. 2004; Song et al. 2004).

As forests mature, canopy height reaches its maximum and species differences in height-growth rate become less important in determining structural development. Species with similar levels of shade tolerance will occupy similar positions in the canopy. Late-successional species that can reach the upper canopy eventually catch up and take the place of less shade-tolerant, early-successional species in the upper canopy (Guldin and Lorimer 1985). However, in regions where there are few physical limitations to canopy height (e.g., typhoons and poor soil conditions), physiological limitations to height growth and crown expansion may reduce competition among the tallest trees, allowing early successional species to coexist in the upper canopy with the late-successional species. For example, in the old-growth Douglas-fir–western hemlock forest at the Wind River Canopy Crane Research, mild climate and abundant rainfall allow canopy trees to reach heights over 60 m. In this forest, upper-canopy trees (tree height >40 m) of all species have attained maximum crown size and show very little crown expansion growth (Ishii et al. 2003). In the absence of crown competition, long-lived pioneer species such as Douglas-fir may be able to coexist with the late-successional species in the upper canopy (Ishii and Ford, 2002). Similarly, persistence of a long-lived, emergent pioneer species, *Eucalyptus regnans*, is observed in southeastern Australia and the Western Pacific where mild climatic conditions allow for development of tall canopies (Lindenmayer et al. 2000).

Stochastic Processes and Development of Structural Complexity

With increasing stand age, stochastic processes such as mortality of individual trees that create gaps in the canopy and small-scale disturbances that cause damage and die-back of crowns play increasingly important roles in creating structural complexity of the forest canopy. As trees reach maximum size, limitations to crown expansion in the upper canopy enhance structural complexity of the canopy surface. In Douglas-fir–western hemlock forests, gaps created in the upper canopy after individual tree mortality are not filled by neighboring trees, and the upper-canopy surface becomes increasingly heterogeneous with increasing stand age (see Figure 5-4). This may be a distinguishing structural characteristic of older stands that develops as a result of stochastic processes and physiological limitation to crown expansion.

Many trees in old-growth forests show evidence of past damage and re-growth, such as forks and crooks in the main stem (Ishii et al. 2000). In tropical forests and monsoon regions that are frequently affected by windstorms and typhoons, crown damage is commonly observed (Putz and Brokaw 1989; Coutts and Grace 1995). Droughts, fungal infections, insect outbreaks, and forest fires can also cause defoliation and die-back of the crown and are often followed by re-growth

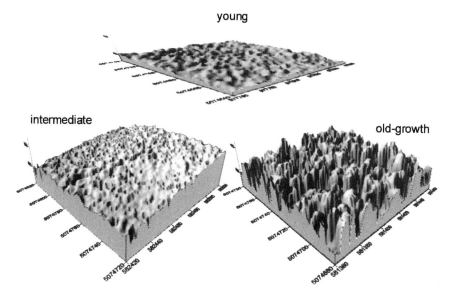

Figure 5-4 Topography of the upper-canopy surface of Douglas-fir forests of different ages. The upper-canopy surface becomes increasingly heterogeneous and rough with increasing stand age. The rugosity, a measure of surface roughness calculated as the standard deviation of outer canopy height, is 4.0 m, 11.5 m, and 15.4 m for young, intermediate, and old-growth stands, respectively.

(e.g., Lanner and Bryan 1981; Piene and Eveleigh 1996; Sharma and Rikhari 1997). Stochasticity of small-scale disturbances followed by re-growth of the crown adds variability to the otherwise deterministic architecture of trees and enhances structural complexity of the forest canopy.

In large, old trees that have reached maximum crown size and foliage amount, the balance between productive and nonproductive organs becomes increasingly important to maintain overall productivity (Remphrey and Davidson 1992). The crowns of large old trees are maintained by "reiteration," where architectural units are duplicated within the tree from suppressed buds (proleptic reiteration) or from growing axes (sylleptic reiteration) (Hallé et al. 1978; Bégin and Filion 1999). Reiteration often occurs by epicormic shoot production, where new units are produced when dormant buds in older tissue are released from suppression. Damage and die-back of the crown due to small-scale disturbances or physiological limitations are often followed by reiteration generating productive organs from existing branching structure, thus efficiently maintaining foliage amount without increasing non-productive biomass. Reiteration contributes to prolonging tree longevity by reproducing dead and dying crown components (Bryan and Lanner 1981). Many long-lived, late-successional species can reiterate architectural units and maintain the crown, in contrast to short-lived early-successional species, which tend to have less ability to reiterate crown components (Millet et al. 1998). Reiteration of various architectural units ranging from shoots and twigs to entire branches and vertical axes (reiterated trunks) has been observed in some large, old trees of long-lived species, including redwoods (*Sequoia sempervirens*, Sillett and Van Pelt 2000) and Douglas-fir (Ishii and Ford 2001). Understory trees of European beech (*Fagus sylvatica*), a shade-tolerant, late-successional species, maintain the crown by means of reiteration when growth is suppressed due to limited light conditions (Nicolini et al. 2001). In old Douglas-fir trees, reiteration enhances structural complexity of the crown by increasing branch size variability (Ishii and Wilson 2001). Reiteration also contributes to enhancing structural complexity by creating specific structural features such as reiterated trunks, fan-shaped clusters of epicormic branches, and platform-shaped forks within branches, making the crown of each tree "highly individualistic and irregular" (Franklin et al. 1981, see Figure 5-5).

Structural Complexity Enhances Ecological Function

One of the most pronounced effects of human activity on the global biosphere is the dramatic change incurred on ecosystem function and the declining biological diversity caused by deforestation (Wilson and Peter 1988; Perry and Maghembe 1989; Kimmins 1997). Humans have simplified ecosystem structure by converting natural forests to young stands, plantation forests, and agricultural fields. Management practices affect the structure (Zenner 2000; Lindenmayer and McCarthy 2002), function (Perry and Amaranthus 1997) and biodiversity (North et al. 1996; Beese and Bryant 1999) of forest ecosystems and simplification of stand structure leads to dimin-

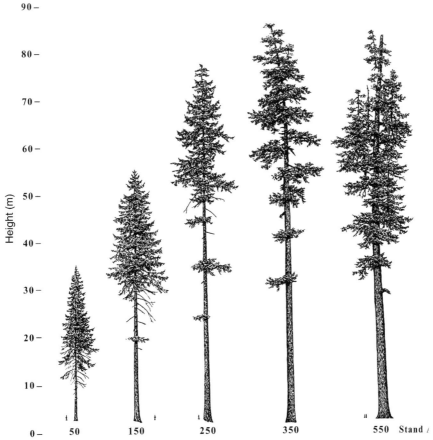

Figure 5-5 Development of structural complexity of the crown of Douglas-fir trees with increasing age. In 50-year-old trees, growth is deterministic and trees show typical conifer tree architecture. Many stands have reached crown closure, and lower branches begin to die. In 150-year-old trees, tree height reaches near 50 m. Further death of lower branches has occurred and the live crown reaches down to about two-thirds of tree height. Branches are reiterated by epicormic branching in the lower crown of some trees. At 250 years, trees are near their maximum height. Epicormic branches can be found in the lower crown of most trees as well as among original branches in the middle to upper crown. At 350 years, epicormic branches in the lower crown have developed into large fan-like clusters, and many of the branches in the main crown are also epicormic. At 550 years, the crown has reached maximum size, suffered several disturbance events, and is being maintained by epicormic branching and reiteration of several vertical axes (reiterated trunks).

ished ecosystem function and biodiversity decline (Bunnell and Huggard 1999). This has raised much debate over future policies for forest ecosystem management. One solution for enhancing ecosystem function and biodiversity of forest ecosystems may be to manage forests to increase stand structural complexity (Önal 1997).

Development of structural complexity, especially that of the forest canopy, enhances stand productivity by promoting complementary resource utilization among species (Ishii et al. 2004). Development of complex canopy structure comprising various species and life forms results in effective packing of biomass in the canopy. This promotes greater spatial and temporal partitioning and more efficient utilization of light resources, leading to increased stand productivity (Kira et al. 1969; Smith and Long 1989; Hartley 2002). Structural complexity of the forest canopy also enhances biodiversity of canopy-dwelling organisms by increasing environmental heterogeneity (e.g., variation of microhabitats and the range of microclimates) (Klopfer and MacArthur 1960; Pianka 1978; Carey et al. 1999a). Structurally complex canopies enhance biodiversity of a variety of organisms, including understory plants (North et al. 1996; McKenzie et al. 2000; Van Pelt and Franklin 2000; Brosofske et al. 2001), epiphytes (Rhoades 1987; McCune 1993; Lyons et al. 2000), birds (MacArthur and MacArthur 1961; Hansen et al. 1995; Beese and Bryant 1999), small mammals (Emmons 1987; Carey 1996), and arthropods (Schowalter 1995; Ozanne et al. 1997; Halaj et al. 2000; Hijii et al. 2001). Although there are other mechanisms that promote biodiversity (e.g., development of complex food webs and sufficient time for colonization), the development of structural complexity with increasing stand age may be especially important for forest ecosystems where trees predominantly produce the structural framework and create a resource-rich template for biodiversity in terms of both food and habitat (Hunter and Price 1992; Power 1992; Jones et al. 1997). Understanding how structural complexity of the forest canopy develops with increasing stand age is important for understanding forest ecosystem functioning as well as for integrative management of forest ecosystems.

References

Aiba, S. and Kohyama, T. (1996). Tree species stratification in relation to allometry and demography in a warm-temperate rain forest. *J. Ecol.* **84**, 207–218.

Bégin, C. and Filion, L. (1999). Black spruce (*Picea mariana*) architecture. *Can. J. Bot.* **77**, 664–672.

Beese, W.J. and Bryant, A.A. (1999). Effect of alternative silvicultural systems on vegetation and bird communities in coastal montane forests of British Columbia, Canada. *For. Ecol. Manage.* **115**, 231–242.

Bicknell, S.H. (1982). Development of canopy stratification during early succession in northern hardwoods. *For. Ecol. Manage.* **4**, 41–51.

Biondi, F., Myers, D.E., and Avery, C.C. (1994). Geostatistically modeling stem size and increment in an old-growth forest. *Can. J. For. Res.* **24**, 1354–1368.

Bravo, F., Hann, D.W., and Maguire, D.A. (2001). Impact of competitor species composition on predicting diameter growth and survival rates of Douglas-fir trees in southwestern Oregon. *Can. J. For. Res.* **31**, 2237–2247.

Brosofske, K.D., Chen, J., and Crown, T.R. (2001). Understory vegetation and site factors: implications for a managed Wisconsin landscape. *For. Ecol. Manage.* **146**, 75–87.

Bryan, J.A. and Lanner, R.M. (1981). Epicormic branching in Rocky Mountain Douglas-fir. *Can. J. For. Res.* **11**, 190–199.

Bunnell, F.L. and Huggard, D.J. (1999). Biodiversity across spatial and temporal scales: problems and opportunities. *For. Ecol. Manage.* **115**, 113–126.

Carey, A.B. (1996). Interactions of canopies and Pacific Northwest mammals. *Northwest Sci.* **69** (Special Issue), 72–78.

Carey, A.B., Kershner, J., Biswell, B., and Domínguez de Toledo, L. (1999a). Ecological scale and forest development: squirrels, dietary fungi, and vascular plants in managed and unmanaged forests. *Wildlife Monographs* **142**, 1–71.

Carey, A. B., Lippke, B.R., and Sessions, J. (1999b). Intentional systems management: managing forests for biodiversity. *Journal of Sustainable Forestry* **9**, 83–125.

Coutts, M.P. and Grace, J. (1995). "Wind and Trees." New York, Cambridge University Press.

Dubrasich, M.E., Hann, D.W., and Tappeiner II, J.C. (1997). Methods for evaluating crown area profiles of forest stands. *Can. J. For. Res.* **27**, 385–392.

Duncan, R.P. (1991). Competition and the coexistence of species in a mixed podocarp stand. *J. Ecol.* **79**, 1073–1084.

Easter, M.J. and Spies, T.A. (1994). Using hemispherical photography for estimating photosynthetic photon flux density under canopies and gaps in Douglas-fir forests of the Pacific Northwest. *Can. J. For. Res.* **24**, 2050–2058.

Emmons, L.H. (1987). Mammals of rain forest canopies. *In* "Forest Canopies" (M.D. Lowman and N.M. Nadkarni, Eds.), pp. 199–223. Academic Press, San Diego.

Ford, E.D. (2000). "Scientific Method for Ecological Research." Cambridge University Press, Cambridge.

Ford, E.D. and Sorrensen, K. A. (1990). Theory and models of inter-plant competition as a spatial process. *In* "Individual Based Models and Approaches in Ecology" (D. DeAngelis and L. Gross, Eds). pp. 363–407. Chapman and Hall, New York.

Franklin, J.F., Cromack, K., Denison, W., McKee, A., Maser, C., Sedell, J., Swanson, F., and Juday, G. (1981). Ecological characteristics of old-growth Douglas-fir forests. *USDA For. Serv. Gen. Tech. Rep.* PNW-118. 48 pages.

Franklin, J.F., Spies, T.A., Van Pelt, R., Carey, A.B., Thornburgh, D.A., Berg, D.R., Lindenmayer, D.B., Harmon, M.E., Keeton, W.S., Shaw, D.C., Bible, K., and Chen, J. (2002). Disturbances and structural development of natural forest ecosystems with silvicultural implications, using Douglas-fir forests as an example. *For. Ecol. Manage.* **155**, 399–423.

Gholz, H.L., Fitz, F.K., and Waring, R.H. (1976). Leaf area differences associated with old-growth forest communities in the western Oregon Cascades. *Can. J. For. Res.* **6**, 49–57.

Guldin, J.M. and Lorimer, C.G. (1985). Crown differentiation in even-aged northern hardwood forests of the Great Lakes region, USA. *For. Ecol. Manage.* **10**, 65–86.

Halaj, J., Ross, D.W., and Moldenke, A.R. (2000). Importance of habitat structure to the arthropod food-web in Douglas-fir canopies. *Oikos* **90**, 139–152.

Hallé, F.R., Oldeman, R.R., and Tomlinson, P.B. (1978). "Tropical Trees and Forests: An Architectural Analysis." Springer-Verlag, Berlin.

Hansen, A.J., McComb, W.C., Vega, R., Raphael, M.G., and Hunter, M. (1995). Bird habitat relationships in natural and managed forests in the west Cascades of Oregon. *Ecol. Appl.* **5**, 555–569.

Hartley, M.J. (2002). Rationale and methods for conserving biodiversity in plantation forests. *For. Ecol. Manage.* **155**, 81–95.

Hijii, N., Umeda, Y., and Mizutani, M. (2001). Estimating density and biomass of canopy arthropods in coniferous plantations: an approach based on a three-dimensional parameter. *For. Ecol. Manage.* **144**, 147–157.

Hix, D.M. and Lorimer, C.G. (1990). Growth-competition relationships in young hardwood stands on two contrasting sites in southwestern Wisconsin. *For. Sci.* **36**, 1032–1049.

Hunter, M.D. and Price, P.W. (1992). Playing chutes and ladders: heterogeneity and the relative roles of bottom-up and top-down forces in natural communities. *Ecology* **73**, 724–732.

Hunter, M.L.J. (1999). Biological diversity. *In* "Maintaining Biodiversity in Forest Ecosystems" (M. L. J. Hunter, Ed.), pp. 3–21. Cambridge University Press, Cambridge, UK.

Ishii, H. and Ford, E.D. (2001). The role of epicormic shoot production in maintaining foliage in old *Pseudotsuga menziesii* (Douglas-fir) trees. *Can. J. Bot.* **79**, 251–264.

Ishii, H. and Ford, E.D. (2002). Persistence of *Pseudotsuga menziesii* (Douglas-fir) in temperate coniferous forests of the Pacific Northwest Coast, USA. *Folia Geobotanica* **37**, 63–69.

Ishii, H., Ford, E.D., and Sprugel, D.G. (2003). Comparative crown form and branching pattern of four coexisting tree species in an old-growth *Pseudotsuga-Tsuga* forest. *Eurasian J. For. Res.* **6**, 99-109.

Ishii, H., Reynolds, J.H., Ford, E.D., and Shaw, D.C. (2000). Height growth and vertical development of an old-growth *Pseudotsuga-Tsuga* forest in southwestern Washington State, USA. *Can. J. For. Res.* **30**, 17–24.

Ishii, H., Tanabe, S., and Hiura, T. (2004). Exploring the relationships among canopy structure, stand productivity and biodiversity of temperate forest ecosystems. *For. Sci.*

Ishii, H. and Wilson, M.E. (2001). Crown structure of old-growth Douglas-fir in the western Cascade Range, Washington. *Can. J. For. Res.* **39**, 1250–1261.

Johnson, E.A., Miyanishi, K., and Kleb, H. (1994). The hazards of interpretation of static age structures as shown by stand reconstructions in a *Pinus contorta–Picea engelmannii* forest. *J. Ecol.* **82**, 923–931.

Jones, C.G., Lawton, J.H., and Shachak, M. (1997). Positive and negative effects of organisms as physical ecosystem engineers. *Ecology* **78**, 1946–1957.

Kenkel, N.C. (1988). Pattern of self-thinning in jack pine: testing the random mortality hypothesis. *Ecology* **69**, 1017–1024.

Kimmins, J.P. (1997). Biodiversity and its relationship to ecosystem health and integrity. *Forestry Chronicle* **73**, 229–232.

Kira, T., Shinozaki, K., and Hozumi, K. (1969). Structure of forest canopies as related to their primary productivity. *Plant and Cell Physiology* **10**, 129–142.

Klopfer, P.H. and MacArthur, R. (1960). Niche size and faunal diversity. *Am. Nat.* **94**, 293–300.

Kohm, K.A. and Franklin, J.F. (1997). "Creating a New Forestry for the 21st Century: The Science of Ecosystem Management." Island Press, Washington DC.

Koike, F. (1989). Foliage-crown development and interaction in *Quercus gilva* and *Q. acuta*. *J. Ecol.* **77**, 92–111.

Krajicek, J.E., Brinkman, K.A., and Gingrich, S.F. (1961). Crown competition: a measure of density. *For. Sci.* **7**, 35–42.

Larson, B.C. (1986). Development and growth of even-aged stands of Douglas-fir and grand fir. *Can. J. For. Res.* **16**, 367–372.

Lanner, R.M. and Bryan, J.A. (1981). Association of rhabdocline needle blight and epicormic branching in Douglas-fir. *Great Basin Naturalist* **41**, 476–477.

Lindenmayer, D.B. and Franklin, J.F. (2003). "Conserving Forest Biodiversity." Island Press, Washington.

Lindenmayer, D. and McCarthy, M.A. (2002). Congruence between natural and human forest disturbance: a case study from Australian montane ash forests. *For. Ecol. Manage.* **155**, 319–335.

Lindenmayer, D.B., Cunningham, R.B., Donnelly, C.F. and Franklin, J.F. (2000). Structural features of old-growth Australian montane ash forests. *For. Ecol. Manage.* **134**, 189–204.

Lyons, B., Nadkarni, N.M. and North, M.P. (2000). Spatial distribution and succession of epiphytes on *Tsuga heterophylla* (western hemlock) in an old-growth Douglas-fir forest. *Can. J. Bot.* **78**, 957–968.

MacArthur, R.H. and MacArthur, J.W. (1961). On bird species diversity. *Ecology* **42**, 594–598.

McCune, B. (1993). Gradients in epiphyte biomass in three Pseudotsuga-Tsuga forests of different ages in western Oregon and Washington. *The Bryologist* **96**, 405–411.

McKenzie, D., Halpern, C.B., and Nelson, C.R. (2000). Overstory influences on herb and shrub communities in mature forests of western Washington, U.S.A. *Can. J. For. Res.* **30**, 1655–1666.

Millet, J., Bouchard, A., and Édelin, C. (1998). Plant succession and tree architecture: an attempt at reconciling two scales. *Acta Biotheoretica* **46**, 1–21.

Moeur, M. (1993). Characterizing spatial patterns of trees using stem-mapped data. *For. Sci.* **39**, 756–775.

Moffett, M.W. and Lowman, M.D. (1995). Canopy access techniques. *In* "Forest Canopies" (M.D. Lowman and N.M. Nadkarni, Eds.), pp. 3–26. Academic Press, San Diego.

Nicolini, E., Chanson, B., and Bonne, F. (2001). Stem growth and epicormic branch formation in understory beech trees (*Fagus sylvatica* L.). *Ann. Bot.* **87**, 737–750.

North, M., Chen, J., Smith, G., Krakowiak, L., and Franklin, J.F. (1996). Initial response of understory plant diversity and overstory tree diameter growth to a green tree retention harvest. *Northwest Sci.* **70**, 24–34.

Önal, H. (1997). Trade-off between structural diversity and economic objectives in forest management. *American Journal of Agricultural Economics* **79**, 1001–1012.

Ozanne, C.M.P., Hambler, C., Foggo, A. and Speight, M.R. (1997). The significance of edge effects in the management of forests for invertebrate biodiversity. *In* "Canopy Arthropods" (N.E. Stork, J. Adis, and R.K. Didham, Eds.), pp. 534–550. Chapman & Hall, London.

Palik, B.J. and Pregitzer, K.S. (1991). The relative influence of establishment time and height-growth rates on species vertical stratification during secondary forest succession. *Can. J. For. Res.* **21**, 1481–1490.

Perry, D.A. and Amaranthus, M.P. (1997). Disturbance, recovery and stability. *In* "Creating a New Forestry for the 21st Century: The Science of Ecosystem Management" (K.A. Kohm and J.F. Franklin, Eds.), pp. 31–56. Island Press, Washington DC.

Perry, D.A. and Maghembe, J. (1989). Ecosystem concepts and current trends in forest management: time for reappraisal. *For. Ecol. Manage.* **26**, 123–140.

Pianka, E.R. (1978). "Evolutionary Ecology." Harper and Row, New York.

Piene, H. and Eveleigh, E.S. (1996). Spruce budworm defoliation in young balsam fir: The 'green' tree phenomenon. *Can. Entomol.* **128**, 1101–1107.

Power, M.E. (1992). Top-down and bottom-up forces in food webs: do plants have primacy? *Ecology* **73**, 733–746.

Putz, F.E. and Brokaw, N.V.L. (1989). Sprouting of broken trees on Barro Colorado Island, Panama. *Ecology* **70**, 508–512.

Putz, F.E., Parker, G.G., and Archibald, R.M. (1984). Mechanical abrasion and intercrown spacing. *American Midland Naturalist* **112**, 24–28.

Remphrey, W.R. and Davidson, C.G. (1992). Spatiotemporal distribution of epicormic shoots and their architecture in branches of *Fraxinus pennsylvanica*. *Can. J. For. Res.* **22**, 336–340.

Rhoades, F.M. (1987). Nonvascular epiphytes in forest canopies: worldwide distribution, abundance, and ecological roles. *In* "Forest Canopies" (M.D. Lowman and N.M. Nadkarni, Eds.), pp. 353–408. Academic Press, San Diego.

Rudnicki, M., Lieffers, V.J., and Silins, U. (2003). Stand structure governs the crown collisions of lodgepole pine. *Can. J. For. Res.* **33**, 1238–1244.

Schowalter, T.D. (1995). Canopy arthropod communities in relation to forest age and alternative harvest practices in western Oregon. *For. Ecol. Manage.* **78**, 115–125.

Sharma, S. and Rikhari, H.C. (1997). Forest fire in the central Himalaya: climate and recovery of trees. *International Journal of Biometeorology* **40**, 63–70.

Sillett, S.C. and Van Pelt, R. (2000). A redwood tree whose crown is a forest canopy. *Northwest Sci.* **74**, 34–43.

Smith, D.M., Larson, B.C., Metthew, J.K., and Ashton, P.M.S. (1997). "The Practice of Silviculture: Applied Forest Ecology." John Wiley & Sons, New York.

Smith, F.W. and Long, J.N. (1989). The influence of canopy architecture on stemwood production and growth efficiency of *Pinus contorta* var. *latifolia. J. Appl. Ecol.* **26**, 681–691.

Solomon, D.S. and Gove, J.H. (1999). Effects of uneven-age management intensity on structural diversity in two major forest types in New England. *For. Ecol. Manage.* **114**, 265–274.

Song, B., Chen, J., and Silbernagel, J.M. (2004). Slicing the 3-D canopies of an old-growth Douglas-fir forest for structural analysis. *For. Sci.*

Spies, T.A. and Franklin, J.F. (1991). The structure of natural young, mature and old-growth Douglas-fir forests in Oregon and Washington. *In* "Wildlife and Vegetation of Unmanaged Douglas-Fir Forests" (L.F. Ruggiero, K.B. Aubry, A.B. Carey, and M.H. Huff, Eds.), pp. 91–109. USDA Forest Service General Technical Report PNW-GTR 285.

Staudhammer, C.L. and LeMay, V.M. (2001). Introduction and evaluation of possible indices of stand structural diversity. *Can. J. For. Res.* **31**, 1105–1115.

Stewart, G.H. (1986). Population dynamics of a montane conifer forest, Western Cascade Range, Oregon, USA. *Ecology* **67**, 534–544.

Stohlgren, T.J. (1993). Intra-specific competition (crowding) of giant sequoias (*Sequiadendron giganteum*). *For. Ecol. Manage.* **59**, 127–148.

Sumida, A., Terazawa, I., Togashi, A., and Komiyama, A. (2002). Spatial arrangement of branches in relation to slope and neighborhood competition. *Ann. Bot.* **89**, 301–310.

Thomas, S.C. (1999). Asymptotic height as a predictor of photosynthetic characteristics in Malaysian rain forest trees. *Ecology* **80**, 1607–1622.

Umeki, K. (1997). Effect of crown asymmetry on size-structure dynamics of plant populations. *Ann. Bot.* **79**, 631–641.

Van Pelt, R. and Franklin, J.F. (2000). Influence of canopy structure on the understory environment in tall, old-growth, conifer forests. *Can. J. For. Res.* **30**, 1231–1245.

Wells, R.W., Lertzman, K.P., and Saunders, S.C. (1998). Old-growth definitions for the forests of British Columbia. *Nat. Areas J.* **18**, 279–292.

Wierman, C.A. and Oliver, C.D. (1979). Crown stratification by species in even-aged mixed stands of Douglas-fir–western hemlock. *Can. J. For. Res.* **9**, 1–9.

Wilson, E.O. and Peter, F.M. (1988). "Biodiversity." National Academy Press, Washington, DC.

Zenner, E.K. (2000). Do residual trees increase structural complexity in Pacific Northwest coniferous forests? *Ecol. Appl.* **10**, 800–810.

CHAPTER 6

A History of Tree Canopies

David L. Dilcher, T. A. Lott, Xin Wang, and Qi Wang

*In the plants of our own day we have the outcome of an age-long series of
experiments, the result of selection of certain designs which, like those
of a good architect, owe their . . . quality to simplicity combined with
efficiency and the absence of features that are unessential.*
—*A. C. Seward,* Plant Life through the Ages, *1931*

Introduction

A history of tree canopies must begin with the history of trees. Tree evolution was not sudden
nor was it restricted to only one group of plants or one way of constructing a tree. There was a
time before there were any trees. This was during the Silurian and Lower Devonian (about 425
to 380 million years ago) when early land plants first appeared and later became common on
exposed land surfaces. The early land plants were small, often consisting of a single or dichotomous branching axis terminated by a sporangium or bearing lateral sporangia (Edwards 1970,
1994; Gensel and Andrews 1984). These plants stood from a few centimeters in the Silurian to
a meter or two in height in the Lower Devonian. They were involved in extensive modifications
of vegetative and reproductive structures as plants adapted to life on land. These early land plants
changed the biological and abiotic environments in ways that prepared the world for the evolution of trees, forests, and the forest canopy.

Early Modifications of the Plant Body

The earliest land plants could be typified by *Cooksonia,* which consists of a simple plant only a
few centimeters tall with dichotomous branching one to four times, and many axes ending in
homosporous sporangia (Edwards 1970; Gerrienne et al. 2001). These and other simple plants
modified their form to increase and maximize their potential for photosynthesis by various modifications, starting from a simple naked round axis. Some flattened the axis into a ribbon-like
form (*Taeniocradia*); some grew protuberances (*Discalis,* Hao 1989) or lateral flap-like tissue out of
their axes (*Asteroxylon*). Others modified their branching systems into condensed lateral profusely
branched organs (*Pertica*). In many of these modifications, the increased photosynthetic tissue was
complemented by an increase in laminar and vascular tissue, resulting in plants with leaves, however derived, that maximized the potential to capture sunlight and promote gaseous exchanges
and carbon fixation. The benefit to the plant of all these changes is an increase in the photosynthetic potential of these plants with leafy shoots and branches. This driving force to increase

118

photosynthetic potential, which also increases reproductive potential, has been important in the evolution of tree canopies throughout the history of trees.

The organization of early land plants was being modified both above and below ground. The requirements for anchoring the plant, nutrients and water, must have increased tremendously from the small *Cooksonia* plant of the Upper Silurian to the Middle Devonian plants standing 2 m tall and having photosynthetic leafy axes or axes with complex lateral branches. A simple axis extending into the ground could no longer manage to absorb the nutrients and water needed for photosynthesis and to grow in an upright stature. These early land plants solved this problem in two basic ways through multiple steps of evolution. One strategy that was adopted was a co-evolution of early land plants with mycorrhizal fungi (Pirozynski and Malloch 1975; Wagner and Taylor 1981; Remy et al. 1994). These fungi increased the absorptive potential of the root, and such associations are still an absolute requirement of most forest trees today.

The second feature of root evolution in many plants is the development of branches and bipolar growth as well as the elaboration of epidermal cells into root hairs (Kenrick 2002). The Lycophytes adopted another strategy of underground stems using numerous leaf-like lateral appendages for absorption rather than root hairs (Rothwell and Erwin 1985).

Biotic Factors Necessary for Tree Evolution

It should be obvious that a lot of plant evolution took place before the advent of the tree habit. This evolution proceeded in multiple directions but always toward a more complex plant body with greater photosynthesis, hence greater reproductive potential, than known for the earliest land plants. It is important to consider the changes that often happen independently in various lineages of early land plants before the evolution of trees took place. Also, it should become clear that various plant groups solved the need for photosynthesis and nutrient water uptake by very different and independent ways. The importance of maximizing the photosynthesis potential of plants can be seen as an early and controlling factor in their evolution, long before trees were a part of the plant form.

The importance of photosynthesis to the success of a plant, both in its potential to grow more vegetative tissue and in its potential to maximize reproduction, cannot be overemphasized. This increase in plant size also required an efficient and well-developed root system to serve as the function of absorption and anchoring the plant. These are the biotic factors that were required and in place before trees could evolve.

Abiotic Factors Important in Tree Evolution

Among the abiotic factors important in the evolution of trees must have been shade, wind, and soil. The initial land plant vegetation was a one-level herbaceous layer near the ground (less than 0.5 m), which could be considered the *ancestral herbaceous canopy* of the Late Silurian world. This was followed, as already discussed, by taller (up to 1 to 2 m) leafy and/or complex branching vegetation representing the Lycopsida, Sphenopsida, Pteridopsida, and perhaps a few progymnosperms.

Light vs. shade was critical to the maximum growth and energy available for reproduction. Because shade was not a typical component of the single canopy of the *ancestral herbaceous canopy*, the photosynthetic system of green land plants would have been adapted to full sunlight exposures. As the herbaceous shrub vegetation grew higher, shade became a factor in plant evolution. Either plants had to modify the physiology of their energy production, thus their photosynthetic mechanism, or they needed to grow taller, finally becoming tall shrubs or trees.

THE EVOLUTION OF RAINFOREST ANIMALS

Dale A. Russell

Six important fossil localities and living tropical rainforests reflect seven stages in the general evolution of rainforest animals. Dates are in millions of years before the present (myr) and can be correlated with time intervals (see Palmer and Geissman 1999).

Near the time of origin of vascular plants (Ludlow locality, UK, 419 myr), small arthropods in low latitudes fed upon organic detritus and microorganisms among the stalks of simple archaic herbs. These animals were in turn consumed by proto-spiders and centipedes (Gensel and Edwards 2001). With the appearance of trees at ~364 myr, photosynthetic surfaces were thrust 20 m above the ground, and canopies began to close (Scheckler 2001). Insects probably colonized the canopy simultaneously. By 311 myr, giant herbivorous insects flew between tree crowns in equatorial forests, feeding on sap and pollen. Large insect predators preyed upon them. On the forest floor, gigantic millipedes ingested decaying wood, as lizard-like vertebrates hunted smaller terrestrial arthropods (Mazon Creek, Illinois and Linton, Ohio localities, Baird et al. 1985; Hook and Baird 1988; Labandeira 2001). Arboreal habitats were later (250–208 myr) invaded by small, *Draco*-like gliding reptiles and peculiar monkey-like forms (drepanosaurs, Frey et. al. 1997; Unwin et al. 2000; Harris and Downs 2002).

Sediments in mid-latitude lakes in China (Jehol localities, 125 myr) contain exquisite skeletons of small vertebrates, as well as evidence of high productivity and warm climates (Si'en 1999; Smith et al. 2001; Zhou et al. 2003). Feathered gliding dinosaurs (ambush predators) and relatively large primitive birds harvested vertebrate and invertebrate prey, as well as seeds in coniferous canopies. A few small herbivorous dinosaurs and insectivorous mammals frequented detritus-based communities at ground level. Large dinosaurs were very rare. The diversification of angiosperms in tropical and paratropical forests (95 myr) coincided with the origin and radiation of wasps, ants, and butterflies (Labandeira et al. 1994; Grimaldi 2000; Grimaldi et al. 2002). Much later (49 myr), leaves and insects preserved in lake deposits in Germany (Messel locality) resembled those in Southeast Asian forests today. Mammals were much more diverse, both taxonomically and morphologically, than their Jehol antecedents, although they were similar in size. A third of them were arboreal. Bats and primates had already made their appearance (Schaal and Ziegler 1992).

Dominican amber contains evidence of bromeliads, orchids, and a diverse insect fauna, indicating the presence of modern tropical rainforests in low latitudes at 20–15 myr (Poinar and Poinar 1999). The Amazon Basin remained covered with rainforests through climatic excursions of the last 2 myr (Colinvaux and De Oliveira 2001). Molecular phylogenies suggest that most species of tropical vertebrates originated before the onset of ice-age climates, and that faunal continuity was undisturbed by them (Moritz et al. 2000).

Modern rainforest animals are more diverse and more abundantly nourished by canopy nuts and fruits than those of the past. Vertebrates of the forest floor are larger, implying richer detrital food chains linked to greater productivity in forest canopies (Potts and Behrensmeyer 1992; Richards 1996; Kingdom 1997). Through geologic time, the canopy has been the site of multiple origins of animal flight. Because biodiversity is correlated with competitive fitness (McCann 2000; Purvis and Hector 2000), living tropical rainforests may be among the most fit of terrestrial ecosystems ever to inhabit the planet. This is suggested by the anthropogenic extermination of savanna and grassland megavertebrates over 10,000 years ago, whereas rainforest faunas probably remained intact until a few centuries ago. The record of animal life indicates that the rainforest canopy has for ~364 myr been a primary driver of terrestrial evolutionary innovation.

References

Baird, G.C., Shabica, C.W., Anderson, J.L., and Richardson, E.S. (1985). Biota of a Pennsylvanian muddy coast: habitats within the Mazonian delta complex, northeastern Illinois. *Journal of Paleontology* **52**, 253–281.

Colinvaux, P.A. and De Oliveira, P.E. (2001). Amazon plant diversity and climate through the Cenozoic. *Palaeogeography, Palaeoclimatology, Palaeoecology* **166**, 51–63.

Frey, E., Sues, H.D., and Munk, W. (1997). Gliding mechanism in the late Permian reptile *Coelurosauravus*. *Science* **275**, 1450–1452.

Gensel, P.G. and Edwards, D. (2001). "Plants Invade the Land: Evolutionary and Environmental Perspectives." Columbia University Press, New York.

Grimaldi, D. (2000). Mesozoic radiations of the insects and origins of the modern fauna. Proceedings of the XXI International Congress of Entomology, Iguassu, Brazil, August 17–26, 2000, Volume 1, pp. xix–xxvii.

Grimaldi, D., Engel, M.S., and Nascimbene, P. (2002). Fossiliferous Cretaceous amber from Myanmar (Burma): its rediscovery, biotic diversity, and paleontological significance. *American Museum Novitates* 3361.

Harris, J.D. and Downs, A. (2002). A drepanosaurid pectoral girdle from the Ghost Ranch (Whitaker) *Coelophysis* quarry (Chinle Group, Rock Point Formation, Rhaetian), New Mexico. *Journal of Vertebrate Paleontology* **22**, 70–75.

Hook, R.W. and Baird, D. (1988). An overview of the Upper Carboniferous fossil deposit at Linton, Ohio. *Ohio Journal of Science* **88**, 143–154.

Kingdom, J. (1997). "The Kingdom Guide to African Mammals." Academic Press, London.

Labandeira, C.C. 2001. Rise and diversification of insects. *In* "Palaeobiology II" (D.E. Briggs and P.R. Crowther, Eds.), pp. 82–88. Blackwell Science, London.

Labandeira, C.C., Dilcher, D.L., Davis, D.R., and Wagner, D.L. (1994). Ninety-seven million years of angiosperm-insect association: paleobiological insights into the meaning of co-evolution. *Proceedings of the National Academy of Sciences USA* **91**, 12278–12282.

McCann, K.S. (2000). The diversity-stability debate. *Nature* **405**, 228–233.

Moritz C., Patton, J.L., Schneider, C.J., and Smith, T.B. (2000). Diversification of rainforest faunas: an integrated molecular approach. *Annual Review of Ecology and Systematics* **31**, 533–563.

Palmer, A.R. and Geissman, J. (1999). "1999 Geologic Time Scale." Geological Society of America, product code CTS004. [http://www.geosociety.org/science/timescale/timescl.pdf]

Poinar, G. and Poinar, R. (1999). "The Amber Forest: A Reconstruction of a Vanished World." Princeton University Press, Princeton, NJ.

Potts, R. and Behrensmeyer, A.K. (1992). Late Cenozoic terrestrial ecosystems. *In* "Terrestrial Ecosystems Through Time" (A.K. Behrensmeyer, J.D. Damuth, W.A. DiMichele, R. Potts, H.D. Sues, and S.L. Wing, Eds.), pp. 419–541. University of Chicago Press, Chicago.

Purvis, A. and Hector, A. (2000). Getting the measure of biodiversity. *Nature* **405**, 212–219.

Richards, P.W. (1996). "The tropical rain forest, Second Edition." Cambridge University Press, Cambridge, UK.

Schaal, S. and Ziegler, W. (1992). "Messel: An Insight into the History of Life and of Earth." Clarendon Press, Oxford.

Scheckler, S.E. (2001). Afforestation: the first forests. *In* "Palaeobiology II" (D.E.G. Briggs and P.R. Crowther, Eds.), pp. 67–71. Blackwell Science, London.

Si'en, W. (1999). Palaeoecology and palaeoenvironment of the Jehol biota: A palaeoecological and palaeoenvironmental reconstruction of conchostracan palaeocommunities in the northern Hebei/western Liaoning area. *Dizhi Xuebao* **73**(4), 289–301 (Chinese, English abstract).

Smith, J.B., Harris, J.D., Omar, G.I., Dodson, P. and You, H.L. (2001). Biostratigraphy and avian origins in northeastern China. *In* "New Perspectives on the Origin and Early Evolution of Birds" (J. Gauthier and L.F. Gall, Eds.), pp. 551–589. Peabody Museum of Natural History, Yale University.

Unwin, D.M., Alifanov, V.R., and Benton, M.J. (2000). Enigmatic small reptiles from the Middle-Late Triassic of Kirgizstan. *In* "The Age of Dinosaurs in Russian and Mongolia" (M.J. Benton, M.A. Shishkin, D.M. Unwin, and E.N. Kurochkin, Eds.), pp. 177–186. Cambridge University Press, Cambridge, UK.

Zhou, Z.H., Barrett, P.M., and Hilton, J. (2003). An exceptionally preserved Lower Cretaceous ecosystem. *Nature* **421**, 807–814.

Therefore, shade vs. sunlight was a major, if not the main, factor in the evolution of taller plants and thus the canopy layers of the forest through time.

The early land plants produced terminal or lateral sporangia on terminal portions of the plants. The plants were homosporous and the spores were small. These early land plant spores were produced only a few centimeters above the ground and dispersed by near ground breezes and/or water droplets or flow. As plants increased in complexity with increased branches, height, and potential for photosynthesis, the potential for spore production increased. More and larger sporangia were complemented by increased dispersal of spores resulting from breezes and wind higher off the ground (Chaloner and Sheerin 1981). We can assume that those plants with the most height provided the best use of the wind for spore dispersal and, as a result, the best potential for reproductive success. So wind as related to spore dispersal may have played a direct role in the evolution of trees. Also, safety factors against wind-induced stem failure are not negligible (Niklas and Speck 2001).

In order for plants to attain any stability as they grow in height, they must be well anchored by a substantial root system. A number of root systems have evolved through time, as already discussed, but common to all roots is the soil in which they develop. Soil is a complex of living and dead organisms and organic matter combined with the inorganic residue of weathered rock material. The evolution of soil and its accumulation in the quantities necessary to anchor and sustain forest-size trees must have required many millions of years. Perhaps this substrate was only available by Middle Devonian time when we find the early evidence for, and early proliferation of, trees. During the Late Devonian, increases in plant size and distribution and in root biomass may result in the depth and volume of soils (Algeo et al. 1995).

Evidence for Ancient Trees

The earliest trees or tree-like forms of plants are found first in the lower Middle Devonian (in the lower Eifelian) about 385 million years ago. *Calamophyton* (see Figure 6-1) is found then and falls just short of a basic forestry definition for a tree, being smaller than 4 m with a stem diameter just short to slightly larger than 7.5 cm, according to Mosbrugger (1990). Others suggest that a tree is any plant with a single stem over 6 m tall (Thomas 2000). Figure 6-1 shows that there was what appears to be a rapid development of tree size as related to stem diameter and estimated height. This covers a period of time from about 385 million years ago with a trunk diameter of 10 cm and estimated height of 4 m for *Calamophyton*, to 360 million years ago when *Callixylon* (trunk of *Archaeopteris*) are known that are 1.5 m in diameter and an estimated height of 30 m (see Figure 6-1). This continues into the Mississippian or Lower Carboniferous (350 million years ago) when *Pitus* is found with a trunk diameter of 2.5 m and an estimated height of 43 m, according to Mosbrugger (1990).

During the Late Devonian (372 to 352 million years ago), numerous forest trees are shorter with canopies that are up to 20 m in height. Actual measurements of *Callixylon* fossils suggest that the tree reached heights of only 18 m (Scheckler, pers. com. 2003) while Long (1979) reports a *Pitus* trunk diameter of 2.5 m and the longest length at just 21 m. Table 6-1 presents a list of fossil trees and their heights as reported in the literature. It is clear from the variations in these measurements that they represent a variety of estimates of the real height of these trees. Mosbrugger (1990) has done the most careful recent research in an attempt to estimate the heights of these Paleozoic trees. Some of the assumptions made in these estimates are clearly seen in the use of "estimated tree height" derived from tree diameter in Figure 6-1.

The heights of fossil trees as discussed in this chapter are based on a variety of facts and assumptions. Fossil plants, by their very nature, are fragmentary. Tree trunks often preserve partially when uprooted and laid down in the sediment and very rarely will a trunk be found

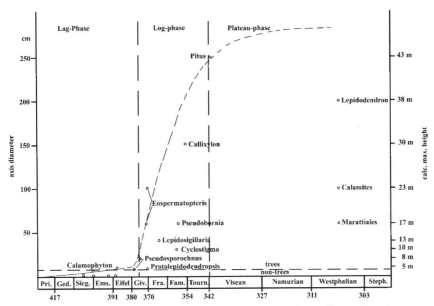

Figure 6-1 The development of trees through the Late Devonian and the Carboniferous. Trees began to occur since the Givetian and quickly increased their height until the end of the Tournaisian. Diameter data from literature in Mosbrugger (1990) and citations in Figure 6–3. Heights of individual trees plotted here are based on the calculations and assumptions given by Mosbrugger (1990). Pri. Pridoli; Ged. Gedinnian; Sieg. Siegenian; Ems. Emsian; Eifel. Eifelian; Giv. Givetian; Fra. Frasnian; Fam. Famennian; Tourn. Tournaisian; Stephan. Stephanian time scale after Haq and Eysinga (1998). This table is modified from Mosbrugger (1990).

that shows roots, trunk, and intact branches. Most often, tree height is estimated from the basal trunk diameter and the nature of the trunk taper. Generally, no allowance is given for the swollen bases of trees that are typical of swamp-grown trees of today. To this can be added reconstructed crown for the total tree height. For this chapter we have searched literature and report the various estimates that several paleobotanists have presented. The general range of canopy height is probably within the range given. Also, the genera reported here consist of many species that may have varied in size. When a size is reported for many species within a genus, we then have a better reconstruction of the various heights and levels within the fossil canopies of the Paleozoic and Mesozoic. Also, fossil plants may be preserved at any growth stage, again adding size variation for that particular tree. Sometimes the ages can be determined when trunk anatomy is preserved. More reports of partial, nearly whole, and complete trees should be considered an important contribution to paleobotany and we hope careful observations will be reported in the future as people find tree trunks often exposed only briefly in mining or natural outcrops.

Modern-day forests consist of a mixture of mono-layer trees, whose branches spread out to form a single leafy plane where the individual leaves do not shade each other, vs. multi-layer trees, whose leafy layers overlap above one another. The mono-layer trees are generally understory forest trees. A plant can use only about 20 percent of the available sunlight energy when exposed to direct sunlight (Thomas 2000). The earliest trees appear to have been multi-layer-type trees in which the light filters through the branches and leaves to those below or reaches the tree by exposing these lateral tree layers to rays of the sun where exposed along the edges of open areas such as rivers. Apparently, about 54 percent full sunlight is needed for a multi-layer tree, while mono-layers can manage with less light (Thomas 2000). This leads to the suggestion that some early shrubby plants such as *Pertica* (see Figure 6-2, B), *Duisbergia* (see Figure 6-2, D), *Arctophyton*

Table 6-1 Data Set for Major Land Plants from the Devonian to the Present

Age	Genus	Height	Spread of Crown	Habitat	Source
D	*Homaephyton* (A)	15-20 cm	8 cm	bog, swamp	Stewart and Rothwell 1993
D	*Pertica* (B)	<1 m	20 cm (middle)		Kasper and Andrews 1972
D	*Pseudosporochnus* (C)	1.5-4 m	3 m	dry upland?	Potonié and Bernard 1904, Taylor 1981, Mosbrugger 1990, Stewart and Rothwell 1993
D	*Duisbergia* (D)	2 m	45 cm	swamp	Schweitzer 1990
D	*Eospermatopteris* (E)	12 m	9 m	dryland	Schweitzer 1990, Stewart and Rothwell 1993
D	*Arctophyton* (F)	<2 m	<1 m		Schweitzer 1968
D	*Calamophyton* (G)	2-3 m	1 m	swamp	Mosbrugger 1990, Schweitzer 1990
D	*Lycopod* (H)	10 m	3.5 m	floodplain	
D	*Archaeopteris* (I)	10-18 m	3.5-11 m	floodplain	Beck 1962, Mosbrugger 1990, Stewart and Rothwell 1993, Schleckler per. com.
D	*Cyclostigma* (J)	8 m	4.5 m	river edge	Schweitzer 1990
D	*Pseudobornia* (K)	15-20 m	1m nr. apex, 4m nr. ground	river edge	Taylor 1981, Schweitzer 1990
C-P	*Medullosa* (L)	3.5-10 m	3-8 m	swamp	Stewart and Delevoryas 1956, Taylor 1981, Stewart and Rothwell 1993
C-P	*Psaronius* (M)	6-10 m	6 m	swamp	Morgan 1959, Taylor 1981, Stewart and Rothwell 1993
Penn	*Diaphorodendron* (N, P, R)	8-20 m	2.5-7 m	swamp	Stewart and Rothwell 1993
Penn	*Lepidodendron* (O)	54 m	24 m	swamp	Stewart and Rothwell 1993

124

Penn	*Lepidophloios*	20 m	6 m	swamp	Stewart and Rothwell 1993
Penn	*Sigillaria* (Q)	<20-34 m	>9-12 m	swamp	Taylor 1981, Stewart and Rothwell 1993
Penn	*Chaloneria*	2 m	0.1 m	wet land?	Stewart and Rothwell 1993
Penn	*Sporangiostrobus*	3 m	0.4 m	swamp	DiMichele and Philips 1994
C-P	*Calamites* (S)	10-20 m	3.5-7 m	lowland+swamp	Taylor 1981, Stewart and Rothwell 1993
Penn	*Cordaites* (T, U)	5-40 m	2.5-22 m	swamp(mangrove)	Taylor 1981, Stewart and Rothwell 1993
P-T	*Glossopteris* (V)	4-6 m	2-3 m	dry upland?	Stewart and Rothwell 1993
T-Pre	*Araucaria* (W)	12-70 m	5-25 m	dry upland?	Krüssman 1995
P-Pre	*Cyas* (X)	12 m	13 m	dry upland	Jones 1993
T-Pre	*Agathis* (Y)	12-60 m	9-47 m	dry upland	Krüssman 1995
K-Pre	*Pinus* (Z)	1-70 m	.4-25 m	dry upland	Elias 1989, Krüssman 1995
P-Pre	*Ginkgo* (AA)	20-40 m	13-26 m	dry upland?	Elias 1989, Krüssman 1995, Zhang et al. 2000
K-Pre	*Pseudotsuga* (BB)	50-90 m	14-25 m	upland	Elias 1989, Krüssman 1995
J	*Williamsonia* (CC)	1.5-2 m	1 m	upland	Sahni 1932
K-Pre	*Sequoiadendron* (DD)	60-105 m	17-30 m	dry upland	Elias 1989, Krüssman 1995
K	*Tempskya* (EE)	4.5-6 m	1 m	dry upland?	Taylor 1981, Stewart and Rothwell 1993
J-Pre	*Sequoia* (FF)	60-110 m	9-17 m	upland	Elias 1989, Krüssman 1995

Age abbreviations: D, Devonian; C-P, Carboniferous-Permian; Penn, Pennsylvanian; P-T, Permian-Triassic; P-Pre, Permian-Present; T-Pre, Triassic-Present; J-Pre, Jurassic-Present; K-Pre, Cretaceous-Present; K, Cretaceous. Order of genera related to Figure 6-2. All height and some canopy data from references; most canopy data estimated from height.

(see Figure 6-2, F), and *Pseudobornia* (see Figure 6-2, K) may have grown in open areas or along stream sides where lateral light was available. This is much like the *Juniperus* in North America that is a much branched tree from the base to the apex and grows well as a pioneer species in open areas. It is often shaded out when the forest trees overtake it.

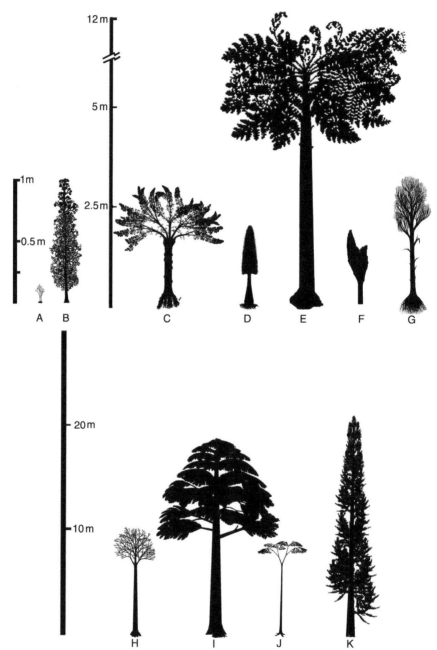

Figure 6-2 Growth habit silhouettes of land plant in relation to canopy heights, from the Devonian to the present. (A) *Horneophyton*, 15–20 cm. (B) *Pertica*, up to 1 m. (C) *Pseudosporochnus*, 3 m. (D) *Duisbergia*, 2 m. (E) *Eospermatopteris*, 12 m. (F) *Arctophyton*, <2 m (not to scale). (G) *Calamophyton*, 2–3 m. (H) *Diaphorodendron*, 10 m. (I) *Archaeopteris*, 18 m (estimated). (J) *Cyclostigma*, 8 m. (K) *Pseudobornia*,

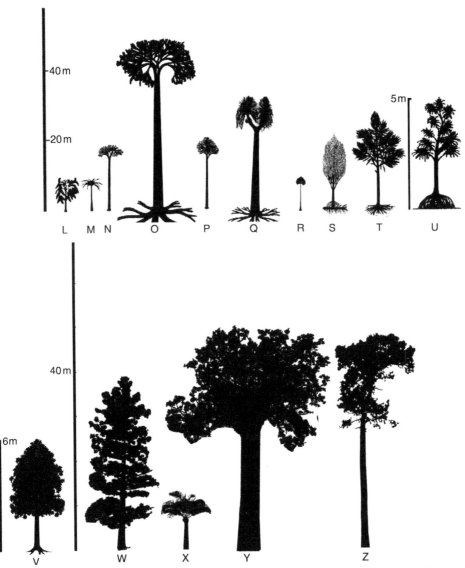

Figure 6-2—*Continued* 15–20 m. (L) *Medullosa*, 10 m. (M) *Psaronius*, 10 m. (N) *Diaphorodendron*, 16–17 m. (O) *Lepidodendron*, 54 m. (P) *Diaphorodendron*, 20 m. (Q) *Sigillaria*, 30–34 m. (R) *Diaphorodendron*, 8–10 m. (S) *Calamites*, 10–20 m. (T) Cordaitean plant, 30 m. (U) Cordaitean plant, 5 m. (V) *Glossopteris*, 4–6 m. (W) *Araucaria*, 70 m. (X) *Cycas*, 12 m. (Y) *Agathis*, 60 m. (Z) *Pinus*, 60 m.

Continued

    Forest canopy and emergent giants are also multi-layered because they receive lateral light and allow some light to filter down to the lower branches, but they do not maintain branches below a crown of terminal branches and leaves. Examples of this type of forest tree are *Archaeopteris* (see Figure 6-2, I) and perhaps some Lycopods (see Figure 6-2, H) from the Late Devonian during the log phase of tree evolution. Numerous trees from the Carboniferous (Tournaisian at 352 million years) have typical multi-layer crowns of forest trees. Thus, we can consider that from Late Devonian time of about 355 million years ago there were some forests

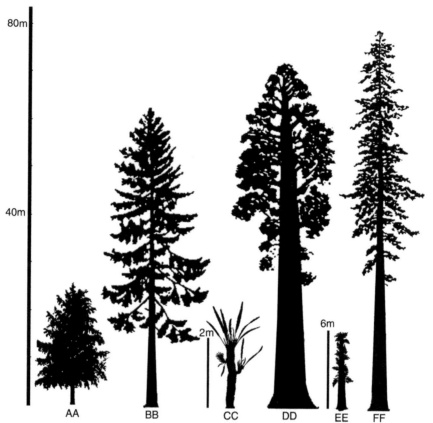

Figure 6-2—Continued   (AA) *Ginkgo*, 40 m. (BB) *Pseudotsuga*, 90 m. (CC) *Williamsonia*, 1.5–2 m. (DD) *Sequoiadendron*, 100 m. (EE) *Tempskya*, 4.5–6 m. (FF) *Sequoia*, 110 m. (A) After Stewart and Rothwell 1993. (B) Modified after Kasper and Andrews 1972. (C) Modified after Leclercq and Banks 1962. (D, G, J) After Schweitzer 1990. (E) After Goldring, in Andrews 1961. (F) From Gensel and Andrews 1984. (H) After Niklas and Speck 2001. (I) After Beck 1962. (K) After Taylor 1981. (L) After Stewart and Delevoryas 1956. (M) After Morgan 1959. (N, P, R) After DiMichele 1981. (O, Q, V) After Stewart and Rothwell 1993. (S) After Hirmer 1927. (T) After Scott 1909. (U) After Cridland 1964. (W, Y, Z, AB, AD, AF) From Gelderen 1996. (X) From Jones 1993. (AA) Photograph by Xin Wang. (AC) After Sahni 1932. (AE) After Andrews and Kern 1947.

with contiguous crowns to form a closed canopy. More attention needs to be paid to the construction of the branching systems of trees from the Late Devonian and Early Carboniferous. It is rare to find whole trunks or branching systems of shrubs and trees in the fossil record when they are preserved and then exposed in mining. It is even rarer when these features are found and recorded in sufficient detail to be useful in reconstructing the nature of a whole fossil plant. Often only fragments of plants exposed by mining or weathering can be collected, and little detailed information can be recorded of the nature of the whole organism.

The evolution of trees began in the late Early Devonian, about 385 million years ago, and stem size and height (often assumed by stem diameter) increased rapidly from about 375 million years ago in a log phase until about 344 million years ago. During this time, several separate lineages (i.e. rhizomorphic lycopsids, cladoxylaleans, and progymnosperms) of evolution developed the tree habit. As a result of these multiple evolutionary events trees were *invented* in several different ways (see Figure 6-3).

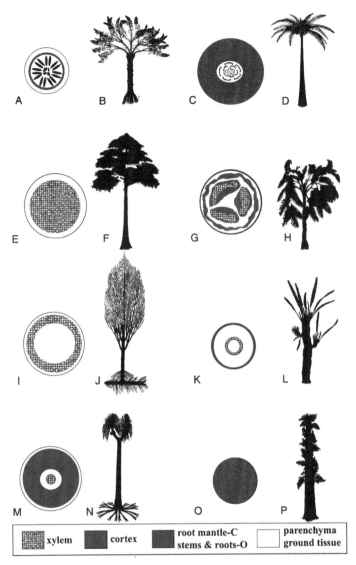

Figure 6-3 Cross-sections and basic growth habits of major land plants. (A, C, E, G, I, K, M, O) Diagrammatic cross-sections. (B, D, F, H, J, L, N, P) Growth habit silhouettes. (A, B) *Pseudosporochnus*. (A) Diameter 5–20 cm. (B) Height 1.5–4 m. (C, D) *Psaronius*. (C) Diameter 6.5 cm. (D) Height 3–10 m. (E, F) *Archaeopteris*. (E) Diameter 1.5 m. (F) Height 10–18 m. (G, H) *Medullosa*. (G) Diameter 45–50 cm. (H) Height 4.5–10 m. (I, J) *Calamites*. (I) Diameter ca. 30 cm. (J) Height 10–20 m. (K, L) *Williamsonia*. (K) Diameter 10 cm. (L) Height 2 m. (M, N) *Sigillaria*. (M) Diameter 1 m. (N) Height ca. 20–34 m. (O, P) *Tempskya*. (O) Diameter 40 cm. (P) Height 6 m. (B) Modified from Leclercq and Banks 1962. (D) After Morgan 1959. (F) After Beck 1962. (H) After Stewart and Delevoryas 1956. (J) After Hirmer 1927. (L) After Sahni 1932. (N) From Stewart and Rothwell 1993. (P) After Andrews and Kern 1947. Cross sections from original data. Diameter and height data from Potonié and Bernard 1904; Hirmer 1927; Sahni 1932; Andrews and Kern 1947; Stewart and Delevoryas 1956; Morgan 1959; Beck 1962; Leclercq and Banks 1962; Taylor 1981; Mosbrugger 1990; Stewart and Rothwell 1993.

How Trees Were Constructed

Basic to most of the different ways of making trees were the common elements: roots, trunk, and a crown of leafy branches. However, the anatomical construction of these trunks varied, and thus the basic support for these various trees was accomplished through a variety of evolutionary pathways.

In some early shrubby plants from the early Middle Devonian such as *Pseudosporochnus* (see Figure 6-3, A, B) the multiple vascular strands and some thickening of the cortex accommodated a modest height of 2.5 to 3 m (Berry and Fairon-Demaret 2002). The central woody tissue was developed in the progymnosperms and later in the gymnopserm lineages as the major support for these trees. During the Mesozoic, a similar support mechanism of central woody tissue developed again in angiosperms. *Archaeopteris* (see Figure 6-3, E, F) was the first woody tree that must have had a well-developed and active vascular cambium. This tree grew to at least 18 m with a well-formed multi-layered crown and a long naked trunk. This tree provides further evidence for a well-developed forest canopy by the Late Devonian and Early Mississippian times.

The lycopods increased trunk height and diameter by an increase in the outer cortex thickness and strengthening of the cortical cells (see Figure 6-3, M, N). This resulted in the lycopods supporting their trees by a tube of cortex rather than by an extensive proliferation of the woody tissue. The Carboniferous lycopod trees have been referred to as pole-trees (DiMichele and Hook 1992). In the Carboniferous swamp forests, various species of these lycopod pole-trees may have stood as narrow linear green poles. Some produced crowns of often deciduous branches containing cones, while other species may have remained in the pole-stage until they were ready to reproduce. Then, with a growth spurt, a crown of branches and cones was produced. Some species dispersed seed-like structures (Phillips 1979) and other megaspores and microspores (Dilcher et al. 1992) into the swamp water. These forest pole-trees allowed sunlight to reach numerous understory plants that were also a part of the Carboniferous swamp forests.

Other trees that are well known from the Carboniferous include *Calamites* (10 to 20 m) supported by a ring of woody tissue derived from the xylem (see Figure 6-3, I, J). *Psaronius* (3 to 10 meters) supported itself in a fashion similar to that of tree ferns living today (see Figure 6-3, D). The adventitious root mantle forms a dense layer surrounding and supporting the stem (see Figure 6-3, C). These trees were swamp-living understory, as demonstrated by the aerenchyma tissue in their roots. They may have formed, more or less, a mono-layer in the canopy because of their large and spreading leaves. This would benefit them, living in the shade of numerous lycopod and *Cordaites* trees. *Medullosa* was a short tree to a large shrub (3.5 to 10 m). It was supported by a combination of weakly developed woody tissue arranged in several large vascular bundles and some thickening of the outer cortex of the stem (see Figure 6-3, G, H). The large leaves formed the crown of the tree and also bore the pollen-producing organs and seeds. *Medullosa* was an understory tree living in floodplains or peat-rich swamps.

The cycads became common during the Mesozoic and formed trees of medium stature that may have resembled palms. The structural support for these cycads and those still living today is derived from a vascular cylinder and a tough outer cortex (see Figure 6-3, K, L). Also, during the Mesozoic, another plant is known from a pseudotrunk (or false trunk). This is the petrified "trunks" of *Tempskya* from the Cretaceous (Galtier and Hueber 2001). It consists of numerous fern stems twisted and tangled together with the leaf petioles and numerous adventitious roots all tightly packed together (see Figure 6-3, O, P). This forms a mass that holds together into a pseudotrunk 4.5 to 6 m tall. Quite independent of the evolution of such a "tree," in the Cretaceous, a similar pseudotrunk formed from the same materials, stems, petioles, and roots was described from the Lower Carboniferous (Hueber and Galtier 2002). This pseudotrunk was developed by zygopterid ferns over 200 millions years earlier than *Tempskya* evolved. This is an excellent example of independent and parallel evolution of ways to make a tree. Similar independent repeated parallel evolutionary events in the development of the tree habit also must have happened in gymnosperms and angiosperms.

Trees have evolved independently and repeatedly, sometimes using the same tissues used by other major clades so that their trees may be considered homologous. Other trees also evolved but used very different tissues to construct the trees, and their trees should be considered analogous. Figure 6-3 demonstrates eight analogous tree types. Because these architectural uses of tissues have been repeatedly used in the development of trees in different phylogenetic lines, we might consider that the tree habit as such has evolved many, many times. Thus, the species in the forest landscape and their contribution to the nature of the forest canopy should be understood as dynamic in nature through time. Certain physical limitations or biological factors may constrain the tree habit and their canopy forming crowns.

The Ecology of Tree Canopies

The world lacked tree canopies until the late Middle Devonian (370 million years ago) to the Late Devonian (345 million years ago) when we find many diverse types of trees (see Figure 6-1 and 6-2). Sufficient numbers of different species of trees and sufficient numbers of individuals were present during the late Devonian so that stands of trees with well-developed crowns would provide a tree canopy environment.

The earliest forest trees probably were gallery or at least floodplain forest trees (DiMichele and Hook 1992; Scheckler, pers. com.). *Archaeopteris* (see Figure 6-2, I and Figure 6-3, E, F) is perhaps the best model we have of an early woody tree. The woody stems (*Callixylon*) found at the Frasnian/Famennian (364 million years ago) are 10 to 50 cm in diameter and probably 10 to 15 m in height (Scheckler, pers. com.). Trivett (1993) and Meyer-Berthaud et al. (1999, 2000) sectioned the petrified trunks of *Archaeopteris* and found evidence of lateral branches that had been shed and subsequently grown over. This is similar to the knots in pinewood sold as knotty pine that are old branches "buried" in or overgrown by the wood after being shaded out by the tree canopy of the forest. Scheckler (pers. com.) estimated that Famennian (Late Devonian, 360 million years ago) floodplain forests consisted of dense stands of *Archaeopteris* trees spaced 1 to 1.5 m. Here, the seasonal floods accommodated the possible aquatic megaspores adapted for outcrossing (Dilcher et al. 1992; Kar and Dilcher 2002) and easily dispersed these fertilized megaspores throughout the floodplain environment as the floodwaters receded. There, new trees could develop with minimal water stress.

Figure 6-1 gives the impression that there was a lag phase of about 40 million years for the early land plants to begin to develop stature, followed by a log phase of about 30 million years during which tall trees evolved, and then a plateau phase. When looked at over long spans of geologic time, the terms lag, log, and plateau may seem appropriate. However, when considered in terms of the evolutionary potential of plants over 40 to 30 or 70 million years, there does not seem to be anything quick or fast about the evolution of trees. There does not seem to be a "sudden" rise of trees by an increase in stem diameter and height.

It is important to understand that the forests of the Late Devonian, Carboniferous, and Permian have no living counterparts, and their forest structure and canopies do not have any extant homologues. The forest tree canopies of the Late Devonian and Carboniferous were poorly developed. The calamite trees were somewhat like bamboo. They were full of branches and small leaves held close to the branches. The lycopod trees had crowns that shed branches, consisted of twigs covered in long narrow leaves, and DiMichele and Hook (1992) suggest that the crowns of some species developed only shortly before reproduction, after which the whole tree died. *Sigillaria* (see Figure 6-2, Q), also a lycopod, branched only once or a few times, so the crown was rather compact. The trees that produced the most extensive crowns were the shorter tree ferns and seed ferns, whose crowns consisted entirely of large leaves and the branching woody trees of the *Cordaites* (see Figure 6-2, T, U). The *Cordaites* had branching crowns that supported long strap-shaped leaves, and some species may have been forest trees (see Figure 6-2, T) while other species may have been mangrove trees (see Figure 6-2, U). The forest may have contained more than one canopy layer. *Psaronius* (see Figure 6-2, M) and *Medullosa* (see Figure 6-2, L) may represent an understory mono-layer vegetation while *Cordaites* and some lycopods (see Figure 6-2, O, Q) were tall trees with other lycopods (see Figure 6-2, N, P, R) and calamites (see Figure 6-2, S) being intermediate.

The tree canopies in the Paleozoic are of medium height. Figure 6-2 is constructed to demonstrate the change and variability in canopy heights during the Paleozoic, and the Triassic and Jurassic periods of the Mesozoic. Canopy heights seldom range over 10 to 20 m, with some ranging upwards of 20 to 40 m. These are quite similar to tree heights of many angiosperms in the temperate areas of the world today. By the Mesozoic time (240 million years ago) we begin to see modern aspects of the forest and forest tree canopy after the recovery phase from the largest extinction event known on Earth at the end of the Paleozoic (Looy et al. 1999). The tree heights of the Triassic and Jurassic are estimated to be similar to many of the extant conifer lineages they are related to. This might appear to be true from the petrified tree trunks known from this time. When trunk length and taper are considered, these trees could equal the heights of 40 to 80 m or more.

Tree canopies were important places for insects to climb up into and hide. The wing venation of one Carboniferous insect has a camouflage wing pattern very similar to that of the venation of the pinnule of a seed fern (Carpenter 1964). This cryptic pattern demonstrates the time spent resting on plant foliage. Retallack and Dilcher (1988) suggest that insects fed upon the pollination droplets of some seed ferns. Insects may have fed on the pollen of these seed ferns (Dilcher 1979; Labandeira 2000). Also, the hollow, broken-off trunks of the lycopods and *Calamites* are known nest sites for some of the early reptiles of the Carboniferous. During the Mesozoic, the conifers, *Ginkgo*, *Glossopteris*, and numerous broad-leaf trees, including the Cretaceous age angiosperms, produced forests with canopies similar to those in the forest today. These were present when flying reptiles and early birds were evolving.

Special Adaptations as a Product of the Canopy

During the Paleozoic, several special adaptations associated with the forest canopy evolved. These include the winged dispersal of seeds, for example *Cordaites* and seed ferns (Mei-tang et al. 1992; Stewart and Rothwell 1993), and the dispersal of megaspores and seed-like organs into water for fertilization and further dispersal (Dilcher et al. 1992; DiMichele and Hook 1992). The hollow branches in the canopy and on the forest floor provide hiding, resting, and nesting places for animals, as already discussed. The biomass of these trees provided food sources for a diversity of animals as well. Examples of such feeding is given by Labandeira (1997, 2000, 2001). The important aspect to note is that trees and their vegetative and reproductive organs in the canopy became an important source of food for animal populations. The plants must have supplied the food for the insects that the early amphibians and reptiles fed upon before they also became herbivorous at a later time. The canopy provided a safe resting and perching place for animals that could climb up the trunks of the trees. It must have been from such high places in the canopy that flight first began. Probably gliding from some elevated canopy perch was important in the evolution of flight. It may have been to escape from a predator or to be successful as a predator.

THE BOTANICAL GHOSTS OF EVOLUTION

Connie Barlow

Many plants, especially those in or beneath a dense and therefore wind-restricted canopy, depend on mobile animals not only for transferring pollen but also for dispersing seeds. We are achingly familiar with the consequences for plants when bird, bat, or insect pollinators dwindle or become extinct. But what about the loss of seed dispersal partners?

Vanishing pollinators is an anthropogenic phenomenon that is happening on our watch (see Buchmann and Nabhan 1995). So, too, with vanishing dispersal partners, but with this distinction: The loss of seed carriers began long before humans learned to till the soil (Barlow 2000). Modern humans are culpable, of course, but so were our distant ancestors— and in a very big way.

Insects and small vertebrates can evolve into superb carriers of pollen, but it takes a much larger creature to swallow and defecate very large seeds in a viable condition. Elephants, rhinos, and other "megafauna" are thus the primary agents of seed dispersal for many plants that produce large fleshy fruits or nutritious, indehiscent pods. Dispersal troubles began for many plants when large mammals were sent into extinction by spear-toting humans entering landscapes in which the inhabitants had not coevolved with projectiles.

This happened in Australia some 60,000 years ago when the first humans arrived, and in North and South America just 13,000 years ago when the Clovis culture of mammoth hunters crossed the Bering land bridge (Martin 1990). Thanks to the longevity of many trees, a capacity to reproduce clonally from roots, and the fact that the human invaders fancied fruit, there are plants alive today whose fruits are markedly anachronistic.

Consider the avocado, *Persea americana*, native to South America. Wild stock of avocado (now rare) from which our plumped-up varieties were bred had only a thin layer of oily flesh to attract dispersers, but the seed was about the same size. Except for the rare jaguar, no animal native to South America today will swallow and defecate the seed of avocado. But there was no shortage of dispersers 15,000 years ago, when the Americas were home to elephantine gomphotheres, rhinolike toxodons, and lumbering ground sloths.

Tropical and semi-tropical species of family Annonaceae (cherimoya, soursop, pawpaw) that are native to the Western Hemisphere are also anachronistic. Omnivores (notably, bears) are capable dispersers, but these plants have lost their herbivore partners, especially the "hindgut fermenters" whose microbial symbionts are located far enough down the line that fruit sugars do not interfere with acid-intolerant microbes.

Avocado, annona fruits, and some large-podded legumes (such as Kentucky coffee tree, *Gymnocladus*, in North America) are all suited for shady environments of the forest understory. Their large seeds contain energy stores sufficient for young plants to establish beneath a mature canopy. But those that would swallow such seeds, enticed by the surrounding pulp, are now gone.

Some small trees of forest-edge and savanna environments, such as honey locust (*Gleditsia* sp.) of both Americas and osage orange (*Maclura pomifera*) of North America, also bear anachronistic fruit—and widely spaced anachronistic thorns. The centimeter-long seeds within the sweet pods of honey locust are so well adapted to withstand the grinding teeth of their extinct dispersal partners that they now must be scarred by knife or acid in order to sprout.

Five hundred years ago, when Spanish invaders returned the horse to its continent of origin, they unknowingly restored a key dispersal partner for osage orange, a close relative of Asian breadfruit. A deep time understanding of the range of osage orange in the late Cenozoic suggests that it truly is native to much of North America even though early

Continued

THE BOTANICAL GHOSTS OF EVOLUTION–*cont'd*

botanists noted that its "natural" range entailed only a bit of eastern Texas and Arkansas. Apparently, once-prolific osage orange was heading toward extinction for lack of dispersers. The return of the horse, along with the tree's appeal as an ornamental, reversed this downward evolutionary trend (Barlow 2001).

An awareness of lost dispersal partners is vital, therefore, not only for determining the extent of native range but also for managing plant diversity with a look toward intact forest canopy.

References

Barlow, C. (2000). "The Ghosts of Evolution: Nonsensical Fruit, Missing Partners, and Other Ecological Anachronisms." Basic Books, New York.

Barlow, C. (2001). Anachronistic fruits and the ghosts who haunt them. *Arnoldia* **61**(2), 14–21.

Buchmann, S.L. and Nabhan, G.P. (1995). "The Forgotten Pollinators." Island Press, Washington, DC.

Martin, P.S. (1990). 40,000 years of extinctions on the 'Planet of Doom.' *Palaeogeography, Palaeoclimatology, Palaeoecology* **82**, 182–201.

Some Carboniferous age plants developed the vine habit (Kerp and Krings 1998; Krings and Kerp 1999, 2000). Various species of these vines evolved sucker-like pads or hooks in order to "climb" up or "climb" over the existing trees, into their canopies. It is interesting that these vines are seed ferns (for example *Mariopteris*) and thus produced seeds with sufficient energy to give the young seedlings a good chance to reach some height after germination. Another group of epiphytes are the fern stems, roots, and petioles that are sometimes found petrified as embedded inclusions in the root mantle of *Psaronius* in coal balls or petrified peat of the Carboniferous. These are a type of epiphyte finding a place to live, much as ferns and orchids do today on tree fern root mantles. Ferns as *Ankyropteris* and *Anachoropteris* are examples of such epiphytes (D. Eggert, pers. com. 2003).

Increase of Woody Biomass

The woody plants and the tall lycopods with a thick strong cortex, the branched crowns, the leaf material, and the roots all contributed huge amounts of organic material to the forest floor. Some of this could have been broken down mechanically and digested by microbes or fungi. However, it does not appear that wood-rotting fungi, important in today's forests, were present until the Permian/Triassic boundary (Visscher et al. 1996; Looy et al. 1999). This accumulation of biomass aided in the deposition of the Carboniferous coals of the world.

Abiotic Importance of Canopies

Today, the importance of forest canopies is well understood because forests act as windbreaks that help disperse the energy of storms, and forest canopies hold huge amounts of moisture that can be recycled within an ecosystem. It has been said that up to one-third of the rainfall in the Amazon comes from the Amazon forest canopy. As the forests are removed in some tropical American areas, this moisture cycle may be lost, and deserts may form. The canopies of trees help to break the force of rain and the roots of forest trees hold the soil in place. These factors have reduced erosion and influenced weathering around the world.

Summary

The search for light was the major factor that influenced the evolution of trees. Plant height and complexity increased from the Late Silurian into the Middle Devonian when they reached tree height. More diverse forests with complex canopy types were probably established by the Lower Carboniferous. Many structural, physiological, and even coevolutionary events needed to evolve before trees could develop height, complex branching systems (canopy), root support, and anchoring systems. The development of supporting tissue such as xylem, outer cortex, adventitious roots, or complex of stems and roots were independently evolved in different lines of plants to form trees. The strong linear trunks of trees held the photosynthetic portions of the trees above the shade of other plants and/or above the ground for the dispersal of spores, seeds, and fruits. They formed a coevolutionary bond with fungi, bacteria, algae, other green plants, and numerous animals. The forests and their canopies are the products of millions of years of evolution. The factors affecting this evolution are both biotic and abiotic that continually change through time.

Acknowledgments

The authors would like to thank Stephen Scheckler and Donald A. Eggert for helpful discussions and suggestions. Financial support for this research was provided by National Science Foundation INT 0074295. This paper is the University of Florida contribution to *Paleobiology* publication no. 557.

References

Algeo, T.J., Berner, R.A., Maynard, J.B., and Scheckler, S.E. (1995). Late Devonian oceanic anoxic events and biotic crises: "rooted" in the evolution of vascular land plants? *GAS Today* **5**, 45, 64–66.

Andrews, H.N. (1961). "Studies in Paleobotany." Wiley, New York.

Andrews, H.N. and Kern, E.M. (1947). The Idaho Tempskyas and associated fossil plants. *Ann. Mo. Bot. Gard.* **34**, 119–186.

Beck, C.B. (1962). Reconstructions of *Archaeopteris*, and further consideration of its phylogenetic position. *Am. J. Bot.* **49**, 373–382.

Berry, C.M. and Fairon-Demaret, M. (2002). The architecture of *Pseudosporochnus nodosus* Leclercq et Banks: a Middle Devonian Cladoxylopsid from Belgium. *Int. J. Plant Sci.* **163**, 699–713.

Carpenter, F.M. (1964). Studies on North American Carboniferous insects. 3. A spilapterid from the vicinity of Mazon Creek, Illinois (Palaeodictyotera). *Psyche* **71**, 117–124.

Chaloner, W.G., and Sheerin, A. (1981). The evolution of reproductive strategies in early land plants. *In* "Evolution Today, Proceedings of the Secondary International Congress of Systematic and Evolutionary Biology" (G.G.E. Scudder, and J.L. Reveal, Eds.), pp. 93–100. Carnegie-Mellon University, Pittsburgh.

Cridland, A.A. (1964). *Amyelon* in American coal-balls. *Palaeontology* **7**, 186–209.

Dilcher, D.L. (1979). Early angiosperm reproduction: an introductory report. *Ann. Rev. Pal. Pal.* **27**, 291–328.

Dilcher, D.L., Kar, R.K., and Dettmann, M.E. (1992). The functional biology of Devonian spores with bifurcate processes—a hypothesis. *In* "Essays in Evolutionary Plant Biology: The Palaeobotanist" (B.S. Venkatachala, D.L. Dilcher, and H.K. Maheshwari, Eds.), pp. 67–74. Birbal Sahni Institute of Palaeobotany, India.

DiMichele, W.A. (1981). Arborescent lycopods of Pennsylvanian age coals: *Lepidodendron*, with description of a new species. *Palaeontographica Abt B* **175**, 85–125.

DiMichele, W.A. and Hook, R.W. (1992). Paleozoic terrestrial ecosystems. *In* "Terrestrial Ecosystems Through Time: Evolutionary Paleoecology of Terrestrial Plants and Animals" (A.K. Behrensmeyer, J.D. Damuth, W.A. DiMichele, R. Potts, H.-D. Sues, and S.L Wing, Eds.), pp. 205–325. University of Chicago Press, Chicago.

DiMichele, W.A. and Philips, T.L. (1994). Paleobotanical and paleoecological constraints on models of peat formation in the Late Carboniferous of Euramerica. *Rev. Pal. Pal.* **106**, 39–90.

Edwards, D. (1970). Fertile Rhyniophytina from the Lower Devonian of Britain. *Palaeontology* **13**, 451–461.

Edwards, D. (1994). Towards an understanding of pattern and process in the growth of early vascular plants. *In* "Shape and Form in Plants and Fungi" (D.S. Ingram, and A. Hudson, Eds.), pp. 39–59. The Linnean Society of London, London.

Elias, T.S. (1989). "Field Guide to North American Trees." Grolier Book Clubs, Danbury, CT.

Galtier, J. and Hueber, F.M. (2001). How early ferns become trees. *Proc. R. Soc. Lond. B* **268**, 1955–1957.

Gensel, P.G. and Andrews, H.N. (1984). "Plant Life in the Devonian." Praeger Publishers, New York.

Gerrienne, P., Bergamaschi, S., Pereira, E., Rodrigues, M.-A., and Steemans, P. (2001). An early Devonian flora, including *Cooksonia*, from the Paraná Basin (Brazil). *Rev. Pal. Pal.* **116**, 19–38.

Hao, S.G. (1989). A new zosterophyll from the Lower Devonian of Yunnan, China. *Rev. Pal. Pal.* **57**, 155–171.

Haq, B.U. and Van Eysinga, F.W.B. (1998). "Geological Time Table." Elsevier Science, Netherlands.

Hirmer, M. (1927). "Handbuch der Paläobotanik. Band 1: Thallophyta-Bryophyta-Pterodophyta." Oldenbourg, München, Berlin.

Hueber, F.M. and Galtier, J. (2002). *Symplocopteris wyattii* n. gen. et n. sp.: a zygopterid fern with a false trunk from the Tournaisian (Lower Carboniferous) of Queensland, Australia. *Rev. Pal. Pal.* **119**, 241–273.

Jones, D.L. (1993). "Cycads of the World." Smithsonian Institution Press, Washington.

Kar, R.K. and Dilcher, D.L. (2002). An argument for the origins of heterospory in aquatic environments. *Palaeobotanist* **51**, 1–11.

Kasper, A.E. and Andrews, H.N. (1972). *Pertica*, a new genus of Devonian plants from northern Maine. *Am. J. Bot.* **59**, 897–911.

Kenrick, P. (2002). The origin of roots. *In* "Plant Roots: The Hidden Half" (Y. Waisel, A. Eshel, and U. Kafkafi, Eds.), pp. 1–13. Marcel Dekker, New York.

Kerp, H. and Krings, M. (1998). Climbing and scrambling growth habits: common life strategies among Late Carboniferous seed ferns. *C. R. Acad. Sci. Paris* **326**, 583–588.

Krings, M. and Kerp, H. (1999). Morphology, growth habit, and ecology of *Blanzyoperis praedentata* (Gothan) nov. comb., a climbing neuropteroid seed fern from the Stephanian of Central France. *Int. J. Plant Sci.* **160**, 603–619.

Krings, M. and Kerp, H. (2000). A contribution to the knowledge of the pteridosperm genera *Pseudomariopteris* Danzé-Corsin nov. emend. and *Helenopteris* nov. gen. *Rev. Pal. Pal.* **111**, 145–195.

Krüssmann, G. (1995). "Manual of Cultivated Conifers." Timber Press, Portland, OR.

Labandeira, C.C. (1997). Insect mouthparts: ascertaining the paleobiology of insect feeding strategies. *Ann. Rev. Earth Plan. Sci.* **26**, 329–377.

Labandeira, C.C. (2000). The paleobiology of pollination and its precursors. *In* "Phanerozoic Terrestrial Ecosystems" (R.A. Gastaldo and W.A. DiMichele, Eds.), pp. 233–269. Paleontological Society Papers 6. Yale University, New Haven.

Labandeira, C.C. (2001). The history of associations between plants and animals. *In* "Plant–Animal Interactions: An Evolutionary Approach" (C. Herrera and O. Pellmyr, Eds.), pp. 26–74, 248–261. Blackwell Science, London.

Leclercq, S. and Banks, H. P. (1962). *Pseudosporochnus nodosus* sp. nov., a Middle Devonian plant with Cladoxylalean affinities. *Palaeontograhica Abt. B* **110**, 1–34.

Long, A.G. (1979). Observations on the Lower Carboniferous genus *Pitus* Witham. *Trans. Roy. Soc. Edinburgh* **70**, 111–127.

Looy, C.V., Brugman, W.A., Dilcher, D.L., and Visscher, H. (1999). The delayed resurgence of equatorial forests after the Permian-Triassic ecologic crisis. *Proc. Nat. Acad. Sci.* **96**, 13857–13862.

Mei-tang, M., Dilcher, D.L., and Zhi-hui, W. (1992). A new seed-bearing leaf from the Permian of China. *Palaeobotanist* **41**, 98–109.

Meyer-Berthaud, B., Scheckler, S.E., and Wendt, J. (1999). *Archaeopteris* is the earliest known modern tree. *Nature* **398**, 700–701.

Meyer-Berthaud, B., Scheckler, S.E., and Bousquet, J.L. (2000). The development of *Archaeopteris*: new evolutionary characters from the structural analysis of an early Famennian trunk of SE Morocco. *Am. J. Bot.* **87**, 456–468.

Morgan, J. (1959). The morphology and anatomy of American species of the genus *Psaronius*. *Illinois Biological Monographs* **27**, 1–108.

Mosbrugger, V. (1990). The tree habit in land plants. *Lecture Notes in Earth Sciences* **28**, 1–161.

Niklas, K.J. and Speck, T. (2001). Evolutionary trends in safety factors against wind-induced stem failure. *Am. J. Bot.* **88**, 1266–1278.

Phillips, T.L. (1979). Reproduction of heterosporous arborescent lycopods in the Mississippian-Pennsylvanian of Euramerica. *Rev. Pal. Pal.* **27**, 239–289.

Pirozynski, K.A. and Malloch, D.W. (1975). The origin of land plants: a matter of mycotropism. *Biosystems* **6**, 153–164.

Potonié, H. and Bernard, C.H. (1904). "Flore Dévonienne de l'étage H-h₁ de Barrande. 86 S., Leipzig.

Remy, W., Taylor, T.N., Hass, H., and Kerp, H. (1994). Four hundred-million-year-old vasicular arbuscular mycorrhizae. *Proc. Natl. Acad. Sci. USA* **91**, 11841–11843.

Retallack, G.J. and Dilcher, D.L. (1988). Reconstructions of selected seed ferns. *Ann. Missouri Bot. Gard.* **75**, 1010–1057.

Rothwell, G.W. and Erwin, D.M. (1985). The rhizomorph apex of *Paurodendron*: implications for homologies among the rooting organs of Lycopsida. *Am. J. Bot.* **72**, 86–98.

Sahni, B. (1932). A petrified *Williamsonia* (*W. sewardiana*, sp. nov.) from the Rajmahal Hills, India. *Palaeontologia Indica N. S.*, **20**, 1–19.

Schweitzer, H.-J. (1968). Pflanzenreste aus dem Devon Nord-westspitzbergens. *Palaeontographica Abt B* **123**, 43–75.

Schweitzer, H.-J. (1990). Pflanzen erobern das land. *Kleine Senckenbert-Reihe* **18**, 1–75.

Scott, D.H. (1909). "Studies in Fossil Botany, Third Edition." Black Publishers, London.

Seward, A.C. (1931). "Plant Life Through the Ages." Macmillan Company, New York.

Stewart, W.N. and Delevoryas, T. (1956). The medullosan pteridosperms. *Bot. Rev.* **22**, 45–80.

Stewart, W.N. and Rothwell, G.W. (1993). "Paleobotany and the Evolution of Plants." Cambridge University Press, New York.

Taylor, T.N. (1981). "Paleobotany: An Introduction to Fossil Plant Biology." McGraw-Hill Book Company, New York.

Thomas, T. (2000). "Trees: Their Natural History." Cambridge University Press, Cambridge, UK.

Trivett, M.L. (1993). An architectural analysis of *Archaeopteris*, a fossil tree with pseudomonopodial and opportunistic adventitious growth. *Bot. J. Linn. Soc.* **111**, 301–329.

Van Gelderen, D.M.. (1996). "Conifers: the Illustrated Encyclopedia." 2 vols. Timber Press, Portland, OR.

Visscher, H., Brinkhuis, H., Dilcher, D.L., Elsik, W.C., Eshet, Y., Looy, C.V., Rampino, M.R., and Traverse, A. (1996). The terminal Paleozoic fungal event: Evidence of terrestrial ecosystem destabilization and collapse. *Proc. Nat. Acad. Sci.* **93**, 2155–2158.

Wagner, C.A. and Taylor, T.N. (1981). Evidence for endomycorrhizae in Pennsylvanian age plant fossils. *Science* **222**, 562–563.

Zhang, Y.-T., Yin, X.-L., Cui F.-Z., Tong, B.-L., Lui, J.-J., Pan, Z.-X., and Guan, C.-Y. (2000). Investigation of the old trees of *Ginkgo* in Tangshan and Qinghuangdao, Heberi Province, China. *Chinese Bulletin of Botany* **17**, 234–236.

II _____

Organisms
in Forest Canopies

In the chronological development of forest canopy biology, the structure of canopies was the first aspect of this emerging science to receive careful investigation (Section I). Many of the methods required to assess the structure of forest canopies were relatively simple to assemble and/or could be initiated from ground level. The second phase in the evolution of this emerging science of canopy science was the investigation of individual organisms (Section II). Population studies require the prerequisite information about forest structure as a baseline. Canopy scientists who study organisms cannot always rely on the conventional methods of sampling employed at ground level so, in many cases, new tools and techniques were required to reach the individual organisms high overhead.

Two challenges arose for scientists to study canopy organisms. First, **sessile organisms** (e.g., mosses, orchids, bromeliads, and insects such as psyllids) require the development of new methods that facilitate marking and returning to the same spot in the treetops over long periods of time. To study organisms in bromeliad tanks, for instance, an investigator needs to access specific viewing spots in the canopy for continuous measurement and observation. If the bromeliads are harvested, the microecosystem in the tank is consequently destroyed. There may be occasions when destructive sampling is useful, but most ecological studies require the bromeliads to remain *in situ*. Similarly, nonvascular epiphytes are almost invisible when viewed with binoculars from ground level; however, access to the canopy, with the subsequent development of tools to mark and return to the same branch over time, often involves a vastly different sampling regime as compared to similar observations at ground level.

Second, **mobile organisms** create obvious challenges to population studies. The notion of tracking birds or even fast-moving monkeys through the treetops is almost inconceivable for us human beings, who are slow and clumsy in our aerial mobility. But the advent of canopy walkways, platforms, and other semi-permanent base camps in the canopy facilitated long-term studies of creatures that would otherwise move away from encroaching scientists. The advent of such tools, in conjunction with more sophisticated sampling techniques, has resulted in a large body of exciting research about organisms in forest canopies.

In Chapter 7, Lynn Margulis offers an exciting microbial perspective on canopy science, emphasizing a theme echoed in many of her recent writings: our minds are incarcerated by words. Ill-defined words can muddle an author's meaning. On the other hand, words too narrowly defined can throttle creative endeavor. Vocabulary for an emerging science such as canopy biology needs to be a middle-ground affair—not too vague, not too specific—to ensure its lasting contribution to society. Steve Sillett and Marie Antoine focus on the ecology of epiphytic lichens and bryophytes in Chapter 8. The former, as mutualistic myco- and photobionts, are an easily recognized association rather than a taxonomic description that initiates biological succession. The latter are low-lying nonvascular plants that often avoid competition in forest habitats by growing epiphytically on their vascular hosts. In Chapter 9, David Benzing discusses vascular epiphytism as a gradation rather than a discreet lifestyle or ecological strategy in subtropical and tropical systems. Exceptionally vulnerable to climatic change, epiphytes may also be used as the proverbial "miner's canary" when studying the threatened habitats of host forests. In Chapter 10, David Watson suggests that mistletoes, as unique constituents of forest canopies worldwide, represent a keystone resource and are often used as indicators of environmental integrity.

In Chapter 11, we begin our review of mobile organisms in forest canopies. Dave Walter distinguishes between decomposition in perched litter and ground litter and then homes in on canopy acarofauna. In Chapter 12, Bruce Rinker examines edaphic microarthropods as a biological frontier and then attempts to link canopy and soil processes, pointing to some of the difficulties with the term "canopy" from a systems perspective. William Miller looks at tardigrades, or water bears, in Chapter 13 as an unmeasured component of nutrient cycling in forests and posits that the lack of basic biological information has hindered the growth of interdisciplinary environmental and ecological studies. In Chapter 14, Terry Erwin argues effectively for a wise expenditure of resources (including time, money, and effort) to answer a question fundamental to most of Earth's biota: How many species of terrestrial arthropods are there? In Chapter 15, Roman Dial and Joan Roughgarden present an original study to document the interaction strengths among canopy anoles, their prey, and their preys' resources using individual tree crowns as replicates; heterogeneity in both the arthropod and lizard communities in forest ecosystems may have interesting effects on community structure. Finally, in Chapter 16, Jay Malcolm provides an overview of some of the main features of mammalian canopy biology; deforestation, and other anthropogenic changes that have devastating effects on canopy populations because of their adaptations to life in the treetops.

These chapters, along with their sidebars of information, highlight some of the latest information about sessile and mobile organisms in forest canopies around the planet. Enhanced by the development of innovative canopy access techniques, studies of sessile organisms such as trees, vines, and epiphytes and of mobile organisms such as insects, birds, reptiles, and mammals can now be conducted with rigorous experimental methodologies. With this critical documentation of species richness in the treetops, we can now move toward their complex interactions in the following section of *Forest Canopies*: that of ecological processes in forest canopies.

CHAPTER 7

What Is Canopy Biology?
A Microbial Perspective*

Lynn Margulis

To be poor and be without trees is to be the most starved human being in the world.
To be poor and to have trees is to be completely rich in ways that money can never buy.
—*Clarissa Pinkola Estes,* The Faithful Gardener, *1995*

The canopy at Tiputini, tributary of the Napo River in Ecuador, is accessed by a magnificent three-tiered tower connected to three sets of hanging bridges constructed by Bart Bouricius some 40 m above the ground. I am especially proud because Bart, biologist and carpenter, is a neighbor and former student of mine in the graduate section of our Environmental Evolution course. I have more than once verified his reputation as the world's premiere designer-builder of canopy walkways, both temperate and tropical.

In addition to the walkways at the Tiputini Biological Diversity Station, two towers of similar height permit views of the treetops. Emergent species are what first strike the eye. The viewer is drawn to a mental tracing of the curving river far below. Greenery of different shades topped by blue sky delight the walkway-climber as far as his or her view can fathom the recognizable.

After a flight from Quito to Coca and a two-hour bus ride to the "canoa" (open-to-the-rain motor-powered river-boat), the visitor realizes that all evidence of people has receded. Roofed huts of fishing folk have given way to sunning birds, turtles, and palms on riverbanks profuse with a tangle of the unknown and unknowable liana, shrub, and tree. Experiencing the river, one wonders how any canopy could be, in principle, even more mysterious. For the next seven hours, down-river noise abounds—clacking, flapping, croaking, splashing, cawing, buzzing—but no human sounds other than the motor and murmurs of fellow boat people are heard. Speeding downstream, we marvel at the rapid currents and total absence of the 21st century—not even an airplane penetrates the wilderness. In this timeless land beside the river, the lush green appears nameless.

Since 1977, in field studies and back in the campus laboratory, my students and I have studied a thriving ecological community of intertidal marine microbes, beholden to photosynthesizers of various types. At Laguna Figueroa in Baja California Norte, Mexico (see Figure 7-1), oxygenic *Microcoleus, Lyngbya,* and *Gloeothece* are underlain by the anoxygenic purple *Thiocapsa* and *Chromatium.* Certain hardy organisms, the resistant forms of *Paratetramitus,* in among the green and purple, are known to survive freezing and desiccation for more than five years. These *Paratetramitus* forms are difficult to distinguish from another hardy, similar-looking organism called *Mychonastes desiccatus.* These two kinds of life are the same size, the same spherical shape, and probably present in many populations in the same abundance. Distinguishing them in the community requires fluorescence for the chlorophyll wavelength. Both fluoresce when placed in

---

A version of this paper appeared in Selbyana **22**(2): 232–235; used with permission.

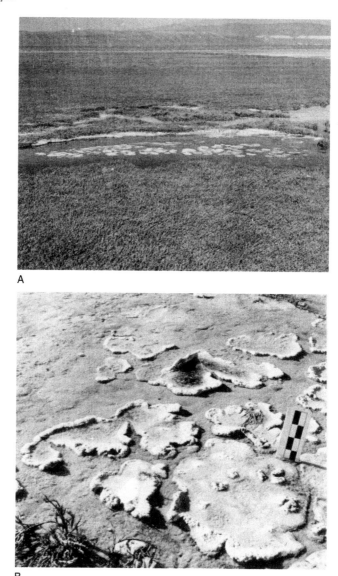

Figure 7-1   Microbial mat communities. (A) An aerial view of Laguna Figueroa in Baja California Norte, Mexico. (B) *Microcoleus* in microbial mats *in situ* (each unit of measure = 10 cm).

the ultraviolet spotlight but give off different colors. *Mychonastes* glows red, as chlorophylls enable it to photosynthesize, unlike *Paratetramitus*, whose fluorescent image reveals a green-glowing cyst wall.

Beneath these two 8-micron spheres, greenish chlorobia of various types abound. Globules of sulfur give rise to hydrogen sulfide and other sulfurous gases, as we descend into the microbial community. Still further down in this vertically laminated ecological wilderness, at the low-tide level, *Desulfovibrio* and, presumably, *Desulfobacter* thrive. So, too, dwell *Spirochaeta* and the viviparous giant serpentine swimmer *Spirosymplokos*. A plethora of diatoms, *Microcoleus*, *Lyngbya*, and *Gloeothece*, form the "canopy," in terms of Moffett (2000); *Chromatium*, *Paratetramitus*, *Mychonaste*s,

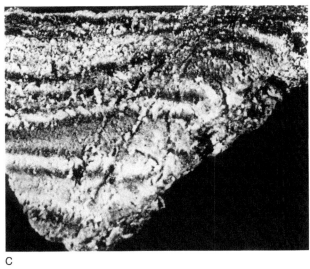

C

Figure 7-1—*Continued* (C) A hand sample (15 cm across) of a microbial mat community dominated by *Microcoleus chthonoplastes* (cyanobacterium).

and their associates comprise the understory. The emergent photosynthesizers (primarily diatoms), those gliding and swimming above the vertically laminated strata, tend to be brown and boat-shaped. Some 60 genera (perhaps 100 species) of these delicate-filigreed "emergents" have been identified at our field site by talented taxonomists.

More than 200 distinguishable species documented in the literature are repeatedly present at our Mexican field site and at comparable cosmopolitan locales. Like the canopy seen from the swinging bridge built with plastic footholds by Bouricius, the vast majority of species in our field samples is unknown. We have an inkling of only those beings that survive mistreatment in the laboratory or greenhouse. The real number of the kinds of inhabitants in our samples is far more likely to be 1000 than the 210 or so that we and our colleagues have tabulated.

A sample containing a thousand life forms cut from a marine microbial mat on the Pacific shore is but a 1 cubic millimeter in size. Similar samples from the delta of the Ebro River (Spain) or the shore at Matanzas (Cuba) are roughly 1 mm high by 1 mm deep by 1 mm millimeter long. How can the glorious canopy sample of Amazonia (say $100 \times 100 \times 50$ km, so vastly larger than the microbial sample) possibly be analyzed by the amateurs who love her or even by the professionals who make her their life's work? Scientists admit to an extremely deficient view of our tiny microbial mat samples after more than two decades of study of this particular microcosm. We hardly know its major components and how the populations change in the most dramatic of temperature, salt concentration, and precipitation swings with the seasons. If so little is known of a cubic millimeter, imagine my awe at the view from the tower of the Amazonian macrocosm (see Figure 7-2).

"Life," sang John Lennon, "is what happens when you're making other plans." "Life" is what you see, smell, hear, and feel at the canopy tower. Words and numbers hardly make an organizational dent when attempting to describe the prodigiosity. Moffett (2000) makes a valiant attempt to regulate the unruly by delineating terms of canopy biology. At least he has ignored the pernicious financial jargon so detrimental to scientific analysis: benefit, cost, fitness, reciprocal altruism, and the like. Here I have an opportunity for a single suggestion, as I applaud his effort to stay as close as he can to the observations themselves.

My suggestion regards terms for physical associations between organisms that are members of different taxa. First, abandon the use of common meaningless words. Purge the prose that uses

Figure 7-2 At the Tiputini Biodiversity Station, this area under the canopy walkway may contain several hundred stable microbial communities that await study. Here we see the lower portion of a tree about 100 feet tall.

"host," "parasite," and other implied taxonomic categories such as "epiphyte" and "endophyte." Replace these with those pernicious terms that convey precisely what is meant. Why? Because the ambiguity intrinsic to these words is unavoidable and obfuscating. Topological, nutritional, and ecological concepts are not distinguished. Identificational, positional, metabolic, genetic, and temporal information becomes so conflated that these terms entirely lack meaning.

All intertaxonomic physical associations that last for most of the life history of at least one of the partners are, by definition, symbioses. Such associations should never be classified by ecological outcome (e.g., beneficial, pathogenic, mutualistic, parasitic) because outcome always depends on particulars of environment and timing. Rather symbioses (whether strangler fig, vesicular-arbuscular zygomycotous fungi in the tissues of dicotyledonous plants, or bromeliads perched on palms) require analysis by level of association. By "levels" I mean whether the association is at the behavioral level (such as the topological relation of the bromeliad with the branch), metabolic level (such as the fungus that derives photosynthate from the dicot), or at the level of shared gene product (such as the nitrogen-fixing rhizobium of the *Acacia* or *Mimosa* root nodule), or even integration at the most intimate genic level (such as the *Agrobacterium* that sends its plasmid-borne genes to be incorporated into the plant cell's chromosomes in the crown gall). Often, in interspecific physical associations that are

ARBOREAL STROMATOLITES: A 210 MILLION YEAR RECORD

Jessica Hope Whiteside

Stromatolites (from Greek for "stony carpets") are lithified, microbially produced arches, spheres, or domes that can reach over one meter in height. Along with their nonlithified forerunners, microbial mats, stromatolites are widely assumed to have been the dominant form of microbial community life in the Precambrian. Stromatolites are dominated by cyanobacteria that shed their polysaccharide sheaths and move up in search of light in an incremental process (often less than 1 mm per year). The polysaccharide sheaths trap and bind sediment and calcium carbonate and the cyanobacteria provide photosynthate and additional compounds for other bacteria. (Another means of stromatolite growth is via the formation by cyanobacteria of their own carbonate network.) Originally known as "cryptozoans" ("hidden animals") because of their mysteriously biogenic shape, the first stromatolites to be correctly identified as containing and requiring bacteria for their precipitation were entirely marine.

Freshwater stromatolites are also fairly common (Winsborough et al., 1994; DeWet et al. 2002; Whiteside et al., 2003). Fossil lacustrine examples include a ~210-million-year-old record of tree encrusting (arboreal) stromatolites (from 0.1 to +1 m in diameter) from the Newark Supergroup of Eastern North America (Whiteside et al. 2003), and in other age strata primarily in Wyoming (~50 Ma) and in California (~16 Ma).

An analogy may be made between modern mats and stromatolites that inhabit hypersaline and otherwise extreme environments in space and lacustrine stromatolites arising during extreme climatic shifts (Olsen 1996) in time. For example, tufa calcite, stromatolitic coatings of carbonate on branches of trees from the Passaic Formation of the Newark basin have been interpreted to result from the transgression of the lake as low-lying vegetation is submerged. Fossilized arboreal stromatolites have been recovered from Triassic, Jurassic, Eocene, and Miocene lacustrine strata in North America. Cheirolepidiaceous conifers, encased by stromatolites and dated as old as ~210 and ~200 million years, have been uncovered from the Passaic and Towaco Formations of the Newark basin (Figure 1), and tree- and stem-encrusting stromatolites have also been retrieved from the East Berlin Formation in Connecticut and from the Scots Bay Formation in Nova Scotia, both about 200 Ma. Eocene fossils (~50 Ma) from the Laney Member of the Green River Formation in Wyoming (Figure 2) and Miocene forms (~16 Ma) from the Middle Miocene Barstow Formation in California provide more recent examples of "tree-hugging" stromatolites. In

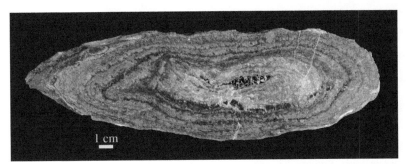

Figure 1   Cross-section of Towaco (~200 Ma) Formation, Newark basin stromatolite from New Jersey. The remnants of the tree in cross section are represented by the dark oblong spearhead in the middle. The concentric stromatolitic rings suggest that this tree, a cheirolepidiaceous conifer, was upright and continued to grow after submergence.

Continued

ARBOREAL STROMATOLITES: A 210 MILLION YEAR RECORD–*cont'd*

cyclical environments, as the climate shifts from arid to humid, the trunks and roots of flooded trees serve as temporary habitable zones for the anciently evolved stromatolites. Autoradiography of East Berlin, Towaco, Passaic (Figure 3), and Barstow stromatolites

Figure 2   Outcrop from Firehole Canyon in the Laney Member (~50 Ma) of the Green River Formation, Green River Basin in Wyoming. Left (panoramic perspective): Contiguous stromatolitic laminations embedded in cliff face. Upper right (close-up): lobate-shaped stromatolite with lithified remnant of tree trunk visible as dark end of cylinder protruding at right end of rock. Lower right (close-up): in situ photograph of wavy laminated stromatolite with two trees visible in cross-section at each end of photograph (light rings surrounded by concentric layers). Rock hammer for scale. Photograph courtesy of Malka Malchus.

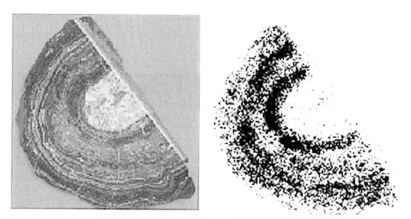

Figure 3   Polished slab of stromatolitic tufa- (porous calcite-) coated branch from the Metlars Member (~210 Ma), Passaic Formation, Newark basin in Pennsylvania. Left: The central area with the light-colored calcite was a branch that provided the nucleus for stromatolite growth. Layers accrued from the center outward. Right:Autoradiograph of polished slab, showing a clear elevated uranium concentration with growth layers (dark areas). Photograph courtesy of Troy Rasbury.

suggests that the bacteria that form the coatings were also involved in uranium mineralization. As the climate becomes more humid, previous desiccated flats are vegetated. As the expanding lake floods these flats, stromatolites encrust the trunks and exposed roots of the drowned vegetation (De Wet et al. 2002) until they too are submerged below the photic zone, knocked down and transported, and/or buried by accumulating muds. These arboreal stromatolites are very similar morphologically to living forms described by Winsborough et al. (1994) in spring-fed lakes and streams of the Cuatre Cienegas basin, Mexico. The 230-milllion-year record of tree-encrusting stromatolites shows that forms of life once thought to be solely marine and mostly ancient can, under extreme environmental conditions, form close associations, in freshwater and in Phanerozoic times, with submerged forest canopies.

References

DeWet, C. B., Mora, C. I., Gore, P. J. W., Gierlowshi-Kordesch, E., and Cucolo, S. J., 2002. Deposition and geochemistry of lacustrine and spring carbonates in Mesozoic rift basins, eastern North America. *SEPM Special Publication* 73: 309-325.

Olsen, P.E., 1986, A 40-million-year lake record of early Mesozoic orbital climatic forcing. *Science* 234: 842-848.

Whiteside, J.H., Olsen, P.E., and Rasbury, E.T. 2003. A 230 million year old record of arboreal stromatolites. *Fourth International Symbiosis Society Congress Schedule and Abstracts*. Etc. Press, Halifax, Nova Scotia, p. 173-174.

Winsborough, B.M., Seeler, J-S., Golubic, S., Folk, R.L. & Maguire, Jr., B. 1994. Recent fresh-water lacustrine stromatolites, stromatolitic mats and oncoids from northeastern Mexico, pp. 71-100. *In*: Bertrand-Sarfati, J. & Monty, C. (Eds.) *Phanerozoic Stromatolites II*. Kluwer Academic Publishers, Amsterdam.

symbioses by definition, the levels of partner association are simply not known, which needs to be explicitly stated. The topological and temporal bases of associations should be reflected in any permanent or casual physical association—at the canopy or below. Does the physical presence of one partner persist during the entire life history of both, as is the case for chloroplasts, mitochondria, and many fungi of plant roots? If impermanent for both, then, by definition, the association must be cyclical for at least one partner as, for example, the fungi that induce germination of orchid seeds. Terms like "endophyte" are particularly egregious since the "phyte" often refers to the fungal partner (and the fungi are, of course, in no way plants, even though "-phyte" is a Greek legacy). This type of rampant confusion and taxonomic ignorance by ecologists has a single major consequence. The field of ecology (which, in principle, extends far beyond the restrictions of academic biology, for example, into climatology, geology, paleontology, stratigraphy, atmospheric and soil sciences) is maligned, disdained, or ignored—except when desperately needed for practical (including financial-aesthetic) purposes like at Disneyland.

If our descriptors employ unambiguous taxonomic information, such as that found in the list of all taxa of all organisms classified from phylum to class (Folch 2000), or for the higher taxa of kingdoms and phyla (Margulis and Schwartz 1998), one serious problem is resolved immediately. When ambiguous terms like "parasite," "endophyte," and "protozoa" are replaced by the actual taxonomic appellations or the metabolic, temporal, and spatial relationships that are actually meant, the canopy and its biology will remain a mystery, but a somewhat less daunting one.

References

Estes, C.D. (1995). "The Faithful Gardener: A Wise Tale about that which Can Never Die." Harpercollins, New York.

Folch, R. (2000). The biosphere concept and index. *In* "Encyclopedia of the Biosphere." Volume 11, pp. 393–399. Gale Group, Detroit, Michigan.

Margulis, L. and Schwartz, K.V. (1998). "Five Kingdoms. An Illustrated Guide to the Phyla of Life on Earth." 3rd Edition. W.H. Freeman, New York.

Moffett, M.W. (2000). What's "up"? A critical look at the basic terms of canopy biology. *Biotropica* **32**(4a), 569–596.

CHAPTER 8

Lichens and Bryophytes in Forest Canopies

Stephen C. Sillett and Marie E. Antoine

*Humanity has so far played the role of planetary killer, concerned only with
its own short-term survival. We have cut much of the heart out of biodiversity.
The conservation ethic, whether expressed as taboo, totemism, aor science,
has generally come too late and too little to save the most vulnerable of life forms.*
—E.O. Wilson, The Future of Life, 2002

Exploration of tree crowns worldwide often reveals rich epiphyte communities composed of lichens, bryophytes, and vascular plants. In very old, wet forests, nearly every tree surface, from outermost twigs to thick branches and the main trunk, is covered by epiphytes. An arboreal botanist faced with this diversity quickly realizes that these plants are not uniformly distributed across tree surfaces and that most of the species seldom, if ever, grow on the forest floor below. Many epiphytes, including all of the lichens and bryophytes, are poikilohydric organisms. Their survival in the canopy depends on an ability to become dormant when humidity is low, with physiological activity resuming only upon rehydration by atmospheric moisture.

In this chapter, we focus on the ecology of epiphytic lichens and bryophytes. After introducing the organisms, we describe their distributions along environmental gradients, summarize biotic interactions, discuss the strong associations between some species and old-growth forests, evaluate their roles in forest ecosystems, and consider their conservation in an uncertain future. Rather than provide an exhaustive literature review, our goal is to inspire future research on lichens and bryophytes by highlighting insights from recent studies in forests around the world.

Major Groups

Although lichens and bryophytes often co-occur as epiphytes in forest canopies, they are easily distinguished. Lichens are fungi with inhabitant populations of extracellularly located photosynthetic partners. This mutualistic association involves a mycobiont, which is usually an ascomycete fungus, and a photobiont, which can be a green alga or cyanobacterium. Lichenization is a nutritional strategy that has arisen independently several times in the course of fungal evolution (Tehler 1996). The mycobiont provides living space (i.e., the thallus) for the photobiont, which, in turn, shares carbohydrates produced through photosynthesis. The type of photobiont has important consequences for epiphytic lichen ecology. Lichens with a green algal photobiont (hereafter "chlorolichens") are physiologically activated by water vapor in humid air, whereas lichens with only a cyanobacterial photobiont remain dormant until they are wetted by liquid water in

Figure 8-1 A rich community of lichens and bryophytes inhabits branches 63 m up in a 75-m-tall Douglas-fir tree (H. J. Andrews Experimental Forest, Oregon, USA). The lettuce-like lobed thalli belong to the old-growth-associated cyanolichen *Lobaria oregana*, which has a green algal primary photobiont and a cyanobacterial secondary photobiont. The pendulous thalli belong to the chlorolichens *Usnea* and *Alectoria*. Photograph by Thomas B. Dunklin.

fog or rain (Green et al. 2002). In the 3 to 4 percent of lichen species containing both types of photobionts (Honegger 1991), it is likely that the green algal component photosynthesizes in humid air while the cyanobacterial component remains dormant until wetted (see Figure 8-1). Lichens with cyanobacterial photobionts (hereafter "cyanolichens") are able to reduce atmospheric nitrogen (i.e., fix N) into forms available for uptake by the mycobiont.

Aside from the type of photobiont, epiphytic lichens are readily distinguished by growth form. Microlichens, or crustose lichens, are so closely appressed to the substrate that their entire lower surface functions as an attachment device. Macrolichens are categorized as foliose (i.e., lobed with distinct dorsal and ventral surfaces) or fruticose (i.e., stalked, tufted, or pendant with no differentiation of upper and lower surfaces). Sizes of epiphytic macrolichens vary greatly. The largest foliose species (e.g., *Lobaria oregana*) can develop lobed thalli several decimeters in width, and the largest fruticose species (e.g., *Usnea longissima*) can develop pendant thalli several meters in length.

Bryophytes are green plants capable of photosynthesis by virtue of intracellular organelles called chloroplasts. They consist of three distantly related phyla—*Anthocerophyta* (hornworts), *Marchantiophyta* (liverworts), and *Bryophyta* (mosses)—that evolved from an aquatic green algal ancestor and were among the first plants to invade the land (Kenrick and Crane 1997; Goffinet 2000). Bryophytes are distinguished from other land plants by a number of retained ancestral characters, including a life cycle dominated by the gametophyte (i.e., haploid) generation and the absence of lignified vascular tissues. In forest environments, many bryophytes avoid competition with lignified vascular plants by growing upon them as epiphytes (see Figure 8-2).

Bryophytes are often distinguished on the basis of their colonial life-forms (Bates 1998). Common epiphytic life-forms include cushions (i.e., dome-shaped clusters of shoots radiating from a single point), turfs (i.e., aggregates of erect shoots), mats (i.e., aggregates of creeping shoots), fans (i.e., highly branched shoots arranged in a horizontal plane), and pendants (i.e., shoots hanging from the substrate). Large cushions, dense turfs, and thick mats hold moisture like a sponge, prolonging activity periods after rain, whereas fans and pendants quickly desiccate when air humidity drops (Zotz et al. 1997). Such effects of life-forms on physiological activity are noteworthy because, like cyanolichens, most bryophytes require liquid water for reactivation following desiccation-induced dormancy (Green and Lange 1994).

Figure 8-2 Extensive bryophyte mats dominated by the moss *Antitrichia curtipendula* cover branches 60 m up in a 75-m-tall Douglas-fir tree (H. J. Andrews Experimental Forest, Oregon, USA). Water stored by these bryophytes likely plays an important role in regulating canopy microclimate. Photograph by Thomas B. Dunklin.

Environmental Gradients

Moisture

Moisture availability is obviously of overriding importance to poikilohydric epiphytes, and lichens and bryophytes are differentially distributed along moisture gradients. Chlorolichens tend to be more tolerant of desiccation than cyanolichens, which tend to be more tolerant than bryophytes. Nearly all lichens require alternating wet and dry periods to maintain symbiosis (Honegger 1998). Prolonged hydration severely limits CO_2 diffusion to the photobiont and, therefore, photosynthesis of lichens (Lange et al. 1993, 2000; Zotz et al. 1998). In contrast, many bryophytes can thrive under continuously wet conditions (Proctor 2000). Thus, bryophyte biomass is generally highest in rainforests, cyanolichens reach their greatest abundance in moist forests with a summer dry season, and chlorolichens are often the only epiphytes in dry forests. There are, of course, exceptions to these general patterns. For example, some epiphytic chlorolichens (e.g., *Hypogymnia duplicata*) are restricted to coastal rainforests (McCune 1997), and some epiphytic bryophytes (e.g., *Porella platyphylla*) are very tolerant of desiccation (Marschall and Proctor 1999). However, the typical abundance sequence (bryophytes—cyanolichens—chlorolichens) recurs along gradients of decreasing moisture availability at multiple spatial scales.

The lichen and bryophyte communities in coniferous forests of northwestern North America have been intensively studied in sites spanning the moisture gradient relevant to epiphytes. Here, *Picea-Tsuga* rainforests along the immediate coast grade into *Pseudotsuga-Tsuga* forests of the coast ranges and the western Cascade Mountains, which grade into dry interior forests of the eastern Cascades and the western Rocky Mountains (Van Pelt 2001). Wet winters and dry summers characterize the climate of this region, but coastal fog ameliorates the summer dry season in the rainforests. Bryophytes reach their greatest abundance in rainforests with individual *Picea* trees supporting up to 36 kg of bryophyte biomass (Ellyson and Sillett 2003). Cyanolichens reach peak abundance and diversity in moist montane forests with individual *Pseudotsuga* trees supporting up to 30 kg of the dominant species, *Lobaria oregana* (Rhoades 1983), along with dozens of other cyanolichen species (Pike et al. 1975; Sillett 1995; Goward and Arsenault 2000b). Chlorolichens are present in forests throughout the moisture gradient, but they are the dominant epiphytes in dry interior forests (Rominger et al. 1994; Rosso and Rosentreter 1999).

Within montane forest watersheds, lower slopes in the vicinity of major streams are more humid than sites located farther upslope away from streams, and epiphytic lichens are distributed accordingly. Cyanolichens tend to be more abundant in riparian forests than in upland forests where chlorolichens dominate epiphyte communities (Sillett and Neitlich 1996; McCune et al. 2002; Peterson and McCune 2003). In some rainforest watersheds, however, epiphytic cyanolichens are scarce in sheltered riparian zones, perhaps because conditions are too consistently wet for positive net photosynthesis (Price and Hochachka 2001). The abundance of epiphytic bryophytes along moisture gradients within montane forest watersheds has yet to be quantified.

A coast to inland moisture gradient affecting epiphytes is also evident in boreal forests. For example, epiphytic cyanolichens are less abundant in inland (Holien 1998) than in coastal (Rolstad et al. 2001) *Picea abies* forests of Norway. Compared to *Pseudotsuga-Tsuga* forests in northwestern North America, however, epiphytic cyanolichens are much less abundant in *Picea abies* forests of Scandinavia. Cold temperatures probably limit epiphytic cyanolichens in boreal forests (see the "Temperature" section below).

The vertical distribution of lichens and bryophytes within forest stands also reflects an underlying moisture gradient, and epiphyte communities are strongly stratified by height in tall forests (McCune 1993). Changes in microclimatic conditions of particular relevance to poikilohydric epiphytes include decreasing humidity and increasing exposure to desiccating wind with increasing height in the canopy (Coxson et al. 1992; Parker 1995; Campbell and Coxson 2001). In old-growth coniferous forests of western North America, bryophytes and chlorolichens tend to be most abundant at opposite ends of an exposure gradient, with cyanolichens occupying an intermediate position (McCune 1993; Sillett 1995). Relative heights of peak abundances for the different epiphyte groups vary according to the prevailing moisture regime. In rainforests, bryophytes are abundant throughout the vertical profile and lichens are only abundant in the upper canopy (Sillett and Goward 1998; Ellyson and Sillett 2003). In moist montane forests, bryophytes dominate the lower canopy, cyanolichens dominate the middle canopy, and chlorolichens dominate the upper canopy (McCune 1993; Sillett and Rambo 2000). In relatively dry montane forests and wet interior forests, bryophytes are restricted to the lower canopy and understory, cyanolichens frequent the lower canopy, and chlorolichens dominate epiphyte communities (Lesica et al. 1991; McCune et al. 1997; Clement and Shaw 1999). In very dry interior forests, bryophytes and cyanolichens are essentially restricted to the forest floor, and epiphytic chlorolichens reign supreme. A few desiccation-tolerant mosses, which normally occur on rocks, occasionally occur as epiphytes on centuries-old dead branches and trunks of *Sequoiadendron giganteum* (pers. obs.).

The relative importance of height as a variable influencing epiphyte communities varies with forest stature. Taller forests obviously have a much stronger vertical component of moisture gradients than shorter forests. Thus, vertical stratification of epiphyte communities is distinct in tall forests (e.g., McCune et al. 2000), while substrate effects and horizontal gradients of moisture availability are more easily distinguished in short forests (e.g., Kantvilas and Minchin 1989).

Within individual tree crowns, lichens and bryophytes are differentially distributed along horizontal gradients of moisture availability. In *Pseudotsuga-Tsuga* forests, for example, chlorolichens tend to be most abundant on twigs in the outer crown, bryophytes often dominate the inner crown, and cyanolichens occupy intermediate positions along branches (Clement and Shaw 1999; Lyons et al. 2000). Much of this horizontal gradient is attributable to successional processes on branches (see the "Succession" section below), but the fact that epiphytes in the outer crown are more exposed to desiccation than those in the inner crown is indicated by the greater downward extension of chlorolichens on twigs than on branches in tall conifers (Ellyson and Sillett 2003).

Epiphytic lichens and bryophytes are also affected by moisture gradients in tropical forest canopies. The horizontal component of the gradient tends to be pronounced in tropical trees due to their broadly domed crowns. In montane rainforests, bryophytes, especially liverworts, dominate inner and middle crowns (Gradstein et al. 2000), and lichens are most abundant in outer crowns (van Leerdam et al. 1990; Wolf 1995). In tall lowland rainforests, lichens are most abundant in outer crowns and upper trunks, while bryophytes are most abundant in inner and middle crowns as well as tree bases (Cornelissen and ter Steege 1989; Montfoort and Ek 1990). As in temperate forests, vertical distributions of epiphytes in tropical forest canopies vary greatly depending on annual precipitation and forest stature. For example, a seasonally dry, tall *Quercus* forest has a more distinct vertical stratification of epiphytes than a shorter forest with persistent cloud cover in Costa Rica (Holz et al. 2002).

In some forests, shifts in the abundance of epiphyte groups during stand development parallel the moisture-related abundance sequence summarized above and may, in fact, reflect a similar underlying gradient (McCune 1993). For example, after catastrophic disturbances (e.g., fire or logging) in *Pseudotsuga-Tsuga* forests, chlorolichens dominate epiphyte communities for at least the first century of stand development (Neitlich 1993). As tree density declines due to self-thinning and gap formation, cyanolichens and bryophytes eventually arrive and occupy the lower canopy and understory (Neitlich and McCune 1997). These epiphyte groups migrate upward over time, displacing chlorolichens and ultimately establishing the vertical stratification seen in old-growth forests (McCune 1993). The rate and extent of this upward migration depends on moisture availability, with wetter forests accumulating bryophytes and cyanolichens more quickly than drier forests (compare mature forests in McCune 1993 and Sillett and McCune 1998). In very old (> 500 years), very wet (> 2 m annual rainfall or very foggy) forests, the upward migration of epiphytes appears to have progressed to its limit: the almost complete domination of epiphyte communities by bryophytes (Sillett and Goward 1998; Ellyson and Sillett 2003). Canopy studies in other forests are needed to evaluate the generality of these vertical distribution patterns during stand development.

The ability of bryophytes to dominate epiphyte communities may be related to the water-storage capacities of colonial life-forms. Dense turfs, thick mats, and large cushions remain wet for extended periods after precipitation events (Zotz et al. 2000). The resulting prolonged hydration periods may give bryophytes a competitive advantage over lichens in rainforest canopies (see the "Succession" section below). The amount of precipitation required to achieve a comparable bryophyte biomass is probably higher in tropical than temperate forests, especially those with winter rainy seasons, because the rate of evaporative water loss is positively correlated with temperature.

Temperature

Temperature gradients can have pronounced effects on the abundance of epiphytic lichens and bryophytes (see Figure 8-3). For example, lowland tropical rainforests are rich in species, especially crustose lichens (Cornelissen and ter Steege 1989; Komposch and Hafellner 2000), but they support a much lower macrolichen and bryophyte biomass than cooler tropical montane rainforests (Pócs 1980; Nadkarni 1984a; Hofstede et al. 1993; Wolf 1993a; Büdel et al. 2000; Freiberg and Freiberg 2000) or lowland rainforests at higher latitudes (Nadkarni 1984b; Ellyson and Sillett 2003). Consistently high temperatures lead to high respiration rates in low-elevation tropical forests (Zotz et al. 1998; Lange et al. 2000). The situation is exacerbated by the fact that lichens and bryophytes have small quantities of chlorophyll per unit area or mass compared to vascular

A

Figure 8-3   Comparison of epiphyte loads in lowland and montane tropical rain forests in Sabah, Malaysia. (A) The heat of the lowland rain forest is unrelenting 70 m up in an 80-m-tall *Shorea gibbosa* tree in the Danum Valley Conservation Area. White crustose lichens cover much of the bark surface, but macrolichens and bryophytes are relatively scarce.

plants (Green and Lange 1994; Martin and Adamson 2001). As a result, many lichens and bryophytes are simply incapable of the high photosynthetic rates required to overcome respiratory energy losses during warm, wet periods (Frahm 1990b). High temperatures during the summer rainy season may also explain why the biomass of epiphytic lichens and bryophytes is so low in deciduous forests of the southeastern United States (Becker 1980). The increasing abundance of epiphytic lichens and bryophytes with altitude in tropical forests is, however, more a consequence of the positive correlation between moisture availability and elevation than it is a reflection of the underlying temperature gradient (Wolf 1993a).

Just as certain species are intolerant of sustained high temperatures, it is plausible that excessively low temperatures might limit the distribution of some epiphytes. For example, N fixation in *Lobaria oregana* virtually ceases below 1°C (Denison 1979; Antoine 2004). This may explain why *L. oregana* and other epiphytic cyanolichens are sparse in high-elevation montane forests even near streams (Sillett and Neitlich 1996). Temperatures during the wet season may simply be too low to allow sufficient N fixation for adequate growth and survival of cyanolichens.

Low temperatures may also affect rates of colonization by epiphytic cyanolichens in cool temperate forests. In northern British Columbia, for example, chlorolichens readily re-colonize regenerating trees in forest gaps, but cyanolichens remain scarce despite abundant propagule sources in adjacent old-growth forest (Benson and Coxson 2002). Similar-aged trees in the old-growth forest understory are colonized by both chlorolichens and cyanolichens. The temperature limitations on N fixation may aggravate the effects of suboptimal light and moisture conditions in densely regenerated forest gaps, thereby severely limiting cyanolichen establishment.

B

Figure 8-3—*Continued* (B) In the montane rain forests of Mt. Kinabalu National Park, cool temperatures allow bryophytes to flourish from the forest floor to the treetops. Epiphytic cyanolichens and vascular plants are also abundant. Photographs by M. E. Antoine.

The dearth of epiphytic cyanolichens in boreal forests and the slow accumulation of chlorolichen biomass on branches in subalpine and alpine boreal forests (Arseneau et al. 1997, 1998) may also be attributable to low temperatures. In addition, temperature gradients may influence the stand-level distribution of epiphytic chlorolichens in boreal forests (Campbell and Coxson 2001). During sunny winter days, snowmelt occurs in the upper canopy, rehydrating resident lichens, but lichens in the lower canopy remain frozen. After summer rains, lichens in the upper canopy dry quickly, but hydration of lichens in the lower canopy is prolonged. *Bryoria*, which dominates the upper canopy, relies on short bursts of physiological activity during the rapid wetting/drying cycles of snowmelts and summer rains. This may reflect *Bryoria*'s intolerance of prolonged hydration periods that characterize the lower canopy after summer rains. In contrast, lower canopy dominants such as *Alectoria* exploit these prolonged hydration periods perhaps because they have high resaturation respiration rates that preclude survival in the rapid wetting/drying cycles of the upper canopy. This complex of interactions among vertical gradients in air temperature, wetting/drying cycles, and species-specific physiological tolerances deserves more research attention.

Light

Light availability varies along exposure gradients in forest canopies (Parker 1995), but photosynthesis in lichens and bryophytes may be limited less by light than by the interrelated effects of temperature and moisture. In upper forest canopies, poikilohydric epiphytes rapidly desiccate and become dormant under the lower humidities and higher temperatures accompanying bright, cloudless conditions. Many lichens and bryophytes become light-saturated at relatively low levels (Demmig-Adams et al. 1990; Green et al. 1997; Zotz et al. 1997), and some species can attain positive net photosynthesis in deep shade (Green et al. 1991). Furthermore, epiphytic bryophytes and lichens growing in exposed habitats often use pigmentation to shield themselves from excessive irradiation (Rikkinen 1995; Solhaug and Gauslaa 1996; Proctor 2000; Gauslaa and Solhaug 2001). The importance of light on the distribution of lichens and bryophytes cannot be ignored, however, because upper-canopy species tend to have higher light-saturation points than lower-canopy species (Hosokawa et al. 1964; Proctor 2000), and growth rates of some species increase with increasing light availability (Rincon and Grime 1989; Hilmo 2002). Light may only be limiting in the shaded lower canopies of dense forests, a hypothesis that is supported by observations of positive effects of hardwood gaps, riparian edges, and silvicultural thinning on epiphyte diversity (Rose 1992; Neitlich and McCune 1997; Peterson and McCune 2003). In order to test this hypothesis, however, the light environment of epiphytes must be experimentally manipulated in ways that do not also affect moisture availability or temperature.

Nutrients

In addition to microclimatic gradients of moisture, temperature, and light, epiphytic lichens and bryophytes are affected by gradients of nutrient availability in forests. Small-scale spatial variation in the nutrient content of soils may account for some of the tree-to-tree variation in cyanolichen abundance. Trees growing on calcium-rich soils have enriched bark from root uptake and support more cyanolichens than nearby trees on relatively poor soils (Gauslaa 1985; Gauslaa and Holien 1998). In nutrient-rich sites, cyanolichens are consistently more abundant on conifer branches within the drip zones of trees (e.g., *Populus, Chamaecyparis*) that leach relatively large quantities of calcium than they are outside these drip zones (Sillett and Goward 1998; Goward and Arsenault 2000a). The calcium in leachates apparently increases the bark pH, allowing cyanolichens to colonize young conifers whose bark is often too acidic for these epiphytes. Indeed, acid deposition may have resulted in the decline of cyanolichens on conifer bark throughout much or Europe, Asia, and eastern North America (Goward and Arsenault 2000b). Conversely, increasing deposition of ammonia has caused increased bark pH and a concomitant decline in acidophytic epiphytes in some parts of Europe (van Herk 2001).

Biotic Interactions

Substrate Effects

Environmental gradients clearly influence the distribution and abundance of lichens and bryophytes in forest canopies, but interactions among tree surfaces and epiphytes have equally profound effects on community structure. From trunk bases to uppermost twigs, trees provide an incredible diversity of substrate types. The expanded, often buttressed bases of trunks harbor distinctive communities where normally terrestrial species can occur as epiphytes (Pike et al. 1975;

Thomas et al. 2001; Glime and Hong 2002). Above the base, a tree's main trunk may be too steep and well drained to support desiccation-sensitive epiphytes, but chlorolichens often thrive here (McCune et al. 2000). However, if the trunk leans substantially, bryophytes may occupy upper bark surfaces where stemflow is concentrated (Pike et al. 1975; Kenkel and Bradfield 1986). In tropical rainforests, the understory is so humid that many bryophytes occupy the main trunk all the way to the crown (Cornelissen and ter Steege 1989; Montfoort and Ek 1990; Wolf 1993b, c).

The vast majority of surface area on a large tree is in its crown above the lowest branches, and most of the epiphyte biomass and diversity in many forests occurs here (McCune 1990; Arseneau et al. 1997). The lower crowns of many tall trees contain few living branches because many branches have died or broken (Ng 1999). The bare xylem of decorticated branches and branch stubs represents an important habitat for many epiphytic lichens (Liu et al. 2000; McCune et al. 2000). On a living branch, bark thickness and roughness often increase dramatically along the age gradient from outermost twigs to the base, and some epiphytes (e.g., *Cladonia*) colonize the rough bark of old branches more readily than the smooth bark of young branches (Sillett et al. 2000a). Tree leaves, and the leaves of vascular epiphytes, also provide substrates for epiphyllous lichens and bryophytes (e.g., Vitt et al. 1973; Coley et al. 1993).

An old tree's original branches represent only a small proportion of the total surface area available to epiphytes in its crown; reiteration, fire, and animal visitation also provide distinctive substrates. Nearly all tree species are capable of reiteration, which is the duplication of architectural units from suppressed buds or growing axes (Hallé et al. 1978). For example, most branch surfaces in large *Pseudotsuga* trees are the product of reiteration (Ishii and Ford 2001; Ishii and Wilson 2001; Ishii et al. 2002). Furthermore, many tree species are capable of reiterating multiple trunks from other trunks and branches within the crown. These are particularly obvious in redwoods (*Sequoia sempervirens*) whose reiterated trunks resemble smaller trees, each with their own branch systems (Sillett 1999). Litter often accumulates on large branches and at the bases of reiterated trunks, fungi process this material, vascular epiphytes colonize the resulting soil, and their fine, interwoven roots greatly increase its water-holding capacity (Sillett and Bailey 2003). During rainy seasons, these sponge-like accumulations of organic material become the source of "canopy waterfalls" (i.e., areas of sustained stemflow) in which desiccation-sensitive epiphytes, especially bryophytes, thrive (Sillett 1999). Many large, old trees have survived repeated crown fires, and the charred bark and wood provide distinctive substrates for some lichens (pers. obs.). Finally, the very tops of tree crowns often support unique epiphyte communities, presumably because perching birds carry propagules on their feet and also increase availability of nutrients, especially N, by defecation (McCune et al. 2000).

Not all tree species provide equally suitable substrates for epiphytes. Some trees (e.g., *Eucalyptus*, *Koompassia*, *Sequoia*) have inhospitable bark relative to other trees in the same forests. Although preferences for certain tree species are common among epiphytic lichens and bryophytes, true specificity rarely occurs because the ecological conditions prevailing on tree surfaces are more important than taxonomic identity in determining substrate characteristics (Wolf 1994). In temperate forests, angiosperm trees, or hardwoods, tend to have cation-enriched, less acidic bark than conifers and often support greater abundances and diversities of epiphytes (Barkman 1958; Kuusinen 1996; Holien 1998). Even among conifers, some species are more suitable for epiphytes than others. For example, epiphytic lichen biomass is higher on *Picea abies* than *Pinus sylvestris* in boreal forests of Finland (Liu et al. 2000), and bryophyte species richness is higher on *Picea sitchensis* than either *Tsuga heterophylla* or *Thuja plicata* in coastal rainforests of British Columbia (Peck et al. 1995).

Perhaps the most striking example of substrate quality differences among conifers occurs in the old-growth *Sequoia sempervirens* rainforests of northern California. Here, Sitka spruce (*Picea sitchensis*) and redwood trees over 90 m tall grow side-by-side along streams. The spruce trees support enormous masses of epiphytes (Ellyson and Sillett 2003). Adjacent redwoods support even greater quantities of vascular epiphytes (Sillett and Bailey 2003), but bryophytes and lichens, especially cyanolichens, are relatively scarce (pers. obs.). Sitka spruce occasionally occurs epiphytically on redwoods (Sillett 1999), and these epiphytic trees harbor cyanolichens not found anywhere else in the crowns of the supporting redwoods! Whether these differences are due to bark chemistry, texture, or stability remains to be determined.

Succession

Epiphytes also interact with each other on tree surfaces. Succession in epiphyte communities involves autogenic (i.e., competition and facilitation) as well as allogenic (i.e., tree crown expansion) processes. New substrates are continuously added to a tree's crown as branches grow. Short-lived pioneer species rapidly colonize twigs along with longer-lived pioneer species that continue to proliferate as branches age, and both are eventually displaced by later successional species, including those that grow over other epiphytes (Stone 1989; John 1992). In dry forests, bryophytes and cyanolichens are often absent, relatively large chlorolichens dominate inner and middle portions of branches, and smaller chlorolichens and crustose lichens are confined to outer portions of branches (Hilmo 1994). In humid forests, the result of succession on branches is a predictable spatial sequence with bryophytes dominating branch bases, cyanolichens mingling with bryophytes in middle regions, and chlorolichens, especially fruticose species, being most abundant on outer twigs (Stone 1989; Lyons et al. 2000). In rainforests, or in the sheltered understory of tall, humid forests, aggressive bryophytes quickly overwhelm lichens on branches (Wolf 1995; Ruchty et al. 2001), forming dense turfs, thick mats, or "moss balls" that accumulate layers of underlying dead organic matter and facilitate colonization by vascular epiphytes (van Leerdam et al. 1990; Jarman and Kantvilas 1995; Heitz et al. 2002; Ellyson and Sillett 2003).

Lichens are not necessarily defenseless against bryophytes and vascular plants during epiphyte succession. Some crustose lichens produce allelopathic chemicals that may enable them to avoid being overgrown. For example, extracts of *Pertusaria* inhibit germination of seeds and moss spores (Frahm et al. 2000), and extracts of *Cryptothecia* inhibit seedling survival and growth of the bromeliad *Tillandsia usneoides* (Callaway et al. 2001). Some crustose lichens (e.g., *Lepraria*) overgrow liverworts in rainforest canopies (pers. obs.). Furthermore, a few cyanolichens (e.g., *Peltigera britannica*, *Pseudocyphellaria rainierensis*) thrive amidst bryophyte shoots (Sillett 1995; Sillett and McCune 1998).

The richness of lichens and bryophytes generally declines during epiphyte succession in rainforests as colonial bryophytes, vascular plants, and their associated layers of dead organic matter exclude smaller species that cannot withstand prolonged hydration (Wolf 1995). In the absence of disturbance, stable communities develop on old branch surfaces in tree crowns. Branch-falls, "epislides," and scientific sampling remove epiphytic material from tree branches, and the resulting disturbed areas are slowly recolonized by lichens, bryophytes, and vascular plants (Nadkarni 2000; Cobb et al. 2001).

As trees grow and their crowns expand, they accumulate lichens and bryophytes, including pioneers and late successional species. However, the sequence of epiphyte colonization during tree growth depends to a large extent on the stand-level context of the tree. For example, in old-growth *Pseudotsuga-Tsuga* forests, young *Tsuga* trees, which dwell in lower canopy, are initially colonized by bryophytes, and lichens do not become prevalent in their crowns until the trees reach the middle canopy (Lyons et al. 2000). Epiphyte communities in the lower canopies of these forests are overwhelmingly dominated by bryophytes, especially *Isothecium myosuroides* (McCune et al. 1997; Sillett and Rambo 2000). An enormous propagule supply, combined with a microenvironment suitable for bryophyte establishment (Sillett et al. 2000a), greatly increases the probability that bryophytes are the first epiphytes to colonize young *Tsuga* in old-growth forests. This sequence contrasts with the situation in forests developing after catastrophic disturbances in which chlorolichens are the initial colonists of young trees (see above).

Associations with Old-Growth Forests

As forests age, they accumulate epiphytic lichens and bryophytes, whose communities develop on ever-expanding tree surfaces in predictable ways, depending on environmental gradients that influence rates of succession. If the moisture regime is not too dry, bryophytes eventually develop large colonies on branches regardless of their height in the canopy (Sillett 1995; Clement and Shaw 1999; Clement et al. 2001). Large, old trees also provide a greater surface area and variety of epiphytic substrates than smaller, younger trees. Thus, old-growth forests tend to support a greater quantity and diversity of epiphytes than young forests (Lang et al. 1980; Lesica et al. 1991; McCune 1993; Neitlich 1993; Selva 1994; Esseen et al. 1996; Dettki and Esseen 1998; Pipp et al. 2001; Price and Hochachka 2001).

Some epiphytic lichens and bryophytes that occur in old-growth forests are seldom found in younger forests (see Figure 8-4; Lesica et al. 1991; Dettki *et al.* 2000; Rosso et al. 2000). On the landscape scale, however, some desiccation-sensitive species may actually be less strongly associated with old-growth forests *per se* than with moist habitats such as riparian zones and forested wetlands (Ohlson et al. 1997; McCune et al. 2002; Peterson and McCune 2003). Others may be associated with old-growth forests because of unsuitable microenvironments (e.g., lack of water-storing bryophytes) or a shortage of suitable substrates (e.g., deeply fissured bark, bare xylem, hardwoods) in young forests (Tibell 1992; Minami and Takahashi 1994; Sillett and Neitlich 1996; Neitlich and McCune 1997). If particular microclimates or substrates are not limiting, slow

Figure 8-4 Old-growth-associated lichens nearly cover a branch 65 m up in a 75-m-tall Douglas-fir tree (H. J. Andrews Experimental Forest, Oregon, USA). The dominant bluish-gray species is the rare cyanolichen *Nephroma occultum*. Photograph by Thomas B. Dunklin.

dispersal may delay the arrival of certain species and greatly decrease their probability of occurrence in young forests.

Transplanting old-growth-associated epiphytes into younger forests and monitoring their growth is one way to determine if particular species require microenvironments found only in old-growth forests. The mat-forming moss *Antitrichia curtipendula* as well as the cyanolichens *Lobaria oregana* and *Pseudocyphellaria rainierensis* grow equally well in young and old *Pseudotsuga-Tsuga* forests (Sillett and McCune 1998; Rosso et al. 2001). Similarly, the cyanolichen *Lobaria scrobiculata* and the chlorolichen *Platismatia norvegica* grow equally well in young and old *Picea abies* forests (Hilmo 2002). *Usnea longissima*, an old-growth-associated chlorolichen, actually grows more rapidly on wooden racks in clearings than it does in forests in coastal Oregon (Keon 2001). These studies demonstrate that suitable microclimates for some old-growth-associated epiphytes are available regardless of stand age, so their scarcity in young forests must be attributable to other factors.

Dispersal and substrate limitations can be evaluated as potential causes of old-growth associations by manipulative experiments. For example, when smooth- and rough-barked branches were inoculated with *Lobaria oregana* propagules in forests of different ages, rates of establishment were dramatically higher on inoculated branches, regardless of bark texture, compared to un-inoculated controls (see Figure 8-5; Sillett et al. 2000a, b). This result, which provides conclusive evidence that this species is dispersal-limited, has been corroborated by independent studies (Peck and McCune 1997; Sillett and Goslin 1999) and has ecological implications because *L. oregana* is the dominant cyanolichen in old-growth *Pseudotsuga-Tsuga* forests (Pike et al. 1975; McCune 1993; Neitlich 1993, Sillett 1995). The extent to which other old-growth-associated epiphytes are dispersal-limited remains to be determined.

Figure 8-5 Experimental branches were lashed to living branches in Douglas-fir forest canopies to determine the cause of *Lobaria oregana*'s strong association with old-growth forests. Rates of establishment were higher on branches inoculated with lichen propagules compared to controls, such as those shown here, which implicated dispersal limitation as the leading cause. Photograph by Thomas B. Dunklin.

Ecological Importance

In forests where they are abundant, epiphytic lichens and bryophytes can greatly modify nutrient cycles, canopy microclimates, and habitats of other arboreal organisms. Nitrogen fixation by cyanolichens is the most obvious contribution of epiphytes to forest nutrient cycles. Most of the N fixed by cyanolichens is used for thallus growth and reproduction (Ahmadjian 1993). This N is ultimately released to the ecosystem via litterfall and decomposition of thalli (Rhoades 1983; Guzman et al. 1990; McCune and Daly 1994; Esseen and Renhorn 1998b). Some of the newly fixed inorganic nitrogenous compounds (e.g., NH_4^+) in cyanolichens are leached by rainfall (Millbank 1982). The enriched throughfall solutions are a source of N for other organisms, including epiphytic chlorolichens and bryophytes (Pike 1978). Bacteria living on their thalli also quickly absorb inorganic N leached from cyanolichens and are, in turn, consumed by arthropods (Carroll 1980). Thus, cyanolichens subsidize food webs in the forest canopy. The total N contribution of epiphytic cyanolichens depends on their stand-level biomass and local patterns of temperature and precipitation (Antoine 2004). In some forests, N fixed by cyanolichens represents a major ecosystem-level input (Sollins et al. 1980). For example, some old-growth *Pseudotsuga-Tsuga* forests support over three tons of cyanolichens per hectare (Neitlich 1993), and these epiphytes may contribute over 16 kg of N per hectare annually (Antoine 2004). Such high N inputs may actually boost ecosystem productivity in these forests, but critical experiments testing this hypothesis have yet to be conducted.

Aside from N fixation, epiphytic lichens and bryophytes also influence forest nutrient cycles by expanding canopy surface areas, thereby increasing atmospheric deposition of nutrients (Nadkarni 1986). For example, the chlorolichen *Ramalina menziesii* increases the deposition of N, Ca, Mg, Na, and Cl in oak woodlands (Knops and Nash 1996), and bryophytes are particularly effective at absorbing inorganic N from precipitation in cloud forests (Clark 1994). Nutrients scavenged from the atmosphere by lichens and bryophytes ultimately reach the forest floor episodically in litterfall (Esseen 1985; Boucher and Nash 1990) or catastrophically when the supporting tree falls. Mineral nutrients and stored carbohydrates of epiphytic bryophytes are also released in pulses during rewetting episodes following desiccation-induced dormancy, enriching throughfall solutions and nourishing canopy biota, including other epiphytes (Coxson 1991b; Coxson et al. 1992). However, mobile forms of some nutrients (e.g., NO_3^-) absorbed by epiphytic bryophytes are transformed into recalcitrant, organic forms that may be retained for long periods in living bryophytes and underlying dead organic matter (Clark et al. 1998).

Epiphytic bryophytes are one of the major contributors to suspended soils that develop on branches in rainforests (Edwards and Grubb 1977; Ingram and Nadkarni 1993; Heitz et al. 2002). These sponge-like accumulations can store enormous quantities of water in the canopy (Pócs 1980; Frahm 1990a; Veneklaas et al. 1990), extending periods of physiological activity for poikilohydric organisms in their vicinity and allowing desiccation-sensitive organisms to flourish high above the ground. Epiphytic bryophytes and their underlying soils are home to fungi (Stone 1996), arthropods (Winchester 1997), annelids (Wolf 1993c), mollusks, and salamanders in rainforest canopies (pers. obs.). Aboveground adventitious roots are found beneath bryophytes on branches of some rainforest trees, and these roots are frequently mycorrhizal (Nadkarni 1981; Ellyson and Sillett 2003). In addition to the importance of their water-storage capacities, branch-dwelling bryophytes on rainforest conifers often form broad platforms suitable for nesting birds, including the endangered marbled murrelet (Hamer and Nelson 1995).

Epiphytic lichens are also an important resource for other arboreal biota. Lichen thalli provide shelter and food for many arthropods, especially oribatid mites (Seyd and Seaward 1984; Stubbs 1989; Prinzing and Wirtz 1997), birds frequently hunt arthropods amidst lichens while foraging in forest canopies (T. Sillett 1994; Pettersson et al. 1995), flying squirrels consume and build nests with fruticose lichens (Hayward and Rosentreter 1994; Rosentreter et al. 1997), and a surprising diversity of parasitic and lichenicolous fungi live on lichens (Stone et al. 1996; Richardson 1999; Lawrey and Diederich 2003). Furthermore, after epiphytic lichens fall from the canopy, they are often consumed by terrestrial animals, including mollusks (Hesbacher et al. 1995) and ungulates (Sharnoff 1994).

Conservation

Given their ecological importance and high diversity, epiphytic lichens and bryophytes are increasingly recognized as worthy of conservation. Reduced impact logging (Romero 1999), the establishment of forest reserves, and certain silvicultural practices have great potential to conserve epiphytic lichens and bryophytes in managed forest landscapes. Maintaining propagule sources in and around cutting units is one of the most effective ways to facilitate epiphyte colonization of regenerating forests because many species are dispersal-limited (Dettki et al. 2000; Sillett et al. 2000a, b) but can find suitable habitats in young forests (Sillett and McCune 1998; Hilmo 2002). Reducing the size of cutting units promotes dispersal by epiphytes from surrounding forests, and retaining large, live trees inside cutting units promotes dispersal by epiphytes surviving on retained trees (Neitlich and McCune 1997; Peck and McCune 1997; Hazell and Gustafsson 1999; Sillett and Goslin 1999; Rosso et al. 2000; Peterson and McCune 2001).

While some old-growth-associated lichens in humid forests acclimate to exposed environments and grow quite well in clearings or along edges (Sillett 1994; Sillett et al. 2000b; Keon 2001), the microclimatic changes accompanying logging (e.g., increased irradiance and desiccation) certainly impact epiphyte communities. For example, isolated trees support a lower abundance and diversity of bryophytes than intact cloud forest trees (Sillett et al. 1995), abundance of desiccation-sensitive lichens is lower along edges than in the interior of boreal forests (Esseen and Renhorn 1998a; Kivistö and Kuusinen 2000), upper canopy epiphytes extend farther down in the canopy along edges than in the interior of *Pseudotsuga-Tsuga* forests (Sillett 1995), cyanolichens are less abundant on remnant trees in gaps than in adjacent old-growth cool temperate forests (Benson and Coxson 2002), and some old-growth associated lichens are damaged by high irradiance (Gauslaa and Solhaug 1996, 2000) and may gradually decline and disappear in response to logging (Gauslaa et al. 2001). Fortunately, the conservation of epiphytic lichens and bryophytes is not limited to modifications of logging practices in older forests; young, regenerating forests can also be manipulated to promote epiphytes.

The lower canopies of dense, young coniferous forests are often inimical to epiphytes, especially chlorolichens, because of an absence of well-developed lower branches (Esseen et al. 1996) and an excessively sheltered microenvironment (Sillett et al. 2000a). Cyanolichens accumulate more rapidly on hardwoods than on conifers (Sillett and Neitlich 1996), and understory shrubs can develop diverse and abundant epiphyte communities within a few decades (Ruchty et al. 2001). Thus, thinning conifers around hardwoods and shrubs will promote epiphyte diversity in these forests and facilitate dispersal of epiphytes to conifers once they develop suitable branch habitats. Uneven thinning will also benefit epiphytes by increasing the structural complexity of regenerating forests and thus providing a greater diversity of microhabitats (Neitlich and McCune 1997; Pipp et al. 2001).

Despite the promise of selective logging and uneven thinning for epiphyte conservation, some species are probably too rare, vulnerable to disturbances, or dispersal-limited to benefit from these silvicultural techniques. For example, the cyanolichen *Erioderma pedicellatum* is known from only a few ravine valleys in Norwegian spruce forests (Holien et al. 1995), and the cyanolichen *Nephroma occultum* is likely to vanish from matrix forests in southern Oregon under current forest management practices (Rosso et al. 2000). The creation of sufficiently large forest reserves may be the only viable way to conserve such species into the foreseeable future.

The effectiveness of conservation measures will likely vary geographically because old-growth associated species may be more vulnerable to disturbances in some portions of their ranges than others. The chlorolichen *Usnea longissima* and the cyanolichen *Lobaria pulmonaria* appear to be more sensitive to the microclimatic changes accompanying forest clearing in boreal forests (Esseen et al. 1981; Gauslaa et al. 2001) than in temperate rainforests (Sillett et al. 2000a; Keon 2001). Such differences may ultimately be related to moisture availability; these species may be more sensitive in drier than wetter portions of their ranges. Similarly, the cyanolichen *Nephroma occultum* is more strongly associated with old-growth forests in drier, inland forests of British Columbia, where it is restricted to the lower canopy, than in wetter, coastal forests, where it occurs in younger forests as well as the upper canopy of old-growth forests (Goward 1995). Thus, conservation measures that may be effective for coastal populations (e.g., live tree retention) may fail to protect inland populations of this species.

In addition to the obvious threats posed by deforestation, epiphytic lichens and bryophytes around the world face the menace of global warming (van Herk et al. 2002) and air pollution (see Farmer et al. 1992; Gries 1996; Bates 2000 for recent reviews). Even seemingly pristine forests that are far removed from pollution sources are often contaminated because of prevailing weather patterns. Acid rain and deposition of toxins (e.g., SO_2, NH_3) alter bark pH, and trace elements may accumulate to deadly levels in lichen and bryophyte thalli. Lichens are especially useful indicators of air quality because of the extreme sensitivity of some species to pollutants. Polluted areas are characterized by depauperate lichen communities and by shifts in species composition that reflect the expansion or invasion of pollution-tolerant and acidophytic species. Epiphytic cyanolichens are particularly vulnerable to air pollution, especially acid deposition (Gauslaa 1995), and have declined in large areas of the Northern Hemisphere (Goward and Arsenault 2000b).

Future Research Directions

Access to forest canopies is now safer and easier than ever before, especially with the use of arborists' rope techniques (see Jepson 2000) and construction cranes. Recent studies have used canopy access to study lichens and bryophytes around the world (see Table 8-1). While our survey of the literature is far from complete (e.g., including English-language publications only), relatively little research on lichens and bryophytes has occurred in the canopies of Australian, Asian, African, and temperate South American forests (see also Rhoades 1995). Floristic surveys of these forest canopies are urgently needed because the majority of lichen and bryophyte species in forests occur exclusively above tree bases (Pike et al. 1975; Cornelissen and Gradstein 1990; Komposch and Hafellner 2000) and are only sporadically detectable as litterfall unless tree crowns are sampled directly.

Most canopy research on epiphytic lichens and bryophytes has been survey work; there have been relatively few experiments (see Table 8-1). Thus, our understanding of epiphyte distribution and abundance along environmental and successional gradients is more developed than our understanding of the causal factors underlying these patterns. Transplant studies and manipulations of substrates and microclimates in tree crowns have great potential to unravel the critical determinants of epiphyte distribution and abundance. Growth rates of transplanted lichens and bryophytes are easy to measure (Russell 1988; McCune et al. 1996; Rosso et al. 2001), and a plethora of techniques are now available to quantify gas exchange rates (Lange *et al.* 1997), chlorophyll fluorescence (Gauslaa and Solhaug 2000), cell osmotic potential (Proctor 1999), and

Table 8-1 Summary of Studies Using Access Methods

Citation	Forest type	Country	No. Stands	No. Trees per stand	Epiphyte focus	Study type	Access method
Clark et al. 1998	tropical	Costa Rica	3	5-12	B	E	R
Cornelissen and Gradstein 1990	tropical	Guyana	3	1-11	mL, B	S	R
Cornelissen and ter Steege 1989	tropical	Guyana	1	11	cL, mL, B	S	R
Freiberg and Freiberg 2000	tropical	Ecuador	4	5	mL, B, V	S	R
Hofstede et al. 1993	tropical	Colombia	1	1	B, V	S	L
Komposch and Hafellner 2000	tropical	Venezuela	1	9	cL, mL	S	C
Montfoort and Ek 1990	tropical	French Guiana	1	28	cL, mL, B	S	R
Nadkarni 1984a	tropical	Costa Rica	1	4	B, V	S	R
Nadkarni 2000	tropical	Costa Rica	1	5	mL, B, V	E	R
Pócs 1980	tropical	Tanzania	2	?	B, V	S	R, F
Sillett et al. 1995	tropical	Costa Rica	2	3	B	S	R
van Leerdam et al. 1990	tropical	Colombia	1	2	mL, B, V	S	L
Veneklaas et al. 1990	tropical	Colombia	1	4	mL, B, V	E	R
Wolf 1993a,b,c	tropical	Colombia	15	4	cL, mL, B, V	S	R
Wolf 1995	tropical	Colombia	7	4	mL, B	S	R
Antoine 2001	temperate	USA	1	36	mL	S	C
Benson and Coxson 2002	temperate	Canada	3	18-21	mL	S	R
Clement et al. 2001	temperate	Chile	1	7	mL, B, V	S	R
Clement and Shaw 1999	temperate	USA	1	9	mL, B	S	R
Ellyson and Sillett 2003	temperate	USA	1	5	mL, B, V	S	R
Lyons et al. 2000	temperate	USA	1	30	mL, B	S	C
McCune 1993	temperate	USA	3	7-22	mL, B	S	F
McCune et al. 1997	temperate	USA	1	14	mL, B	S	C
McCune et al. 2000	temperate	USA	1	40	cL, mL, B	S	C
Nadkarni 1984b	temperate	USA	1	3	B, V	S	R
Pike et al. 1975	temperate	USA	1	20	cL, mL, B	S	R
Pike et al. 1977	temperate	USA	1	1	mL, B	S	R
Rhoades 1981	temperate	USA	2	5	mL, B	S	R
Rhoades 1983	temperate	USA	1	4	mL	S	R
Rosso et al. 2000	temperate	USA	10	1-4	mL	S	R
Rosso et al. 2001	temperate	USA	3	4	B	E	R
Sillett 1994	temperate	USA	2	2	mL	E	R
Sillett 1995	temperate	USA	2	2	mL, B	S	R
Sillett and McCune 1998	temperate	USA	10	2	mL	E	R
Sillett and Rambo 2000	temperate	USA	5	2	mL, B	S	R
Sillett et al. 2000a,b	temperate	USA	10	4	mL	E	R, L
Stone 1989	temperate	USA	1	6	mL, B	S, E	P
Arseneau et al. 1997	boreal	Canada	5	9-12	mL	S	F
Arseneau et al. 1998	boreal	Canada	5	9-12	mL	S	F
Campbell and Coxson 2001	boreal	Canada	8	8	mL	S	R
Liu et al. 2000	boreal	Finland	1	4	mL	S	F
Solhaug et al. 1995	boreal	Norway	1	1	cL	E	F

(R = ropes, L = ladders, C = crane, F = tree felling, P = platforms) to study epiphytic lichens or bryophytes in forest canopies. Studies accessing epiphytes from the ground are excluded. The epiphyte focus (B = bryophytes, mL = macrolichens, cL = crustose lichens, V = vascular plants) and Study Type (E = experiment, S = survey) are also indicated. Citations are listed alphabetically by forest type.

thallus moisture content (Coxson 1991a). Such measurements are especially powerful when combined with field monitoring of canopy microclimates (e.g., Renhorn et al. 1997; Campbell and Coxson 2001; Green *et al.* 2002). Moreover, solar-powered sensor arrays connected to dataloggers now permit long-term monitoring of epiphytic habitats even in the tallest forest canopies.

In conclusion, the loss and modification of epiphytic habitats caused by logging and air pollution have already eliminated much of the lichen and bryophyte biomass and diversity on Earth. The ecological importance of these poikilohydric organisms certainly necessitates their consideration in global forest conservation efforts. Studies that focus on methods to promote and maintain epiphytic lichens and bryophytes in managed forest canopies are becoming increasingly important. Rapid progress in this area can be made if lichenologists and bryologists collaborate with other scientists studying forest canopies.

Acknowledgments

The Global Forest Society provided financial support during the writing of this chapter (GF-18-1999-52). We also thank Bruce McCune and three anonymous reviewers for constructive comments on the manuscript.

References

Ahmadjian, V. (1993). "The Lichen Symbiosis." John Wiley and Sons, New York.

Antoine, M. E. (2004). An ecophysiological approach to quantifying nitrogen fixation by *Lobaria oregana*. *Bryologist* **107**, in press.

Arseneau, M.-J., Ouellet, J.-P., and Sirois, L. (1998). Fruticose arboreal lichen biomass accumulation in an old-growth balsam fir forest. *Canadian Journal of Botany* **76**, 1669–1676.

Arseneau, M.-J., Sirois, L., and Ouellet, J.-P. (1997). Effects of altitude and tree height on the distribution and biomass of fruticose arboreal lichens in an old growth balsam fir forest. *Ecoscience* **4**, 206–213.

Barkman, J.J. (1958). "Phytosociology and Ecology of Cryptogamic Epiphytes." Van Gorcum, Assen, The Netherlands.

Bates, J.W. (1998). Is 'life-form' a useful concept in bryophyte ecology? *Oikos* **82**, 223–237.

Bates, J.W. (2000). Mineral nutrition, substratum ecology, and pollution. *In* "Bryophyte Biology" (A.J. Shaw and B. Goffinet, Eds.), pp. 248–311. Cambridge University Press, Cambridge, UK.

Becker, V.E. (1980). Nitrogen fixing lichens in forests of the southern Appalachian Mountains of North Carolina. *Bryologist* **83**, 29–39.

Benson, S. and Coxson, D.S. (2002). Lichen colonization and gap structure in wet-temperate rainforests of northern interior British Columbia. *Bryologist* **105**, 673–692.

Boucher, V.L. and Nash III, T. H. (1990). The role of the fruticose lichen *Ramalina menziesii* in the annual turnover of biomass and macronutrients in a blue oak woodland. *Botanical Gazette* **151**, 114–118.

Büdel, B., Meyer, A., Salazar, N., Zellner, H., Zotz, G., and Lange, O.L. (2000). Macrolichens of montane rainforests in Panama, province Chiriquí. *Lichenologist* **32**, 539–551.

Callaway, R.M., Reinhart, K.O., Tucker, S.C., and Pennings, S.C. (2001). Effects of epiphytic lichens on host preference of the vascular epiphyte *Tillandsia usneoides*. *Oikos* **94**, 433–441.

Campbell, J. and Coxson, D. (2001). Canopy microclimate and arboreal lichen loading in subalpine spruce-fir forest. *Canadian Journal of Botany* **79**, 537–555.

Carroll, G.C. (1980). Forest canopies: complex and independent subsystems. *In* "Forests: Fresh Perspectives from Ecosystem Analysis" (R.H. Waring, Ed.), pp. 87–107. Oregon State University Press, Corvallis, OR.

Clark, K.L. (1994). The role of epiphytic bryophytes in the net accumulation and cycling of nitrogen in a tropical montane cloud forest. Ph. D. dissertation, University of Florida, Gainesville, FL.

Clark, K.L., Nadkarni, N.M., and Gholz, H.L. (1998). Growth, net production, litter decomposition, and net nitrogen accumulation by epiphytic bryophytes in a tropical montane forest. *Biotropica* **30**, 12–23.

Clement, J.P., Moffett, M.W., Shaw, D.C., Lara, A., Alarçon, D., and Larrain, O.L. (2001). Crown structure and biodiversity in *Fitzroya cupressoides*, the giant conifers of Alerce Andino National Park, Chile. *Selbyana* **22**, 76–88.

Clement, J.P. and Shaw, D.C. (1999). Crown structure and the distribution of epiphyte functional group biomass in old-growth *Pseudotsuga menziesii* trees. *Ecoscience* **6**, 243–254.

Cobb, A.R., Nadkarni, N.M., Ramsey, G.A., and Svoboda, A.J. (2001). Recolonization of bigleaf maple branches by epiphytic bryophytes following experimental disturbance. *Canadian Journal of Botany* **79**, 1–8.

Coley, P.D., Kursar, T.A., and Machado, J.-L. (1993). Colonization of tropical rain forest leaves by epiphylls: effects of site and host plant leaf lifetime. *Ecology* **74**, 619–623.

Cornelissen, J.H.C. and Gradstein, S. R. (1990). On the occurrence of bryophytes and macrolichens in different lowland rain forest types at Mabura Hill, Guyana. *Tropical Bryology* **3**, 29–35.

Cornelissen, J.H.C. and ter Steege, H. (1989). Distribution and ecology of epiphytic bryophytes and lichens in dry evergreen forest of Guyana. *Journal of Tropical Ecology* **5**, 131–150.

Coxson, D.S. (1991a). Impedance measurement of thallus moisture content in lichens. *Lichenologist* **23**, 77–84.

Coxson, D.S. (1991b). Nutrient release from epiphytic bryophytes in tropical montane rain forest (Guadeloupe). *Canadian Journal of Botany* **69**, 2122–2129.

Coxson, D.S., McIntyre, D.D., and Vogel, H.J. (1992). Pulse release of sugars and polyols from canopy bryophytes in tropical montane rain forest (Guadeloupe, French West Indies). *Biotropica* **24**, 121–133.

Demmig-Adams, B., Máguas, C., Adams III, W.W., Meyer, A., Kilian, E., and Lange, O.L. (1990). Effect of high light on the efficiency of photochemical energy conversion in a variety of lichen species with green and blue-green phycobionts. *Planta* **180**, 400–409.

Denison, W.C. (1979). *Lobaria oregana*, a nitrogen-fixing lichen in old-growth Douglas-fir forests. *In* "Symbiotic Nitrogen Fixation in the Management of Temperate Forests" (J.C. Gordon, C.T. Wheeler, and D.A. Perry, Eds.), pp. 266–275. Forest Research Laboratory, Oregon State University, Corvallis, OR.

Dettki, H. and Esseen, P.-A. (1998). Epiphytic macrolichens in managed and natural forest landscapes: a comparison at two spatial scales. *Ecography* **21**, 613–624.

Dettki, H., Klintberg, P., and Esseen, P.-A. (2000). Are epiphytic lichens in young forests limited by local dispersal? *Ecoscience* **7**, 317–325.

Edwards, P.J. and Grubb, P.J. (1997). Studies of mineral cycling in a montane rain forest in New Guinea. 1. The distribution of organic matter in the vegetation and soil. *Journal of Ecology* **65**, 943–969.

Ellyson, W.J.T. and Sillett, S.C. (2003). Epiphyte communities on Sitka spruce in an old-growth redwood forest. *Bryologist* **106**, 197–211.

Esseen, P.-A. (1985). Litter fall of epiphytic macrolichens in two old *Picea abies* forests in Sweden. *Canadian Journal of Botany* **63**, 980–987.

Esseen, P.-A., L. Ericson, H. Lindström, and O. Zackrisson. 1981. Occurrence and ecology of *Usnea longissima* in central Sweden. Lichenologist 13, 177–190.

Esseen, P.-A. and Renhorn, K.-E. (1998a). Edge effects on an epiphytic lichen in fragmented forests. *Conservation Biology* **12**, 1307–1317.

Esseen, P.-A. and Renhorn, K.-E. (1998b). Mass loss of epiphytic lichen litter in a boreal forest. *Annales Botanici Fennici* **35**, 211–217.

Esseen, P.-A., Renhorn, K.-E., and Pettersson, R.B. (1996). Epiphytic lichen biomass in managed and old-growth boreal forests: effect of branch quality. *Ecological Applications* **6**, 228–238.

Farmer, A.M., Bates, J.W. and Bell, J.N.B. (1992). Ecophysiological effects of acid rain on bryophytes and lichens. In "Bryophytes and Lichens in a Changing Environment" (J.W. Bates and A.M. Farmer, Eds.), pp. 284–313. Clarendon Press, Oxford, UK.

Frahm, J.-P. (1990a). The ecology of epiphytic bryophytes on Mt. Kinabalu, Sabah (Malaysia). *Nova Hedwigia* **51**, 121–132.

Frahm, J.-P. (1990b). The effect of light and temperature on the growth of the bryophytes of tropical rain forests. *Nova Hedwigia* **51**, 151–164.

Frahm, J.-P., Specht, A. Reifenrath, K. and Vargas, Y.L. (2000). Allelopathic effect of crustaceous lichens on epiphytic bryophytes and vascular plants. *Nova Hedwigia* **70**, 245–254.

Freiberg, M. and Freiberg, E. (2000). Epiphyte diversity and biomass in the canopy of lowland and montane forests in Ecuador. *Journal of Tropical Ecology* **16**, 673–688.

Gauslaa, Y. (1985). The ecology of *Lobarion pulmonariae* and *Parmelion caperatae* in *Quercus* dominated forests in south-west Norway. *Lichenologist* **17**, 117–140.

Gauslaa, Y. (1995). The *Lobarion*, an epiphytic community of ancient forests threatened by acid rain. *Lichenologist* **27**, 59–76.

Gauslaa, Y. and Holien, H. (1998). Acidity of boreal *Picea abies* canopy lichens and their substratum, modified by local soils and airborne acidic deposition. *Flora* **193**, 249–257.

Gauslaa, Y., Ohlson, M., Solhaug, K.A., Bilger, W., and Nybakken, L. (2001). Aspect-dependent high-irradiance damage in two transplanted foliose forest lichens, *Lobaria pulmonaria* and *Parmelia sulcata*. *Canadian Journal of Forest Research* **31**, 1639–1649.

Gauslaa, Y. and Solhaug, K.A. (1996). Differences in susceptibility to light stress between epiphytic lichens of ancient and young boreal forest stands. *Functional Ecology* **10**, 344–354.

Gauslaa, Y. and Solhaug, K.A. (2000). High-light-intensity damage to the foliose lichen *Lobaria pulmonaria* within a natural forest: the applicability of chlorophyll fluorescence methods. *Lichenologist* **32**, 271–289.

Gauslaa, Y., and Solhaug, K.A. (2001). Fungal melanins as a sun screen for symbiotic green algae in the lichen *Lobaria pulmonaria*. *Oecologia* **126**, 462–471.

Glime, J.M. and Hong, W. S. (2002). Bole epiphytes on three conifer species from Queen Charlotte Islands, Canada. *Bryologist* **105**, 451–464.

Goffinet, B. (2000). Origin and phylogenetic relationships of bryophytes. *In* "Bryophyte Biology" (A.J. Shaw and B. Goffinet, Eds.), pp. 124–149. Cambridge University Press, Cambridge, UK.

Goward, T. (1995). *Nephroma occultum* and the maintenance of lichen diversity in British Columbia. *In* "Conservation Biology of Lichenised Fungi" (C. Scheidegger, P.A. Wolseley, and G. Thor, Eds.), pp. 93–101. Herausgeber, Birmensdorf, Germany.

Goward, T. and Arsenault, A. (2000a). Cyanolichen distribution in young managed forests: a dripzone effect? *Bryologist* **103**, 28–37.

Goward, T. and Arsenault, A. (2000b). Cyanolichens and conifers: implications for global conservation. *Forest, Snow, and Landscape Research* **75**, 303–318.

Gradstein, S.R., Griffin III, D., Morales, M.I., and Nadkarni, N.M. (2000). Diversity and habitat differentiation of mosses and liverworts in the cloud forest of Monteverde, Costa Rica. *Caldasia* **23**, 203–212.

Green, T.G.A., Büdel, B., Meyer, A., Zellner, H., and Lange, O.L. (1997). Temperate rainforest lichens in New Zealand: light response of photosynthesis. *New Zealand Journal of Botany* **35**, 493–504.

Green, T.G.A., Kilian, E., and Lange, O.L. (1991). *Pseudocyphellaria dissimilis*: a desiccation-sensitive, highly shade-adapted lichen from New Zealand. *Oecologia* **85**, 498–503.

Green, T.G.A. and Lange, O.L. (1994). Photosynthesis in poikilohydric plants: a comparison of lichens and bryophytes. In "Ecophysiology of Photosynthesis" (E.-D. Schulze and M.M. Caldwell, Eds.), pp. 319–341. Springer-Verlag, New York, NY.

Green, T.G.A., Schlensog, M., Sancho, L.G., Winkler, J.B., Broom, F.D., and Schroeter, B. (2002). The photobiont determines the pattern of photosynthetic activity within a single lichen thallus containing cyanobacterial and green algal sectors. *Oecologia* **130**, 191–198.

Gries, C. (1996). Lichens as indicators of air pollution. *In* "Lichen Biology" (T.H. Nash III, Ed.), pp. 240–254. Cambridge University Press, Cambridge, UK.

Guzman, G., Quilhot, W., and Galloway, D.J. (1990). Decomposition of species of *Pseudocyphellaria* and *Sticta* in a southern Chilean forest. *Lichenologist* **22**, 325–331.

Hallé, F., Oldeman, R.A.A., and Tomlinson, P.B. (1978). "Tropical Trees and Forests: An Architectural Analysis." Springer-Verlag, New York.

Hamer, T.E. and Nelson, S.K. (1995). Characteristics of marbled murrelets nest trees and nesting stands. *In* "Ecology and Conservation of the Marbled Murrelet" (C.J. Ralph, G.L. Hunt Jr., M.G. Raphael, and J.F. Piatt, Eds.), pp. 69–82. Pacific Southwest Research Station, Albany, CA.

Hayward, G.D. and Rosentreter, R. (1994). Lichens as nesting material for northern flying squirrels in the northern Rocky Mountains. *Journal of Mammology* **75**, 663–673.

Hazell, P. and Gustafsson, L. (1999). Retention of trees at final harvest—evaluation of a conservation technique using epiphytic bryophyte and lichen transplants. *Biological Conservation* **90**, 133–142.

Heitz, P., Wanek, W., Wania, R., and Nadkarni, N. (2002). Nitrogen-15 natural abundance in a montane cloud forest canopy as an indicator of nitrogen cycling and epiphyte nutrition. *Oecologia* **131**, 350–355.

Hesbacher, S., Baur, B., Baur, A., and Proksch, P. (1995). Sequestration of lichen compounds by three species of terrestrial snails. *Journal of Chemical Ecology* **21**, 233–246.

Hilmo, O. (1994). Distribution and succession of epiphytic lichens on *Picea abies* branches in a boreal forest, central Norway. *Lichenologist* **26**, 149–169.

Hilmo, O. (2002). Growth and morphological response of old-forest lichens transplanted into a young and an old *Picea abies* forest. *Ecography* **25**, 329–335.

Hofstede, R.G.M., Wolf, J.H.D., and Benzing, D.H. (1993). Epiphytic biomass and nutrient status of a Colombian upper montane rain forest. *Selbyana* **14**, 37–45.

Holien, H. (1998). Lichens in spruce forest stands of different successional stages in central Norway with emphasis on diversity and old growth species. *Nova Hedwigia* **66**, 283–324.

Holien, H., Gaarder, G., and Håpnes, A. (1995). *Erioderma pedicellatum* still present, but highly endangered in Europe. *Graphis Scripta* **7**, 79–84.

Holz, I., Gradstein, S.R., Heinrichs, J., and Kappelle, M. (2002). Bryophyte diversity, microhabitat differentiation, and distribution of life forms in Costa Rican upper montane *Quercus* forest. *Bryologist* **105**, 334–348.

Honegger, R. (1991). Functional aspects of the lichen symbiosis. *Annual Review of Plant Physiology and Plant Molecular Biology* **42**, 553–578.

Honegger, R. (1998). The lichen symbiosis—what is so spectacular about it? *Lichenologist* **30**, 193–212.

Hosokawa, T., Odani, N., and Tagawa, H. (1964). Causality of the distribution of corticolous species in forests with special reference to the physio-ecological approach. *Bryologist* **67**, 396–411.

Ingram, S.W. and Nadkarni, N.M. (1993). Composition and distribution of epiphytic organic matter in a neotropical cloud forest, Costa Rica. *Biotropica* **25**, 370–383.

Ishii, H. and Ford, E.D. (2001). The role of epicormic shoot production in maintaining foliage in old *Pseudotsuga menziesii* trees. *Canadian Journal of Botany* **79**, 251–264.

Ishii, H., Ford, E.D., and Dinnie, C.E. (2002). The role of epicormic shoot production in maintaining foliage in old *Pseudotsuga menziesii* trees II. Basal reiteration from older branch axes. *Canadian Journal of Botany* **80**, 916–926.

Ishii, H. and Wilson, M.E. (2001). Crown structure of old-growth Douglas-fir in the western Cascade Range, Washington. *Canadian Journal of Forest Research* **31**, 1250–1261.

Jarman, S.J. and Kantvilas, G. (1995). Epiphytes of an old Huon pine tree (*Lagarostrobos franklinii*) in Tasmanian rainforest. *New Zealand Journal of Botany* **33**, 65–78.

Jepson, J. (2000). "The Tree Climber's Companion." Beaver Tree Publishing, Longville, MN.

John, E. (1992). Distribution patterns and interthalline interactions of epiphytic foliose lichens. *Canadian Journal of Botany* **70**, 818–823.

Kantvilas, G, and Minchin, P.R. (1989). An analysis of epiphytic lichen communities in Tasmanian cool temperate rainforest. *Vegetatio* **84**, 99–112.

Kenkel, N.C. and Bradfield, G.E. (1986). Epiphytic vegetation on *Acer macrophyllum*: a multivariate study of species-habitat relationships. *Vegetatio* **68**, 43–53.

Kenrick, P. and Crane, P.R. (1997). The origin and early evolution of plants on land. *Nature* **389**, 33–39.

Keon, D.B. (2001). Factors limiting the distribution of the sensitive lichen *Usnea longissima* in the Oregon Coast Range: habitat or dispersal? M. Sc. thesis, Oregon State University, Corvallis, OR.

Kivistö, L. and Kuusinen, M. (2000). Edge effects on the epiphytic lichen flora of *Picea abies* in middle boreal Finland. *Lichenologist* **32**, 387–398.

Knops, J.M.H. and Nash III, T.H. (1996). The influence of epiphytic lichens on the nutrient cycling of an oak woodland. *Ecological Monographs* **66**, 159–179.

Komposch, H. and Hafellner, J. (2000). Diversity and vertical distribution of lichens in a Venezuelan tropical lowland rain forest. *Selbyana* **21**, 11–24.

Kuusinen, M. (1996). Epiphyte flora and diversity on basal trunks of six old-growth forest tree species in southern and middle boreal Finland. *Lichenologist* **28**, 443–463.

Lang, G.E., Reiners, W.A., and Pike, L.H. (1980). Structure and biomass dynamics of epiphytic lichen communities of balsam fir forests in New Hampshire. *Ecology* **61**, 541–550.

Lange, O.L., Büdel, B., Heber, U., Meyer, A., Zellner, H., and Green, T.G.A. (1993). Temperate rainforest lichens in New Zealand: high thallus water content can severely limit photosynthetic CO_2 exchange. *Oecologia* **95**, 303–313.

Lange, O.L., Büdel, B., Meyer, A., Zellner, H., and Zotz, G. (2000). Lichen carbon gain under tropical conditions: water relations and CO_2 exchange of three *Leptogium* species of a lower montane rainforest in Panama. *Flora* **195**, 172–190.

Lange, O.L., Reichenberger, H., and Walz, H. (1997). Continuous monitoring of CO_2 exchange of lichens in the field: short-term enclosure with an automatically operating cuvette. *Lichenologist* **29**, 259–274.

Lawrey, J.D. and Diederich, P. (2003). Lichenicolous fungi: interactions, evolution, and biodiversity. *Bryologist* **106**, 80–120.

Lesica, P., McCune, B., Cooper, S.V., and Hong, W.S. (1991). Differences in lichen and bryophyte communities between old-growth and managed second growth forests in Swan Valley, Montana. *Canadian Journal of Botany* **69**, 1745–1755.

Liu, C., Ilvesniemi, H., and Westman, C.J. (2000). Biomass of arboreal lichens and its vertical distribution in a boreal coniferous forest in central Finland. *Lichenologist* **32**, 495–504.

Lyons, B., Nadkarni, N.M., and North, M.P. (2000). Spatial distribution and succession of epiphytes on *Tsuga heterophylla* in an old-growth Douglas-fir forest. *Canadian Journal of Botany* **78**, 957–968.

Marschall, M. and Proctor, M.C.F. (1999). Desiccation tolerance and recovery of the leafy liverwort *Porella platyphylla*: chlorophyll-fluorescence measurements. *Journal of Bryology* **21**, 257–262.

Martin, C.E. and Adamson, V.J. (2001). Photosynthetic capacity of mosses relative to vascular plants. *Journal of Bryology* **23**, 319–323.

McCune, B. (1990). Rapid estimation of abundance of epiphytes on branches. *Bryologist* **93**, 39–43.

McCune, B. (1993). Gradients in epiphyte biomass in three *Pseudotsuga-Tsuga* forests of different ages in western Oregon and Washington. *Bryologist* **96**, 405–411.

McCune, B. (1997). Biogeography of rare lichens from the coast of Oregon. *In* "Conservation and Management of Native Plants and Fungi" (T.N. Kaye, A. Liston, R.M. Love, D.L. Luoma, R.J. Meinke, and M.V. Wilson, Eds.), pp. 234–241. Native Plant Society of Oregon, Corvallis, OR.

McCune, B., Amsberry, K.A., Camacho, F.J., Clery, S., Cole, C., Emerson, C., Felder, G., French, P., Greene, D., Harris, R., Hutten, M., Larson, B., Lesko, M., Majors, S., Markwell, T., Parker, G.G., Pendergrass, K., Peterson, E.B., Peterson, E.T., Platt, J., Proctor, J., Rambo, T., Rosso, A., Shaw, D., Turner, R., and Widmer, M. (1997). Vertical profile of epiphytes in a Pacific Northwest old-growth forest. *Northwest Science* **71**, 145–152.

McCune, B. and Daly, W.J. (1994). Consumption and decomposition of lichen litter in a temperate coniferous rainforest. *Lichenologist* **26**, 67–71.

McCune, B., Derr, C.C., Muir, P.S., Shirazi, A.S., Sillett, S.C., and Daly, W.J. (1996). Lichen pendants for transplant and growth experiments. *Lichenologist* **28**,161–169.

McCune, B., Hutchinson, J., and Berryman, S. (2002). Concentration of rare epiphytic lichens along large streams in a mountainous watershed in Oregon, USA. *Bryologist* **105**, 439–450.

McCune, B., Rosentreter, R., Ponzetti, J.M., and Shaw, D.C. (2000). Epiphyte habitats in an old conifer forest in western Washington, U.S.A. *Bryologist* **103**, 417–427.

Millbank, J.W. (1982). The assessment of nitrogen fixation and throughput by lichens. III. Losses of nitrogenous compounds by *Peltigera membranacea*, *P. polydactyla*, and *Lobaria pulmonaria* in simulated rainfall episodes. *New Phytologist* **97**, 229–234.

Minami, Y. and Takahashi, K. (1994). Environmental factors affecting the growth of epiphytic bryophytes and lichens in *Cryptomeria japonica* forest. *National Environmental Science Research (Japan)* **7**, 1–8.

Montfoort, D. and Ek, R. (1990). Vertical distribution and ecology of epiphytic bryophytes and lichens in a lowland rain forest in French Guyana. Institute of Systematic Botany, Utrecht, The Netherlands.

Nadkarni, N.M. (1981). Canopy roots: convergent evolution in rainforest nutrient cycles. *Science* **214**, 1023–1024.

Nadkarni, N.M. (1984a). Epiphyte biomass and nutrient capital of a neotropical elfin forest. *Biotropica* **16**, 249–256.

Nadkarni, N.M. (1984b). Biomass and mineral capital of epiphytes in an *Acer macrophyllum* community of a temperate moist coniferous forest, Olympic Peninsula, Washington State. *Canadian Journal of Botany* **62**, 2223–2228.

Nadkarni, N.M. (1986). The nutritional effects of epiphytes on host trees with special reference to alteration of precipitation chemistry. *Selbyana* **9**, 44–51.

Nadkarni, N.M. (2000). Colonization of stripped branch surfaces by epiphytes in a lower montane cloud forest, Monteverde, Costa Rica. *Biotropica* **32**, 358–363.

Neitlich, P.N. (1993). Lichen abundance and biodiversity along a chronosequence from young managed stands to ancient forest. M. Sc. thesis, University of Vermont, Burlington, VT.

Neitlich, P.N. and McCune, B. (1997). Hotspots of epiphytic lichen diversity in two young managed forests. *Conservation Biology* **11**, 172–182.

Ng, F.S.P. (1999). The development of the tree trunk in relation to apical dominance and other shoot organisation concepts. *Journal of Tropical Forest Science* **11**, 270–285.

Ohlson, M., Söderström, L., Hörnberg, G., Zackrisson, O., and Hermansson, J. (1997). Habitat qualities versus long-term continuity as determinants of biodiversity in boreal old-growth swamp forests. *Biological Conservation* **81**, 221–231.

Parker, G.G. (1995). Structure and microclimate of forest canopies. In "Forest Canopies" (M.D. Lowman and N.M. Nadkarni, Eds.), pp. 73–106. Academic Press, San Diego.

Peck, J.E., Hong, W.S., and McCune, B. (1995). Diversity of epiphytic bryophytes on three host tree species, Thermal Meadows, Hotspring Island, Queen Charolotte Islands, *Canada. Bryologist* **98**, 123–128.

Peck, J.E. and McCune, B. (1997). Remnant trees and canopy lichen communities in western Oregon: a retrospective approach. *Ecological Applications* **7**, 1181–1187.

Peterson, E.B. and McCune, B. (2001). Diversity and succession of epiphytic macrolichen communities in low-elevation managed conifer forests in western Oregon. *Journal of Vegetation Science* **12**, 511–524.

Peterson, E.B. and McCune, B. (2003). The importance of hotspots for lichen diversity in forests of western Oregon. *Bryologist* **106**, 246–256.

Pettersson, R.B., Ball, J.P., Renhorn, K.-E., Esseen, P.-A., and Sjöberg, K. (1995). Invertebrate communities in boreal forest canopies as influenced by forestry and lichens with implications for passerine birds. *Biological Conservation* **74**, 57–63.

Pike, L.H. (1978). The importance of epiphytic lichens in mineral cycling. *Bryologist* **81**, 247–257.

Pike, L.H., Denison, W.C., Tracy, D.M., Sherwood, M.A., and Rhoades, F.M. (1975). Floristic survey of epiphytic lichens and bryophytes growing on old-growth conifers in western Oregon. *Bryologist* **78**, 389–402.

Pike, L. H., Rydell, R.A., and Denison, W.C. (1977). A 400-year-old Douglas-fir tree and its epiphytes: biomass, surface area, and their distributions. *Canadian Journal of Forest Research* **7**, 680–699.

Pipp, A.K., Henderson, C., and Callaway, R.M. (2001). Effects of forest age and forest structure on epiphytic lichen biomass and diversity in a Douglas-fir forest. *Northwest Science* **75**, 12–24.

Pócs, T. (1980). The epiphytic biomass and its effect on the water balance of two rain forest types in the Uluguru Mountains (Tanzania, East Africa). *Acta Botanica Academiae Scientiarum Hungaricae* **26**, 143–167.

Price, K. and Hochachka, G. (2001). Epiphytic lichen abundance: effects of stand age and composition in coastal British Columbia. *Ecological Applications* **11**, 904–913.

Prinzing, A., and Wirtz, H.-P. (1997). The epiphytic lichen, *Evernia prunastri*, as a habitat for arthropods: shelter from desiccation, food-limitation and indirect mutualism. *In* "Canopy Arthropods" (N.E. Stork, J. Adis, and R.K. Didham, Eds.), pp. 477–494. Chapman and Hall, London, UK.

Proctor, M.C.F. (1999). Water-relations parameters of some bryophytes evaluated by thermocouple psychrometry. *Journal of Bryology* **21**, 263–270.

Proctor, M.C.F. (2000). Physiological ecology. *In* "Bryophyte Biology" (A.J. Shaw and B. Goffinet, Eds.), pp. 225–247. Cambridge University Press, Cambridge, UK.

Renhorn, K.-E., Esseen, P.-A., Palmqvist, K., and Sundberg, B. (1997). Growth and vitality of epiphytic lichens. I. Responses to microclimate along a forest edge-interior gradient. *Oecologia* **109**, 1–9.

Rhoades, F. M. (1981). Biomass of epiphytic lichens and bryophytes on *Abies lasiocarpa* on a Mt. Baker lava flow, Washington. *Bryologist* **84**, 39-47.

Rhoades, F.M. (1983). Distribution of thalli in a population of the epiphytic lichen *Lobaria oregana* and a model of population dynamics and production. *Bryologist* **86**, 309–331.

Rhoades, F.M. (1995). Nonvascular epiphytes in forest canopies: worldwide distribution, abundance, and ecological roles. In "Forest Canopies" (M.D. Lowman and N.M. Nadkarni, Eds.), pp. 353–408. Academic Press, San Diego.

Richardson, D.H.S. (1999). War in the world of lichens: parasitism and symbiosis as exemplified by lichens and lichenicolous fungi. *Mycological Research* **103**, 641–650.

Rikkinen, J. (1995). What's behind the pretty colours? A study on the photobiology of lichens. *Bryobrothera* **4**, 1–239.

Rincon, E. and Grime, J.P. (1989). Plasticity and light interception by six bryophytes of contrasted ecology. *Journal of Ecology* **77**, 439–446.

Rolstad, J., Gjerde, I., Storaunet, K.O., and Rolstad, E. (2001). Epiphytic lichens in Norwegian coastal spruce forest: historic logging and present forest structure. *Ecological Applications* **11**, 421–436.

Romero, C. (1999). Reduced-impact logging effects on commercial non-vascular pendant epiphyte biomass in a tropical montane forest in Costa Rica. *Forest Ecology and Management* **118**, 117–125.

Rominger, E.M., Allen-Johnson, L., and Oldemeyer, J.L. (1994). Arboreal lichen in uncut and partially cut subalpine fir stands in woodland caribou habitat, northern Idaho and southeastern British Columbia. *Forest Ecology and Management* **70**, 195–202.

Rose, F. (1992). Temperate forest management: its effects on bryophyte and lichen floras and habitats. *In* "Bryophytes and Lichens in a Changing Environment" (J.W. Bates and A.M. Farmer, Eds.), pp. 211–233. Clarendon Press, Oxford, UK.

Rosentreter, R., Hayward, G.D., and Wicklow-Howard, M. (1997). Northern flying squirrel seasonal food habits in the interior conifer forests of central Idaho, USA. *Northwest Science* **71**, 97–102.

Rosso, A.L., McCune, B., and Rambo, T.R. (2000). Ecology and conservation of a rare, old-growth-associated canopy lichen in a silvicultural landscape. *Bryologist* **103**, 117–127.

Rosso, A.L., Muir, P.S., and Rambo, T.R. (2001). Using transplants to measure accumulation rates of epiphytic bryophytes in forests of western Oregon. *Bryologist* **104**, 430–439.

Rosso, A.L. and Rosentreter, R. (1999). Lichen diversity and biomass in relation to management practices in forests of northern Idaho. *Evansia* **16**, 97–104.

Ruchty, A., Rosso, A.L, and McCune, B. (2001). Changes in epiphyte communities as the shrub, *Acer circinatum*, develops and ages. *Bryologist* **104**, 274–281.

Russell, S. (1988). Measurement of bryophyte growth. 1. Biomass (harvest) techniques. In "Methods in Bryology" (J.M. Glime, Ed.), pp. 249–257. Hattori Botanical Laboratory, Nichinan, Japan.

Selva, S.B. (1994). Lichen diversity and stand continuity in the northern hardwoods and spruce-fir forests of northern New England and western New Brunswick. *Bryologist* **97**, 424–429.

Seyd, E.L. and Seaward, M.R.D. (1984). The association of oribatid mites with lichens. *Zoological Journal of the Linnean Society* **80**, 369–420.

Sharnoff, S. (1994). Use of lichens by wildlife in North America. *Research and Exploration* **10**, 370–371.

Sillett, S.C. (1994). Growth rates of two epiphytic cyanolichen species at the edge and in the interior of a 700-year-old Douglas-fir forest in the western Cascades of Oregon. *Bryologist* **97**, 321–324.

Sillett, S.C. (1995). Branch epiphyte assemblages in the forest interior and on the clearcut edge of a 700-year-old forest canopy in western Oregon. *Bryologist* **98**, 301–312.

Sillett, S.C. (1999). Tree crown structure and vascular epiphyte distribution in *Sequoia sempervirens* rain forest canopies. *Selbyana* **20**, 76–97.

Sillett, S.C. and Bailey, M.G. (2003). Effects of tree crown structure on biomass of the epiphytic fern *Polypodium scouleri* in redwood forests. *American Journal of Botany* **90**, 255–261.

Sillett, S.C. and Goslin, M.N. (1999). Distribution of epiphytic lichens in relation to remnant trees in a multiple-age Douglas-fir forest. *Canadian Journal of Forest Research* **29**, 1204–1215.

Sillett, S.C. and Goward, T. (1998). Ecology and conservation of *Pseudocyphellaria rainierensis*, a Pacific Northwest endemic lichen. *In* "Lichenographia Thomsoniana" (M.G. Glenn, R.C. Harris, R. Dirig, and M.S. Cole, Eds.), pp. 377–388. Mycotaxon Ltd., Ithaca, NY.

Sillett, S.C., Gradstein, S.R., and Griffin III, D. (1995). Bryophyte diversity of *Ficus* tree crowns from cloud forest and pasture in Costa Rica. *Bryologist* **98**, 251–260.

Sillett, S.C. and McCune, B. (1998). Survival and growth of cyanolichen transplants in Douglas-fir forest canopies. *Bryologist* **101**, 20–31.

Sillett, S.C., McCune, B., Peck, J.E., and Rambo, T.R. (2000a). Four years of epiphyte colonization in Douglas-fir forest canopies. *Bryologist* **104**, 126–132.

Sillett, S.C., McCune, B., Peck, J.E., and Rambo, T.R., and Ruchty, A. (2000b). Dispersal limitations of epiphytic lichens result in species dependent on old-growth forests. *Ecological Applications* **10**, 789–799.

Sillett, S.C. and Neitlich, P.N. (1996). Emerging themes in epiphyte research in westside forests with special reference to cyanolichens. *Northwest Science*, **70** (Special Issue), 54–60.

Sillett, S.C. and Rambo, T.R. (2000). Vertical distribution of dominant epiphytes in Douglas-fir forests of the central Oregon Cascades. *Northwest Science* **74**, 44–49.

Sillett, T.S. (1994). Foraging ecology of epiphyte-searching insectivorous birds in Costa Rica. *Condor* **96**, 863–877.

Solhaug, K. A., Gauslaa, Y., and Haugen, J. 1995. Adverse effects of epiphytic crustose lichens upon stem photosynthesis and chlorophyll of *Populus tremula*. *Botanica Acta* **108**, 233–239.

Solhaug, K.A. and Gauslaa, Y. (1996). Parietin, a photoprotective secondary product of the lichen *Xanthoria parietina*. *Oecologia* **108**, 412–418.

Sollins, P., Grier, C.C., McCorison, F.M., Cromack Jr., K., and Fogel, R. (1980). The internal element cycles of an old-growth Douglas-fir ecosystem in western Oregon. *Ecological Monographs* **50**, 261–285.

Stone, D.F. (1989). Epiphyte succession on *Quercus garryana* branches in the Willamette Valley of western Oregon. *Bryologist* **92**, 81–94.

Stone, J.K., Sherwood, M.A., and Carroll, G.C. (1996). Canopy microfungi: function and diversity. *Northwest Science*, **70** (Special Issue), 37–45.

Stubbs, C.S. (1989). Patterns of distribution and abundance of corticolous lichens and their invertebrate associates on *Quercus rubra* in Maine. *Bryologist* **92**, 453–460.

Tehler, A. (1996). Systematics, phylogeny and classification. *In* "Lichen Biology" (T.H. Nash III, Ed.), pp. 217–239. Cambridge University Press, Cambridge, UK.

Thomas, S.C., Liguori, D.A., and Halpern, C.B. (2001). Corticolous bryophytes in managed Douglas-fir forests: habitat differentiation and responses to thinning and fertilization. *Canadian Journal of Botany* **79**, 886–896.

Tibell, L. (1992). Crustose lichens as indicators of forest continuity in boreal coniferous forests. *Nordic Journal of Botany* **12**, 427–450.

van Herk, C.M. (2001). Bark pH and susceptibility to toxic air pollutants as independent causes of changes in epiphytic lichen composition in space and time. *Lichenologist* **33**, 419–441.

van Herk, C.M., Aptroot, A., and van Dobben, H.F. (2002). Long-term monitoring in the Netherlands suggests that lichens respond to global warming. *Lichenologist* **34**, 141–154.

van Leerdam, A., Zagt, R.J., and Veneklaas, E.J. (1990). The distribution of epiphyte growth-forms in the canopy of a Colombian cloud-forest. *Vegetatio* **87**, 59–71.

Van Pelt, R. (2001). "Forest Giants of the Pacific Coast." University of Washington and Global Forest Press, Seattle, WA.

Veneklaas, E.J., Zagt, R.J., van Leerdam, A., Van Ek, R., Broekhoven, A.J., and van Genderen, M. (1990). Hydrological properties of the epiphytic mass of a montane tropical rain forest, Colombia. *Vegetatio* **89**, 183–192.

Vitt, D.H., Ostafichuk, M., and Brodo, I.M. (1973). Foliicolous bryophytes and lichens of *Thuja plicata* in western British Columbia. *Canadian Journal of Botany* **51**, 571–580.

Wilson, E.O. (2002). "The Future of Life." New York, Knopf.

Winchester, N.N. (1997). Canopy arthropods of coastal Sitka spruce trees on Vancouver Island, British Columbia, Canada. *In* "Canopy Arthropods" (N.E. Stork, J. Adis, and R.K. Didham, Eds.) pp. 151–168. Chapman and Hall, London, UK.

Wolf, J.H.D. (1993a). Diversity patterns and biomass of epiphytic bryophytes and lichens along an altitudinal gradient in the northern Andes. *Annals of the Missouri Botanical Gardens* **80**, 928–960.

Wolf, J.H.D. (1993b). Epiphyte communities of tropical montane rain forests in the northern Andes. I. Lower montane communities. *Phytocoenologia* **22**, 1–52.

Wolf, J.H.D. (1993c). Epiphyte communities of tropical montane rain forests in the northern Andes. II. Upper montane communities. *Phytocoenologia* **22**, 53–103.

Wolf, J.H.D. (1994). Factors controlling the distribution of vascular and non-vascular epiphytes in the northern Andes. *Vegetatio* **112**, 15–28.

Wolf, J.H.D. (1995). Non-vascular epiphyte diversity patterns in the canopy of an upper montane rain forest (2550-3670 m), Central Cordillera, Colombia. *Selbyana* **16**, 185–195.

Zotz, G., Budel, B., Meyer, A., Zellner, H., and Lange, O.L. (1997). Water relations and CO_2 exchange of tropical bryophytes in a lower montane rain forest in Panama. *Botanica Acta* **110**, 9–17.

Zotz, G., Budel, B., Meyer, A., Zellner, H., and Lange, O.L. (1998). *In situ* studies of water relations and CO_2 exchange of the tropical macrolichen *Sticta tomentosa*. *New Phytologist* **139**, 525–535.

Zotz, G., A. Schweikert, W. Jetz, and H. Westerman. 2000. Water relations and carbon gain are closely related to cushion size in the moss *Grimmia pulvinata*. *New Phytologist* **148**, 59–67.

CHAPTER 9

Vascular Epiphytes

David H. Benzing

*Plants are superb opportunists, making the most of different combinations of water, air,
soil, and climate. Thier grip on the planet, their capacities for colonization, and
their intergration with the environment are due to an astounding
diversification and variety.*
—Anthony Huxley, Plant and Planet, *1975*

Introduction

Botanists based in the temperate zone see little evidence that bark is one of the most intensively exploited of the many kinds of substrates used by higher plants. Numerous algae and bryophytes inhabit the canopies of mid and higher latitude woodlands, but few vascular types do. Tropical forests tell quite a different story, especially those that support many tons per hectare of suspended vegetation, including up to one-third of the local vascular flora (e.g., Gentry and Dodson 1987a; Freiberg and Freiberg 2000; Hsu et al. 2002; see Figure 9-1). In fact, about 25,000 species, or somewhere between 8 and 10 percent of the vascular plants, routinely spend at least part of their lives as epiphytes—in effect, they root in the crowns of larger plants rather than on the ground (see Table 9-1; Benzing 1990).

Epiphytes as defined here use their botanical hosts or "phorophytes" (literally meaning "supporting plant") solely for mechanical support. Lacking the ground-based plant's access to earth soil and the mistletoe's parasitic habit (see Chapter 10), they instead draw on less conventional alternatives for moisture and nutrients, such as ant wastes; sodden, rotting litter; and precipitation as it flows over shoots and roots, either or both of which can be appropriately absorptive (Benzing 1990).

Fossils indicate that vascular epiphytism has existed since at least the Carboniferous period, and for many modern species, ancient patterns persist. Hundreds of extant lycopods and ferns, for example, root in mats of humus, much like *Botryopteris forensis* did on the trunks of extinct tree fernlike *Psaronius* about 250 million years ago (Rothwell 1991; see Figure 9-2; see Chapter 7). Judging from the taxonomic scatter of epiphytism (see Table 9-1; Benzing 1990) and the variety of ways that modern flora accomplish this lifestyle (see Table 9-2), bark has been colonized many times and along numerous evolutionary pathways. Paradoxically, epiphytism is rare or absent in some of the largest families of flowering plants such as Asteraceae, Fabaceae, and Poaceae.

So why devote a chapter in a text dedicated to forest canopy biology to a group of plants that predominately inhabit just the humid tropics? Perhaps most compelling is the 8 to 10 percent statistic just mentioned, but there are plenty of additional reasons. Vascular epiphytes also demonstrate extraordinary adaptive variety, some undetermined portion of which represents evolutionary response to constraints and opportunities more or less unique to the forest canopy habitat. No less important are the ways that these plants influence the ecosystems that host them.

Figure 9-1 Wet montane forest densely populated by *Tillandsia* spp. (Bromeliaceae) in Chiapas State, Mexico.

Table 9-1 Salient Facts about Epiphyte Diversity and Taxonomic Scatter

1	About two thirds of the ferns are epiphytic.
2	Epiphytism is common among the homosporous lycopods (i.e., *Lycopodium sensu lato*) but infrequent among the heterosporous types (*Selaginella* and *Isoetes*).
3	Only one cycad and a couple of gnetophytes among the gymnosperms are epiphytic.
4	More than 80 families of flowering plants contain at least one epiphytic species.
5	The orchids alone account for more than half of the vascular epiphytes, and the family is over two-thirds epiphytic.
6	The remaining most heavily epiphytic angiospermous families are scattered through Magnoliophyta (Araceae and Bromeliaceae of Liliopsida and Asclepiadaceae, Cactaceae, Ericaceae, Gesneriaceae, Melastomataceae, Moraceae, Piperaceae, and Rubiaceae of Magnoliopsida).

Finally, the vascular epiphytes require inclusion if only because they occur in such great abundance in threatened ecosystems like the tropical montane cloud forest (see Figure 9-1).

The following discussion is organized around several topics and related questions. Foremost are the growing conditions experienced by epiphytes and those aspects of plant life history, morphology, and physiology that enable their aerial lifestyle. How, for example, do the more stress-tolerant types within this group cope with the combined effects of drought, physical disturbance, high irradiance, and nutritional deprivation? And is it the specialization required to accommodate this punishing suite of constraints that as well confines so many of them to canopy-based

Figure 9-2 *Tmesipteris* sp. (Psilotophyta) anchored in the mantle of adventitious roots covering the trunk of a tree fern on North Island, New Zealand.

Table 9-2 Criteria Used to Distinguish and Label the Types of Epiphytes

1	Fidelity to bark as a substratum: obligate, facultative, and accidental epiphytes
2	Timing during the life cycle: holo-epiphytes versus hemi-epiphytes (primary versus secondary hemi-epiphytes)
3	Light requirement: exposure, sun, and shade-tolerant epiphytes (Pittendrigh 1948)
4	Type of substrate used: twig, naked bark, suspended humus-soil, ant nest, knothole or rotten wood epiphytes
5	Relationship with ants: nonmyrmecophytic and myrmecophytic (ant house, ant nest epiphytes)
6	Nutritional mode: atmospheric, tank/trash basket, soil-rooted (in suspended soil), myrmecotrophic (ant-fed), carnivorous

substrates? Might we expect these plants to be extraordinarily sensitive to climate change because they are so specialized (Benzing 1998)? Another major topic concerns how epiphytes influence community structure, process, and character. Are these plants, for instance, as important to the ecosystems that host them as some authors suggest (e.g., Wilson 1987; Stork 1987; Paoletti et al. 1991; Stuntz et al. 1999, 2001; Merwin et al. 2003)?

How Distinct Is the Epiphyte Lifestyle?

Truth be told, epiphytism, rather than constituting a discrete lifestyle or ecological strategy, grades into terrestrialism (the soil-rooting habit) and even more so into lithophytism (the utilization of rocky substrates; see Figure 9-3). For example, every member of a population of an "obli-

Figure 9-3 Lithophytic *Alcantarea* sp. (Bromeliaceae) on Precambrian granite in Rio de Janeiro State, Brazil.

gate" epiphyte roots on bark; but for the "accidental" epiphyte, only the exceptional individual does this (see Table 9-2). Still other epiphytes known as "facultative" types grow on the ground and on bark seemingly indifferent to which one of these two media supports them. Not surprisingly, the diversity and abundance of the facultative epiphytes peak where growing conditions within and below the canopy converge, being either exceptionally wet or dry. Bennett (1991) identified a suite of vegetative and reproductive characters that distinguish types within a mixed group of lithophytic and epiphytic species of *Tillandsia* (Bromeliaceae), but for the most part, the biology that underlies substrate preference among the epiphytes remains obscure.

The fact that so many of the widely cultivated epiphytes grow about as well in manufactured containers and in a variety of garden soils as on phorophytes suggests that only one or a few characteristics or life cycle stages (germination and establishment?) dictate substrates in nature. However, most of these popular cultigens are derived from relatively unspecialized wild types, and when compared to many of the less commonly grown forms, epiphytism again reveals it graded nature. These latter, less amenable plants survive in cultivation only if maintained on specific natural substrates or close facsimiles (e.g., naked bark and twigs, as illustrated below). A growing number of reports describe the fates of fallen epiphytes and the possible reasons why the obligate types eventually, if not always promptly, succumb (Matelson et al. 1993; Pett-Ridge and Silver 2002; Mondragón et al. in press).

For some species, ecological habit shifts during the normal course of growth. These so-called hemi-epiphytes routinely spend only part of their lives rooted above ground (Figures 9-4 and 9-5 and Table 9-2). Secondary hemi-epiphytes (e.g., many Araceae) begin as vines, and after the

Figure 9-4 Hemi-epiphytic *Raphidiophora* sp. (Araceae) at the Marie Selby Botanical Gardens, Sarasota, Florida. This specimen has already lost stem and most root contact with the ground.

Figure 9-5 Hemi-epiphytic *Ficus* sp. (Moraceae) in Tamil Nadu State, India.

STRANGLER FIG TREES: DEMONS OR HEROES OF THE CANOPY?

Timothy G. Laman

High in the canopy, a bird dropping splats on a branch. A tiny seed from the dropping gets lodged there, and the next rain washes it down amongst the moss growing in the tree's main crotch. Soon the seed germinates, and a strangler fig seedling begins its unlikely life. Assuming the tree crotch where it grows has a large enough deposit of soil and retains enough water to keep the seedling alive, the young fig will soon begin to send roots down its host tree trunk toward the ground. These figs are called "stranglers" because roots that descend and wrap the host trunk fuse together, forming a basket-like network. They appear to strangle, but actually are forming rigid rings around the host's trunk that restrict its further growth. Sometimes, this may kill the host, and the fig may grow to stand on its own as the dead tree rots away.

This strategy seems to be an effective one for taking a shortcut to the canopy. As discussed by Lowman (this volume), the chance of a tree seedling making it to the canopy from the forest floor is incredibly small. Strangler figs in the genus *Ficus*, as well as other hemi-epiphytes such as *Clusia*, have evolved their unique growth habits as an alternative way to get to the high-light zone of the canopy. So are these strangler figs with their demonic strategy killing trees and taking over the rain forest canopy? This hardly seems to be the case, since I know of no forests where figs are dominant. The reason is that their lifestyle also has numerous bottlenecks and limitations. First of all, the number of suitable establishment sites in the canopy is low, since a position such as a large knot-hole or tree crotch with adequate nutrients and water retention is needed (Laman 1995a, 1995b). Second, the problem of seed dispersal to these small canopy sites is also a formidable obstacle. Figs have evolved to produce large crops of fruits with many small seeds that attract birds, bats, and other arboreal mammals to maximize their seed dispersal. Still, as one moves away from the fruiting fig, the density of seed rain drops off very quickly (Laman 1996a). Even after seeds manage to land in a tree crown, there are other complications, such as seed-harvesting ants (Laman 1996b). Ants may consume a large proportion of seeds, but they may also move some seeds from bare branches to their nests in tree crotches where, if abandoned, the seeds are in a better position to get established. Finally, canopy trees vary greatly in their likelihood of having good fig establishment sites because of differences in architecture, bark type, and epiphyte load, and this further limits the density of good establishment sites (Laman 1996c).

With all these obstacles to overcome, it does seem rather heroic when a tiny seed in a bird dropping grows into a giant strangler fig. Their hemi-epiphytic growth habit itself is highly specialized, but interestingly, different strangler fig species show even further specialization. At my research site in Borneo, there are 56 species in the genus *Ficus*, and 28 of them are stranglers in the subgenus *Urostigma* (Laman and Weiblen 1998). But among these 28 species growing in the same forest, there are clear differences in growth habits and in such parameters as preferred height of establishment in the canopy. Only seven species regularly become freestanding trees, while the rest invest less in supportive roots and rely on their host tree for continued support (Laman and Weiblen 1998). Some species show preference for establishment in the highest, sunniest canopy positions, whereas others do best growing on shady mid-canopy trees (Laman 1996c). We have much left to learn about how such specialization can evolve in a complex ecosystem like a rainforest.

Strangler figs may have some negative impact on host trees, but their most significant role in rainforest communities is providing year-round food resources for frugivorous vertebrates. Although individual trees ripen their crops of figs in synchrony, fruiting is

Figure 1 Researcher Cheryl Knott climbs a rope past the roots of a strangler fig (*Ficus stupenda*). This fig exhibits a single main descending root (on the front of host bole) with branching roots wrapping around the host trunk. The point of initial establishment of the fig on its host at 32 m above the ground is visible at the top of the frame where the pale fig trunk begins to ascend. Photograph by Tim Laman.

dispersed throughout the year at the population level. This has to do with the fig's relationship with mutualistic pollinator wasps, but its consequence for the vertebrate community is very significant since animals can reliably find fruiting figs even during seasons when other fruit is scarce. Thus, figs have been identified as "keystone species" in rainforests throughout the tropics (Lambert and Marshall 1991; Leighton and Leighton 1983; Spencer et al. 1996). Demons or heroes? Of course they are neither. Strangler figs are just another group of organisms that has found a unique way to make a living in the forest canopy.

Continued

STRANGLER FIG TREES: DEMONS OR HEROES OF THE CANOPY?–*cont'd*

References

Laman, T.G. (1995a). *Ficus stupenda* germination and seedling establishment in a Bornean rain forest canopy. *Ecology* **76**, 2617–2626.

Laman, T.G. (1995b). The ecology of strangler fig seedling establishment. *Selbyana* **16**, 223–229.

Laman, T.G. (1996a). *Ficus* seed shadows in a Bornean rain forest. *Oecologia* **107**, 347–355.

Laman, T.G. (1996b). The impact of seed harvesting ants (*Pheidole* sp. nov.) on *Ficus* establishment in the canopy. *Biotropica* **28**, 777–781.

Laman, T.G. (1996c). Specialization for canopy position by hemiepiphytic *Ficus* species in a Bornean rain forest. *J. Trop. Ecol.* **12**, 789–803.

Laman, T.G. and Weiblen, G.D. (1998). Figs of Gunung Palung National Park (West Kalimantan, Indonesia). *Tropical Biodiversity* **5**, 247–299.

Lambert, F.R. and Marshall, A.G. (1991). Keystone characteristics of bird-dispersed *Ficus* in a Malaysian lowland rain forest. *J. Ecol.* **79**, 793–809.

Leighton, M. and Leighton, D.R. (1983). Vertebrate responses to fruiting seasonality within a Bornean rain forest. *In* "Tropical Rain Forest: Ecology and Management" (S.L. Sutton, T.L. Whitmore, and A.C. Chadwick, Eds.), pp. 181196. Blackwell, Oxford, UK.

Spencer, H., Weiblen, G. and Flick, B. (1996). Phenology of *Ficus variegata* in a seasonal wet tropical forest at Cape Tribulation, Australia. *J. Biogeogr.* **23**, 467–475.

oldest portion of the shoot dies, they lose contact with soil (Figure 9-4). Primary hemi-epiphytes do the opposite, germinating on bark and later, with descending roots, access the forest floor. Most notable among the primary hemi-epiphytes are the stranglers of genus *Ficus* (see sidebar, "Strangler Fig Trees," and Figure 9-5). Provided sufficient time, a strangler can kill its host, but usually not before becoming robust enough to stand on its own roots. Hemi-epiphytes, and even more so the complete epiphytes (holo-epiphytes), fall into additional recognized subtypes based on criteria like growth habit, the type (if any) of their relationship with ants, tolerance for exposure, the presence or absence of phytotelmata (water-filled cavities), and so on (see Table 9-2; Benzing 1990).

Epiphytes native to the wettest forests deviate least in form and function from their terrestrial counterparts. Shoot and root systems develop more or less proportionally, and they exhibit conventional anatomy and morphology. Quite different arrangements prevail among the epiphytes suited for more stringent growing conditions. Here, appropriately modified shoots or root systems at once capture moisture and nutrients in addition to performing photosynthesis (see Figures 9-6, 9-7, 9-8, and 9-9 and the sidebar, "Orchid Adaptations to an Epiphytic Lifestyle"). Morphology promotes this capacity by allowing plants to accumulate debris and moisture in lieu of access to earth soil, or they possess special tissues that prolong contact with flowing precipitation, as illustrated by Orchidaceae with its velamentous roots (see the sidebar) and *Polypodium aureum* with its densely scaled rhizome (see Figure 9-9).

At the adaptive extreme, the plant is essentially "shootless" (e.g., *Taeniophyllum*; see Figure 9-10) or rootless (e.g., Spanish moss; see Figure 9-11), consistent with the need for pronounced material economy (see below) and the way the bodies of these highly abbreviated plants are almost completely exposed to the atmosphere (Benzing and Ott 1981). For reasons also elaborated below, few lineages beyond some bromeliads, ferns, and orchids have evolved the requisite qualities for growth where canopy-based substrates and climate challenge vascular epiphytism most severely. Even though the rest of the epiphytes are relegated to generally more permissive environments, many of them nevertheless utilize narrowly proscribed

Figure 9-6 *Neoregelia* sp. illustrating the bromeliad phytotelmata in Rio de Janeiro State, Brazil.

substrates (e.g., the ant-garden species), or they grow only where specific temperatures and humidity regimens prevail (e.g., Araceae in humid, warm forests).

Other aspects of form, physiology, and natural history definitely do not distinguish epiphytic from terrestrial vegetation, including the woody plants that mechanically support them. For example, the same coterie of fauna carry pollen for both kinds of flora. Tendencies differ, but usually only coincidentally. For example, epiphytes are often bird-dependent or reliant on some long-range insect for seed-set, owing to the over-representation of families like Gesneriaceae and Orchidaceae in forest canopy habitats (e.g., Nadkarni and Matelson 1989). Self-compatibility may over-occur among the epiphytes, but broader sampling is needed to confirm what remains still speculative (Bush and Beach 1995).

Epiphytes also require aerial vectors to disperse their seeds—seeds that tend to be smaller than those of related terrestrials and better able to defy gravity by more tenaciously adhering to aerial substrates (Benzing 1990). But here again the same groups of animals serve plants of both descriptions. Likewise, the epiphytes and terrestrials share the same devices and mechanisms to maintain water balance and perform photosynthesis, although important details often distinguish the ways that many of the epiphytes conduct these processes compared to their hosts (as described below).

All of this ambiguity about definitions and defining characteristics simply underscores what was previously stated: No single attribute identified so far consistently differentiates epiphytes from all of the other types of vegetation beyond their anchorage on bark. Nor does the epiphyte's habitat impose growing conditions that consistently differ from those encountered by plants of

Figure 9-7 *Crytopodium punctatum* on *Taxodium distichum* in south Florida, illustrating how a mass of negatively geotropic roots produces a "trashbasket" alternative to soil. Both the phorophyte and its epiphytes are in leafless winter condition.

other descriptions. But even though epiphytism fails in most respects to qualify as a discrete plant lifestyle, this fact in no way reduces its biological importance or lessens the justification for including the plants that practice this way of life in this volume.

The Forest Canopy as Plant Habitat

Chemical and physical variety, widely dispersed, relatively ephemeral substrates, and droughtiness better describe the conditions under which most epiphytes grow than those experienced by the terrestrial plants that support them.

Bark is an exceptionally heterogeneous rooting medium because its chemical composition and capacities to secure mechanically seeds and roots and sequester moisture differ among kinds of trees (Nicolai 1986). Moreover, all of these features change as weathering and aging progress. Bark also exposes the plants that use it to virtually all of the many microclimates that occur from the top of the canopy to its base (see Chapters 3 and 5; Bohlman et al. 1995; Freiberg 1997). It also assures that epiphytes more than terrestrial flora experience rapid change as dry and humid weather alternate. Last, its patchy distribution through the canopy and relatively high turnover further challenge the plants that require bark as their rooting medium (Benzing 1981).

Figure 9-8   Rosulate fern with impounded litter in Amazonas State, Venezuela.

Constraints imposed by substrates and climate powerfully elevate the material cost of epiphytism. Every life cycle stage is affected, although not evenly. First, bark being extraordinarily dispersed and aerial assures that most of the seeds produced to utilize it end up useless on the ground. Attrition continues every time a patch of bark upon which the exceptional seed managed to settle fails before the embryo inside can mature. Recruitment is further constrained by stress, particularly drought, through its suppression of photosynthesis and, ultimately, fecundity. The marked morphological reductions illustrated in Figures 9-9 and 9-10 plausibly constitute extreme responses to the high costs of maintaining populations on bark (Benzing and Ott 1981).

Co-occurring epiphytes often exhibit different adaptation for their shared lifestyle, and they sort themselves in space accordingly. Individuals rooted to twigs even in humid forests frequently experience desert-like conditions. Not so their counterparts rooted in nearby knotholes, on arboreal ant nests, or in soil-like mantles of lichens and bryophytes (see Figures 9-12 and 9-13). Some of these better-provisioned epiphytes deploy strategies (e.g., drought deciduousness, Figure 9-9) unsustainable for the plant obliged by stress to be more materially conservative (e.g., Figure 9-14), or they possess impoundments and other devices that enhance their access to H_2O and nutrients (e.g., Figures 9-6, 9-7, 9-8, and 9-9). Accommodation to what are on average at least moderately impoverished rooting media is evident in the multiple ways that so many epiphytes reduce their reliance on bark for all services save holdfast.

Many authors have documented how populations comprising mixed communities of epiphytes employ divergent means to counter stress and acquire resources from often unconventional

Figure 9-9 Deciduous *Polypodium aureum* (Polypodiaceae) in leafless winter condition in west central Costa Rica. Attached scales obscure the rhizome.

sources in order to partition forest canopies and even the crowns of individual trees (e.g., Pittendrigh 1948; Davidson 1988; Kernan and Fowler 1995; Ibisch 1996; Hietz and Briones 1998; Vandunne 2001). Patterns reminiscent of dense humid forests sometimes also occur on considerably shorter phorophytes in open, more arid woodlands where environmental gradients must be less pronounced (DeSouza Werneck and Do Espírito-Santo 2002).

Little is known about the quantities of nutrients and moisture available to epiphytes. The few "suspended soils" examined to date, although modest in volume, proved to be at least as fertile as many of those on the ground (e.g., Lesica and Antibus 1991; Nadkarni and Matelson 1991; Ingram and Nadkarni 1993; Hietz et al. 1999; Benzing 2000). Conversely, the "atmospheric" epiphyte anchored on naked bark or to a twig has only precipitation, mostly only while it flows by, and dry deposition for sustenance (see Figures 9-11 and 9-14). For these plants, essential resources arrive in short pulses instead of the more continuous streams afforded by a fertile, moist suspended soil or the contents of a capacious tank-shoot (Sheridan 1994; Zotz and Thomas 1999). Accordingly, epiphytes, because they experience a greater variety of growing conditions, are likely to augment more impressively the functional diversity of a forest than add to its biomass (Benzing 1995). Likewise, epiphytes are likely to provide associated biota novel goods and services as described in more detail below.

Even less is known about the bioactivity of canopy-based substrates than their fertility or moisture-holding capacities. Compared to mineral soil, bark is both organic and perhaps sufficiently

ORCHID ADAPTATIONS TO AN EPIPHYTIC LIFESTYLE

Wesley E. Higgins

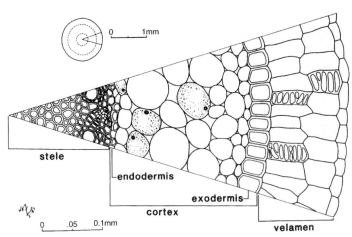

Figure 1   Transection of a *Restrepiella ophiocephala*. Drawing courtesy of Dr. Wendy Zomlefer.

The Orchidaceae originated as terrestrial forest understory herbs approximately 100 million years ago (Chase et al. 2001). The transition to an epiphytic canopy habitat required adaptations in plant morphology. Orchids have specialized adaptations in the roots, stems, leaves, and seed. Epiphytic orchids (e.g., *Restrepiella ophiocephala*) have no vascular connection to the host tree. The host supplies support only in a habitat that has more sunlight than the forest floor. Orchids absorb required nutriments from the surface of the host and rainwater.

Orchid roots serve as anchorage for the plant and take part in photosynthesis and water and nutrient uptake and storage. These adventitious roots typically arise from the rhizome. Orchid roots have a spongy layer of cells outside the exodermis known as the *velamen* that functions for temporary water storage (Pridgeon 1987). These cells rapidly absorb rainwater (and nutrients) and hold it until it can be translocated across the exodermis into the vascular system. Roots of epiphytic orchids are exposed to the light and the cells in the roots contain functioning chloroplasts. This is why wet orchid roots appear green in color. Velamen can also be found in Aroids that are adapted to an epiphytic habit.

Epiphytic orchids often have enlarged portions of the stem, called *pseudobulbs*, which are used for water and carbohydrate storage. The pseudobulb may form in one internode or it can consist of several internodes. The pseudobulbs swell or shrink as moisture is stored or withdrawn. This adaptation allows orchids to flourish in areas with seasonal rainfall where the plants experience months without rainfall. The pseudobulbs and leaves have a thick cuticle to reduce moisture loss.

The leaves of a plant are the primary photosynthetic organs that are sometimes modified for water storage. Some orchids have thick succulent leaves and no pseudobulbs. Orchids have a modified photosynthetic pathway as an adaptation to the dry canopy habitat. The opening of the stomata to take up carbon dioxide is always connected with large losses of water. To inhibit this loss, Crassulacean Acid Metabolism (CAM) includes a process that allows the uptake of carbon dioxide during the night when relative humidity is higher than during daylight periods. The prefixed carbon dioxide is stored in vacuoles and is used during the daytime for photosynthesis.

Continued

ORCHID ADAPTATIONS TO AN EPIPHYTIC LIFESTYLE–*cont'd*

Orchid seeds are adapted for wind dispersal. The dust-like seed consists of a tiny embryo and a net like testa. The seeds lack endosperm, the 3N tissues that typically feed a developing embryo. In orchids, when germination occurs, a mycorrhizal fungus penetrates the testa and feeds the embryo. This symbiotic relationship also occurs in the seed germination of terrestrial orchid species.

Although Orchidaceae is a member of the monocotyledons, the embryo lacks a cotyledon. However, orchids do have the general characteristics of other members of the lower Asparagales: mycorrhizal relationships, simultaneous microsporogenesis, sympodial growth, inferior ovary, sepal nectaries, and lateral inflorescences. Orchidaceae is one of the 84 families of vascular plant families that have species adapted to the epiphytic lifestyle (Kress 1986).

References

Chase, M.W., Freudenstein, J.V., and Cameron, K.M. (2001). DNA data and Orchidaceae systematics: a new phylogenetic classification. *The First International Orchid Conservation Congress Incorporating the 2nd International Orchid Population Biology Conference* **22**. Perth, Australia.

Kress, W.J. (1986). The systematic distribution of vascular epiphytes: an update. *Selbyana* **9**, 2-22.

Pridgeon, A.M. (1987). The velamen and exodermis of orchid roots. In "Orchid Biology: Reviews and Perspectives" (J. Arditti, Ed.), pp. 139–192. Cornell University Press, Ithaca, New York.

Figure 9-10   *Taeniophyllum* sp. a "shootless" orchid in Papua, New Guinea. The flat green roots are 3 to 4 mm wide.

Figure 9-11   Dense festoons of "rootless" *Tillandsia usneoides* (Bromeliaceae) growing on *Casuarina* sp. in southern Florida.

unaltered by processes similar to those that build soil to still contain nonnutritive substances that influence epiphyte welfare, especially during germination (e.g., Frei and Dodson 1972). These compounds might affect epiphytes directly as cytotoxins or secondarily by enhancing or inhibiting beneficial microbes like mycorrhizal fungi. Mycorrhizae of all types appear to be less routine among the epiphytes than soil-based flora, with the possible exception of Orchidaceae (e.g., Bermudes and Benzing 1989; Lesica and Antibus 1990; Janos 1993; Schmid et al. 1999). On the other hand, diverse fungi of mostly undocumented lifestyle associate with epiphytes (Richardson and Currah 1995). A final point concerns the possibility that epiphytism obliges exceptional physiology even where growing conditions don't always seem to warrant it.

Not surprisingly, epiphytes native to the most exposed micro-sites mature slowly, possess long-lived foliage, use water sparingly, and react sensitively to certain pollutants, indicating pronounced capacity to scavenge scarce nutrients and robust preparation for drought (Benzing 1990). But epiphytes that grow under more generally favorable circumstances tend to perform similarly on several counts. They, too, gain carbon and grow slowly, despite access to what appear to be relatively lavish supplies of moisture and nutrients, as if epiphytism per se mandates a stress-adapted phenotype (see the next section). In any case, more inquiry is needed to appreciate fully the forest canopy as a living space for epiphytes and to understand how conditions there have influenced the evolution of these plants and dictate how and where they grow today.

Figure 9-12   Ant nest supporting an ant-garden made up of members of at least five families in Amazonian Ecuador. The carton portion of the nest is about 40 cm wide at its broadest part.

Ecophysiology

Aspects of basic metabolism that vary in ways that adapt organisms to their environments constitute ecophysiology. All green higher plants, for example, contain the same pigments, electron carriers, and the enzyme rubisco; but how they use them to fix CO_2, how much H_2O is expended in the process, and whether PEP carboxylase is also involved varies among species. In effect, the ubiquitous apparatus for capturing energy and carbon has been fine-tuned to match the many conditions under which autotrophic plants grow. Certain features of the mechanisms that protect the plant's light-harvesting apparatus from excess excitation energy and those that allow it to accumulate mineral nutrients and much more also fall under the heading of ecophysiology.

Epiphyte ecophysiology, especially that of Bromeliaceae, is better known than for the members of just about any other category of vegetation (Benzing 2000). Particularly well studied are the most stress-tolerant types, especially those adapted to endure prolonged drought. It can now be said that CAM metabolism is more pervasive among the epiphytes than any other ecologically defined group of plants. It occurs in many degrees and under a variety of growing conditions, including unexpectedly wet ones. It permits the epiphyte so equipped to achieve high water use economy, to reduce photo-injury, and perhaps also to enhance N-use efficiency and carbon

Figure 9-13 Diverse humus epiphytes in wet forest in central Costa Rica.

gain where high humidity might otherwise reduce plant access to CO_2 (e.g., Holtum and Winter 1999; Pierce et al. 2001, 2002a, b; Fernandes et al. 2002; Haslam et al. 2002; Maxwell 2002; Zotz 2002).

One bromeliad in particular has received exceptional attention because it shifts between C_3 and CAM photosynthesis. Tank-forming *Guzmania monostachia* undergoes rapid transformations between high irradiance and shade-adapted phenotypes coincident with the flushing and shedding of foliage by its often deciduous hosts (Maxwell et al. 1995). Capacity for photochemical quenching and chlorophyll content and the amounts of xanthophyll cycle pigments present, as well as reliance on CAM versus C_3 photosynthesis also quickly adjust to prevailing light intensities and plant moisture status. Deeply shade-tolerant species may react to intense light in still other ways. The stomata of *Aspasia principissa* (Orchidaceae), for example, respond too slowly to allow sunflecks to account for as much of the carbon budget as some other understory flora, perhaps reflecting the overriding importance of water economy even to this trunk epiphyte (Zotz and Mikona 2003).

Another modest contingent of epiphytic bromeliads, ferns, and orchids shed their foliage during the dry season, i.e., are drought-deciduous (see Figures 9-7 and 9-9), and a still smaller group made up entirely of ferns tolerate frequent severe desiccation, i.e., are poikilohydrous (see Figure 9-15). It is the ferns more than any other vascular plant clade that warrant recognition for diverse and often novel modes of water balance, much of it associated with, if not adopted to, facilitate

Figure 9-14   *Tillandsia pruinosa*, an "atmospheric" bromeliad growing on unembellished bark provided by *Quercus virginiana* in southern Florida. Note the dense, light-reflecting cover of absorptive foliar scales that if wet too long kill the plant by preventing gas exchange.

epiphytism. Hietz and Briones (1998), for example, noted that co-occurring populations of Mexican cloud forest ferns partitioned the crowns of shared trees in part according to which mechanism for water balance they deployed of the many distinct types serving one or more member species of that single community.

Mounting evidence obtained primarily from bromeliads and orchids identifies the epiphytes as deeply stress-tolerant. Inherently slow growth is typically part of this syndrome, but perhaps more so here than usual (Schmidt and Zotz 2002). A decade or more is required to mature, even for the tank bromeliads with their substantial reservoirs of leaf-impounded H_2O and decaying, nutrient-laden debris (see Figure 9-6). Water seems to be the resource that limits growth most commonly, but the fact that subjects with thin, mesomorphic foliage and ample supplies of moisture also exhibit exceptionally low photosynthetic capacity suggests that additional factors account for their conservative performances. Physiology that shifts during the life cycle may further distinguish at least some epiphytes from much other flora, as may some important physical properties of leaves. Helbsing et al. (2000) reported the lowest cuticular permeabilities to water on record for a group of epiphytic aroids, peperomias, and orchids, even though the plants sampled inhabited a moist lowland Panamanian forest.

Additional study is needed to determine whether the inverse relationship between body size and growth rate, specifically relative growth rate, demonstrated for the tank bromeliad *Vriesea*

Figure 9-15   Poikilohydrous *Pleopeltis polypodioides* (Polypodiaceae) in dry condition on *Quercus virginiana* in central Florida.

sanguinolenta (Schmidt and Zotz 2001; Zotz et al. 2002) indeed distinguishes it and perhaps many other epiphytes from otherwise similar herbaceous, non-epiphytic flora. Whether comparable to other plants or not, the fact remains that physiology and development are coupled in some bromeliads in ways that foster "ontogenetic drift." Diminishing ratios of green to heterotrophic tissue as the shoot matures may account for this behavior. Any major change in the allocation of current photosynthate from the production of additional foliage to storage preparatory to flowering and sympodial branching would also slow growth. Many epiphytes, including most Bromeliaceae and Orchidaceae, produce determinate (monocarpic) ramets (shoots) by regular sympodial branching. Each ramet in these arrangements is programmed to produce a set number of leaves followed by one, usually apical inflorescence, and one or more axillary ramets after which it dies.

Heterophylly obliges ontogenetic drift in certain tank-forming Bromeliaceae (e.g., Adams and Martin 1986). In this case, water relations, accompanying shifts in CO_2 exchange, and leaf anatomy in tank-forming *Tillandsia deppeana* were the variables recorded. The authors noted that change in shoot form from non-impounding to impounding capacity increased the amount and reliability of the H_2O supply. Improved access to moisture in turn explained the observed transition to less xeromorphic leaf anatomy and greater vulnerability of photosynthesis to drought. Additional non-heterophyllic epiphytes may alter their resource supplies and likewise adjust their structure and function to match that reality. Given the rather spotty occurrence of canopy-suspended media that provide streams of moisture and nutrients as continuous as many earth soils can, we can reasonably expect more epiphytes to exhibit pronounced ecophysiological adjustments as they mature.

Mineral nutrition more conspicuously distinguishes the epiphytes from terrestrial flora than their modes of carbon and water balance or light relations. Moreover, environmental supply, not plant demand, makes the greater difference. Alternatives to earth soil in canopy habitats are numerous and widely exploited by epiphytes (see Figures 9-6, 9-7, 9-8, 9-9, 9-12, 9-13, and 9-16 and Table 9-2). Impoundments fashioned from roots or overlapping foliage allow hundreds of species representing at least five families to intercept litter and house the detritivores and sapro-phytes necessary to process it for plant use (see Figures 9-6, 9-7, and 9-8). Another group of epiphytes specializes on ant nests—seldom, if ever, rooting anywhere else (see Figure 9-12). Still other members of Asclepiadaceae, Bromeliaceae, Orchidaceae, Rubiaceae, and additional fami-lies entice ants to nest in plant organs provisioned with cavities (mymecodomatia) that permit them to use ant wastes for nutrition (Huxley 1978; see Figure 9-16). Carnivory, on the other hand, is rare perhaps because only the wettest forests afford the epiphyte the requisite high expo-sure and abundant moisture to render the use of prey cost-effective (Givnish et al. 1984).

Recent assessments using mass spectrometry have revealed that some epiphytes obtain N from sources other than those used by co-occurring terrestrial flora. Isotopic signatures specific to this element in tree litter and precipitation, and that fixed biologically allow determinations of how and from where different types of epiphytes obtain it (Steward et al. 1995; Hietz et al. 1999;

Figure 9-16   Ant-house (myrmecotrophic) *Tillandsia caput-medusae* (Bromeliaceae) in central Costa Rica. Dead members of the resident ant colony are also displayed.

TANK BROMELIADS: FAUNAL ECOLOGY

Barbara A. Richardson

Bromeliad plants are complex structures that provide a variety of compartments and ecological gradients for their animal communities. They are, therefore, true microcosms and not simple phytotelmata as they have often been regarded.

The bromeliads considered here have leaves in basal rosettes that retain water in the leaf axils. These rosettes contain a wide range of animals, from typical litter organisms to truly aquatic ones. The drier, outer area of the plant traps canopy litter, the resource base of the ecosystem. The litter is first processed by an army of terrestrial shredders and chewers that, in Puerto Rico (Richardson 1999), include isopods, diplopods, and cockroaches. In Dominica, however, they are largely replaced by iridescent blue worms (*Eutrigaster* sp., Octochaetidae) that may be used as a food source by an opportunist omnivorous bird, the trembler (*Cinclocerthia ruficauda*, Mimidae), often observed tearing bromeliads apart. Small particles of canopy debris are washed down between the closely appressed leaves, an area of increased humidity above the water meniscus, and a habitat for amphibious detritivores such as scirtid beetle larvae and adult hydrophilid beetles (*Omicrus ingens*, see Figure 1) and their carnivorous larvae (see Figure 2). The plant pools in the expanded leaf bases contain free-swimming organisms, mainly filter feeders such as ostracods, and culicid larvae that have important implications for human health (Frank 1983; Machado-Allison et al. 1985). Fine organic soil, faecal pellets, and bacteria collect at the base of the pools and the end-users of the system are mainly enchytraeid worms, harpacticoid copepods, and a variety of dipteran larvae, characteristically tipulids (*Trentepohlia* spp.), chironomids, and ceratopogonids.

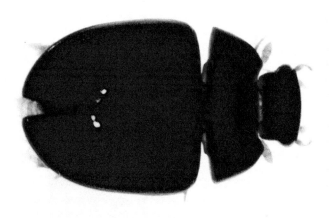

Figure 1 *Omicrus ingens*, Hydrophilidae, Tribe Omicrini. 2.5-2.8 mm. Black, glabrous, and shining dorsally, flattened ventrally. Endemic to bromeliads and to Puerto Rico.

Continued

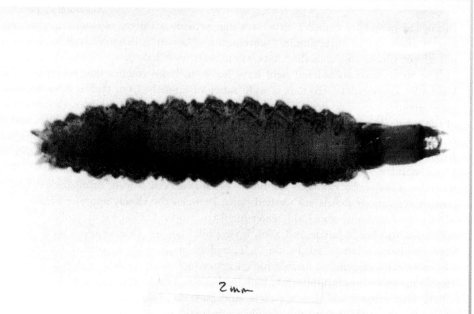

Figure 2   *Omicrus ingens* larva, 3.9-5.6 mm. pale yellowish gray, dorsal tergites brown.

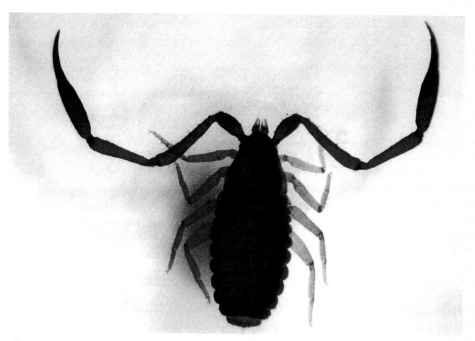

Figure 3   *Macrochernes attenuatus*, Pseudoscorpion, 2-5 mm. Endemic to bromeliads and to Puerto Rico.

In such a detritus-based system, there are few predators, mainly spiders and opilionids in the outer areas, pseudoscorpions and predatory beetle larvae (Elateridae, Staphylinidae, Pselaphidae) in the humid areas, free-swimming Chaoboridae larvae and occasional predatory mosquito larvae in the plant pools, and Tanypodinae larvae (chironomids) in the organic soil. Large predatory Odonata nymphs, mainly *Mecistogaster* spp., have been reported from mainland South and Central America (Picado 1913; Armbruster et al. 2002; Richardson and Srivastava, pers. obs.) and damselfly nymphs, *Diceratobasis macrogaster*, have been found in Jamaican bromeliads (Laessle 1961), but have not yet been found in other Caribbean islands. Picado (1913) reported bromeliad frogs in Costa Rica, dorso-ventrally flattened to fit into the leaf axils, but these have not been reported from the Caribbean. Many species, such as scirtid beetle larvae, show this same morphological adaptation, allowing them to move quickly between leaves, as do predators such as larvae of *Platycrepidius* sp. (Elateridae) and pseudoscorpions (*Macrochernes attenuatus*). This pseudoscorpion is endemic to bromeliads and to Puerto Rico and has specialized long pedipalps, allowing it to reach for prey (see Figure 3).

Bromeliad microcosms are discrete units, and animal communities and their resource base (canopy detritus and water) can be sampled and quantified (Richardson 1999); such complete sampling, not possible in larger ecosystems, can be used to answer a variety of ecological questions. Further, since bromeliads occur at all elevations and provide a constant habitat, they can be used to compare α- and β-diversity of their faunal communities within and among forests (Richardson 1999; Richardson et al. 2000).

Bromeliad plants behave as islands in that, within a particular forest, species richness and abundance of their animal communities are significantly and positively correlated with plant size, as is the amount of canopy debris collected by each plant. Species richness is unrelated to the total area of bromeliads available for colonization in that forest.

In Puerto Rico, α-diversity reflects forest net primary productivity (NPP) and mineral content of the debris, all declining with increasing elevation, when animal abundance measures such as Simpson's Diversity Index are used. In particular, abundance of large predators, such as spiders and elaterid larvae, is reduced at high elevation, consistent with the inability of the food chain to support them.

Animal species richness peaks in the mid-elevational palo colorado (*Cyrilla racemiflora*) forest, and is independent of abundance, declining NPP, and declining mineral concentrations with elevation. This unimodal pattern at mid-elevation is found in plant and animal assemblages in other forests (Rosenzweig 1995). It suggests that factors other than nutrient concentrations are important in determining animal species assemblages, such as the diversity of plant species, spatial heterogeneity of the forest, and historical and evolutionary factors. Many of the species with larvae in bromeliads spend their adult lives in the larger forest ecosystem, come under its influences, and contribute to its food chains. Mid-elevational forests are usually benign, drying out less than lower forests and not subject to the extreme water-logging, leaching, and lower temperatures of cloud forests. Survivorship may be higher and extinction of small populations less likely.

Bromeliad plants are long-lived and can support many generations of invertebrates that pass into forest food chains. In forests where they are found in high density, their contribution may be considerable. They contribute to forest diversity, as the constraints of bromeliad architecture and competition for standing water has favored the evolution of bromeliad specialists, some of which are endemic to particular localities or islands.

Continued

TANK BROMELIADS: FAUNAL ECOLOGY–*cont'd*

References

Armbruster, P., Hutchinson, A., and Cotgrave, P. (2002). Factors influencing community structure in a South American tank bromeliad fauna. *Oikos* **96**, 225–234.

Frank, J. H. (1983). Bromeliad phytotelmata and their biota, especially mosquitoes. In "Phytotelmata: Terrestrial plants as hosts for aquatic insect communities" (J.H. Frank and L.P. Lounibos, Eds.), pp. 101-128. Plexus Publishing Inc., Medford, New Jersey.

Laessle A.M. (1961). A micro-limnological study of Jamaican bromeliads. *Ecology* **42**, 499–517.

Machado-Allison, C.E., Barrera, R.R., Frank, J.H., Delgado, L., and Gomez-Cova, C. (1985). Mosquito communities in Venezuelan phytotelmata. In "Ecology of Mosquitoes" (L.P. Lounibos, J.R. Rey, and J.H. Frank, Eds.), pp. 79-93. Proceedings of a Workshop, Florida Medical Entomology Laboratory, Vero Beach, Florida.

Picado, M.C. (1913). "Les broméliacées épiphytes considérées comme milieu biologique." Doctoral Thesis, faculty of Sciences, University of Paris.

Richardson, B.A. (1999). The bromeliad microcosm and the assessment of faunal diversity in a neotropical forest. *Biotropica* **31**, 321–336.

Richardson, B.A., Richardson, M.J., Scatena, F.N., and McDowell, W.H. (2000). Effects of nutrient availability and other elevational changes on bromeliad populations and their invertebrate communities in a humid tropical forest in Puerto Rico. *Journal of Tropical Ecology* **16**, 167–188.

Rosenzweig, M.L. (1995). "Species Diversity in Space and Time." Cambridge University Press, Cambridge.

Wania et al. 2002). Nitrogenase activity prevails in the phytotelmata of certain bromeliads and aroids (see sidebar; Bermudes and Benzing 1991; Sheridan 1994). Ant-gardened species may benefit from more intimate associations with the same microbes if the discovery by Cedeño et al. (1999) that cyanobacteria form dense sheaths around the fine roots of nest-dwelling *Philodendron deflexum* is representative and that this prokaryote indeed fixes N_2.

Answers to questions about the importance of mycorrhizae in canopy habitats, aspects of mineralization there, and whether certain epiphytes possess nutrition to match the chemical peculiarities of their substrate are not available yet. Circumstantial evidence, however, indicates some interesting possibilities. Consider, for example, the over-representation of families like Ericaceae in the tropical forest canopy. Being predominantly woody certainly does not predispose this taxon for epiphytism (see Table 9-3), but perhaps penchants for certain growing conditions do. Suspended soils in wet, cool montane forests where the diversity of epiphytic Ericaceae peaks may feature low pH and considerable N in organic form much as terrestrial flora experience in certain boreal habitats where this family is also well represented.

Phytosociology and Demography

Epiphyte diversity ranges from zero to hundreds of species per hectare across the tropics, with climate serving as its most reliable predictor (e.g., Pittendrigh 1948; Gentry and Dodson 1987b; Andreade and Park 1997; Griffiths and Maxwell 1999; Vandunne 2001). Humidity more than any other climate variable favors species richness, up to 107 species (Valdivia 1977) on a single tree where dense evergreen canopies feature multiple microclimates and numerous alternatives to earth soil. Specializations for narrowly defined substrates (e.g., ant carton, suspended humus, naked bark; see Table 9-2) and the distributions of these media and microclimates as described

Table 9-3 Traits, Many of Which Are Homoplasious (Evolved Repeatedly),
that Characterize Epiphytes More Often Than They Occur Among Plants in General

I. Vegetative morphology	1. Herbaceous and evergreen
	2. Sympodial branching with, short determinant ramets
	3. Sclerophylly and succulence (xeromorphy)
	4. Anatomy and morphology that prolong contact with nutritive fluids and solids (see sidebars)
II. Ecophysiology	1. High water use efficiency
	2. Slow growth
	3. Robust photo protective capacity
	4. CAM photosynthesis
	5. Low inherent photosynthetic capacity
	6. Mycorrhizae (vescicular at least) uncommon
III. Life Cycle	1. Perennial with prolonged juvenile phase
	2. Polycarpy (iteroparity)
	3. Numerous small seeds
	4. Autogamous breeding systems?

below often structure communities of epiphytes and allow them to accommodate extraordinary high concentrations of species.

Epiphytes contribute to overall floristic diversity differently depending on geographic scale. Species-area curves rise steeply compared to those for co-occurring terrestrial flora, perhaps more so in montane compared to lower elevation forests (Ibisch 1996; Nieder et al. 2000). Epiphytes can comprise fully half of the vascular flora in small plots (<0.1 ha.), but representation falls precipitously as the censused area expands. Modest body size explains in part the disparity between fine scale (alpha-diversity) and regional scale (beta-diversity), but other factors must explain why, on average, the ranges of species tend to expand as elevation decreases below about 400 to 600m.

Features that help vascular epiphytes recruit substrates in broadly three-dimensional habitats include seeds that contain ant-attracting chemicals (the ant-nest garden epiphytes; Seidel et al. 1990) and fleshy fruits, some of which explain peculiar fine-grain distributions, such as the fidelity of several Brazilian *Billbergia* species to the sheathing bases of palm fronds that also provide daytime refuges for frugivorous bats. Exclude the orchids, virtually all of which produce minute "dust-type" seeds, and the epiphytes constitute a largely animal-dispersed group. How many additional epiphytes also rely on behavioral peculiarities of their seed disperser to utilize narrowly circumscribed substrates is unknown.

Many epiphytes require high exposure and others, like certain filmy ferns, cannot endure either as much sun or the associated aridity (e.g., Hietz and Briones 1998). Consequently, epiphytes segregate along environmental gradients as well as among different kinds of substrates. At the same time, few species exhibit pronounced specificity for the kinds of trees that support them (but see Díaz Santos 2000 and Merwin et al. 2003). Instead, a phorophyte that hosts one species of epiphyte generally accommodates many of the other local populations (i.e., epiphytes tend to be gregarious) (Benzing 1990). Other trees within range of the same dispersing seeds remain completely epiphyte-free. The basis for this kind of discrimination can be obvious, such as smooth or unstable bark. Additional, more subtle possibilities include allelopathy and the absence of the fungi that all orchids need to germinate.

Spacings exhibited by the members of mixed populations of epiphytes argue against competition as a major determinant of community structure and composition (e.g., Benzing 1981; Hietz and Briones 1998). Instead, the combined effects of disturbance, scattered substrates, and high seed mobility probably foster coexistence and perhaps also allow greater ecological equivalence than usual among populations that share rooting media (e.g., several *Tillandsia* spp. on naked bark, Benzing 1981; Catling and Lefkovitch 1989). In effect, minimal antagonism among residents with similar requirements for growth and lottery-style recruitment of patchy, relatively ephemeral living spaces may militate against structure within many communities of epiphytes. Noninterference seems to be most pronounced where little of the local rooting medium is particularly hospitable (e.g., dry vs. wetter forest) or endures long enough to permit epiphytes to reach densities that would force competition.

Species-specific needs for specific kinds of rooting media and tolerances for narrowly defined combinations of exposure and humidity definitely impose coarse-grained structure (e.g., vertical stratification) within many epiphyte communities, but what about more subtle agencies that might impose finer-scale patterns? Differences in the wetability of bark, the patchwork distributions of rain shadows and stem flow channels, and uneven chemical/physical conditioning by weather and previous occupants may translate into less conspicuous, but still decisive, habitat heterogeneity for epiphytes, Whatever the cause, Catling and Lefkovitch (1989) reported early and later developing assemblages of bromeliads and orchids in citrus orchards in Belize. Ibisch (1996) noted serial occupancy by epiphytes on *Alnus acuminata* in wet montane forest in Peru as did Merwin et al. (2003) in a Costa Rican wet forest. Regular progressions of species also characterize at least some ant-nest gardens in tropical America (Catling 1995).

Hietz and Hietz-Seifert (1995a, b), among others, used nearest neighbor analysis to demonstrate that many epiphytes distribute non-randomly within their communities. Sorting out the operative agencies(s) from the many possibilities responsible for aggregation (e.g., low seed mobility and the nurse and seed-trapping effects of existing plants, in addition to substrate preferences) poses a more daunting challenge. Efforts to quantify the impacts of disturbance on community structure have yielded similarly equivocal results (e.g., Robertson and Platt 1991; Rosenberger and Williams 1999) and not unexpectedly. Like most constraints on populations, disturbance— or more specifically, its biological consequence—is target-defined, hence variable. Put more directly, perturbation of a given frequency and spatial scale will affect a population more or less adversely depending on certain life history characteristics (e.g., seed mobility, growth rate) of its membership, all of which range widely among epiphytes.

Epiphytes also vary in their capacities to moderate canopy substrates for other users. Tank-forming epiphytes, especially bromeliads, accumulate much wet humus in their leaf axils, thereby creating rooting media for seedlings more drought-sensitive than their own (see Figures 9-6 and 9-7). Water seeping from leaky phytotelmata increases the area improved for other biota, particularly from species with exceptionally leafy shoots (see Figure 9-3). The root mass of *Immatophyllum palmatum* (Orchidaceae) attracts gravid queens seeking nest sites for new ant colonies in Central America (Catling 1995). As her brood and nest increase in number and size, additional ant-nest garden epiphytes appear. Arboreal ants may also influence the welfare of epiphytes that lack evident myrmecophytism. Catling (1997) demonstrated high associations between some members of a diverse community of epiphytes and the presence of arboreal ant nests in trees in Belize.

Much effort has been invested to characterize the demography of epiphytes, mostly bromeliads and orchids (e.g., Tremblay 1997; Mondragón et al. 1999; Winkler and Hietz 2001; Hietz et al. 2002). Growth rates of individuals and populations generally accord with the findings on water balance, photosynthesis, and light responses described in the section devoted to ecophysiology. Just about every epiphyte examined so far exhibited high stress-tolerance and low vigor, but again, are these findings characteristic? Are the mostly bromeliads and orchids studied to date both representative for epiphytes in general and truly exceptional for their sluggish growth? Might comparisons with similarly constructed (herbaceous perennials) flora native to oligotrophic bogs, for example, dispel the growing impression that epiphytes per se exhibit unusually modest capacities for carbon gain combined with unusually robust mechanisms for photo-protection? Probably not, but this condition may describe more of the epiphytes than flora native to most other kinds of habitats.

Taxonomists suggest positive relationship between speciation and epiphytism. Big families with substantial epiphytic and terrestrial memberships like Araceae, Bromeliaceae, and Orchidaceae include numerous genera, many of the largest of which contain primarily epiphytes (e.g., *Anthurium* in Araceae, *Tillandsia* and *Vriesea* in Bromeliacae, and *Pleurothallus* and *Bulbophyllum* in Orchidaceae). These associations between ecological habit and clade size accord with extensive, recent radiations favored by epiphytism. Exuberant speciation unaccompanied by divergence in the characters that dictate both medium and climate conditions for plant growth would produce large groups of closely related epiphytes with similar body plans and close ecological equivalency. This possibility draws circumstantial support from several parts of the angiosperm complex, for example, from the roughly 4000 species-strong orchid Subtribe Pleurothallidinae, most members of which inhabit cool humid Andean habitats as fly-pollinated epiphytes and lithophytes.

The possibility that epiphytism encourages speciation has provoked considerable published discourse (e.g., Hietz and Hietz-Siefert 1995a; Ibisch et al. 1996; Tremblay 1997; Nieder et al. 2000), although nothing definitive yet. Epiphytes are clearly more sensitive to climate than soil-rooted vegetation—witness their precipitous declines in diversity along geographic rainfall gradients compared to co-occurring trees, shrubs, and terrestrial herbs (Gentry and Dodson 1987b). Moreover, herbarium records suggest, and plant collectors often report, that many epiphytes exhibit narrow geographic ranges and that closely related populations often occupy different parts of the same multiply dissected life zones (e.g., in Andean South America; Nieder et al. 2000).

A comparison of montane epiphytes with related terrestrial flora by Ibisch et al. (1996) in Peru, however, does not support epiphytism as impetus for speciation, nor does Kessler's (2002) analysis of range sizes (larger for epiphytes than terrestrial species) and ecological correlates among Bolivian Bromeliaceae. Perhaps conditions that promote radiations among canopy-based flora exceed the norm only where certain climates prevail or local substrates exhibit characteristics conducive to the same outcomes. More comprehensive collecting and analyses of population genetics should help resolve this issue and assist conservation efforts (see sidebar in Chapter 5 this volume).

Interactions with Fauna

Epiphytes interact with animals in all of the ways that promote reproductive success among seed plants generally, plus some additional ones. The relevant question for us is whether any of these relationships are novel or important enough to warrant mention.

It seems unlikely that the epiphytes on average depend any more or less than other plants on animals to disperse their pollen and seeds. Pollination syndromes exhibited by certain epiphytic orchids rank among the most elaborate of all, but quite a few terrestrial members of the same family engage in similar sorts of arrangements. As mentioned above, the same broad collection of insects, birds, and bats seems to provide reproductive services for the epiphytes, as for the trees upon which they grow and the herbs that root below them on the forest floor. Still, some surprises are possible. Virtually nothing is known about pollination in some sizable collections of generally little studied species (e.g., *Peperomia*). No epiphyte reportedly sheds wind-borne pollen, but then anemophily is generally uncommon in the humid tropical forest.

Epiphytes also experience herbivory, although whether above or below levels sustained by adjacent vegetation remains poorly studied (e.g., Stuntz et al. 1999; Schmidt and Zotz 2000). Epiphytes, as relatively stress-tolerant flora, tend to produce long-lived leaves equipped with enough mechanical and water storage tissue to render them possibly less nutritive than much of rest of the forage present in the forest. Stuntz and Zotz (2001) report exceptionally low concentrations of N in the foliage of diverse epiphytes, consistent with their equally modest capacities for photosynthesis. Virtually nothing is published on the presence of feeding deterrents or more powerful toxins. Epiphytes are spared the attention of large grazing animals, few of which climb tress, but they sometimes incur considerable damage from primates and other arboreal mammals seeking animal prey (personal observation).

Epiphytes increase opportunities for canopy-based fauna simply by providing space and by humidifying the canopy atmosphere (Freiberg 2001; Stuntz et al. 2002). Sometimes they contribute especially high-quality habitat, such as the moist humus that accumulates in the tanks of bromeliads and the shoots of bird's nest ferns and aroids (see sidebar). A substantial literature documents the high abundance and diversity of the fauna that populate bromeliad phytotelmata, and the other cavities that many other epiphytes offer (e.g., Stork 1987; Paoletti et al. 1991). Morphology and behavior indicate that some of these plant users (e.g., some dendrobatid frogs and crustaceans) have long been dependent on epiphytes for breeding and hiding places (e.g., Diesel and Schuh 1993; Richardson 1999; Stuntz et al. 2001).

Plants and ants engage in a variety of symbioses across a wide range of habitats, but nowhere are these relationships more diverse and refined than in the tropical forest. Virtually all of the ant-fed (myrmecotrophic) plants are epiphytic (see Figure 9-16 and Table 9-2), and the same ants that these plants house sometimes also protect adjacent vegetation from herbivores (e.g., Dejean et al. 1995). The ant-nest garden phenomenon is also exclusively arboreal, and at some sites all of the dominant tree-dwelling species cultivate epiphytes (Wilson 1987; see Figure 9-12). If plants are as essential to maintain the physical integrity of the ant-nest, as one study suggests (Yu 1994), then this small subset of ant-cultivated epiphytes alone plays an inordinately important role in the biology of many neotropical forests (see Figure 9-17).

Epiphytes occur at high enough densities in certain wet forests to blur substantially distinctions between the canopy and understory (Paoletti et al. 1991; see Figures 9-1, 9-13, and 9-18). Enough living and dead plant material sometimes accumulates on bark to accommodate earthworms and other invertebrates that more commonly inhabit the upper humic horizons of terrestrial soil (e.g., Fragoso and Rojas-Fernandez 1996). In a real biological sense, epiphytes, including nonvascular types, at the densities promoted by continuous high humidity add a third dimension to a medium usually viewed as largely two dimensional and confined to the base of the forest.

Epiphytes definitely support forest-dwelling fauna, but to what extent their users partition the resources they provide and how these plants compare with other forest flora on this basis remains obscure. Stuntz et al. (in press) noted in a Panamanian forest that more individuals, but not more ant species, inhabited *Annona glabra* hosts when these trees also supported vascular epiphytes. Another study by Stuntz et al. (2001) that focused on the epiphytes themselves revealed that they attract numerous kinds of arthropods (e.g., beetles, isopods, larval Diptera, spiders), many with less-flexible lifestyles than ants. Moreover, species representing distinct guilds strongly associated with one or another of the two bromeliads and the single orchid surveyed. Should this pattern of plant preference be broadly characteristic, we have even greater reason to believe that epiphytes not only promote biodiversity, but do so in varied and complicated ways.

Influences on Basic Forest Processes

Epiphytes influence several processes fundamental to forest function because of how and where they grow. Location is important because they straddle the biogeochemical pathway connecting the compartment that contains most of the nutrients tied up in biomass (the trees) at a site

Figure 9-17 Twig specialist *Ionopsis utricularioides* (Orchidaceae) in Ecuador. The supporting axis is about 1.5 cm in diameter.

with the compartment (soil) that contains most of the nutrients awaiting redistribution by plants through the same forest ecosystem (see Figure 9-19). Diverse by nutritional mode, epiphytes can extract elements like N and P from all of the vehicles—specifically, precipitation, canopy washes, and litter—that convey them from the largest pool of plant-incorporated nutrients in the typical tropical forest back to the soil.

Nadkarni (1981) posited (now well confirmed, e.g., Clark et al. 1998) that a well-developed community of epiphytes can enhance the nutrient trapping capacity of a forest ecosystem, increase its storage volume, and can act as a "nutrient capacitor," steadily releasing growth-limiting ions for use by other flora. Additionally, she noted that this arrangement has fostered a conspicuous evolutionary response by competing vegetation. Certain trees in some temperate and tropical humid forests produce aerial roots that ramify through the thick mats of epiphytes and supporting humic soils suspended in their crowns. Presumably, this arrangement allows them to absorb incoming and recycling nutrients more directly, if not more efficiently, compared to uptake later from the forest floor. Exactly how epiphytes impact co-occurring biota, however, depends on a variety of site-specific circumstances.

Should a forest ecosystem be exceptionally infertile, large masses of epiphytes might tie up large enough fractions of one or more scarce nutrients to stress co-occurring vegetation, particularly the trees that support them (Benzing and Seemann 1978). Should the resident epiphytes reduce carbon gain by casting shade and denying nearby, more productive vegetation scarce

Figure 9-18 Dense colonies of bryophytes in the canopy of a subtropical cloud forest tree in North Island, New Zealand.

nutrients, whole-forest photosynthesis will fall below potential. Compared to terrestrial flora, with its more continuously available soil-based source of moisture, epiphytes use water sparingly, consistent with their less reliable supply. Consequently, they accumulate carbon more slowly. Although N and P invested in the relatively long-lived foliage characteristic of most epiphytes remain in place to support photosynthesis longer, the same nutrients, even if incorporated for shorter intervals in more productive foliage elsewhere in the forest, would foster higher productivity.

On the other hand, diverse and abundant epiphytes may allow the heavily colonized forest to deploy certain scarce resources more effectively than possible if it were epiphyte-free. Recall that epiphytes, often including residents of the same forests, differ substantially among themselves and from their hosts relative to water use-economy, shade-tolerance, and mineral-use efficiency. Appropriate measurements could determine the effectiveness with which a canopy laden with diverse epiphytes uses light and deploys key nutrients to support photosynthesis compared to systems of similar chemical/physical organization (e.g., comparable by canopy height, leaf area index, investments of N in foliage), but without epiphytes.

Epiphytes also humidify the canopy, and not always solely by transpiration. Fish (1983) reported that the impounding shoots of epiphytic bromeliads in montane forest in Colombia held 50,000 l of H_2O per canopy hectare (see Figure 9-6). Reservoirs of this magnitude and probably smaller ones must significantly mitigate the effects of the inevitable, if brief, droughts on particularly sensitive flora like the bryophytes that so densely inhabit the canopies of many cloud

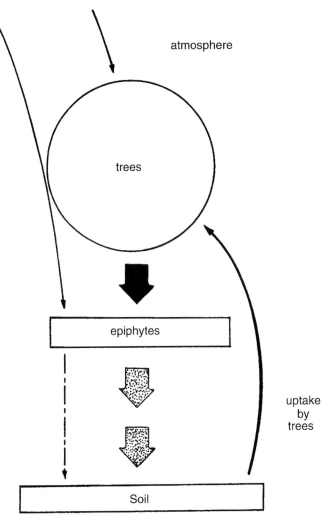

Figure 9-19 Nutrient cycling in forest ecosystems with abundant epiphytes. Arrows indicate directions of nutrient flux, the circle and rectangles pools of the same nutrients (see text for additional details).

forests (see Figure 9-18; Freiberg 1997; Stuntz et al. 2002). Stomata also sense atmospheric humidity and during dry weather can depress diffusive conductance enough to reduce photosynthesis below rates possible for unaffected foliage.

In essence, enough is already known about epiphytic vegetation to conclude that these plants capture and use water, light, and nutrients in ways distinct from co-occurring soil-rooted vegetation. For reasons related in part to these differences, they potentially humidify the canopy, modify rates of nutrient flux, and increase the number of connections among nutrient-containing compartments enough to influence forest structure and function. Viewed somewhat differently, epiphytes add mechanisms for photosynthesis and water management to many forests that most biologists associate with distinctly more arid ecosystems. Without doubt, epiphytes can vastly increase the functional diversity and complexity of the ecosystems they inhabit, but determining how much they influence hydrology, biogeochemical cycling, and energetics will require a systems approach using data of the type obtained from the more narrowly targeted inquiry largely pursued until now.

Conservation and Global Change

Considering what is known about epiphyte ecophysiology and demography, nonvascular and vascular types alike should respond sensitively to climate change (Benzing 1998). Less predictable are the fertilizing effects of elevated CO_2 and the growing supply of reactive nitrogen also released by fossil fuel combustion. Many vascular epiphytes are indisputably threatened by the widespread destruction of tropical ecosystems, particularly those that inhabit especially vulnerable types like rainforests and montane cloud forests (Benzing 1998; Foster 2001). Quite likely, hundreds of species have already disappeared unrecorded as casualties of human activity, and prospects for many of the survivors are dim.

Epiphytes are probably exceptionally sensitive to climate for complex reasons, not all of which relate to death by desiccation. Recall that many epiphytes conduct CAM, and while this mechanism promotes economical water use and other benefits, it increases vulnerability to perturbation, including climate change, by limiting growth rate and reducing tolerance to shade. Other adaptations to aridity narrow tolerance in a different direction. Some of the most specialized xerophytic species (e.g., "atmospheric" Bromeliaceae; see Figure 9-15; Benzing et al. 1978) suffocate if provided excess moisture because they possess dense layers of hydrophilic foliar trichomes. The velamen of the orchid root can act similarly where moisture is too plentiful (Benzing and Ott 1981; sidebar).

Interest in epiphyte conservation is increasing partly for practical reasons, particularly the desire to develop strategies to harvest sustainable plants for sale. Recent inquiries have also dealt with the roles as refugia of forest fragments and isolated trees spared to provide shade for pastured animals (e.g., Williams-Linera et al. 1995; Hietz-Seifert et al. 1996; Wolf and Konings 2001; Nkongmeneck et al. 2002). At issue is whether the epiphytes that survive land use conversion on these phorophytes embody enough of the genetic diversity formerly present among the occupants of intact forest to reestablish viable populations should woodlands be encouraged to regenerate. In essence, do the relatively arid, sunny conditions that prevail in the crowns of isolated individuals and small clusters of trees favor only the most stress-tolerant members of epiphyte populations that generally inhabit darker more humid spaces? Or perhaps the individuals remaining on isolated trees possess sufficient phenotypic plasticity to inhabit all of the kinds of spaces formerly occupied by their relatives in undisturbed forest.

The good news about the future of vascular epiphytes concerns their invasiveness, more specifically, the absence of this tendency beyond the rare exception (e.g., several hemi-epiphytic South American *Ficus* in southern Florida). None of the thousands of widely traded ornamental and epiphytic aroids, bromeliads, cacti, Gesneriaceae, orchids, and members of many additional families has escaped large-scale from cultivation. Routinely animal- or wind-dispersed seeds suggest that something other than insufficient mobility accounts for the near complete absence of invaders in the forest canopy. Perhaps another shared life history characteristic or something about the suitability of aerial habitats for free-living plants precludes the same invasiveness exhibited by so much terrestrial flora.

Summary

Vascular epiphytism is essentially a subtropical/tropical phenomenon, but often a highly significant one in humid regions. In addition to being notably influential where they occur abundantly, many epiphytes display biology consonant with climates that slow photosynthesis and substrates that impose high mortality. The most specialized species compensate for their lack of access to earth soil with devices and mechanisms that in the case of nutrition often involve joint enterprise between plants and symbionts, particularly ants. The vascular epiphytes also exhibit well-developed capacities to tolerate drought and avoid photo-injury, sometimes demonstrating more competence in these performances than circumstances seem to warrant.

Most of the adaptations that foster epiphytism also occur among terrestrial flora, and many of them are broadly homoplasious. Biological novelty, in so far as it distinguishes the epiphytes, seems to involve resource acquisition, the use of fauna in this endeavor, and unusual structure and function that promote high material economy biased to favor reproduction. Growing conditions in the canopies that epiphytes inhabit range from relatively permissive to intolerable for higher plants. Epiphytes that occur where canopy-based substrates and climate most severely challenge the capacities of vascular plants to maintain populations belong to few families. These mostly bromeliads and orchids succeed because they possess features that maximize material economy and promote exceptional access to scarce resources. The ferns that root alongside them do so in part because they can also tolerate desiccation lethal to most vascular plants.

Epiphytes provide incompletely inventoried and sometimes unique products and services to other forest-dwelling biota. As distinctly stress-tolerant plants that utilize unconventional sources of moisture and nutrients, they also contribute adaptive counterpoint to the communities they inhabit, the more so the more mesophytic the hosting forest. Additionally, epiphytes are well disposed by structure and function and location in the canopy to influence the nature and dynamics of entire forest ecosystems. To what extent they do this remains largely speculative. Finally, vascular epiphytes should receive more attention if only because they grow abundantly in several kinds of threatened habitats and are probably exceptionally vulnerable to climate.

References

Adams, III, W.W. and Martin, C.E. (1986). Physiological consequences of changes in life form of the Mexican epiphyte *Tillandsia deppeana* (Bromeliaceae). *Oecologia* **70**, 298–304.

Andrade, J.L. and Park, N. (1997). Microhabitats and water relations of epiphytic cacti and ferns in a lowland neotropical forest. *Biotropica* **29**, 261–270.

Bennett, B. (1991). Comparative biology of neotropical and saxicolous *Tillandsia* species: population structure. *J. Trop. Ecol.* **7**, 361–371.

Benzing, D.H., (1981). Bark surfaces and the origin and maintenance of diversity among angiosperm epiphytes: an hypothesis. *Selbyana* **5**, 248–255.

Benzing, D.H. (1990). "Vascular Epiphytes." Cambridge University Press, Cambridge, UK.

Benzing, D.H. (1995). The physical mosaic and plant variety in forest canopies. *Selbyana* **16**, 159–168.

Benzing, D.H. (1998). Vulnerabilities of tropical forests to climate change: the significance of resident epiphytes. *Climate Change* **39**, 519–540.

Benzing, D.H. (2000). "Bromeliaceae: Profile of an Adaptive Radiation." Cambridge University Press, Cambridge, UK.

Benzing, D.H. and Ott, D. (1981). Vegetative reduction in epiphytic Bromeliaceae and Orchidaceae: its origin and adaptive significance. *Biotropica* **13**, 131–140.

Benzing, D.H. and Seemann, J. (1978). Nutritional piracy and host decline: a new perspective on the epiphyte-host relationship. *Selbyana* **2**, 133–148.

Benzing, D.H., Seemann, J., and Renfrow, A. (1978). The foliar epidermis in Tillandsioideae (Bromeliaceae) and its role in habitat selection. *Am. J. Bot.* **65**, 359–365.

Bermudes, D. and Benzing, D.H. (1989). Fungi in neotropical epiphyte roots. *Biosystems* **23**, 65–73.

Bermudes, D. and Benzing, D.H. (1991). Nitrogen fixation in association with Ecuadorian bromeliads. *J. Trop. Ecol.* **7**, 531–536.

Bohlman, S.A., Matelson, T.J., and Nadkarni, N.M. (1995). Moisture and temperature patterns of canopy humus and forest floor soil of a montane cloud forest, Costa Rica. *Biotropica* **27**, 13–19.

Bush, S.P. and Beach, J.H. (1995). Breeding systems of epiphytes in a tropical montane wet forest. *Selbyana* **16**, 155–158.

Catling, P.M. and Lefkovitch, L.P. (1989). Associations of vascular epiphytes in a Guatemalan cloud forest. *Biotropica* **21**, 35–40.

Catling, P.M. (1995). Evidence of partitioning of Belizean ant nest substrates by a characteristic ant flora. *Biotropica* **27**, 535–537.

Catling, P.M. (1997). Influence of aerial *Azteca* nests on the epiphyte community of some Belizean orange orchards. *Biotropica* **29**, 237–242.

Cedeño, A. Mérida, T., and Zegarra, J. (1999). Ant gardens of Surumoni, Venezuela. *Selbyana* **20**, 125–132.

Clark, K.L., Nadkarni, N.M., Schaefer, D.S. and Gholz, H.L. (1998). Atmospheric deposition and net retention of ions by the canopy of a tropical montane forest, Monteverde, Costa Rica. *J. Trop. Ecology* **14**, 27–45.

Davidson, D.W. (1988). Ecological studies of neotropical ant gardens. *Ecology* **69**, 1138–1152.

Dejean, A., Olmsted, I., and Snelling, R.R. (1995). Tree-epiphyte-ant relationships in low inundated forest of Sian Ka'an Biosphere Reserve, Quintana Roo, Mexico. *Biotropica* **27**, 57–70.

DeSouza Werneck, M. and Do Espírito-Santo, M.M. (2002). Species diversity and abundance of vascular epiphytes on *Vellozia piresiana* in Brazil. *Biotropica*. **34**, 51–58.

Díaz Santos, F. (2000). Orchid preferences for host tree genera in a Nicaraguan tropical rain forest. *Selbyana* **21**, 25–29.

Diesel, R. and Schuh, M. (1993). Maternal care in the bromeliad crab *Metopaulias depressus* (Decapoda): maintaining oxygen, pH, and calcium levels optimal for larvae. *Behav. Ecol. Sociobiol.* **32**, 1–15.

Fernandes, J., Chaloub, R.M., and Reinert, F. (2002). Influence of nitrogen supply on the photosynthetic response of *Neoregelia cruenta* (Bromeliaceae) under high and low light intensities. *Functional Plant Biology* **29**, 757–762.

Fish, D. (1983). Phytotelmata: flora and fauna. *In* "Phytotelmata: Terrestrial Plants as Hosts for Aquatic Insect Communities" (J.H. Frank and L.P. Lounibos, Eds.), pp. 1–27. Plexus, Medford, New Jersey.

Fragoso, C. and Rojas-Fernandez, H. 1996. Earthworms inhabiting bromeliads in Mexican tropical rain forests: ecological and historical determinants. *J. Trop Ecol.* **12**, 729–734.

Frei, Sister John Karen and Dodson, C.H. (1972). The chemical effect of certain bark substrates on the germination and early growth of epiphytic orchids. *Bull. Torrey Bot Club*. **99**, 301–307.

Freiberg, M. (1997). Spatial and temporal pattern of temperature and humidity of a tropical premontane rein forest tree in Costa Rica. *Selbyana* **18**, 77–84.

Freiberg, M. (2001). The influence of epiphyte cover on branch temperatures in a tropical tree. *Plant Ecol.* **153**, 241–250.

Freiberg, M. and Freiberg, E. (2000). Epiphyte diversity and biomass in the canopy of low and montane forest in Ecuador. *J. Trop. Ecol.* **16**, 673–688.

Foster, P. (2002). The potential negative impacts of global change on tropical montane cloud forests. *Earth-Science Reviews* **55**, 580–586.

Gentry, A.H. and Dodson, C.H. (1987a). Contribution of non-trees to species richness of a tropical rain forest. *Biotropica* **19**, 149–156.

Gentry, A.H. and Dodson, C.H. (1987b). Diversity and biogeography of neotropical vascular epiphytes. *Ann. Missouri Bot. Gard.* **69**, 557–593.

Givnish, T.J., Burkhardt, E.L., Happle, R., and Weintraub, J. (1984). Carnivory in the bromeliad *Brocchinia reducta*, with a cost/benefit model for the general restriction of carnivorous plants to sunny, moist, nutrient-poor habitats. *Amer. Nat.* **124**, 479–497.

Griffiths, H. and Maxwell, K. (1999). In memory of C. S. Pittendrigh: does exposure in forest canopies relate to photo protective strategies in epiphytic bromeliads? *Functional Ecology* **13**, 15–23.

Haslam, P.H., Borland, A.M. and Griffiths, H. (2002). Short-term plasticity of crassulacean acid metabolism expression in the epiphytic bromeliad *Tillandsia usneoides*. *Functional Plant Biol.* **29**, 749–756.

Helbsing, S., Riederer, M., and Zotz, G. (2000). Cuticles of vascular epiphytes: Efficient barriers for water loss after stomatal closure? *Ann. Bot.* **86**, 765–770.

Hietz, P., Ausserer, J., and Schindler, G. (2002). Growth, maturation and survival of epiphytic bromeliads in a Mexican humid montane forest. *J. Trop. Ecol.* **18**, 177–191.

Hietz, P. and Briones, O. (1998). Correlation between water relations and within-canopy distributions of epiphytic ferns in a Mexican cloud forest. *Oecologia* **114**, 305–316.

Hietz, P. and Hietz-Seifert, U. (1995a). Structure and ecology of epiphyte communities of a cloud forest in central Vera Cruz, Mexico. *J. Veg. Sci.* **6**, 719–728.

Hietz, P. and Hietz-Seifert, U. (1995b). Intra and interspecific relations within an epiphyte community in a Mexican humid montane forest. *Selbyana* **16**, 135–140.

Hietz, P., Wanek, W., and Popp, M. (1999). Stable isotope composition of carbon and nitrogen and nitrogen content in vascular epiphytes along an altitudinal transect. *Plant Cell Environ.* **22**, 1435–1447.

Hietz-Seifert, U., Hietz, P., and Guevara, S. (1996). Epiphyte vegetation and diversity on remnant trees after forest clearance in southern Vera Cruz, Mexico. *Biological Conservation* **75**, 103–111.

Holtum, J.A.A. and Winter, K. (1999). Degrees of crassulacean acid metabolism in tropical epiphytic and lithophytic ferns. *Australian J. Plant Physiol.* **26**, 749–757.

Hsu, C., Horng, F., and Kao, C. (2002). Epiphyte biomass and nutrient capital of a moist subtropical forest in north-eastern Taiwan *J. Trop. Ecol.* **18**, 659–670.

Huxley, C.R. (1978). The ant-plants *Myrmecodia* and *Hydnophytum* (Rubiaceae) and the relationships between their morphology, ant occupants, physiology and ecology. *New Phytol.* **80**, 231–268.

Ibisch, P.L. (1996). "Neotropische Epiphytendiversität-das Beispiel Bolivien." Martina Galunder-Verlag,Wiehl.

Ibisch, P.L., Boegner, A., Nieder, J., and Barthlott. W. (1996). How diverse are neotropical epiphytes? An analysis based on the "catalogue of the flowering plants and gymnosperms of Peru" *Ecotropica* **2**, 13–28.

Ingram, S.W. and Nadkarni, N.M. (1993). The composition and distribution of epiphytic organic matter in neotropical cloud forest, Costa Rica. *Biotropica* **25**, 370–383.

Janos, D.P. (1993). Vesicular-arbuscular mycorrhizae of epiphytes. *Mycorrhiza* **4**, 1–4.

Kernan, C. and Fowler, N. (1995). Differential substrate use by epiphytes in Corcovado National Park, Costa Rica: a source of guild structure. *J. Ecol.* **83**, 65–73.

Kessler, M. (2002). Environmental patterns and ecological correlates of range size among bromeliad communities of Andean forests in Bolivia. *Bot. Rev.* **68**, 100–127.

Lesica, P. and Antibus, R.K. (1990). The occurrence of mycorrhizae in vascular epiphytes of two Costa Rican rain forests. *Biotropica* **22**, 250–258.

Lesica, P. and Antibus, R.K. (1991). Canopy soils and epiphyte richness. *Natl. Geogr. R. Explor.* **7**, 156–165.

Matelson, T.J., Nadkarni, N.M., and Longino, J.T. (1993). Longevity of fallen epiphytes in a neotropical montane forest. *Ecology* **74**, 265–269.

Maxwell, K. (2002). Resistance is useful: diurnal patterns of photosynthesis in C_3 and crassulacean acid metabolism epiphytic bromeliads. *Functional Plant Biol.* **29**, 679–687.

Maxwell, K., Griffiths, H., Borland, A.M., Broadmeadows, S.J., and Fordham, M.C. (1995). Short-term photosynthesis responses of the C_3-CAM bromeliad *Guzmania monostachia* var. *monostachia* to tropical seasonal transitions under field conditions. *Aust. J. Pl. Physiol.* **22**, 771–781.

Merwin, M.C., Rentmeester, S.A., and Nadkarni, N.M. (2003). The influence of host tree species on the distribution of epiphytic bromeliads in experimental monospecific plantations, La Selva, Costa Rica. *Biotropica* **35**, 37–47.

Mondragón, D., Calvo-Irabien, L., and Benzing, D.H. (In press). The basis for obligate epiphytism in *Tillandsia brachycaulos* (Bromeliaceae) in a Mexican dry tropical forest. *J. Trop Ecol.*

Mondragón, D., Durán, R., Ramírez, I., and Olmsted, I. (1999). Population dynamics of *Tillandsia brachycaulos* Schltdl. (Bromeliaceae) in Dzibilchaltun National Park, Yucatán. *Selbyana* **20**, 250–255.

Nadkarni, N.M. (1981). Canopy roots: convergent evolution in rain forest nutrient cycles. *Science* **214**, 249–256.

Nadkarni, N.M. and Matelson, T.J. (1989). Bird use of epiphyte resources in neotropical montane forest and pasture tree crowns. *Condor* **91**, 891–897.

Nadkarni, N.M. and Matelson, T.J. (1991). Litter dynamics within the canopy of a neotropical cloud forest, Monteverde, Costa Rica. *Ecology* **72**, 849–860.

Nadkarni, N.M. and Solano, R. (2002). Potential effects of climate change on canopy communities in a tropical forest: an experimental approach. *Oecologia* **131**, 580–586.

Nieder, J., Engwald, S., Klawun, M., and Barthlott, W. (2000). Patterns of Neotropical epiphyte diversity. *Selbyana* **20**, 66–75.

Nicolai, V. (1986). The bark of trees: thermal properties, microclimate, and fauna. *Oecologia* **69**, 148–160.

Nkongmeneck, B., Lowman, M.D., and Atwood, J.T. (2002). Epiphyte diversity in primary and fragmented forest of Cameroon, Central Africa: a preliminary survey. *Selbyana* **23**, 121–130.

Paoletti, M.G., Taylor, R.A.J., Stinner, B.R., Stinner, D.D.H., and Benzing, D.H. (1991). Diversity of soil fauna in the canopy and forest floor of a Venezuelan cloud forest. *J. Trop. Ecology* **7**, 373–383.

Pett-Ridge, J. and Silver, W.L. (2002). Survival, growth and ecosystem dynamics of displaced bromeliads in a montane tropical forest. *Biotropica* **34**, 211–224.

Pierce, S., Maxwell, K., Griffith, H., and Winter, K. (2001). Hydrophobic trichome layers and epicuticular wax powders in Bromeliaceae. *Amer. J. Bot.* 1371–1389.

Pierce, S., Winter, K., and Griffiths, H. (2002a). The role of CAM in high rainfall cloud forests: an *in situ* comparison of photosynthetic pathways in Bromeliaceae. *Plant Cell Environ.* **25**, 1181–1189.

Pierce, S., Winter, K., and Griffiths, H. (2002b). Carbon isotope ratios and the extent of daily CAM use by Bromeliaceae. *New Phytol.* **156**, 75–83.

Pittendrigh, C.S. (1948). The bromeliad-*Anopheles*-malaria complex in Trinidad. *Evolution* **2**, 58–89.

Richardson, B.A. (1999). The bromeliad microcosm and the assessment of faunal diversity in a neotropical forest. *Biotropica* **31**, 321–336.

Richardson, K.A. and Currah, R.S. (1995). The fungal community associated with the roots of some rain forest epiphytes of Costa Rica. *Selbyana* **16**, 49–73.

Robertson, K. and Platt, W.J. (1991). Effects of fire on bromeliads in subtropical hammocks of Everglades National Park, Florida. *Selbyana* **13**, 39–49.

Rosenberger, T. and Williams, K. (1999). Responses of vascular epiphytes to branch fall gap formation in *Clusia* trees in a montane rain forest. *Selbyana* **20**, 49–58.

Rothwell, G.W. (1991). *Botryopteris forensis* (Botryopteridaceae), a trunk epiphyte of the tree fern *Psaronius*. *Amer. J. Bot.* **64**, 1254–1262.

Schmid, E., Oberwinkler, I., and Gomez, L.D. (1999). Light and electron microscopy of host-fungus interactions in the roots of some epiphytic ferns from Costa Rica. *Can. J. Bot.* **73**, 991–996.

Schmidt, G. and Zotz, G. (2000). Herbivory in the epiphyte *Vriesea sanguinolenta* Cogn. and Marchal (Bromeliaceae). *J. Trop. Ecol.* **16**, 829–839.

Schmidt, G. and Zotz, G. (2001). Ecophysiological consequences of differences in plant size, *in situ* carbon gain, and water relations of the epiphytic bromeliad *Vriesea sanguinolenta*. *Plant, Cell and Envir.* **24**, 101–112.

Schmidt, G. and Zotz, G. (2002). Inherently slow growth in two Caribbean epiphyte species: a demographic approach. *J. Veg. Science* **13**, 527–534.

Seidel, J.L., Epstein, W.W., and Davidson, D.W. (1990). Neotropical ant gardens. I. Chemical constituents. *J. Chem. Ecol.* **16**, 1791–1816.

Sheridan, R.P. (1994). Adaptive morphology of a tropical wet montane epiphyte *Anthurium hookeri*. *Selbyana* **15**, 18–23.

Steward, G.R., Schmidt, S.C. Handley, L.L., Turnbull, M.H., Erskine, P.D., and Joly, C.A. (1995). [15]N natural abundance in rain forest epiphytes: implications for nitrogen source and acquisition. *Plant Cell Environ.* **18**, 85–90.

Still, C.J., Foster, P.N., and Schneider, S.H. (1999). Simulating the effects of climate change on tropical montane cloud forests. *Nature* **398**, 608–610.

Stork, N.E. (1987). Arthropod faunal similarity of Bornean rain forest trees. *Ecol. Entomol.* **12**, 69–80.

Stuntz, S., Linder, C., Linsenmair, K.E., Simon, U., and Zotz, G. (In press). Do non-myrmecophytic epiphytes influence community structure of arboreal ants? *Basic and Applied Ecology*.

Stuntz, S., Simon, U., and Zotz, G. (1999). Assessing the potential influence of vascular epiphytes on arthropod diversity in tropical tree crowns: hypotheses, approaches, and preliminary data. *Selbyana* **20**, 276–283.

Stuntz, S., Simon, U., and Zotz, G. (2002). Rain forest air-conditioning: the moderating influences of epiphytes on the microclimate in tropical tree crowns. *Int. J. Biometeorol.* **46**, 53–59.

Stuntz, S., Ziegler, C., Simon, U., and Zotz, G. (2001). Diversity and canopy structure of the arthropod fauna within three canopy epiphyte species in Central Panama. *J. Trop. Ecol.* **18**, 161–176.

Stuntz, S. and Zotz, G. (2001). Photosynthesis in vascular epiphytes: A survey of 27 species of diverse taxonomic origin. *Flora* **196**, 132–141.

Tremblay, R. (1997). Distributions and dispersions of individuals in nine species of *Lepanthes* (Orchidaceae). *Biotropica* **29**, 38–45.

Valdivia, P.E. (1977). Estudio botanico y ecológico de la región del Rio Uxpanapa, Vera Cruz. IV. Las Epifitas. *Biotica* **2**, 55–81.

Vandunne, H. (2001). Effects of the spatial distribution of trees and conspecific epiphytes and geomorphology on the distributions of epiphytic bromeliads in a secondary montane forest (Cordillera Central, Colombia). *J. Trop. Ecol.* **18**, 193–213.

Wania, R., Heitz, P., and Wanek, W. (2002). Natural ^{15}N abundance of epiphytes depends on the position within the forest canopy: source, signals, and isotope fractionation. *Plant, Cell Environ.* **25**, 581–589.

Williams-Linera, G., Sosa, V., and Platas, T. (1995). The fate of epiphytic orchids after fragmentation of a Mexican cloud forest. *Selbyana* **16**, 36–40.

Wilson, E.O. (1987). The arboreal ant fauna of Peruvian Amazon forest: a first assessment. *Biotropica* **19**, 245–282.

Winkler, M. and Hietz, P. (2001). Population structure of three epiphytic orchids (*Lycaste aromatica*, *Jacquiniella leucomelana*, and *J. teretifolia*) in a Mexican humid montane forest. *Selbyana* **22**, 27–33.

Wolf, J.H.D. and Konings C.J.F. (2001). Toward a sustainable harvesting of epiphytic bromeliads: a pilot study from the highlands of Chiapas, Mexico. *Biological Conservation* **101**, 23–31.

Yu, W. (1994). The structural role of epiphytes in ant gardens. *Biotropica* **26**, 217–221.

Zotz, G. (1995). How fast does an epiphyte grow? *Selbyana* **16**, 150–154.

Zotz, G. (2002). Categories and CAM: blurring divisions, increasing understanding? *New Phytol.* **156**, 4–5.

Zotz, G. and Mikona, C. (2003). Photosynthetic induction and leaf carbon gain in the tropical understory epiphyte *Aspinasia principissa*. *Ann. Bot.* **91**, 353–359.

Zotz, G., Reichling, P., and Valladares, F. (2002). A simulation study of the importance of size-related changes in leaf morphology and physiology for carbon gain of an epiphytic bromeliad. *Ann. Bot.* **90**, 437–443.

Zotz, G. and Thomas, V. (1999). How much water is in the tank? Model calculations for two epiphytic bromeliads. *Ann. Bot.* **87**, 183–192.

CHAPTER 10

Mistletoe: A Unique Constituent of Canopies Worldwide

David M. Watson

As the mistletoe is disseminated by birds, its existence depends on them; and it may methodically be said to struggle with other fruit-bearing plants, in tempting the birds to devour and thus disseminate its seeds. In these several senses, which pass into each other, I use for convenience' sake the general term of struggle for existence.
—*Charles Darwin,* The Origin of Species by Means of Natural Selection, *1859*

Introduction

Mistletoes are a unique group of plants that have long fascinated people, inspiring a prominent role in folklore and mythology. Lacking roots, remaining green and leafy year-round, and even bearing fruit during freezing winters, the plant was revered as sacred by the Druids and was the focus of midwinter fertility rites—customs still evident today in the use of mistletoe in Christmas celebrations. The ultimate goal of Aeneus in Virgil's epic poem "The Aeneid" was to find the golden bough (a mistletoe)—needed to gain entry into the world of souls (see Figure 10-1). The "burning bush" parable related in the Bible probably referred to a mistletoe common on acacias throughout the Middle East (Polhill and Wiens 1998), the fiery red and yellow flowers appearing during summer when the shrub has lost its leaves. Mistletoes were highly esteemed by alchemists and figured prominently in early herbals and pharmacœpias and, more recently, these plants have yielded promising compounds for the treatment of cancer, epilepsy, and other diseases.

This fascination with mistletoe is reflected in the biological literature, starting with Pliny the Elder and Theophrastos around 2,000 years ago, with Linnaeus and Darwin both contributing to early mistletoe research. Describing species, documenting the physiological and anatomical adaptations associated with their parasitic habit, and evaluating the relationship between mistletoe and host has dominated mistletoe research, and mistletoes have been used as a model system for understanding fruit dispersal (McKey 1975; Howe and Smallwood 1982) and the spread of parasites generally (Martinez del Rio et al. 1996). More recently, associations with fruit dispersers, pollinators, and herbivores have motivated studies throughout the world with an increasing body of literature focusing on mistletoe–animal interactions (Hawksworth and Wiens 1996; Watson 2001).

In this chapter, I synthesize our current understanding of mistletoe ecology in forests around the world, highlighting the importance of these plants to animals and forest communities generally. Having established the key attributes of mistletoe and summarized their distribution, diversity, and life history, I focus on recent developments in the study of these plants. Although mistletoe has

Figure 10-1   Leaves from the golden bough. (Illustrated by H. M. Brock). From Frazer, L. (Ed.) (1924). "Leaves from the Golden Bough." Grosset and Dunlap, New York.

reached the greatest diversity in tropical forests, most mistletoe research has been conducted in temperate forests and woodlands and semi-arid shrub-lands and savannahs. Hence, unlike many other chapters in this book that deal primarily with tropical ecosystems, most of the research discussed here is from temperate regions. Whether these patterns apply to rainforests and other tropical habitats is unclear and is a productive area for future research by canopy biologists.

Distribution, Diversity, and Attributes of Mistletoe

There are approximately 1,400 species of mistletoe, growing in a broad range of habitats throughout the world, including tropical rainforests, subarctic coniferous forests, deserts, and coastal heathlands across all continents except Antarctica, and many isolated oceanic islands (Calder and Bernhardt 1983; Watson 2001). Like mangroves and succulents, mistletoes are poly-phyletic—grouped together on the basis of common growth-form rather than shared ancestry. Mistletoes differ from epiphytes in that they are obligate hemiparasites: reliant on their host for water and minerals but synthesizing their own carbohydrates by photosynthesis. A range of plants acts as host with trees, shrubs, ferns, herbs, grasses, and even cacti, orchids, and other mistletoes known to support mistletoes (Kuijt 1969). Most mistletoe species, however, parasitize trees and shrubs. Accordingly, the group is most prominent in forests and woodlands. They range in size from the diminutive *Arceuthobium minutissimum*, a specialist on Himalayan conifers that rarely exceeds 10 mm in overall size, to the spectacular Western Australian Christmas Tree *Nuytsia flori-bunda*—one of three root-parasitic mistletoes—that can grow to heights of 10 m or more. Most species, however, are small to medium-sized shrubby plants with semi-succulent leaves and stems (see Figure 10-2). In addition to the plant itself, mistletoe infection can cause a reaction in the host, ranging from a slight swelling on the branch to a large mass of thickened and twisted branches (known as a witch's broom; see Figures 10-2 and 10-5).

Of the five families comprising the group, true mistletoes (Viscaceae) and showy mistletoes (Loranthaceae) are the most diverse and widespread. Recent systematic research suggests that the aerial parasitic habit evolved independently in these groups and at least five times in mistletoes as a whole (Nickrent 2001). The three root parasitic species are considered basal, consistent with the suggestion that aerial parasitism evolved from root-parasitic ancestors. Mistletoes are known from as long ago as the late Cretaceous with the Loranthaceae originating in Gondwanaland and the Viscaceae in Laurasia, both groups subsequently radiating throughout the world (Calder and Bernhardt 1983; Polhill and Wiens 1998).

While many plants are parasitic, mistletoes are the only group of woody plants to be obligate aerial parasites and exhibit a range of adaptations associated with their growth form, the most obvious being the complete absence of roots. Instead, they have developed a specialized holdfast, or haustorium, that anchors the mistletoe to its host and surrounds the vascular connection between host and mistletoe (see Figure 10-2). In addition to diffuse xylem–xylem connections, several groups of mistletoe also tap the phloem of their host, either supplementing the carbohy-drates they synthesize or obtaining most their carbohydrates directly from the host (Press and Graves 1995). In order to draw water and minerals from their host, mistletoe plants concentrate a range of soluble organic compounds and cations in their tissues, maintaining a low water poten-tial relative to host tissue. As parasites, the distribution and life-history of mistletoes is not con-strained by soil fertility and water availability to the same extent as other plants. Instead, mistletoe occurrence depends primarily on the presence of suitable hosts and effective dispersal agents (Reid et al. 1995; Restrepo et al. 2002). Several groups of mistletoe use wind or insects for pollination, but most rely on birds to disperse both pollen and seeds. The advantage of birds is that they are highly mobile and, unlike wind and water, are selective and have a higher like-lihood of depositing the pollen or seed in the right place (Sargent 1995; Restrepo et al. 2002).

As with other bird-pollinated plants, mistletoes display a range of characteristics to maximize the likelihood of birds visiting and pollinating their flowers. Flowers are typically brightly colored (yellow or red) with a fused corolla tube (see Figure 10-3), borne in large inflorescences on short stems, and producing abundant nectar rich in available carbohydrates. To enhance pollination success, mistletoes exhibit pronounced complementarity of flowering within a particular region, with only one species in peak flower at any time—minimizing the chances of the pollen of one

Figure 10-2 Clockwise from top left: young *Amyema miquelii*, Australia, showing haustorium (photograph by author); Witches' broom associated with *Arceuthobium douglasii* growing on fir tree in New Mexico (dwarf mistletoe shoots visible at center of clump; photograph by D.L. Nickrent); *Atkinsonia ligustrina*, one of three root parasitic mistletoes, Australia (photograph by author); *Viscum crassulae*, a parasite on succulent shrubs from South Africa. Photograph by D.L. Nickrent.

species being wasted on flowers of another. While some mistletoe species in India and New Zealand are pollinated by a small number of bird species (Davidar 1985; Robertson et al. 1999), most mistletoes are pollinated by a broad range of nectar-feeding birds, and no bird species can be considered a mistletoe-flower specialist. While mammals frequently visit the flowers, no mammals are known to function as pollinators.

In contrast, many of the birds that disperse mistletoe fruits are highly specialized—their diet is composed almost entirely of mistletoe fruits. The fruits are small—typically less than 10 mm in diameter—and brightly colored (often red or white), with an extremely sticky layer of pulp surrounding the seed (viscin). Birds either swallow the fruit whole, peel off the exocarp and ingest the seed and pulp, or pick out the seed and ingest only the pulp (Snow and Snow 1988; Reid 1991). Once ingested, the seeds are either regurgitated or defecated; in either case, the seed is still surrounded by viscin and readily adheres to the perch. The sticky seeds often adhere to the birds' feathers, and this may be an important means of dispersal for some groups in the Viscaceae, considered by Restrepo et al. (2002) to be the most likely mechanism explaining the widespread occurrence of these groups on isolated oceanic islands (whereas ingested seeds would be voided in transit).

Figure 10-3 Flowers of mistletoes in the Loranthaceae, clockwise from top left: *Psittacanthus ramiflorus*, Mexico (photograph by D.L. Nickrent); *Macrosolen crassus*, Malaysia (photograph by D.L. Nickrent); *Amyema cambagei*, Australia (photograph by author); *Atkinsonia ligustrina*, Australia (photograph by author). All flowers are pollinated by birds except *Atkinsonia*, the small fragrant flowers, which are insect-pollinated.

In most continents, mistletoes are dispersed by one or two species in a given region, most of them mistletoe specialists that depend on mistletoe fruit as their primary (or in some cases sole) food source (see Table 10-1). These mistletoe specialists represent many families and orders, but all are small and highly mobile, foraging in pairs or small parties. As their food source is often highly dispersed and temporally variable, vagility is an important attribute. Thus, mistletoe specialists are typically capable of long-distance movement, predisposing them to be successful colonists of outlying areas like islands and mountains (Restrepo et al. 2002).

Table 10-1   Dispersal Agents for Mistletoe Worldwide

Region	Main Mistletoe Dispersers	Other Mistletoe Dispersers
Indian subcontinent	Dicaeidae	Zosteropidae
Southeast Asia	Dicaeidae	Paridae
Eurasia	Muscicapidae	Sylvidae, Bombycillidae
Africa	Lybiidae	Coliidae, Paridae
North America	Ptilogonatidae	Picidae, Bombycillidae, Corvidae, Muscicapidae
Mesoamerica	Fringillinae, Tyrannidae, Ptilogonatidae	Columbidae
South America	Fringillinae, Cotingidae, Tyrannidae, Mimidae, Microbiotheriidae* (Andes)	Thraupini, Columbidae
New Zealand	Meliphagidae	Columbidae, Pachycephalidae
Australia	Meliphagidae, Dicaeidae	Oriolidae

*This taxon is a mammal, but all others are bird families/subfamilies. Many other groups are known to consume mistletoe fruit, often regularly, but these groups comprise the main dispersers throughout the world After Reid 1991; Watson 2001; Restrepo et al. 2002.

Some of the most detailed studies of mistletoe dispersal have been carried out in England, where the main disperser of the sole species of mistletoe, *Viscum album,* is the aptly named Mistle Thrush, *Turdus viscivorus* (literally, "thrush that eats mistletoe" in Latin). This large thrush eats a variety of fruit but feeds primarily on mistletoe berries in winter, when little else is available (Snow and Snow 1984, 1988). As well as acting as sole disperser, this large thrush actively defends mistletoe clumps against opportunistic fruit eaters, guarding mistletoe clumps just as some hummingbirds vigorously guard nectar-rich plants in the neotropics. Other species have been recorded feeding on the seeds and hunting for insects within the mistletoe clump. In some areas, the Blackcap *Sylvia atricapilla,* a small warbler, may also act as an important disperser. Unlike the larger thrush, this species squeezes out the seed onto a twig, ingesting only the skin and pulp. This process leads to a high success rate of dispersal, and the Blackcap is considered to be the main mistletoe disperser in some areas of Europe (Snow and Snow 1988).

Some research on diversity patterns in New World mistletoes found that mistletoe clades dispersed by birds exhibited far greater diversities than those that use wind or hydrostatic expulsion to disperse seeds (Restrepo et al. 2002), suggesting that birds (specifically, suboscine passerines) and Loranthaceous mistletoes may have undergone reciprocal diversification during the late Cretaceous and early Tertiary. While many mammals eat mistletoe fruits, most are considered fruit predators—the plant not gaining any benefit from the interaction. It was recently discovered, however, that a small marsupial acts as the sole disperser for a mistletoe in the southern Andes (Amico and Aizen 2000). It has been suggested that ancient lineages of marsupial may have been the original dispersal agents in Gondwanaland, but birds subsequently became the dominant dispersers.

Value and Popularity as a Food Source

In addition to pollinators and fruit dispersers, many other animals visit mistletoe clumps, taking advantage of the abundant high-quality nutritional resources supplied by the plants. Unlike pollinators and dispersers that participate in mutually beneficial interactions, obtaining food and ensuring the continued survival of mistletoes, these opportunists are of no direct benefit to the plant. As the nectar and fruit are provided as an incentive to potential dispersers, they are reliable, abundant, and of high nutritional quality. Accordingly, they attract an amazing variety of animals. Cassowaries and cuscuses, lemurs and lories, gibbons and hornbills, hummingbirds and woodpeckers—the diversity of birds and mammals recorded feeding on the nectar and fruits is immense. There is even a small fish (*Triportheus angulatus*) that has been recorded feeding on mistletoe fruits during seasonal floods in the Amazon basin (Watson 2001).

As pointed out earlier, mistletoe tissue in general (including flowers, leaves, and branches) is rich in minerals and, as such, is highly sought after by browsing mammals. Elephants and rhinoceroses actively search out mistletoe clumps in the African savannah, while at higher altitudes, mistletoe is preferentially eaten by mountain gorillas and other primates (Watson 2001). Mammalian herbivory of mistletoes is so prevalent that it is considered to be one of the key selective forces driving mistletoe-host mimicry in leaf shape and growth form (Canyon and Hill 1997; see Figure 10-4). Some of the best examples come from the eucalypt-dominated forests and woodlands of Australia where a range of folivorous marsupials are known to favor mistletoe foliage over the secondary compound-rich leaves of the eucalypt host. A notable exception to this pattern occurs in mangrove woodlands—a habitat in which there are no mammalian herbivores (see Figure 10-4). Other factors are doubtless involved, however, since mistletoes in New Zealand also resemble their hosts but evolved in the complete absence of browsing mammals.

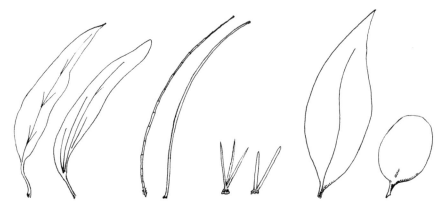

Figure 10-4 Diagrams of host (left) and mistletoe (right) leaves in the Australian genus *Amyema* (from left to right): *Eucalyptus meliodora* and *Amyema miquellii, Casuarina cunninghamiana* and *Amyema cambagei, Acacia tetragonophylla* and *Amyema preissii, Avicennia maritima* and *Amyema mackayense*. All display a high degree of mimicry except for the mangrove example—the only system in which there are no mammalian folivores. Courtesy of Maggie J. Watson.

Mistletoe is also an important food source for a range of invertebrates, although there hasn't been an equivalent amount of research conducted on these interactions. Some work on a species of mistletoe in Texas noted more than 200 species of insect acting as pollinators (Whittaker 1984). Overall, a large number of specialist insect herbivores are known from at least eight taxonomic orders, living and feeding only on mistletoe leaves. The best studied of these groups is the Lepidoptera, and mistletoe is the sole host plant for many butterflies and moths, including several rare and endangered species. Other animals capitalize on this diverse and abundant insect community, with mistletoe clumps being the favored foraging substrate for many insectivorous birds. This may constitute an example of Gentry's (1978) anti-pollinator hypothesis, insectivorous birds driving away pollen-bearing insects from mass flowering plants, further ensuring outcrossing and successful pollination.

Effects on Host Growth and Survivorship

The effect of mistletoes on hosts (especially trees) has been the subject of considerable research, especially for those species of tree used commercially for fruit and timber production. Although popularly viewed as the cause of death and declining vigor in trees, mistletoes don't necessarily have a deleterious effect on their host. Nick Reid and colleagues have investigated the effects of mistletoe infection on a range of acacia and eucalypt species in Australian forests and woodlands (Reid et al. 1995). While infected hosts often had reduced radial growth rates, there was little change in mortality. Several African mistletoes are facultatively deciduous, and some studies suggest mistletoe infection may afford greater resilience to severe water stress during drought through a decreased incidence of xylem cavitation (Ehleringer and Marshall 1995).

The most extensive research conducted on the effects of mistletoe on host survivorship and growth rates has been on the dwarf mistletoes (Viscaceae: Arceuthobium) of North America. In addition to direct effects on host survivorship, the many indirect effects of mistletoe infection are substantially greater. Infected trees are more susceptible to infection by various fungi and are preferentially attacked by bark beetles (Hawksworth 1983). The response of many coniferous trees to mistletoe infection is to produce a dense clump of thickened branches at the site of infection, known as a witch's broom (Figure 10-5). While of immense value to native mammals and birds,

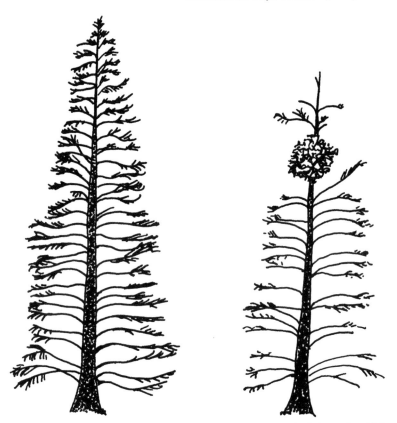

Figure 10-5   Examples of white spruce trees (*Picea glauca*) with and without dwarf mistletoes. In addition to the witch's broom, note the reduced number of branches and dead top—often used by woodpeckers and other cavity-nesting species. Courtesy of Maggie J. Watson.

these brooms represent a dangerous accumulation of resin-rich dead wood, leading to a dramatic increase in fire intensity in infected stands. They can function as "fire-ladders," drawing a ground fire up into the crown, leading to dramatically increased damage to trees and altering successional dynamics of the overall stand. Finally, mistletoe infection and the resultant witch's brooms lead to an overall change in growth form of the tree—often shorter, with clumped foliage and irregular branch structure (Figure 10-5). While considered a natural part of forest succession in this ecosystem (Bennetts et al. 1996), this process has a major negative impact on forestry in the region—in North America alone, dwarf mistletoe infection is estimated to account for annual wood losses of approximately 12 million cubic meters (Hawksworth 1983 and references cited therein).

Role in Promoting Diversity

Given the importance of mistletoe to a variety of animals, it has been proposed that mistletoe functions as a keystone resource throughout the world, having a disproportionate influence on the distribution of diversity in a range of forests and woodlands (Watson 2001). In addition to the breadth of known interactions between mistletoe and various animals, several studies have described positive associations between species richness and mistletoe abundance—areas with

MISTLETOE—A HOMEMAKER'S DELIGHT!

David M. Watson

In addition to providing a concentrated and reliable food source, mistletoe clumps are also used extensively by animals as a preferred location to shelter. Whether used as a nest site, a nocturnal roost, a place to hibernate, or a refuge from the heat, species from over 45 families of birds and 7 families of mammals have been recorded living in them (Watson 2001). Strategies vary with larger birds like vultures, plantain-eaters, hawks, and owls using the clump as a platform for their large stick nest (see Figure 1); smaller animals like squirrels, thrushes, pigeons, and possums building nests within them; and small birds like finches, honeyeaters, wrens, and tyrant flycatchers hanging their nest from mistletoe twigs. The haustorium connecting mistletoe and host is often enlarged; and some birds, including hummingbirds, warblers and the endangered marbled murrelet, build their nest atop this

Figure 1   Illustration of a grey go-away bird *Corythaixoides concolor* building a nest within a mistletoe clump in an Acacia. The evergreen foliage and robust branches provide an ideal nest-site, with a diverse range of species nesting and roosting preferentially in mistletoes. Courtesy Maggie J. Watson.

convenient platform. In cold climates, a variety of mammals (including porcupines, martens, squirrels, and pack-rats) have been recorded using dwarf mistletoes and associated witch's brooms as hibernaculae (Parks et al. 1999), while in the hot Australian summer, koalas, possums, and owls escape the daytime heat by roosting within the clumps. Mistletoe is so important as shelter that it has been used effectively to predict the distribution of particular animals. Australian ringtail possums and the highly endangered northern spotted owl both nest preferentially in/on mistletoe clumps in parts of their range, and the presence of mistletoe has been used in predictive models to identify suitable habitat for them. Finally, a variety of raptors have been recorded lining their nest with fresh sprigs of mistletoe (Viscaceae), often replacing them daily—a strategy that may exploit mistletoe's known immuno-stimulant properties to prevent nest-borne infections (Watson 2001).

In addition to these direct uses, mistletoe has an important indirect effect in many habitats, being the main mechanism for hollow formation in trees (Bennetts et al. 1996; Parks et al. 1999). Many trees react to mistletoes by restricting vascular flow to the infected branch, often subsequently shedding the branch and mistletoe, leaving a hollow or spout (referred to as self-pruning). In many forests throughout the world, a range of animals depend on tree hollows for shelter (Gibbons and Lindenmayer 2002), and research from the Rocky Mountains found a close relationship between numbers of cavity-nesting birds and mistletoe occurrence (Bennetts et al. 1996).

References

Bennetts, R.E., White, G.C., Hawksworth, F.G., and Severs, S.E. (1996). The influence of dwarf mistletoe on bird communities in Colorado ponderosa pine forests. *Ecol. Appl.* **6**, 899–909.

Gibbons, P. and Lindenmayer, D. (2002). "Tree Hollows and Wildlife Conservation in Australia." CSIRO Publishing, Collingwood, Australia.

Parks, C.G., Bull, E.L., Tinnin, R.O. Shepherd, J.F., and Blumton, A.K. (1999). Wildlife use of dwarf mistletoe brooms in Douglas-fir in northeast Oregon. *West. J. Appl. For.* **14**, 100–105.

Watson, D.M. (2001). Mistletoe—a keystone resource in forests and woodlands worldwide. *Ann Rev Ecol. System.* **32**, 219–249.

more mistletoe plants have higher diversities. This is not a linear relationship, however, with mistletoe sometimes occurring in great numbers in perturbed or highly disturbed areas.

In order to evaluate the effect of mistletoe occurrence on biodiversity, several studies have compared the species richness of areas with varying mistletoe densities. Bennetts et al. (1996) conducted the most extensive study, focusing on bird communities in the coniferous forests of the Rocky Mountains. The intensity of dwarf mistletoe occurrence was found to be an important predictor of bird distribution, having significant positive effects on 24 of 28 species, as well as relating to the number of snags and number of bird nests. Similar findings have been reported from elsewhere (summarized by Watson 2001), lending broad support to the keystone resource hypothesis.

The trouble with this sort of study is that, while suggestive of a relationship between mistletoe occurrence and biodiversity, it cannot demonstrate a direct causative effect. Hence, areas with more mistletoe may simply support greater plant diversity generally, and any relationship with animal richness is purely coincidental. Some work in African savannahs has demonstrated that mistletoes occur in greater numbers in areas with more fertile soils (Dean et al. 1994; Polhill and Wiens 1998)—thus any relationship between mistletoes and richness may have nothing to do with mistletoes directly, but may merely be an expression of the widely supported diversity-productivity association.

If there really is a causative effect of mistletoe on biodiversity, the most definitive way to demonstrate it is to conduct a removal experiment, comparing areas with no mistletoe (treatment areas) to areas where mistletoe has been left (control areas). The only published example comes from eucalypt-dominated forests of eastern Australia where two adjacent habitat fragments were compared directly (Watson 2002). Mistletoe plants had been removed from one patch for five years prior to the study while the other patch had not been manipulated. Otherwise, the fragments were comparable in area, habitat structure, floristic composition, and grazing history. Based on extensive inventories, the bird community was found to differ significantly between the two fragments, with 20 percent higher richness of birds in the control patch. There were also significant differences for woodland bird species and those species known to feed on mistletoe—both were recorded more frequently in the control patch. While based on a one-to-one comparison, this study provides the first direct evidence that mistletoe does have a positive influence on diversity, and ongoing research is exploring the repercussions of mistletoe removal at the catchment scale.

In addition to promoting diversity, mistletoe is a useful indicator of environmental integrity and habitat health. Mistletoes rely on animal vectors for both pollination and fruit dispersal, are the preferred diet of many herbivores, and depend on their hosts for all mineral and water requirements. As such, an interconnected web of interactions mediates their distribution, so that a change in any one of these other organisms can alter the distribution patterns of mistletoe. In fragmented woodlands of southeastern Australia, fire frequency and the density of many arboreal marsupials have been reduced, resulting in expansion of mistletoes, particularly along edges of woodlands. In the southwestern region of Australia, however, habitat fragmentation has had the reverse effect. Associated sheep grazing has reduced the understory, removing cover for small birds which function as mistletoe pollinators and dispersers, resulting in a dramatic decrease in mistletoe occurrence (Norton and Reid 1997). Accordingly, mistletoe has been included in restoration plans for degraded habitats (e.g., Rice et al. 1981), and is seen as an integral constituent of a functioning woodland or forest.

Summary

Mistletoes are a polyphyletic group of approximately 1,400 species of obligate hemiparasites, distributed worldwide. While they obtain all of their minerals and water from their host (typically a tree or shrub), they are green plants that generally manufacture their own carbohydrates by photosynthesis. They maintain low water potentials relative to their hosts by concentrating soluble cations and organic compounds in their tissues, and typically have higher rates of transpiration than their hosts. They are pollinated primarily by birds, and birds are also the main vector for fruit dispersal. In addition to these mutualistic interactions, mistletoe is a popular food source for a wide variety of animals, often providing nutritional resources when little else is available (midwinter in temperate areas, midsummer in arid environments). They are also highly favored roost and nest sites for a wide range of animals and are considered one of the main causes of hollow formation in many regions. Mistletoe can therefore be considered a keystone resource, having a disproportionate influence on the ecosystem as a whole, affecting distribution patterns of organisms both directly and indirectly. While some mistletoe species can have deleterious effects on their host, most mistletoes are best regarded as water-parasites, having a negligible effect on host survivorship. Given the range of interactions mistletoes have with pollinators, dispersers, herbivores, and hosts, they are a sensitive assay of environmental integrity and have been used as indicators of overall habitat health.

References

Amico, G. and Aizen, M.A. (2000). Mistletoe seed dispersal by a marsupial. *Nature* **408**, 929–930.

Bennetts, R.E., White, G.C., Hawksworth, F.G., and Severs, S.E. (1996). The influence of dwarf mistletoe on bird communities in Colorado ponderosa pine forests. *Ecol. Appl.* **6**, 899–909.

Calder, M. and Bernhardt, P. (Eds.) (1983). "The Biology of Mistletoes." Academic Press, Sydney, Australia.

Canyon, D.V. and Hill, C.J. (1997). Mistletoe host-resemblance: a study of herbivory, nitrogen and moisture in two Australian mistletoes and their host trees. *Aust. J. Ecol.* **22**, 395–403.

Darwin, C. (1859). "The Origin of Species by Means of Natural Selection." J. Murray, London.

Davidar, P. (1985). Ecological interactions between mistletoes and their avian pollinators in south India. *J. Bombay Nat. Hist. Soc.* **82**, 45–60.

Dean, W.R.J., Midgley, J.J., and Stock, W.D. (1994). The distribution of mistletoes in South Africa: patterns of species richness and host choice. *J. Biogeogr.* **21**, 503–510.

Ehleringer, J.R. and Marshall, J.D. (1995). Water relations. In "Parasitic Plants" (M.C. Press and J.D. Graves, Eds.), pp. 125–140. Chapman and Hall, London.

Gentry, A. (1978). Anti-pollinators for mass-flowering plants? *Biotropica* **10**, 68–69.

Gibbons, P. and Lindenmayer, D. (2002). "Tree Hollows and Wildlife Conservation in Australia." CSIRO Publishing, Collingwood, Australia.

Hawksworth, F.G. (1983). Mistletoes as forest parasites. *In* "The biology of mistletoes" (M. Calder and P. Bernhardt, Eds.), pp. 317–334. Academic Press, Sydney, Australia.

Hawksworth, F.G. and Wiens, D. (1996). "Dwarf Mistletoes: Biology, Pathology, and Systematics, Agriculture Handbook 709, Second Edition" United States Department of Agriculture Forest Service, Washington DC.

Howe, H.F. and Smallwood, J. (1982). Ecology of seed dispersal. *Ann. Rev. of Ecol. System.* **13**, 201–228.

Kuijt, J. (1969). "The Biology of Parasitic Flowering Plants." University of California Press, Berkeley.

Martinez del Rio, C., Silva, A., Medel, R., and Hourdequin, M. (1996). Seed dispersers as disease vectors: bird transmission of mistletoe seeds to plant hosts. *Ecology* **77**, 912–921.

McKey, D. (1975). The ecology of coevolved seed dispersal systems. *In* "Coevolution of Plants and Animals" (L.E. Gilbert and P. Raven, Eds.), pp. 159–191. Austin TX: University of Texas Press.

Nickrent, D.L. (2001). Mistletoe phylogenetics: current relationships gained from analysis of DNA sequences. Proceedings of the Western International Forest Disease Work Conference, August 14–18, Kona, Hawaii.

Norton, D.A. and Reid, N. (1997). Lessons in ecosystem management from management of threatened and pest loranthaceous mistletoes in New Zealand and Australia. *Conserv. Biol.* **11**, 759–769.

Parks, C.G., Bull, E.L., Tinnin, R.O. Shepherd, J.F., and Blumton, A.K. (1999). Wildlife use of dwarf mistletoe brooms in Douglas-fir in northeast Oregon. *West. J. Appl. For.* **14**, 100–105.

Polhill, R. and Wiens, D. (1998). "Mistletoes of Africa." Royal Botanic Gardens, Kew.

Press M.C. and Graves, J.D. (Eds.) (1995). "Parasitic Plants." Chapman and Hall, London.

Reid, N. (1991). Coevolution of mistletoes and frugivorous birds? *Aust. J. Ecol.* **16**, 457–469.

Reid, N., Stafford Smith, M., and Yan, Z. (1995). Ecology and population biology of mistletoes, *In* "Forest Canopies" (M.D. Lowman and N.M. Nadkarni, Eds.), pp. 285–310. Academic Press, San Diego.

Restrepo, C., Sargent, S., Levey, D., and Watson, D.M. (2002). The role of vertebrates in the diversification of New World mistletoes. *In* "Seed Dispersal and Frugivory; Ecology, Evolution and Conservation" (D.J. Levey, W.R. Silva, and M. Galetti, Eds.), pp 83–98. CABI Publishing, Wallingford, UK.

Rice, J., Ohmart, R.D., and Anderson, B. (1981). Bird community use of riparian habitats: the importance of temporal scale in interpreting discriminant analysis. *U.S.D.A. Forest Service, General Technical Report* **RM-87**, 186–196.

Robertson, A.W., Kelly, D., Ladley, J.J., and Sparrow, A.D. (1999). Effects of pollinator loss on endemic New Zealand mistletoes (Loranthaceae). *Conserv. Biol.* **13**, 499–508.

Sargent S. (1995). Seed fate in a tropical mistletoe: the importance of host twig size. *Func. Ecol.* **9**, 197–204.

Snow, B.K. and Snow, D.W. (1984). Long-term defense of fruit by mistle thrushes *Turdus viscivorus. Ibis* **126**, 39–49.

Snow, B.K. and Snow, D.W. (1988). "Birds and Berries: A Study of an Ecological Interaction." T. and A.D. Poyser, Calton, UK.

Watson, D.M. (2001). Mistletoe—a keystone resource in forests and woodlands worldwide. *Ann Rev Ecol. System.* **32**, 219–249.

Watson, D.M. (2002). Effects of mistletoe on diversity: a case-study from southern New South Wales. *Emu* **102**, 275–281.

Whittaker, P.L. (1984). The insect fauna of mistletoe (*Phoradendron tomentosum*, Loranthaceae) in southern Texas. *Southwest. Natur.* **29**, 435–444.

CHAPTER 11

Hidden in Plain Sight: Mites in the Canopy

David Evans Walter

"What scientists know about tropical rainforests serves above all to convince them they are deeply ignorant about them."
—Catherine Caufield, In the Rainforest, *1984*

What Are Mites?

Pick a leaf, stem, epiphyte, beetle, bird or handful of perched litter from a forest canopy and you are likely to be holding a swag of mites—but these tiny denizens of even the highest reaches of the canopy are rarely noticed. This is because mites scurry along the margins of our perceptions, too small to see easily without a microscope, but just too large for microbiological investigations. To many, this absence would be little cause for comment because even those observers looking for mites tend to search down-to-earth habitats like soil, ground-dwelling animals, or crops. Forest canopies, however, from cool-temperate through tropical regions, are infused with an extraordinary diversity and abundance of mites (Walter and O'Dowd 1995a, b; Walter and Proctor 1998; Walter et al. 1998; Walter and Behan-Pelletier 1999; Winchester et al. 1999; Basset 2001).

What are these animals and what are they doing in the canopy? The first question is most easily answered, at least in generalities. Mites are chelicerate arthropods with a unique body organization (see Sidebar on "Mites") that indicates a long evolutionary separation from other arachnids. More than 50,000 species of Acari have been described, and perhaps a million await description. Those we know are distributed across three Superorders, six Orders, more than two dozen taxonomic suborders and "Cohorts" (~Infrasuborders), >400 Families, and over 5,000 Genera (see Sidebar, "What are Mites"). Devonian fossils demonstrate that mites were among the earliest colonists of land (Walter and Proctor 1999; Hallan 2003), but when these mites began to climb onto living vegetation is more obscure (O'Dowd et al. 1991).

Origins of the Canopy Acarofauna

The ancestors of mites probably colonized early soils directly from the ocean, and soil is still the ecosystem stratum with the highest acarine densities and diversities (Norton 1994; Walter and Proctor 1999). Even a handful of forest litter may have hundreds of individual mites representing dozens of species; this detritus-based diversity extends into the canopy in the suspended soils that collect in mosses, treeholes, crotches, and among the leaves and holdfasts of ferns, orchids, and other epiphytes (Nadkarni and Longino 1990; Paoletti et al. 1991; Walter et al. 1998; Shaw

WHAT ARE MITES?

David Evans Walter

Mites are chelicerate arthropods. That is, like insects and spiders, mite bodies are covered by a chitinous exoskeleton composed of specialized body regions that develop through the expression and modification of serially repeated segments, each of which originally had the potential to produce a pair of jointed appendages (Telford and Thomas 1998a, b). In contrast to insects and their relatives (but the same as spiders and their relatives), the first pair of mite appendages are feeding structures (chelicerae) rather than antennae and the appendages of the third segment are ambulatory (legs I) rather than feeding structures (mandibles) (Thomas and Telford 1999). Unlike any other kind of arthropod, including other arachnids, mites have a unique organization of the body into a headlike feeding region called the capitulum (also, gnathosoma), bearing the chelicerae and a pair of palps, and a body (idiosoma) with up to four pairs of legs (three pairs in the larval stage and only two pairs in some plant-parasites), zero to three pairs of eyes, the brain, digestive tract, and reproductive organs.

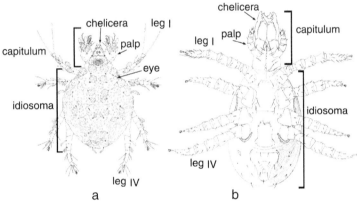

Figure 1 An acariform mite (a), *Cheletogenes ornatus*, dorsal view; (b) a parasitiform mite, *Blattisocius* sp., ventral view.

References

Telford, M.J. and Thomas, R.H (1998a). Of mites and *zen*, Expression studies in a chelicerate arthropod confirm *zen* is a divergent Hox gene. *Dev. Genes Evol.* **208**, 591–594.

Telford, M.J. and Thomas, R.H. (1998b). Expression of homeobox genes show chelicerate arthropods retain their deutocerebral segment. *Proc. Na. Acad. Sci., USA* **95**, 10671–10675.

Thomas, R.H. and Telford, M.J. (1999). Appendage development in embryos of the oribatid mite *Archegozetes longisetosus* (Acari, Oribatei, Trhypochthoniidae). *Acta Zoologica* **80**, 193–200.

and Walter 2003). Mites of both suspended and forest-floor soils have a high degree of overlap at the generic and family levels, but the species composition is largely complementary. That is, mites of perched litter differ from, but are close relatives of, those found in forest litter (Walter et al. 1998; Beaulieu and Walter, in review). Patterns at the family and generic levels indicate that relatively few lineages have made the transition from soil to living on the surface of plants or animals, but many of these have undergone spectacular radiations. Thus, although the acarofauna of suspended soils is little changed from that in more basal soils, a history of repeated

A SYSTEMATIC SYNOPSIS OF CANOPY MITES

David Evans Walter

Class Arachnida, Subclass Acari (Acarina)—mites and ticks
    Superorder Opilioacariformes—Order Opilioacarida—none arboreal
    Superorder **Acariformes**
        Order Sarcoptiformes—ca. 230 families, >15,000 described species
            Suborder **Oribatida** (also known as Cryptostigmata)
                These mites feed on algae, lichens, fungi, and detritus on vegetation (e.g., Cymbaeremaeidae, Adhaesozetidae, Camisiidae, Oribatulidae, Plateremaeidae) or inhabit tree trunks, epiphytes (especially mosses) and suspended soil.
            Suborder **Astigmata**
                Astigmatans live on leaves, under bark, in fungal sporocarps, in sap fluxes, in nests, and as symbionts of arboreal arthropods, birds, mammals, and reptiles.
        Order Trombidiformes—ca. 125 families, >22,000 described species
            Arboreal taxa belong to the Suborder **Prostigmata** and include the plant-parasitic spider mites and their relatives (Tetranychoidea); gall and rust mites (Eriophyoidea); and broad mites and their relatives (Tarsonemidae); also, many predators on vegetation and in nests (e.g., Anystidae, Bdellidae, Cheyletidae, Cunaxidae, Eupallopsellidae, and Mecognathidae); fungivores or scavengers on leaves, bark, or nests (e.g., Tarsonemidae, Tydeidae); chiggers and velvet mites (Parasitengona); and a variety of symbionts of arboreal animals (Cheyletoidea).
    Superorder **Parasitiformes**
        Order Holothyrida—holothyrans—none arboreal
        Order **Ixodida** (Metastigmata)—ticks—3 families, <900 described species
            All ticks are obligate blood-feeding parasites of vertebrates.
        Order **Mesostigmata** (Gamasida)—ca. 70 families, >12,000 described species,
            On vegetation, only two families of predatory Mesostigmatans are common (Phytoseiidae, Ascidae), but suspended soils contain a diversity of predators in numerous families. The flower mites (Ascidae, Ameroseiidae) and bee, bird and rat mites (Dermanyssoidea) also are Mesostigmatans.

invasions and diversifications on vegetation, fungal sporocarps, animals, and other canopy habitats has produced a bewildering diversity of arboreal mites (see Sidebar, "A Systematic Synopsis of Canopy Mites").

Because of their importance to agriculture, mites that are parasitic on vascular plants and their predators are the best known of the canopy mites and are useful models of the evolutionary history of mite colonization of the canopy or at least of its most extreme section—the leaves that form the outermost canopy. The inhabitants of leaf surfaces (the phylloplane) occupy a habitat as different from soil as any in the canopy and the harshest and most difficult of canopy habitats to colonize. Of the six Orders of living mites (see Sidebar), only one (Trombidiformes, Prostigmata) has produced plant-parasites that subsist entirely on leaf cell contents, but this lifestyle has arisen at least seven times within the Order (Walter and Proctor 1999) and twice resulted in extensive radiations of canopy mites: over 1,200 described species of spider mites (Tetranychoidea) and nearly 3,000 described species of gall mites (Eriophyoidea) are all obligate plant-parasites. Additionally, a more insidious kind of plant parasitism—flower mites that feed on nectar and pollen and move from host to host on pollinators—has arisen at least four times in the

Mesostigmata, once in the Ameroseiidae, and at least three times in the Ascidae (OConnor et al. 1991; Colwell and Naeem 1994; Colwell 1995; Seeman and Walter 1995; Seeman 1996; Naskrecki and Colwell 1998).

A similar pattern of repeated independent colonizations of leaves, sometimes followed by spectacular radiations, is found among the predatory mites that followed the parasites onto vegetation. For example, the predatory Prostigmatans that can be found on leaf surfaces represent a subset of predatory families that inhabit soil but are characterized by genera found only or primarily on vegetation. The same is even more true of the Mesostigmata, where only one family has spectacularly succeeded in colonizing leaves—the Phytoseiidae, with almost 2,000 described species (Walter and Proctor 1999). Smaller radiations of foliar mesostigmatans also have occurred among two unrelated genera of Ascidae in tropical forests (Walter et al. 1993; Walter and Lindquist 1997).

Thus, the plant-parasitic and predatory acarofauna of the outermost canopy is derived from at least 20 independent colonization events. These mites, however, are only part of the functional complexity of the mite fauna inhabiting the outermost canopy. Australian rainforest canopies often have densities of 1,000 to 10,000 mites per square meter of leaf area, but the vast majority of these mites are neither plant-parasites nor their predators but rather mites that graze on microbial growth (fungi, algae, lichens) or scavenge detritus (Walter and O'Dowd 1995a). These microbivore-scavengers represent dozens of other independent colonizations of and radiations on vegetation. Below, some of the functional complexity in these canopy mite assemblages is reviewed.

Functional Diversity

Only the insects rival mites for species diversity, so not surprisingly, the roles played by mites in forest canopies are diverse. The three primary roles of leaf-inhabiting mites were introduced above: plant-parasites, microbivore-scavengers, and predators. These groups will be considered in more detail below, along with associations with canopy animals.

Plant-Parasites

Most plant-parasites (Figure 11-1) use modified stylet-like chelicerae to puncture individual host cells and drain their contents, but the relative effects of feeding by the two major groups, Tetranychoidea and Eriophyoidea, have different implications for canopy dynamics. The association between eriophyoid mites and vascular plants appears to be an ancient one (Lindquist 1998). About 2,900 species of eriophyoid mites are described, but three-fourths of them are from temperate regions and perhaps less than 5 percent of the tropical fauna has been described (Amrine and Stasny 1994). All of these mites are tiny, most less than 0.2 mm in length, and have worm-like bodies with only two pairs of legs. Most woody dicots and many herbaceous dicots, monocots, conifers, and ferns are attacked by eriophyoids, and the vast majority of these mites are host-specific (Amrine and Stasny 1994; Lindquist et al. 1996). Gall, erinose, blister, witch's-broom, and big bud mites cause galling and deformation of leaves, petioles, stems, buds, flowers, and fruits that are often confused with microbial diseases. Perhaps the most unusual of these cancer-like growths are the pelts of hair-like processes called erinea (from their fleece-like structure). Other galls have more discrete forms and are variously called bladder, finger, nail, and pocket galls. The exact mechanism causing the galling response is not well understood, but appears to result from the release of some component that the mite introduces during feeding. Mites galling flowers and fruit may have a strong effect on plant fitness, but in most cases the effect of galls on the host is not clear. Unlike feeding by spider mites and most other plant

Figure 11-1   Plant-parasitic mites: (a) a fig mite, *Paratarsonemella giblindavisi* Walter (Prostigmata, Tarsonemidae); (b) a peacock mite, *Tuckerella* sp. (Prostigmata, Tuckerellidae); (c) a male and (d) a female spider mite (Prostigmata, Tetranychidae); (e) broad mites, *Polyphagotarsonemus latus* (Banks) (Prostigmata, Tarsonemidae), phoretic on a whitefly.

parasites, host cells often survive feeding by gall mites. This is the most likely explanation for the ability of eriophyoid mites to transmit plant diseases (Lindquist et al. 1996). Since eriophyoid mites migrate on air currents, their potential to spread disease is high.

The ability to gall plant tissue has evolved several times in acariform mites, but it is common only within the Eriophyoidea. However, more than half of all known eriophyoid species do not live within deformed plant tissues (Amrine and Stasny 1994). These mites are leaf vagrants that cause a silvering as cells die and fill with air or a rust-like pattern of dead cells. Other mites may cause russeting of leaves, but the common name is applied correctly only to Eriophyoidea.

All of the 1,800-plus known species in the five families of Tetranychoidea are parasites of vascular plants (Helle and Sabelis 1985). Because of the damage caused to economically important plants, the spider mites (Tetranychidae) are the best studied. Although only one species has been demonstrated to vector a plant virus, spider mite feeding is often extremely damaging and can result in death, especially in herbaceous hosts. Generation times are usually short (10 to 14 days), and populations rapidly reach damaging levels. As with eriophyoid mites, spider mites migrate on the wind and readily colonize new plants.

Spider mites (see Figure 11-1c, d) get their common name from the silken webs spun by some species. Others, however, are silkless or use silken strands only to strap their eggs to the plant and feed, unprotected, on either side of the leaf. Master-class weavers occur only within the tribe Tetranychini, and their webs vary from seemingly random arrays of lines to tightly woven, tent-like sheets. Species of *Tetranychus* produce the largest and most complex webs and often form dense entanglements of many layers (Saito 1985). Some tetranychines have subsocial colonies where young are protected and predators are attacked, driven off, and sometimes killed (Saito 1986a, b, 1997, 2000).

Some spider mites are highly polyphagous, but of the approximately 1,200 described species in 70 genera, only a few dozen species are known from many hosts and most appear to be relative specialists (Bolland et al. 1998). This is also true of the more than 600 known species in 22 genera of false spider mites (Tenuipalpidae): most appear to be host specialists. Of the remaining three families of tetranychoid mites, only peacock mites (Tuckerellidae) are canopy inhabitants—but spectacular ones (see Figure 11-1b). Peacock mites have a long tail of plumose posterior setae that trails out behind them and is used to flick away approaching predators; the body is further protected by large, leaf-like setae.

Several genera of Tarsonemidae feed on leaves, but the most interesting canopy tarsonemid mites are cryptic and rarely noticed. Fig mites (see Figure 11-1a) belong to the tribe Tarsonemellini, and develop inside fig fruits (*Ficus* spp.) on galled ovaries, and move from fig to fig on female fig wasps. Only a handful of fig mites have been described to date, but they have been found in South America, Africa, Southeast Asia and Australia; a high diversity of host-specific mites appears to be present (Compton 1993; Ho 1994; Lindquist 1998; Walter 2000).

A final, and much better studied, special case of plant parasitism deserves mention: flower mites. Rob Colwell (University of Connecticut, Storrs), who knows the neotropical representatives of this group better than anyone, has referred to flower mites as a venereal disease of plants: they infest flowers (plant genitalia), consume significant amounts of nectar and pollen (reproductive juices and gametes), and ride pollinators from flower to flower (Colwell 1986; Heyneman et al. 1991; OConnor et al. 1991; Ohmer et al. 1991; Colwell and Naeem 1994; Paciorek et al. 1995; OConnor et al. 1997). The ascid *Rhinoseius–Tropicoseius* lineage appears to have arisen in the montane neotropics and to have diverged in association with different families of plant hosts (Naskrecki and Colwell 1998). A quick dash down the hummingbird's bill is necessary for prompt disembarkation, and the decision to exit is probably mediated by the olfactory information in the stream of air generated by the birds as they breathe. An Australian system analogous to the neotropical hummingbird–flower mites has evolved between species of the genus *Hattena* (Ameroseiidae) and birds that visit flowers. The mites feed on pollen and nectar in the flowers of mangroves, mistletoes, and forest trees and shrubs, and they use honeyeaters, spinebills, and lorikeets to move from plant to plant (Halliday 1996; Seeman 1996). In Africa, Australasia, and Asia, other ameroseiid mites (e.g., *Neocypholaelaps*, *Afrocypholaelaps*) live in flowers, feed on pollen and nectar, and use bees and butterflies for phoretic transport (Eickwort 1994; Seeman and Walter 1995; Halliday 1996).

Scavenger-Microbivores and Detritivores

Outside of the known plant-parasitic mite groups, reports of mites feeding on the leaves of canopy plants are rare and mostly caused by misidentifications, facultative leaf-feeding by predators (probably to replenish fluids), or rare reports of microbivore-scavengers feeding on living leaves. Feeding on dead or dying vegetation, however, is a common behavior in canopy Oribatida *(Figure 11-2)*. Even small dead leaf patches (often caused by leaf fungi) can be colonized by oribatids (typically members of the Oribatuloidea), which bore through dead, fungus-ridden leaf or stem material like tiny beetles. Other oribatids graze on detritus and decomposers (see Figure 11-2c, g) in suspended soils or on mildews, leaf fungi, algae, or lichens on vegetation or stems (see Figure 11-2a, b, f). Technically, algae and lichens are primary producers so algivorous and lichenivorous mites could be considered true herbivores.

Predators

Prostigmatans in the family Tydeidae (see Figure 11-2i) are one of the most commonly encountered arboreal mites, inhabiting leaves, stems, trunks, suspended soils, and specialized habitats such as sporocarps and nests. Algae, fungi, and pollen are known sources of food for tydeids, but at least some species are facultative predators (Figure 11-3) (Walter and O'Dowd 1995a). The extent of facultative predation among canopy scavenger-microbivores has not been well explored,

Figure 11-2 Microbivore-scavenger mites: (a) cepheid nymph, a grazer on stems (Oribatida, Cepheidae); (b) *Scapheremaeus* sp., a grazer on leaves (Oribatida, Cymbaeremaeidae); (c) box mite from litter (Oribatida, Phthiracaroidea); (d) microdipsid, a litter fungivore (Prostigmata, Pygmephoridae); (e) insect-phoretic stage (hypopus) of acarid (Astigmata, Acaridae); (f) *Zygoribatula* sp., a grazer on leaves (Oribatida, Oribatuloidea); (g) *Notophthiracarus* sp., a box mite (Oribatida, Phthiracaroidea); (h) *Asperolaelaps rotundus* Womersley, a fungivore associated with decaying wood (Mesostigmata, Ameroseiidae); (i) a tydeid mite, (Prostigmata, Tydeidae).

but many more typical predators often take nonprey foods, especially pollen, honeydew, extrafloral nectar, and fungi (McMurtry and Croft 1997; Walter and Proctor 1999).

Ambush or sit-and-wait predators can often be found at the juncture of the petiole and the leaf blade, lying in wait for small arthropods moving along the stem-petiole highway system that connects one leaf to another. Generally, ambush predators take active prey and ignore those that are sessile or slow-moving. Ambush predators are capable of taking prey much larger than themselves, and some make use of toxins or silken strands to ensnare their victims (Walter and Proctor 1999). Prostigmatans (see Figure 11-3c, d) dominate the lie-in-wait predators on the phylloplane; however, in suspended soils, some Mesostigmatans (e.g., species of *Euepicrius*) use a probe-and-pounce mode of prey capture.

In contrast, cruise predators (see Figure 11-3a, b, e) actively seek out prey, although their pace varies from a constant and deliberate searching of every nook and cranny to the dizzy, spiraling

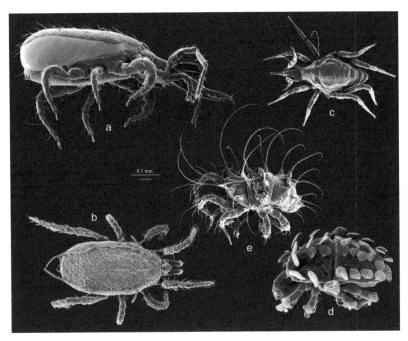

Figure 11-3   Predatory mites: (a) *Athiasella* sp., a predator in suspended soils (Mesostigmata, Ologamasidae); (b) *Phytoseius oreillyi* Walter and Beard, a predator on leaf surfaces; (c) a cunaxid mite predatory on leaves (Prostigmata, Cunaxidae); (d) a cheyletid mite from treeholes and nests (Prostigmata, Cheyletidae); (e) a stem-inhabiting predator, *Mecognatha hirsuta* Wood.

dash across the leaf characteristic of "whirligig" or "footballer" mites (Anystidae) that overrun mites and small insects, stab them with their mouthparts, and paralyze them with a toxin (Walter and Proctor 1999). Thus, cruise predators are capable of taking both passive prey (e.g., eggs, scale insects, molting arthropods) and actively moving prey. Passive prey can be much larger than the predators and may be fed on in an almost parasite-like manner. In other cases, the smaller stages of much larger animals fall victim. For example, an erythraeid mite (*Balaustium* sp.) has been reported to be a significant cause of egg mortality in a butterfly (Hayes 1985).

On plants, Mesostigmatans in the family Phytoseiidae (see Figure 11-3b) and Prostigmatans in the families Bdellidae, Cheyletidae (see Figure 11-3d), Cunaxidae (see Figure 11-3c), Stigmaeidae, Mecognathidae (see Figure 11-3e), Eupallopsellidae, and Erythraeidae are the most common cruise predators, but in suspended soils and nests a much greater diversity of predatory families is represented. The feeding habits of phytoseiid mites are the best known of any group of canopy mites. Most appear to be general and opportunistic predators of mites and insects, but are capable of subsisting on nonprey foods (McMurtry and Croft 1997).

Since little is known about the feeding habits of many canopy mites, relegation to feeding guilds often depends on taxonomic inference from what is known about related mites in other systems. This type of trophic classification needs to be treated with caution, as an example from the biocontrol literature illustrates. The first mite reported in the literature as a biological control agent was not a member of a group usually thought to be predatory. Instead, *Hemisarcoptes malus*, a member of the Astigmata—free-living members of which are typically scavenger-microbivores—was found feeding on oyster scale on apple in 1868 (Gerson and Smiley

1990). Subsequent research has demonstrated that all species of *Hemisarcoptes* are predators of armored scales (Homoptera: Diaspididae). Mite eggs are laid under scales and larvae, nondispersing nymphs and adults feed on scales, especially ovipositing female scales. Deutonymphs are phoretic under the elytra of coccinellid beetles (*Chilocorus* spp.) (Gerson and Izraylevich 1997) and may also be parasitic on their phoretic carriers (Houck 1994).

Arthropod Associates

Any arthropod that forms a burrow, den, or nest is likely to be associated with a diversity of mites (Figure 11-4). This is especially true of beetles, millipedes, and social Hymenoptera (Eickwort 1994; Walter and Proctor 1999). Many Astigmata have a specialized dispersal morph (the hypopus) with modified suckers for hitching rides on insects (see Figures 11-2e, 11-4d). Ants (see Figure 11-4g) and bees, in particular, are associated with extraordinary diversities of mites in their nests that often hitch rides on the bodies of workers and reproductives (Walter and Behan-Pelletier 1999; Walter et al. 2002). Honeybees, for example, have more than 100 species of mites associated with them. However, even non-social insects that spend their lives on leaves as predators (Hurst et al. 1995) or herbivores (Husband and de Moraes 1999) can have mite parasites living on them.

Nest-Associates, Ectocommensals, and Parasites of Vertebrates

Every canopy vertebrate species, and probably most individuals, have symbiotic mites associated with them. That may seem an extreme statement, but it is certainly true of all canopy mammals (see Figure 11-4a, b) and birds (see Figure 11-4i-k) and many canopy reptiles (see Figure 11-4h). Amphibians, alone among terrestrial vertebrates, seem to have few mite associates (Walter and Proctor 1999), perhaps because few are fully terrestrial (other than plethodontid salamanders, which often eat mites). In most cases, vertebrate-symbiotic mites are as fully and completely canopy inhabitants as their hosts, but I know of no study of arthropod canopy diversity that has considered what contribution these mites make to canopy biodiversity.

Hosts, Habitats, and Having a Good Time

The term "canopy" may seem self-explanatory, but in reality, its definition varies with worker and study. In this chapter, the forest canopy is considered to be represented by the woody vegetative cover of a site and all the living and dead organic material that it supports. Within this concept of canopy are a diversity of "habitats" that contain mites or, rather, one can find subunits that seem somehow discrete to people. Whether or not mites partition the canopy in this manner is one of the more intriguing questions in canopy research (Walter 2001). Next, I review some of the major habitat assemblages that can be recognized in the canopy and the types of mites associated with them.

Tree Trunks: Epicorticolous Habitats

The trunks of the trees that support forest canopies link the upper canopy with the ground, but rather than having an intermediate acarofauna, they tend to be dominated by characteristic trunk taxa. For example, sap fluxes, where sap oozes out of a wound, are colonized by Astigmatans, particularly histiostomatid mites, wading filter-feeders in many wet habitats (OConnor 1982). Certain families of Mesostigmata, Prostigmata, and Oribatida also live on tree trunks, especially in mosses and lichens (Walter et al. 1998; Walter and Behan-Pelletier 1999). Indeed, oribatid mites are often called "moss mites" and are typically the most abundant and diverse mites on cryptogam-covered tree boles. Larger epiphytes such as ferns and orchids can also be found on the main trunks.

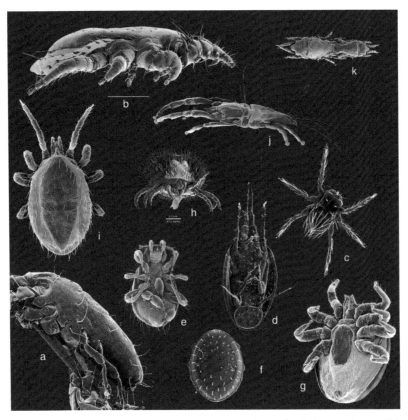

Figure 11-4 Symbiotic mites: (a) koala fur mite, *Koalachirus perkinsi* (Domrow) (Astigmata, Atopomelidae), on a koala hair; (b) a possum fur mite, *Trichosurolaelaps striatus* Domrow; (c) insect-parasitic stage (larva) of a trombidioid (Prostigmata, Trombidioidea); (d) insect-phoretic stage (hypopus) of a histiostomatid (Astigmata, Histiostomatidae); (e) ologamasid mite with phoretic hypopi; (f) an ant associated uropodid (Mesostigmata, Uropodidae); (g) *Myrmozercon* sp., a parasite of leaf-nesting *Polyrachis* ants; (h) a gecko mite, *Geckobia* sp. (Prostigmata, Pterygosomatidae); (i) a tropical fowl mite, *Ornithonyssus bursa* (Berlese); (j) a pigeon feather mite, *Falculifer* sp. (Astigmata, Falculiferidae); (k) mate-guarding in a rosella feather mite, *Rhytidelasma punctata* Mironov et al. (Astigmata, Pterolichidae). Scale bars, a-d at b; e-k at h, both = 0.1 mm.

Basidiomycete fungi often produce relatively large, fleshy to woody sporocarps on tree boles that soon become home to numerous insects and mites (Wheeler and Blackwell 1984). Mushrooms, conks, shelf fungi, and bracket fungi grow from standing dead wood or sprout from the heartwood of infected living trees. All are inhabited by distinct mite faunas that include Oribatida, Astigmata (especially Histiostomatidae), Prostigmata (especially Tarsonemidae), and Mesostigmata (especially Ascidae, Digamasellidae). One of the most common groups of mites associated with mushrooms are Astigmatans in the Histiostomatidae. These mites crawl through rotting sludge, filter-feeding on bacteria, yeasts, and very fine particulate organic matter (Walter and Kaplan 1990). Most mushrooms are an ephemeral resource, but shelf fungi (conks) tend to be woodier and persist much longer than mushrooms. Most conks belong to the Polyporales and contain many thousands of narrow spore tubes on their undersides. Some of them have highly elongate mites that live within the spore tubes (e.g., Lindquist 1965, 1995).

Perched Litter, Suspended Soil, and Hanging Humus

A rain of dead and dehisced plant material, dead and molted animal matter, and windblown soil is trapped by canopy epiphytes, crotches and crevices of trees, and the deeper cavities discussed below. Alliteratively described as suspended soil or hanging humus, this pool of perched nutrients and organisms is one of the least studied of canopy habitats (Basset 2001). Ongoing investigations of the Mesostigmatan fauna in Australian rainforests indicates that perched litter is inhabited by a diverse and distinct fauna (Walter et al. 1998). Although the generic composition of the perched and forest floor litter habitats are similar, the species assemblages are highly complementary. For example, at least seven species of ologamasid mites in the genus *Athiasella* (see Figure 11-3a) inhabit litter at Lamington National Park in Queensland, but three of these have been found only in the perched litter captured by epiphytic crow's nest ferns, three only in forest floor samples, and only one species in both habitats (Beaulieu and Walter, in review).

Suspended soils also have a diverse fauna of oribatid mites burrowing in decaying vegetation or grazing on fungi and other microbes growing on the detritus (see Figure 11-2c, g). The taxa found in suspended soils, however, are quite different from those found on leaves, stems, and trunks. This is also true of scavenger-microbivore Prostigmata (see Figure 11-2d) and Astigmata (see Figure 11-2e): species that graze on fungi in pockets of soil are not the same as those found wandering on the surfaces of plants. Other specialized faunas are those that inhabit fungal sporocarps, the galleries of wood-boring insects, and nests (Walter and Behan-Pelletier 1999; Shaw and Walter 2003).

Under Bark: Subcorticolous Habitats

Few mites burrow into the bark of living trees, but many are found in the galleries of insects, under dead bark, or in fungus-infected wounds. Galleries constructed under bark by wood-boring beetles are home to numerous Astigmata, Mesostigmata, and Prostigmata (Walter and Behan-Pelletier 1999). Tarsonemid mites sucking beetle eggs or fungi (and sometimes vectoring their spores) are especially common in this set of habitats (Lindquist 1986; Moser et al. 1989).

Dead wood on living trees forms a significant component of aboveground biomass and diversity, including nest sites for many vertebrates and insects. Colonization and decomposition may proceed for some time before the stem or trunk crashes to the forest floor. This standing dead wood is home to a variety of oribatid mites and other detritivores that invade from the bark or in association with wood-boring insects. Predatory mites follow the scavenger-microbivores and other prey (e.g., nematodes) into the wood. Areas of dead wood or damaged and healed-over trunk that become tree hollow nests accumulate their own special mite faunas (Shaw and Walter 2003).

Suspended Water: Phytotelmata

Pools of water that collect in cavities in plants, from the leaf axils of bromeliads to bamboo internodes to large tree holes, are referred to as phytotelmata (Kitching 2000, 2001). Except for those epiphytes that can stand direct exposure to the sun, most of these plant-borne habitats are in the middle to lower levels of the canopy, even down to water trapped in cavities at the base of trunks or in the axils of understory shrubs and herbs. Astigmatans in the families Acaridae, Algophagidae, and Histiostomatidae are the dominant phytotel acarofauna (Fashing 1994, 1998; Fashing et al. 2000), but some Mesostigmatans, oribatids, and occasionally a true water mite (Hydracarina) can be found in tree holes (Walter and Behan-Pelletier 1999; Kitching 2000, 2001). Bromeliads in the neotropics appear to have an especially rich, if understudied, fauna (Armbruster et al. 2002) and may represent an important pool of canopy mite diversity (Basset 2001).

Branches and Stems: The Rafters of the Canopy

Most of the woody surface area in forest canopies is occupied by branches and stems. Their bark varies from very smooth and relatively featureless to deeply fissured and from limited epiphytic growth to deep coverings of cryptogams. Large epiphytes with their accumulations of suspended soils perch on the larger branches that ramify near the main bole. In contrast, the smaller stems lack the deep crevices and thick mats of epiphytes often present on large stems, dry out quickly after rain, and are a relatively exposed and hostile habitat. During rains, stems channel water and resemble tiny ephemeral streams—with different faunas active during wet and dry periods. Leaf mites in transit can be found on small stems, but many others are stem specialists (Momen 1987; Walter et al. 1994; Walter 1995). For example, in Australian rainforests ornate false spider mites (Tuckerellidae, see Figure 11-1b), certain cruising predators (Eupallopsellidae, Mecognathidae), and many species of *Scapheremaeus* (Oribatida) are found almost exclusively on stems.

Leaves: The Phylloplane

The phylloplane (leaf surface) in a forest canopy is regularly exposed to intense sunlight, driving winds, and pouring rain. Intuitively, this does not seem a suitable habitat for oribatid mites— denizens of soils, bark crevices, and mosses. Species in 17 families, however, have been recorded from leaf surfaces (Behan-Pelletier and Walter 2000). Most of these mites graze on phylloplane fungi or scavenge dead plant material. Leaves with a tomentum often have large numbers of oribatid mites, perhaps because this microhabitat provides higher relative humidity than surrounding air and accumulates food such as pollen and fungi (Spain and Harrison 1968).

The most successful radiations of phylloplane mites are those that obtain fluids, nutrition, and often shelter from the leaf itself (see the "Plant-Parasites" section above), but the most abundant of all mites on rainforest canopy leaves are the microbivore-scavengers. The Tydeidae is perhaps the most ubiquitous of all foliar mite families: almost any tree, shrub, or liana that has not been treated by pesticides will have these mites. Tarsonemidae also are common phylloplane mites with about a dozen genera consisting of phylloplane specialists (Lindquist 1986). Among the most bizarre are species of *Fungitarsonemus* that glue pollen grains, spores, and other small bits of detritus to their backs or in patches on leaves where eggs and molting individuals are secreted.

Since mites are small, they tend to seek out small crevices in which to shelter. Being small also means that things such as a bee or a leaf may seem more like a densely forested mountain or vast plain to a mite, depending on the host's surface architecture. For example, smooth, rainshedding leaves are often veritable acarine deserts with little other than the occasional tydeid wandering the wasteland or crouching under whatever shelter the midvein provides. Some mites (e.g., spider mites, some rust mites) have adapted to this hostile environment by constructing web nests, cementing together small piles of debris (*Fungitarsonemus*) or hiding in empty insect exuviae or leaf spot fungi. In general, however, a smooth leaf is an empty one.

In contrast, leaves covered by a forest of hairs, at least if the hairs have space for mites to move among them, often have high densities and diversities of mites. The differences in mite populations between adjacent trees with differences in pubescence can be striking. Hairy leaves in tropical rainforests in Australia typically average three times as many species and five times as many individuals as smooth leaves at the same site. (Walter 1992; Walter and O'Dowd 1995b). Factors such as the density of hairs (Karban et al. 1995) or leaf domatia (see below) may strongly influence the distribution of predatory mites. Leaves with dense layers of appressed hairs or scales, however, have no place for mites to hide and are equivalent to smooth leaves as mite habitat.

Canopy Animals: A Rich Supply of Acarine Habitats

Canopy mammals, birds, snakes, lizards, amphibians, snails, myriapods, spiders, and all manner of insects are relatively large and rich arrays of habitats for symbiotic mites (Walter and Proctor 1999; Proctor and Owens 2000; Walter and Shaw, in press). The proportion of living acarine diversity that these commensals, parasites, and mutualists represent is difficult to quantify—or even to delimit. For example, mites live within the tracheae of bees; in the stink glands of bugs; within the vein cells on grasshopper wings; under the elytra of beetles; between the barbules of bird feathers (see Figure 11-4j-k); within feather quills; in the fur of mammals (see Figure 11-4a, b); in the nares, lungs, or airsacs of snails, frogs, birds, mammals, and reptiles; in the skin pores of frogs, birds, and mammals; in the eyes of bats, cloacas of turtles, and gills of crabs; and so on (Walter and Proctor 1999). The diversity of mites on a single host can be staggering. For example 18 species of mites have been collected from a single species of Australian passalid beetle (Seeman 2002) and more than 25 species of feather mites are known from a single species of neotropical rainforest parrot (Pérez 1997). Nests of vertebrates and social insects also are filled with mite inquilines and parasites, for example, about 50 genera of obligate myrmecophiles are known from ant nests (Eickwort 1990).

Accommodating Mites: Hair, Holdfasts, Acarinaria, and Acarodomatia

One indication of the importance of mite associations with other canopy organisms over evolutionary time is the production of plant or animal structures that accommodate mites. Perhaps the best known of these are the small "mites houses," or acarodomatia, that occur on the leaves of plants from more than 30 different families (Walter 1996). Even more bizarre are the mite pockets or acarinaria that occur on some insects and, perhaps, on some lizards (Walter and Proctor 1999).

Investigations into the responses of mites to leaf architecture are more than a century old (O'Dowd and Willson 1989), but only recently has it become clear that one particular class of leaf architecture, the domatium, is of critical importance to understanding the complexity of the relationship between mites and plants. Acarodomatia are plant structures (i.e., not mite-induced) that provide refugia for mites and small insects, almost always predators or scavenger-microbivores and only rarely plant-parasites. Plants with intact or well-developed leaf domatia usually have significantly more predators and scavenger-microbivores and fewer acarine, insect, and fungal plant parasites (Agrawal et al. 2000; Norton et al. 2000; Norton et al. 2001; Walter 1996; Walter and O'Dowd 1999). Domatia allow plants to retain potentially beneficial mites on leaves that would otherwise be attractive only to parasites, and support populations of scavenger-microbivores and predators characteristic of leaves with an open tomentum.

The same can be true for insect hosts. Although sparse to moderate setation allows mites to amble over an insect's surface and to grab a hair to increase their purchase, densely hairy bees may be mite-proofed. This is especially true for phoretic stages such as the Astigmatan deutonymph (hypopus) that usually relies on suckers (see Figures 11-2e, 11-4d) to hold onto the host's cuticle: suction requires a smooth surface to be effective. When densely hairy bees are host to mites, however, it is usually because of two phenomena: either the mites are able to penetrate the dense tomentum or the bees have kindly provided refugia in the form of mite pockets or hairless regions where mites can cluster (Walter et al. 2002).

Although few mite-bee associations have been studied in detail, those that are known cover the full range of possible interactions. Some mites feed on nest detritus, have no strong positive or negative effect on their "hosts," and are best considered commensals. In contrast, other mites are mutualists, essential for the successful development of bee brood. One way that wasps and bees can assure their nests contain these beneficial mutualists is to provide a secure hiding place on dispersing females: the smooth platforms or pocket-like recesses called acarinaria (OConnor 1993; Walter and Proctor 1999; Walter et al. 2002). Many bees and wasps produce acarinaria that have no obvious function other than to facilitate mite travel to the next nest: a pattern that strongly supports the idea of mutually beneficial interactions. Similar mutualisms may exist between mites and birds (Proctor and Owens 2000), but there is little evidence that canopy mammals benefit from their mite associates (Kim 1985).

Pondering the Imponderable: How Many Kinds of Mites?

May (1978, 1988) speculated that insects were highly diverse because they were small and that very small species such as mites may be even more diverse than insects. Since mites are smaller than most other canopy animals, and many live symbiotically on them, mite diversity is largely a function of the diversity and suitability of hosts. For example, most species of bees (Apoidea) host many species of symbiotic mites, and at least 32 lineages of mites have independently colonized the bee-habitat (Eickwort 1994; Walter et al. 2002). Individual bee-mite species tend to be geographically restricted in their distribution and to be host-specific at the species, species-complex, or generic levels. About 25,000 species of bees have been described, and perhaps 40,000 currently exist; therefore, the diversity of bee-mites alone is likely to exceed the currently described diversity of the Acari. As detailed above, repeated colonization and diversification by lineages of symbiotic mites has occurred in association with many species-rich groups of both plants and animals. The canopy is the biotic frontier where these intimate relationships between mites, other animals, plants, and microbes are least known.

Future Directions

Almost a decade ago in the first version of this chapter, the authors posed four questions about canopy mites:

1. What factors limit the populations of plant parasitic mites in rainforest canopies?
2. Who eats whom in the rainforest canopy?
3. Are scavenging mites good for plants?
4. How important are mites to canopy biodiversity?

This chapter is a partial answer to the last question: mites are a major component of forest canopy diversity and they interact strongly with the other organisms that share this complex of habitats. Unfortunately, in most cases, the first three questions—those that deal with understanding the ecology and evolution of canopy mites—remain largely unanswered. Fortunately, there is one exception: the importance of microbivore-scavengers to plant health.

Although the earliest work on mites and leaf domatia was inspired by a belief that mites protected trees from fungal disease (O'Dowd and Willson 1989), mite–fungus–plant interactions received experimental attention only recently. Norton et al. (2000) examined the influence of domatia on the abundance of a tydeid mite on wild grape and its impact on a key fungal pathogen, grape powdery mildew. In two field experiments, they found that:

1. Mite densities were significantly higher on plants with intact domatia than on plants with blocked domatia;
2. Grape clones with larger domatia had significantly more mites than clones with smaller domatia; and,
3. In two of three experiments plants with intact or larger domatia had significantly less mildew.

Norton et al. (2001) found that domatia on wild grape protected beneficial mites from larger predaceous insects but gave no protection against low humidity under laboratory conditions. Thus, at least one aspect of mite canopy ecology has become better understood: microbivore-scavengers can protect plants from pathogens, and leaf domatia are integral to this interaction. The challenge for future canopy acarophiles is to develop experiments to illuminate other interactions between mites and other canopy inhabitants.

Acknowledgments

Over the years, many people have contributed to my understanding of canopy mites, but two deserve special recognition: Dennis O'Dowd, who first lifted my eyes out of the soil, and Heather Proctor, who has kept me synergistically enthused about mites wherever we found them. Thanks also to the students and technicians who have helped along the way, especially Catherine Harvey, Caroline Meacham, Anthony O'Toole, and Matthew Shaw for help with some of the figures. Special thanks are also due to the Australian Research Council and the Australian Biological Resources Study for funding much of my research.

References

Agrawal, A.A, Karban, R., and Colfer, R.G. (2000). How leaf domatia and induced plant resistance affect herbivores, natural enemies and plant performance. *Oikos* **89**, 70–80

Amrine, J.W.J. and Stasny, T.A. (1994). "Catalog of the Eriophyoidea (Acarina: Prostigmata) of the World." Indira Publishing House, West Bloomfield, Michigan.

Armbruster, P., Hutchinson. R.A., and Cotgreave, P. (2002). Factors influencing community structure in a South American tank bromeliad fauna. *Oikos* **96**, 225–234.

Basset, Y. (2001). Invertebrates in the canopy of tropical rain forests: How much do we really know? *Plant-Ecology* **153**, 87–107.

Beaulieu, F. and Walter, D.E. (In review). Habitat specificity and predatory mites: the genus *Athiasella* (Mesostigmata: Ologamasidae) of Australian rainforest canopies. Proceedings of the XIth International Congress of Acarology, Merida, Mexico, July 2003.

Behan-Pelletier, V. and Walter, D.E. (2000). Biodiversity of oribatid mites (Acari: Oribatida) in Tree Canopies and Litter. *In* "Invertebrates as Webmasters in Ecosystems" (D.C. Coleman and P. Hendrix, Eds.), pp. 187–202. CABI, Wallingford.

Bolland, H.R., Gutierrez, J., and Flechtmann, C.H.W. (1998). "World Catalogue of the Spider Mite Family (Acari: Tetranychidae)." Brill, Boston.

Colwell, R.K. (1986). Community biology and sexual selection: lessons from hummingbird flower mites. *In* "Community Ecology" (J. Diamond and T.J. Case, Eds.), pp. 406–424. Harper and Row, New York

Colwell, R.K. (1995). Effects of nectar consumption by the hummingbird flower mite *Proctolaelaps kirmsei* on nectar availability in *Hamelia patens*. *Biotropica* **27**, 206–217

Colwell, R.K. and Naeem, S. (1994). Life-history patterns of hummingbird flower mites in relation to host phenology and morphology. *In* "Mites: Ecological and Evolutionary Analysis of Life-History Patterns" (M Houck, Ed.), pp. 23–44. Chapman and Hall, New York.

Compton, S.G. (1993). One way to be a fig. *African Entomol.* **1**, 151–158

Eickwort, G.C. (1990). Associations of mites with social insects. *Annu. Rev. Entomol.* **35**, 469–488

Eickwort, G.C. (1994). Evolution and life-history patterns of mites associated with bees. *In* "Mites: Ecological and Evolutionary Analysis of Life-History Patterns" (M Houck, Ed.), pp. 218–251. Chapman and Hall, New York.

Fashing, N.J. (1994). Life-history patterns of astigmatid inhabitants of water-filled treeholes. *In* "Mites: Ecological and Evolutionary Analysis of Life-History Patterns," (M Houck, Ed.), pp. 160–185. Chapman and Hall, New York.

Fashing, N.J. (1998). Functional morphology as an aid in determining trophic behaviour: the placement of astigmatic mites in food webs of water-filled tree-hole communities. *Exper. Appl. Acarol.* **22**, 435–453.

Fashing, N.J., OConnor, B.M., and Kitching, R.L. (2000). *Lamingtonacarus*, a new genus of Algophagidae (Acari: Astigmata) from water-filled treeholes in Queensland, Australia. *Invert. Taxon.* **14**, 591–606

Gerson, U. and Izraylevich, S. (1997). A review of host utilization by *Hemisarcoptes* (Acari: Hemisarcoptidae) parasitic on scale insects. *Syst. Appl. Acarol.* **2**, 33–42.

Gerson, U. and Smiley, R.L. (1990). "Acarine Biocontrol Agents: An Illustrated Key and Manual." Chapman and Hall, Melbourne.

Hallan, J. (2003). Genera of Acari. Texas A&M University, Department of Entomology, Insect Collection. http://insects.tamu.edu/research/collection/hallan/acarallgen.html

Halliday, R.B. (1996). Revision of the Australian Ameroseiidae (Acarina: Mesostigmata). *Invert. Tax.* **10**, 179–201

Hayes, J.L. (1985). The predator–prey interaction of the mite *Balaustium* sp. and the pierid butterfly *Colais alexandra*. *Ecology* **66**, 300–303.

Helle, W. and Sabelis, M. W. (Eds.) (1985). *In* "Spider Mites, Their Biology, Natural Enemies, and Control, Volume 1A." Elsevier, New York

Heyneman A.J., Colwell, R.K., Naeem, S., Dobkin, D., and Hallet, B. (1991). Host plant discrimination: experiments with hummingbird flower mites. *In* "Plant Animal Interactions: Evolutionary Ecology in Tropical and Temperate Regions" (P.W. Price, T.M. Lewisohn, G.W. Fernandes, and W.W. Benson, Eds.), pp. 455–485. John Wiley and Sons, New York.

Ho, C-C. (1994). A new genus and two new species of Tarsonemidae from *Ficus* spp. (Acari: Heterostigmata). *Int. J. Acarol.* **20**, 189–197.

Houck, M.A. (1994). Adaptation and transition into parasitism from commensalism, a phoretic model. *In* "Mites: Ecological and Evolutionary Analysis of Life-History Patterns" (M Houck, Ed.), pp. 252–281. Chapman and Hall, New York.

Hurst, G.D.D., Sharpe, R.G., Broomfield, A.H., Walker, L.E., Majerus, T.M.O., Zakharov, I.A., and Majerus, M.E.N. (1995). Sexually transmitted disease in a promiscuous insect, *Adalia bipunctata. Ecol. Entomol.* **20**, 230–236.

Husband, R.W. and de Moraes, G.J., (1999). A new species of *Chrysomelobia* (Acari, Podapolipidae) from *Platyphora testudo* (Demay) (Coleoptera, Chrysomelidae) from Peru, with a key to known species of the genus. *Int. J. Acarol.* **25**, 309–315.

Karban, R., English-Loeb, G., Walker, M.A., and Thaler, J. (1995). Abundance of phytoseiid mites on *Vitis* species, effects of leaf hairs, domatia, prey abundance and plant phylogeny. *Exper. Appl. Acarol.* **19**, 189–197.

Kim, K.C. (Ed.) (1985). "Coevolution of Parasitic Arthropods and Mammals." John Wiley & Sons, Brisbane.

Kitching, R.L. (2000). "Food Webs and Container Habitats: The Natural History and Ecology of Phytotelmata." Cambridge University Press, Cambridge.

Kitching, R.L. (2001). Food webs in phytotelmata: "bottom-up" and "top-down" explanations for community structure. *Annu. Rev. Ecol. Syst.* **46**, 729–760.

Lindquist, E.E. (1965). An unusual new species of the genus *Hoploseius* Berlese (Acarina, Blattisociidae) from Mexico. *Can. Entomol.* **97**, 1121–1131.

Lindquist, E.E (1986). The world genera of Tarsonemidae (Acari, Heterostigmata), a morphological, phylogenetic, and systematic revision, with a reclassification of the family-group taxa in the Heterostigmata. *Mem. Entomol. Soc. Can.* **136**, 1–517.

Lindquist, E.E. (1995). Remarkable convergence between two taxa of ascid mites (Acari, Mesostigmata) adapted to living in pore tubes of bracket fungi in North America, with description of *Mycolaelaps* new genus. *Can. J. Zool.* **73**, 104–28.

Lindquist, E.E. (1998). Evolution of phytophagy in trombidiform mites. *Exp. Appl. Acarol.* **22**, 81–100

Lindquist, E.E., Sabelis, M.W., and Bruin, J. (1996). "Eriophyoid Mites, Their Biology, Natural Enemies and Control." *World Crop Pests* Volume 6, Elsevier, Amsterdam.

May, R.M. (1978). The dynamics and diversity of insect faunas. *In* "Diversity of Insect Faunas" (L.A. Mound and N. Waloff, Eds.), pp. 188–204. Blackwell Scientific Publications, Oxford.

May, R.M. (1988). How many species are there on earth? *Science* **241**, 1441–1449.

McMurtry, J.A. and Croft, B.A. (1997). Life-styles of phytoseiid mites and their roles in biological control. *Annu. Rev. Entomol.* **42**, 291–321.

Momen, F.M. (1987). The mite fauna of an unsprayed apple orchard in Ireland. *Zeitschrift fuer Angewandte Zoologie* **74**, 417–434.

Moser, J.C., Perry, T.J., and Solheim, H. (1989). Ascospores hyperphoretic on mites associated with *Ips typographus. Mycol. Res.* **93**, 513–517

Nadkarni, N.M. and Longino, J.T. (1990). Invertebrates in canopy and ground organic matter in a neotropical montane forest, Costa Rica. *Biotropica* **22**, 286–289.

Naskrecki, P. and Colwell, R.K. (1998). Systematics and host plant affiliations of hummingbird flower mites of the genera *Tropicoseius* Baker and Yunker and *Rhinoseius* Baker and Yunker (Acari, Mesostigmata, Ascidae). Thomas Say Foundation Monographs, Entomological Society of America, Washington DC.

Norton, A.P., English-Loeb, G., and Belden, E. (2001). Host plant manipulation of natural enemies, Leaf domatia protect beneficial mites from insect predators. *Oecologia* **126**, 535–542.

Norton, A.P., English-Loeb, G., Gadoury, D., and Seem, R.C. (2000). Mycophagous mites and foliar pathogens, Leaf domatia mediate tritrophic interactions in grapes. *Ecology* **81**, 490–499.

Norton, R.A. (1994). Evolutionary aspects of oribatid mite life histories and consequences for the origin of the Astigmata. *In* "Mites: Ecological and Evolutionary Analysis of Life-History Patterns" (M Houck, Ed.), pp. 99–135. Chapman and Hall, New York.

OConnor, B.M. (1982). Evolutionary ecology of astigmatid mites. *Annu. Rev. Entomol.* **27**, 385–409

OConnor, B.M. (1993). The mite community associated with *Xylocopa latipes* (Hymenoptera, Anthophoridae, Xylocopinae) with description of a new type of acarinarium. *Intern. J. Acarol.* **19**, 159–166

OConnor, B.M., Colwell, R.K, and Naeem, S. (1991). Flower mites of Trinidad II. The genus *Proctolaelaps* (Acari, Ascidae). *Gr. Basin Naturalist* **51**, 348–376

OConnor, B.M., Colwell, R.K., and Naeem, S. (1997). The flower mites of Trinidad III. The genus *Rhinoseius* (Acari, Ascidae). *Misc. Publ. Mus. Zool. Univ. Michigan* **184**, 1–32

O'Dowd, D.J. and Willson, M.F. (1989). Leaf domatia and mites on Australian plants, ecological and evolutionary implications. *Biol. J. Linn. Soc.* **37**, 191–236.

O'Dowd, D.J., Brew, C.R., Christophel, D.C., and Norton, R.A. (1991). Mite-plant associations from the Eocene of southern Australia. *Science* **252**, 99–101

Ohmer C., Fain, A., and Schuchmann, K.L. (1991). New ascid mites of the genera *Rhinoseius* Baker and Yunker, (1964, and *Lasioseius* Berlese 1923, Acari, Gamasida, Ascidae) associated with hummingbirds or hummingbird-pollinated flowers in southwestern Columbia. *J. Nat. Hist.* **25**, 481–497

Paciorek, C.J., Moyer, B.R., Levin, R.A., and Halpern, S.L. (1995). Pollen consumption by the hummingbird flower mite *Proctolaelaps kirmsei* and possible fitness effects on *Hamelia patens*. *Biotropica* **27**, 258–262

Paoletti, M.G., Stinner, B.R., Stinner, D. Benzig, D., and Taylor, R. (1991). Diversity of soil fauna in the canopy and forest floor of a Venezuelan cloud forest. *J. Trop. Ecol.* **7**, 373–383

Pérez, T.M. (1997). Eggs of feather mite congeners (Acarina, Pterolichidae, Xolalgidae) from different species of New World Parrots (Aves, Psittaciformes). *Int. J. Acarol.* **23**, 103–106

Proctor, H. C., and Owens, I. (2000). Mites and birds. *Tr. Ecol Evol.* **15**, 358–364.

Saito, Y. (1985). Life types of spider mites. *In* "Spider Mites: Their biology, Natural Enemies and Control. Volume 1a." (W. Helle and M.W. Sabelis, Eds.), pp. 253–264. Elsevier, New York.

Saito, Y. (1986a). Prey kills predator, counter-attack success of a spider mite against its specific phytoseiid predator. *Exp. Appl. Acarol.* **2**, 47–62.

Saito, Y. (1986b). Biparental defence in a spider mite (Acari, Tetranychidae) infesting Sasa bamboo. *Behav. Ecol. Sociobiol.* **18**, 377–386.

Saito, Y. (1997). Sociality and kin selection in Acari. *In* "Evolution of Social Behaviour in Insects and Arachnids" (J. C. Choe and B. Crespi, Eds.), pp. 443–457. Cambridge University Press, Cambridge.

Saito, Y. (2000). Do kin selection and intra-sexual selection operate in spider mites? *Exp. Appl. Acarol.* **24**, 351–363.

Seeman, O.D. and Walter, D.E. (1995). Life history of *Afrocypholaelaps africana* (Evans) (Parasitiformes, Ameroseiidae), a mite inhabiting mangrove flowers and phoretic on honeybees. *J. Aust. Entomol. Soc.* **34**, 45–50

Seeman, O.D. (1996). Flower mites and phoresy, the biology of *Hattena panopla* Domrow and *Hattena cometis* Domrow. *Aus. J. Zool.* **44**, 193–203.

Seeman, O.D. (2002). "Mites and Passalid Beetles, Diversity, Taxonomy and Biogeography." PhD thesis, University of Queensland.

Shaw, M. and Walter, D.E. (2003). Hallowed hideaways: tree hollows and allied habitats as "hotspots" for early derivative mites. *In* "Arthropods of Tropical Forests, Spatio-Temporal Dynamics and Resource Use in the Canopy" (Y. Basset, R. Kitching, S. Miller, and V. Novotny, Eds.), pp. 291–303. Cambridge University Press, Harvard.

Spain, A.V. and Harrison, R.A. (1968). Some aspects of the ecology of arboreal Cryptostigmata (Acari) in New Zealand with special reference to the species associated with *Olearia colensoi* Hoof. f. *N.Z. J. Sci.* **11**, 452–458.

Walter, D.E. (1992). Leaf surface structure and the distribution of *Phytoseius* mites (Acarina, Phytoseiidae) in south-east Australian forests. *Aust. J. Zool.* **40**, 593–60.

Walter, D.E. (1995). Dancing on the Head of a Pin, Mites in the Rainforest Canopy. *Rec. W. Aust. Mus. Suppl.* **52**, 49–53.

Walter, D.E. (1996). Living on leaves, Mites, tomenta, and leaf domatia. *Annu. Rev. Entomol.* **41**, 101–114.

Walter, D.E. (2000). First record of a fig mite from the Australian Region, *Paratarsonemella giblindavisi* (Acari, Tarsonemidae). *Aust. J. Entomol.* **39**, 229–232.

Walter, D.E. (2001). Achilles and the mite, Zeno's Paradox and rainforest mite diversity. *In* "Acarology, Proceedings of the 10th International Congress" (R.B. Halliday, D.E. Walter, H.C. Proctor, R.A. Norton, and M.J. Colloff, Eds.), pp. 113–120. CSIRO Publishing, Melbourne.

Walter, D.E. and Behan-Pelletier, V. (1999). Mites in forest canopies, filling the size distribution shortfall? *Annu. Rev. Entomol.* **44**, 1–19.

Walter, D.E. and Kaplan, D.T (1990). Feeding observations on two astigmatic mites, *Schwiebea rocketti* Woodring (Acaridae) and *Histiostoma bakeri* Hughes and Jackson, associated with citrus feeder roots. *Pedobiologia* **34**, 281–286.

Walter, D.E. and Lindquist, E.E. (1997). Australian species of *Lasioseius* (Acari, Mesostigmata, Ascidae), the *porulosus* group and other species from rainforest canopies. *Invert. Taxon.* **11**, 525–547.

Walter, D.E. and O'Dowd, D.J. (1995a). Life on the forest phylloplane, hairs, little houses, and myriad mites. *In* "Forest Canopies" (M.D. Lowman and N. Nadkarni, Eds.), pp. 325–351. Academic Press, New York.

Walter, D.E. and O'Dowd, D.J. (1995b). Beneath biodiversity, factors influencing the diversity and abundance of canopy mites. *Selbyana* **16**, 12–20.

Walter, D.E., and O'Dowd, D.J. (1999). The good, the bad, & the ugly: Which really inhabit leaf domatia? *In* "Acarology IX: Symposia" (G.R. Needham, R. Mitchell, D.J. Hornand, C.A. Welbourn, Eds.), pp. 215–220. Ohio Biological Survey, Columbus, Ohio.

Walter, D.E. and Proctor, H.C. (1998). Predatory mites in tropical Australia, local species richness and complementarity. *Biotropica* **30**, 72–81.

Walter, D.E. and Proctor, H.C. (1999). "Mites, Ecology, Evolution and Behaviour." University of NSW Press, Sydney and CABI, Wallingford.

Walter, D.E. and Shaw, M. (In press). Mites and disease. *In* "Biology of Disease Vectors, Second Edition" (W.C. Marquardt, W.C. Black, J. Freier, H. Hagedorn, J. Hemingway, S. Higgs, A.A. James, and B. Kondratieff, Eds.). Academic Press, New York.

Walter, D.E., Halliday, R.B., and Lindquist, E.E. (1993). A review of the genus *Asca* (Acarina, Ascidae) in Australia, with descriptions of three new leaf-inhabiting species. *Invert. Taxon.* **7**, 1327–1347.

Walter, D.E., O'Dowd, D.J., and Barnes, V. (1994). The forgotten arthropods, Foliar mites in the forest canopy. *Mem. Qld. Mus.* **36**, 221–226.

Walter, D.E., Seeman, O., Rodgers, D., and Kitching, R.L. (1998). Mites in the mist: How unique is a rainforest canopy knockdown fauna? *Aust. J. Ecol.* **23**, 501–508.

Walter, D.E., Beard, J.J., Walker, K., and Sparks, K. (2002). Of mites and bees, A review of mite-bee associations in Australia and a revision of *Raymentia* (Acari, Mesostigmata, Laelapidae), with the description of two new species of mites from *Lasioglossum* (*Parasphecodes*) (Hymenoptera, Halictidae). *Aus. J. Entomol.* **41**, 128–148.

Wheeler, Q. and Blackwell, M. (Eds.) (1984). "Fungus-Insect Relationships: Perspectives in Ecology and Evolution." Columbia University Press, New York.

Winchester, N.N., Behan-Pelletier, V., and Ring, R.A. (1999). Arboreal specificity, diversity and abundance of canopy-dwelling oribatid mites (Acari, Oribatida). *Pedobiologia* **43**, 391–400.

CHAPTER 12

Soil Microarthopods: Belowground Fauna that Sustain Forest Systems

H. Bruce Rinker

Even among the tiniest creatures of the insect world, as well as among the larger forms of life, there are mysteries which all science is at a loss to explain.
—*Edwin Way Teale,* Grassroot Jungles: A Book of Insects, *1944*

Forest Soil Fauna: A Gauntlet for Erwin's Speculations

Erwin (1982, 1983) suggested a total of about 30 million insect species worldwide, based on a number of broad assumptions, including a laconic note in both of his articles about soil fauna: "My own observations indicate that the canopy is at least twice as rich in species as the forest floor." This number has been challenged in recent decades (e.g., see Stork 1988; Lowman et al. 1991; André et al. 1994) because of new insights on tree phenologies and soil processes. André et al. (1994) applied Erwin's assumptions to their analysis of sand dune fauna in southern France and then extrapolated to an impractical estimate between 29 million and 366 million species of arthropods in the tropics! Clearly, we do not know—even to the nearest order of magnitude—how many brethren species accompany us on the planet (Wilson 1992). Forest soils, along with forest canopies, represent two of the most promising frontiers of biological diversity now under scientific investigation.

The distribution, abundance, and diversity of edaphic (or soil) fauna appear to exceed all previous expectations. No natural terrestrial environments exist that exclude soil invertebrates (Anderson 1987). They include microfauna, defined as organisms <100 μm in length (e.g., bacteria, nematodes, fungi, and protozoans); mesofauna such as mites and springtails that are typically 100 μm to 2 mm long; and macrofauna >2 mm in length (Anderson 1988). Collectively, soil biota are divided into two ecological categories: hemiedaphon (organisms in the surface organic layer, historically emphasized in the literature) and euedaphon (organisms in the underlying mineral soils, historically neglected by field researchers). Mites and springtails, also called microarthropods (but not exclusively), are significant mesofauna because of their moderate impacts on energy flows and great influence on other fauna in the surface layers of forest soils (Paoletti et al. 1991). Though taxonomic evaluations are far from complete for microarthropod groups, 30,000 species of mites are known and more than 2,000 types of springtails described. The former can reach hundreds of thousands of individuals per m^3 of surface soil, the latter up to 100,000 springtails per m^3 (see Luxton 1972, 1975, 1981a, 1981b for comprehensive studies of mite taxa in Europe; Wallwork 1976, 1983; Anderson 1978; Petersen and Luxton 1982; Schenker 1984). That said, at most 10 percent of soil microarthropod populations has been explored and 10 percent of species described (André et al. 2002). Their prodigious numbers, along with speculations about their species richness, corroborate the belief that microarthropods are vital to the health and integrity of forest ecosystems.

As biodiversity is a component of the ecological health (or the autopoietic environmental well-being) of forests, so is health a subset of the biological integrity (or the wholeness) of those same living systems (see Pimentel et al. 2000). A patient can have a missing part—a finger, a kidney, or an eye—and still be healthy; but, undoubtedly, she has lost some of her integrity. A forest can lose one or several species—the passenger pigeon, the American chestnut, a mite or springtail species—and still maintain itself; but no one would argue that the system retains its original integrity. Though the autotrophs are the primary drivers for forest ecosystems, determining carbon inputs, heterotrophs such as microarthropod decomposers are ultimately responsible for governing the availability of nutrients required for primary productivity (Wardle 2002). Plants and decomposers are then obligate mutualists for the long-term maintenance of forest ecosystems. Is functional diversity or taxonomic diversity of decomposers, however, crucial for the forest's well-being and wholeness?

A Broad Spectrum of Queries for Canopy and Soil Scientists

A review of the recent literature on microarthropods provides an initial list of latter-day unknowns for forest ecologists as we ponder the functional diversity vs. taxonomic diversity of soil fauna. Aspects of each question are addressed throughout this chapter on ground decomposers:

- How many species of soil microarthropods exist worldwide (e.g., André et al. 1994, 2002)? Soils, including the deepest horizons and the rhizosphere, constitute a huge reservoir of under-studied biodiversity, yet our sampling techniques may be inadequate to assess the diversity and abundance of soil fauna. André et al. (2002) provided a thorough review of current extraction methods.
- How do the diversity and function of soil microarthropods in temperate areas, showing marked seasonality, compare to those in the less seasonal tropical regions (e.g., Heneghan et al. 1999; Reynolds and Hunter 2001; Rinker et al. 2001)? Decomposition in temperate regions proceeds more slowly than in the humid tropics. Thus, seasonal climatic variability acts as a constraint on biotic activity. In the tropics, however, decomposition is strongly influenced by site-specific characteristics of the biota, such as faunal assemblage structure (Heneghan et al. 1999). This has significant implications for land-use practices and for various scenarios for global climate change.
- What is the explicit role of soil microarthropods in the regulation of nutrient fluxes among litter, soil, and plant roots (e.g., Anderson 1987, 1988) or between forest canopy and forest soils (e.g., Reynolds et al. 2000)? As integrating variables, the three processes of decomposition, soil respiration, and nutrient cycling are important in our understanding of soil systems (Reynolds and Hunter 2001). After initial colonization of soil material by bacteria and fungi, microarthropods and other invertebrates physically break down the litter (comminute). Thus, invertebrates affect directly and indirectly the processes of carbon and nitrogen mineralization over a wide range of spatial/temporal scales (Anderson 1988; Lovett and Rusink 1995). Recent studies have provided strong evidence for a top-down ecological link between canopy herbivores and soil processes (Reynolds and Hunter 2001).
- Could more studies integrating the work of canopy ecologists and soil microbiologists help to assess whether these ecological roles have manipulative potential or can be regarded as a source

of variation or "noise" within extremely heterogeneous soil systems (e.g., Anderson 1987)? Rather than disconnected autecologies, can these roles be tested experimentally (e.g., Anderson 1978; Seastedt 1984; Seastedt and Crossley 1984; Crossley et al. 1992)? Interdisciplinary studies involving botanists, palynologists, entomologists, and chemists might prove the most beneficial to the spatial/temporal analysis of intricate soil processes.

- Is biodiversity per se vital for belowground ecological functioning (e.g., Crossley et al. 1992; Henghan et al. 1999; Laakso and Setälä 1999; Oldeman 2001)? How do stratified diversity and function compare belowground and aboveground (e.g., Moore 1988; Moore et al. 1988)? How are they linked (e.g., van Breemen 1992; Strong et al. 1996)? Microarthropods play a major role in the regulation of microbial populations and the decomposition of belowground detritus (Moore 1988). Thus, they are strong mediators for food web stability (Moore et al. 1988). Is the apparent ecological redundancy among soil invertebrates a primary or secondary aspect of ecosystem performance?

These are just a few of the pressing issues for forest canopy and forest soil scientists. Addressing them will clarify the ongoing row over functional diversity vs. taxonomic diversity in forest systems, especially regarding the ecological links between canopy herbivory and soil fauna (see Wardle 2002).

Functional Diversity vs. Taxonomic Diversity in Forest Systems: More Unknowns

Though documentation is scant, a number of authors (e.g., Crossley et al. 1992; Heneghan et al. 1999; Laakso and Setälä 1999; Oldeman 2001) hint that ecological function, rather than biological diversity, among edaphic fauna may be their most crucial contribution to the complex food webs of forest soils. Unlike the strict delineations typical in aboveground trophic pyramids, belowground biota may commonly exhibit a fluctuating feeding process called switching or different-channel omnivory (Moore et al. 1988). Switching may introduce longer-than-expected food chains into forest systems, making them, ironically, more stable in the long run (Moore 1988). Soil invertebrates consume 20 to 100 percent of annual litter input and, by doing so, produce an immense amount of excrement (Webb 1977) that then enters the forest system. Thus, the precise species of forest soil biota, presently a troubling uncertainty among taxonomists, may be a vital though secondary aspect of ecosystem performance compared to their collective trophic assemblages. "Species are carriers of ecological functions, but many redundant species or species webs each ensure one and the same ecological function. Species counts hence are not false, but they are impractical as a foundation for canopy theory" (Oldeman 2001). It's not what they are, but what they do that counts.

Over time, soils develop horizons that result from the interactions of climate, organisms, parent material, and topology (Jenny 1980; Coleman and Crossley 1996). Like the forests above them, soil profiles eventually exhibit pronounced stratification, each layer showing its own complex of microhabitats such as litter layers and a hierarchical matrix of aggregates. Also like layers of forest vegetation, components and processes in belowground strata remain understudied. Unlike forests, however, the constituents of soil food webs appear ambiguous at times regarding their numerous ecological associations. Microarthropods live on litter surfaces and in air-filled macropores in small-scale soil environments (Anderson 1987), comminuting litter and associating with other soil fauna. Some litter-living species, specialists as long as leaf matter remains discreet, become generalists after litter is comminuted and reduced by microbes and fungi (Anderson 1978). Others can fluctuate among competition, predation, and mutualism in ways that warrant closer study (Moore 1988). Unlike the more easily definable food web above ground, soil microarthropods often manifest indistinctness and redundancy in their ecological functions (Laakso and Setälä 1999). "Paradoxically, most soil animals show little evidence of trophic specialization and large numbers of species apparently co-exist while utilizing rather similar food resources" (Anderson 1978). These tiny organisms then represent a promising mosaic of lithe interactions that, undoubtedly, will enhance our understanding of forest health and integrity.

The forest canopy includes both the structural and the functional components of the aboveground strata of forest vegetation (even the epiphytes and vines), their attending fauna, and, importantly, all their ecological interactions and processes (Nadkarni 1995; Parker 1995; Moffett 2000). An expanded meaning of the term "canopy" may eventually include soils as part of the forest canopy if, indeed, measurable links are established definitively between aboveground and belowground strata to show these layers as part of a single system. In time, "canopy" may have to be abandoned altogether in the scientific literature because of its ambiguity in favor of a term as inelegant as biological column or biocolumn to describe the multi-layered, almost fractal profile of forest architecture and composition sandwiched between atmosphere and bedrock. As it has become more generally used, "canopy" has lost meaning like other words that once had specific significance in ecology (e.g., "stability," "equilibrium," and "keystone species"). Further, multi-trophic analysis of ecosystem function, examining possible links between canopy and soils, may be the most pragmatic approach toward understanding the full nature of forests, given that (1) soil biodiversity within trophic groups may be poorly related to ecosystem functioning, (2) redundancy may exist at the species level in edaphic food webs, and (3) the belowground compartment of forest systems may be influenced mainly by the physiological attributes of whatever species happen to be present (Wardle 1999). This kind of analysis is also the most complex approach, requiring long-term ecological study in multiple spatial/temporal settings.

Reynolds and Hunter: A Temperate Model for Top-Down Linkage

Conceptual gaps remain in our knowledge about the linkages of aboveground and belowground organisms. For example, detrital food webs show a high degree of aggregation of taxa, yet little is known about the diets and physiology of many edaphic species (Hunt et al. 1987). Despite much speculation, little data exist concerning the role played by microarthropods in nutrient transformation in soil (Ineson et al. 1982). Except for recent work in the mountains of western North Carolina (Reynolds and Hunter 2000, 2001; Reynolds et al. 2003) and in the tropical moist forest of eastern Puerto Rico (Rinker, in preparation), few studies have quantified the linkages between canopy herbivory and soil decomposition in forest ecosystems.

Reynolds et al. (2000) and Reynolds et al. (2003) tested whether inputs from canopy herbivores affected critical soil processes such as respiration, nutrient cycling, and decomposition. Their results, and early results from the parallel study in Puerto Rico, shed much light on the ecological linkages between the top and bottom of the biological column in forests. Corroborating a few other studies (Kolb 1991; Schowalter et al. 1991; Lovett and Rusink 1995; Chen and Wise 1997, 1999; Stadler et al. 2001), Reynolds and Hunter found that forest canopies and soils are indeed linked quantitatively through the process of insect herbivory. They observed increases in four groups of soil fauna (collembolans, prostigmatid mites, bacterial-feeding nematodes, and fungal-feeding nematodes), following the addition of herbivore-derived inputs to the forest floor. For the collembolans, at least, the addition of insect frass had a strong, positive effect on their abundance, suggesting that herbivore-derived inputs from the forest canopy can play a role in the spatial and temporal dynamics of microarthropods that then directly affect soil processes (Reynolds et al. 2003). Litter provides a major source of carbon for soil respiration and may contribute to a microclimate that favors decomposers (Reynolds and Hunter 2001). Increased soil respiration indicates greater microbial activity that then increases decomposition. Microarthropods often track this decomposition, especially as soil fungi become established, thereby increasing the rate of decomposition even more through the comminution of litter.

Undoubtedly, parallel studies in tropical and subtropical forests will provide additional evidence for the importance of inputs from canopy herbivores on soil decomposition. For example, a project proposed for Fall 2003 in the coastal Panamanian rain forest of San Lorenzo Protected Area used the San Lorenzo canopy crane and the French-sponsored canopy raft (Yves Basset, pers. comm.; see Rinker et al. 1995 for applicable aspects of the mission's operations) to study vertical stratification and β-diversity of arthropods including edaphic Acari and Collembola similar to work by Hammond et al. (1997) in Sulawesi. Though this large-scale study focused on the abundance and diversity of tropical species in forest soil vs. canopy habitats (but not on ecological linkages), it promises to be an important parallel to the work of Reynolds et al. (2003) in western North Carolina and Rinker (in preparation) in eastern Puerto Rico. Both analyses measured the ecological links between soil and canopy strata in temperate and tropical forests.

Trickles and Cascades in Forest Systems

The scientific literature is replete with acknowledgments that bemoan our meager grasp of the linkages between forest canopies and forest soils. Aboveground and belowground compartments in forest ecosystems are coupled intuitively because of the connections between the processes of production (carbon input) and decomposition (carbon loss). Based on the limited data available, the extent of those connections seems to be case-specific, especially regarding trophic cascades (Wardle 1999).

Trophic cascades are sequences of resource consumption along a food chain induced by top predators that eventually influence lower trophic levels (Halaj and Wise 2001). Strong (1992) labeled these as "archetypical trophic cascades." Strong (1992) also identified them as runaway consumption exerting a downward dominance through a food chain. They are well documented in aquatic systems (e.g., Carpenter 1985; Strong 1992), but the evidence for such chains in terrestrial systems is equivocal (Strong et al. 1996; Pace et al. 1999; Wardle 1999; Holt 2000; Polis et al. 2000; Power 2000; Halaj and Wise 2001) because these tend to be speciose systems with suppressing forces, such as herbivory and interspecific competition, set in a stochastic habitat matrix. In some cases, there is little evidence to suggest that top predators induce a chain of effects with ecosystem-level consequences. In fact, for these situations, lower trophic levels in soil food webs seem much more important than higher ones in regulating key ecosystem functions, such as plant productivity and decomposition processes (Wardle 1999). Strong (1992; see Figure 12-1) suggested a continuum between trophic cascades in unstable systems (with low complexity and diversity) and trophic trickles (with high complexity and diversity). Most trophic cascades are aquatic (e.g., some tropical and temperate streams, trout streams, temperate lakes, and kelp forests), depauperate of species, and have algae at the base. At the other end of Strong's continuum, trophic trickles can occur in speciose terrestrial environments with top forces that peter out between levels. Trophic trickles are thus a more appropriate metaphor for many communities on land and in diverse aquatic systems than are trophic cascades (Strong 1992).

To investigate the possible existence of trophic cascades in terrestrial systems, Strong et al. (1996) examined root herbivory of bush lupine (*Lypinus arboreus*) in a natural shrub community on the northern California coast. Similarly, Heneghan et al. (1999) studied the decomposition of *Quercus prinus* in two humid tropical forests (viz., La Selva, Costa Rica, and Luquillo Experimental Forest, Puerto Rico). Both groups of researchers found forceful links (Strong et al. 1996) reminiscent of aquatic trophic cascades, especially at tropical sites, but also indications that such links are site-specific. For instance, Strong et al. (1996) established a positive spatial correlation

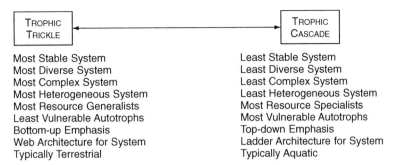

Figure 12-1 A spectrum of trophic interactions where stability implies a balanced reciprocity between consumers and consumed. (See Strong 1992.)

between high death rates of bush lupine and high densities of herbivorous moth caterpillars in roots and a negative spatial correlation between bush death rate and incidence of entomopathogenic nematodes in the soil. A third link in this trophic chain exists between the suppression of nematodes and the abundance of nematophagous fungi. Heneghan et al. (1999) found that tropical microarthropods have pronounced influence on decomposition, whereas temperate species have little effect. If trophic cascades are important in soil food webs, then the question emerges as to what effect this has on overall ecosystem function, especially on those tasks provided by lower trophic levels affected by the cascade (Wardle 1999). In other words, is the trophic ecology of soils necessarily a top-down force for temperate and tropical forests?

Hunter and Price (1992; see Figure 12-2) argued for a model of ecological interactions in terrestrial systems that synthesizes the fundamental disagreements over (1) whether bottom-up or top-down forces predominate in populations and communities and (2) whether little things or big things run the world: "From every intermediate level in a trophic web there are 'ladders' going up and 'chutes' going down, and the major players in the game are not restricted to the top or the

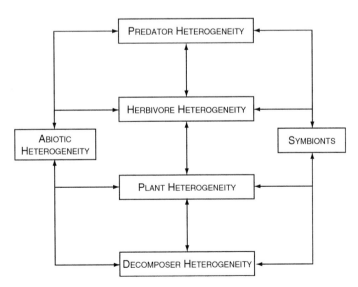

Figure 12-2 Bottom-up and top-down trophic forces in a forest system, showing cascading feedback loops. Note the interplay between abiotic and biotic factors throughout the model. Adapted from Hunter and Price 1992.

bottom of the web" (Hunter and Price 1992). An example of a bottom-up force in such an eco-logical model includes a biotic interaction that limits resources, the cycling of nutrients from forest litter to primary producers; instances of a top-down force may involve mutualism, facilitation, or the influence of herbivores by their predators. Menge (1992) made a convincing case that bottom-up and top-down forces are *joint* determinants of community structure. Compellingly, the use of the oxymoron "cascading upward" by Hunter and Price (1992) captures the fundamental reality that with the removal of higher trophic levels, lower levels remain present, whereas the removal of pri-mary producers allows no system at all. In other words, plants have primacy (Power 1992). Plants provide a bottom-up template for communities, shaping the herbivore diversity above in the trophic pyramids of forests. On the other hand, heterogeneity in the plant community is influenced directly—perhaps ultimately—by decomposer heterogeneity, including an assortment of microarthropods active in the associated soil horizons. Feedback loops at every trophic level can influence diversity throughout the pyramid by cascading effects both up and down the system. Fieldwork by researchers highlights the ongoing debate among forest ecologists over the role of trophic interactions in determining distributions and abundances of organisms (e.g., Hunter and Price 1992; Menge 1992; Power 1992; Strong et al. 1996; Wardle 1999; Halaj and Wise 2001). In the food webs of temperate and tropical forests, the bottom-up and top-down ecological roles—rather than species diversity—of soil and canopy fauna appear to share center stage for the health and integrity of these systems.

Conclusion: Soil Fauna—Another Last Biotic Frontier

André et al. (1994) called soil fauna "the other last biotic frontier" as a playful rebuke of Erwin's assertion that forest canopies are the remaining biological opportunity for discovery. Erwin's work in neotropical forests emphasized the biological diversity of canopies without sufficient analysis of soil fauna. Inadvertently, his work also underscored taxonomy rather than ecological function as fundamental to forest health and integrity. Though connected through carbon gain and loss, the aboveground and belowground compartments of forest systems seem to move carbon through different suites of trophic interactions with arthropods as the most abundant herbivores and detri-tivores in many of these terrestrial associations (Seastedt and Crossley 1984).

The richness and abundance of edaphic fauna may rival that of any other terrestrial ecosys-tem. Forest soils, along with forest canopies, are two promising frontiers for scientific discovery. Though investigation of their ecological links is still in its infancy (e.g., Reynolds and Crossley 1997; Reynolds et al. 2000; Reynolds and Hunter 2001; Reynolds et al. 2003), the emerging pic-ture of these two biological strata is an integrated column of life sandwiched between atmosphere and bedrock with top-down and bottom-up processes actively influencing that integration.

References

Anderson, J.M. (1978). Inter- and intra-habitat relationships between woodland cryptostigmata species diversity and the diversity of soil and litter microhabitats. *Oecologia* **32**, 341–348.

Anderson, J.M. (1987). Interactions between invertebrates and microorganisms: Noise or necessity for soil processes? *In* "Ecology of Microbial Communities" (M. Fletcher, T.R.G. Gray, M. Scourfield, and J.G. Jones, Eds.), pp. 125–145. Society for General Microbiology Symposium 41, Cambridge University Press, Cambridge.

Anderson, J.M. (1988). Spatiotemporal effects of invertebrates on soil processes. *Biology and Fertility of Soils* **6**, 216–227.

André, H.M., Noti, M-I., and Lebrun, P. (1994). The soil fauna: the other last biotic frontier. *Biodiversity and Conservation* **3**, 45–56.

André, H.M., Ducarme, X., and Lebrun, P. (2002). Soil biodiversity: myth, reality, or conning? *Oikos* **96**, 3–24.

Carpenter, S.R., Kitchell J.F., and Hodgson J.R., (1985). Cascading trophic interactions and lake productivity. *BioScience* 35(10):634–539.

Chen, B., Wise, D.H., (1997). Responses of forest-floor fu givores to experimental food enhancement. *Pedobiologia* 41:316–326.

Chen, B., and Wise, D.H., (1999). Bottom-up limitation of predaceous arthropods in a detritusbased terrestrial food web. *Ecology* 80(3):761–772.

Coleman, D.C. and Crossley, D.A. (1996). "Fundamentals of Soil Ecology." Academic Press, San Diego.

Crossley, D.A., Mueller, B.R., and Perdue, J.C. (1992). Biodiversity of microarthropods in agricultural soils: Relations to processes. *Agriculture, Ecosystems, and Environment* **40**, 37–46.

Erwin, T. (1982). Tropical forests: Their richness in Coleoptera and other arthropod species. *The Coleopterists Bulletin* 36(1), 74–75.

Erwin, T. (1983). Tropical forest canopies: The last biotic frontier. *Bulletin of the Entomological Society of America* **29**, 14–19.

Halaj, J. and Wise, D. H. (2001). Terrestrial trophic cascades: how much do they trickle? *The American Naturalist* 157(3), 262–281.

Hammond, P.M., Stork, N.E., and Brendell, M.J.D. (1997). Tree-crown beetles in context: A comparison of canopy and other ecotone assemblages in a lowland forest in Sulawesi. *In* "Canopy Arthropods" (N.E. Stork, J. Adis, and R.K. Didham, Eds.), pp. 184–223. Chapman and Hall, London.

Heneghan, L., Coleman, D.C., Zou, X., Crossley, D.A., and Haines, B.L. (1999). Soil microarthropod contributions to decomposition dynamics: Tropical-temperate comparisons of a single substrate. *Ecology* 80(6), 1873–1882.

Holt, R.D. (2000). Trophic cascades in terrestrial ecosystems. Reflections on Polis et al. *Trends in Ecology and Evolution* 15(11), 444–445.

Hunt, H.W., Coleman, D.C., Ingham, E.R., Ingham, R.E., Elliott, E.T., Moore, J.C., Rose, S.L., Reid, C.P.P., and Morley, C.R. (1987). The detrital food web in a shortgrass prairie. *Biology and Fertility of Soils* **3**, 57–68.

Hunter, M.D. and Price, P.W. (1992). Playing chutes and ladders: Heterogeneity and the relative roles of bottom-up and top-down forces in natural communities. *Ecology* **73**, 724–732.

Ineson, P., Leonard, M.A., and Anderson, J.M. (1982). Effect of collembolan grazing upon nitrogen and cation leaching from decomposing leaf litter. *Soil Biology and Biochemistry* **14**, 601–605.

Jenny, H. (1980). "The Soil Resource: Origin and Behavior: Ecological Studies 37." Springer-Verlag, New York.

Kolb, T.E., Dodds, K.A., and Clancy, K.M. (1991). Effects of Western spruce budworm defoliation on the physiology and growth of potted Douglas-Fir seedlings. *Forest Science* 45(2), 280–291.

Laakso, J, and Setälä, H. (1999). Sensitivity of primary production to changes in the architectures of belowground food webs. *Oikos* **87**, 57–64.

Lovett, G.M. and Rusink, A.E. (1995). Carbon and nitrogen mineralization from decomposing gypsy moth frass. *Oecologia* **104**, 133–138.

Lowman, M.D., Moffett, M.W., and Rinker, H.B. (1991). Insect sampling in forest canopies: A new method. *In* "Biologie d'Une Canopée de Forêt Équatoriale II" (F. Hallé and O. Pascal, Eds.), pp. 41–44. Opération Canopée, Lyon, France.

Luxton, M. (1972). Studies on the oribatid mites of a Danish beech wood soil. I. Nutritional biology. *Pedobiologia* **12**, 434–463.

Luxton, M. (1975). Studies on the oribatid mites of a Danish beech wood soil. II. Biomass, calorimetry, and respirometry. *Pedobiologia* **15**, 161–200.

Luxton, M. (1981a). Studies on the astigmatic mites of a Danish beech wood soil. *Pedobiologia* **22**, 29–38.

Luxton, M. (1981b). Studies on the prostigmatic mites of a Danish beech wood soil. *Pedobiologia* **22**, 277–303.

Menge, B.A. (1992). Community regulation: under what conditions are bottom-up factors important on rocky shores? *Ecology* 73(3), 755–765.

Moffett, M.W. (2000). What's "up"? A critical look at the basic terms of canopy biology. *Biotropica* 32(4a), 569–596.

Moore, J.C. (1988). The influence of microarthropods on symbiotic and non-symbiotic mutualism in detrital-based below-ground food webs. *Agriculture, Ecosystems, and Environment* **24**, 147–159.

Moore, J.C., Walter, D.W., and Hunt, H.W. (1988). Arthropod regulation of micro- and mesobiota in below-ground food webs. *Ann. Rev. Entom.* **33**, 419–439.

Nadkarni, N. (1995). Good-bye, Tarzan. *The Sciences* 35(1), 28–33.

Oldeman, R.A.A. (2001). Canopies in canopies in canopies. *Selbyana* 22(2): 235–238.

Pace, M.L., Cole, J.J., Carpenter, S.R., and Kitchell, J.F. (1999). Trophic cascades revealed in diverse ecosystems. *Trends in Ecology and Evolution* 14(12), 483–488.

Paoletti, M.G., Favretto, M.R., Stinner, B.R., Purrington, F.F., and Bater, J.E. (1991). Invertebrates as bioindicators of soil use. *Agriculture, Ecosystems, and Environment* **34**, 341–362.

Parker, G.G. (1995). Structure and microclimate of forest canopies. *In* "Forest Canopies" (M.D. Lowman and N.M. Nadkarni, Eds.), pp. 73–106. Academic Press, San Diego.

Petersen, H and Luxton, M. (1982). A comparative analysis of soil fauna populations and their role in decomposition processes. *Oikos* **39**(3), 290–388.

Pimentel, D., Westra, L. and Noss, R.F. (Eds.) (2000). "Ecological Integrity: Integrating Environment, Conservation, and Health." Island Press, Washington, DC.

Polis, G.A., Sears, A.L.W., Huxel, G.R., Strong, D.R., and Maron, J. (2000). When is a trophic cascade a trophic cascade? *Trends in Ecology and Evolution* **15**(11), 473–475.

Power, M.E. (1992). Top-down and bottom-up forces in food webs: Do plants have primacy? *Ecology* **73**(3), 733–746.

Power, M.E. (2000). What enables trophic cascades? Commentary on Polis *et al. Trends in Ecology and Evolution* **15**(11), 443–444.

Reynolds, B.C., Crossley, D.A., and Hunter, M.D. (2003). Response of soil invertebrates to forest canopy inputs along a productivity gradient. *Pedobiologia* **47**, 127–139.

Reynolds, B.C. and Hunter, M.D. (2001). Responses of soil respiration, soil nutrients, and litter decomposition to inputs from canopy herbivores. *Soil Biology and Biochemistry* **33**(12, 13), 1641–1652.

Reynolds, B.C., Hunter, M.D., and Crossley, D.A. (2000). Effects of canopy herbivory on nutrient cycling in a northern hardwood forest in western North Carolina. *Selbyana* **21**(1,2), 74–78.

Rinker, H.B., Lowman, M.D., and Moffett, M.W. (1995). Africa from the treetops. *American Biology Teacher* **57**(7), 393–401.

Rinker, H.B., Lowman, M.D., Hunter, M.D., Schowalter, T.D., and Fonte, S.J. (2001). Literature review: Canopy herbivory and soil ecology—the top-down impact of forest processes. *Selbyana* **22**(2), 225–231.

Rinker, H.B. (In preparation). The effects of canopy herbivory on soil microarthropods in a tropical rainforest.

Schenker, R. (1984). Spatial and seaonal distribution patterns of oribatid mites (Acari: Oribatei) in a forest soil ecosystem. *Pedobiologia* **27**, 133–149.

Schowalter, T.D., Sabin, T.E., Stafford, S.G., and Sexton, J.M. (1991). Phytophage effects on primary production, nutrient turnover, and litter decomposition of young Douglas-fir in western Oregon. *Forest Ecology and Management* **42**, 229–243.

Seastedt, T.R. (1984). The role of microarthropods in decomposition and mineralization processes. *Annual Review of Entomology* **29**, 25–46.

Seastedt, T.R. and Crossley, D.A. (1984). The influence of arthropods on ecosystems. *BioScience* **34**, 157–161.

Stadler, B., Solinger, S., and Michalzik, B. (2001). Insect herbivores and the nutrient flow from the canopy to the soil in coniferous and deciduous forests. *Oecologia* **126**(1), 104–113.

Stork, N.E. (1988). Insect diversity: Facts, fiction and speculation. *Biological Journal of the Linnaean Society* **35**, 321–337.

Strong, D.R. (1992). Are trophic cascades all wet? Differentiation and donor-control in speciose ecosystems. *Ecology* **73**(3), 747–754.

Strong, D.R., Maron, J.L., and Connors, P.G. (1996). Top down from underground? The underappreciated influence of subterranean food webs on aboveground ecology. *In* "Food Webs: Integration of Patterns and Dynamics" (G.A. Polis and K.O. Winemiller, Eds.), pp. 170–175. Chapman and Hall, New York.

Teale, E.W. (1944). "Grassroot Jungles: A Book of Insects." Dodd, Mead, and Company, New York.

van Breemen, N. (1992). Soils: biotic constructions in a Gaian sense? *In* "Responses of Forest Ecosystems to Environmental Changes" (European Symposium on Terrestrial Ecosystems: Forests and Woodland). pp. 189–207. Elsevier Applied Science, New York.

Wallwork, J.A. (1976). "The Distribution and Diversity of Soil Fauna." Academic Press, New York.

Wallwork, J.A. (1983). Oribatids in forest ecosystems. *Ann. Rev. Entom.* **28**, 109–130.

Wardle, D.A. (1999). How soil food webs make plants grow. *Trends in Ecology and Evolution* **14**(11), 418–420.

Wardle, D.A. (2002). "Communities and Ecosystems: Linking the Aboveground and Belowground Components." Princeton University Press, Princeton, NJ.

Webb, D.P. (1977). Regulation of deciduous forest litter decomposition by soil arthropod feces. *In* "The Role of Arthropods in Forest Ecosystems" (W.J. Mattson, Ed.), pp. 57–69. Springer, New York.

Wilson, E.O. (1992). "The Diversity of Life." Harvard University Press, Cambridge, Massachusetts.

CHAPTER 13

Tardigrades: Bears of the Canopy

William R. Miller

A lifetime can be spent in a Magellanic voyage around the trunk of a single tree.
—*E.O. Wilson,* Naturalist, *1994*

Introduction

Tardigrades, or water bears, are aquatic animals not usually associated with the canopy. In fact, they are generally associated with the ground, or ponds, or even the ocean. To equate them with the canopy stretches the imagination.

Tardigrades are micro-invertebrates that are less than half a millimeter in length. They are located systematically above the nematodes and below the arthropods though the actual affinities are still under debate (Kinchin 1994). Goeze (1773) described the first "Kleiner Wasser Bär" and Spallanzani (1776) named them "il Tardigrado," translated as "slow-stepper" in reference to their slow, deliberate movements. Since then, they have been commonly called water bears because of their bear-like appearance, their legs with claws, and their slow lumbering gait.

These bilaterally symmetrical micro-metazoans actually look more like a miniature caterpillar. Adults average 250 to 500 μm in length. Their body is composed of five somewhat indistinct segments: a cephalic segment and four trunk segments. Each body segment supports a telescopic pair of legs. The first three pairs of legs are directed ventrolaterally and are the primary means of locomotion. The fourth pair of legs is directed posteriorly on the terminal segment of the trunk and is used for grasping the substrate. Each leg terminates in either claws (see Figures 13-1, 13-2, and 13-3) or digits with suction discs secreted by claw glands (Dewel et al. 1993).

Tardigrades have a ventral nervous system but a dorsal brain that supports many sensory inputs, including light-sensitive "eyes." They have a complete digestive system with a complex mouth and pharynx structure. They eat bacteria, detritus, algae, nematodes, rotifers, and each other. Their sexes are separate and they lay spectacularly ornate eggs free in the environment (see Figure 13-4) or smooth eggs left in their shed skin. They have well-developed muscles and are flexible and articulate in all directions (Smith 2001).

Two characteristics that help make tardigrades a unique phylum are the lack of either a circulatory system or a respiratory system. They achieve circulation with a hemocoel-type body cavity, moving fluid throughout the body by muscular movement. They obtain oxygen through their cuticle and gut walls (Nelson 2001).

The animals are best known for cryptobiosis, a deep dormancy they enter when their habitat dries out; they simply desiccate to about 1/3 their normal size and look like a flake of dust. While in cryptobiosis, they can survive almost any environmental extreme on the planet. They have been exposed to −272.95°C (functional absolute zero) for 20 hours, to 130°C, to 6,000 atmospheres of

251

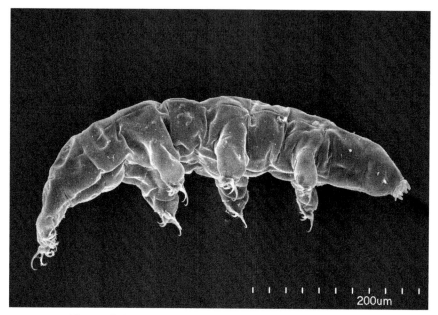

Figure 13-1   Scanning electron microscope image of eutardigrada.

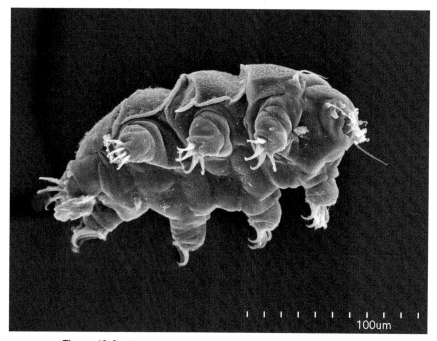

Figure 13-2   Scanning electron microscope image of heterotardigrada.

pressure, and to excessive concentrations of most common gases and x-rays, and were still able to return to active life. They have remained cryptobiotic for many years (some say a century) and re-activated (Kinchin 1994).

Cryptobiosis is defined as a reversible cessation of metabolism. It is a scientific paradox because when any animal stops metabolizing completely, it is biologically dead, yet tardigrades may return to an active state. Imbedded within cryptobiosis is yet another paradox, the ability to survive freezing (Miller 1997).

Working with Tardigrades

Tardigrades are easily collected by putting a handful of moss or a scraping of lichen into a small paper bag. The animals go into cryptobiosis and simply wait for processing, months or years later. In the laboratory, the sample is re-hydrated in a small bowl, the debris on the bottom of the bowl is examined the next day under a dissecting microscope at about 30-power, and the animals are transferred to either a preservative or a microscope slide (Miller 1997).

The taxonomy of tardigrades is based on morphological characteristics visible under light microscopy. The nature and structure of the cuticle and the presence/absence of sensory cirri separate the two major classes: Eutardigrada (see Figure 13-1) and Heterotardigrada (see Figure 13-2). Each class is further divided by the structure of their cuticle; the size, shape, and location of sensory appendages; and the toes, pads, claws, spines, and papillae on the legs. Families and genera are based on the shape of claws (see Figure 13-3), the organization of the feeding apparatus, and the characteristics of eggs (see Figure 13-4). Identification to family and genus is straightforward for the non-taxonomist while identification to species generally takes an expert (Nelson 2001).

A recent study (Garey et al. 1999) showed that 18S rRNA gene-based phylogenies of major tardigrade groups are congruent with morphologically-based phylogenies, suggesting that the characters used for current morphological studies are being interpreted correctly and that molecular-based approaches will be a useful tool for understanding tardigrade taxonomy.

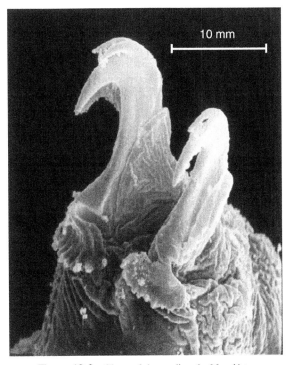

Figure 13-3   Claws of the tardigrade *Macrobiotus*.

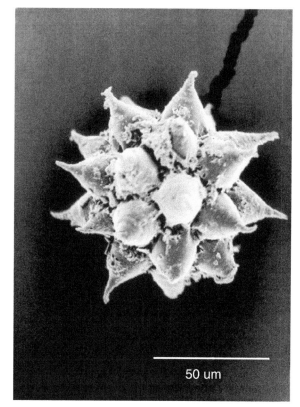

Figure 13-4   Tardigrade Egg.

Value of Tardigrades

Tardigrades are a component of the planetary biodiversity and food web that remains unmeasured. Limno-terrestrial tardigrades consume bacteria, plants, and other microscopic animals. They are direct food for mites, collembolans, and insect larvae (Kinchin 1994). Grazing animals such as deer, reindeer, cattle, and birds are indirect tardigrade consumers. Tardigrades make resources available to larger feeders by being consumed themselves or by increasing the activity of microbes in the environment. Thus, tardigrade biomass represents a transfer of energy to higher trophic levels in the food web for many important species. Our knowledge of the ecological role of this animal is inadequate. Tardigrades are part of this unquantified path in the nutrient cycle of the planet.

Tardigrades have recently been used as bio-indicators of environmental status. Hohl et al. (2001) looked at the tardigrades on trees upwind and downwind from a source of atmospheric pollution. They found significant differences in the tardigrade diversity and density. Steiner (1994a, b) made similar observations near roadways in Germany. Neither of these studies extended into the canopy; but perhaps as three-dimensional models of tardigrade populations are developed, it will be possible to identify cause-and-effect relationships.

The physiological process of cryptobiosis, which is the reversible suspension of metabolism in response to environmental conditions such as heat, cold, and humidity, is of interest for applications in medicine, aging, and space travel. Krantz et al. (1999) recently demonstrated

ROTIFERS IN THE WATER FILM

Howard L. Taylor

Rotifers are microscopic animals belonging to the phylum Rotifera, a small group of some 1,800 species of mainly freshwater, pseudocoelomate metazoans. Two features distinguish all rotifers: a ciliated anterior end, called the corona, and a set of jaws called trophi. The corona is used in locomotion and the collection of food particles; the jaws process the food.

The phylum is divided into two classes:

1. Class Digononta with two orders: Seisonidea, which has only two epizootic marine species; and Bdelloidea, with fewer than 400 described species distributed among four families; and
2. Class Monogononta, which contains three orders: Collothecacea (with two families), Flosculariacea (with six families), and Ploimida (with 18 families).

Altogether, about 2,000 rotifer species have been reported.

The name Bdelloid (with only the "d" pronounced) is derived from the Greek word *bdelloida*, or leech. The name was chosen for these rotifers because of the similar way by which they and leeches move. They attach the apical end to a surface, pull their bodies forward, attach their posterior end, extend their bodies, and repeat the sequence. Most bdelloids also swim freely by means of their apical coronas, which are retractable and have two trochal disks of metachronal beating cilia. Males are unknown among the bdelloids; consequently, they reproduce asexually by means of parthenogenesis.

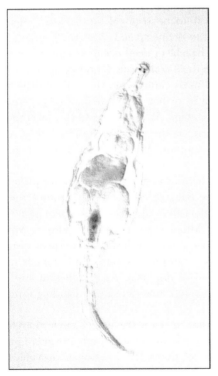

Figure 1 Typical Bdelloid, free swimming with corona opened. Scale bar: 50 μm. (Photo Bigorajski.)

Figure 2 Typtical Bdelloid, corona withdrawn and body extended for crawling. Scale bar: 50 μm (Photo Bigorajski.)

(Continued)

ROTIFERS IN THE WATER FILM—Cont'd

Bdelloids are of particular interest in forest canopy studies because their habitats include the water films that cover lichens, mosses, and liverworts. Live rotifers can be found easily in moist samples of these tiny habitats transported to the laboratory. Adult bdelloids and their eggs have the capacity of surviving slow desiccation, a process called anhydrobiosis. When dried samples are transported to the laboratory and moistened, bdelloids will be found after a few hours because some of the adults have become re-hydrated or because their eggs have hatched.

Rotifers are the most cosmopolitan of the freshwater zooplankton and are a major food source for some animals in freshwater communities. They also aid in soil decomposition. Not surprisingly, taxonomic studies of bdelloids from such collections will add to our understanding of this species-rich ecosystem called the forest canopy.

that tardigrades can carry plant pathogenic bacteria in their gut and suggested that they may be an unrecognized vector, even in cryptobiosis.

Tardigrades in the Canopy

Tardigrades are found all over the world, in high mountains, in deserts, and even in Antarctica. Of the 1,000 described species, 22 are considered cosmopolitan and can be found anywhere (McInnes 1994). The remaining 900+ species appear to be regional, insular, or endemic, but because so few distributional records exist, it is hard to define spatial patterns. Tardigrades have been found to live between the grains of sand on a beach, in deep ocean sediments, in freshwater algae, in soil, and in leaf litter. However, they are most commonly found living in the droplets of water that become trapped between the leaves of moss clumps and the thali of lichens. Here, in this sponge-like micro-world, it is not hard to envision that, as the habitat is soaked by rain, the animals emerge from cryptobiosis into the active stage. When active, they move about, eat, grow, shed, reproduce, and defend themselves. Then, as the moss slowly dries, they return to their cryptobiotic state to await the next rain (Ramazzotti and Maucci 1983). The cycle may happen many times a year or once every few years.

It is.thought that during cryptobiosis, when the habitat is dry, tardigrades are picked up by the winds as a flake of dust and dispersed a few feet, a few miles, or around the world. Windborne distribution has not been proven but offers the only plausible explanation for their collection on small, remote islands (McInnes 1995; Miller et al. 1995). In addition, Miller and Heatwole (1996) were able to use wind patterns to explain the insular pattern of distribution of tardigrade in ice-free areas of Antarctica. Passive transport by birds and animals may also occur. Of course humans are the newest mechanism of dispersal. As ornamental and decorative plants are imported for gardens and homes, the micro-invertebrates, including tardigrades, may also be introduced.

"Tardigrade rain," the cryptobiotic tardigrades that float on the winds, circle the globe like volcanic ash, smoke, or dust, and settle wherever they run out of wind, often over a forest. The presence of mosses and lichens in the upper reaches of many forest canopies provides a ready habitat. Thus, it is not surprising that tardigrades can be found in the canopy.

There are no published studies of tardigrades from the canopy. However, Voegtlin (1982) collected arthropods from an old-growth Douglas-fir canopy and reported observing many tardigrades and rotifers on the bottom of his sieves. Nalini Nadkarni, of Evergreen State University, provided a glimpse of tardigrade habitat when she climbed moss-covered limbs to demonstrate canopy exploration on television. She later collected samples from which a dozen different species of tardigrades were extracted. A new frontier for tardigrades was opened.

Hal Heatwole, of North Carolina State University, participated in a French balloon expedition to the Masoala National Park, Baie d'Amanizana, near Tampolo, Madagascar, in the fall of 2001. He and his assistant used an individual helium balloon to access mosses and lichens along a horizontal transect at the very top of the canopy and later they walked a parallel transect on the ground.

From these Madagascaran samples, 179 tardigrades of 10 species in seven genera were extracted. The upper canopy had the highest species richness with all 10 species present, whereas the moss and lichens from the rocks, dead wood, and soil on the forest floor had only six of the species. There were both juveniles and adults, indicating healthy and established populations. There were gravid females and freely deposited eggs, some with visible embryos, all evidence of actively breeding populations. These results were presented by W. R. Miller at the Third International Canopy Conference in Cairns, Australia, during the summer of 2002 and are being prepared for publication.

Two students at Chestnut Hill College have chosen to conduct canopy studies of micro-invertebrates in temperate deciduous forests for their senior research projects. Both found greater diversity and greater density of tardigrades the higher they went into the trees. Both projects identified differences in diversity on different species of tree, suggesting host specificity. The projects were presented as posters at the Cairns conference. These pilot projects are indicative of what may be found when the micro-invertebrates in the canopy are studied further.

At the Cairns meeting, Neville Winchester presented a talk on invertebrates from temperate coniferous rainforests. He described the millions of mites he had extracted from 200-foot tall Douglas-fir trees on Vancouver Island. He lamented that most were undescribed species, and that no one could help him identify them. He looked into the audience and said, "Maybe I should have used tardigrades." Neville's tardigrades are presently being analyzed.

Conclusions

The animals of the phylum Tardigrada remain a little-known, little-studied group despite their overall abundance. Beyond our basic physiological knowledge, what we know about tardigrades is in a primitive state. The ecological requirements of tardigrades are virtually unknown, and the contribution of the phylum to the biodiversity of the planet is under-documented. The lack of basic biological information has hindered the growth of interdisciplinary environmental and ecological studies.

In terms of phylogeny, biomass, economic contribution, vectors, trophic-level energy transfer, food-chain relationships, and keystone species, our knowledge of the role of tardigrades in the canopy is inadequate. At the same time, tardigrades present a unique opportunity for research. They are easy to work with, forgiving about collection, and their removal does little environmental damage. However, they must be gathered by direct sampling. Their study can be added onto other projects for little or no increase in investment.

The Academy of Natural Sciences in Philadelphia is developing a new website about tardigrades (http://tardigrade.acnatsci.org). It will post both educational and scientific information. The site already has "How to" sections, a bibliography, and list of species. In the near future, it will add geographic information system mapping, an interactive and illustrated key for identification. Eventually. the site will be a digital monograph with images. They will offer an identification service to researchers as they build a voucher collection.

E.O. Wilson (1994) must have had animals like tardigrades in mind when he wrote, "A lifetime can be spent in a Magellanic voyage around the trunk of a single tree." I now look up and wonder how many lifetimes we will need for a three-dimensional voyage into the canopy.

References

Dewel, R.A., Nelson, D.R., Dewel, W.C. (1993). *In* "Microscopic Anatomy of Invertebrates, Volume 12: Onychophora, Chilopoda, and Lesser Protostomia." pp. 143–183. Wiley-Liss, Inc, Chichester.

Garey, J.R., Nelson, D.R., Mackey, L.Y., and Li, J. (1999). Tardigrade phylogeny: congruency of morphological and molecular evidence. *Zoologischer Anzeiger* **238**, 205–210.

Goeze, J.A.E. (1773). Über den Kleinen Wasserbär. *In* "Abhandlungen aus der Insektologie, aus d. Französ. Übers." (H.K. Bonnets, Ed.), p. 367–375. Usw, 2. Beobachtg, Halle.

Hohl, A.M., Miller, W.R., and Nelson, D.R. (2001). The distribution of tardigrades upwind and downwind of a Missouri coal-burning power plant. *Zoologischer Anzeiger* **240**, 395–402.

Kinchin, I.M. (1994). "The Biology of Tardigrades." Portland Press, London, pp. 186.

Krantz, S.L., Benoitt, T.G., and Beasley, C.W. (1999). Phyropathogenic bacteria associated with tardigrada. *Zoologischer Anzeiger* **238**, 259–260.

McInnes, S.J. (1994). Zoogeographic distribution of terrestrial/freshwater tardigrades form current literature. *Journal of Natural History* **28**, 257–352.

McInnes, S.J. (1995). Tardigrades from Signy Island, South Orkney Islands, with particular reference to freshwater species. *Journal of Natural History* **29**, 1419–1445.

Miller, W.R. (1997). Tardigrades: bears of the moss. *The Kansas School Naturalist*. Emporia State University, **43**, 1–16.

Miller, W.R. and Heatwole, H.F. (1996). Tardigrades of the Australian Antarctic Territories: the Northern Prince Charles Mountains, East Antarctica. *Proceedings of the Linnean Society of New South Wales* **116**, 245–260.

Miller, W.R., Horning, D.S., and Dastych, H. (1995). Tardigrades of the Australian Antarctic: the description of two new species from Macquarie Island, Subantarctica. Entomologische Mitteilungen aus dem Zoologishen Museum, Hamburg, **11**, 231–240.

Nelson, D.R. (2001). 15. Tardigrada. *In* "Ecology and Classification of North American Freshwater Invertebrates" (J.H. Thorp, and A.P. Covich, Eds.) pp. 501–521. Academic Press, San Diego.

Ramazzotti, G. and Maucci, W. (1983). Il Philum Tardigrada (Third edition), Memorie dell'Istituto Italiano di Idrobiologia Dott. Marco de Marci. **41**, 1–1012. Istituto Italiano di Idrobiologia, Verbania Pallanza.

Smith, D.G. (2001). "Pennak's Freshwater Invertebrates of the United States." John Wiley & Sons, New York, p. 638.

Spallanzani, L. (1776). "Opuscolui di Fisica animale e vegetabile, Bd 2, il Tardigrado etc." *Oposcula Zoologica,* 4:222. Modena.

Steiner, W. (1994a). The influence of air pollution on moss-dwelling animals: 2. Aquatic fauna with emphasis on Nematoda and Tardigrada. *Rev. Suisse Zool.* **101**, 699–724.

Steiner, W. (1994b). The influence of air pollution on moss-dwelling animals: 4. Seasonal and long term fluctuations on rotifer, nematode and tardigrade populations. *Rev. Suisse Zool.* **101**, 1017–1031.

Voegtlin, D.J. (1982). "Invertebrates of the H.J. Andrews Experimental Forest, Western Cascade Mountains, Oregon: A Survey of the Arthropods Associated with the Canopy of Old-Growth *Pseudotsuga menziesii*." Forest Research Laboratory, Oregon State University, Special Publ. #4.

Wilson, E.O. (1994). "Naturalist." Island Press, Washington, pp. 380.

CHAPTER 14

The Biodiversity Question: How Many Species of Terrestrial Arthropods Are There?

Terry L. Erwin

I believe future generations will find it blankly incomprehensible
that we are devoting so little money and effort to the study . . .
[of] how many species are there on Earth . . . [how many]
in one representative hectare in the tropical rain forest . . .
how many individual organisms . . . in a given environment?
—*Sir Robert M. May*, How many species are there on Earth? Science 241, *1988*

Introduction

The question posed first by Raven in 1985 and then by May in 1988 is very important because those species we live with as humans on this planet are our resources, our connection, our dependency that keeps the planet healthy, or not. This is our only long-term equilibrium as a species that must manage the planet from now on. Our technological status as a species demands this, nothing less.

An answer to Raven and May's question is that an equatorial rainforest hectare in the western Amazon Basin, the most species-rich place on the planet, has 6.03×10^{12} individual organisms, representing more than 60,000 arthropod species; moreover, species turnover (β-diversity) of beetles, the most species-rich taxon, between two such hectares on the equator a mere 21 km apart is 50 to 70 percent, depending on beetle family (Erwin and Pimienta, in press). Yet, even after eight years of intensive fieldwork to obtain the necessary comparative samples where the universe of local species is completely known (see Figure 14-1), an answer to May's core biodiversity question remains out of reach. Why? Because, what we do know now is the magnitude of effort it will take to gather the data in finding an acceptable scientifically supported answer (Erwin and Lacher, in prep.) is not receiving the necessary funding to find answers.

This chapter reviews the path leading to our present knowledge and the questions raised along the way by both fieldwork and theoretical approaches. Then, I offer plans that will lead directly to a resolution of the biodiversity question, if money and effort are seriously applied to the endeavor. If life were to be discovered on Mars, or nearby Alpha Centuri, money and effort would be the least of the worries for an inventory project of ALL alien life forms. What about ALL of Earth's species, those we live with, those that support humanity now and into the future—i.e., those that keep this planet healthy, those that supply commodities, medicinal bases, reprieves from deleterious genes, beneficial ecological services, molecules for suggesting new types

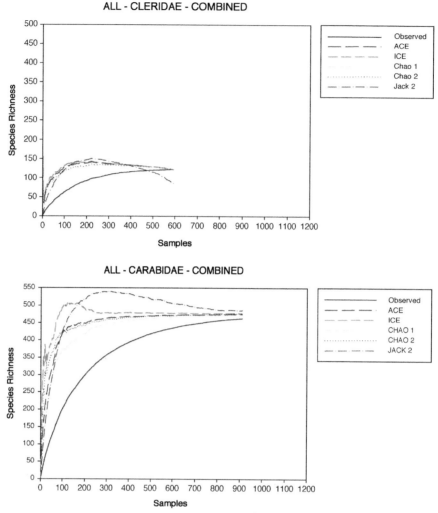

Figure 14-1   Species accumulation curve for clerid and carabid beetles in the canopy at two western Amazon sites. Note observed (Obs) and the five estimators from R. Colwell's EstimateS Program.

of drugs and perfumes, general knowledge about where and with whom we live? Our knowledge of these is pathetically little and is in a little-growth phase. Our only hope of solving the problem is adequate expenditure of interest, money, time, and effort. Is this effort in our future? "Short-term" planning is normal in the life cycle of humans, versus the long-term timing of bio-evolution, the changing world, upon which we mortally depend.

Previous Methods Employed to Acquire an Answer

After two and a half centuries of inquiry, we still do not have the crucial answer regarding most of Earth's biota: How many species of terrestrial arthropods are there (see Table 14-1)? As E.O. Wilson (1988) pointed out, we are not even close to knowing the order of magnitude of species on Earth. Recently, at various meetings both in the systematics and conservation

Table 14-1   History of Guess-timates, Anecdotes, and Failed Extrapolations

Estimator	Date	Methods	Total Species
Ray (in Westwood)	<1833	"induced to consider"	20,000
Kirby (in Westwood)	<1833	? -no quote given	4 – 600,000
Sabrosky	1952	Catalog extrapolation	2.5 – 10 million
Erwin	1982	Canopy samples, Panamá	30 million
Raven	1985	Taxon ratios, vertebrates	2.5 million
May	1988	Size ratios > 1.0 mm	10 – 30 million
Stork	1988	Tweaking Erwin's formula	7 – 80 million
Stork & Gaston	1990	Taxon ratios, butterflies	5 – 6.6 million
May	1990	Size ratios > 0.2 mm	10 million
Hodkinson & Casson	1991	Taxon ratios, true bugs	1.8 – 2.6 million
Gaston	1991	Anecdotal from taxonomists	>10 million
Hammond	1992	Counting descriptions	1.7 million
Gaston	1992	Herbivore/plant extrapolations	6.8 – 8.7 million
Ødergaard	1999	Herbivore/plant extrapolations	5 – 10 million
Novotny, et al.	2002	Herbivore/plant extrapolations	4.8 – 6.6 million

communities, some researchers have argued that the question is not important, but rather our energy must go to conserving whatever is out there. Others have argued that the question itself is not helpful scientifically. Those who oppose those two arguments have marshaled various arguments of their own. For example, if we don't know it and where it lives, we can't save it. Others believe that the actual number does not matter; the academic pursuit of all knowledge is important. Several previous papers have added to the discussion in some detail (May 1988, 1990; Stork 1988, 1993; Erwin 1991b; Hammond 1992; Stork and Gaston 1993; Miller et al. 2002).

Divine Insight

From the time of Aristotle (384 to 322 BCE) until the late 17th century, there does not seem to be any recorded effort at guessing how many species the world contained. At the very end of this period, discoveries were being made on newly explored continents. John Ray (cited in Westwood 1833) was "induced to consider" there were some 20,000 species. Since religion tempered much of the thinking during the 16 or so centuries leading up to Ray's guess, one might think that the size of the Ark was the limiting factor here. By the early 19th century, specimens from many continents were pouring into Europe; many taxonomists realized what the world might contain, so the Reverend William Kirby ventured a guess at 400,000 to 600,000 species (cited in Westwood 1833). Miller et al. (2002) pointed out that others (e.g. Sharp 1895; Frost 1942) held to the notion that at least two million was closer to the mark. Some modern scientists and conservationists have adopted a commercial version of Ray's "divine insight" that might be called the "Ouija Board" technique, or "Second-Order Guess-timate" approach—that is, using some unknown mechanism to point to a previously published number and then using it as "probably in the neighborhood. . . ." Why these methods have passed editorial oversight in scientific and conservation journals is a scandal.

Catalog Counts

Metcalf (1940), Sabrosky (1952), and later Hammond (1992) reported the number of species actually described (640,000 in 1940, 686,000 in 1952, and 1.7 million in 1992). From those real numbers, they then used the previous technique to "guess-timate" that there are 2.5 to 10 million species, and 12.5 million, respectively. While their contributions at giving us numbers of

INSECT ZOOS AS WINDOWS INTO FOREST CANOPIES

Nathan Erwin

Forest canopies, particularly rainforest canopies, teem with life. The challenge for an exhibit such as an insect zoo is to provide a space in which museum visitors experience the movement, sound, color, and real drama of life in the treetops.

The National Museum of Natural History's (NMNH) O. Orkin Insect Zoo in Washington, DC, presents exhibits that show the diversity of insects and related arthropods. It also highlights adaptations that have insured arthropod survival for about 500 million years, the crucial roles insects play in terrestrial ecosystems, and some of the research on insects that goes on behind-the-scenes at NMNH (Love 1996). One of the ecosystem exhibits reveals some of the myriad facets of rainforests and their canopies.

"Rain Forest Riches" features live arthropods to show visitors some of the key players in rainforest ecology. A live colony of leaf-cutter ants set in the base of a fabricated kapok tree demonstrates the importance of the nutrient cycling conducted by these formicine farmers that take pieces of leaves from the canopy and grow fungus on them underground. A fabricated strangler fig tree, with five small exhibit cases lodged in it, shows a variety of arthropods that can be found in the cavities and crevices of this hemi-epiphytic tree species. One exhibit case displays tank bromeliads that have "fallen" to the forest floor, revealing a glimpse of what insect life is found in the canopy of a rainforest.

NMNH education and exhibit staff uses the 4-MAT Learning System (McCarthy 1987). McCarthy identified four learning styles: imaginative, analytic, common sense, and dynamic. The Insect Zoo answers the "why" questions that imaginative learners have by providing docents in addition to the label text. Analytic learners asking "what" questions find layered text throughout; layered text begins at the top of a panel stating the broad concept(s) in large type and progresses down the panel with more and more detailed information at each layer of smaller type. Common-sense learners find interactive exhibits to answer their "how does it work" questions about insects. Dynamic learners discover interactive exhibits along with docents at carts filled with live specimens and other objects to answer their "what if . . ." questions. Engaging visitors by appealing to multiple learning styles will lead them to take advantage of the full experience provided by the exhibit.

About 65 percent of NMNH visitors are family groups, with school-group visitors coming in at about 10 percent (Bielick and Pekarik 1995). With this in mind, "Amazing Arthropods," "Dealing With Danger," and "Thriving Through Change" are more engaging exhibit titles than "Arthropod Biology," "Batesian vs. Mullerian Mimicry," and "Contrasting Holometabolous and Paurametabolous Metamorphosis." Knowing the audience allows exhibit developers to use layered text to appeal to visitors while explaining more technical terms.

Museum staff must also consider what visitors want from a museum visit. McLean (1993) says, "Entertainment and social interaction make up a large portion of a [visitor's] museum experience." Tough as that statement might be to accept by museum staff, it is why exhibits such as the Insect Zoo are crowded on weekends and holidays by individuals and family groups. School groups must now consider learning standards developed by the federal and state governments. Does this mean museums must abandon their educational missions in order to attract visitors just for entertainment purposes or not present objects and topics to satisfy standardized tests? No. It does mean, however, that exhibits need to engage and not bore or tire visitors. Teacher materials and websites developed by the museum must address teachers' concerns about national and state learning standards.

Rainforests and their canopies are complex ecosystems. Insect zoos and similar exhibits, now numbering over 100 in the United States alone (Corley 2002), have the opportunity to give visitors a glimpse of the daily lives of some of the amazing and essential inhabitants of forest canopies in an engaging and lasting way.

References

Bielick, S. and Pekarik, A. (1995). "Beyond the Elephant." Institutional Studies Office. Smithsonian Institution, Washington, DC.

Corley, T. (2002). "Let's Go Buggy: The Ultimate Guide to Insect Zoos and Butterfly Houses." Corley Publications, Encino, CA.

Love, S. (1996). "Curators as agents of change: an insect zoo for the nineties" *In* "Exhibiting Dilemmas: Issues of Representation at the Smithsonian" A. Henderson and A. Kaeppler, Eds.) Smithsonian Press.

McCarthy, B. (1987). "The 4MAT System: Teaching to Learning Styles with Right/Left Mode Techniques." Excel, Inc., Barrington, Illinois.

McLean, K. (1993). "Planning for People in Museum Exhibitions." Association of Science-Technology Centers. Washington, DC.

described species is lauded, milestones by year, their resulting "guess-timates" venture onto the next category below as they extend from the milestone into what is really extant, and those figures become part of the anecdotal literature.

Anecdotal Approach

The phrase "trading in anecdotes" in the search for the "near-imponderable" (Gaston, 1991b) readily summed up the exchange more than a decade ago between Gaston (1991a, 1991b) and Erwin (1991a). While Gaston legitimately cast his net looking for "kinds of information that may or may not contribute to debate" on the question of how many species there are, Erwin focused on direct answers from field data (Erwin and Scott 1981; Erwin and Pimienta, in press), two very different approaches. Erwin, being a taxonomist, was not confident that a survey of responding taxonomists would contribute anything of note to the debate. Some 12 years have passed and taxonomists are still taxonomists, but one thing has changed dramatically: more of them have had an opportunity to see the results of mass collecting efforts beginning in the 1980s from sustained countrywide light traps (Costa Rica), extensive malaise, intercept, and sifting sampling, and multinational canopy and understory fogging programs. A new Gastonesque "taxonomist's survey" that takes into account the participants' sample and field knowledge would, I believe, be instructive for a follow-up on the anecdotal approach. I would suggest the following categories: small museum taxonomists with only local field experience; large museum taxonomists with only museum experience; taxonomists with extensive but traditional field and museum experience; and taxonomists with access to mass sampling results. Perhaps using ordination analyses of participants with respect to who they are in the categories mentioned above and where the taxonomists are located geographically would shed light on how well taxonomists can see nature's reality. Regardless of the perhaps predictable results, this method still will result in only anecdotal kinds of information. An additional item that Gaston (1991b) left out of his original paper is the number of non-responding taxonomists that felt the survey question was patently impossible to answer. Regardless, Gaston's efforts helped others both in and out of science to focus on the main point; i.e., we do not know much about biodiversity on our planet.

Extrapolation

Extrapolations of various kinds and based on various types of pre-existing data have been under-taken seeking a reasonable and supportable estimate of the number of species on Earth. The ear-liest of these, following the introduction of the idea by Erwin (1982), perhaps was Morse et al. (1985), who used individuals versus physical size. Erwin (1982) used known beetles from a known tree species in Panamá. This study was the first extrapolation from field-gathered data. Although the presented hypothesis of 30 million species in the world was focused upon and has been roundly debated over the past 20 years, an overlooked element included in this paper has not been tested until recently. The main purpose of the paper was to provide an estimate of how many species there are in an acre of tropical rainforest (Raven and May's biodiversity question). From the Panamanian data, Erwin (1982) hypothesized that there were 41,389 arthropod species in a hectare of "scrubby" seasonal tropical forest in an area of Panamá long inhabited by peo-ple. After an exhaustive canopy fogging regime over a recent period of eight years in an equa-torial lowland tropical rainforest in eastern Ecuador with no people in which nine million + specimens were recovered from 1,800 canopy samples at two sites about 21 km apart, it is now known that there are more than 60,000 species per hectare (Erwin and Pimienta, in press).

Extrapolation from Rate of Taxonomic Descriptions

White (1975), following an idea of Steyskal (1965), attempted to project how many species there might be in North America in 29 various families of Coleoptera using "trend curves." His sup-position was that the rate of species description follows a sigmoid curve. Therefore, the point at which the curve begins to level toward the asymptote one might fit the lower ascendancy por-tion of the curve, flipped, onto the upper end of the curve. Since the basis of data is human activity rather than actual species, the effort fails from the start (cf. Erwin, 1994).

Extrapolation from Known Faunas

Raven (1985) initiated this method of looking at data in hand (i.e., well-known faunal elements such as mammals and birds) and working out ratios based on what is known in the temperate regions and then estimating what should occur in tropical regions. He was followed by Hodkinson and Casson (1991), Stork (1993), and Stork and Gaston (1993) in which these authors adapted the method to insect groups.

Extrapolation from Samples Approach

Scaling up to Gaston's idea of the "near imponderable" from virtually nothing has been the method employed from players across the board—taxonomists to ecologist to mathematicians. Erwin (1982) began this play with beetles from 19 tree samples in Panamá. Controversy and debate ensued for nearly 20 years before serious field testing of the main underlying assumption of Erwin's hypothesis (host specificity) were made (Ødegaard 2000; Novotny et al. 2002). While various players in Project Wallace (a year-long entomological expedition to Sulawesi) used sam-ple data to address the biodiversity question—How many species are there?—they did so not by rigorously testing Erwin's assumptions, but by looking at the data in a variety of different ways.

Theoretical Constructs

May (1988) fully reviewed several attempts at answering his own biodiversity question includ-ing analysis of the structure of food webs, patterns in relative abundance of species, patterns in the number of species or number of individuals in different categories of physical size dis-tribution, and general observations about trends in the commonness or rarity of organisms, and other ecological approaches. Have these methods he reviewed been successful? May (1988) wrote: ". . . I will not trust any estimate of the global total of species . . . [until] "a rough list of all the species found in one representative hectare in the tropical rain forest" [is made]. This is close to being available (Erwin and Pimienta, in press). Later, May (1990) used a theoretical distribution function based on known British beetles and associations with trees species and came to the conclusion that Erwin's use of 20 percent host specificity was too high.

Fatal Flaws in Previous Approaches

Divine insight, guess-timating, and anecdotes are all unscientific: they are not scientifically testable. Their positive value comes to play within the lay public by making the **quest** for an answer very exciting in a humanistic way.

Catalog counts and rates of descriptions tell us what we know and how fast we are knowing it, but do not contribute to a final resolution to the question until the very last organism is catalogued. Their positive value is that they help keep us (scientists and public) abreast of what is known.

Extrapolation and the theoretical constructs techniques of various designs are interesting in their own right as scientific endeavors, yet those applied thus far have not taken into account enough of biodiversity across space, time, and size scales (fractal universes, or sets of species within smaller and smaller-sized sets of species) to arrive at satisfactory biodiversity estimates even within a magnitude of order.

The fatal flaw in all that has come before is the absence of β-diversity data on a grand spatial and temporal scale that included enough taxonomic groups to detect the diverse taxonomic and geographic patterns that exist in nature (Erwin and Pimienta, in press).

A Synthetic Approach Offered

Thus far, all extrapolations are parochial. It is well known that different parts of the planet have different floral and faunal components (Primack, pers. comm.) and that scaling up from single sites or even single areas does not lead to valid results applicable globally.

Colwell and Coddington (1994) provided a highly useful overview of the tools necessary for good biodiversity inventories—namely, estimates of species richness and measures of complementarity (see Figure 14-2). These authors, riding the front of the wave, also connected their tools to acquiring data for conservation purposes. To date, few have undertaken tests of their suggestions on any meaningful scale. The ALAS (Arthropods of La Selva) Project in Costa Rica is heading in that direction, although its design focuses mainly on a comparison of sampling techniques; it is a large-scale inventory of arthropod diversity in a lowland tropical rainforest. Erwin's studies in Perú and Ecuador directly address the question of spatial and temporal turnover, but with only two sites over six years in Ecuador (Erwin and Pimienta, in press; Lucky et al. 2002) and six microhabitat samplings in Perú over two years (unpublished data). In the Ecuador study, tests of complementarity (see Table 14-2) resulted in values from 0.48 to 0.69 for the seven families of beetles thus far analyzed, the species-rich families having the higher values. Species accumulation curve asymptotes (for example, see Figure 14-1) were reached, or nearly so, for all families in 600 to 1200 samples (see results in Table 14-2). These results suggest a markedly high degree of species turnover across a forest mosaic within short distances even in similar forests in the western Amazon Basin, which in turn suggests that a new inventory protocol is necessary for terrestrial arthropods.

Calculating Complementarity

Total species richness for both sites combined: $Sab = Sa + Sb - Vab$

Number of species unique to either list: $Uab = Sa + Sb - 2Vab$

Complementarity Index: $Cab = Uab / Sab$

Sa   = species richness at first site.
Sb   = species richness at second site.
Vab = # species shared between the two sites.

Figure 14-2   Formula for calculating complementarity.

Erwin and Lacher (in prep.) have designed a new inventory program for the Conservation International hotspots (go to *www.biodiversityscience.org* to understand Conservation International's definition of hotspots) and the Smithsonian Institution's 50 hectare plots based on discovery and documentation of all species in 10 ha permanent plots at each of the sites using mobile labs and the richness estimates (Colwell 1997) and complimentary tools (Colwell and Coddington 1994) in novel ways. In the long term, it will be necessary to establish these plots in all plant formations within each 1° by 1° grid of the Earth's surface if we are truly going to know all life on the planet. Imagine a blitz of taxonomists and their students plus a battalion of parataxonomists rapidly working a focused inventory in each grid quadrate for one year to obtain the first significant cut of biodiversity knowledge. As these plots are established and inventoried, their data may lead to even faster ways to project an answer to the biodiversity question not now available, or at least arrive at the correct order of magnitude of species on Earth. This could be accomplished within 25 years.

Conclusions

If the goal of conservation strategies is to provide reserved areas for ***most*** of biodiversity, and that ***should*** be the goal, knowledge is needed from across the spectrum of local fractal universes from the world of mites and tardigrades to the world of large vertebrates, those in our own human fractal universe. As early as 1991, a significant call went out to include terrestrial arthropods in conservation planning (Erwin 1991b; Vane-Wright et al. 1991; Colwell and Coddington 1993; Kremen et al. 1993). The ideas put forth in all papers regarding terrestrial arthropods indicated that knowledge of arthropod biodiversity is central to strategic planning for conservation areas in knowing species occurrences and distributions at both local and regional scales. If we consider this at a multitude of fractal universes at each inventoried site, as well as the phylogenetic relationships of species across the landscape matrix, we cannot but conclude that saving most of biodiversity means paying attention to all the small species. The very heart of biodiversity resides in tropical forest canopies (Erwin 1988), yet as pointed out by Lowman and Wittman (1996) and Basset (2001), we know little thus far about what lives there, let alone their patterns of occurrence (Wilson 2000). Thus, the question arises, how does one plan areas for conserving most of the species occurring locally or in a given region? Areas for attention may or may not contain the same species, while those not attended to may have a completely different fauna and flora (Erwin and Pimienta, in press).

Table 14-2 Summary of Data for 7 Families of Beetles Sampled from the Tropical Forest Canopy in the Yasuni Area of Eastern Ecuador between 1994 and 2002

	Samples	Occurrence	Sobs	Individuals	Singletons	Doubletons	ACE	ICE	Chao 1	CI	Occurrence in samples	Feeding guild
BUPR	600	282	210	750	86	90	152	231	401	0.69	47%	plant tissues
CARA	1200	911	462	3536	58	64	486	492	549	0.57	76%	predators
CCAN	600	132	31	208	0	3	52	36	36	0.48	22%	fungivores
CLER	1200	600	122	1202	0	36	122	122	122	0.48	50%	predators
ELAT	600	447	157	1930	35	39	192	177	197	0.69	75%	scavengers
EROT	600	362	168	1130	44	49	167	185	231	0.59	60%	fungivores
TENE	600	168	92	309	33	34	94	104	209	0.67	28%	fungivores
Totals			1242	9065	256	315						
Percent total species as singletons, doubletons					20.6%	25.4%						

Ideally, a 90 percent (or more) knowledge of species occurrence in every 1° by 1° grid of the Earth's surface would provide a solid footing of information for conservation strategies (Erwin and Lacher, in prep.). And, for purposes of answering May's core biodiversity question, the data would be available. We now know that even for the incredible diversity of tropical beetles in the richest place on Earth, the western Amazon Basin, we can know 95 percent or more in the forest canopies with about 1,200 samples taken over a trans-annual period of a mere three years (Erwin and Pimienta, in press), the comments of Bartlett et al. (1999) regarding the difficulty of knowing all the species at a site, notwithstanding. The key is the use of species accumulation curves (Colwell 1997) and the Complementarity Index (Colwell and Coddington 1994) as discussed in Erwin and Pimienta, in press.

Considering potential benefits for humanity, not accomplishing an inventory of life on Earth has been the greatest failing of the human race thus far. Or, perhaps, not even ***recognizing*** that such knowledge is fundamental to our long-term survival on Earth as a species is a greater failing. Captured in a commercial, religious, technological-dependent, military-political, nationalistic, Keynesian economics box, humankind at the decision-making level cannot see the underlying significance that excellent health of the global environment keeps us, and all other species upon which we depend, in nature's fine balance through the ages. Myopic institutional and political leadership concerning the environment likely will be our downfall as a species more than anything else. It is important to always keep in mind that Nature bats last. Paul Ehrlich (1968) provided an excellent analogy of the rivets popping from an airplane's wing during flight (species going extinct); which will be the rivet (species) that brings the plane (ecosystem) down? May's dismay that so little attention is given to this serious world-wide problem will find its way into the history books, but then, after Nature bats last, who will be left to read history?

References

Bartlett, R., Pickering, J., Gauld, I., and Windsor, D. (1999). Estimating global biodiversity, tropical beetles and wasps send different signals. *Ecological Entomology* **24**, 118–121.

Basset, Y. (2001). Invertebrates in the canopy of tropical rain forests, how much do we really know? *Plant Ecology* **153**, 87–107.

Colwell, R.K. (1997). EstimateS, Statistical estimation of species richness and shared species from samples. Version 6.01b Beta. User's Guide and application published at *http//viceroy.eeb.uconn.edu/estimates*.

Colwell, R.K. and Coddington, J.A. (1994). Estimating terrestrial biodiversity through extrapolation. *Philosophical Transactions of the Royal Society (Series B)* **345**, 101–118.

Ehrlich, P.R. (1968). "The Population Bomb." Ballantine Books, New York.

Erwin, T.L. (1982). Tropical forests, their richness in Coleoptera and other arthropod species. *The Coleopterists Bulletin* **36**(1), 74–75.

Erwin, T.L. (1988). The tropical forest canopy, the heart of biotic diversity. *In* "Biodiversity" (E.O. Wilson and F.M. Peters, Eds.), pp. 123–129. National Academy Press, Washington, DC.

Erwin, T.L. (1991a). How many species are there? Revisited. *Conservation Biology* **5**(3), 330–333.

Erwin, T.L., (1991b). An evolutionary basis for conservation strategies. *Science* **253**, 750–752.

Erwin, T.L. (1994). Arboreal beetles of tropical forests, the Xystosomi group, Subtribe Xystosomina (Coleoptera, Carabidae, Bembidiini). Part I. Character analysis, taxonomy, and distribution. *The Canadian Entomologist* **126**(3), 549–666.

Erwin, T.L. and Scott, J.C. (1981). Seasonal and size patterns, trophic structure, and richness of Coleoptera in the tropical arboreal ecosystem: the fauna of the tree *Luehea seemannii* Triana and Planch in the Canal Zone of Panamá. *The Coleopterists Bulletin* **34**(3), 305–322.

Erwin, T.L. and Pimienta, M.C. *in press*. Mapping patterns of β-diversity for beetles across the western Amazon Basin: a preliminary case for improving conservation strategies. *WWF Assessment Book*, 2003.

Erwin, T.L. and Lacher, T. (In prep). The quest to discover and document all species: a new terrestrial biodiversity inventory protocol. To be submitted to TREE.

Frost, S.W. (1942). "General Entomology." Dover Publications, New York.

Gaston, K.J. (1991a). The magnitude of global insect species richness. *Conservation Biology* **5**(3), 283–296.

Gaston, K.J. (1991b). Estimates of the near-imponderable: a reply to Erwin. *Conservation Biology* **5**(4), 564–565.

Hammond, P. (1992). Species inventory. *In* "Global Biodiversity: Status of Earth's Living Resources" (B. Groombridge, Ed.), pp. 17–39. Chapman and Hall, London.

Hodkinson, I.D. and Casson, D. (1991). A lesser predilection for bugs, Hemiptera (Insecta) diversity in tropical forests. *Biological Journal of the Linnean Society* **43**, 101–109.

Kremen, C., Colwell, R.K., Erwin, T.L., Murphy, D.D., Noss R.F., and Sanjayan, M.A. (1993). Terrestrial arthropod assemblages: their use in conservation planning. *Conservation Biology* **7**(4), 796–808.

Lowman, M.D. and Wittman, P.K. (1996). Forest canopies, methods, hypothesis, and future directions. *Annual Review of Ecology and Systematics* **27**, 55–81.

Lucky, A., Erwin, T.L., and Witman, J.D. (2002). Temporal and spatial diversity and distribution of Arboreal Carabidae (Coleoptera) in a Western Amazonian rain forest. *Biotropica* **34**(3), 376–386.

May, R.M. (1988). How many species are there on Earth? *Science* **241**, 1441–1448.

May, R.M. (1990). How many species? *Philosophical Transactions of the Royal Society of London B.* **330**, 293–304.

Metcalf, Z.P. (1940). How many insects are there in the world? *Entomological News* **51**, 219–222.

Miller, S., Novotny, V., and Basset, Y. (2002). Case studies of arthropod diversity and distribution. *In* "Foundations of Tropical Forest Biology: Classic Papers with Commentaries" (R.L. Chazdon, and T.C. Whitmore, Eds.) pp. 407–413. University of Chicago Press, Chicago.

Morse, D.R., Lawton, J.H., Dodson, M.M., and Williamson, M.H. (1985). Fractal dimension of vegetation and the distribution of arthopod body length. *Nature* **314**, 731–733.

Novotny, V., Basset, Y., Miller, S.E., Welbens, G.D., Bremer, B., Cizek, L., and Drozd, P. (2002). Low host specificity of herbivorous insects in a tropical forest. *Nature* **416**, 841–844.

Ødegaard, F. (2000). How many species of arthropods? Erwin's estimate revised. *Biological Journal of the Linnean Society* **71**, 583–597.

Raven, P.H. (1985). Disappearing species. A global tragedy. *Futurist* **19**, 8–14.

Sabrosky, C.W. (1952). How many insects are there? *In* "The Yearbook of Agriculture 1952: Insects" (A. Stefferud, Ed.). Washington, Government Printing Office.

Sharp, D. (1895). "Insects." Macmillan, New York.

Steyskal, G.C. (1965). Trend curves of the rate of species description in zoology. *Science* **149**(3686), 880–882.

Stork, N.E. (1988). Insect diversity: facts, fiction, and speculation. *Biological Journal of the Linnean Society* **35**, 321–337.

Stork, N.E. (1993). How many species are there? *Biodiversity and Conservation* **2**, 215–232.

Stork, N.E. and Gaston, K.J. (1993). Counting species one by one. *New Scientist* **1729**, 43–47.

Vane-Wright, R.I., Humphries, C.J., and Williams, P.H. (1991). What to protect? Systematics and the agony of choice. *Biological Conservation* **55**, 235–254.

Westwood, J.O. (1833). On the probable number of species of insects in the creation; together with descriptions of several minute Hymenoptera. *The Magazine of Natural History and Journal of Zoology, Botany, Mineralogy, Geology, and Meteorology* **6**, 116–123.

White, R.E. (1975). Trend curves of the rate of species description for certain North American Coleoptera. *The Coleopterists Bulletin* **29**(4), 281–295.

Wilson, E.O. (1988). The current state of biological diversity. *In* "Biodiversity" (E.O. Wilson and F.M. Peters, Eds.), pp. 3–18. National Academy Press, Washington, DC.

Wilson, E.O. (2000). A global biodiversity map. *Science* **289**, 2279.

CHAPTER 15

Physical Transport, Heterogeneity, and Interactions Involving Canopy Anoles

Roman Dial and Joan Roughgarden

*One lesson from affiliating with a tree—perhaps the greatest moral lesson anyhow from
earth, rocks, animals is that same lesson
of inherency, of what is, without the least regard to what the
looker on supposes or says, or whatever he likes or dislikes.*
—*Walt Whitman,* Specimen Days, *1892*

Introduction

Like marine science, forest canopy science is a collection of disciplines sharing a seemingly discrete physical environment. Both sciences consider phenomena unique to their focal ecosystem as well as more general processes that may apply across ecosystems. This chapter reviews one particular canopy ecosystem, the tropical rainforest canopy of eastern Puerto Rico, in the context of general principals. The system of *Anolis* lizards, their arthropod prey, and the surrounding canopy has proven particularly informative for investigating the interactions between consumers, their resources, and the physical environment. The interactions can be characterized in the context of a physical transport process, in this case wind, that subsidizes local consumers with non-local resources. The apparent outcome of this subsidy on both the resources and the consumer species who compete for these resources is the subject of this review.

In general, the prospect of a physical transport process leads to concepts of *local* and *non-local*. *Local consumers* are consumers whose entire life cycle occurs near their natal habitat. These consumers feed on both *local resources* that originate from the same habitat as the consumers and on *non-local resources* that originate from beyond the consumer's natal habitat. If the consumers are predators, then the dispersal phase of prey with complex life cycles often constitutes the non-local resources, whereas local prey often have simple (but sometimes complex) life cycles. Consumers that feed on non-local resources are considered *subsidized.*

Subsidized communities include, among others, riparian zones, cave communities, coral reefs, hard bottom intertidal communities, and near-shore deserts (reviewed by Polis et al., 1997). While forest canopies clearly subsidize forest floors with carbon, this chapter investigates the reverse—a canopy community that appears to be subsidized from below—and asks the following questions:

- How important to local canopy consumers are non-local resources, resources whose source originates outside the canopy?
- How might local-source prey differ from non-local prey in their response to predation within the canopy?

- Does the coexistence of potentially competing local consumers depend on the local resources or non-local resources?
- What possible impact do non-local prey have on local prey as mediated by their shared predators?
- How are non-local resources transported to the canopy habitat anyway?
- Do the consumers respond to the patchiness of this transport?

It is for these more general questions that the specific field of canopy science might be useful, both to formulate hypotheses and to test them.

In this chapter, a forest canopy is defined as a collection of adjacent tree crowns. Each crown has a set of organisms—the local community—that were born within or near to that tree and recruited there. These tend to be arthropods with limited dispersal ability—crickets, cockroaches, Homoptera, Psocoptera, and jumping spiders—plus several species of *Anolis* lizards. Members of this local community spend most if not all of their life cycle within individual tree crowns. The non-local prey of the lizards are flying insects, mostly Diptera, whose larval stages are principally terrestrial and whose dispersing adult stages are found within the forest canopy. The community members of canopy arthropods, their resources, and their consumers have been examined both through observation (Reagan 1986, 1991, 1992, 1995, 1996; Dial 1992; Dial et al. 1994; Dial and Roughgarden 1995, 1996; Garrison and Willig 1996; Reagan et al. 1996) and experimental manipulation (Dial 1992; Dial and Roughgarden 1995; Lawton 1995).

The study of canopy anole ecology in Puerto Rico began with Rand's influential work on anoline "ecotypes" (Rand 1964), followed by Turner and Gist (1970), Schoener and Schoener (1971), Williams (1972), Moll (1978), and Lister (1981). These early studies provided an empirical basis for classical niche theory (e.g., MacArthur and Levins 1964) by documenting vertical use of forest habitat by anoles; however, their observations were ground-based, not canopy-based, and thus incomplete. In a series of papers based on canopy-level observations, Reagan (1986, 1991, 1992, 1995, 1996) documented the abundance of arboreal anoles and their use of canopy habitat and resources. Reagan repeatedly visited two sites in the Caribbean National Forest of Puerto Rico's Luquillo Mountains where he accessed the canopy using three towers. More recently, the ecological relationships of canopy anoles in Puerto Rico were investigated in a series of papers by Dial and coworkers (Dial 1992; Dial et al. 1994; Dial and Roughgarden 1995, 1996). The purpose of the Dial study was to measure interaction strengths among canopy anoles, their prey, and resources of the prey using individual tree crowns as replicates. By applying arborist methods facilitating canopy access and movement (Dial and Tobin 1994; Jepson 2000), the effects of lizards on their prey and their prey's resources were determined by lizard removal and exclosure from individual tree crowns. Individuals of all arboreal lizards observed in seven treatment crowns were removed and excluded using bole collars for six months. These seven crowns and seven additional control crowns were monitored monthly for lizard and arthropod abundances, and for herbivory on leaves at the end of the experiment. Any lizards observed during arthropod censuses were also removed from manipulated crowns. The experiment is detailed in Dial (1992) and Dial and Roughgarden (1995). Dial and coworkers also documented canopy anole parasitism (Dial and Roughgarden 1996), predation (Dial and Roughgarden 1995), community structure (Dial et al. 1994), and competition between canopy species (Dial 1992), as well as the abundance, distribution, and wind transport of canopy arthropods (Dial 1992). The goal of this review is to develop a conceptual view of a canopy ecosystem based on observational and experimental data.

Section I describes the study system in general. Section II reviews autecology for *Anolis stratulus* and *A. evermanni*, the focal, possibly competing, consumers in the rainforest canopy of eastern Puerto Rico. Section III reviews arthropod data from gaps, experimental, and control crowns in the context of lizards as consumers of arthropod resources. Section IV attempts to synthesize the observations linking physical transport, canopy heterogeneity, and competition between canopy anoles.

Section I: Tabonuco Canopy: Description, Access, and Sampling

Forest Description

Tabonuco forest is a 20- to 30-m tall montane, wet forest type common to the windward sides of Caribbean islands and dominated by the broadleaf evergreen tabonuco tree, *Dacryodes excelsa* (Burseraceae). Understory trees and shrubs include *Cecropia*, *Prestoea* palms, *Piper* shrubs and *Urera* nettles. These forests are structurally simpler and less diverse than mainland forests of similar climate and more strongly influenced by hurricanes (Laurence 1996). The second-growth tabonuco forests of the Caribbean National Forest in Puerto Rico's Luquillo Mountains have been extensively studied for more than three decades (Odum and Pigeon 1970; Brown et al. 1983; Reagan and Waide 1996). Most of the data (Reagan 1986, 1991, 1992, 1995, 1996; Dial 1992; Dial et al. 1994; Dial and Roughgarden 1995, 1996; Reagan et al. 1996; Garrison and Willig 1996) reviewed here were collected near the El Verde and Bisley Watersheds in the Luquillo Long Term Ecological Research (LTER) site, Caribbean National Forest, Puerto Rico, approximately 300 to 500 m above sea level. About half the stems > 20 cm diameter in tabonuco forest are tabonuco trees. Most arthropod and lizard sampling and all lizard manipulations took place within tabonuco crowns. Crowns used in these studies, 7 to 60 m^2 in projected area and 5 to 13 m deep, are exposed to trade winds. Hurricanes defoliate the canopy at approximately half-century intervals. Two years after Hurricane Hugo (Walker 1991; Walker and Brokaw 1991), when many studies reviewed here took place, large gaps remained typical of the forest (see Figure 15-1). About half the stems supported lianas. Dead limbs, pockets of arboreal soil, bromeliads, orchids, and other epiphytes, both vascular and non-vascular, were rare or absent from the study crowns as a result, in part, of hurricanes.

Canopy Sampling

Reagan (1986, 1991, 1992) measured anole abundances along three canopy towers at two sites using the variable-width transect measure (Overton 1971). This method requires capturing and marking lizards. The Heckel and Roughgarden (1979) method used by Dial and colleagues for lizard abundance estimates may be more suitable for canopy work. The Heckel and Roughgarden technique uses paint guns to mark animals and, since lizards do not have to be caught, can be marked up to 7 m away.

Several canopy arthropod studies in Puerto Rico have been conducted from within the canopy. Garrison and Willig (1996) suspended sticky traps alongside a canopy tower and Schowalter (1994; Schowalter and Ganio 1999, 2003) sampled by clipping branches with a long-handled pruner. These studies were done independently of lizard censuses. Dial and coworkers visited tabonuco crowns monthly for eight months to sample arthropods in concert with lizard abundance estimates, removal, and gut content analysis (Dial 1992; Dial et al. 1994; Dial and Roughgarden 1995). Using non-injurious arborist methods (Dial and Tobin 1994), they gained access and movement to within 3 m of the crown tops. Within 14 study crowns, resident populations of two species of *Anolis* lizards were censused (Dial et al. 1994), airborne arthropod abundances were sampled using sticky traps (21 per tree, 2 × 15 cm in area), foliage arthropods were counted on leaves (525 leaflets per tree), and orb spiders were counted monthly; ant colonies were sampled under bark flakes, and leaves were measured for herbivory at the end of the experiment (Dial 1992; Dial and Roughgarden 1995). All arthropods in sticky traps and leaf samples were identified to order and sized into 1 mm size classes while most individual Diptera, Homoptera, Hymenoptera, Orthoptera, and spiders were resolved to family (Dial 1992). Using several kinds of sticky traps (see Section III), Dial (1992) also sampled arthropods in the airspace of two forest gaps 12 to 15 m above ground to determine arthropod abundance in the air column between crowns. Figure 15-1 shows the locations of these samples as dashed lines.

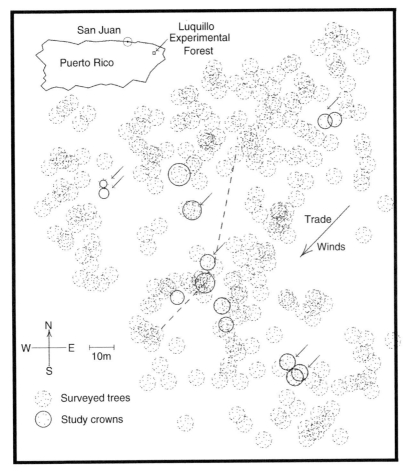

Figure 15-1 Map of study site showing locations of mapped stems > 20 cm (shading without solid circles) and experimental trees (shading with solid circles indicating crown area). Dashed lines show locations of suspended lines supporting gap sticky traps. Arrows indicate study crowns exposed to trade winds blowing over gaps. (From Dial and Roughgarden, unpublished manuscript.)

Section II: Autecology of Crown-Dwelling *Anolis* Lizards

Lizards of the genus *Anolis* are common, conspicuous, diurnal elements of Caribbean forest canopies (Rand 1964; Schoener and Schoener 1971; Lister 1981; Reagan 1992, 1996). In Puerto Rico's tabonuco forests, there are four species of canopy-dwelling anoles: the crown giant *Anolis cuvieri*, the primitive twig-dwarf *A. occultus*, the uniformly green-colored *A. evermanni*, and the black-spotted, gray *A. stratulus*. All four lay their eggs in ground soil, then spend most of their adult lives foraging arboreally, usually in well-defended territories. Dial et al. (1994) reported on two individual *A. cuvieri* observed over nine months and showed that their home range sizes (350 m²) were similar to that reported by Losos et al. (1990), who used radio telemetry on one individual. All studies reported that *A. cuvieri* (body weight ~40 g) includes a substantial amount of snails, plant material, and smaller anoles in its diet (Losos et al. 1990; Dial et al. 1994; Reagan 1996). The insectivorous *A. stratulus* (body weight ~2 g) and *A. evermanni* (body weight ~4 g), orders of magnitude more abundant than *A. cuvieri* (Reagan 1992; Dial et al. 1994), are considered quite closely related to each other (Williams 1972). This section reviews known consumers, body size distributions, microhabitat use, gut content, and abundance of canopy *A. stratulus* and *evermanni*, with the objective of determining the strength of present-day competition between the two.

Predators and Parasites of Crown Dwelling *A. evermanni* and *stratulus*

Besides *Anolis cuvieri*, predators of canopy anoles include pearly-eyed thrashers (Reagan 1996), Puerto Rican lizard cuckoos (Dial and Roughgarden 1995; Reagan 1996), red-tailed hawks (Reagan 1996), broad winged hawks (Dial and Roughgarden 1995; Reagan 1996); *Alsophis* (Reagan 1996), and *Epicrates* snakes (Dial and Roughgarden 1995; Reagan 1996); *Eleuthrodactylus* tree frogs (Dial and Roughgarden 1995); and several large invertebrates (Reagan 1996), including tailless whip scorpions (*Phrynus longipes*), arboreal tarantulas (*Avicularia laeta*), and centipedes (*Scolopendra alternans*). Dial and Roughgarden (1996) noted mortality of *A. evermanni* and *stratulus* by a sarcophagid (Diptera) parasitoid. The infection rate by the sarcophagid larvae was three times higher in *A. evermanni* than in *stratulus*, a difference attributed by the authors to the presence of black spotting on *stratulus*. These black spots, lacking on *evermanni*, were located where sarcophagid-inflicted wounds would be present (Dial and Roughgarden 1996), perhaps duping flies and preventing them from larvipositing on spotted lizards. Dial and Roughgarden (1995, 1996) observed more total canopy anole deaths due to the sarcophagid parasites than due to vertebrate predators.

Body Size Distributions

Classical niche theory (e.g., MacArthur and Wilson 1967) suggests that body length ratios of 1.3 to 1.5:1 are a crude estimate of how different two species must be to coexist syntopically. Reagan (1996) gives mean snout vent lengths (SVL) for *Anolis evermanni* as 55 mm and for *stratulus* as 41 mm, a ratio of 1.34:1. The mean SVL from seven tabonuco crowns sampled by Dial and Roughgarden (unpublished data) were similar to those reported by Reagan: mean SVL for *evermanni* was 51.1 mm (n = 44) and for *stratulus* mean SVL was 38.6 mm (n = 42), a ratio of 1.33:1. Sexual dimorphism in the two species led to substantial overlap in SVL between female *evermanni* and male *stratulus*, with mean body sizes differing by 2 mm (see Figure 15-2). While the mean SVL among the species differentiated by sex differed significantly (ANOVA, $F_{3,84}$ = 86.0, P < 0.001), Scheffe's post hoc comparisons (Ramsey and Schafer 1997) showed that female *A. evermanni* and male *stratulus* did not differ (P = 0.34) and shared very similar body size distributions (see Figure 15-2).

Microhabitat Use

Rand (1964) and Williams (1972) were the first to identify microhabitat variables as structural niche axes for Caribbean *Anolis*. Reagan (1995, 1996) reviewed data on microhabitat use by canopy-dwelling *A. evermanni* and *stratulus* at two tabonuco sites in Puerto Rico's Luquillo Mountains using canopy towers as vertical transects. He found that *stratulus* perched 15 to 20 m higher (modal height 18 to 20 m) than *evermanni* (modal height 0 to 2 m) and used smaller perches (median perch diameter = 0 to 10 cm) than *evermanni* (median perch diameter = 10 to 20 cm). Vertical distributions of *A. stratulus* paralleled the distribution of small perch availability (Reagan 1992). Reagan (1986) considered *stratulus* more tolerant of low humidity than *evermanni* and found that *stratulus* foraged over a wider range of heights during the dry season than during the wet. He did not provide data by season on perch diameters. Reagan's (1996) review suggested that *stratulus* is a crown-dwelling species and *evermanni* a trunk species.

Dial (1992) reported perch diameters (see Figure 15-3), heights (see Figure 15-4), and solar exposure collected from 305 observations. Dial (1992 unpublished data) found that modal perch diameters during both seasons were greater for males than for females, and greater for *evermanni* than for *stratulus*, observations consistent with larger bodied animals using larger perches. During January, the modal class of perch diameters for *evermanni* females was 8 to 16 cm, equal to that

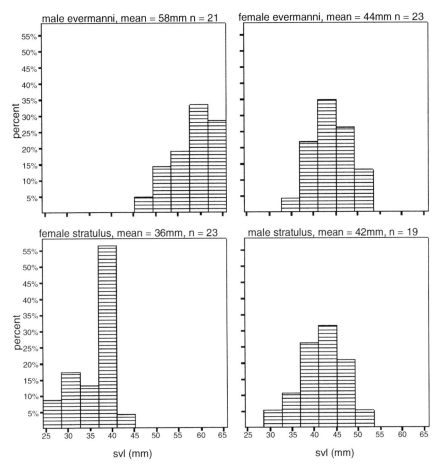

Figure 15-2 Body size distributions of male and female *Anolis evermanni* and *A. stratulus* collected from seven tabonuco crowns. Number of individual lizards indicated by *n*. (From Dial 1992.)

for *stratulus* males (see Figure 15-3). During January, female *evermanni* did not differ significantly from *stratulus* in mean perch diameter (Scheffe's post hoc, two-way ANOVA); however, mean perch diameters of both *stratulus* sexes and female *evermanni* differed significantly from mean perch diameter of male *evermanni* (ANOVA $F_{3,282} = 6.64$, $P < 0.001$), a pattern consistent with the idea that anole morphology matches microhabitat use (Irschick and Losos 1999). In contrast, mean perch diameters during August did not differ for any species-sex comparisons (ANOVA $F_{3,85} = 2.18$, $P = 0.10$). Thus, while there is a tendency for smaller lizards to use smaller perches, the difference was not always significant.

Mean perch heights (see Figure 15-4) were not significantly different among sexes and species during January ($F_{3,282} = 1.62$, $P = 0.19$) or August ($F_{3,84} = 0.90$, $P = 0.45$; August samples were taken above 10 m only), although mean perch height of *evermanni* (mean $= 16.5$ m, s.d. $= 5.4$ m, n $= 215$) was approximately 1.5 m lower than mean perch height of *stratulus* (mean $= 17.9$ m, s.d. $= 4.6$ m, n $= 159$). Dial (1992 unpublished data) found both species were essentially crown anoles using similar perches in both diameter and height above ground.

Dial (1992 unpublished data) classified solar exposure for each lizard observed as "sun," "shade," "cloudy," or "rainy" following Rand (1964). Because lizards could choose to perch in

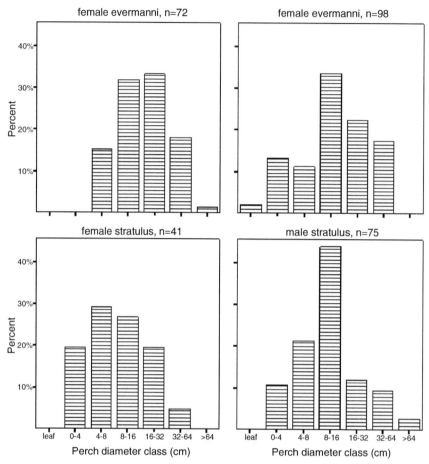

Figure 15-3 Distributions of perch diameters by species and sex from 14 tabonuco crowns during January. Number of observations indicated by *n*. (From Dial 1992.)

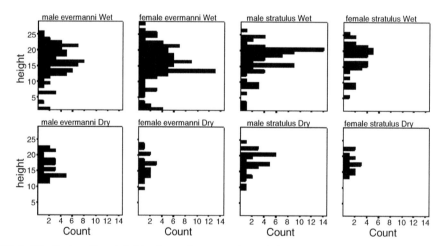

Figure 15-4 Distribution of perch heights by species, sex, and month from 14 tabonuco crowns. Number of observations indicated by *n*. Top row shows January observations, bottom row shows August. (From Dial 1992.)

sun or shade, but not in rainy or cloudy conditions, only sun vs. shade observations are considered important in classifying the niche of each species. Sample sizes of each species–sex combination are too small to test adequately for differences in solar preferences. Pooling sexes by species suggests that during January, *evermanni* preferred sun to shade (sun:shade observations = 32:20) and *stratulus* preferred the reverse (sun:shade observations = 2:16). During August, each species was observed more frequently in the shade (*evermanni* sun:shade observations =15:19; *stratulus* sun:shade observations = 7:17).

In summary, body size distributions, perch diameters, and perch heights were essentially the same for female *evermanni* and male *stratulus* in Dial's (1992) data, but not in Reagan's (1996). It appears that the structural niche and body size distributions of these two species can overlap substantially, but do not always. In particular, post- and pre-hurricane conditions provide striking contrasts in the structural niche overlap of these two lizard species.

Diets: Prey Size and Taxa

Anole diets as inferred from gut contents of crown-dwelling *A. evermanni* and *stratulus* have been reviewed by Reagan (1996), both as number of items and as gut volume. Those data showed Homoptera as the most abundant arthropod order by volume in *evermanni* stomachs, followed by spiders, Blattodea, Coleoptera, and non-ant Hymenoptera. Homoptera were also the most abundant order by volume in *stratulus* stomachs, followed by ants, Orthoptera, and Coleoptera. Reagan (1996) also reported that *stratulus* stomachs averaged 11.5 items/individual during the wet season and 8.7 items/individual during the dry season; *evermanni* averaged 17.3 items/individual (no distinction made as to season). The mean prey volume of *evermanni* stomachs was identified as twice that of *stratulus* stomachs. Due to high variances in prey size metrics, no significant differences were detected among species, sexes, or seasons. Reagan (1996) concluded that *evermanni* was a "true generalist" species and that there was a potential for competition between *stratulus* and *evermanni*, particularly during the dry season, although no data on prey availability were presented.

Here we present unpublished data from the gut contents of 40 individual *evermanni* and 37 individual *stratulus* providing information on the size distributions, biomass, and identity of 931 prey items. Descriptive statistics of the number of prey items and their biomass by lizard species and sex are provided in Table 15-1 and are comparable to that reported in Reagan (1996) with 9 to 17 items weighing 6 to 10 mg total per lizard. ANOVA of the four species-by-sex groups showed a significant difference among the groups for mean items per gut ($F_{3,73} = 3.21$, P = 0.03); however, post hoc comparisons could not identify any pair-wise differences (P > 0.08), despite a tendency for more items in the smaller lizards. Total gut biomass was not significantly different among groups ($F_{3,73} = 1.65$, P = 0.186), although there was a tendency for mean biomass of gut contents to be higher for larger lizards (see Table 15-1).

Dial and Roughgarden (1995) described the expected biomass distributions by body size for aerial and for leaf arthropods based on measured crown abundances. Their curves were products of biomass allometry (Rogers et al. 1976) and size distributions of sampled arthropods and represent resource availability (measured as biomass) as a function of arthropod body size. For crowns with lizards, peaks in biomass occurred at 2.9 mm for aerial arthropods and at 3.3 mm for leaf arthropods. If lizards are maximizing their energy intake, then these peaks could be considered the optimal arthropod size for lizards to prey upon. Gut contents of lizards (Dial unpublished data) showed that mean prey length for individual lizards was closest to optimal for *evermanni*; individual *stratulus* on average consumed arthropod prey about 0.5 to 1.0 mm smaller than the arthropod body size of maximal biomass (see Figure 15-5).

Aerial arthropods Diptera, non-ant Hymenoptera, and Coleoptera make up 90 percent of individuals caught in canopy sticky traps (Dial and Roughgarden 1995; Garrison and Willig

THE COLOR OF POISON: FLAMBOYANT FROGS IN THE RAINFOREST CANOPY

Donna J. Krabill

Neotropical rainforests of Central and South America are filled with a spectacular diversity of living organisms and are home to some of the most exquisite amphibians in the world. Poison dart frogs incite curiosity with their lethal toxins, brilliant colors, and fascinating behaviors. Unlike many poisonous animals, dart frogs do not manufacture their own toxins. Their flaunting colors warn visual predators, such as birds and small mammals, of their toxicity. It is generally believed the toxin, released from pores on their backs, is sequestered from their natural diet of flies, ants, and other small insects. The bad-tasting toxin serves as a defense mechanism and also prevents fungal and bacterial growth on their moist skin (Heselhaus 1992). Conversely, captive dart frogs lose their toxicity due to a typical diet of fruit flies and baby crickets.

Poison dart frogs belong to the family Dendrobatidae with more than 65 species comprising the four genera considered poisonous: *Dendrobates, Epipedobates, Minyobates,* and *Phyllobates* (Walls 1994). These small diurnal frogs range in size from a human thumbnail to the thumb itself. They range from the forest floor to the canopy. Some species, such as *D. biolat* and *D. reticulatus,* spend most of their lives aboveground. Epiphytic bromeliads collect and hold water that houses the eggs, tadpoles, and adult frogs. In some species, the male transports newly hatched tadpoles on its back, traveling high into the treetops where it drops one tadpole per bromeliad tank (Walls 1994). The female provides nourishment by making regular visits to each canopy nursery, depositing an unfertilized egg until each colorful froglet emerges from the water two to three months later.

In the early 1970s, scientists John Daly and Charles Meyers began unlocking the mysteries of some poison dart frog toxins (Wilson 2002). They found that the Golden Poison frog (*Phyllobates terriblis*), used by the Choco Indians of Colombia on their blowgun darts, was the most toxic frog (see Figure 1). A single frog produces enough batrachotoxin to kill

Figure 1 The lethal golden poison frog (*Phyllobates terriblis*) of Colombia produces enough toxin to kill 10 adults. Photograph by R.D. Bartlett.

20,000 mice or approximately 10 adult humans (Heselhaus 1992). These researchers also found that the tiny *Epipedobates tricolor* of Ecuador produces epidatidine, a toxin that has biopharmaceutical potential as a non-addictive painkiller. This compound is similar to morphine but is 200 times stronger with fewer side effects (Plotkin 2000).

Many poison dart frogs are now at risk of extinction due to human destruction of the rainforest. For example, when Daly and Meyers returned to the Ecuadorian rainforest in search for *Epipedobates* to collect more epidatidine, they found that one of two prime sites had been replaced with banana plantations (Wilson 2002). Wilson (2002) further warned that, "The search for natural medicine is a race between science and extinction and will become critically so as more forests fall." The extinction of these flamboyant amphibians would be an irreparable loss to both global biodiversity and medical research.

References

Heselhaus, R. (1992). "Poison-Arrow Frog: Their Natural History and Care in Captivity." Blanford, London.
Plotkin, M. (2000). "Medicine Quest." Penguin Books, New York.
Walls, J.G. (1994). "Jewels of the Rainforest: Poison Frogs of the Family Dendrobatidae." T.F.H. Publications, Neptune City, New Jersey.
Wilson, E.O. (2002). "The Future of Life." Vintage Books, New York.

1996). Spiders, Orthoptera, Blattodea, Homoptera, and Psocoptera made up more than 85% percent of all arthropod individuals in leaf arthropods (Dial and Roughgarden 1995, Schowalter and Ganio 2003). These eight orders made up 89 percent of gut biomass in *evermanni* and 85 percent in *stratulus*. Figure 15-6 shows size distributions by biomass for aerial arthropods collected from lizard gut contents (Dial unpublished data) compared to the expected biomass distributions of arthropods caught in sticky traps published by Dial and Roughgarden (1995). For aerial arthropods, both species together consumed a modal size class (by gut biomass) of 2 to 3 mm, the size class of maximal expected biomass for aerial arthropods in crowns with lizards (see Figure 15-6). The conformance statistic, B, $B = 1 - \sum |U_i - A_i| / 2$, where U_i = proportion total biomass utilized in class i and A_i = the environmental availability of class i (proportion of total expected biomass from Dial and Roughgarden 1995), provides a measure of how closely diet matches resource availability (MacNally 1995). If diet matches availability exactly, then $U_i = A_i$ for all i, and $B = 1$ (perfect generalist); if the organism specializes far out of proportion to availability, then $B = 0$ (specialist). Using only the 1 mm size classes between 0 mm and 6 mm for each species (these are size

Table 15-1   Arthropod Prey of *Anolis* from Gut Contents

Species and Sex	Mean Number of Items (standard deviation)	Mean Biomass (mg) (standard deviation)
Anolis evermanni	9.0	10.1
male (n = 19)	(7.1)	(10.6)
A. evermanni	9.1	6.1
female (n = 21)	(6.1)	(5.3)
A. stratulus	16.5	6.2
male (n = 16)	(13.4)	(5.0)
A. stratulus	14.6	5.7
female (n = 21)	(9.2)	(5.5)

From Dial, unpublished data.

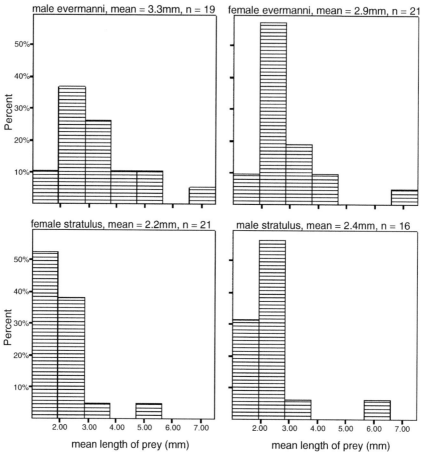

Figure 15-5 Distributions of mean prey body lengths in lizard gut contents by species and sex. Number of individual lizards indicated by *n*. (From Dial, unpublished data.)

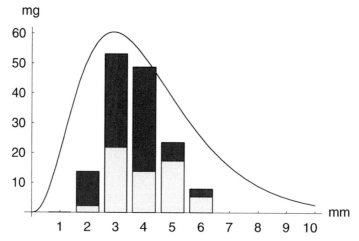

Figure 15-6 Total biomass of aerial arthropods from gut contents of 77 lizards (stacked histogram; white = *evermanni*, black = *stratulus*) compared to expected distribution of biomass from sticky traps (curve). (From Dial, unpublished data; Dial and Roughgarden 1995.)

classes with > 2 items in pooled gut contents) yields conformance values of $B = 0.79$ for evermanni and $B = 0.70$ for stratulus, supporting the contention that evermanni is more generalist in choice of prey by size than stratulus.

The gut contents of 40 *evermanni* and 37 *stratulus* (Dial unpublished data) were used to estimate the importance of each of eight prey taxa based on percent gut biomass (see Figure 15-7). Using ANOVA on arcsin-square root transformed proportions and Scheffe's post hoc pair-wise comparisons, spiders ($F_{3,73} = 3.8$, $P = 0.01$) and Psocoptera ($F_{3,73} = 3.9$, $P = 0.01$) showed significant differences among species-by-sex comparisons; Diptera ($F_{3,73} = 2.4$, $P = 0.08$) and Homoptera ($F_{3,73} = 2.6$, $P = 0.06$) showed marginally significant differences; the other taxa showed no significant differences ($P > 0.10$). Diptera made up over half the gut biomass of aerial arthropods for each species and sex. Orthoptera were more important for *evermanni* than *stratulus* and equally important for male and female *evermanni;* Hymenoptera were important for *stratulus* females and Homoptera for *stratulus* males. Gut biomass of all aerial arthropods (= Diptera + Hymenoptera + Coleoptera) and for all leaf arthropods (= spiders + Orthoptera + Blattodea + Homoptera + Psocoptera) were summed for each individual lizard. Proportions of both aerial ($F_{3,73} = 2.5$, $P = 0.07$) and leaf arthropods ($F_{3,73} = 3.4$, $P = 0.02$) were significantly different between female *evermanni* and female *stratulus* (see Figure 15-7).

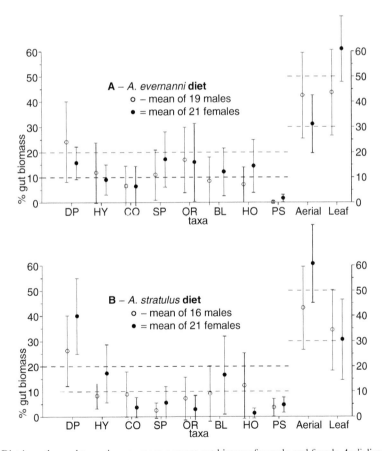

Figure 15-7 Diet by arthropod taxa given as mean percent gut biomass for male and female *Anolis* lizards. Circles give means and error bars are ± 0.5 standard deviations. DP = Diptera, HY = Hymenoptera, CO = Coleoptera, HO = Homoptera, AR = Araneae, PS = Psocoptera, OR = Orthoptera, BL = Blattodea. Dashed lines indicate 10%, 20%, 30% and 50% biomass. "Aerial" and "Leaf" are defined in the text. (From Dial, unpublished data.)

The implications of these data are: (1) There is substantial dietary overlap between the two species with biomass of Diptera making up the highest or second highest proportion of each species-by-sex diet, and (2) females differ most between species, with *evermanni* specializing more on leaf arthropods and *stratulus* females specializing more on aerial arthropods. These observations differ from Reagan (1996), whose data indicate similarity between females and males of both species and a lesser importance for Diptera and greater one for Homoptera in the diets of both species. In summary, dietary analysis from two studies (Dial unpublished data, Reagan 1996) suggests that the two species consume very similar-sized prey and that similar taxa are important for satisfying energy demands. Both Reagan (1996) and Dial (unpublished data) find 5 to 15 items per lizard gut, equal to 5 to 10 mg of arthropod biomass per lizard. When environmental abundances of food resources are known, *evermanni* conforms slightly more than *stratulus*. Overlap in prey sizes, as with lizard body sizes and perch diameters, was greatest between male *stratulus* and female *evermanni*. Some partitioning of prey taxa was evident along aerial vs. foliage groups, particularly with females. When prey differences between species were apparent, *stratulus* seemed to prefer primarily smaller-sized, aerial prey and *evermanni* larger-sized, leaf prey, and both species relied principally on 2 to 3 mm or larger sized aerial prey to satisfy a substantial portion of their energy requirements. The question is posed, then: do these species respond to spatial and temporal variability in the abundance of aerial arthropods?

Distribution and Abundance

Reagan (1996) reviewed pre-Hurricane Hugo density estimates of *Anolis* taken from canopy towers in the Luquillo Mountains. Density estimates of *stratulus* ranged from 2.1 lizards/m^2 to 2.8 lizards/m^2. Using relative abundance estimates of *evermanni* and the absolute density estimates of *stratulus,* Reagan (1996) estimated *evermanni* densities at 0.15 lizards/m^2. Only about 35 percent of the *evermanni* observations were above 10 m, suggesting that crown densities were 0.05 lizards/m^2. The high densities of *stratulus* were attributed to the vertical layering of home ranges. Prey limitation was suggested, although direct comparisons of lizard abundance with prey availability were not presented. Reagan (1996) also reported that for *stratulus* "in areas of severe forest damage, populations of this species remained low four years following" hurricane Hugo. The reason was most certainly reduced habitat availability following canopy defoliation.

Dial and colleagues' study (Dial 1992; Dial et al. 1994; Dial and Roughgarden 1995, 1996; Dial unpublished data) began 18 months after Hurricane Hugo. Lizard censuses were made in 14 experimental trees at Bisley during January and August to determine densities (see Figure 15-8). Unlike pre-Hurricane Hugo studies (Reagan 1996), mean densities of the two species (0.2 to 0.3 lizards/m^2) were not significantly different from each other (see Figure 15-8). Dial (1992) also reports on a census from eight crowns near El Verde (< 10 km from Bisley), which were not defoliated during the hurricane. In these crowns, mean *stratulus* density was 0.15 lizards/m^2 and densities of *evermanni were* 0.04 lizards/m^2, the latter figure quite close to Reagan's (1996). Clearly, post-hurricane densities of *stratulus* appear to be as much as an order of magnitude less than pre-hurricane, while *evermanni* densities were approximately the same at El Verde pre- and post-hurricane, and substantially higher at Bisley post-hurricane.

Anolis stratulus, A. evermanni, and number of eggs per female lizard were each positively and significantly correlated with aerial arthropods captured in sticky traps (see Figure 15-9). Extrapolation of the *stratulus* regression to the arthropod abundance axis suggests that aerial arthropods may be more important for *stratulus* (248 arthropods m^{-2} 12 hours^{-1}) than for *evermanni* (0 arthropods m^{-2} 12 hours^{-1}). This is consistent with dietary analysis (see Section II) showing a greater proportion of aerial arthropods in *stratulus* diets than in *evermanni* diets (see Figure 15-7). The regression coefficients for lizard density on arthropods were not significantly different

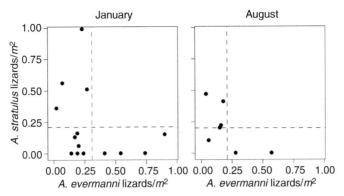

Figure 15-8 Densities of crown-dwelling *Anolis evermanni* and *A. stratulus* censused in January (left, n = 14 crowns) and August (right, n = 7 crowns). Each circle represents densities for a single tabonuco crown. Dashed lines represent mean densities for each species averaged across crowns. Mean *A. evermanni* density (number of lizards/projected area of tree crown) in January was 0.31 lizards/m² (s.d. = 0.25) and mean *stratulus* density was 0.21 lizards/m² (s.d. = 0.30); mean density of both species together was 0.50 lizards/m² (s.d. = 0.32). In the seven crowns sampled both months, mean density of *evermanni* in January was significantly higher (2-tailed paired sample t-test, t = 2.55, P =0. 04) than mean density in August (mean = 0.21 lizards/m²). Mean *stratulus* density in January was not significantly different (2-tailed paired sample t-test, t = 0.26, P =.80) from density in August (mean = 0.20 lizards/m²). Mean density of *evermanni* did not differ significantly from mean density of *stratulus* during either August (df = 12, t = 0.094, P = 0.926) or January (df = 26, t = 0.954, P = 0.349). (From Dial 1992 and unpublished data).

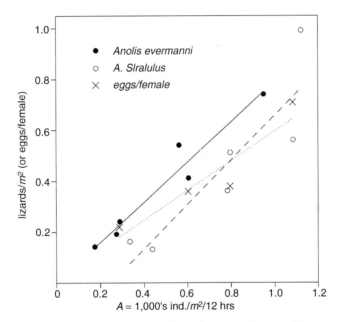

Figure 15-9 Population measures of *Anolis evermanni* and *A. stratulus* regressed against sticky trap catches of arthropods in thousands (10^3) per m² per 12 hours, **A**. Solid line and bullets give density of *A. evermanni* individuals /m², **E**, regressed against arthropods from n = 6 crowns with no *A. stratulus* present (Pearson's r² = 0.94; **E = 0.79 A + 0.00**). Dashed line and open circles give *A. stratulus* individuals /m², **S**, regressed against arthropods from n = 6 crowns with 3 or more *A. stratulus* present (Pearson's r² = 0.81; **S = 0.88 A − 0.22**). Dotted line and X's give the number of eggs per female, **G**, regressed against aerial arthropods from n = 4 crowns with 10 or more lizards collected and dissected (Pearson's r² = 0.88; **G = 0.58 A + 0.13**). Regression slopes in each case are significantly different from zero (F test, P < 0.06), and the intercepts are not significantly different from zero (t-test, P > 0.25). (From Dial 1992 and unpublished data.)

between the two species ($F_{1,8} = 1.07$, $P = 0.45$). Lizard densities did not correlate significantly with leaf sampled arthropods (Pearson's $r^2 = 0.02$).

In summary, and in contrast to Reagan (1996), Dial (1992, unpublished data) found that densities of *Anolis evermanni* and *A. stratulus* in tabonuco canopy were not significantly different. While Reagan (1996) describes *evermanni* as a generalist species, his data show clearly that it is not primarily a crown species. Dial's observations (1992, unpublished data) indicate that microhabitat, density, diet, and the relationship between aerial arthropods and lizard densities are essentially the same for both *Anolis evermanni* and *A. stratulus*. These results suggest the potential for strong competition that is exacerbated by hurricane defoliation.

Competition Between *Anolis evermanni* and *A. stratulus*

Taken together, the canopy data here show substantial overlap in lizard morphology (see Figure 15-2), microhabitat use (see Figures 15-3 and 15-4), and diet between *A. evermanni* and *A. stratulus* (see Table 15-1, Figures 15-5–7), with very similar population responses to aerial arthropod abundance (see Figure 15-9). Moreover, Dial and Roughgarden (1995) have shown that lizards in this system deplete (Section III) similar sizes and taxa of arthropods from tabonuco crowns reported here as food resources. Given such similar structural and trophic niches, classic competition theory (e.g., May 1973) and a model tailor-made for this system (Dial unpublished) suggest that these two species may coexist only if within-species interactions are stronger than between-species. In essence, models for two species that share the same niche can coexist only if they each limit themselves more than the other species. If between-species interactions are stronger than within-species, coexistence is not possible. Behavioral interactions between canopy anole species and between sexes are suggestive in this regard. Reagan (1984) reported agonistic interactions between *evermanni* and *stratulus* during staged encounters. Dial (1992) reports that negative between-sex interactions were observed only between species; within a given species males fought with males and females with females, but males and females of the same species were never seen interacting antagonistically. This contrasts with female–male agonistic interactions observed between species (Dial 1992). These anecdotal observations were not quantified; however, competition theory (May 1973) predicts "priority effects" between competitors with stronger between-species than within-species interactions. At equilibrium, any given habitat patch holds only one or the other species, depending on initial conditions, but not both. During both January and August, no crowns were observed with above-average densities of both lizards; only crowns with low densities of lizards supported both species in relatively equal abundances (see Figure 15-8). Trees with high lizard densities (> 0.3 lizards/m^2) supported relative abundances of *stratulus* > 70 percent or relative abundances of *evermanni* > 80 percent. Thus, the checkerboard pattern in distribution and abundance of crown-dwelling populations in this study is consistent with competitive priority effects where behavioral interference—driving individual lizards away from pair-inhabited territories—is the mechanism of exclusion.

The checkerboard pattern may well be the result of Hurricane Hugo having "reset" the system by stripping all foliage from the canopy and relocating individual lizards to the ground. Then, as lizards re-colonized the canopy on a crown-by-crown basis, competitive priority effects came into play, leading to "*evermanni*-only" and "*stratulus*-only" crowns. Ultimately, as the canopy regrows fully, it is likely that the abundance of small *stratulus* perches increases sufficiently so that *stratulus* may exclude *evermanni* from crowns, as seen in the data reviewed by Reagan (1995, 1996). Alternatively (or additionally), the apparently greater susceptibility of *evermanni* to lethal sarcophagid fly infestations (Dial and Roughgarden 1996) may provide *stratulus* with a parasitoid-mediated advantage. In any event, data on relative abundances of crown-dwelling anoles from four towers at two pre-hurricane sites (Reagan 1996) and eight crowns at one undamaged site (Dial 1992) suggest that *stratulus* ultimately establishes itself as the canopy lizard of greatest relative abundance.

In summary, pre-hurricane densities and microhabitats of each lizard species differed (Reagan 1995, 1996), with *A. stratulus* a very abundant (~2 lizards/m^2) canopy lizard and *evermanni* less common there (~0.05 lizards/m^2). In contrast, 1 to 2 years after Hurricane Hugo, very similar densities (0.2 to 0.3 lizards/m^2) and microhabitat distributions for the two species occurred at the scale of several groves (Dial 1992). In contrast, similar densities of both species *in the same crown* were observed only for crowns with low total densities (see Figure 15-8). These differences in density across scales are attributed to within-crown interference competition between the species, since each species appeared to forage in similar microhabitats (see Figures 15-3 and 15-4) for the same (see Figures 15-5–7) limiting aerial arthropod resources (see Figure 15-9). These resources were determined experimentally to be depleted by lizards (Section III; Dial and Roughgarden 1995). The checkerboard distribution of lizard abundance (high densities of one species if and only if the other species was in low abundance) appears to result from competition for limiting food resources within the same microhabitats, combined with stronger between-species behavioral interactions than within-species. While pre-hurricane and undamaged trees suggest that this checkerboard pattern is not stable, what factors may lead to stability in community structure?

Stability of Crown-Dwelling Lizard Communities

Community stability can be defined as the temporal constancy of relative abundances of community members. In the case of two-species communities, the relative abundance of one species in a given habitat patch can be followed, and the standard deviation of relative abundance over time used as an index of community instability. A greater standard deviation in relative abundance over time would suggest weaker community stability. During the seven-month experimental period for Dial and colleagues' study, the number of each lizard species observed per hour was recorded during monthly visits to each of the 14 crowns (mark-remark methods for absolute abundances were not used to avoid disturbing the system). Using seven control crowns with un-manipulated lizard abundances (Dial 1992) provides estimates of the standard deviation of relative abundance. Figure 15-10 shows the time series plots for these seven crowns, the standard deviation of the relative abundance of *stratulus* in each crown, and the mean catch of sticky trapped 2- to 3-mm arthropods—the latter shown to be abundant in guts (see Figure 15-6) and depleted by lizards (Dial and Roughgarden 1995)—in each crown over the same time period. Using non-parametric, rank order statistics for the correlation of standard deviation of relative abundance of *stratulus* with abundance of aerial arthropods 2 to 3 mm long (Spearman's, *P = 0.01*) suggests that lizard community stability may depend on resource abundance. In particular, more resources increase lizard community stability, while fewer resources appear to lead to community turnover. Thus, the question of what factors determine the abundance of lizard resources seems appropriate.

Section III: Determinants of Canopy Arthropod Abundance

Figure 15-1 shows the location of crowns and gaps at the study site of Dial and colleagues (Dial 1992). A crown was considered exposed to northeasterly trade winds (i.e., windward), if a gap lay immediately to its northeast. A crown was considered sheltered from the trades (i.e., leeward), if it was contiguous with forest canopy immediately to its northeast. Pre-manipulation abundances of aerial arthropods caught in sticky traps in windward crowns were significantly greater ($F_{1,12}$ = 6.4, P = 0.026) than in leeward crowns (Dial 1992): the mean number of insects caught within windward crowns (776 arthropods m^{-2} 12hrs^{-1}) was nearly twice that in leeward crowns (424

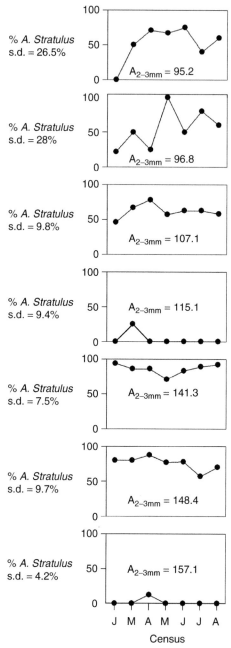

Figure 15-10   Lizard community stability and resource abundance in seven control crowns. Each stylized plot shows the relative abundance of *Anolis stratulus* (measured as % of all crown lizards on vertical axis) over time (horizontal axis, measured in monthly census periods) in a single tree crown. s.d. = standard deviation of *A. stratulus* relative abundance (%) over time for each crown and $A_{2\text{-}3mm}$ = mean number of 2-3 mm arthropods caught in sticky traps averaged over all census periods. J, M, A, M, J, J, and A represent months from January through August (there was no census in February). (From Dial 1992.)

arthropods m^{-2} 12hrs^{-1}). Dial (1992) also showed that the mean density of anoles (both species combined) in windward crowns (mean = 0.76 lizards/m^2, s.e. = 0.10) was nearly three times higher than in leeward crowns (mean = 0.26 lizards/m^2, s.e. = 0.03).

Gap Arthropods

Dial (1992) reported on the results of suspending small, cylindrical traps (25 cm^2) on horizontal lines stretched at mid-canopy height (12 to 15 m) across two gaps (see Figure 15-1). The overall mean number of insects captured in small gap traps (1,945 arthropods m^{-2} 12hrs^{-1}; n = 23 traps, s.e. = 60.4) was significantly greater than the catch from small within-crown traps (1,241 arthropods m^{-2} 12 hrs^{-1}; n = 6 traps, s.e. = 244.4). Using the approach of Dial and Roughgarden (1995) to calculate the expected biomass of gap arthropods as a function of body length using gap arthropod body size distribution and biomass allometry (Rogers et al. 1976) gives a peak biomass for arthropods in the 2 to 3 mm size class (Dial 1992). Many of the same taxonomic orders of insects and spiders captured in crowns (Diptera, Thysanoptera, Hymenoptera, Coleoptera, Homoptera, Araneae, Hemiptera, Psocoptera, Neuroptera, Lepidoptera, Blattodea, Orthoptera) were also captured in gaps (see Figure 15-11). The numerical majority of individuals in sticky traps were Diptera in both crowns (Dial and Roughgarden 1995) and gaps (see Figure 15-11). Diptera captured in crown and gap traps included many families (Phoridae, Chironomidae, Sciaridae, Scatopsidae, Psychodidae, Chloropidae, Drosophilidae, Tephritidae, Neriidae) with larval habitats in litter, soil, fungi, and water (Ferrar 1987; Borrer et al. 1989), although larvae of Syrphidae and Sarcophagidae were observed in crowns (Dial 1992). Because debris and epiphytes in the canopy had not yet accumulated following Hurricane Hugo, sticky trapped Diptera most likely originated from the ground and other non-arboreal sources.

The greater abundance of aerial arthropods over gaps than within forest canopy may be due to several factors. Two possibilities suggested by the literature are: (1) greater food resources for dipteran larvae (including detrital as well as living material) from gap plants than from later successional plants (Janzen 1973; Coley et al. 1985; La Caro and Rudd 1985) and (2) insects are

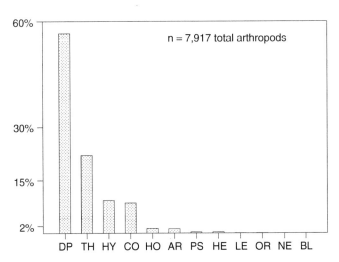

Figure 15-11 Taxonomic distribution of arthropods captured using cylindrical sticky traps suspended 11-15 m over two forest gaps shown in Figure 1. DP = Diptera, TH = Thysanoptera, HY = Hymenoptera, CO = Coleoptera, HO = Homoptera, AR = Araneae, PS = Psocoptera, HE = Hemiptera, LE = Lepidoptera, OR = Orthoptera, NE = Neuroptera, BL = Blattodia. (From Dial and Roughgarden, unpublished manuscript.)

attracted to increased light levels within gaps (Craig and Bernard 1990). Dial (1992) reports on the results of a wind vane trap (0.5 m sticky trap-coated rod), where samples showed a significant, positive correlation of insect catch per cm to windward position along the rod ($r^2 = 0.81$, $P = 0.01$); during one sample period, the number of insects captured on the first, most windward cm was an order of magnitude more than on the last, most leeward cm of the rod. These results support the conclusion that insects arrive from the instantaneous wind direction. Dial (1992) used two large, fixed-cylinder traps ($\sim$425 cm^2) to estimate the circular mean direction of arthropod catch and circular statistics (Stephens 1962; Batschelet 1965, 1981) and found that the mean direction differed significantly from random (V-test, $P = 0.005$). Arthropod catches on the large, fixed cylinders indicated a general southeasterly origin of arthropods during those months (see Figure 15-12).

In summary, five lines of evidence suggest that wind transports aerial arthropods into forest canopy from adjacent gaps. First, sticky traps caught more arthropods within windward crowns than in leeward crowns. Second, arthropods existed in the air within crowns and over gaps with Diptera predominating in both habitats. Third, significantly more arthropods were captured in gap traps than in crown traps. Fourth, arthropods arrived from the instantaneous wind direction. And fifth, the mean direction of arthropod arrival coincided with an easterly trade-wind direction. It appears that the observed spatial distribution of canopy anoles within the forest canopy at this site may represent the local consumers' response to variation in non-local resource abundance, variation determined by canopy structure and wind direction. These conclusions follow from: (1) arthropods in observed lizard diets (see Figure 15-7), (2) lizard depletion of these arthropods (Section III), and (3) evidence from gap catches of arthropods.

Arthropod Dispersal and Response to Predator Removal

Dial (1992) and Dial and Roughgarden (1995) documented the effects of removing anoles from tabonuco crowns. Only arthropods > 2 mm increased significantly following lizard removal. Dial (1992) compared arthropod response following removal of *evermanni* lizard to removal of *stratulus*. The comparison used two control crowns and two removal crowns for each lizard species (n = 8 total crowns). The results suggested that *evermanni* suppressed larger arthropods more than did *stratulus*, which suppressed smaller arthropods more (see Figure 15-13a, b). Abundances from these same eight experimental crowns also suggested that *evermanni* suppressed Homoptera,

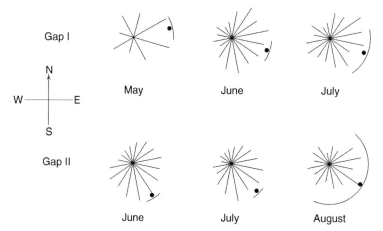

Figure 15-12 Compass rose distributions for number of arthropods captured 11-15 m over two gaps shown in Figure 15-1 using large, fixed cylinder sticky traps. (From Dial and Roughgarden, unpublished manuscript.)

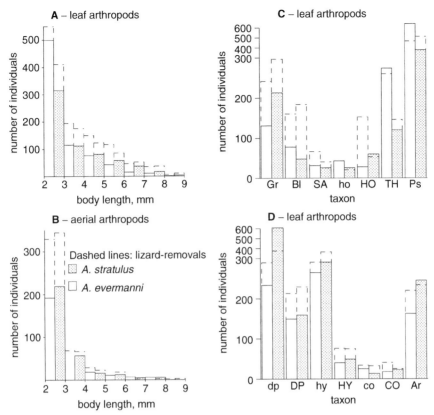

Figure 15-13 Pooled samples for arthropod abundance (sampled with sticky traps for Diptera, Hymeoptera, and Coleoptera; and with visual inspections of foliage for other taxa) from eight crowns and six censuses used in a canopy-level lizard removal experiment. Histograms outlined in dashed lines show arthropod abundances from lizard removal crowns (n = 2 crowns for each lizard species). Histograms outlined in solid lines give arthropod abundances from trees with the specified lizard species present (n = 2 control crowns for each lizard species). Shaded histograms are crowns with > 70% relative abundance of *A stratulus*, unshaded are crowns with > 70% relative abundance of *A evermanni* (in the case of lizard removal crowns the relative abundance figures are pre-lizard removal). A and B compare arthropod size distributions and suppression (defined as the difference between arthropod abundances in lizard removal crowns and controls) for leaf arthropods and aerial arthropods respectively. C and D compare arthropod taxonomic distributions and suppression. Gr = Gryllidae, Bl = Blattodea, SA = salticid spiders > 2 mm, ho = Homoptera < 2 mm, HO = Homoptera > 2 mm. TH = green theridiid spiders > 2 mm, Ps = Psocoptera, dp = Diptera < 2mm, DP = Diptera > 2mm, hy = parasitic hymenoptera < 2 mm, HY = parasitic hymenoptera > 2 mm, co = Colepotera < 2mm, CO = Coleoptera > 2mm, and Ar = web spiders/m². (From Dial 1992.)

gryllids, and orb and salticid spiders more than did *stratulus*, while *stratulus* suppressed Blattodea and Diptera more than did *evermanni* (see Figure 15-13c, d). These observations on prey size and taxa are consistent with species-specific lizard gut content analysis (see Figure 15-7). Dial and Roughgarden (1995) also reported weak negative effects on arthropods < 2 mm following lizard removal, effects attributed to increased predation by foliage-dwelling spiders preyed upon by lizards. Dial (1992) also found that certain leaf-dwelling arthropod taxa > 2 mm that were green in color showed no responses to lizard removal: Psocoptera, tropiduchid and issid Homoptera, and theridiid spiders (see Figure 15-14), a result similar to that found by Van Bael et al. (2003) with insectivorous birds and their prey in a Panamanian rainforest canopy. Finally Dial and Roughgarden (1995) suggested that the strong positive response to lizard removal shown by large

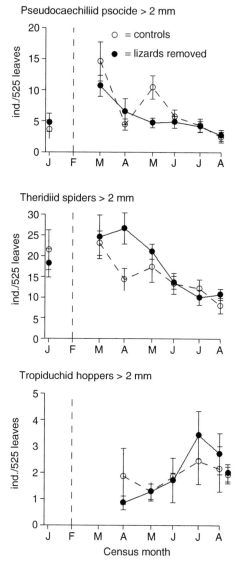

Figure 15-14 Numerical responses of green-colored arthropods > 2mm found on foliage surfaces. The vertical dashed line indicates the date of lizard removal. Data points are means (solid circles and lines for seven lizard removal crowns and open circles and dashes for seven control crowns) 1 s.e. J, M, A, M, J, J, and A represent months from January through August (there was no census in February). (From Dial 1992.)

(>2 mm), abundant, chewing foliovores (Blattodea and Gryllidae) cascaded downward as increased herbivory and a doubling of leaf damage (see Figure 15-15a, b). Removals may also have encouraged trophic compensation upward as increased abundance of parasitic hymenoptera (see Figure 15-15c). In summary, many local arthropod prey of lizards showed strong responses to lizard removal (Dial 1992; Dial and Roughgarden 1995; see Figure 15-16).

If arthropod resources originate upwind of a local anole population consuming the resources, then few surviving adult arthropods may actually return to their original, upwind gap. This could decouple non-local prey populations from local predator populations, weakening the prey–predator interaction. Crown populations of lizards might limit, but not regulate, non-local arthropods.

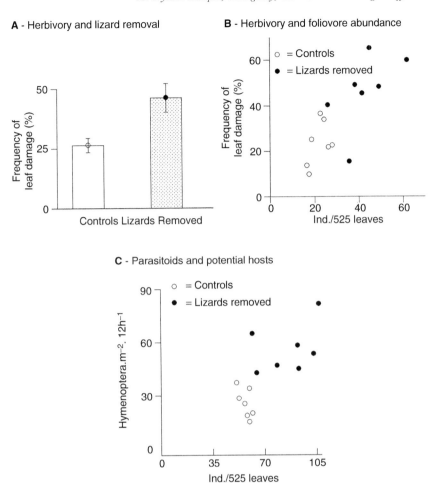

Figure 15-15 Indirect responses to lizard removal. (A) Frequency of tabonuco leaf damage by treatment on new leaves produced by apical meristems marked before lizard removal. (B) Frequency of tabonuco leaf damage in relation to Blattodea + Gryllidae abundance in each of fourteen study crowns. Each point represents % of leaflets showing any herbivore damage within a given crown plotted in relation to the sum of the overall mean abundances of Blattodea and Gryllidae for the same crown. (C) Abundance of parasitic Hymenoptera > 2 mm caught in sticky traps plotted against potential host arthropods in each of fourteen study crowns. Each point represents the mean abundance of Hymenoptera for a crown plotted with the overall mean abundance of leaf arthropods > 2mm in the same crown. (From Dial and Roughgarden 1995.)

In contrast, local prey populations that inhabit crowns throughout their life cycle (such as Gryllidae, Blattodea, Homoptera, and Salticidae) may not be decoupled from local predator populations. Local prey might well be regulated by local predators. Experimental detection of these effects may be suggested by predator removals. Local prey populations with the predator present would show dynamics out of step with the dynamics of the same prey with the predator experimentally removed, if the predator regulates the prey. Non-local prey abundance with the predator present would most likely fluctuate in tandem with prey abundance with the predator removed, if some other factor regulated the prey. These interpretations are supported by the lizard removal experiment (see Figure 15-16; Dial 1992; Dial and Roughgarden 1995). Data from that experiment showed that Diptera, the taxon with greatest gap abundance, was more abun-

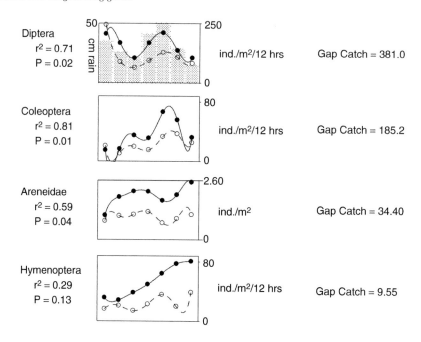

Figure 15-16 Stylized temporal dynamics of arthropod taxa from a lizard removal experiment. Only taxa showing significant differences between crowns with lizards present and lizards removed are used. Each plot provides the time series for the means of seven lizard removal and seven control crowns for a different arthropod taxon. Horizontal axes give time as monthly censuses. J, M, A, M, J, J, and A represent months from January through August (there was no census in February). Solid circles (means of seven crowns) connected by solid polynomial curves represent dynamics for lizard removal crowns. Open circles (means of seven crowns) and dashed polynomial curves represent dynamics for control crowns. Vertical axes in each graph represent arthropod abundance in units given on right axis. The horizontal dashed line extending across the figure separates sticky trap catches (upper) from foliage samples (lower).

Continued

dant in lizard removal crowns than in controls, but the dynamical trajectories were unchanged by lizard removal, and in fact paralleled rainfall in both lizard removal and control crowns (see Figure 15-16). Thus lizards likely limited Diptera, but dynamics of Diptera may be regulated by rainfall, or at least by some factor correlated with rainfall (Didham and Springate 2003). In contrast, taxa poorly represented in gaps, such as gryllids, salticids > 2 mm, and Hymenoptera > 2 mm differed both in numbers and dynamics between removals and control (see Figure 15-16). Indeed, gryllid dynamics in crowns without lizards paralleled leaf growth of tabonuco, while gryllid dynamics in crowns with lizards did not (see Figure 15-16). This suggests that population regulation of canopy arthropods can be conditional on predator presence: top down in their presence and bottom up in their absence.

One way to compare the influence of a predator on a variety of its prey is to consider the correlation of prey abundances with and without the predator over time, to see if the abundances change together. Figure 15-16 shows stylized polynomial plots of arthropod abundances with and without lizards for eight taxa that demonstrated significant responses to lizard removal (Dial 1992; Dial and Roughgarden 1995). The census points were used to calculate the Spearman's rank correlation coefficients for the mean abundance of each taxa between crowns with and without lizards. These rank correlation statistics themselves were significantly and positively correlated with ln(A), where A = mean abundance in gap sticky traps (Pearson's r = 0.76, P = 0.02, n = 7,

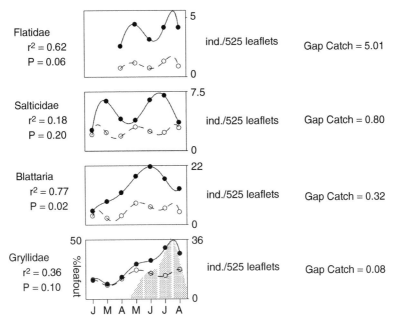

Figure 15-16—*Continued* Spearman's rank correlation coefficient, r, equals correlation of arthropod abundance in control crowns with arthropod abundance in lizard removal crowns (p-value gives significance of difference from zero for r). "Gap Catch" = number of individuals caught in gap sticky traps normalized as individuals m^{-2} 12 hrs^{-1}. Shading in plot for Diptera (topmost plot) gives monthly rainfall during the experiment. Census means of Diptera > 2mm correlated with monthly rainfall in removals ($r^2 = 0.73$, P = 0.01) but not in controls ($r^2 = 0.17$, P = 0.36). Shading in plot for Gryllidae (lowermost) gives % of tabonuco leaves newly flushed one month previous. Mean number of Gryllidae in each census correlated well with percent of new leaf growth one month earlier in removals ($r^2 = 0.99$, P < 0.001) but not in controls ($r^2 = 0.15$, P = 0.52). Blattodea showed a similar response to new leaf growth during the same month. (From Dial 1992; Dial and Roughgarden unpublished data; and Dial and Roughgarden 1995.)

omitting Blattodea). In other words, well-dispersing taxa (those with large gap catches as indicated in Figure 15-16) showed similar dynamics in lizard removal and control crowns (Diptera > 2 mm: Spearman's $r^2 = 0.71$, P = 0.02; all Coleoptera: $r^2 = 0.81$, P = 0.01; Areneidae: $r^2 = 0.59$, P = 0.04), while poorly dispersing taxa did not (Gryllidae: $r^2 = 0.36$, P = 0.10; Salticidae > 2 mm: $r^2 = 0.18$, P = 0.20; Hymenoptera > 2 mm: r = 0.29, P = 0.13). Therefore, while airborne arthropods appear to regulate anole population densities (see Figures 15-9 and 15-10), anoles may not be important in regulating aerial arthropod numbers.

The canopy experiments of Dial (1995) and Dial and Roughgarden (1995) suggest several generalizations that likely extend to other non-canopy systems: (1) Prey body size and crypsis may determine the predation pressure exerted by visually orienting predators on their potential prey, with visible, energetically abundant prey preferred (see Figure 15-14 and 15-16). (2) Removal of abundant top level predators can lead to trophic cascades even in terrestrial systems (see Figure 15-15). (3) Indirect effects tend to be weaker than direct effects (see Figures 15-15 and 15-16). (4) Resource rich communities can be more stable than resource poor ones (see Figure 15-10). (5) Physical transport and dispersal of prey can determine predator distribution and abundance (see Figures 15-1 and 15-9) and response to predation (see Figure 15-16). (6) Factors that regulate canopy arthropods can be conditional on predator presence/absence (see Fig 15-16 Gryllidae), or not (see Fig 15-16 Diptera). (7) Pair-wise interactions may often not be commutative: non-local prey may limit and regulate local predators, but local predators may not regulate these same non-local prey. Similarly, local prey may neither limit nor regulate local predators, but local predators may both limit and regulate these local prey (see Figures 15-7, 15-8, 15-9 and 15-16). Weak predator control of limiting-prey suggests that dynamic models of predator-prey interactions that predict predator-mediated extinction of prey, or other food web properties sensitive to interaction magnitudes (Pimm 1982; Lawton 1992), can be irrelevant at certain scales. Coexistence and stability of predator-prey dynamics may result simply from the spatial decoupling of prey from predator populations, as in the case of non-local prey and local predators. The strong regulatory and limiting role exerted by anoles on local prey may be the selective agent for green-colored crypsis in foliage-dwelling arthropods, a strategy apparently less prominent among non-local prey.

Section IV: Synthesis

This chapter has set out to describe a community of two consumers and their resources in a tropical rainforest canopy. Additional canopies may show similar processes, particularly where abundant predators are local to individual crowns exposed to persistent winds. Systems involving other arboreal insectivores, such as tree frogs, lizards such as flying lizards (*Draco*) in Asia and day geckos (*Phelsuma*) in Madagascar, insectivorous birds (e.g. Van Bael *et al.*, 2003), and even ants may show analogous function and process. In particular, a hypothesized view of competing consumers whose spatial abundance is regulated by non-local resources has been advanced. The individual crowns of this ecosystem subject to hurricane disturbance are seen as habitat patches for two anole species that scramble for the same limiting prey and interfere in establishing territories within very similar microhabitats. The limiting prey, whose body size corresponds to the size of maximal expected biomass in the environment, include a preponderance of potentially non-local taxa, arthropods that appear to be delivered by trade winds blowing across large gaps. Differential exposure to these winds, combined with priority effects from competition, may lead to the observed spatial heterogeneity in lizard community structure as a micro-checkerboard distribution. This distribution depends on variation in arthropod arrival. Trees with higher abundance of limiting prey for consuming lizards seem to remain more stable in their community structure, while trees with low abundance may turn-over in their community structure more quickly. Abundance of consumers shows little response to spatial variation in local resource abundance but a strong response, both temporally and spatially, to non-local prey abundance. In contrast and highlighting the asymmetry of trophic relationships, the effect of consumers on local prey seems stronger than the effect on non-local prey. In this system, several of the local prey are herbivores. Thus, the subsidy of local consumers by non-local resources may have indirect effects on local producers through the local herbivore pathway. For instance, increased abundance of non-local prey may increase local predator populations, thereby reducing local herbivore populations and herbivory.

This chapter has demonstrated that forest canopies can serve as model ecosystems for general ecological processes. Indeed, the discrete crowns of forest canopies may be as amenable to experimental investigations as the tide pools and boulders of intertidal systems. Many canopy animals neither fly nor glide, but climb and jump, and thus may be controlled with stem and bole collars to restrict them from canopy elements. Indeed, one contribution of canopy science may well be elucidating community interactions. As issues of canopy access and movement are increasingly addressed, the reduced dimensionality of canopy surfaces will be seen to present a terrestrial habitat well suited for manipulative investigations, and the lizards and arthropods living there seen as fruitful and informative experimental subjects.

References

Batschelet, E. (1965). "Statistical Methods for the Analysis of Problems in Animal Orientation and Certain Biological Rhythms." American Institute of Biological Science, Washington, D.C.

Batschelet, E. (1981). "Circular Statistics in Biology." Academic Press, London, England.

Borrer, D. J., Triplehorn, C. A. and Johnson, N. F. (1989). "An Introduction to the Study of Insects." Saunders, New York, New York.

Brown, S., Lugo, A. E. Silander, S., and Liegel, L. (1983). Research history and opportunities in the Luquillo Experimental Forest. United States Department of Agriculture, Forest Service General technical report SO-44. Southern Forest Experimental Station, New Orleans, Louisiana.

Coley, P. D., Bryant, J. P. and Chapin, F. S. (1985). Resource availability and plant-herbivore defense. *Science* **230**, 895–899.

Craig, C. and Bernard, G. D. (1990). Insect attraction to ultraviolet-reflecting spider webs and web decorations. *Ecology* **71**, 616–623.

Dial, R. (1992). A food web for a tropical rain forest: The canopy view from Anolis. Ph.D. Dissertation, Stanford University, Stanford, California.

Dial, R. and Roughgarden, J. (1995). Experimental removal of insectivores from a tropical rain forest canopy: Direct and indirect effects. *Ecology* **76**, 1821–1834.

Dial, R. and Roughgarden, J. (1996). Natural history observations of *Anolisomyia rufianailis* (Diptera: Sarcophagidae) infesting Anolis lizards in a rain forest canopy. *Environ. Entom.* **25**, 1325–1328.

Dial, R., Roughgarden, J. and Tobin, S. C. (19940. Notes on the absolute abundances of canopy anoles, Anolis *cuvieri, A. stratulus*, and *A. evermanni* (Lacertillia: Polychridae) in eastern Puerto Rico. Carib. J. Sci. **30**, 278–279.

Dial, R. and Tobin, S. T. (1994). Description of arborist methods for forest canopy access and movement. *Selbyana* **15**, 24–37.

Didham, R. K. and Springate, N. D. (2003). Determinants of temporal variation in community structure. *In* "Arthropods of Tropical Forests: Spatio-temporal Dynamics and Resource Use in the Canopy" (Y. Basset, V. Novotny, S. E. Miller and R. L. Kitching, Eds.), pp. 28–39. Cambridge University Press, Cambridge, UK.

Ferrar, P. (19870. A Guide to the Breeding Habits and Immature Stages of Diptera Cyclorrapha. Entomonograph 8. Brill/Scandanavian Science Press, Copenhagen, Denmark.

Garrison, R. W. and Willig, M. R. (1996). Arboreal arthropods. *In* "The Food Web of a Tropical Rain Forest" (D.P. Reagan and R. B. Waide, Eds.), pp. 183–245. The University of Chicago Press, Chicago, IL.

Heckel, D.G. and Roughgarden, J. (1979). A technique for estimating the size of Lizard Populations. *Ecology* **60**, 966–975.

Irschick, D.J. and Losos, J. B. (1999). Do lizards avoid habitats in which performance is suboptimal? The relationship between sprinting capabilities and structural habitat use in Caribbean anoles. *Amer. Natur.* **154**, 293–305.

Janzen, D. H. (1973). Sweep samples of tropical foliage insects: Effects of seasons, vegetation types, elevation, time of day, and insularity. *Ecology* **54**, 659–708.

Jepson, J. (2000). The tree climber's companion. Beaver Tree Publishing, Longville, Minnesota, USA.

La Caro, F. and Rudd, R. L. (1985). Leaf litter disappearance rates in Puerto Rican montane rain forest. *Biotropica* **17**, 269–273.

Laurence, W. T. (1996). Plants: The food base. *In* "The Food Web of a Tropical Rain Forest" (D.P. Reagan and R.B. Waide, Eds.), pp. 17–51. The University of Chicago Press, Chicago, IL.

Lawton, J. H. (1992). Feeble links in food webs. *Nature* **355**, 19–20.

Lawton, J. H. (1995). Tabonuco trees, trophic manipulations and Tarzan: lizards in a tropical rain forest canopy. *Trends Ecol. Evol.* **10**, 392–393.

Lister, B. C. (1981). Seasonal niche relationships of rain forest anoles. *Ecology* **62**, 1548–1560.

Losos, J. B., M. R. Gannon, W. F. Pfeiffer, and Waide, R. B. (1990). Notes on the ecology and behavior of *Anolis cuvieri* (Lacertillia: Iguanidae) in Puerto Rico. *Carib. J. Sci.* **26**, 65–66.

MacArthur, R. H. and Levins, R. (1964). Competition, habitat selection and character displacement in a patchy environment. *Proc. Natl. Acad. Sci. U.S.* **51**, 1207–1210.

MacArthur, R. H. and Wilson, E. O. (1967). The Theory of Island Biogeography. Princeton University Press, Princeton, NJ.

MacNally, R. C. (1995). "Ecological Versatility and Community Ecology." Cambridge University Press, Cambridge, UK.

May, R. M. (1973). "Stability and Complexity in Model Ecosystems." Princeton University Press, Princeton, NJ.

Moll, A. G. (1978). Abundance studies on the *Anolis* lizards and insect populations in altitudinally different tropical forest habitats. CEER-11. Center for Energy and Environmental Research, Rio Piedras, Puerto Rico.

Odum, H. T. and Pigeon, R. F. (1970). A Tropical Rain Forest: A Study of Irradiation and Ecology at EL Verde, Puerto Rico. U.S. Atomic Energy Commission, Oak Ridge, TN.

Overton, W.S. Estimating the numbers of animals in wildlife populations. *In* "Wildlife Management Techniques" (R. H. Giles, Ed.), pp. 403–456. The Wildlife Society, Washington, DC.

Pimm, S. L. (1982). Food Webs. Chapman and Hall, London, England.

Polis, G. A., Anderson, W. B. and Holt, R. D. (1997). Toward and integration of landscape and food web ecology: The dynamics of spatially subsidized food webs. *Annu. Rev. Ecol. Syst.* **28**, 289–316.

Rand, A. S. (1964). Ecological distribution in anoline lizards of Puerto Rico. *Ecology* **45**, 745–752.

Ramsey, F. L. and Schafer, D. W. (1997). "The Statistical Sleuth: A Course in Methods of Data Analysis." Duxbury Press, New York, NY.

Reagan, D. P. (1984). Competitive interactions between rain forest lizards: field observations and experimental evidence. *Bull. Ecol. Soc. Am.* **65**, 233.

Reagan, D. P. (1986). Foraging behavior of *Anolis stratulus* in a Puerto Rico rain forest. *Biotropica* **18**, 157–160.

Reagan, D. P. (1991). The response of *Anolis* lizards to hurricane-induced habitat changes in a Puerto Rican rain forest. *Biotropica* **23**, 468–474.

Reagan, D. P. (1992). Congeneric species distribution and abundance in a three-dimensional habitat: The rain forest anoles of Puerto Rico. *Copeia* **1992**, 392–403.

Reagan, D. P. (1995). Lizard ecology in the canopy of an island rain forest. *In* "Forest Canopies" (M. D. Lowman and N. M. Nadkarni, Eds.), pp. 149–164. Academic Press, New York, NY.

Reagan, D. P. (1996). Anoline lizards. *In* "The Food Web of a Tropical Rain Forest" (D.P. Reagan and R. B. Waide, Eds.), pp. 321–345. The University of Chicago Press, Chicago, IL.

Reagan, D. P. and Waide, R. B. (1996). The Food Web of a Tropical Rain Forest. The University of Chicago Press, Chicago, IL.

Reagan, D.P. Camilo, G. R., and Waide, R. B. (1996). The community food web: Major properties and patterns of organization. *In* "The Food Web of a Tropical Rain Forest" (D.P. Reagan and R. B. Waide, Eds.), pp. 461–510. The University of Chicago Press, Chicago, IL.

Rogers, L. E., Hinds, W. T. and. Buschbom, R. L (1976). A general weight vs. length relationship for insects. *Annal. Ent. Soc. Am.* **69**, 387–389.

Schoener, T.S. and Schoener, A. (1971). Structural habitats of West Indian Anolis lizards II. Puerto Rico uplands. *Breviora* **375**, 1–39.

Schowalter, T. D. (1994). Invertebrate community structure and herbivory in a tropical rain forest canopy in Puerto Rico following Hurricane Hugo. *Biotropica* **26**, 312–319.

Schowalter, T. D., and Ganio, L. M. (1999). Invertebrate communities in a tropical rain forest canopy in Puerto Rico following Hurricane Hugo. *Ecol. Ent.* **24**, 191–201.

Schowalter, T. D., and Ganio, L. M. (2003). Diel, seasonal, and disturbance-induced variation in invertebrate assemblages. *In* "Arthropods of Tropical Forests: Spatio-temporal Dynamics and Resource Use in the Canopy" (Y. Basset, V. Novotny, S. E. Miller and R. L. Kitching, Eds.), pp. 315–328. Cambridge University Press, Cambridge, UK.

Stephens, M.A. (1962). Methods for estimating circular statistics. *Biometrika* **49**, 463–477.

Turner, F. B. and Gist, C. S. (1970). Observations of lizards and tree frogs in an irradiated Puerto Rican forest. *In* "A Tropical Rain Forest: A Study of Irradiation and Ecology at EL Verde, Puerto Rico" (H. T. Odum, and R. F. Pigeon, Eds.), pp. E29-E49. U.S. Atomic Energy Commission, Oak Ridge, TN.

Van Bael, S. A., Brawn, J. D., and Robinson, S. K. (2003). Birds defend trees from herbivores in a Neotropical forest canopy. *Proc. Natl. Acad. Sci.* **100**, 1804–1807.

Walker, L.R. (1991). Tree damage and recovery from Hurricane Hugo in Luquillo Experimental Forest, Puerto Rico. *Biotropica* **23**, 379–385.

Walker, L. R. and Brokaw, N. V. L. (1991). Summary of the effects of Caribbean hurricanes on vegetation. *Biotropica* **23**, 442–447.

Whitman, W. (1950). "Leaves of Grass and Selected Prose." Random House, New York.

Williams, E. E. (1972. The origin of faunas: Evolution of lizard congeners in a complex island fauna: A trial analysis. *In* "Evolutionary Biology" (T. M. Dobzhansky, M. Hecht, and W. Steere, eds.), pp. 47–89. Appleton-Century-Crofts, New York, NY.

CHAPTER 16

Ecology and Conservation of Canopy Mammals

Jay R. Malcolm

---

*If disturbed during the daytime they seem to be in no way disconcerted by the light,
but issue forth from the holes, careen about the surface of the trunk and branches,
faster than mice on a level floor, take to the air in clouds, floating away among
the neighbouring trees like bits of soot from a chimney . . .*
—*Ivan Sanderson,* Mammals of the North Cameroons, Transactions of the
Zoological Society of London 24, *1940*

Introduction

Canopy mammals include not only some of the most charismatic and visible forest animals, but as a group, they are notably diverse from morphological and ecological viewpoints. The earliest eutherian mammals, and perhaps the earliest therians as well, appear to have been scansorial (i.e., they used both the ground and arborescent vegetation), which facilitated their radiation into diverse niches in the Cretaceous (Szalay 1994; de Muizon 1998; Argot 2001; Ji et al. 2002). Of 26 present-day orders of mammals, 12 include species that use the forest canopy and 39 of 79 families of non-flying mammals include canopy-feeding taxa (Emmons 1995). In size, non-flying canopy mammals span nearly five orders of magnitude, from the diminutive arboreal rice rats (*Oecomys* spp.) of the Amazon Basin to the orangutan (*Pongo pygmaeus*) of Southeast Asia. Methods of locomotion range from the agonizingly slow-motion movements of three-toed sloths (*Bradypus* spp.) to the acrobatic antics of primates, and feeding specializations in the group span nearly the full set shown by mammals, including adaptations for feeding on leaves, nectar, fruits, tree exudates, and insects. With the exception of Australia, almost all canopy mammals are found in tropical rainforests (Emmons 1995), and the pinnacle of diversity and abundance in the group is in that biome, where as many as 60 percent of non-volant species may be at least partly arboreal (Emmons et al. 1983) and where canopy folivores alone may comprise more than 70 percent of the total non-volant mammalian biomass (Eisenberg and Thorington 1973).

Because of their often secretive and nocturnal habits, canopy mammals also include some of the planet's least-known animals. For example, among 10 neotropical sites judged to be best sampled for mammals, only one was thought to have had sufficient canopy trapping (Voss and Emmons 1996). In some cases, rare species turn out to be common once the canopy is sampled; for example, canopy trapping in the central Amazon revealed that the arboreal wooly opossum *Caluromys philander* was the most abundant mammal in the forest (Malcolm 1991a, b). On the other hand, because of their diurnal habits and close phylogenetic relationship with humans, primates are some

297

of the best-known canopy vertebrates (Kays and Allison 2001). Unfortunately, because of rampant deforestation and other anthropogenic influences in the tropics, many species of canopy mammals are in considerable peril.

In this chapter, I provide an overview of some of the main features of mammalian canopy biology and review key concepts that have contributed to our understanding of the ecology, evolution, and conservation of the group. I include information from both relatively well-known groups, such as primates, and lesser-known groups, such as tropical small mammals. My purpose in part is to stimulate further research, especially in the tropics, where the fauna is not only incredibly diverse, but understudied and endangered. This overview complements the earlier, excellent overview by Emmons (1995): it focuses on different aspects of canopy biology and reviews some of the key concepts in the conservation of canopy mammals. In common with Emmons (1995), I exclude bats, which because of their species richness and ecological importance could be the subject of an entire chapter in their own right.

The Physics of Canopy Living: Adaptations to Arboreality

Cartmill (1974) identified three main physical differences between the arboreal and terrestrial worlds. As discussed below, these differences represent challenges that canopy mammals have overcome in a variety of ways. Adoption of canopy living in turn influences subsequent selective pressures, with implications for further evolution in the group.

Branch Orientation and Diameter

On sloping surfaces, friction must be used to maintain a mammal's position. Most arboreal mammals produce the necessary friction by grasping the support, with either opposing digits or limbs, in some cases using a pincer-like grip (for example, sloths, porcupines, anteaters, and pangolins). However, when the support is large in diameter relative to the grip size, part of the grasping force is squeezing the support away from the animal and hence opposing the frictional force.

Several consequences follow. First, for grasping species, it makes sense for individuals to choose support diameters that match their body plans. For example, when released on a large-diameter tree trunk in the tropics, arboreal marsupials make their way slowly and laboriously along the trunk to the nearest small-diameter vine where they quickly ascend into the canopy. Cunha and Vieira (2002) used spools of thread to track scansorial small mammals in the Brazilian Atlantic rainforest and found that all species used the smallest diameters most frequently (despite body sizes that ranged across two orders of magnitude), but that the average diameters of supports used followed the body size gradient. Presumably, this reflects the fact that all species can grasp small supports better than large supports, but that larger-diameter supports can be used efficiently only by relatively large species. Similar limitations with respect to support diameter are evident when humans climb trees using the traditional Amazonian "foot-belt" (*peconha*), a device which in essence transforms the feet into a pincer-like grasping appendage (see Moffett and Lowman 1995). Trees above a certain diameter cannot be climbed and the larger the trunk, the more laborious the climb.

A second consequence of a grasping strategy is that because long digits provide a greater surface area for friction than shorter ones, the digits of grasping appendages tend to be relatively long among arboreal mammals. Indeed, Ji et al. (2002) used the proportions of the phalanges to argue that the earliest-known eutherian mammal (genus *Eomaia*) was at the very least scansorial, and possibly fully arboreal. They noted that the intermediate phalanges of arboreal species tended to be long relative to the proximal phalange, and that in this respect, *Eomaia* was intermediate between the fully arboreal *Micoureus* and *Caluromys*, and the scansorial *Didelphis* and fully terrestrial *Metachirus* (see Figure 16-1). Similarly, relative arm lengths of primates increase with muscle mass, which is

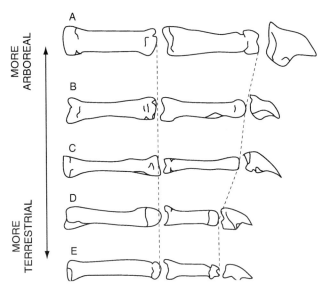

Figure 16-1 Schematic illustration of manual phalanges and claws comparing the early eutherian *Eomia scansoria* (C) with the didelphid marsupials *Micoureus* (A, fully arboreal), *Caluromys* (B, fully arboreal), *Didelphis* (D, scansorial), and *Metachirus* (E, fully terrestrial). The proximal phalanges are standardized to the same length. Modified from Ji et al. (2002).

thought to correspond to the challenges of grasping and climbing vertical supports (Preuschoft et al. 1998). At the same time, long digits require strong (and hence heavy) muscles to compensate for the longer load arms. Arboreal species tend to have more robust forearm bones than non-arboreal species, conferring greater strength to the bone and providing a greater area for muscle attachments (Iwaniuk et al. 1999). Not surprisingly, primates, especially the large ones, tend to have heavy limbs and appendages relative to other mammals (Preuschoft et al. 1998).

Friction can be increased not only through longer appendages, but also through changes to the surface of the foot (for example, by having large pads, elastic pad surfaces, and/or friction ridges that interlock with the surface of the support). An example is provided by the palm civet (*Nandinia binotata*), which has large, soft, and finely ridged palms on both the fore and hind feet. In addition, the heels of the hind feet bear deeply carunculated, transversely ridged pads of thick skin that serve as friction pads (Rosevear 1974).

As discussed by Cartmill (1974), an alternate strategy to grasping is to use claws. Because claws penetrate into the support, clawed species are able to use a broader range of support diameters than grasping species. A good example is provided by Cartmill (1974), who showed photographs of a gray squirrel (*Sciurus carolinensis*) climbing a cinder block wall. Claws are especially useful for small mammals, which cannot achieve the broad grip required for large supports except by an inefficient increase in body weight. The advantage of claws, or at least something similar in function, is illustrated by the traditional climbing techniques of the BaAka pygmies of central Africa (unpubl. obs.). For the smallest trees, they grasp the tree with their hands and walk up the trunk with their feet. As tree diameter increases, they increase the diameter of their grasp by using a stiff vine loop that encircles both the tree trunk and their torso (just under the arms). Again, they walk up the trunk, periodically flipping the vine loop upward by thrusting their weight toward the trunk. However, as tree diameter increases even further, their feet are unable to attain sufficient purchase to allow an upward flip of the vine; here, they resort to claws of a sort. A small hatchet is used to create small notches in the tree surface, which provide their feet the extra bit of purchase to flip the unwieldy vine loop upward.

BODY MASS OF GLIDING MAMMALS: AN ENERGETIC APPROACH

Roman Dial

Together with epiphytes, gliding vertebrates are among the quintessential forest canopy organisms. While there are 60 species of gliding mammals known worldwide (Jackson 1999), gliding is known to have evolved at least seven and possibly eight independent times (Stafford et al. 2002) in the class Mammalia. Curiously, the largest gliders on three continents (Asia, Australia, and Africa) from four separate evolutionary histories are each around 2 kg (Emmons 1995). This constancy in maximum observed weight may reflect an upper limit to the whole animal efficiency of gliding over quadrupedal locomotion. Schmidt-Nielsen (1986) reviewed mammal energetics and allometry to find that the whole animal energy expenditure, E_D (in ml O_2), for an animal of mass m (grams) traveling distance D (meters) is given generally as:

$$E_D = \alpha\, m^b D, \tag{1}$$

($b \sim 0.67$ and $2.45 \times 10^{-3} \leq \alpha \leq 8.61 \times 10^{-3}$), while the energetic cost (ml O_2) of vertical climbing E_V for an animal of size m (grams) climbing a vertical distance H (meters) is:

$$E_V = \beta\, mH, \tag{2}$$
$$= (1.36 \times 10^{-3} \leq \beta \leq 2.0 \times 10^{-3}).$$

These two equations show that climbing is more costly than walking, but what about gliding? Trigonometry suggests that a glider can glide horizontal distance D by climbing to height $H = D \tan \theta$ where θ is the glide angle measured over the entire glide. The whole animal energetic difference between walking (Equation 1) and gliding (Equation 2) for a mammal of size m is found by subtracting the cost of climbing to glide from walking:

$$S(D,m) = E_D - E_v = D\left[\alpha m^b - \beta m \tan(\theta)\right] \tag{3}$$

The linear relationship of energetic savings to D supports the Emmons and Gentry (1983) hypothesis that open forests encourage the evolution of gliders, if gliding evolves as an energetically favorable mode of travel. The non-linearity in body mass and glide angle implies both upper and lower bounds for—as well as an optimal—body weight. While studies have shown that gliders can reach angles as shallow as 11 degrees for some part of their glide, overall angle from takeoff to landing for most gliders most of the time lies between 20 and 30 degrees (Scholey 1986; Scheibe and Robins 1998; Jackson 2000; Vernes 2001; Stafford et al. 2002). Using a mid-range value of $\theta = 25°$ allows a comparison of the cost of climbing-to-glide horizontal distance D with walking distance D. Using the allometric constants provided by Schmidt-Nielsen (1986) in Equation (3) to find the mass where walking and gliding are equally costly (that is, when $S(D,m) = 0$, yields $m_{max} = 2{,}573$ g). This suggests that mammalian gliders should be less than 2.5 kg. A bigger mammal would find it more costly to climb and glide than simply to walk to the destination. Further analysis (Dial 2003) shows that solving for the body mass that maximizes the energetic savings in $S(D,m)$ gives $m_{opt} \sim 420$ g while setting $\theta \geq 45°$ to find the mass where jumping is sufficient gives $m_{min} \leq 19$ g. Mean, minimum, and maximum body mass for three glider assemblages with independent evolutionary histories (see Table) support the hypothesis that body mass in mammal gliders is determined, in part, by energetic considerations. Mammalian gliders may have fairly straightforward energetic constraints on their body mass.

Body Mass Statistics for Mammalian Glider Assemblages from Jackson (1999)

	Borneo	Australia	Africa
Number of gliders	15	6	6
Mean mass (g)	544	419	560
Minimum mass (g)	13	12	16
Maximum mass (g)	1,985	1,175	1,417

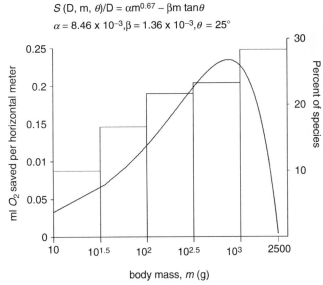

$$S\,(D, m, \theta)/D = \alpha m^{0.67} - \beta m \tan\theta$$

$$\alpha = 8.46 \times 10^{-3}, \beta = 1.36 \times 10^{-3}, \theta = 25°$$

Figure 1 Energetic savings for a mammalian glider superimposed on glider body mass distributions. The curve shows the energetic savings per horizontal meter traveled as a function of mass using Equation (3) with parameters given in the figure. The histogram shows the percent of all known mammalian gliders in $\log_{10}$ size classes (data from Jackson 1999). At a body mass of ~2500 g there is no savings over walking for a glide of 25 degrees.

References

Dial, R. (2003). "Energetic Savings and The Body Size Distributions of Gliding Mammals." *Evolutionary Ecology Research* **5**, 1–12.

Emmons, L.H. (1995). Mammals of rain forest canopies. *In* "Forest Canopies" (M.D. Lowman and N.M. Nadkarni, Eds.), pp. 199–223. Academic Press, San Diego.

Emmons, L.H. and Gentry, A.H. (1983). Tropical forest structure and the distribution of gliding and prehensile vertebrates. *Am. Nat.* **121**, 513–524.

Jackson, S.M. (2000). Glide angle in the genus Petaurus and a review of gliding in mammals. *Mammal Review* **30**, 9–30.

Scheibe, J.S. and Robins, J.H. (1998). Morphological and performance attributes of gliding mammals. In "Ecology and Evolutionary Biology of Tree Squirrels" (M.A. Steele, J.F. Merritt, and D.A. Zegers, Eds.), pp. 131–144. Special Publication 6, Virginia Museum of Natural History.

Schmidt-Nielsen, K. (1986). "Scaling: Why Is Animal Size So Important?" Cambridge University Press, Cambridge, UK.

Continued

BODY MASS OF GLIDING MAMMALS: AN ENERGETIC APPROACH–*cont'd*

Scholey, K. (1986). The climbing and gliding locomotion of the giant red flying squirrel *Petaurista petaurista* (Sciuridae). In "Bat Flight—Fledermausflug" (W. Nauchtigall, Ed.), pp. 187–204. Gustav-Fischer Verlag, Stutgart, Germany.

Stafford, B.J., Thorington, R.W., and Kawamichi, T. (2002). Gliding behavior of Japanese giant flying squirrels. *J. Mammal.* **83**, 553–562.

Vernes, K. (2001). Gliding performance of the northern flying squirrel (*Glaucomys sabrinus*) in mature mixed forests of eastern *Canada. J. Mammal.* **82**, 1026–1033.

It becomes especially useful to be able to change the angle of claw purchase by changing the orientation of the foot. Supination (reversal) of the hind foot in particular allows an animal to use its claws effectively when descending a tree headfirst, a feature that has evolved numerous times and is used, for example, by squirrels, kinkajous (*Potos flavus*), bassariscus (*Bassariscus* spp.), palm civets (*Nandinia binotata*), Wiedi's cat (*Felis weidii*; the only known example of a cat), and galagos (*Galago* spp.). Because of hind foot reversal, the arboreal Wiedi's cat is able to grab branches equally well with its fore and hind paws and is capable of arboreal acrobatics not seen among the other Felidae (Ewer 1973). Some primarily terrestrial species that use trees show partial hind-foot reversal, such as, raccoons (*Procyon* spp.) and coatis (*Nasua* spp.) (McClearn 1992).

The physical challenges imposed by sloping and vertical supports are exacerbated for large mammals. As size increases, the oxygen consumption demanded by climbing comprises a relatively larger fraction of resting metabolism; that is, the efficiency of climbing goes down as body size increases (Cartmill 1974). Bakker and Kelt (2000) found that, compared to terrestrial species, both scansorial and arboreal mammals did not attain the largest sizes, which may partly reflect these energetic constraints. Arboreal mammals also did not attain the smallest sizes of terrestrial species, which may be due to the inability of small mammals to subsist on an energetically poor diet of leaves and fruits (McNab 1986; Eisenberg 1987) and the greater success of folivory/frugivory than insectivory as a canopy lifestyle (Emmons 1995). Support diameter and strength also may limit the body size of arboreal mammals (Cristoffer 1987; Bakker and Kelt 2000). Differences in vegetation structure among the rainforests of South America, Asia, and Africa have been implicated in influencing not only body size, but also the distribution of gliding mammals. The prevalence of gliders in Southeast Asian dipterocarp forests may reflect the lack of canopy connections between the widely spaced emergent trees of the canopy (Emmons and Gentry 1983; Emmons 1995).

Substrate Discontinuity and Complexity

The arboreal environment also differs from the terrestrial one in that it is discontinuous, complex, and three-dimensional. However, at small enough scales, the terrestrial substrate also is discontinuous and three-dimensional; hence, the anatomical differences between scansorial and arboreal species become insignificant for very small mammals (Ji et al. 2002).

Crossing of gaps between canopy trees is a particularly difficult challenge for canopy mammals. Three basic solutions have evolved: reaching across, leaping across, and gliding across (Emmons 1995). Many canopy-dwelling species can only cross gaps that they can reach across; otherwise, they must descend to the ground, which has important implications for the types of food resources that they can utilize effectively (Emmons 1995). When crossing a gap and moving from a known to an unknown support, a grasping hind foot (and a prehensile tail) provides insurance against falling, which may explain why a grasping pollex tends to be more common than a grasping halux (Cartmill 1974). The optic and orbital convergence characteristic of most primates may be an adaptation to estimate distances across gaps (Mittermeier and Fleagle 1976), although, as pointed out by Cartmill (1974), although laterally directed eyes also are found among arboreal species that customarily take long leaps (such as squirrels). Relatively large eyes and brains tend to be characteristic of arboreal species, presumably to allow the collection and processing of the large amounts of information required for effective locomotion in the complex three-dimensional environment (McNab and Eisenberg 1989; Bernard and Nurton 1993). Large brain size also may be useful to keep track of widely dispersed and seasonal food resources such as canopy fruits; for example, spatial memory can increase the efficiency of random searching by up to 300 percent (Janson 1998). Perhaps not surprisingly, the most arboreal and frugivorous South American marsupial, the woolly opossum, also has the largest relative brain (Eisenberg and Wilson 1981).

An example of the relatively large eyes of arboreal species is provided by the small mammal fauna of the Juruã River of the western Amazon (see Figure 16-2). In both the Muridae and Didelphidae, the most arboreal genera have the largest eye-lens sizes relative to body size, including arboreal rice rats (*Oecomys* spp.), climbing rats (*Rhipidomys* spp.), the woolly mouse opossum (*Micoureus demerarae*), and the woolly opossum (*Caluromys lanatus*). In contrast, the most terrestrial species, such as the gray spiny mouse (*Scolomys juruaense*), the short-tailed opossum (*Monodelphis emiliae*), and the brown four-eyed opossum (*Metachirus nudicaudatus*), have the smallest relative eye-lens weights. Interestingly, a third family, the Echimyidae, fails to show these relationships. For unknown reasons, all species in this family have relatively large eye lenses, including both arboreal and terrestrial species. In this family, more variation is shown among species in the strictly terrestrial genus *Proechimys* than between arboreal and terrestrial taxa.

It is has been argued that the lower peak vertical reaction forces of fore compared to hind limbs may have improved the ability of early primates to reach and grasp branches (and food) in a discontinuous environment, which in turn may have facilitated the evolution of other locomotory styles such as brachiation in gibbons (*Nomascus* spp.) and spider monkeys (*Ateles* spp.) and bipedalism in humans (references in Schmitt and Lemelin 2002). Glazier and Newcomer (1999) found that, compared to terrestrial species, arboreal species spend a greater percentage of their maturation time in the juvenile period, which they attributed to the greater amount of learning required to forage in three compared to two dimensions.

Low Support Stability

Narrow-diameter supports in the arboreal environment provide low stability, which means that a low center of gravity is advantageous. A relatively short radius to humerus length appears to be associated with arboreality, presumably to keep the center of gravity close to the surface (Cartmill 1985; Iwaniuk et al. 1999). Similarly, belly vibrissae may assist tree squirrels in keeping their center of gravity low (Cartmill 1974).

As discussed by Schmitt and Lemelin (2002), walking gaits of arboreal species may be designed in part to increase stability in the arboreal environment. The four-footed walking gate of primates has several features that distinguish it from the gaits of most other mammals, including diagonal- vs. lateral-sequence walking gaits, more protracted forelimb postures at forelimb touchdown, and lower peak vertical substrate reaction forces on forelimbs relative to hind limbs. These same characteristics are shown in the gait of the arboreal woolly opossum compared to the strictly terrestrial short-tailed opossum (Schmitt and Lemelin 2002). The first two characteristics appear to be adaptations to living in a fine branch environment—the first because it improves balance and the second because it increases the relative length of the forelimb, allowing a longer stride length and hence greater speed without increasing branch oscillations (references in Schmitt and Lemelin 2002).

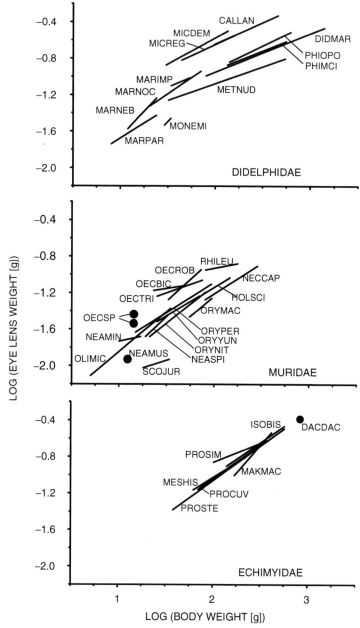

Figure 16-2 Eye-lens weight plotted against bodyweight for small mammals of the Juruá River Basin in Brazil. For each species, lines are best-fit linear regressions and are shown across the range of body weights. For infrequently captured species, points represent single individuals. Acronyms represent the first three letters of the genus and the first three letters of the species (see Patton et al. 2000).

Adaptations to Arboreality Per Se?

The set of adaptations shown by a species reflects the various and often conflicting selection pressures that individuals have experienced through time. As a result, one must be very cautious in attributing features to arboreality per se; in many cases, a feature may have evolved in response to a suite of factors, of which arboreality was only one. In other cases, alternative explanations may make more sense.

A good example of the difficulty in sorting out the selective pressures that led to a particular adaptation is provided by several of the characteristics that are common to primates, which often are found in marsupials as well (Cartmill 1974). Despite huge ecological and evolutionary diversity in the group, most species tend to share a set of features that are distinctive relative to those of other mammals. These include relatively large brains, well-developed stereoscopic vision, grasping hands and feet with nails rather than claws, small litters, and reduced developmental rates (Rasmussen 1990). A grasping hind foot is probably the most widespread adaptation in the group (Cartmill 1974), and four principal hypotheses have been put forward to explain its evolution:

1. It is an arboreal adaptation per se.
2. It permitted vertical clinging and leaping habits (as shown by *Tarsius* and *Propithecus*, for example).
3. It facilitated visual insect predation.
4. It permitted effective access to terminal branches (Cartmill 1974; Rasmussen 1990).

The first was raised by Elliot and Smith (1913, cited by Cartmill 1974), who hypothesized that grasping hands and feet (along with stereoscopic vision, lack of olfaction, and large brains) were specific adaptations to the demands of locomotion and information processing in the complex, three-dimensional arboreal habitat. However, Cartmill (1974) argued that arboreality per se did not make sense as an explanation given that claws would serve just as well. Indeed, all primates could easily have had squirrel-like feet, which would serve them even better on large-diameter supports. Cartmill (1974) argued that claws also can function effectively in a fine-branch environment (although perhaps not as effectively as grasping digits). As a result, he favored the idea that the grasping hind foot had evolved because early primates needed to hold on while they identified arthropods by sight and then caught them by hand. Sussman (cited in Rasmussen 1990) disagreed with Cartmill (1974), noting that primate species as a group tend to rely on fruits and flowers on terminal branches rather than on insect predation. Sussman hypothesized instead that terminal branch exploitation of flowers and fruits was the critical step in the early Eocene radiation of primates, which occurred at a time when angiosperms were undergoing taxonomic upheaval and when large fruits first appeared.

Rasmussen (1990) revisited these hypotheses using the woolly opossum (*Caluromys*) as a model for early primates, proposing that *Caluromys* had diverged from other didelphids in many of the same ways that early primates had diverged from the earliest euprimates. *Caluromys* shares many features with primates (including a relatively large brain and short snout, small litters, slow development, and a long life span), although it also shows important differences (including claws vs. nails, no broad tactile pads at the fingertips, and relatively little orbital convergence). Rasmussen (1990) concluded that all four hypotheses probably played a role in the evolution of the grasping hind foot in both *Caluromys* and early primates. He noted that all authors had started with a grasping foot as one possible solution to the challenges imposed by arboreality and that the earliest primates were arboreal leapers and climbers. Perhaps most significantly, however, Rasmussen (1990) found that *Caluromys* was both a terminal-branch fruit/flower feeder and a visually oriented predator (for example, grabbing moths out of the air). He suggested that visually oriented predation was associated with terminal branches because at the same time that the angiosperm radiation was unfolding, a radiation of pollinating and fruit-feeding insects was occurring.

Other authors have noted the importance of feeding behaviors in determining the suite of adaptations shown by a species and the importance of locomotory adaptations in influencing foraging opportunities. Emmons (1995) developed a classification scheme of morphological adaptations to arboreality in which levels of increased mobility were associated with increased foraging opportunities. For example, compared to the relatively limited foraging opportunities available to non-leaping species, highly mobile gliding and flying species may be better able to specialize on certain resources or to make use of otherwise marginal ones. Another example of the interplay between locomotory adaptations and feeding behaviors are suspensory postures such as hanging from the hind feet or by the tail, which may have evolved to increase access to feeding areas (Mittermeier and Fleagle 1976). McClearn (1992) examined the relationship between feeding behaviors and arboreal adaptations for three species of Procyonids, examining the degree to which they were in conflict, were integrated, or had no apparent relationship with one other. She concluded that food-gathering and handling behaviors precluded the use of fine branches for coatis and raccoons, and hence that the location of a major food source had indirectly influenced the range of arboreal microhabitats that was accessible to them.

Arboreal life in turn may lead to new sets of selective pressures. For example, Holmes and Austad (1994) argued that because of their gliding, hole-living, and nocturnal habits, North American flying squirrels (*Glaucomys* spp.) may suffer lower rates of predation than other tree squirrels, which may have influenced the evolution of their life history characteristics. Compared to other tree squirrels, they have relatively low basal metabolic rates and a set of K-selected life history traits, including a longer period of maternal investment, slower rates of growth to maturation, larger relative body weight at birth, lower annual fecundity, and longer life spans. Stapp (1992 cited by Holmes and Austad 1994) attributed these differences to the different microclimates and energetic costs associated with nocturnality and gliding; however, Holmes and Austad (1994) suggested that, as predicted by evolutionary senescence theory, low intrinsic mortality may have contributed to the evolution of slow reproductive maturity and late senescence. Malcolm et al. (in press) suggested that compared to terrestrial mammals, the relatively K-selected features of arboreal mammals has led to lower speciation rates in the group.

In summary, the physical challenges of arboreality have been solved in various ways by canopy mammals, resulting in the evolution of distinctive morphological and life-history characteristics. Grasping digits and/or limbs and claws are two main strategies used to climb; however, other features of the arboreal environment, such as its complex, three-dimensional, and discontinuous nature and the low-stability of its supports, also have importantly influenced evolution in the group. At the same time, the suite of adaptations shown by a species reflects the full suite of selective pressures influencing it, making the study of adaptations to arboreality a challenging undertaking.

Ecology of Canopy Mammals

Field Methods

Arboreal mammals often are difficult to study because of their small size and nocturnal habits; as a result, our knowledge of the entire canopy fauna has depended on the development of suitable study methods. Primates are the best-studied group in part because they usually are large and diurnal. It is relatively straightforward, although laborious, to undertake detailed observational studies of habituated primate groups; indeed, detailed single-species studies have now been undertaken for the vast majority of primate genera (Peres and Janson 1999). In addition, accurate density and community composition estimates can be obtained easily via diurnal strip-censuses. For example, only 16 h of censuses will usually reveal 50 percent or more of local primate species and a complete inventory usually can be accomplished in less than 50 h (Emmons 1999; Peres and Janson 1999). In contrast, inventories of the entire mammalian fauna at a site may require several years, if not decades (Emmons 1999). In part, this reflects the fact that sampling techniques targeting small, elusive, nocturnal species have not been developed, or are not easily undertaken.

The paucity of even basic information for small and nocturnal species is well illustrated by a recent study on the Juruã River in the western Amazon (see Patton et al. 2000). Prior to the study, reasonably detailed and accurate distributional information was available at the subspecific level for primates (e.g., Emmons and Feer 1990); in contrast, it turned out that a large portion of the small mammal fauna had not even been described: a year of intensive terrestrial and canopy trapping resulted in 10 new small mammal species. Nocturnal canopy species are rarely seen at night, even those with bright eye-shine, and some species rarely frequent the lower (and hence more human-accessible) areas of the canopy. An example is provided by Malcolm (1990), who reported that densities of several canopy species estimated from terrestrial trapping were only a small fraction of those from canopy trapping and that strip transects even more seriously underestimated densities (see also Smith and Phillips 1984; Lindenmayer et al. 2001).

A variety of techniques has been used to study the ecology of canopy species, including observational studies, spotlighting, hair-tubing, trapping, radio telemetry, and interviewing of local peoples. Several live-trapping techniques are now available and are increasingly incorporated into census work. A major challenge in these techniques is access to the canopy, which tends to be very laborious and/or very expensive. Many access techniques are now available, each of which has advantages and disadvantages with respect to time, energy, safety, and expense (e.g., Moffett and Lowman 1995). Because of the difficulty of canopy access, rather than directly affixing traps to trunks or branches (e.g., Smith and Phillips 1984; Meggs et al. 1991) and climbing repeatedly to service the traps, it is advantageous to use some sort of a pulley system that allows repeated raising and lowering of traps without repeated canopy access. Perhaps the most elaborate scheme is Malcolm's (1991a), in which pulleys permitted repeated hoisting of traps to an upper frame affixed to the tree (see also Kays 1999). Vieira's (1998) technique is potentially an improvement in that tree climbing is not required at all; instead, access is provided by shooting lines into the canopy with a slingshot or such. Lambert et al. (in prep.) have developed an even simpler technique in which only a single line is shot into the canopy and no wooden box is required.

As yet, studies comparing trap success among the various trapping methods have not been undertaken. Malcolm's (1991a) technique offers some advantages in that animals can readily enter the traps and higher trap heights can be attained (because the tree climber can remove obstructing vegetation). However, it is quite possible that many canopy species will just as readily enter less-accessible traps (for example, by climbing along the side of the trap to find the door) and that trap height does not matter much (provided that one reaches the lower canopy at least). If true, trapping techniques such as Vieira's (1998) may be of particular value because they avoid tree climbing.

Vertical Stratification

Given their dependence on trees, not surprisingly, arboreal mammals reach their greatest diversity and abundance in the complex, multi-storied canopies of tropical rainforests. In a comparison of mammal faunas among boreal, temperate, and tropical forests, Fleming (1973) found that arboreal species increased from just a few species in boreal and temperate forests to >10 percent of the total fauna in Central American tropical rainforests. Many taxonomic groups are more or less restricted to tropical rainforests; for example, some 90 percent of primates worldwide are restricted to tropical forest habitats (references in Peres and Janson 1999). One of the main hypotheses explaining this variation with latitude, and indeed variation over smaller spatial scales

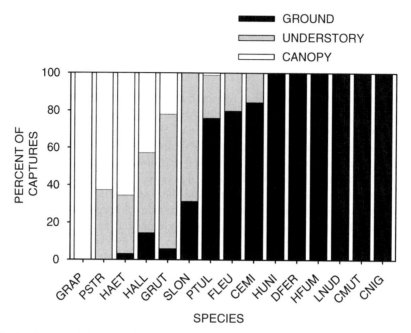

Figure 16-3 Small mammal abundance by forest stratum (ground, understory, or canopy) at a rainforest site in the Central African Republic. Acronyms are the first letter of the genus and the first three letters of the species (see Malcolm and Ray 2000).

as well, is the physical structure of the arborescent vegetation. One possibility is that the complex rainforest canopy offers more axes for niche partitioning, and hence the possibility of more species being packed into the same area. This implies that suitable food resources are available; presumably, both forest structure and availability of food are intimately related in determining axes for niche partitioning. Numerous authors have pointed out that it is the year-round availability of diverse food resources in rainforests that makes the third dimension available as habitat (e.g., August 1983; Emmons 1995; Bakker and Kelt 2000). In support of this idea, despite a climate that does not support a significant canopy fauna elsewhere, the evergreen habits of eucalypts, coupled with the ability to use them as a primary dietary item, have led to the development of a significant canopy fauna in the temperate and dry forests of Australia (Emmons 1995). In forests that lack year-round canopy resources, canopy-feeding species tend to either hibernate or feed on the ground for part of the year (Emmons 1995).

Many studies have shown that arboreal mammalian species tend to be most common at certain heights than others; for example, Cunha and Vieira (2002) cite 15 such studies in the South American tropics (recent additions include Patton et al. 2000 and Voss et al. 2001). Examples from other tropical localities include McClearn et al. (1994) and Malcolm and Ray (2000). Typically, these studies demonstrate diverse use of vertical space among species, with many species being strictly terrestrial; others displaying significant use of both ground and the understory; others using both the understory and the canopy; and still others being more or less restricted to the canopy (see Figure 16-3). In general, the small mammal fauna of the canopy tends to be less species-rich than that of the ground; however, differences in abundances between the two faunas are less consistent, with some studies showing approximately equal abundances in the two faunas and others showing overall lower abundances in the canopy (Malcolm et al. in press). Species that utilize both the forest floor and the canopy tend to be very rare, perhaps

VERTICAL STRATIFICATION OF SMALL MAMMALS IN LOWLAND RAINFOREST OF THE AUSTRALIAN WET TROPICS

Romina Rader

Conducting a mammal study in a rainforest canopy is an exciting and immensely challenging task. This is not only because formidable obstacles and dangers hinder access to the environment, but also because those lofty green layers do not surrender their secrets easily. Indeed, accumulating data and coaxing hypotheses from those secretive worlds is a complex and time-consuming enterprise.

Traditional sampling methods have included hoistable arboreal traps (Malcolm 1991; Viera 1998; Kays 1999), single rope techniques to access fixed traps (Barker and Sutton 1997) and, more recently, the use of canopy towers and cranes (Allen 1996; Kyoto University 2000). Accessing the treetops using a crane is advantageous as setting and checking traps is not only safer but considerably less labor intensive. This is no small advantage when a comprehensive study requires hundreds of traps to be set every night and checked every morning!

Nine canopy cranes currently exist around the globe as tools to aid scientific research. This research was conducted using the canopy crane in the lowland rain forests of Cape Tribulation, Australia (see Figures 1 and 2).

Figure 1   Using the Australian canopy crane to access the treetops. Photograph by Michael Cermak.

Canopy use by small mammals in Australia is poorly known, primarily due to problems with access. Several species, particularly rodents, have been known to use the canopy, but the extent of this use has never been quantified. The aim of this study was to investigate which species utilize this relatively unexplored habitat, how often, and for what purpose. Folding aluminum mammal traps and cage traps were tied to timber platforms wired to tree branches. Traps were placed at four different height levels from the ground to the canopy (0 m, 1 to 3 m, 10 to 15 m, and 25 to 30 m) for 5,300 trap nights. Trapped animals were tagged, measured, weighed, and released at the location of capture.

Continued

VERTICAL STRATIFICATION OF SMALL MAMMALS IN LOWLAND RAINFOREST OF THE AUSTRALIAN WET TROPICS–*cont'd*

Figure 2   Checking a mammal trap at 35 meters. Photograph by Michael Cermak.

Several interesting results emerged from this study, the most significant being the trapping of the prehensile-tailed rat (*Pogonomys mollipilosus*), a rodent that had never been trapped prior to this study. However, while its ecology is practically unknown, failure to trap it in the past is probably more due to its elusive arboreal nature than its being particularly rare. It was originally identified on the basis of specimens captured by cats and in the fecal pellets of Lesser Sooty Owls (*Tyto multipunctata*).

In addition, two rodent species, the giant white-tailed rat (*Uromys caudimaculatus*) and the mosaic-tailed mouse (*Melomys cervinipes*), used the canopy more extensively than previously thought and were most commonly trapped above the ground at the 1 to 3 m level. The marsupial Antechinus (*Antechinus flavipes rubeculus*) was also trapped more frequently at the 1 to 3 m level yet was not trapped at the higher level, even though it is considered to be semi-arboreal. The Cape York rat (*Rattus leucopus*) was found to be primarily terrestrial with most captures on the ground.

Species distributions are illustrated in Figure 3. The gradation from darker red squares to lighter red squares represents the locations where the species was most often captured to least often captured, respectively. Squares lacking color indicate that the species was not captured at that height.

The results from this study have implications for any future sampling methodology involving mammals. Considering most estimates of mammal species diversity are derived from ground trapping, it is possible that populations of arboreal and scansorial species may be underestimated or overlooked. In addition, perceived inter-specific competitive interactions may not represent the true dynamics of the system if one species actually occupies vertical strata unknown to the researcher. For example, in the present study, the literature describes the possible occurrence of inter-specific competition between *Rattus leucopus* and *Melomys cervinipes* where the system is negatively driven by *Rattus leucopus* (Heinsohn and Heinsohn 1999). It was thought that if *Rattus* numbers were high, the abundance of *Melomys*

Figure 3 Distribution of canopy mammals at different heights. The gradation from darker to lighter red squares represents the locations where a species was most often captured to least often captured, respectively. Photograph by Michael Cermak.

would decrease. However, if vertical strata are sampled, another scenario becomes evident whereby *Melomys* is using vertical strata more than previously expected. Such relationships would have to be tested further with removal experiments. However, at the very least, vertical sampling provides a new perspective on arboreal and terrestrial mammal interactions.

In conclusion, accessing the canopy offers a unique opportunity to discover new relationships as well as allow more efficient sampling of arboreal and scansorial mammal populations. Failing to sample the canopy would result not only in underestimates of species abundances and distributions, but significant relationships may remain unknown. Considering that estimates of species richness are used to set conservation priorities, it is important that they are as closely aligned to reality as possible, and herein lies a practical application for this research.

References

Allen, W.H. (1996). Travelling across the treetops. *Bioscience* **46**, 796–799.

Barker, M.G. and Sutton, S.L. (1997). Low-tech methods for forest canopy access. *Biotropica* **29**, 243–247.

Heinsohn, G. and Heinsohn, R. (1999). Long term dynamics of a rodent community in an Australian tropical rainforest. *Wildlife Research* **26**, 187–198.

Kays, W.R. (1999). A hoistable arboreal mammal trap. *Wildlife Society Bulletin* **27**, 298–300.

Kyoto University. (2000). List of publications of canopy biology program (1994–1999). Centre for Ecological Restoration, Kyoto University, Japan.

Malcolm, J.R. (1991). Comparative abundances of neotropical small mammals by trap height. *Journal of Mammalogy* **72**, 188–192.

Vieira, E.M. (1998). A technique for trapping small mammals in the forest canopy. *Mammalia* **62**, 306–310.

because effective locomotory adaptations to one stratum preclude efficient use of (and access to) the other. An example of vertical stratification of scansorial species is provided by Cunha and Vieira (2000; see also Charles-Dominique et al. 1981), who used spool-and-line devices to study three marsupial species of the Brazilian Atlantic rainforest. The slender mouse opossum, *Marmosops incana*, was an understory species, whereas the black four-eyed opossum (*Philander frenata*) and the common opossum (*Didelphis aurita*) were more terrestrial, with the former ranging into the understory and the latter sometimes into the canopy.

To date, most arboreal trapping has sampled at most two arborescent strata, including one close to the forest floor; hence, vertical partitioning within the small mammals' canopy is poorly known. Horizontal partitioning also has been rarely investigated. A possibility is that the distribution of tree fall gaps in primary forest will influence both the vertical and horizontal partitioning of space. For example, Malcolm (1995) found in primary forest that horizontal variability in understory foliage density was a strong correlate of terrestrial small mammal species richness, which he attributed to the fact that areas with more tree fall gaps offered a broader set of habitat types (especially for understory insectivores) (see also Cork and Catling 1996). Similarly, if gaps are a focus of some arboreal species, then abundances and height distributions of arboreal species also can be expected to vary as a function of gap density.

As this last example suggests, variation in forest vertical structure can be expected to influence the richness and abundance of the canopy fauna not only as a function of latitude, but locally as well. Malcolm (1995) sampled a gradient of edge-influenced forests at the Biological Dynamics of Forest Fragments site in the central Amazon and found that the best correlate of canopy small mammal species richness was a measure of the grain (semivariance) of the canopy cover. In canopies that were finely dissected by gaps, arboreal species tended to be rarer than in canopies that were more coarsely dissected, which was attributed to the relative difficulties and dangers of near-ground travel for canopy species. Several studies have found that as the forest canopy becomes more open, proportionately less small mammal abundance and biomass is comprised of arboreal species (e.g., Malcolm 1997). Similar to edge effects, logging can also result a shift in small mammal biomass toward the ground as the forest canopy is opened (Struhsaker 1997; Malcolm 1997; Malcolm and Ray 2000), presumably because of more frequent near-ground movements by canopy-specializing species and increased abundances of understory-specializing and terrestrial species. As the intensity of such disturbances increases, the remaining patches of high-stature forest will become increasingly isolated from each other, with the net effect that canopy-loving species may disappear from these forests.

Unfortunately, although several studies have examined the effects of local changes in forest structure on canopy mammals, studies comparing such effects among continents have not been undertaken, at least for small mammals. Two intensive live-trapping studies, one in the Amazon (Malcolm 1991a, b) and the other in central Africa (Malcolm and Ray 2000), offer some interesting contrasts in this regard. In central Africa, most canopy-dwelling species were captured frequently in the understory as well; indeed, overall, small mammal species composition and abundance in the understory was quite similar to that in the canopy. This appeared to contrast markedly with Malcolm (1991a, b), who found in the central Amazon that some abundant canopy species tended to be very rare in the understory and hence that the overall similarity between the understory and canopy fauna was low. This difference may relate to differences in forest structure. In the central African forest, canopy gaps tended to be relatively common with the net effect that the functional canopy layer (i.e., the layer that intercepts the majority of light) often dipped close to the ground. As a result, vertical use of space in these African forests may be more reminiscent of that in logged or edge-modified forests in the Amazon. Similar conclusions may apply in lower-rainfall rainforests at the edge of Amazon Basin, which can have large areas of relatively open-canopy forest. Fa and Peres (2001) used a similar argument in explaining the large difference in arboreal relative to terrestrial biomass between the neotropics and Africa. Arboreal biomass averaged approximately 50 percent of the total across eight neotropical sites (range: 12 to 92 percent), but only 10 percent for nine African sites (range: 1 to 76 percent) (see Figure 16-4). They hypothesized that in the closed canopy forests more typical of the neotropics, most primary productivity occurs in the tree canopy, but due to higher gap frequencies in Africa, more primary productivity occurs in the understory. Another contributory factor may be the greater abundance of large terrestrial herbivores in Africa (such as elephants), which keep gaps open, and the relative lack of aggressive colonizing tree species in Africa, again contributing to longer-lived gaps (Fa and Peres 2001; Cristoffer and Peres 2003). The shift in productivity toward the forest floor in Africa can support a higher terrestrial biomass, which because of arboreal constraints on body size (Bakker and Kelt 2000), can be comprised of larger-bodied species. The large and paleoecological-persistent areas of savanna in Africa also may contribute to this difference by providing a ready supply of woodland and large herbivorous species that can invade the forest and potentially speciate there depending on the amount of canopy disturbance.

Additional support for the importance of forest structure in contributing to the species richness of tropical mammal faunas comes from bodyweight distributions. If rainforest mammals partition resources vertically and if competition structures local body-size distributions, then bodyweight distributions within a forest height stratum should be "over-dispersed" relative to those across strata; i.e., species within a stratum should be more dissimilar in weight than among strata (Bakker and Kelt 2000). Many authors have noted that species within a stratum tend to differ from each other in body size (e.g., Charles-Dominique et al. 1981; Woodman et al. 1995). For example, Emmons (1980) reported that squirrel species in Gabon within the same or neighboring height strata differed in body size from each other. In the most comprehensive test of the idea, using the neotropical fauna Bakker and Kelt (2000) found that, as predicted, bodyweight distributions within strata (terrestrial, scansorial, or canopy) tended to be more uniform than among strata. Moreover, they found that bodyweight distributions within a stratum tended to be similar to those in North American biomes, where the arborescent stratum is largely unavailable for partitioning. Emmons (1995) also argued for the importance of competitive interactions in mammalian canopy communities in her comparison of five rainforest sites, where she found striking convergence among sites with respect to species number, bodyweight distributions, and taxonomic balance. She hypothesized that such competition was especially important among arboreal mammals because of the energetic demands associated with endothermy and the generalized, and hence overlapping, diets demanded by canopy living.

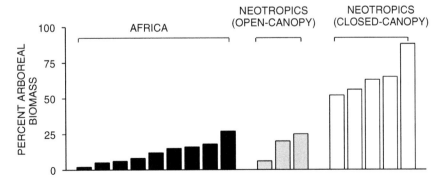

Figure 16-4 Percent of total mammal biomass that is arboreal for sites in Africa and the neotropics. For the latter, percentages are given for both open-canopy and closed-canopy sites. Modified from Fa and Peres (2001).

Feeding Ecology of Canopy Species

Canopy resources are particularly abundant in tropical rainforests, where the canopy is the site of greatest primary productivity and where conditions for plant growth are relatively equitable year round. Unlike in temperate forests, fruits become a highly significant resource for mammals in the tropics; for example, in Fleming's (1973) comparisons among forest types, frugivorous mammals species comprised approximately 30 percent of the Panamanian rainforest fauna, but were absent from temperate and boreal forests. Not surprisingly, given that most of the light energy is intercepted in the canopy, folivory is especially important in rainforest canopies, especially for relatively large-bodied groups such as primates. In neotropical forests, arboreal folivores may comprise more than 70 percent of the mammal biomass (Eisenberg and Thorington 1973), and in Africa, folivorous columbines account for approximately 60 percent of the primate biomass (Oates et al. 1990). By contrast, carnivory tends to be relatively rare in tropical rainforests (Fleming 1973), and exclusively canopy-feeding mammals that utilize only vertebrate prey are unknown (Emmons 1995). Canopy-frequenting mammals from carnivorous groups tend to either frequently forage on the ground or be frugivorous (such as palm civets; binturongs, *Arctictis binturong*; kinkajous; and olingos, *Bassaricyon*). Presumably, specializations on vertebrate prey in the canopy are unknown because stealth is difficult and because smaller prey can always retreat to smaller branches. It should be noted, however, that some large carnivores (such as leopards, *Panthera pardus*) can feed extensively on arboreal taxa (for example, primates accounted for 24 to 34 percent of their diet in three rainforest studies; Ray and Sunquist 2001). Interestingly, large cats in neotropical and Asian tropical forests do not appear to include arboreal primates in their diet to the same extent as African leopards (Ray and Sunquist 2001).

Folivory   In their recent overview of primate feeding ecology, Janson and Chapman (1999) note that although leaves are the most abundant food source in the canopy, and their nutritional characteristics are relatively good, they represent a considerable digestive challenge for mammals. Leaf cellulose can be digested only with the help of bacterial or protozoan symbiots and, most seriously, leaves have considerable chemical defenses. The importance of both factors is demonstrated by the fact that researchers have had considerable success predicting the biomass of folivorous primates using indices of leaf quality, such as the ratio of protein to fiber or of protein to fiber plus tannins (references in Janson and Chapman 1999).

Mammals have evolved various strategies for dealing with these challenges. One response is to focus on the soluble carbohydrates and proteins and avoid cellulose by selecting low fiber leaves. The problem here is that such high-quality leaves tend to be rare and hard to find, being available only during a short window of leaf flushing, and they still may be heavily defended. A second response is through the evolution of digestion chambers, either through specialization of the foregut (a rare evolutionary feature) or the hindgut (Janson and Chapman 1999). Foregut digestion is advantageous compared to hindgut digestion in that materials released from the foregut (including the microbes themselves) can be digested in the rest of the digestive tract; however, it also is disadvantageous in that soluble nutrients are lost to the microbes themselves and that further feeding is inhibited during the digestive process. This slow through-time for microbial digestion limits food intake and hence affects energetics and social behavior (Janson and Chapman 1999).

In general, folivory is less viable as a feeding strategy for small compared to large mammals. The high leaf selectivity required for a high-quality diet requires high mobility, which is problematic for small mammals. A low-quality diet is similarly problematic: gut capacity declines in direct proportion to bodyweight, whereas mass-specific metabolic rates increase as bodyweight decreases, which means that obtaining sufficient energy from a low-quality diet is difficult for small mammals. These problems can be circumvented to some extent by lowering metabolic needs and, where possible, by focusing on high-quality leaves.

Plant defenses can be expected to vary in certain predictable situations. In his classic paper, Janzen (1974) hypothesized that plants growing on especially nutrient-poor soils would vigorously defend any growth that they could achieve and hence that these areas would have low densities of folivorous animals. Conversely, if plants can easily replace lost tissues, such as on rich soils, then they should invest less in secondary compounds. The resulting higher-quality leaves can be expected to maintain a higher folivore biomass. As a possible example, compared to the relatively infertile upland (terra firme) forests of the Amazon, the seasonally enriched floodplain (varzea) forests of the Amazon's white-water rivers have a higher folivore density (Peres 1997a). Low leaf defenses also may give rise to abundant insect populations and high biomasses of insectivorous mammals (Malcolm et al. in press). Other situations in which plants can be expected to spend fewer resources on leaf defense include during early succession, when growth is at a premium in the race for light, and perhaps in more seasonal forests. For example, Malcolm (1995, 1997) suggested that the high small mammal productivity of secondary and edge-modified forests compared to primary forest was due to lower investments by plants in defensive compounds and resulting higher insect abundances. The biomass of both folivorous lemurs in Madagascar (Ganzhorn 1992) and howler monkeys in the neotropics (Peres 1997b) increases with increasing seasonality. However, as the period of drought increases, a larger number of trees will show deciduous behavior, and declines in folivore diversity can be expected (e.g., Woinarski et al. 1992).

In addition to these implications for secondary compounds, certain nutrients may be in especially short supply on impoverished soils, which may lead to nutrient limitations affecting the abundance of canopy folivores (Janson and Chapman 1999). Much of the phosphorous has been lost from the old soils of the tropics, or is chemically bound and unavailable to plants, and hence tends to be especially limiting for herbivorous mammals. The phosphorous-to-calcium ratio in leaves is 1:5 or less, whereas in mammal bones it is 1:1 or more. The higher abundance of folivorous primates in Africa vs. the neotropics may be related to the prevalence of foregut fermenters in the former and their greater ability to extract phosphorous (Terborgh and van Schaik 1987; Janson and Chapman 1999). Braithwaite et al. (1983) and Cork and Catling (1996) hypothesized that nutrient limitation was especially problematic for folivorous mammals in Australia. The latter authors proposed that if nutrient supplies were too low to combat the toxic effects of secondary compounds, then folivorous gliders would be absent. Above that threshold, however, other factors such as food quality and abundance and cavity abundance would determine folivore density (see Figure 16-5). This hypothesis may explain the conclusions of Oates et al. (1990) that soil chemistry in Africa tends to be less important in determining abundance of columbines than forest age, structural heterogeneity, taxonomic composition, and history.

Frugivory   Other abundant resources for canopy mammals in the tropics are the fruits and seeds produced by canopy trees. Trees have been selected to provide fruits that result in effective dispersal, but protect the nutritionally rich seed from predation. Unlike leaves, aside from the seed, fruits tend to be deficient in protein; hence, most frugivores supplement their diets with leaves and/or insects (Janson and Chapman 1999). Presumably, trees provide the minimum possible payoff to the disperser, or perhaps by providing an incomplete diet, ensure that the dispersers actually leave the tree (Janzen 1983).

Some of the same factors that influence leaf quality and quantity also influence fruit quality and quantity. Fruit productivity appears to decrease as site productivity decreases; for example, the extremely low mammalian biomass of the impoverished soils of the Guyana Shield has been attributed to low soil and fruit productivity (Emmons 1984; see also Malcolm 1990). Frugivore

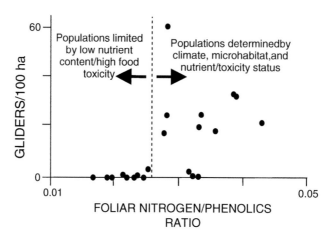

Figure 16-5 Proposed hierarchical model of habitat determinants for the greater glider (*Petauroides volans*) in eucalypt forests of southeastern Australia. Modified from Cork and Catling (1996).

and folivore biomass tend to decrease with rainfall, which Janson and Chapman (1999) ascribe to relatively low site productivity under high rainfall. In some cases, however, leaf and fruit resources do not vary in parallel. For example, early successional plants often produce relatively small seeds, making them of less utility to large-bodied frugivores. Some evidence exists of lower fruit biomass and fewer fleshy fruit species in floodplain forests compared to upland forests, which may result in relatively low abundances of frugivores compared to folivores and contribute to the relatively low primate species richness of floodplain forests (Peres 1997a).

The low nutritional quality of fruits is exacerbated by seasonality. Although some aspects of the tropical rainforest environment are remarkably invariant, rainfall is not, often varying markedly from month to month, from year to year, and from site to site. Most tropical forests have a distinct dry season that is severe enough to limit plant growth, resulting in a period of marked resource scarcity that tends to be synchronous among the various resources available to canopy mammals (e.g., Terborgh 1986). In general, frugivores switch to other foodstuffs during the dry season when fruit availability decreases. The convergence of many frugivore species to a few resources led Terborgh (1986) to hypothesize that these were "keystone" resources; i.e., they played a disproportionate role in supporting the vertebrate community (and, indirectly, forest composition because of the seed predation and dispersal undertaken by vertebrates). Inter-annual variation in food resources can sometimes be even more dramatic. For example, El Niño events can lead to synchronous mass fruiting events (Curran et al. 1999) or to synchronous fruit failure (Windsor 1990) depending on the region. In the central Amazon, the strong El Niño of 1982/83 was associated with a peak in the populations of virtually all small mammal species (Malcolm 1991b; see also Struhsaker 1997). Foster (1982) noted on Barro Colorado Island in Panama that unusually wet dry seasons were correlated with periods of fruiting failure and high mortality in the frugivore community. Emmons (1995) noted that a consequence of spatial and temporal variability in fruit availability was high mobility among frugivores. For example, because of their high mobility, some bats are able to travel long distances to feed exclusively on fruit. They make up for nutritional deficiencies by intake of calories far in excess of metabolic requirements (references in Emmons 1995).

In order to optimize dispersal, plants have evolved ways to restrict seed dispersal to a fraction of the frugivore community. In turn, this provides possibilities for niche differentiation in the frugivore community with certain frugivores being restricted to certain fruits, and the possibility that the set of available frugivores in turn will influence fruit size spectra (e.g., Emmons 1995). Although vertebrate specializations have been observed with respect to pulp richness and seed size, fruit morphology, fruit coloration, fruit phenology, and fruit position (references in Charles-Dominique 1993), narrow partitioning of the fruit resource does not appear to be generally true for canopy mammals or for tropical mammals in general. Instead, most fruits tend to be consumed by a relatively diverse assemblage of vertebrates; for example, the "bird-monkey syndrome" or the "ruminant-rodent-elephant syndrome" (Gautier-Hion et al. 1985; Shanahan and Compton 2001). Charles-Dominique (1993) hypothesized that early in the radiation of a plant group, plant species tend to be fed upon by a relatively small number of vertebrates that have specializations that allow them first access. In this case, selective pressures exerted by the frugivores will be similar and can result in the development of specializations within the plant group, providing a potential example of what Janzen (1980) termed diffuse coevolution (the selective pressures exerted between two groups of species linked by mutualistic relationships). However, as other vertebrates begin to focus on the resource and the frugivore assemblage becomes progressively composed of phylogenetically unrelated organisms, selective pressures will sometimes work in opposition. This dilution of selective pressures (Herrera 1985) will slow the coevolutionary process.

Some evidence exists of niche differentiation among canopy frugivores according to foraging habitats. For example, the nonspecialized (*sensu* Charles-Dominique 1993) frugivorous woolly opossum and kinkajou (*Potos flavus*) showed broad dietary overlap with fruit selection apparently being driven primarily by pulp accessibility rather than by fruit color or nutrient composition (Atramentowicz 1982; Julien-Laferriere 1999). Partitioning by these species seemed to occur primarily with respect to fruit patch productivity. Compared to the woolly opossum, the larger kinkajou tended to use more productive patches, but stayed for a shorter period of time in each and showed longer, faster movements between the more widely dispersed productive patches (Julien-Laferriere 1999). Another example of partitioning of fruit resources by space appears to be vertical partitioning in figs, although spatial constraints to mobility within a stratum may limit both the potential for partitioning within the dispersal assemblage and the effectiveness of dispersal (Shanahan and Compton 2001). As discussed below for insects, an inability to partition fruit resources according to size instead may result in microhabitat differentiation. Alternatively, the lack of fruit partitioning may be an example of broad niche overlap on common resources (fruit), but divergence in use of fitness-enhancing ones (such as insects) (Robinson and Wilson 1998; Janson and Chapman 1999).

Insectivory   Insects are a high-quality resource; however, they are a scarce one. As pointed out by Janson and Chapman (1999) for frugivorous primate species, time searching for insects increases with body size. However, beyond a certain body size (roughly capuchin size), enough protein cannot be obtained solely from insects, and individuals must rely instead on leaves. Small canopy mammals have lower total metabolic demands because of their small size and hence can rely upon this scare, high-quality food resource. Emmons (1995) reported that the relatively few species that fed almost entirely on arboreal insects were all small, nocturnal forms. Not surprisingly, the proportion of insects in the diet increases with decreasing body size (Redford et al. 1984). At the same time, most arboreal species, even those considered frugivores, will exploit insects opportunistically (Redford et al. 1984).

Dickman (1988) hypothesized that niche partitioning with respect to insect size is unlikely for mammals because even the smallest mammals can handle quite large insects. He hypothesized that niche differentiation among insectivorous mammals therefore would be with respect to foraging microhabitats, an idea that finds support from Charles-Dominique (1977) and Terborgh (1983), who found that canopy species tended to differ in the heights that they frequented and the sizes of substrates that they used while searching for insects.

Variation in Species Composition

Not only may different sites have different numbers of canopy species, but the identity of the species may vary as well, leading to a turnover in species composition from site to site. For example, some species may specialize on certain habitat types (β-diversity), or as sites become increasingly distant within the same habitat type, a turnover in species composition may be observed (γ-diversity). Unfortunately, relatively little work has been undertaken in this area, in part because species distributions and habitat associations are poorly known. Even for primates, studies at the assemblage level are rare (Peres and Janson 1999).

The increase in species richness with decreasing latitude often is accompanied by a decrease in the sizes of geographic ranges, in what is known as Rapport's Rule. The implication is that, all things being equal, one can expect a higher turnover in species composition from site to site (and higher endemism) at low than high latitudes. This rule appears to apply for primates, although more strongly in the neotropics than Africa (Cowlishaw and Hacker 1997; Eeley and Lawes 1999). Many explanations have been put forward for this phenomenon, including:

1. Selection to favor broader tolerances at higher latitudes because of greater environmental variability, which translates into an ability to use larger geographic areas;
2. More prevalent biotic interactions at low latitudes, which as a result limit ranges more often;
3. Relatively low extinction probabilities for the small-range species at low latitudes because of lower environmental heterogeneity through time; and
4. Higher relative severity of elevation barriers at low latitudes and hence higher rates of allopatric speciation (Eeley and Lawes 1999).

Eeley and Lawes (1999) found a correlation between dietary diversity and geographic range size among primates, providing some evidence for the first hypothesis. An interesting parallel for arboreal species, as yet untested, would be an increase in range size with increased breadth of vertical habitat usage.

In some cases, variation among sites in species composition will reflect both γ- and β-variation. For example, although upland (terra firme) and floodplain (varzea) forests in the Amazon Basin can be very similar physiognomically, they both contain species that are unique to one forest type or the other (e.g., Malcolm et al., in press). In headwaters regions, where the floodplain is narrow and less regularly flooded, both sets of species can be found at one site, leading to high local diversity (Robinson and Terborgh 1990; Malcolm et al., in press). Another potential example concerns the savanna/rainforest distinction in Africa where, for example, Eeley and Lawes (1999) reported a pronounced turnover in primate species composition between the two forest types, with genera often being more or less restricted to one biome or the other. However, in some cases, savanna taxa also use rainforest, with the result that variation in species composition among rainforest sites may to a certain extent reflect proximity to, and connectivity among, savanna and savanna-like habitats (see Malcolm and Ray 2000).

The magnitude of habitat-specific variation will depend on a host of factors, including the size of the habitats through time. Smaller areas of habitat can be expected to be subject to higher extinction rates, which may explain the lower species richness of primate faunas in Southeast Asia compared to Africa or the neotropics (Eeley and Lawes 1999). Similarly, instead of at the equator, Peres and Janson (1999) reported highest primate diversity in the Amazon where the area of the tropical rainforest was most extensive, which they attributed to area effects through time. Body size and life history characteristics also may help to explain variation in γ- and β-diversity. All things being equal, the same-sized area will contain fewer individuals of large species, suggesting that larger species are more likely to go extinct than smaller ones. Although smaller-bodied species may be more susceptible to fluctuations in resources and temperature than large species, they may be more effective at colonizing new areas and hence replenishing locally rare or extirpated populations. The net effect may be a triangular relationship between body size and geographic range size, with larger-bodied species being restricted to larger range sizes and smaller species covering the full spread of range sizes (Eeley and Lawes 1999; Brown and Mauer 1987). Similarly, Malcolm et al. (in press) speculated that lower γ-diversity of canopy compared to terrestrial mammals along the Juruá River in western Amazonia reflected the more K-selected traits of canopy mammals. One might similarly expect larger range sizes for canopy compared to terrestrial mammals. The same phenomenon may explain the generally lower species richness of canopy compared to terrestrial small mammal faunas. An alternative possibility is fewer possibilities for niche differentiation in the canopy compared to the ground.

Conservation of Canopy Mammals

Deforestation and Forest Fragmentation

The outright loss of the forest canopy through deforestation has devastating consequences for arboreal mammals, even the least specialized ones. Extremely high levels of forest loss in the tropics, which continues at high rates, can be expected to result in the extinction of many species, especially if contiguous blocs of rainforest are completely lost (Kinzig and Harte 2000).

In deforested landscapes, some of the key factors influencing the persistence of canopy species include fragment size and shape, fragment connectivity, the nature of the habitat types in the matrix, and the degree to which fragments represent the geographic ranges and habitats of the original faunal. If fragments are small enough, they may not provide the minimum space requirements for large and wide-ranging canopy species, especially if the resources that they use are patchy in space and time (Lovejoy et al. 1986). All things being equal, species at low densities are expected to show higher extinction rates than more common species (MacArthur and Wilson 1967). In some cases, possibly for arboreal folivores, densities may be higher in fragments due to a lack of predators and flushes of foliage along fragment edges. Because of their comparatively K-selected life history characteristics, based solely on fragment size, one might expect canopy species to be less vulnerable to extinction than terrestrial ones (MacArthur and Wilson 1967); however, to counter any such lower vulnerability, canopy species presumably require larger home ranges and the human-modified landscapes of the matrix are typically more hostile to arboreal than terrestrial species because of the relative rarity of arboreal habitat.

Connectivity of arboreal habitats will be particularly important in influencing metapopulation persistence if fragments are small. Connectivity for canopy species can be increased in several ways; for example, by placing fragments in close proximity to each other, by increasing the amount and quality of the arborescent vegetation in the matrix, or by providing forested corridors between habitats. An example of the first two is provided by Malcolm (1991b, 1997). Apparently, the combination of abundant arborescent vegetation in pastures and proximity to nearby large areas rainforest (100 to 800 m away) meant that small mammal populations in fragments varied in response to variation in habitat quality, but not fragment isolation or size. Malcolm (1991b, 1997) and Laurance (1991) hypothesized that tolerance to the matrix was a key factor influencing persistence in fragmented landscapes. Interestingly, most small mammals in both the central Amazon and central Africa were at higher densities in young secondary habitats than in primary forest habitats (Malcolm 1991b, 1995; Malcolm and Ray 2000). Even arboreal species such as *Oecomys paricola* and *Micoureus demerarae* had higher densities in secondary than in primary forests (Malcolm 1991b), indicating that they were fully capable of utilizing the low-stature secondary forests than might have been suspected from trapping in primary forest. Possible explanations include the importance of insects in their diets, which can often be super-abundant in secondary forests; another is a pre-adaptation to secondary habitats because of preferential use of gap margins even in primary forest. In araucarian vine forests in Australia, Bentley et al. (2000) found that habitat generalists tended to be more tolerant of fragmentation than specialists, whereas fecundity, diet, body mass, natural rarity, and tolerance to matrix were poor predictors of tolerance. Concerning the latter, however, they noted that habitat generalists may be more likely to tolerate the modified vegetation of the matrix than specialists. As a result, they were unable to reject matrix tolerance as a predictor of tolerance to fragmentation. Large, frugivorous canopy species can be expected to fare poorly in secondary forests, which offer little in the way of large-diameter supports and large-sized fruit. A good example is provided by brown capuchin monkeys in the central Amazon. They abandoned a 10-ha fragment shortly after it was isolated from the surrounding forest, and only revisited it approximately five years later when the surrounding secondary forest had reached a height of nearly 10 m (unpubl. obs.). The value of arborescent vegetation in the matrix is well established for small arboreal mammals, with abundances and diversity of forest taxa increasing with the amount of arborescent vegetation, and with highly modified pastures providing habitat for few, if any, rainforest species, and no habitat for arboreal species (Estrada et al. 1994; Malcolm 1995; Bentley et al. 2000; Gascon et al. 1999).

The value of corridors in promoting connectivity has been examined in a few cases. At the smallest scale, Estrada et al. (1994) found that live fences retained approximately one-quarter of the rainforest species in contrast to the adjoining pasture, which had almost none. Laurance and Laurance (1999) found that linear remnants of moderate width were used by five of six species of marsupials, but that the lemuroid ringtail occurred only in remnants that were > 200 m wide and that were connected to large tracts of continuous forest. For species that avoid edge-modified habitats, corridors evidently must be of considerable width to provide suitable interior habitat, especially if they are to maintain a population rather than just support occasional dispersal movements.

Because of the huge contrast between clearings and closed-canopy rainforest, the creation of clearings has a powerful impact on adjacent rainforest communities. In response to increased light levels, vines and secondary plants proliferate, and increased exposure to winds and vine loads leads to high mortality among canopy trees, further increasing the light available for secondary plant species. The net effect is a loss of habitat for many canopy species, although as noted earlier, many scansorial and arboreal small mammal species can increase in densities in this highly modified secondary environment, at least in the short term. In the central Amazon, the effects of fragmentation on the small mammal fauna could be entirely attributed to these edge effects (Malcolm 1991b, 1997). Laurance and Yensen (1991) also noted the importance of edge effects in influencing mammal communities in tropical Queensland. These edge effects can extend considerable distances into the forest. For example, at 5 to 8 years after edge formation, Malcolm (1994) found that they extended nearly 100 m into the forest. Pardini and Malcolm (unpubl.; see also Pardini 2001) estimated that they extended 330 m into the edges of fragments of Atlantic rainforest of Bahia. Because of drying, these edges may leave the forest more vulnerable to fires, with devastating effects for many canopy tree species and canopy-dwelling mammals. For example, relative to interior forest, Cochrane and Laurance (2002) found elevated fire return intervals within 2,400 m of forest edges in the eastern Amazon.

A final note should be made on the topic of representation; namely, the set of species and habitats represented in a reserve/fragment network. Emmons (1999) recently argued that because primate species richness could be used to predict the species richness of other mammalian groups (notably rodents and marsupials), a reasonable strategy for choosing sites for protected areas would be to focus on areas of high species richness and endemicity of primates. Because of high γ diversity in tropical regions, it is apparent that many such protected areas will be needed to provide representative populations of all species. Unfortunately, canopy species distributions are often poorly known, as are the long-term effects of fragmentation, so it is at present difficult to assess the conservation value of a few large contiguous reserves compared to a more geographically comprehensive network of smaller, interconnected reserves.

Logging

Tree-fall gaps create conditions that are reminiscent of edge effects, albeit on a smaller scale. Indeed, by using appropriate additive edge models, it may be possible to extend results from studies of fragmentation to the effects of logging (Malcolm 2001). In support of this hypothesis, bird species responded in similar ways to logging and fragmentation at two neotropical sites (Malcolm 2001) and Malcolm and Ray (2000) found edge-induced increases in small mammal abundance in both central Amazon and Africa, the former due to fragmentation and the latter due to logging. According to this hypothesis, the most relevant measure of logging damage is the density and distribution of openings created in the canopy, a result supported by Malcolm and Ray (2000; see also Webb 1997). Surprisingly, apparently only one study has examined responses of the canopy fauna as a function of variation in the number of canopy gaps. In that study, Struhsaker (1997) found striking impacts of high-volume logging on biodiversity, hypothesizing that a combination of increased elephant usage and high rodent densities in high-volume logged areas led to suppressed tree regeneration and possible long-term transitions from rainforest to herbaceous/shrub communities. Presumably, the negative effects of this transition to conditions more typical of secondary forests would be strongest on the most frugivorous and arboreal species. The nature of the relationship between biodiversity impacts and the volume of timber harvested is not known in any detail. Some authors (e.g., Johns 1997) have suggested that biodiversity impacts will increase only slowly with the amount of timber removed (see Figure 16-6c). Others (e.g., Malcolm 1998, 2001) have suggested that strong impacts will be felt at even relatively low off-takes (see Figure 16-6a). Various other possibilities include a threshold response (see Figure 16-6b). An understanding of the impacts of logging on canopy species has the potential to lead not only to better designs for careful logging, but also to better understanding of edge effects during fragmentation.

Additional effects of logging on canopy mammals include the felling of important fruiting species, increased access for hunters, and a loss of forest structural complexity. The latter is of

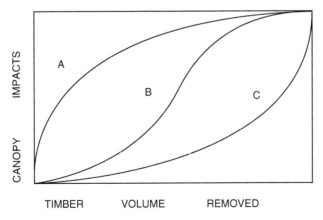

Figure 16-6   Models of canopy impacts as a function of the volume of timber harvested per unit area. In curve A, strong impacts are caused even at low harvest volumes, whereas in curve C, impacts are low until volumes are quite high. Curve B shows a threshold response wherein impacts increase rapidly over a narrow range of harvest volumes.

ORANGUTANS: THE LARGEST CANOPY DWELLERS

Cheryl D. Knott

As I walk through the rainforest of Borneo at dusk, I am startled by a loud "snap" as a tree limb is broken off and inserted into a nest stretching over 2 m wide. Although I can barely make out a glimpse of orange fur, this is unmistakably an adult orangutan bedding down for the night. With adult males weighing 86.3 kg on average and females 38.5 kg (Markham and Groves 1990), orangutans are the largest habitually arboreal animal (Rodman and Mitani 1987; Knott 1999a). My data from Gunung Palung in Borneo show that orangutans spend, on average, 99.6 percent of their waking hours in the canopy. This is in stark contrast to the African apes—chimpanzees and gorillas—that are much more terrestrial. Chimpanzees, smaller than orangutans, climb into trees to eat fruit, but normally travel and rest on the ground. At the Tai forest in the Ivory Coast, Doran (1993) found that they spent close to 50 percent of their time on the ground. The larger mountain gorillas are almost exclusively terrestrial, relying on abundant low-lying vegetation. Recent studies of western lowland gorillas indicate that they often climb trees to obtain fruit, but they rest and travel on the ground like the other African apes. Orangutans are thus unique among the large-bodied hominoids in their essentially exclusive use of the canopy.

Figure 1   Adult male orangutan traveling through the canopy in Gunung Palung National Park, Indonesia. Photograph by Tim Laman.

Their niche as large-bodied canopy dwellers raises intriguing questions about how they can use the canopy efficiently. These large animals are cautious climbers, moving through the forest by quadrumanual clambering—using all four hands and feet to grasp and pull themselves along (Rodman and Mitani 1987). This spreads their bodyweight over several different arboreal supports, lessening the risk that one will give way. Essentially, this is the same method humans use for climbing, although without as much strength or flexibility! Orangutans have extremely shallow hip joints, allowing them almost 360 degrees of rotation and giving them the ability to reach tree limbs in any direction. Smaller individuals exhibit brachiation on occasion.

Size does place constraints on canopy usage, but heavy bodyweight also allows orangutans, particularly the big males, to use unique forms of locomotion. One of their common techniques is to sway a slender tree from side to side, utilizing their bodyweight to bend it. They then grab a branch of an adjacent tree and pull it toward them before scrambling onward. As the new tree swings toward vertical, they quickly move to the other side and use their weight to bend the tree toward the next objective. They are thus effectively using a tree as a spring to propel themselves across a gap until they can reach the next tree. This mode of locomotion is particularly common among the males (Rijksen 1978). They normally come down to lower canopy levels to use this travel method, and males tend to travel lower in the canopy than do females (Setiawan et al. 1996). Pole trees in the mid-canopy are smaller in diameter, easier to bend over, and spaced more frequently. Using the substrate itself to help them locomote is one of the unique possibilities in the canopy that distinguishes the movement of orangutans from terrestrial locomotion.

The preferred food of orangutans is fleshy fruit; however, they also eat bark, leaves, piths, and insects when times are tough (Knott 1998). These foods are primarily located in the canopy and often, such as the case with some ripe fruit and young leaves, are situated at the terminal ends of branches. Comparing orangutan males to females that are half their size provides an excellent test of the way size may constrain the evolution of large-bodied canopy animals. Adult males spend more time on the ground than females do, especially while traveling long distances (Galdikas 1988). Females may be able to travel and feed higher in the canopy than the males because their lighter weight enables them to travel to smaller branches while feeding and to access higher canopy pathways that may be difficult for adult males to use (Cant 1987a, 1987b; Rodman and Mitani 1987). However, males (as well as females) may also just break off a large fruit-bearing branch and take it to a more comfortable perch for consumption, such as a tree crotch. Whether males really suffer a foraging cost because of their greater heft has not been demonstrated. Other sex differences also exist: males travel significantly slower in the canopy than do females, are more likely to build their nests close to their last food tree, and tend to build their nests in trees that are lower in the canopy and smaller in diameter than do females (Setiawan et al. 1996).

An arboreal lifestyle may also place constraints on orangutan reproduction. Orangutans give birth on average only once every 8 years (Galdikas and Wood 1990; Knott 2001)— the longest inter-birth interval of any mammal. This is compared to a mean of only 4 years between births in gorillas and approximately 5 years in chimpanzees (Knott 2001). Juvenile orangutans at least occasionally ride on their mothers until they are as old as 7 years. Mothers spend significant amounts of time helping juveniles negotiate the canopy, and often have to wait for slower juveniles to catch up. This may place significant energetic costs on orangutan females. We now know that being in poor energetic status lowers ovarian function in female orangutans and is a main contributor to long inter-birth intervals (Knott 1999b, 2001). Thus, the added energetic cost of arboreality on orangutan females may be one of the factors that have had a significant effect on orangutan inter-birth intervals and thus their evolution.

Why then are orangutans habitual arboreal travelers when the African apes are not? Avoiding terrestrial predators seems unlikely to be the reason as there are records of leopard predation on chimpanzees (Boesch and Boesch-Achermann 2000), but no records of

Continued

ORANGUTANS: THE LARGEST CANOPY DWELLERS–*cont'd*

predation on wild orangutans. Borneo, in fact, has no cats large enough to be a threat to orangutans, although Sumatra has tigers that have occasionally killed ex-captive orangutans. The answer may lie in differences in canopy structure and food availability and distribution between the Southeast Asian rainforests and those of Africa. Forests inhabited by orangutans potentially provide a more continuous canopy for travel. These forests also tend to harbor taller trees. It may simply be more energetically efficient for orangutans to travel arboreally rather than descend to the ground between feeding bouts. Additionally, orangutans, which are predominantly solitary, may feed in trees that are smaller and closer together whereas more group-living chimpanzees feed on large fruit trees that are relatively far apart, favoring the use of faster terrestrial travel. Cross-continent comparisons will, undoubtedly, shed more light on the evolution and ecology of the great apes. Seeing orangutans so effectively negotiate the canopy surely challenges our notion of what a canopy dweller can be.

References

Boesch, C. and Boesch-Achermann, H. (2000). "The Chimpanzees of the Tai Forest: Behavioural Ecology and Evolution." Oxford University Press, New York.

Cant, J.G.H. (1987a). Effects of sexual dimorphism in body size on feeding postural behavior of Sumatran Orangutans (*Pongo pygmaeus*). *Am. J. Phys. Anthropol.* **74**, 143–148.

Cant, J.G.H. (1987b). Positional behavior of female Bornean orangutans (*Pongo pygmaeus*). *Am. J. Primatol.* **12**, 71–90.

Doran, D. (1993). Sex differences in adult chimpanzee locomotor behavior: The influence of body size on locomotion and posture. *Am. J. Phys. Anthropol.* **91**, 99–116.

Galdikas, B.M.F. (1988). Orangutan diet, range, and activity at Tanjung Puting, Central Borneo. *Int. J. Primatol.* **9**, 1–35.

Galdikas, B.M.F. and Wood, J.W. (1990). Birth spacing patterns in humans and apes. *Am. J. Phys. Anthropol.* **83**, 185–191.

Knott, C.D. (1998). Changes in orangutan diet, caloric intake and ketones in response to fluctuating fruit availability. *Int. J. Primatol.* **19**, 1061–1079.

Knott, C.D. (1999a). Orangutan Behavior and Ecology. *In* "The Nonhuman Primates" (P. Dolhinow and A. Fuentes, Eds.), pp. 221–235. Mayfield Press, Mountain View, CA.

Knott, C.D. (1999b). "Reproductive, Physiological and Behavioral Responses of Orangutans in Borneo to Fluctuations in Food Availability." Ph.D. Thesis. Harvard University.

Knott, C.D. (2001). Female reproductive ecology of the apes: implications for human evolution. *In* "Reproductive Ecology and Human Evolution" (P. Ellison, Ed.), pp. 429–463. Aldine de Gruyter, New York.

Markham, R.J. and Groves, C.P. (1990). Brief communication: weights of wild orangutans. *Am. J. Phys. Anthropol.* **81**, 1–3.

Rijksen, H.D. (1978). "A Field Study on Sumatran Orang-utans (Pongo pygmaeus abelii, Lesson 1827): Ecology, Behavior, and Conservation." H. Veenman and Zonen, Wageningen, the Netherlands.

Rodman, P.S. and Mitani, J.C. (1987). Orangutans: sexual dimorphism in a solitary species. *In* "Primate Societies" (B.B. Smuts, D.L. Cheney, R.M. Seyfarth, R.W. Wrangham, and T.T. Struhsaker, Eds.), pp. 146–152. University of Chicago Press, Chicago.

Setiawan, E., Knott, C.D., and Budhi, S. (1996). Preliminary assessment of vigilance and predator avoidance behavior of orangutans in Gunung Palung National Park, Indonesia. *Tropical Biodiversity.* **3**, 269–279.

particular concern in temperate regions, where logging typically results in losses of old-growth forests and/or a reduction in forest structural complexity. In plantations, many of the resources required by canopy mammals may be lacking or at lower abundances, including cavity trees, flowers, pollen, nectar, fruit, and invertebrates (e.g., Lindenmayer et al. 2000).

Hunting

Hunting can have as dramatic an effect on some canopy mammal species as habitat loss (Redford 1992; Robinson and Bennett 2000). Although hunting is a traditional activity in many human societies, it has become a serious problem in many areas of the world because of high human populations, improved hunting technologies, and increased commercialization of the hunt (Redford and Robinson 1987; Wilkie and Carpenter 1999; Fa and Peres 2001). The introduction of firearms in particular has had an important impact on canopy species. Hunters target larger-bodied species, which tend to be most susceptible to over-hunting because of lower densities and reproductive rates (Fa and Peres 2001). As larger species are hunted out, hunters shift to smaller species. Frequently, extraction rates exceed population productivity (Robinson and Bodmer 1999), leading to local extirpations (e.g., Peres 1990). An indirect effect of logging is increased access for hunters. As a result, in some cases, the ecological impacts of hunting may exceed those of logging (e.g., Wilkie et al. 2001). Hunting is an especially serious problem in African rainforests, where the biomass of the bushmeat harvest is an order of magnitude higher than in neotropical rainforests. In Africa, because of dense human populations, a small forest area, improved technology, and commercialization of the bushmeat trade, hunting poses one of the most insidious threats to the survival of many canopy species (Fa and Peres 2001).

Conclusions and Summary

Canopy mammals have evolved various morphological adaptations to meet the challenges of arboreal living, including features to increase friction during climbing, to traverse the gaps between tree crowns and between branches, to reconnoiter the complex, three-dimensional canopy environment, and to maintain stability in a fine-branch environment. Large body size has important implications for all of these solutions, with increasing body size increasing the dangers of weak supports and the metabolic costs associated with climbing. The set of morphological adaptations shown by a canopy species influences foraging possibilities and vice versa; indeed, adoption of canopy living has implications not only for morphology and feeding behavior, but for the evolution of life history features as well.

Ecological study of canopy species is relatively easy for diurnal and large-bodied species such as primates, but has lagged for the smaller and nocturnal forms, which are poorly known from taxonomic, distributional, and ecological perspectives. Although techniques are being developed that make comprehensive censuses of canopy mammals more routine, scientific knowledge about the entire canopy community remains rudimentary, especially in the tropics, where canopy mammals are at their most diverse and abundant. The combination of a complex and multi-strata canopy combined with a diverse resource base has provided abundant opportunities for niche differentiation in the tropics, with various species specializing on certain height strata and micro-habitats within the canopy. Variation in the development of the canopy, including its extent, spatial configuration, and physical structure, importantly influences the characteristics of the canopy community, both locally and regionally and in both ecological and evolutionary time.

Food resources of the canopy also offer particular challenges for canopy mammals. Although a valuable resource from a nutritional perspective, leaves are heavily defended both mechanically and chemically. The ability of canopy mammals to breach these defenses is constrained by a variety of factors, including body size, the evolution of appropriate fermenting chambers, and species mobility. In certain situations, such as during early succession and on rich soils, plants can be expected to devote fewer resources to leaf defenses, which can lead to high biomasses of folivores and, indirectly, insectivores. Frugivory becomes especially important in tropical regions and is of particular interest because of the mutualistic relationships between trees and their dispersers and because of the potential for coevolution. As a resource, fruits again offer unique challenges, in particular because they tend to be nutritionally incomplete, widely dispersed, and show strong variation in abundance through time. The net effect of the various constraints and challenges offered by canopy food resources appears to have resulted in generalized feeding habits for many canopy mammal species, which may contribute to the importance of competitive interactions in influencing key features of the canopy community.

Many canopy species are in considerable peril due to human influences, especially in the tropics, where deforestation is rampant. Anthropogenic habitats tend to have structurally and taxonomically impoverished tree communities, with the net result that they are an unfriendly environment for many canopy species and offer few prospects for movement and even fewer for the maintenance of viable habitats. Some smaller canopy species may utilize and even flourish in secondary forests, but many attain their highest densities in complex, old-growth forests. The protection of the canopy fauna may thus depend to a large extent on forest fragments and protected areas. Relatively small fragments can be expected to support populations of large and wide-ranging canopy species in the long term only if they are broadly connected to other larger areas.

Other important anthropogenic influences include logging and hunting, each of which can have devastating effects on canopy populations. The long-term effects of logging on canopy communities are poorly understood, but can be expected to be considerable if harvest volumes are high. Under increasing human population densities, improved hunting technologies, and increased trade, many large-bodied canopy species are facing extirpations from hunting.

Acknowledgments

Helpful comments on an earlier draft of this chapter were provided by Justina Ray. I would also like to thank Carlos Peres, who pointed me in the direction of some valuable primate literature.

References

Argot, C. (2001). Functional-adaptive anatomy of the forelimb in the Didelphidae, and the paleobiology of the paleocene marsupials *Mayulestes ferox* and *Pucadelphys andinus*. *Journal of Morphology* **247**, 51–79.

Atramentowicz, M. (1982). Influence du milieu sur l'activité locomotrice et la reproduction de *Caluromys philander* (L.). *Revue d'Ecologie (Terre et Vie)* **36**, 376–395.

August, P.V. (1983). The role of habitat complexity and heterogeneity in structuring tropical mammal communities. *Ecology* **64**, 1495–1513.

Bakker, V.J. and Kelt, D.A. (2000). Scale-dependent patterns in body size distributions of Neotropical mammals. *Ecology* **81**, 3530–3547.

Bentley, J.M., Catterall, C.P., and Smith, G.C. (2000). Effects of fragmentation of araucarian vine forest on small mammal communities. *Conservation Biology* **14**, 1075–1087.

Bernard, R.T.F. and Nurton, J. (1993). Ecological correlates of relative brain size in some South-African rodents. *South African Journal of Zoology* **28**, 95-98.

Braithwaite, L.W., Dudzinski, M.L., and Turner, J. (1983). Studies on the arboreal marsupial fauna of eucalypt forests being harvested for woodpulp at Eden, New South Wales. II. Relationship between fauna density, richness and diversity, and measured variables of the habitat. *Australian Wildlife Research* **10**, 231–247.

Brown, J.H. and Mauer, B.A. (1987). Evolution of species assemblages: effects of energetic constraints and species dynamics on the diversification of the North American avifauna. *American Naturalist* **130**, 1–17.

Cartmill, M. (1974). Pads and claws in arboreal locomotion. *In* "Primate Locomotion" (F.A. Jenkins, Jr., Ed.), pp. 45–83. Academic Press, New York.

Cartmill, M. (1985). Climbing. *In* "Functional Vertebrate Morphology" (M. Hidelbrand, D.M. Bramble, Liem K.F., and D.B. Wake, Eds.), Harvard University Press, Cambridge, Mass.

Charles-Dominique, C. (1977). "Ecology and Behaviour of Nocturnal Primates: Prosimians of Equatorial West Africa." Columbia University Press, New York.

Charles-Dominique, P. (1993). Speciation and coevolution: an interpretation of frugivory phenomena. *Vegetatio* **107/108**, 75–84.

Charles-Dominique, P.M., Atramentowicz, M., Charles-Dominique, M., Gerard, H., Hladik, C.M., and Prevost, M.F. (1981). Les mammifères frugivores arboricoles nocturnes d'une forêt Guyanaise: inter-relations plantes-animaux. *Revue d'Ecologie (Terre et Vie)* **35**, 341–435.

Cochrane, M.A. and Laurance, W.F. (2002). Fire as a large-scale edge effect in Amazonian forests. *Journal of Tropical Ecology* **18**, 311–325.

Cork, S.J. and Catling, P.C. (1996). Modelling distributions of arboreal and ground-dwelling mammals in relation to climate, nutrients, plant chemical defences and vegetation structure in the eucalypt forests of southeastern Australia. *Forest Ecology and Management* **85**, 163–175.

Cowlishaw, G. and Hacker, J.E. (1997). Distribution, diversity, and latitude in African primates. *American Naturalist* **150**, 505–512.

Cristoffer, C. (1987). Body size differences between New World and Old World, arboreal, tropical vertebrates: causes and consequences. *Journal of Biogeography* **14**, 165–172.

Cristoffer, C. and Peres, C.A. (2003). Elephants vs. butterflies: the ecological role of large herbivores in the evolutionary history of two tropical worlds. *Journal of Biogeography* **30**, 1–24.

Cunha, A.A. and Vieira, M.V. (2002). Support diameter, incline, and vertical movements of four didelphid marsupials in the Atlantic forest of Brazil. *Journal of Zoology* **258**, 419–426.

Curran, L.M., Caniago, I., Paoli, G.D., Astianti, D., Kusneti, M., Leighton, M., Nirarita, C.E., and Haeruman, H. (1999) Impact of El Nino and logging on canopy tree recruitment in Borneo. *Science* **286**, 2184–2188.

de Muizon, C. (1998). *Mayulestes ferox*, a borhyaenoid (Metatheria, Mammalia) from the early palaeocene of Bolivia. Phylogenetic and palaeobiological implications. *Geodiversitas* **20**, 19–142.

Dickman, C.R. (1988). Body size, prey size, and community structure in insectivorous mammals. *Ecology* **69**, 569–580.

Eeley, H.A.C. and Lawes, M.J. (1999). Large-scale patterns of species richness and species range size in anthropoid primates. *In* "Primate Communities" (J.G. Fleagle, C.H. Janson, and K.E. Reed, Eds.), pp. 191–219. Cambridge University Press, Cambridge, UK.

Eisenberg, J.F. (1987). The evolution of arboreal herbivores in the Class Mammalia. *In* "The Ecology of Arboreal Folivores" (G.G. Montgomery, Ed.), pp. 135-152. Smithsonian Institution Press, Washington, DC.

Eisenberg, J.F. and Thorington, Jr., R.W. (1973). A preliminary analysis of a neotropical mammal fauna. *Biotropica* **5**, 150–161.

Eisenberg, J.F. and Wilson, D. (1981). Relative brain size and demographic strategies in didelphid marsupials. *American Naturalist* **118**, 1–15.

Emmons, L.H. (1984). Geographic-variation in densities and diversities of non-flying mammals in Amazonia. *Biotropica* **16**, 210–222.

Emmons, L.H. (1980). Ecology and resource partitioning among nine species of African rain forest squirrels. *Ecological Monographs* **50**, 31–54.

Emmons, L.H. (1995). Mammals of rain forest canopies. *In* "Forest Canopies" (M.D. Lowman and N.M. Nadkarni, Eds.), pp. 199–223. Academic Press, San Diego.

Emmons, L.H. (1999). Of mice and monkeys: primates as predictors of mammal community richness. *In* "Primate Communities" (J.G. Fleagle, C.H. Janson, and K.E. Reed, Eds.), pp. 171–188. Cambridge University Press, Cambridge, UK.

Emmons, L.H. and Feer, F. (1990). "Neotropical Rain Forest Mammals: A Field Guide." University of Chicago Press, Chicago.

Emmons, L.H., Gautierhion, A., and Dubost, G. (1983). Community structure of the frugivorous folivorous forest mammals of Gabon. *Journal of Zoology* **199**, 209–222.

Emmons, L.H. and Gentry, A.H. (1983). Tropical forest structure and the distribution of gliding and prehensile-tailed vertebrates. *American Naturalist* **121**, 513–524.

Ewer, R.F. (1973). "The Carnivores." Cornell University Press, Ithaca.

Estrada, A., Coatesestrada, R., and Meritt, D. (1994). Non-flying mammals and landscape changes in the tropical rain-forest region of Los-tuxtlas, Mexico. *Ecography* **17**, 229–241.

Fa, J.E. and Peres, C.A. (2001). Game vertebrate extraction in African and Neotropical forests: an intercontinental comparison. *In* "Conservation of Exploited Species" (J.D. Reynolds, G.M. Mace, K.H. Redford, and J.G. Robinson, Eds.), pp. 203–241. Cambridge University Press, Cambridge, UK.

Fleming, T.H. (1973). Numbers of mammal species in North and Central American forest communities. *Ecology* **54**, 555–563.

Foster, R.B. (1982). Famine on Barro Colorado Island. *In* "The Ecology of a Tropical Forest: Seasonal Rhythms and Long-Term Changes" (E.G. Leigh, Jr., A.S. Rand, and D.M. Windsor, Eds.), pp. 201–212. Smithsonian Institution Press, Washington, DC.

Ganzhorn, J.U. (1992). Leaf chemistry and the biomass of folivorous primates in tropical forests: test of a hypothesis. *Oecologia* **91**, 540–547.

Gascon, C., Lovejoy, T.E., Bierregaard, R.O., Malcolm, J.R., Stouffer, P.C., Vasconcelos, H.L., Laurance, W.F., Zimmerman, B., Tocher, M., and Borges, S. (1999). Matrix habitat and species richness in tropical forest remnants. *Biological Conservation* **91**, 223–229.

Gautier-Hion, A., Duplantier, J.M., Quris, R., Feer, F., Sourd, C., Decoux, J.P., Dubost, G., Emmons, L., Erard, C., Hecketsweiler, P., Moungazi, A., Roussilhon, C., and Thiollay, J.M. (1985). Fruit characters as a basis of fruit choice and seed dispersal in a tropical forest vertebrate community. *Oecologia* **65**, 324–337.

Glazier, D.S. and Newcomer, S.D. (1999). Allochrony: a new way of analysing life histories, as illustrated with mammals. *Evolutionary Ecology Research* **1**, 333–346.

Herrera, C.M. (1985). Determinants of plant–animal coevolution: the case of mutualistic dispersal of seeds by vertebrates. *Oikos* **44**, 132–141.

Holmes, D.J. and Austad, S.N. (1994). Fly now, die later: life-history correlates of gliding and flying in mammals. *Journal of Mammalogy* **75**, 224–226.

Iwaniuk, A.N., Pellis, S.M., and Whishaw, I.Q. (1999). The relationship between forelimb morphology and behaviour in North American carnivores (Carnivora). *Canadian Journal of Zoology* **77**, 1064–1074.

Janson, C.H. (1998). Experimental evidence for spatial memory in wild brown capuchin monkeys (*Cebus apella*). *Animal Behaviour* **55**, 1129–1143.

Janson, C.H. and Chapman, C.A. (1999). Resources and primate community structure. *In* "Primate Communities" (J.G. Fleagle, C.H. Janson, and K.E. Reed, Eds.), pp. 55–74. Cambridge University Press, Cambridge, UK.

Janzen, D.H. (1974). Tropical blackwater rivers, animals, and mast fruiting by the Dipterocarpacea. *Biotropica* **6**, 69–103.

Janzen, D.H. (1980). When is it coevolution? *Evolution* **34**, 611–612.

Janzen, D.H. (1983). Dispersal of seeds by vertebrate guts. *In* "Coevolution" (D.J. Futuyma and M. Slatkin, Eds.), pp. 232–262. Sinauer, Sunderland, Mass.

Ji, Q., Luo, Z.X., Yuan, C.X., Wible, J.R., Zhang, J.P., and Georgi, J.A. (2002). The earliest known eutherian mammal. *Nature* **416**, 816–822.

Johns, A.G. (1997). "Timber Production and Biodiversity Conservation in Tropical Rain Forests." Cambridge University Press, Cambridge, UK.

Julien-Laferriere, D. (1999). Foraging strategies and food partitioning in the neotropical frugivorous mammals *Caluromys philander* and *Potos flavus*. *Journal of Zoology* **247**, 71–80.

Kays, R. and Allison, A. (2001). Arboreal tropical forest vertebrates: current knowledge and research trends. *Plant Ecology* **153**, 109–120.

Kays, R.W. (1999). A hoistable arboreal mammal trap. *Wildlife Society Bulletin* **27**, 298-300.

Kinzig, A.P. and Harte, J. (2000). Implications of endemics-area relationships for estimates of species extinctions. *Ecology* **81**, 3305–3311.

Lambert, T.D, Malcolm, J.R., and Zimmerman, B.L. (in prep). Variation in small mammal richness by trap height and trap type at a south-eastern Amazonian site, with notes on a new method for of canopy trapping in the canopy.

Laurance, S.G. and Laurance, W.F. (1999). Tropical wildlife corridors: use of linear rain forest remnants by arboreal mammals. *Biological Conservation* **91**, 231–239.

Laurance, W.F. (1991). Ecological correlates of extinction proneness in Australian tropical rain forest mammals. *Conservation Biology* **5**, 79–89.

Laurance, W.F. and Yensen, E. (1991). Predicting the impacts of edge effects in fragmented habitats. *Biological Conservation* **55**, 77–92.

Lindenmayer, D.B., Cunningham, R.B., Donnelly, C.F., Incoll, R.D., Pope, M.L., Tribolet, C.R., Viggers, K.L., and Welsh, A.H. (2001). How effective is spotlighting for detecting the greater glider (*Petauroides volans*)? *Wildlife Research* **28**, 105–109.

Lindenmayer, D.B., Mccarthy, M.A., Parris, K.M., and Pope, M.L. (2000). Habitat fragmentation, landscape context, and mammalian assemblages in southeastern Australia. *Journal of Mammalogy* **81**, 787–797.

Lovejoy, T.E., Bierregaard, Jr., R.O., Rylands, A.B., Malcolm, J.R., Quintela, C.E., Harper, L.H., Brown, Jr., K.S., Powell, A.H., Powell, G.V.N., Shubart, H.O.R., and Hays, M.B. (1986). Edge and other effects of isolation on Amazon forest fragments. *In* "Conservation Biology: The Science of Scarcity and Diversity" (M.E. Soulé, Ed.), pp. 257–285. Sinauer, Sunderland, Massachusetts.

MacArthur, R.H. and Wilson, E.O. (1967). "The Theory of Island Biogeography." Princeton University Press, Princeton.

Malcolm, J.R. (1990). Mammalian densities in continuous forest north of Manaus, Brazil. *In* "Four Neotropical Rain Forests" (A. Gentry, Ed.) pp. 339–357. Yale University Press, New Haven, Connecticut.

Malcolm, J.R. (1991a). Comparative abundances of Neotropical small mammals by trap height. *Journal of Mammalogy* **72**, 188–192.

Malcolm, J.R. (1991b). The small mammals of Amazonian forest fragments: pattern and process. Unpublished Ph.D. Dissertation, University of Florida, Gainesville, Florida.

Malcolm, J.R. (1994). Edge effects in central Amazonian forest fragments. *Ecology* **75**, 2438–2445.

Malcolm, J. R. (1995). Forest structure and the abundance and diversity of neotropical small mammals. *In* "Forest Canopies" (M.D. Lowman and N.M. Nadkarni, Eds.), pp. 179–197. Academic Press, San Diego.

Malcolm, J.R. (1997). Biomass and diversity of small mammals in Amazonian forest fragments. *In* "Tropical Forest Remnants: Ecology, Management and Conservation of Fragmented Communities" (W.F. Laurance and R.O. Bierregaard, Jr., Eds.), pp. 207–221. University of Chicago Press, Chicago.

Malcolm, J.R. (1998). A model of conductive heat flow in forest edges and fragmented landscapes. *Climatic Change* **39**, 487–502.

Malcolm, J.R. (2001). Extending models of edge effects to diverse landscape configurations. *In* "Lessons from Amazonia: The Ecology and Conservation of a Fragmented Forest" (R.O. Bierregaard Jr., C. Gascon, T.E. Lovejoy, and R. Mesquita, Eds.), pp. 346–357. Yale University Press, New Haven, Connecticut.

Malcolm, J.R., Patton, J.L., and da Silva, M.N.F. (In press). Small mammal communities in upland and floodplain forests along an Amazonian white-water river. *In* "Mammalian Diversification: From Population Genetics to Biogeography" (E. Lacey and M. Myers, Eds.). University of California Press, Berkeley.

Malcolm, J.R. and Ray, J.C. (2000). Influence of timber extraction routes on central African small-mammal communities, forest structure, and tree diversity. *Conservation Biology* **14**, 1623–1638.

McClearn, D. (1992). Locomotion, posture, and feeding-behavior of kinkajous, coatis, and raccoons. *Journal of Mammalogy* **73**, 245–261.

McClearn, D., Kohler, J., Mcgowan, K.J., Cedeno, E., Carbone, L.G., and Miller, D. (1994). Arboreal and terrestrial mammal trapping on Gigante Peninsula, Barro Colorado Nature Monument, Panama. *Biotropica* **26**, 208–213.

McNab, B.K. (1986). The influence of food habits on the energetics of eutherian mammals. *Ecological Monographs* **56**, 1–19.

McNab, B.K. and Eisenberg, J.F. (1989). Brain size and its relation to the rate of metabolism in mammals. *American Naturalist* **133**, 157–167.

Meggs, R.A., Lindenmayer, D.B., Linga, T., and Morris, B.J. (1991). An improved design for trap brackets used for trapping small mammals in trees. *Wildlife Research* **18**, 589–591.

Mittermeier, R.A. and Fleagle, J.G. (1976). The locomotory and postural repetoires of *Ateles geoffroyi* and *Colobus guereza*, a reevaluation of the locomotory category of semibrachiation. *American Journal of Physical Anthropology* **45**, 235–256.

Moffett, M.W. and Lowman, M.D. (1995). Canopy access techniques. *In* "Forest Canopies" (M.D. Lowman and N.M. Nadkarni, Eds.), pp. 3–26. Academic Press, San Diego.

Oates, J.F., Whitesides, G.H., Davies, A.G., Waterman, P.G., Green, S.M., Dasilva, G.J., and Mole, S. (1990). Determinants of variation in tropical forest primate biomass: new evidence from West Africa. *Ecology* **71**, 328–343.

Pardini, R. (2001). Pequenos mamíferos e a fragmentação da Mata Atlântica de Una, Sul da Bahia -Processos e Conservação. Unpublished Ph.D. Dissertation, University of São Paulo, São Paulo, Brazil.

Patton, J.L., Da Silva, M.N.F., and Malcolm, J.R. (2000). Mammals of the Rio Juruá and the evolutionary and ecological diversification of Amazonia. *Bulletin of the American Museum of Natural History* **244**, 1-306.

Peres, C.A. (1990). Effects of hunting on western Amazonian primate communities. *Biological Conservation* **54**, 47–59.

Peres, C.A. (1997a). Primate community structure at twenty western Amazonian flooded and unflooded forests. *Journal of Tropical Ecology* **13**, 381–405.

Peres, C.A. (1997b). Effects of habitat quality and hunting pressure on arboreal folivore density in Neotropical forests: a case study of howler monkeys (*Alouatta* spp.). *Folia Primatologica* **68**, 199–222.

Peres, C.A., and Janson, C.H. (1999). Species coexistence, distribution and environmental determiniants of Neotropical primate richness: a community-level zoogeographic analysis. *In* "Primate Communities" (J.G. Fleagle, C.H. Janson, and K.E. Reed, Eds.), pp. 55–74. Cambridge University Press, Cambridge, UK.

Preuschoft, H., Gunther, M.M., and Christian, A. (1998). Size dependence in prosimian locomotion and its implications for the distribution of body mass. *Folia Primatologica* **69**, 60–81.

Rasmussen, D.T. (1990). Primate origins: lessons from a neotropical marsupial. *American Journal of Primatology* **22**, 263–277.

Ray, J.C. and Sunquist, M.E. (2001). Trophic relations in a community of African rainforest carnivores. *Oecologia* **127**, 395–408.

Redford, K.H. (1992). The empty forest. *Bioscience* **42**, 412–422.

Redford, K.H., da Fonseca, G.A.B., and Lacher Jr., T.E. (1984). The relationship between frugivory and insectivory in primates. *Primates* **25**, 433–440.

Redford, K.H. and Robinson, J.G. (1987). The game of choice: patterns of Indian and colonist hunting in the neotropics. *American Anthropologist* **89**, 650-667.

Robinson, B.W. and Wilson, D.S. (1998). Optimal foraging, specialization, and a solution to Liem's paradox. *American Naturalist* **151**, 223–235.

Robinson, J.G. and Bennett, E.L. (2000). "Hunting for Sustainability in Tropical Forests." Columbia University Press, New York.

Robinson, J.G. and Bodmer, R.E. (1999). Towards wildlife management in tropical forests. *Journal of Wildlife Management* **63**, 1–13.

Robinson, S.K. and Terborgh, J. (1990). Bird communities of the Coaha Cashu biological station in Amazonian Peru. *In* "Four Neotropical Rain Forests" (A.H. Gentry, Ed.), pp. 199–216. Yale University Press, New Haven.

Rosevear, D.R. (1974). "The Carnivores of West Africa." British Museum of Natural History, London.

Sanderson, I.T. (1940). The mammals of the North Cameroons Forest Area being the results of the Percy Sladen Expedition to the Mamfe Division of the British Cameroons. *Transactions of the Zoological Society of London* **24**, 623-725.

Schmitt, D. and Lemelin, P. (2002). Origins of primate locomotion: gait mechanics of the woolly opossum. *American Journal of Physical Anthropology* **118**, 231–238.

Shanahan, M. and Compton, S.G. (2001). Vertical stratification of figs and fig-eaters in a Bornean lowland rain forest: how is the canopy different? *Plant Ecology* **153**, 121–132.

Smith, A.P. and Phillips, K. (1984). A systematic technique for census of sugar gliders and other small arboreal mammals. *Australian Wildlife Research* **11**, 83–87.

Struhsaker, T.T. (1997). "Ecology of an African Forest: Logging in Kibale and the Conflict Between Conservation and Exploitation." University Presses of Florida, Gainesville, Florida.

Szalay, F.S. (1994). "Evolutionary History of the Marsupials and an Analysis of Osteological Characters." Cambridge University Press, Cambridge, UK.

Terborgh, J. (1986.) Community aspects of frugivory in tropical forests. *In* "Frugivores and Seed Dispersal" (A. Estrada and T.H. Fleming, Eds.), pp. 371–384. Dr. W. Junk Publishers, Dordrecht, The Netherlands.

Terborgh, J.W. (1983). "Five New World Primates." Princeton University Press, Princeton, New Jersey.

Terborgh, J.W. and van Schaik, C.P. (1987). Convergence vs. nonconvergence in primate communities. *In* "Organization of Communities, Past and Present" (J.H.R. Gee and P.S. Giller, Eds.), pp. 205–226. Blackwell Science, Oxford.

Vieira, E.M. (1998). A technique for trapping small mammals in the forest canopy. *Mammalia* **62**, 306–310.

Voss, R.S. and Emmons, L.H. (1996). Mammalian diversity in Neotropical lowland rain forests: a preliminary assessment. *Bulletin of the American Museum of Natural History* **230**, 3-115.

Voss, R.S., Lunde, D.P., and Simmons, N.B. (2001). The mammals of Paracou, French Guiana: a neotropical lowland rain forest fauna, Part 2: Nonvolant species. *Bulletin of the American Museum of Natural History* **263**, 3-236.

Webb, E.L. (1997). Canopy removal and residual stand damage during controlled selective logging in lowland swamp forest of northeastern Costa Rica. *Forest Ecology and Management* **95**, 117–129.

Wilkie, D.S. and Carpenter, J.F. (1999). Bushmeat hunting in the Congo basin: an assessment of impacts and options for mitigation. *Biodiversity and Conservation* **8**, 927–955.

Wilkie, D.S., Sidle, J.G., Boundzanga, G.C., Blake, S., and Auzel, P. (2001). Deforestation or defaunation: commercial logging and market hunting in northern Congo. *In* "Wildlife Logging Interactions in Tropical Forests" (R. Fimbel, A. Grajal, and J.G. Robinson, Eds.), Columbia University Press, New York.

Windsor, D.M. (1990). Climate and moisture variability in a tropical forest: long-term records from Barro Colorado Island, Panama. *Smithsonian Contributions to the Earth Sciences* **29**, 1-145.

Woinarski, J.C.Z., Braithwaite, R.W., Menkhorst, K.A., Griffin, S., Fisher, A., and Preece, N. (1992). Gradient analysis of the distribution of mammals in stage III of Kakadu National Park, with a review of the distribution patterns of mammals across north-western Australia. *Wildlife Research* **19**, 233–262.

Woodman, N., Slade, N.A., Timm, R.M., and Schmidt, C.A. (1995). Mammalian community structure in lowland, tropical Peru, as determined by removal trapping. *Zoological Journal of the Linnean Society* **113**, 1–20.

III _____

Ecological Processes
in Forest Canopies

When the first edition of *Forest Canopies* was published, few scientists had attempted to study ecological processes in this out-of-reach habitat. Most studies were focused on measuring forest structure or trying to assess the biological diversity overhead. In the past decade, however, the integration of data on both structure and populations within forest canopies has led to more collaborative work on ecological processes. The dynamics of photosynthesis, nutrient cycling, herbivory, decomposition, reproductive biology, and other forces that drive ecosystems represent exciting new thresholds of discovery. Such processes necessarily take into careful account a range of spatial and temporal variables that make forest canopies challenging:

1. The differential use over time of this space by communities of organisms,
2. The vast heterogeneity of substrate,
3. The variability in ages for components within the canopy (e.g., soil/plant communities, sun/shade cohorts of leaves),
4. The variability in microclimates from top to bottom, including the interface between atmosphere and upper canopy,
5. The high levels of organismal diversity, and
6. The development of replicable sampling protocols for quantifying the processes themselves.

Section III illustrates some of the incredible progress in canopy biology made during the past decade via a plethora of new studies and a growing army of researchers who are attempting to quantify ecological processes in this aboveground realm. In Chapter 17, Jennifer Funk and Manuel Lerdau provide an overview of many aspects of scaling and measuring canopy photosynthesis; as ecologists move forward in studies of treetops processes, long-term responses of terrestrial ecosystems to anthropogenic environmental change at stand, regional, and global levels underscore the importance of carbon balance for forest systems.

In Chapter 18, Bruce Rinker and Meg Lowman travel "downstream" in the energy circuit to examine the effects of canopy herbivory as a key regulatory process on healthy stands, providing historical perspective on the science, a comparison of tropical herbivory studies, and an evaluation of various sampling methodologies.

In Chapter 19, Kitti Reynolds and Mark Hunter address nutrient inputs to the canopy, nutrient cycling within the canopy, and nutrient inputs from the canopy to the forest floor in both temperate and tropical forests, illustrating a complex energy circuit that requires critical, long-term examination. Michelle Zjhra and Beth Kaplin discuss in Chapter 20 the reproductive biology and genetics of tropical forests where high biodiversity is the driving force that frames ecological and evolutionary studies in this habitat; plant-pollinators and seed-dispersers allow for the complex gene flow through the system so that access to flowers and fruits is essential for continued analysis.

In Chapter 21, Steve Fonte and Tim Schowalter travel even further "downstream" in the energy circuit to examine the critical role played by decomposition in the functioning of forest ecosystems; whether on the ground or in the treetops, this process has a significant influence on nutrient dynamics aboveground and belowground. Concluding the section on ecological processes in forest canopies, in Chapter 22, Steve Madigosky details mimicry, aposematic coloration, mutualism, and other types of symbiotic interactions vital to survival strategies.

Faced with a Herculean task, canopy scientists attempt to analyze the complex anatomy (or structure and diversity) and physiology (or function) of forest systems that change constantly over time. Unfortunately, the structure, diversity, and functions of both temperate and tropical systems are affected dramatically by anthropogenic changes. The final section of *Forest Canopies* will address issues of stewardship and conservation of the High Frontier.

CHAPTER 17

Photosynthesis in Forest Canopies

Jennifer L. Funk and Manuel T. Lerdau

*If you can keep your leaves when all about you, Are losing theirs
and blaming it on drought; If you can photosynthesize when others shut their Stomates
and stop transpiring; Yours will be the canopy and all beneath it
And—which is more—you'll be a tree, my seed.*
—*Author's digression from Rudyard Kipling, "If" in Songs from Books, 1912*

Introduction

The need to predict long-term responses of terrestrial ecosystems to anthropogenic environmental change at stand, regional, and global levels underscores the importance of understanding the carbon balance of forest ecosystems. One major element of forest carbon balance is photosynthesis, which has typically been studied at the cell to leaf levels. Extrapolating from small-scale measurements of photosynthesis to estimates of canopy and stand-level photosynthesis is difficult for a variety of reasons: ecological and physiological differences among species; leaf-to-leaf variation within a species through the canopy; influences of canopy architecture on leaf microclimate; and nonlinearities in the response of photosynthesis to resource levels. Efforts to model and measure canopy photosynthesis directly have vastly improved over the last decade, and comparisons of model results with direct measures of canopy CO_2 flux provide insight into the mechanisms by which biological sources of variation combine with environmental variables to determine overall canopy photosynthesis. This chapter presents a general overview of many aspects of scaling and measuring canopy photosynthesis. Comprehensive reviews of photosynthesis models (von Caemmerer 2000), canopy structure and micrometeorology (Jones 1992; Baldocchi 1993; Parker 1995; Schulze and Caldwell 1995; Mulkey et al. 1996a), scaling of physiological processes (Ehleringer and Field 1993; Jarvis 1995), and measurement techniques (Baldocchi et al. 1988; Field et al. 1995) can be found elsewhere.

Biochemical Considerations

Mechanistic Photosynthesis Models

Photosynthesis in plants occurs through C_3, C_4, and crassulacean acid metabolism (CAM) pathways. Among trees, the C_3 pathway predominates with C_4 and CAM occurring rarely (e.g., Lerdau et al. 1992). Common to all pathways is the diffusion of CO_2 from air into chloroplasts through stomata. This diffusion of CO_2 to the sites of carboxylation involves resistance at multiple stages: the leaf boundary layer, stomata, and mesophyll. The enzyme rubisco (ribulose-1,5-bisphosphate carboxylase-oxygenase) catalyzes the carboxylation (and oxygenation) of ribulose-1,5-bisphosphate

335

(RuBP), yielding 2 molecules of 3-phosphoglycerate (PGA) (see Figure 17-1). The rate of carboxylation depends on the concentrations of CO_2 and O_2, the competing substrates for rubisco, as well as the ratio of RuBP concentration to enzyme active sites.

All mechanistic (or process-based) models of leaf C_3 photosynthesis are based on the kinetic properties of rubisco (von Caemmerer 2000). This fundamental similarity at the enzyme level allows the development of carbon exchange models that can be applied across forest types and that are ideal for predicting plant, canopy, and ecosystem responses to changes in climate and other ecological conditions. The most widely used photosynthesis model is that of Farquhar et al. (1980), which has been subsequently modified (von Caemmerer and Farquhar 1981; Kirschbaum and Farquhar 1984; Harley et al. 1992). In this model, net leaf CO_2 assimilation (A) is represented as a function of photosynthesis, photorespiration, and respiration:

$$A = V_c - 0.5V_o - R_d = V_c \left(1 - 0.5 \frac{V_o}{V_c}\right) - R_d, \tag{1}$$

where V_c and V_o are the rates of carboxylation and oxygenation and R_d is day respiration, the rate of CO_2 evolution in the light from processes other than photorespiration (i.e., mitochondrial respiration).

The rates of carboxylation and oxygenation can be expressed in terms of enzyme kinetic parameters such that:

$$\phi = \frac{V_o}{V_c} = \left(\frac{V_{o\,max}}{K_o} \frac{K_c}{V_{c\,max}}\right) \frac{O}{C_i} \tag{2}$$

where $V_{c\,max}$ and $V_{o\,max}$ are maximum carboxylation and oxygenation rates, K_c and K_o are Michaelis-Menten constants for carboxylation and oxygenation, and C_i and O are intercellular concentrations of CO_2 and O_2.

Γ^*, the CO_2 compensation point in the absence of mitochondrial respiration (Laisk and Oja 1998), is calculated by setting A and R_d equal to zero in Equation (1), yielding $\phi = 2$, and using Equation (2) to find the corresponding value of C_i (Farquhar and von Caemmerer 1982).

$$\Gamma^* = 0.5 \frac{V_{o\,max}}{K_c} \frac{K_c}{V_{c\,max}} O, \text{ and} \tag{3}$$

$$\phi = \frac{2\Gamma^*}{C_i}. \tag{4}$$

Substituting Equations (3) and (4) into Equation (2) yields:

$$A = V_c \left(1 - \frac{\Gamma^*}{C_i}\right) - R_d. \tag{5}$$

V_c shows three stages of limitation depending on the availability of RuBP (see Figure 17-2). When rubisco is RuBP-saturated, V_c will be determined by the kinetic properties of the enzyme and the concentrations of CO_2 and O_2. Under RuBP-limiting conditions, V_c is determined by the rate of RuBP regeneration, which will depend either on the supply of ATP and NADPH from electron transport or the availability of inorganic phosphate (P_i) (see Figure 17-1). Limitation of

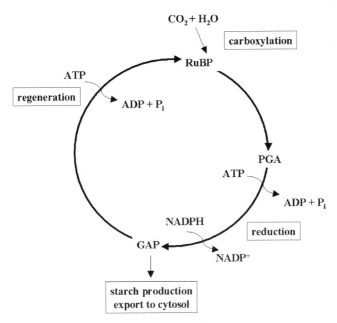

Figure 17-1 The reductive pentose phosphate, or Calvin, cycle involves the carboxylation of ribulose 1,5-bisphosphate (RuBP), the reduction of 3-phosphoglycerate (PGA) to glyceraldehyde 3-phosphate (GAP, or triose phosphate), and the regeneration of RuBP.

carboxylation by P_i has been found to occur under conditions of high CO_2, low temperature, and starch accumulation (reviewed in Sharkey 1985; Harley and Tenhunen 1991).

This yields a general version of the model:

$$A = \min\{W_c, W_j, W_p\} \left(1 - \frac{\Gamma^*}{C_i}\right) - R_d \qquad (6)$$

where W_c, W_j, and W_p are the rates of carboxylation when limited by rubisco, RuBP, and P_i, respectively. For more detail about W_c, W_j, and W_p and their derivations, see von Caemmerer (2000).

Figure 17-2 is an extension of the hypothesis put forth by Blackman that, at a given time, photosynthesis is limited by a single factor. However, under conditions of simultaneously varying light, CO_2 and temperature, it is most likely that A is co-limited by multiple factors (reviewed in Jones 1992).

Values of model parameters for numerous species and functional groups can be found in the literature (Kirschbaum and Farquhar 1984; Wullschleger 1993; reviewed in Evans and Loreto 2000; von Caemmerer 2000). Some parameters can be predicted from empirical relationships with other leaf properties (e.g., V_{cmax} and leaf N; Harley et al. 1992). However, other studies have found substantial and often unexplained variation in parameters within species resulting from differences in light environment (Harley and Baldocchi 1995) and seasonal changes in resources and leaf age (Harley et al. 1992; Wilson et al. 2000a, 2000b, 2001; see Figure 17-3). Small amounts of uncertainty associated with parameter values can lead to large model error. Evans and Loreto (2000) found that small changes in Γ^* and R_d substantially altered the prediction of stomatal conductance in unstressed plants. Model users should be aware of the potential error associated with uncertainty in published values of model parameters as well as their dynamic

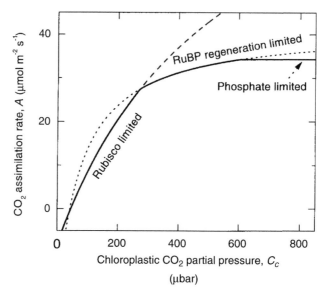

Figure 17-2 Photosynthetic rate as a function of chloroplast CO_2 partial pressure, depicting three phases of potential limitation. The solid curve is the minimum rate of photosynthesis. Copyright © CSIRO. Reproduced by permission of CSIRO Publishing from von Caemmerer, S. (2000). "Biochemical models of leaf photosynthesis. Techniques in Plant Sciences No. 2." CSIRO Publishing, Collingwood, Australia.

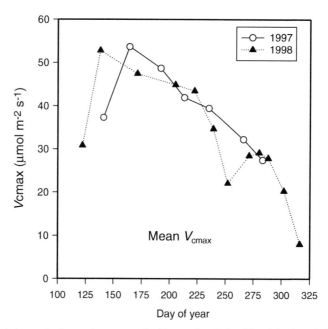

Figure 17-3 Seasonal changes in the maximum rate of rubisco carboxylation (V_{cmax}) for four deciduous species (*Quercus alba, Quercus prinus, Acer rubrum, Acer saccharum*). Symbols denote means from all four species collected within a two-week period. Reprinted from Wilson et al. (2001) with permission from Blackwell Publishing.

nature. This uncertainty increases the confidence intervals surrounding estimates of carbon exchange at the canopy scale and limits attempts to translate these estimates into recommendations in terms of conservation and land-use policy.

Perhaps the greatest challenge facing efforts to model canopy photosynthesis across large spatial scales is accounting for the variation that occurs among co-occurring taxa (e.g., Wilson et al. 2000a). Attempts to model photosynthesis at the stand level typically use parameter values that average across the trees present. An obvious complication of this approach involves the incorporation of realistic stand dynamics (e.g., changes in species composition and abundance due to succession or management practices). When the goal is to model carbon accumulation at the annual scale, such problems tend to be small because the turnover of plants is small at such short time scales. On the other hand, when studies attempt to model carbon balance across decades or centuries, solving the challenge of inter-specific variability becomes paramount because changes in species abundance become the dominant source of variation. This inter-specific variability becomes particularly important when one attempts to predict long-term effects of changes in land use, forest management, and climate.

Stomatal Conductance Models

In order to predict carbon assimilation (A) using Equation (6), an estimate of C_i must be derived. This is usually accomplished by coupling a mechanistic photosynthesis model with a model describing stomatal conductance to water vapor (g_s) (Collatz et al. 1991; Harley et al. 1992). At the time of this writing, there is no generally accepted mechanistic model for stomatal conductance (Farquhar et al. 2001; but see Friend 1991; Mott and Parkhurst 1991; Williams et al. 1996), and the most widely used models of stomatal function are empirical. Despite the lack of biological mechanism in these models, they tend to work remarkably well across environmental conditions (e.g., McMurtrie et al. 1992) and taxa, suggesting a fundamental unity in the underlying control processes. Beginning with work by Wong et al. (1979), the general relationship between g_s and A across light, temperature, humidity, and CO_2 gradients led to the following model, developed by Ball et al. (1987):

$$g_s = kA \frac{h_s}{C_s}, \tag{7}$$

where h_s and C_s are relative humidity and CO_2 concentration at the leaf surface, and k is a constant representing stomatal sensitivity to these factors. C_i is dependent on both g_s and A and is calculated iteratively using Fick's law:

$$C_i = C_a - 1.6 \frac{A}{g_s}, \tag{8}$$

where C_a is the concentration of CO_2 external to the leaf, and the factor 1.6 accounts for the different diffusivities of water vapor and CO_2 in air.

Since the publication of the Ball-Berry model (Equation 7), numerous empirical models of stomatal conductance have been developed (see review in Berry et al. 1997). Berry et al. (1997) tested five empirical models with four large datasets and found that the formulation of Leuning (1995) provided the best overall fit to the data:

$$g_s = g_o + a_1 \frac{A}{(C_s - \Gamma)(1 + D_s/D_o)}, \tag{9}$$

where g_o is a residual stomatal conductance at the light compensation point (or cuticular conductance), Γ is the CO_2 compensation point, D_s is leaf to air vapor pressure deficit, and a_1 and D_o are empirically determined coefficients. Leuning (1995) argues that the use of humidity deficit at the leaf surface rather than relative humidity (used in the Ball-Berry model) is more physiologically appropriate to stomatal response. By replacing C_s with $(C_s\text{-}\Gamma)$, the model of Leuning (1995) also performs better at low values of C_s. However, such physiological realism always carries with it a price in terms of generality or ease of measurement. In this case, relative humidity can be estimated from ambient measures of humidity, such as those available at weather stations. On the other hand, humidity deficit at the leaf surface is a term that can be very difficult to measure or estimate.

Leaf-Level Responses to the Environment

As others have noted, canopies are composed of populations of leaves, and understanding leaf-level processes is the first step to discerning the complexities of whole canopies (Harper 1989). Leaves respond in predictable and consistent ways to major environmental factors, including light, nutrients, temperature, and water availability. By considering leaf-level photosynthetic responses across taxa, one has the basis for developing canopy photosynthesis estimates that can be applied across broad temporal and spatial scales.

Light

Under saturating CO_2 conditions, A increases with irradiance as ATP and NADPH are produced through photochemical reactions (Bjorkman 1981). This increase attenuates as other limitations to carboxylation predominate (e.g., CO_2 diffusion, phosphate limitation), resulting in a nonlinear response of A to light. Differences in light response curves can be found among species, particularly with respect to the light level at which carboxylation saturates (see Figure 17-4). Within an individual species, the shape of the response curve depends on several factors, including leaf light environment and leaf N. In order to maximize carbon assimilation under conditions of low light, Evans (1989) suggested that shade leaves partition more N to thylakoid-related protein, which functions in the electron transport chain (light harvesting portion), than to protein and enzymes associated with carboxylation. However, a more recent study suggested that, by increasing surface area for photon capture, changes in leaf mass per unit area (LMA) have a greater impact on assimilation per unit nitrogen than does N redistribution (Evans and Poorter 2001). Yet others found increased chlorophyll per unit nitrogen with decreasing light, which suggests that shade leaves increase the number of light-harvesting complexes relative to other components of photosynthetic apparatus (Kull and Niinemets 1998). All of these modifications increase assimilation in low light with the cost of lower light-saturated photosynthetic rates relative to sun leaves in high light.

The observation that shade leaves allocate more resources to light capture is consistent with the *optimization* hypothesis (also referred to as functional convergence hypothesis, Field 1991, or acclimation hypothesis, Meir et al. 2002) originally proposed by Mooney and Gulmon (1979): N investments are regulated to maximize CO_2 assimilation. The optimization hypothesis is invoked in canopy photosynthesis models and will be discussed in more detail later in this chapter.

Nitrogen

Of all plant nutrients, nitrogen most often limits carbon exchange and is the most substantially affected by human activities on a global scale (Vitousek et al. 1997). Nitrogen deposition is occurring across temperature Europe, North Asia, and North America, and is becoming important in

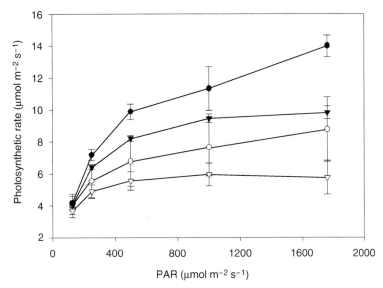

Figure 17-4 The relationship between photosynthesis and light in two tropical tree species: *Dussia munda* (*closed circles*, sun leaves; *open circles*, shade leaves) and *Brosimum utile* (*closed triangles*, sun leaves; *open triangles*, shade leaves). Data from Lerdau and Throop 2000.

South America as well (Bouwman et al. 2002). Fortunately for canopy biologists, there is a well-established relationship between maximum photosynthetic rate, A_{max}, and leaf nitrogen (e.g., Field and Mooney 1986; Reich et al. 1997). The strength of this relationship stems from the essential role of N in photosynthetic proteins and enzymes, which total roughly 75 percent of leaf nitrogen content (Field and Mooney 1986). The A_{max}-N relationship is linear in wild plants, only showing saturation with artificially high nitrogen availability. Schulze et al. (1994) express this relationship empirically as:

$$A_{max} = \alpha_N(N - N_t),\qquad(10)$$

where α_N is a constant, N is leaf nitrogen, and N_t is a threshold below which there is no photosynthesis and is a function of leaf structure and longevity. The A_{max}-N relationship was originally observed to be more robust when A_{max} and N were expressed on a mass basis (Evans 1989; Field and Mooney 1986); however, recent studies have found strong relationships between A_{max} and nitrogen (N_a) on an area basis (e.g., Reich and Walters 1994; Niinemets and Tenhunen 1997). There is a substantial amount of variation in the slope of this relationship across species and functional groups (Reich et al. 1995). For example, crop species tend to have higher photosynthetic rates per unit N than do rainforest and evergreen trees (see Figure 17-5). Thus, the exact nature of the relationship must be determined for specific model applications.

The strong relationship between A_{max} and leaf N documented by the studies above was, in most cases, generated across a diverse range of taxa and across large gradients in leaf N. Therefore, this relationship may not be evident given less subtle gradients in leaf N within a canopy (Wilson et al. 2000a). There is also the possibility that the relationship may deteriorate with leaf age (Sims and Pearcy 1991), plant stress (e.g., Wilson et al. 2000a), and across growth environment. As noted above, A_{max} for a given leaf N can differ between leaves grown in low and high light environments (Evans 1989). In addition, inter-specific differences in the flexibility of N

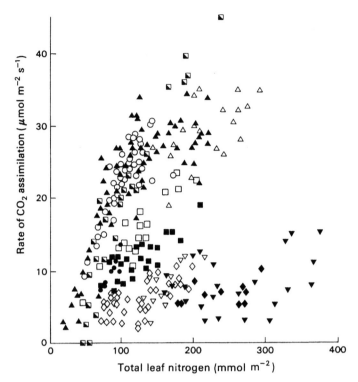

Figure 17-5   Variation in the relationship between the rate of photosynthesis and leaf N content, expressed per unit leaf area. Data represent a wide range of species, including crop species (*closed triangles, open circles*), desert annuals and shrubs (*open triangles*), California trees and shrubs, and rainforest species (*open diamonds, closed diamonds, closed squares*). Reprinted from Evans 1989, copyright Springer-Verlag.

redistribution between light harvesting and carboxylation (Field 1991) may alter the generality of the A_{max}-N relationship in response to changes in light environment and environmental stress.

Temperature

In evaluating the response of terrestrial ecosystems to human activities, one of the most important factors to consider is temperature because of the strong evidence that temperature on a global scale is increasing in response to fossil fuel emissions (reviewed in Walther et al. 2002). Gross photosynthesis increases with temperature, as the CO_2 binding capacity of rubisco increases, until rubisco becomes substrate limited (e.g., electron transport, CO_2 diffusion). Net photosynthesis may not increase, however, because respiration increases with temperature and photorespiration increases as a result of decreases in the specificity factor of rubisco and the relative solubility of CO_2 compared to O_2. At very high temperature (> 40°C, Larcher 1995), photosynthetic system components, including chloroplast membranes and key protein complexes, may become damaged, sometimes irreversibly (Bjorkman 1981). The temperature optimum of photosynthesis varies across species and depends on environmental conditions such as CO_2 concentration and irradiance (von Caemmerer 2000).

Water Availability

Water stress is ubiquitous in forest communities. Even wet forests regularly experience periods of decreased rainfall, and understanding how water stress influences carbon exchange across various time scales is crucial for any effort to estimate canopy photosynthesis. Water stress events over various temporal scales decrease the photosynthetic capacity of leaves. A direct biochemical link between A and water stress has not yet been established. There is controversy over the relative importance of two mechanisms by which water stress causes photosynthetic depression: (1) decreased CO_2 diffusion resulting from stomatal closure and (2) disruption of CO_2 metabolism. A recent experiment by Tezara et al. (1999) concluded that both processes occur but that metabolic impairment via decreased ATP synthesis played a bigger role in depressing A. Some researchers, using gas exchange methods to assess the physiological response of plants to water stress, have observed large C_i values in stressed plants and concluded that stomatal limitation to CO_2 diffusion is not a limiting factor in regulating A (e.g., Schulze 1986). However, water-stressed plants are susceptible to patchy stomatal closure that results in inflated C_i values (Mott and Buckley 2000; but see Wilson et al. 2000b).

One common process in forest species is midday photosynthetic depression. Collatz et al. (1991) suggested two mechanisms for midday photosynthetic depression based on leaf boundary layer conductance (g_b). At high g_b (small boundary layer), humidity at the leaf surface declines, resulting in stomatal closure, decreased C_i and, ultimately, lower A. In contrast, low g_b restricts the exchange of sensible and latent heat between air and leaf, increasing leaf temperature above the optimum. This inhibition of A leads to a decrease in g_s. This process is common in many temperate (Funk et al. 2003) and tropical (Mulkey et al. 1996b) broadleaf forests and may decrease daily photosynthesis by as much as 50 percent.

Canopy Photosynthesis

While the biochemical and leaf-level knowledge outlined above is essential for understanding canopy photosynthesis, one must also consider the entire canopy as a unit, including how the canopy itself influences the leaf environment and how to incorporate these influences in a single comprehensive model.

The Influence of Canopy Structure

Canopy structure influences the transmission of light and thus, indirectly, the temperature profile in which leaves are found. Canopy structure also influences the allocation of resources to leaves at different positions within the canopy, thus altering the photosynthetic potential of these leaves. As discussed below, the impacts of canopies on light and temperature gradients are well understood, but the regulation of resource allocation throughout plant canopies remains to be elucidated.

Light is the single most important environmental parameter driving canopy photosynthesis and varies dramatically throughout the canopy, depending on canopy architecture. The vertical distribution of light can be predicted by a modified version of Beer's law:

$$Q(z) = Q(0)\exp^{-k*LAI(z)}, \tag{11}$$

where $Q(z)$ is the light level at depth z in the canopy, $Q(0)$ is the light level at the top of the canopy, k (dimensionless) is an empirically determined extinction coefficient, and $LAI(z)$ is the cumulative leaf area from the canopy top to depth z. The magnitude of k depends on leaf angle, leaf distribution, and the spectral properties of leaves and typically ranges between 0.3 and 0.8 (Jarvis and Leverenz 1983).

Leaf area index (LAI, m^2 leaf m^{-2} ground) generally increases with precipitation and soil fertility; however, other factors such as species composition, temperature, and wind damage will affect LAI (Gholz 1982). High LAI is commonly found in forests with overlapping generations of leaves (e.g., evergreen) and rarely exceeds 7 in forests dominated by broadleaf species (see Table 17-1, Jarvis and Leverenz 1983). LAI can also show strong seasonal patterns, even in evergreen forests (Jarvis and Leverenz 1983), and will change with stand age (e.g., Parker 1995).

Incoming irradiance is comprised of direct and diffuse components. While diffuse radiation will be distributed to leaves at all angles, direct-beam radiation will be incident to specific leaf angles as solar position changes diurnally and seasonally. Thus, realistic canopy photosynthesis models must account for canopy architecture to estimate radiation absorption by leaves of various orientations. Generally, sun leaves in the upper canopy are vertically oriented (erect), whereas those lower in the canopy are horizontal (planar) (Hollinger 1989; Parker 1995). Orienting leaves away from direct radiation at canopy tops prevents over-excitation of photosynthetic reaction centers and excessive leaf temperature (Bjorkman and Demmig-Adams 1995). Using canopy photosynthesis models, Baldocchi and Harley (1995), Law et al. (2001), and others suggest that the combination of leaf clumping and vertical variation in leaf angle leads to optimal canopy radiation absorption in broadleaf, deciduous forests. Leaf clumping results in canopy gaps that create sunflecks for understory leaves. Modeling the photosynthesis of these leaves is difficult due to the unrepeatable and unpredictable nature of sunflecks (Pearcy and Pfitsch 1995). Both the internal structure and external shape of needles result in different patterns of canopy carbon assimilation and light interception in coniferous forests compared to broad-leaved canopies. Small needle size, clustering of needles on shoots, and deep canopies result in lower intercepted light and more diffuse radiation when compared to a broad-leaved canopy of similar LAI (Stenberg et al. 1995).

The photosynthetic capacity of leaves acclimates to strong light gradients in forest canopies. Variation in LMA with canopy height occurs as plants (1) increase the surface area of shade leaves for maximum light capture and (2) increase the biochemical capacity in sun leaves for maximum assimilation. It is generally thought that the vertical distribution of LMA drives patterns of A and N_a. Indeed, LMA, V_{cmax}, and N_a have all been shown to scale with light as light attenuates with canopy depth (e.g., Ellsworth and Reich 1993; Wilson et al. 2000a; see Figure 17-6). These observations extend the *optimization* hypothesis to the canopy level: whole plant carbon gain will be maximized when N is preferentially allocated to leaves that have the highest light interception (Mooney and Gulmon 1979; Field 1983; Field and Mooney 1986; Hirose and Werger 1987; Hollinger 1989; Field 1991). Following this prediction, N is "optimally" distributed when sun leaves at the canopy top have higher leaf N than shaded leaves at the canopy bottom. Model simulations of canopy photosynthesis with both optimal and uniform N distributions have

Table 17-1 Some Forest Characteristics.

Forest Type	Average Nitrogen Content (mg g^{-1})	Specific Leaf Area (m^2 kg^{-1})	LAI (m^2 m^{-2})
Broad-leaf crop	38.4 ± 1.8	23.6 ± 1.7	3.4
Cereals	33.6 ± 3.2	25.3 ± 1.9	6.1
Deciduous conifer	20.7 ± 1.7	11.3 ± 1.4	1.5
Evergreen conifer	11.0 ± 0.6	4.1 ± 0.4	7.4
Sclerophyllous shrubland	11.4 ± 0.6	6.9 ± 0.7	4.2
Deciduous broad-leaf	19.6 ± 2.7	11.5 ± 2.4	5.6
Evergreen broad-leaf	13.4 ± 1.0	5.7 ± 0.8	6.4
Temperate grassland	25.5 ± 1.6	16.9 ± 1.3	1.8
Tropical rainforest	16.5 ± 1.6	9.9 ± 1.4	6.3
Tundra	20.5 ± 1.1	NA	0.7

Data are means ± 1 standard error. Data compiled by Schulze et al. 1994.

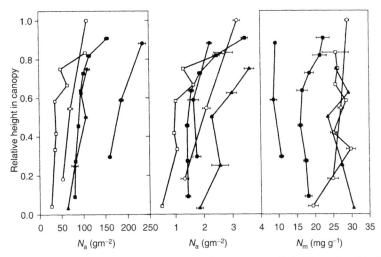

Figure 17-6 Variation with relative canopy height in (a) leaf mass per area (L$_a$), (b) area-based leaf N content (N$_a$), and (c) mass-based leaf N content (N$_m$). The five species examined are birch (*open circles*, n = 31 leaves), oak (*closed squares*, n = 50 leaves), beech (*open squares*, n = 19 leaves), tropical species (*closed triangles*, n = 42 leaves), and spruce (*closed circles*, n = 23 leaves). Leaves were sampled from lower and upper canopy. Error bars are ± 1 SE. Reprinted from Meir et al. 2002, with permission from Blackwell Publishing.

yielded conflicting results. Hirose and Werger (1987) found optimal N distribution to increase photosynthesis by 27 to 89 percent, with larger increases expected in denser forests. However, other studies have found significantly lower increases in canopy photosynthesis (< 10 percent) with optimal N distribution (Field 1983; Leuning et al. 1991; Friend 2001).

Although many studies report non-uniform patterns of N distribution within canopies, the idea of optimal N distribution has been challenged based on discrepancies between modeled optimal distributions and actual distributions found in natural canopies. Numerous studies have found that the actual distribution of N within canopies is more uniform than optimal (Hirose and Werger 1987; Leuning et al. 1991; Hollinger 1996; Anten et al. 1998; Anten and Hirose 2001). Several explanations have been proposed to explain this deviation, including errors associated with the modeling of optimal distribution (e.g., inadequate description of canopy light profile, A$_{max}$-N relationship); failure to account for acclimation of photosynthesis at low light levels; the inseparable effects of leaf age and light on leaf N; the ratio of structural N to total leaf N (Kull and Jarvis 1995); effects of various tree heights (Anten et al. 1998); and increased cost of allocating N to sun leaves, which experience photoinhibition, wind and herbivory (Hollinger 1996). Other concerns regarding the generality of optimal N distribution include inter-specific differences in vertical patterns of N (Kull and Niinemets 1998; Meir et al. 2002) and the complete absence of a relationship between leaf N and light in some forests (Livingston et al. 1998). Finally, Anten and Hirose (2001) argued that it is misguided to expect optimal behavior at the canopy level because evolutionary processes operate at the level of individual plants. In fact, individuals show a higher degree of optimal N distribution than do canopies (Anten et al. 1998).

At the time of this writing, no mechanistic basis for optimal resource distribution within canopies has been described. An alternative, mechanistic explanation of non-uniform N patterns in forest canopies is derived from *coordination* theory proposed by Chen et al. (1993). According to this theory, individual plants redistribute N in order to balance limitations imposed on carboxylation by rubisco (W$_c$) and electron transport (W$_j$). Thus, whereas optimization theory selects a canopy N distribution from all possible profiles to maximize A, ignoring any mechanistic basis, coordination theory redistributes N between leaves in order to balance W_c and W_j based on available N, light, C_i, and leaf temperature. Chen et al. (1993) simulated canopy photosynthesis using uniform, coordinated, and optimal N distributions and found that values generated using coordination theory fall between those generated from uniform and optimal distributions, closely matching actual values found in natural canopies.

Canopy Photosynthesis Models

The *sine qua non* of efforts to understand canopy photosynthesis is the development and testing of models that can be applied to entire canopies and that are broadly useful across a range of taxa and forest types. These models can then be used to estimate canopy photosynthesis within and across landscapes and across large temporal scales. Such models represent the integration of biochemical, leaf-level, and canopy structural effects on carbon exchange and can be used to examine the effects of large-scale environmental changes, such as changes in climate, forest management, or land use.

Several different approaches to modeling canopy photosynthesis (A_{can}) have been developed and are concurrently used at the present time. Mechanistic "bottom-up" models dominate current efforts to predict canopy photosynthesis. While empirical "top-down" models do exist (see Jarvis 1993; Holbrook and Lund 1995), advances in physiology, micrometeorology, and remote sensing have allowed the development of bottom-up models that can be driven by parameters mensurable at large spatial scales. There are many types of mechanistic models, spanning a wide range of detail and complexity. These models typically consist of two submodels: (1) microclimatic and (2) physiological. Note that most of these models, to date, do not yet include population or community dynamics and so are limited in the temporal extent over which they can be used to estimate stand-level photosynthesis.

The general goal of any canopy photosynthesis model is a realistic integration of photosynthesis from all leaves in the canopy, incorporating nonlinear relationships between leaf processes and environmental variables as well as the influence of canopy structure on these processes. Canopy scaling models fall into two broad categories based on their treatment of canopy microclimate: (1) big-leaf and (2) multi-layer models. Big-leaf models treat a forest canopy as a single unit having one set of specified physiological parameters. The simplest form of big-leaf scaling model is as follows:

$$A_{can} = \frac{A_o * f\text{PAR}}{k},$$ \hfill (12)

where A_o is the rate of photosynthesis of a sun leaf, fPAR is the fraction of incident photosynthetically active radiation (PAR, μmol photon m^{-2} s^{-1}) absorbed by the canopy, and k is the light extinction coefficient (Sellers et al. 1992).

Big-leaf models have been evaluated and criticized by many (Sellers et al. 1992; Amthor 1994; Kull and Jarvis 1995; de Pury and Farquhar 1997; Kruijt et al. 1997; Raulier et al. 1999; Friend 2001). In practice, big-leaf models are simple, as they require a limited number of parameters, and typically approximate canopy photosynthesis to within 20 percent (Baldocchi and Amthor 2001). However, big-leaf models neglect important gradients in physiological and microclimatic variables within the canopy. Perhaps the biggest limitation of this approach is the assumption that, across canopy layers, photosynthetic rates will saturate at the same level of incident PAR, an assumption not always supported by data (Kull and Kruijt 1998; see Figure 17-4). De Pury and Farquhar (1997) developed a big-leaf model that accounts for this variation in saturation intensity by creating a sun and shade fraction. These two-layer models are reasonable for open forest stands, but do not accurately represent the complexity of multi-storied canopies (Baldocchi et al. 2002).

Several multi-layer models have been developed (McMurtrie et al. 1992; Baldocchi and Harley 1995; Leuning et al. 1995; Williams et al. 1996; Baldocchi and Meyers 1998; Larocque 2002). In contrast to big-leaf models, multi-layer models integrate leaf-level photosynthesis models over multiple canopy layers and often separate sunlit and shaded fractions within layers. This type of approach requires a detailed knowledge of leaf distribution, canopy architecture, and radiation profiles. Typically, canopies are divided into 20 to 30 layers with an LAI of 0.5 or less per layer (Baldocchi et al. 2002, but see Williams et al. 1996). Detailed characterization of A in canopy layers can be simplified by invoking the optimization hypothesis. For example, assuming that the vertical profiles of V_{cmax} and N track that of light distribution, these parameters can be estimated at any height in the canopy from measured radiation profiles. Multi-layer models can incorporate varying levels of detail, including leaf angle distribution and leaf clumping (Baldocchi and Harley 1995), leaf energy budgets to calculate leaf temperature, the effect of wind on leaf boundary layer, and sunflecks. Baldocchi (1993) provides an excellent review of various methods used to describe micrometeorological conditions within and above the forest canopy (e.g., material and radiative transfer), as these topics are beyond the scope of this chapter.

The physiological submodel derives A using algorithms based either on biological mechanisms or on empirically observed relationships. Biochemical-based submodels, such as the coupled photosynthesis–stomatal conductance model described above (e.g., Collatz et al. 1991), require the measurement of numerous physiological parameters but provide flexibility when these parameters are expected to change through time and space. Generally, mechanistic submodels parameterized with leaf-level data are restricted to multi-layer models. Parameters for mechanistic models are commonly generated from measurements made with leaf and branch enclosures. Current enclosure systems are portable and allow the user to control a number of environmental variables, facilitating the calculation of quantum use efficiency (A vs. light) and V_{cmax} (A vs. C_i). Limitations to this approach include disturbance to a leaf or plant by the enclosure, the artificial environment of the enclosure, and poor spatial and temporal coverage of whole canopies. Big-leaf models often build nonlinearities of physiological responses into model parameters, allowing the input of course-scale, canopy-level data (e.g., parameters determined by remote sensing techniques; see following section).

Alternatively, A can be derived from empirical relationships. An example of this approach includes work by Kull and Jarvis (1995) and Aber et al. (1996), who acquired A_{max} from the dependence of photosynthesis on leaf tissue nitrogen. However, as discussed above, this relationship is subject to spatial and temporal variation and may be difficult, or impossible, to extrapolate across species. For example, canopy photosynthesis will be overestimated if one uses A/N ratios generated for sun leaves across the entire canopy. Similarly, species-rich forests are likely to have component taxa with very different physiological capacities, and deriving mean values for leaf characteristics can be a difficult and error-filled task. A has also been obtained empirically from sap-flow rates using the relationship between A and stomatal conductance (e.g., Catovsky et al. 2002).

Model complexity need only address the goal of a particular application. For example, big-leaf models may be preferred for long-term, large spatial scale predictions when including detailed parameters is unrealistic (Amthor 1994). Detailed multi-layer models are best suited for predictions on short time scales and in response to environmental perturbation (Baldocchi and Harley 1995). Sensitivity analysis can be used to determine which features of a model may be removed without creating serious uncertainty (Jarvis 1993).

Sensitivity analysis can also be used to investigate the sources of variation in canopy photosynthesis estimates. Baldocchi et al. (2002) conducted a sensitivity analysis for various physiological, structural, and environmental parameters using the biophysical model CANOAK. While not a comprehensive parameter analysis, they found that uncertainty in leaf angle and V_{cmax} resulted in large changes in net ecosystem exchange (>20 percent, see Table 17-2, Figure 17-7a), while LAI had a small effect on system CO_2 exchange. Baldocchi et al. (2002) argue that CO_2 fluxes approach a saturating value once the canopy closes (generally LAI > 3) and that A_{can} will only be affected by changes in LAI when the canopy is not closed (see Figure 17-7b). Other studies support the idea that LAI is not a dominant parameter in photosynthesis models for closed canopies (e.g., Larocque 2002), which contradicts assumptions of older models (e.g., Running and Coughlin 1988). Of course, it also possible that these sensitivity analyses reveal more about model structure than they do about underlying biology.

LAI is likely to be an important driver of A_{can} when different canopy types are considered. When Baldocchi and Meyers (1998) applied a multi-layer model (CANVEG) to five different canopies (wheat, potato, soybean, broad-leaf temperate forest, and boreal conifer forest) ranging in LAI from 2 to 5 with V_{cmax} between 40 and 80 $\mu mol\ m^{-2}\ s^{-1}$, they found that LAI and V_{cmax} had strong influences on A_{can} across the different canopies. Most efforts to evaluate model sensitivity have focused on climatic variability and the influence of structural parameters. However, as models are used to derive estimates of longer-term impacts such as climate and land-use change, the sensitivity analyses will have to include explicit consideration of genetic effects both within and across species.

Measurements at the Canopy Scale

Canopy scale measurements of CO_2 flux serve two main roles. First, and most often, they provide direct information on the magnitude of fluxes at any one particular site. Second, and just as important, they are useful for parameterizing and testing the above-described models. As such, canopy measurements are essential for flux estimations across landscape and continental scales. Two types of measurement techniques are frequently used: direct measurements of matter and energy combined with micrometeorological algorithms; and remote sensing approaches that document patterns of spectral reflectance.

Micrometeorological methods are commonly used to measure canopy photosynthesis. The eddy covariance method measures canopy fluxes by the simultaneous determination of vertical wind speed and gas concentration at a fixed height above the canopy (Baldocchi et al. 1988) and can be represented as the time-averaged vertical flux of CO_2 (F_c):

Table 17-2   Sensitivity of Net Ecosystem Exchange to Various Model Parameters Using the Multilayer Canopy Photosynthesis Model CANOAK

Parameter	Values Tested	Max Change in NEE (%)
Leaf dimension (m)	0.001-0.1	1.9
V_{cmax} ($\mu mol\ m^{-2}\ s^{-1}$)	50-73	21.3
Varying V_{cmax} with height	Constant or Variable	0.2
LAI (m^2 leaf m^{-2} ground)	4-6	6.4
Leaf area profile	Skewed or Uniform	1.5
Leaf angle	Planar (10°) or Erect (80°)	80.1
k, stomatal coefficient	8-12	16.2
Reduced direct radiation	-20%	13.5

Data from Baldocchi et al. 2002.

$$F_c = \rho_c \, w'C', \tag{13}$$

where ρ_c is the density of CO_2, w' is the instantaneous deviation of vertical wind speed from mean wind speed, and C' is the instantaneous deviation from the mean CO_2 concentration measured at a reference height.

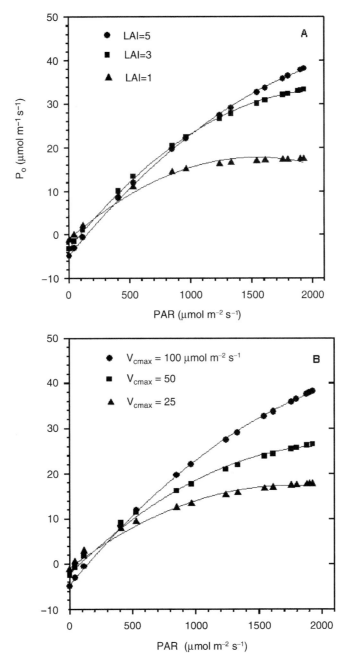

Figure 17-7 The relationship between photosynthesis and light with varying (a) V_{cmax} (constant LAI of 5) and (b) LAI (constant V_{cmax} of 100 μmol m^{-2} s^{-1}). Reprinted from Baldocchi and Amthor 2001, with permission from Elsevier.

This technique does not disturb canopy dynamics and allows continuous measurement over large areas (100 to 2000 m) (Baldocchi and Amthor 2001). Recent collaborations among research groups working at numerous eddy covariance measurement sites have generated large datasets spanning a broad range of biomes, climates, and time scales (e.g., FLUXNET, Baldocchi et al. 2001). Weaknesses of this approach include the restriction of sampling flat terrain during steady atmospheric conditions, the differential effect of disturbance on system components (e.g., soil versus plants), and a lack of mechanistic data regarding the contribution of system elements to total flux. In particular, one must recall that such above-canopy techniques measure the *net flux* of CO_2, and this *net flux* results from the sum of net canopy photosynthesis, bole respiration, root respiration, and soil respiration. Deriving A_{can} requires equations describing the relationships among these terms. Any uncertainty in these equations or in the parameters that are used in them will lead to uncertainty in the estimation of A_{can}. Specifically, total system respiration must be partitioned into autotrophic (R_A) and heterotrophic (R_H) components. This usually involves an estimate of R_A during nighttime respiration. However, the rate of dark respiration in plants differs between light and dark periods and the magnitude of this difference is currently under debate (Graham 1980; Brooks and Farquhar 1985; Atkin et al. 1998). Thus, estimates of canopy photosynthesis from micro-meteorological methods and models are not exactly comparable. Whereas models that scale fluxes from leaves to whole canopies will yield estimates of net primary production (NPP), system CO_2 flux (including fluxes from soil and plants) yields net ecosystem production (NEP). Net ecosystem exchange (NEE) is comparable to NEP but differs in the way it treats atmospheric storage of CO_2 (Ruimy et al. 1995).

As noted above, direct measurements of canopy assimilation are used to evaluate the performance of canopy photosynthesis models and to constrain model estimates (Baldocchi and Harley 1995; Aber et al. 1996; Williams et al. 1996; Baldocchi and Meyers 1998; Baldocchi and Wilson 2001; Wilson et al. 2001). Using this approach, the multi-layer canopy photosynthesis models CANOAK (Baldocchi and Harley 1995; Baldocchi 1997; Baldocchi and Wilson 2001) and CAN-VEG (Baldocchi and Meyers 1998) were found to capture a large percentage of variance in canopy CO_2 flux. Comparisons between models and eddy covariance data can also direct model development and, ultimately, address the mechanistic basis of canopy photosynthesis. For example, as discussed above, it is unclear whether water stress decreases photosynthesis through reductions in g_s or direct impairment of CO_2 metabolism. By manipulating a coefficient in the empirical stomatal conductance submodel of the multi-layer CANOAK model, Baldocchi (1997) successfully fit model results to data from water-stressed oaks. If alterations in CO_2 metabolism were the dominant cause of reduced photosynthesis, the adjustment of stomatal conductance alone would not have improved the fit between model results and observed data. Comparisons between models and eddy flux data can also highlight model limitations. In another case, the CANVEG model required tuning for old forests, which experienced reduced A and g_s possibly as a result of hydraulic limitation (Baldocchi and Meyers 1998). Also, the failure of CANOAK to predict diurnal patterns of carbon assimilation during drought suggests that the empirical model of stomatal conductance does not adequately describe stomatal function during drought periods (Wilson et al. 2001).

Remote sensing techniques are used to create parameters for canopy photosynthesis models by measuring the contrast between the reflectance of red and near infrared wavelengths from a forest canopy, generating a vegetation index. One common index is the normalized vegetation index (NDVI):

$$NDVI = \frac{R_{IR} - R_R}{R_{IR} + R_R}, \tag{14}$$

where R_{IR} is the reflectance in the near infrared and R_R is the reflectance in the red. Empirical relationships are derived between indices and vegetation parameters such as *f*PAR (Sellers et al.

1992), LAI, and total canopy N (reviewed in Field et al. 1995; Parker 1995), which can be used to determine A_{can}. These parameters are often used in top-down models, such as those predicting annual forest and agricultural production from intercepted radiation (e.g., Monteith 1977). In another case, Waring et al. (1995) couple remotely sensed values of PAR and *f*PAR with cuvette-derived, species-specific measures of maximum quantum use efficiency to determine canopy CO_2 assimilation. In addition, reflectance data can estimate levels of chlorophyll and other pigments, which are valuable indicators of environmental stress.

One major limitation of the remote sensing approach is the insensitivity of NDVI to changes in canopy photosynthesis at small temporal and spatial scales. For example, while remote sensing techniques are likely to identify seasonal changes in LAI, small changes in conductance or V_{cmax}, which can greatly alter model output, are likely to remain undetected (Wilson et al. 2001), and these changes may occur in response to brief but severe temperature or water stress events. Like micro-meteorological techniques, remote sensing signals are a complex integration of physical, biochemical, and ecological processes and are not directly linked to biological mechanisms. Remote sensing measurements remain, however, a useful way to estimate fluxes across the spatial scales necessary for conservation and policy efforts.

Flux Patterns Across Scales

There is enough information to begin to make some tentative generalizations regarding which parameters are most important in regulating canopy photosynthesis across broad spatial and temporal scales. Eddy covariance data corroborates many of the flux patterns observed at the leaf level. Higher canopy CO_2 assimilation can occur on cloudy days due to increased diffuse radiation (e.g., Baldocchi 1997; Law et al. 2002), although productivity declines in response to prolonged periods of cloudiness (Goulden et al. 1996). In ecosystems where moisture is supra-abundant (e.g., tropical rainforests), decreases in precipitation are correlated with increases in productivity because of the positive relationship between precipitation and cloudiness and the inverse relationship between cloudiness and light availability. Midday depressions in CO_2 uptake are witnessed at the canopy level (Williams et al. 1996; Baldocchi 1997; Jarvis et al. 1997; Baldocchi and Meyers 1998); however, it is not clear whether this pattern is a result of depressed A or some other process (e.g., increased soil respiration with elevated midday temperature). Also, soil moisture deficits result in depressed A_{can}, although the magnitude of the reduction can vary between 10 and over 50 percent (Goulden et al. 1996; Baldocchi 1997).

Studies have also found a significant effect of length of growing season on annual estimates of canopy assimilation. In particular, Baldocchi and Wilson (2001) found system assimilation to increase by 6 g C m^{-2} day^{-1} in a temperate broad-leaf forest. Jackson et al. (2001) predict that a 5- to 10-day lengthening of the growing season associated with global climate change could increase global NPP by as much as one-third. In addition to changes in growing season length, global increases in atmospheric CO_2 concentration and increased temperature in some ecosystems are expected in the future. While the degree to which elevated CO_2 alters NPP is currently under debate (Ward and Strain 1999), increases in CO_2 flux from forests (e.g., plant and soil respiration) due to increased temperature in some systems may negate any increases in NPP (Canadell et al. 2000).

Taxonomic differences in photosynthesis remain the largest source of variation that is not currently captured well in modeling efforts. Deciduous broad-leaf trees (12.5 ± 4.8, μmol m^{-2} s^{-1} ± 1 SD) tend to have higher leaf level A_{max} than evergreen broad-leaf (6.5 ± 2.2), deciduous needle-leaf (6.7 ± 3.0), or evergreen needle-leaf (7.0 ± 3.1) trees (Reich et al. 1998). However, taxonomic variation within these functional categories is large (e.g., Bassow and Bazzaz 1998; Lerdau and Throop 1999; Catovsky et al. 2002). Canopy CO_2 flux data collected by eddy covariance techniques support these leaf-level patterns (see Table 17-3),

Table 17-3 Seasonal Maximum and Annual Estimates of Gross Primary Production from Multiple Forest Types Collected Using Eddy Covariance Techniques

Forest Type	Number of Forests Measured	Seasonal Max GPP (mmol CO_2 m^{-2} s^{-1})	Annual GPP (g C m^{-2} yr^{-1})
Temperate coniferous	7	21.4 ± 3.8	1487 ± 339
High altitude coniferous	1	15.1	831
Boreal coniferous	5	18.7 ± 8.1	1046 ± 441
Temperate deciduous	5	25.3 ± 3.0	1327 ± 161
Cold temperate deciduous	2	24.0 ± 6.2	1034 ± 185
Mediterranean evergreen	5	21.7 ± 3.6	1583 ± 171
Rainforest	1	29.5	3249
Grassland	2	26.8 ± 17.5	1129 ± 829
Crop	4	38.3 ± 15.1	1142 ± 396

Data from Falge et al. 2002.

although there is much variation between forests within a given forest type. Large differences in the seasonal maximum GPP between forest types are evident in the data presented in Table 17-3. However, when corrected for variation in duration of the growing season between evergreen and deciduous forests, annual GPP is similar among most ecosystems.

A fundamental limitation on using micrometeorological methods in estimating forest NEE involves the vast difference in time scales between those periods when rates of carbon accumulation are positive (i.e., canopy photosynthesis is greater than the sum of all respiratory losses) and periods when ecosystems are net sources of carbon to the atmosphere. Periods of net accumulation tend to last a long time and to dominate the time span of any particular study, and the rate of accumulation is typically very small. In contrast, carbon loss periods occur as a result of disturbance events that typically last a very short period of time, but the loss rates can be high (e.g., a hurricane or fire causes a large increase in respiration over a short time scale; Keller et al. 1996). With micrometeorological towers sampling only a fraction of the forest area, it is extremely unlikely that a tower study of even five years would capture the variability necessary to estimate NEE. Using towers to estimate NEE accurately mandates the use of a spatially explicit disturbance model that can be run for the appropriate time interval.

Summary and Future Directions

As ecologists move forward in their study of canopy photosynthesis and in their efforts to communicate results to conservationists and policy-makers, several challenges remain. Complex canopy structure, differences in leaf and species composition, and nonlinear physiological response to environmental variables complicate efforts to quantify rates of canopy photosynthesis. Despite the complex nature of forest canopies, improved models and direct measurements of system CO_2 flux can now successfully estimate canopy photosynthesis. The incorporation of mechanistic, biochemical photosynthesis submodels into canopy photosynthesis models greatly improves the ability to predict canopy dynamics in response to changing environmental conditions. This effort may be facilitated by the development of a mechanistic model of stomatal conductance. Currently, the effect of parameter uncertainty on bottom-up models has not been thoroughly explored. The relatively large sensitivity of the mechanistic, multi-layer model CANOAK (Baldocchi et al. 2002) to V_{cmax} underscores the importance of accurate model parameterization.

One major obstacle facing efforts to simplify canopy models is the issue of resource distribution within canopies. The optimization of leaf and resource distribution within a canopy to maximize plant carbon gain is a major assumption in many canopy photosynthesis models. This approach decreases the dependence on leaf-level measurements at multiple heights in a canopy, allowing the use of non-intrusive remote sensing techniques. However, while some data sets support the existence of optimal leaf and resource distribution in canopies, this is not true of all systems. The failure of natural canopies to conform to an optimal N distribution and the decoupling of individual plant and canopy evolutionary pressures may warrant the complete removal of the optimal canopy approach from canopy photosynthesis models. Also, the potential breakdown of empirical relationships with stand age and environmental stress is a major concern. For example, Wilson et al. (2000a) found that water stress reduced V_{cmax} with little or no change in N_a, invalidating the commonly employed relationship.

Our ability to model canopy photosynthesis will be instrumental in predicting the effects of global climate change and anthropogenic disturbance on terrestrial ecosystems and global carbon balance. Understanding the influence of elevated atmospheric CO_2 concentration, increased nitrogen availability, atmospheric pollution (e.g., ozone), and increased temperature on physiological characteristics (e.g., V_{cmax}, leaf N content), stand biomass, and canopy architecture is essential in achieving this goal. In addition, species composition in forests can be expected to change in response to shifts in temperature and precipitation, disturbance from development and agriculture, and the inadvertent introduction of exotic species. In relatively undisturbed systems, changes in species composition can alter rates of canopy photosynthesis independently of changes in canopy size, structure, and disturbance. Bassow and Bazzaz (1998) found that A_{max} from maple and oak growing next to each other at Harvard Forest differed by 100 percent. Thus, understanding taxonomic differences in physiology should be included in canopy photosynthesis models. Recent studies on species-specific growth trajectories over decadal scales highlight the potential for species effects to dominate patterns of carbon flux on these longer time scales (Clark and Clark 1999).

Acknowledgments

We thank Peter Harley for valuable comments on the manuscript and Missy Holbrook and Chris Lund for the opportunity to revise this work. Jennifer L. Funk was supported by NSF and EPA pre-doctoral fellowships. Manuel T. Lerdau was supported by grants from NSF and the EPA and by a Hrdy Fellowship from Harvard University.

References

Aber, J.D., Reich, P.B., and Goulden, M.L. (1996). Extrapolating leaf CO_2 exchange to the canopy: a generalized model of forest photosynthesis compared with measurements by eddy correlation. *Oecologia* **106**, 257–265.

Amthor, J. S. (1994). Scaling CO_2-photosynthesis relationships from the leaf to the canopy. *Photosynthesis Research* **39**, 321–350.

Anten, N.P.R. and Hirose, T. (2001). Limitations on photosynthesis of competing individuals in stands and the consequences for canopy structure. *Oecologia* **129**, 186–196.

Anten, N.P.R., Miyazawa, K., Hikosaka, K., Nagashima, H., and Hirose, T. (1998). Leaf nitrogen distribution in relation to leaf age and photon flux density in dominant and subordinate plants in dense stands of a dicotyledonous herb. *Oecologia* **113**, 314–324.

Atkin, O.K., Evans, J. R., and Siebke, K. (1998). Relationship between the inhibition of leaf respiration by light and enhancement of leaf dark respiration following light treatment. *Australian Journal of Plant Physiology* **25**, 437–443.

Baldocchi, D.D. (1993). Scaling water vapor and carbon dioxide exchange from leaves to a canopy: rules and tools. *In* "Scaling Physiological Processes" (J.R. Ehleringer and C.B. Field Eds.), pp. 77–114. Academic Press, San Diego.

Baldocchi, D.D. (1997). Measuring and modelling carbon dioxide and water vapour exchange over a temperate broad-leaved forest during the 1995 summer drought. *Plant, Cell and Environment* **20**, 1108–1122.

Baldocchi, D.D. and Amthor, J.S. (2001). Canopy photosynthesis: history, measurements, and models. *In* "Terrestrial Global Productivity" (J. Roy, B. Saugier, and H.A. Mooney, Eds.), pp. 9–31. Academic Press, San Diego.

Baldocchi, D.D. and Harley, P.C. (1995). Scaling carbon dioxide and water vapour exchange from leaf to canopy in a deciduous forest. II. Model testing and application. *Plant, Cell and Environment* **18**, 1157–1173.

Baldocchi, D.D. and Meyers, T. (1998). On using eco-physiological, micrometeorological and biogeochemical theory to evaluate carbon dioxide, water vapor and trace gas fluxes over vegetation: a perspective. *Agricultural and Forest Meteorology* **90**, 1–25.

Baldocchi, D.D. and Wilson, K.B. (2001). Modeling CO_2 and water vapor exchange of a temperate broadleaved forest across hourly to decadal time scales. *Ecological Modelling* **142**, 155–184.

Baldocchi, D.D., Hicks, B.B., and Meyers, T.P. (1988). Measuring biosphere-atmosphere exchanges of biologically related gases with micrometeorological methods. *Ecology* **69**, 1331–1340.

Baldocchi, D.D., Falge, E., Gu, L., Olson, R., Hollinger, D., Running, S., Anthoni, P., Bernhofer, C., Davis, K., Evans, R., et al. (2001). FLUXNET: a new tool to study the temporal and spatial variability of ecosystem-scale carbon dioxide, water vapor, and energy flux densities. *Bulletin of the American Meteorological Society* **82**, 2415–2434.

Baldocchi, D.D., Wilson, K.B., and Gu, L. (2002). How the environment, canopy structure and canopy physiological functioning influence carbon, water and energy fluxes of a temperate broad-leaved deciduous forest -an assessment with the biophysical model CANOAK. *Tree Physiology* **22**, 1065–1077.

Ball, J.T., Woodrow, I.E., and Berry, J.A. (1987). A model predicting stomatal conductance and its contribution to the control of photosynthesis under different environmental conditions. *In* "Progress in Photosynthetic Research, Vol. IV" (Biggens, J., Eds.), pp. 221–224. Martinus Nijhoff Publishers, The Netherlands.

Bassow, S.L. and Bazzaz, F.A. (1998). How environmental conditions affect canopy leaf-level photosynthesis in four deciduous tree species. *Ecology* **79**, 2660–2675.

Berry, J.A., Collatz, G.J., Denning, A.S., Colello, G.D., Fu, W., Grivet, C., Randall, D.A., and Sellers, P.J. (1997). SiB2, a model for simulation of biological processes within a climate model. *In* "Scaling-Up from Cell to Landscape" (P.R. van Gardingen, G.M. Foody, and P.J. Curran, Eds.), pp. 347–369. Cambridge University Press, Cambridge, UK.

Bjorkman, O. (1981). Responses to different quantum flux densities. *In* "Encyclopedia of Plant Physiology" (O.L. Lange, P.S. Nobel, C.B. Osmond, and H. Ziegler, Eds.), pp. 57–107. Springer-Verlag, Heidelberg, Germany.

Bjorkman, O. and Demmig-Adams, B. (1995). Regulation of photosynthetic light energy capture, conversion, and dissipation in leaves of higher plants. *In* "Ecophysiology of Photosynthesis" (E.-D. Schulze and M.M. Caldwell, Eds.), pp. 17–47. Springer-Verlag, Heidelberg, Germany.

Bouwman, A.F., Van Vuuren, D.P., Derwent, R.G., and Posch, M. (2002). A global analysis of acidification and eutrophication of terrestrial ecosystems. *Water, Air, and Soil Pollution* **141**, 349–382.

Brooks, A. and Farquhar, G.D. (1985). Effect of temperature on the CO_2/O_2 specificity of ribulose-1,5-bisphosphate carboxylase/oxygenase and the rate of respiration in the light. *Planta* **165**, 397–406.

Canadell, J.G., Mooney, H.A., Baldocchi, D.D., Berry, J.A., Ehleringer, J.R., Field, C.B., Gower, S.T., Hollinger, D.Y., Hunt, J.E., Jackson, R.B., et al. (2000). Carbon metabolism of the terrestrial biosphere: a multitechnique approach for improved understanding. *Ecosystems* **3**, 115–130.

Catovsky, S., Holbrook, N.M., and Bazzaz, F.A. (2002). Coupling whole-tree transpiration and canopy photosynthesis in coniferous and broad-leaved tree species. *Canadian Journal of Forest Research* **32**, 295–309.

Chen, J.-L., Reynolds, J.F., Harley, P.C., and Tenhunen, J.D. (1993). Coordination theory of leaf nitrogen distribution in a canopy. *Oecologia* **93**, 63–69.

Clark, D. A., and Clark, D. B. (1999) Assessing the growth of tropical rain forest trees: issues for forest modeling and management. *Ecological Applications* **9**, 981–997.

Collatz, G.J., Ball, J.T., Grivet, C., and Berry, J.A. (1991). Physiological and environmental regulation of stomatal conductance, photosynthesis and transpiration: a model that includes a laminar boundary layer. *Agricultural and Forest Meteorology* **54**, 107–136.

de Pury, D.G.G. and Farquhar, G.D. (1997). Simple scaling of photosynthesis from leaves to canopies without the errors of big-leaf models. *Plant, Cell and Environment* **20**, 537–557.

Ehleringer, J.R. and Field, C.B. (1993). "Scaling Physiological Processes." Academic Press, San Diego.

Ellsworth, D.S. and Reich, P.B. (1993). Canopy structure and vertical patterns of photosynthesis and related leaf traits in a deciduous forests. *Oecologia* **96**, 169–178.

Evans, J.R. (1989). Photosynthesis and nitrogen relationships in leaves of C_3 plants. *Oecologia* **78**, 9–19.

Evans, J.R. and Loreto, F. (2000). Acquisition and diffusion of CO_2 in higher plant leaves. *In* "Photosynthesis: Physiology and Metabolism" (R.C. Leegood, T.D. Sharkey, and S. von Caemmerer, Eds.), pp. 321–351. Kluwer Academic Publishers, The Netherlands.

Evans, J.R. and Poorter, H. (2001). Photosynthetic acclimation of plants to growth irradiance: the relative importance of specific leaf area and nitrogen partitioning in maximizing carbon gain. *Plant, Cell and Environment* **24**, 755–767.

Falge, E., Baldocchi, D., Tenhunen, J., Aubinet, M., Bakwin, P., Berbigier, P., Bernhofer, C., Burba, G., Clement, R., Davis, K.J., et al. (2002). Seasonality of ecosystem respiration and gross primary production as derived from fluxnet measurements. *Agricultural and Forest Meteorology* **113**, 53–74.

Farquhar, G.D. and von Caemmerer, S. (1982). Modelling of photosynthetic responses to environmental conditions. *In* "Encyclopedia of Plant Physiology" (O.L. Lange, P.S. Nobel, C.B. Osmond, and H. Ziegler, Eds.), pp. 550–587. Springer-Verlag, Heidelberg, Germany.

Farquhar, G.D., von Caemmerer, S., and Berry, J.A. (1980). A biochemical model of photosynthetic CO_2 assimilation in leaves of C_3 species. *Planta* **149**, 78–90.

Farquhar, G.D., von Caemmerer, S., and Berry, J.A. (2001). Models of photosynthesis. *Plant Physiology* **125**, 42–45.

Field, C. (1983). Allocating leaf nitrogen for the maximization of carbon gain: leaf age as a control on the allocation program. *Oecologia* **56**, 341–347.

Field, C.B. (1991). Ecological scaling of carbon gain to stress and resource availability. *In* "Response of Plants to Multiple Stresses" (H.A. Mooney, W.E. Winner, and E.J. Pell, Eds.), pp. 35–65. Academic Press, San Diego.

Field, C.B. and Mooney, H.A. (1986). The photosynthesis-nitrogen relationship in wild plants. *In* "On the Economy of Plant Form and Function" (T.J. Givnish, Ed.), pp. 25–55. Cambridge University Press, Cambridge, UK.

Field, C.B., Gamon, J.A., and Penuelas, J. (1995). Remote sensing of terrestrial photosynthesis. *In* "Ecophysiology of Photosynthesis" (E.-D. Schulze and M.M. Caldwell, Eds.), pp. 511–527. Springer-Verlag, Heidelberg, Germany.

Friend, A.D. (1991). Use of a model of photosynthesis and leaf microenvironment to predict optimal stomatal conductance and leaf nitrogen partitioning. *Plant, Cell and Environment* **14**, 895–905.

Friend, A.D. (2001). Modeling canopy CO_2 fluxes: are 'big-leaf' simplifications justified? *Global Ecology and Biogeography* **10**, 603–619.

Funk, J.L., Jones, C.G., Baker, C.J., Fuller, H.M., Giardina, C.P., and Lerdau, M.T. (2003). Diurnal variation in the basal emission rate of isoprene. *Ecological Applications* **13**, 169–178.

Gholz, H.L. (1982). Environmental limits on aboveground net primary production, leaf area, and biomass in vegetation zones of the Pacific Northwest. *Ecology* **63**, 469–481.

Goulden, M.L., Munger, J.W., Fan, S.-M., Daube, B.C., and Wofsy, S.C. (1996). Exchange of carbon dioxide by a deciduous forest: response to interannual climate variability. *Science* **271**, 1576–1578.

Graham, D. (1980). Effects of light on "dark" respiration. *In* "The Biochemistry of Plants, Vol. 2" (D.D. Davies, Ed.), pp. 525–579. Academic Press, New York.

Harley, P.C. and Baldocchi, D.D. (1995). Scaling carbon dioxide and water vapour exchange from leaf to canopy in a deciduous forest. I. Leaf model parameterization. *Plant, Cell and Environment* **18**, 1146–1156.

Harley, P.C. and Tenhunen, J.D. (1991). Modeling the photosynthetic response of C3 leaves to environmental factors. *In* "Modeling Crop Photosynthesis—from Biochemistry to Canopy," pp. 17–39. American Society of Agronomy and Crop Science Society of America, Madison, WI.

Harley, P.C., Thomas, R.B., Reynolds, J.F., and Strain, B.R. (1992). Modeling photosynthesis of cotton grown in elevated CO_2. *Plant, Cell and Environment* **15**, 271–282.

Harper, J. (1989). Canopies as populations. *In* "Plant Canopies: Their Growth, Form, and Function" (G. Russell, B. Marshall, and P.G. Jarvis, Eds.), pp. 105–128. Cambridge University Press, Cambridge, UK.

Hirose, T. and Werger, M.J.A. (1987). Maximizing daily canopy photosynthesis with respect to the leaf nitrogen allocation pattern in the canopy. *Oecologia* **72**, 520–526.

Holbrook, N. M. and Lund, C. P. (1995). Photosynthesis in forest canopies. *In* "Forest Canopies" (M.D. Lowman and N.M. Nadkarni, Eds.), pp. 411–430. Academic Press, San Diego.

Hollinger, D.Y. (1989). Canopy organization and foliage photosynthetic capacity in a broad-leaved evergreen montane forest. *Functional Ecology* 3, 53–62.

Hollinger, D.Y. (1996). Optimality and nitrogen allocation in a tree canopy. *Tree Physiology* 16, 627–634.

Jackson, R.B., Lechowicz, M.J., Li, X., and Mooney, H.A. (2001). Phenology, growth, and allocation in global terrestrial productivity. In "Terrestrial Global Productivity" (J. Roy, B. Saugier, and H.A. Mooney, Eds.), pp. 61–82. Academic Press, San Diego.

Jarvis, P.G. (1993). Prospects for bottom-up models. In "Scaling Physiological Processes" (J.R. Ehleringer and C.B. Field, Eds.), pp. 115–126. Academic Press, San Diego.

Jarvis, P. (1995). Scaling processes and problems. *Plant, Cell and Environment* 18, 1079-1089.

Jarvis, P.G. and Leverenz, J.W. (1983). Productivity of temperate, deciduous and evergreen forests. In "Encyclopedia of Plant Physiology" (Lange, O. L., Nobel, P. S., Osmond, C. B., and Ziegler, H., eds.), pp. 233–280. Springer-Verlag, Heidelberg, Germany.

Jarvis, P.G., Massheder, J.M., Hale, S.E., Moncrieff, J.B., Rayment, M., and Scott, S.L. (1997). Seasonal variation of carbon dioxide, water vapor, and energy exchanges of a boreal black spruce forest. *Journal of Geophysical Research* 102, 28953–28966.

Jones, H. G. (1992). "Plants and Microclimate: A Quantitative Approach to Environmental Physiology, second edition." Cambridge University Press, Cambridge, UK.

Keller, M., Clark, D.A., Clark, D.B., Weitz, A.M., and Veldkamp, E. (1996). If a tree falls in the forest... *Science* 273, 201–201.

Kirschbaum, M.U.F. and Farquhar, G.D. (1984). Temperature dependence of whole-leaf photosynthesis in *Eucalyptus pauciflora* Sieb. ex Spreng. *Australian Journal of Plant Physiology* 11, 519–538.

Kruijt, B., Ongeri, S., and Jarvis, P.G. (1997). Scaling of PAR absorption, photosynthesis and transpiration from leaves to canopy. In "Scaling-Up from Cell to Landscape" (P.R. van Gardingen, G.M. Foody, and P.J. Curran, Eds.), pp. 79–104. Cambridge University Press, Cambridge, UK.

Kull, O. and Jarvis, P.G. (1995). The role of nitrogen in a simple scheme to scale up photosynthesis from leaf to canopy. *Plant, Cell and Environment* 18, 1174–1182.

Kull, O. and Kruijt, B. (1998). Leaf photosynthetic light response: a mechanistic model for scaling photosynthesis to leaves and canopies. *Functional Ecology* 12, 767–777.

Kull, O. and Niinemets, U. (1998). Distribution of leaf photosynthetic properties in tree canopies: comparison of species with different shade tolerance. *Functional Ecology* 12, 472–479.

Laisk, A. and Oja, V. (1998). "Dynamics of Leaf Photosynthesis. Rapid-Response Measurements and Their Interpretations. Techniques in Plant Sciences No. 1." CSIRO Publishing, Melbourne, Australia.

Larcher, W. (1995). Photosynthesis as a tool for indicating temperature stress events. In "Ecophysiology of Photosynthesis" (E.-D, Schulze M.M. and Caldwell, Eds.), pp. 261–277. Springer-Verlag, Heidelberg, Germany.

Larocque, G.R. (2002). Coupling a detailed photosynthetic model with foliage distribution and light attenuation functions to compute daily gross photosynthesis in sugar maple (*Acer saccharum* Marsh.) stands. *Ecological Modelling* 148, 213–232.

Law, B.E., Cescatti, A., and Baldocchi, D.D. (2001). Leaf area distribution and radiative transfer in open-canopy forests: implications for mass and energy exchange. *Tree Physiology* 21, 777–787.

Law, B.E., Falge, E., Gu, L., Baldocchi, D.D., Bakwin, P., Berbigier, P., Davis, K., Dolman, A.J., Falk, M., Fuentes, J.D., Goldstein, A., Granier, A., Grelle, A., Hollinger, D., Janssens, I.A., Jarvis, P., Jensen, N.O., Katul, G., Mahli, Y., Matteucci, G., Meyers, T., Monson, R., Munger, W., Oechel, W., Olson, R., Pilegaard, K., Paw, U.K.T., Thorgeirsson, H., Valentini, R., Verma, S., Vesala, T., Wilson, K., and Wofsy, S. (2002). Environmental controls over carbon dioxide and water vapor exchange of terrestrial vegetation. *Agricultural and Forest Meteorology* 113, 97–120.

Lerdau, M. and Throop, H. (1999). Isoprene emission controls in a tropical wet forest canopy: implications for model development. *Ecological Applications* 9, 1109–1117.

Lerdau, M. and Throop, H.L. (2000). Sources of variability in isoprene emission and photosynthesis in two species of tropical wet forest trees. *Biotropica* 32, 670–676.

Lerdau, M.T., Holbrook, N.M., Mooney, H.A., Rich, P.M., and Whitbeck, J.L. (1992). Seasonal patterns of acid fluctuations and resource storage in the arborescent cactus *Opuntia excelsa* in relation to light availability and size. *Oecologia* 92, 166–171.

Leuning, R. (1995). A critical appraisal of a combined stomatal-photosynthesis model for C3 plants. *Plant, Cell and Environment* 18, 339–355.

Leuning, R., Kelliher, F. M., De Pury, D. G. G., and Schulze, E.-D. (1995). Leaf nitrogen, photosynthesis, conductance and transpiration: scaling from leaves to canopies. *Plant, Cell and Environment.* 18, 1183-1200.

Leuning, R., Wang, Y.P., and Cromer, R.N. (1991). Model simulations of spatial distributions and daily totals of photosynthesis in *Eucalyptus grandis* canopies. *Oecologia* 88, 494–503.

Livingston, N.J., Whitehead, D., Kelliher, F.M., Wang, Y.-P., Grace, J.C., Walcroft, A.S., Byers, J.N., McSeveny, T.M., and Millard, P. (1998). Nitrogen allocation and carbon isotope fractionation in relation to intercepted radiation and position in a young *Pinus radiata* D. Don tree. *Plant, Cell and Environment* **21**, 795–803.

McMurtrie, R.E., Leuning, R., Thompson, W.A., and Wheeler, A.M. (1992). A model of canopy photosynthesis and water use incorporating a mechanistic formulation of leaf CO_2 exchange. *Forest Ecology and Management* **52**, 261–278.

Meir, P., Kruijt, B., Broadmeadow, M., Barbosa, E., Kull, O., Carswell, F., Nobre, A., and Jarvis, P.G. (2002). Acclimation of photosynthetic capacity to irradiance in tree canopies in relation to leaf nitrogen concentration and leaf mass per unit area. *Plant, Cell and Environment* **25**, 343–357.

Mooney, H.A. and Gulmon, S.L. (1979). Environmental and evolutionary constraints on photosynthetic characteristics of higher plants. *In* "Topics in Plant Population Biology" (O.T. Solbrig, S. Jain, G.B. Johnson, and P.H. Raven, Eds.), pp. 316–337. Columbia University Press, New York.

Monteith, J.L. (1977). Climate and the efficiency of crop production in Britain. *Philosophical Transactions of the Royal Society of London Series B* **281**, 277–294.

Mott, K.A. and Buckley, T.N. (2000). Patchy stomatal conductance: emergent collective behaviour of stomata. *Trends in Plant Science* **5**, 258–262.

Mott, K.A. and Parkhurst, D.F. (1991). Stomatal responses to humidity in air and helox. *Plant, Cell, and Environment* **14**, 509–515.

Mulkey, S.S., Chazdon, R.L., and Smith, A.P. (1996a). "Tropical Forest Plant Ecophysiology." Chapman and Hall, New York.

Mulkey, S.S., Kitajima, K., and Wright, S.J. (1996b). Plant physiological ecology of tropical forest canopies. *Trends in Ecology and Evolution* **11**, 408–413.

Niinemets, U. and Tenhunen, J.D. (1997). A model separating leaf structural and physiological effects on carbon gain along light gradients for the shade-tolerant species *Acer saccharum*. *Plant, Cell and Environment* **20**, 845–866.

Parker, G.G. (1995). Structure and microclimate of forest canopies. *In* "Forest Canopies" (M.D. Lowman and N.M. Nadkarni, Eds.), pp. 73–106. Academic Press, San Diego.

Pearcy, R.W. and Pfitsch, W.A. (1995). The consequences of sunflecks for photosynthesis and growth of forest understory plants. *In* "Ecophysiology of Photosynthesis" (E.-D. Schulze, and M.M. Caldwell, Eds.), pp. 343–359. Springer-Verlag, Heidelberg, Germany.

Raulier, F., Bernier, P. Y., and Ung, C.-H. (1999). Canopy photosynthesis of sugar maple (*Acer saccharum*): comparing big-leaf and multilayer extrapolations of leaf-level measurements. *Tree Physiology*. **19**, 407-420.

Reich, P. and Walters, M. (1994). Photosynthesis-nitrogen relations in Amazonian tree species II. Variation in nitrogen vis-a-vis specific leaf area influences mass- and area-based expressions. *Oecologia* **97**, 73–81.

Reich, P.B., Kloeppel, B.D., Ellsworth, D.S., and Walters, M.B. (1995). Different photosynthesis-nitrogen relations in deciduous hardwood and evergreen coniferous tree species. *Oecologia* **104**, 24–30.

Reich, P.B., Walters, M.B., and Ellsworth, D.S. (1997). From tropics to tundra: global convergence in plant functioning. *Proceedings of the National Academy of Science* **94**, 13730–13734.

Reich, P.B., Ellsworth, D.S., and Walters, M.B. (1998). Leaf structure (specific leaf area) modulates photosynthesis-nitrogen relations: evidence from within and across species and functional groups. *Functional Ecology* **12**, 948–958.

Ruimy, A., Jarvis, P.G., Baldocchi, D.D., and Saugier, B. (1995). CO_2 fluxes over plant canopies and solar radiation: a review. *Advances in Ecological Research* **26**, 1–63.

Running, S.W. and Coughlan, J.C. (1988). A general model of forest ecosystem processes for regional applications I. Hydrologic balance, canopy gas exchange and primary production processes. *Ecological Modelling* **42**, 125–154.

Schulze, E.-D. (1986). Whole-plant responses to drought. *Australian Journal of Plant Physiology* **13**, 127–141.

Schulze, E.-D. and Caldwell, M.M. (1995). "Ecophysiology of Photosynthesis." Springer-Verlag, Heidelberg, Germany.

Schulze, E.-D., Kelliher, F.M., Korner, C., Lloyd, J., and Leuning, R. (1994). Relationships among maximum stomatal conductance, ecosystem surface conductance, carbon assimilation rate, and plant nitrogen nutrition: a global ecology scaling exercise. *Annual Review of Ecology and Systematics* **25**, 629–660.

Sellers, P.J., Berry, J.A., Collatz, G.J., Field, C.B., and Hall, F.G. (1992). Canopy reflectance, photosynthesis, and transpiration. III. A reanalysis using improved leaf models and a new canopy integration scheme. *Remote Sensing of Environment* **42**, 187–216.

Sharkey, T.D. (1985). Photosynthesis in intact leaves of C_3 plants: physics, physiology and rate limitations. *The Botanical Review* **51**, 53–105.

Sims, D.A. and Pearcy, R.W. (1991). Photosynthesis and respiration in *Alocasia macrorrhiza* following transfers to high and low light. *Oecologia* **86**, 447–453.

Stenberg, P., DeLucia, E.H., Schoettle, A.W., and Smolander, H. (1995). Photosynthetic light capture and processing from cell to canopy. *In* "Resource Physiology of Conifers: Acquisition, Allocation, and Utilization" (W.K. Smith and T.M. Hinckley, Eds.), pp. 3–38. Academic Press, San Diego.

Tezara, W., Mitchell, V.J., Driscoll, S.D., and Lawlor, D.W. (1999). Water stress inhibits plant photosynthesis by decreasing coupling factor and ATP. *Nature* **401**, 914–917.

Vitousek, P. M., Aber, J. D., Howarth, R. W., Likens, G. E., Matson, P. A., Schindler, D. W., Schlesinger, W. H., and Tilman, D. G. (1997). Human alteration of the global nitrogen cycle: sources and consequences. *Ecological Applications*. **7**, 737-750.

von Caemmerer, S. (2000). "Biochemical Models of Leaf Photosynthesis. Techniques in Plant Sciences No. 2." CSIRO Publishing, Collingwood, Australia.

von Caemmerer, S. and Farquhar, G.D. (1981). Some relationships between the biochemistry of photosynthesis and the gas exchange of leaves. *Planta* **153**, 376–387.

Walther, G.-R., Post, E., Convey, P., Menzel, A., Parmesan, C., Beebee, T.J.C., Fromentin, J.-M., Hoegh-Guldberg, O., and Bairlein, F. (2002). Ecological responses to recent climate change. *Nature* **416**, 389–395.

Ward, J.K. and Strain, B.R. (1999). Elevated CO_2 studies: past, present and future. *Tree Physiology* **19**, 211–220.

Waring, R.H., Law, B.E., Goulden, M.L., Bassow, S.L., McCreight, R.W., Wofsy, S.C., and Bazzaz, F.A. (1995). Scaling gross ecosystem production at Harvard Forest with remote sensing: a comparison of estimates from a constrained quantum-use efficiency model and eddy correlation. *Plant, Cell and Environment* **18**, 1201–1213.

Williams, M., Rastetter, E.B., Fernandes, D.N., Goulden, M.L., Wofsy, S.C., Shaver, G.R., Melillo, J.M., Munger, J.W., Fan, S.-M., and Nadelhoffer, K.J. (1996). Modelling the soil-plant-atmosphere continuum in a *Quercus-Acer* stand at Harvard Forest: the regulation of stomatal conductance by light, nitrogen and soil/plant hydraulic properties. *Plant, Cell and Environment* **19**, 911–927.

Wilson, K.B., Baldocchi, D.D., and Hanson, P.J. (2000a). Spatial and seasonal variability of photosynthetic parameters and their relationship to leaf nitrogen in a deciduous forest. *Tree Physiology* **20**, 565–578.

Wilson, K.B., Baldocchi, D.D., and Hanson, P.J. (2000b). Quantifying stomatal and non-stomatal limitations to carbon assimilation resulting from leaf aging and drought in mature deciduous tree species. *Tree Physiology* **20**, 787–797.

Wilson, K.B., Baldocchi, D.D., and Hanson, P.J. (2001). Leaf age affects the seasonal pattern of photosynthetic capacity and net ecosystem exchange of carbon in a deciduous forest. *Plant, Cell and Environment* **24**, 571–583.

Wong, S.C., Cowan, I. R., and Farquhar, G.D. (1979). Stomatal conductance correlates with photosynthetic capacity. *Nature* **282**, 424–426.

Wullschleger, S.D. (1993). Biochemical limitations to carbon assimilation in C_3 plants: a retrospective analysis of the A/Ci curves from 109 species. *Journal of Experimental Botany* **44**, 907–920.

CHAPTER 18

Insect Herbivory in Tropical Forests

H. Bruce Rinker and Margaret D. Lowman

The degree to which insects regulate ecosystem parameters remains a key issue and one that significantly broadens the scope and value of insect ecology.
—*T.D. Schowalter,* Insect Ecology: An Ecosystem Approach, *2000*

Introduction: The Little Things that Run the World

In an address to inaugurate the invertebrate exhibit at the National Zoo, Washington, DC on May 7, 1987, E.O. Wilson called insects "the little things that run the world" (Wilson 1987a). In terms of their diversity, distribution, and abundance, insects are unrivaled as a global ecological and evolutionary force. "It needs to be repeatedly stressed that invertebrates as a whole are even more important in the maintenance of ecosystems than are vertebrates" (Wilson 1987a). In a single tree in the Peruvian Amazon, Wilson found 43 species of ants (Wilson 1987b). Erwin (1982) discovered more than 1,000 kinds of beetles on 19 individual trees of a single species, *Luehea seemannii,* in a seasonal lowland forest in Panama. From this discovery, Erwin extrapolated an estimate of 30 million species of organisms on the planet (Erwin 1982, 1983).[1] In terms of biomass, insects in tropical forests constitute several tons per hectare compared to a few kilograms per hectare for birds and mammals (reviewed in Dajoz 2000). Eight million ants and one million termites per hectare make up more than one-third the animal biomass in Amazonian *terra firma* rainforest (Hölldobler and Wilson 1990). If we limit this discussion only to their bulk in the treetops, these humble hymenopterans and associates still represent a significant portion of the total animal biomass for tropical forests. For nearly a quarter-century, we have known that canopy arthropods are key regulators of ecosystem processes (Reynolds et al. 2000). Insects then are the little things that run our forests—indeed most terrestrial ecosystems—from top to bottom.

Tropical Insects: Hypotheses for Their High Biodiversity in Equatorial Forests

Why do so many kinds of insects occur in the tropics? Numerous reasons offered in the literature (e.g., MacArthur 1969; Lowman and Nadkarni 1995; Kricher 1997; Dajoz 2000) decant into four prominent hypotheses that deal with history, structure, dynamics, and energetics. Yet each one has problematic aspects that point toward a combination of reasons for insect abundance and diversity in the tropics. As DeVries (1987) aptly wrote, "There have been many theories put forth, but none has satisfactorily answered the question, because almost all of them rely on the

---

[1] See appendix on p.381 for details about Erwin's extrapolations.

circular reasoning: greater diversity generates greater diversity." Researchers may be uncertain about the reasons for the large diversity of equatorial insects, but they are confident that tropical regions are the richest reservoir of arthropod species worldwide.

The historical hypothesis focuses on long-term environmental stability, thus fostering long-term diversification (Terborgh, 1992; Kricher 1997). In tropical forests, where climatic variation is low and predictable, opportunities for specialization exist, such as frugivory and unique pollination strategies (e.g., euglossine bees and canopy orchids). On the other hand, tropical areas have not been immune to climatic change. Amazonia may have endured tremendous biological upheaval during the Pleistocene (but see Colinvaux 1997). Further, the Andes Cordillera—notorious for its geological and climatic fluctuations—manifests extremely high endemism and biological diversity among certain taxa.

A second reason offered for high insect abundance and diversity in the tropics is the structural hypothesis (Dajoz 2000). This hypothesis states that large leaf-surface area and narrow niches promote high levels of specialization among insects, especially among phytophagous species. The breadth of niches for herbivorous insects may be measured by their degree of host-specificity (Janzen 1983). For example, half the species of Lepidoptera in a Costa Rican forest feed on a single plant species (Janzen 1983; DeVries 1987). On the other hand, contradictory data from Borneo, North America, and other locations suggest that there is a lower host-specificity in the tropics than in temperate regions (Mawdsley and Stork 1997)! Thus, we cannot assert whether the elevated biodiversity of tropical arthropods is due to host-specificity that might be higher in the tropics than in temperate regions.

The dynamics hypothesis addresses the intense competition, predation, and parasitism in the tropics that force biological radiation among insect fauna over time (see Macarthur 1972; Kricher 1997). These factors regulate the coexistence of species via ecological principles such as competitive displacement (or competitive exclusion) and put a lower ceiling on the abundance of any given species, thus allowing more species to fit in (MacArthur 1969; 1972). Yet competition seems rare among phytophagous insects (Schowalter 2000), least observed among free-living, chewing species and more prevalent among internal feeders (e.g., miners and borers). If competition does not structure phytophagous insect communities, then perhaps it is not the robust Darwinian principle as traditionally believed.

Finally, the energetics hypothesis holds that high energy diversity and resource allocation favor the emergence of different feeding strategies (Wilson 1992; Kricher 1997). In other words, net primary productivity and species richness are highly correlated. High amounts of productivity in equatorial regions, along with the synthesis of secondary compounds in plants that deter herbivory, may drive selection toward rapid species diversification among surviving types of insects. Once again, the data are inconclusive and suggest a combination of reasons for the abundance of arthropods in the tropics.

Speculation about the basis for invertebrate diversity, however, may narrow into one simple explanation—their small size and accompanying small niches (Wilson 1987a). An insect is *necessarily* small because of two physiological constraints on its body size. First, though water-soluble, its brittle chitinous exoskeleton can support only a limited mass of muscle before collapsing inward. Second, unlike the circulatory system of a vertebrate that delivers both nutriment and oxygen throughout the body, the circulatory system of an insect carries only food molecules. Oxygen is delivered separately and directly to cells via a complex system of minute tracheae. The shell of an elephant-sized insect would shatter from gravity and stress, and its innermost tissues would die of oxygen starvation as soon as the animal exerted itself. The upper limit for body size of terrestrial arthropods seems to have been reached by the world's largest species of beetle (*Titanus giganteus*), moth (*Thysania agrippina*), wasp (*Pepsis heros*), scorpion (*Pandinus imperator*), and spider (*Teraphosa leblondii*)—all found in Amazonia, except for the African scorpion. "We don't know with certainty why invertebrates are so diverse, but a community held opinion is that the key is their small size," (Wilson 1992). Their Lilliputian size, allowing insects to divide up the environment into little domains where specialist can co-exist, seems to have guaranteed them broad ecological and evolutionary success in the tropics—and in just about every other global environment.

Borrowing from MacArthur (1969), the tropical environment is not like a box that will hold only so many eggs, but like a balloon that resists further invasion proportionally to its present contents but can always hold a little bit more if necessary. The data clearly suggest that ecological history, structure, dynamics, and energetics are all important explanations of arthropod diversity in the tropics on some level and that small size allows the metaphorical balloon to distend with Lilliputian richness.

Tropical Insects: Hypotheses for Their High Biodiversity in the Rainforest Canopy

But what accounts for the high species diversity of insects suspected in the upper canopy of equatorial forests? Basset et al. (2003) examined vertical stratification, temporal distribution, resource use, and host specificity of arthropods in tropical rainforests. The high illumination and temperature in the treetops encourage foraging and ovipositioning among insects. Leaf flush also provides a supply of young nutritious leaves that have not yet developed the defensive toughness of older leaves. On the other hand, plants often sequester secondary compounds in these same young leaves that deter herbivory. As an evolutionary barrier, phytochemicals inadvertently promote biological diversification among insects, if these arthropods metabolize the compounds and utilize the byproducts for their own survival. Canopy insects are residents of an aerial continent of sugars held aloft by stems that connect heaven to earth. Forests, like gigantic stands of lollipops, provide nutrients to those arthropods that are able to defeat the trees' attendant defensive poisons.

Intense fogging of one species of tree in Panama revealed a rich insect diversity in the rainforest canopy and led to lively speculations that this fauna is more diverse in the treetops than any other environments, including forest soils (Erwin 1982, 1983). These data are controversial, however, with speculation on sampling biases and with comparisons to recent fieldwork in other habitats (André et al. 1994; Kricher 1997; Linsenmair et al. 2001). Data from Cameroon, Guyana, and French Guiana strongly assert that invertebrate densities obtained from foliage samples in the canopy are higher than in the understory (Basset et al. 1991; Lowman et al. 1998; Linsenmair et al. 2001). Whether those densities are higher in the forest canopy than in forest soils, or in any other terrestrial habitat, is also under investigation (Reynolds et al. 2003; Rinker, in prep.). Many invertebrate herbivores, in particular, have specific food requirements that render most rainforest foliage unsuitable. Spatial/temporal phenology such as leaf flush, nectar availability, and fruiting may be the reason for aggregations of phytophages in the canopy rather than simple forest structure. Rinker et al. (2001) reviewed the ecological linkages between canopy herbivory and soil ecology, shifting emphasis away from a strict autecology of individual species toward a more comprehensive ecosystem approach. Speculations about the high insect diversity in forest canopies may be overstated, however, due to faulty assumptions by Erwin and others, in sampling biases, and by recent fieldwork in other habitats (André et al. 1994; Krichen 1997; Linsenmair et al. 2001).

Plant and Insect Interactions: An Ongoing CoEvolutionary Dance for Forest Survival

In a complex co-evolutionary dance, insects influence, and are influenced by, plant phytochemistry. Though there is much controversy about the magnitude and setting of global species-richness on earth, the bulk of this biodiversity is found in the canopy arthropods of tropical forests (Stork et al. 1997). Tropical vegetation is renowned for its high diversity and incidence of alkaloids, latex, and other secondary metabolites, and also for a diversity of counter-adaptations by phytophagous insects (Novotny et al. 2003). Of the 32 orders of insects, only nine have members that feed on living plants such as Coleoptera and Lepidoptera (see Figure 18-1). Plants are formidable evolutionary barriers against herbivory because of their impressive arsenal of physical (e.g., toughness, trichomes, and stinging hairs) and chemical (e.g., vanilla, salicylic acid, nicotine, caffeine, tannins, and pyrethrum) defenses. Research in plant/animal interactions has been dominated by investigations into the role of secondary plant compounds in determining the distribution, abundance, and evolution of phytophagous insects (Hunter 1992b). For example, herbivory sometimes induces foliar changes that render leaves less suitable for the development of certain herbivores (Schultz and Baldwin 1982; Huffaker and Gutierrez 1999). Once the barriers are surmounted, however, phytophagous insect groups radiate extensively (see Romoser and Stoffolano 1998). Most of this complex radiation has occurred in the treetops of tropical rainforests.

Even if they survive the phytochemical barriers, insects are beleaguered by predators and parasites. (e.g., Romoser and Stofolano 1998, Huffaker and Gutierrez 1999; Schowalter 2000). Predation pressure and parasitic interactions, subject to local microclimate and structure of canopy foliage, modify the behaviors of prey species and, thereby, influence their evolutionary development (e.g., Koike and Nagamitsu 2003); however, the latter tend to be more effective than predators in responding to and controlling eruptions of their host populatons (Schowalter 2000). In the flow of nutrients in forest ecosystems, herbivores are "midstream" components. Pressed above and below in the trophic pyramid, insects bear enormous ecological and evolutionary weight in temperate and tropical forests. The physical and chemical defenses of plants against herbivory, plus the limiting influence of parasites and predators on insect expansion, are two important components for the long-term health of forests.

But a third element exists in the ever-fluctuating trophic equation. Ground measurements historically showed that phytophages typically consume about 5-10 percent of the total net primary productivity in forest ecosystems (Mattson and Addy 1975). Coincidentally, this figure makes an interesting comparison to the estimated amount of energy that reaches successive levels in a trophic pyramid, also 10 percent (Raven and Johnson 1996). Insects in the tropics munch through an estimated 680 kg ha^{-1}y^{-1} of leaves compared to 100 to 300 kg ha^{-1}y^{-1} of leaves by vertebrates (Dajoz 2000). Janzen (1983) examined patterns of herbivory among insects and vertebrates and concluded that leaf-eating insects are a much greater threat as defoliators of forests than vertebrate megafauna. Although 5-10 percent has been a rule of thumb for forest herbivory, the degree of defoliation can vary greatly between sites (Lowman 1987, 1995b; Schowalter 2000). Causes for this variation include phenology, age class of leaves, vegetation strata, forest type, and the natural history of local arthropod herbivores, including the demographics of their predators and parasites. With 10 percent as a good average for the amount of organic matter that reaches a given trophic level, the remainder is dissipated as heat production and waste (Raven and Johnson 1996). And, given their great abundance and diversity, herbivorous insects generate a substantial amount of waste.

In addition to the phytochemicals and the predators/parasites, the third component of the trophic equation is frass or insect droppings. The excreta of terrestrial arthropods are "upstream" from microbes and other decomposers waiting on the forest floor. Though work has just begun to quantify arthropod feces in terrestrial ecosystems, it is already clear that frass plays a significant role in nutrient cycling. "The standing crop of arthropod feces has not been quantified in any terrestrial ecosystem; however, this component is much larger than any of those mentioned previously. The standing crop of fecal pellets from macroarthropod detritivores such as millipedes may locally exceed annual litterfall inputs" (Seastedt and Crossley 1984; see also Ohmart et al.

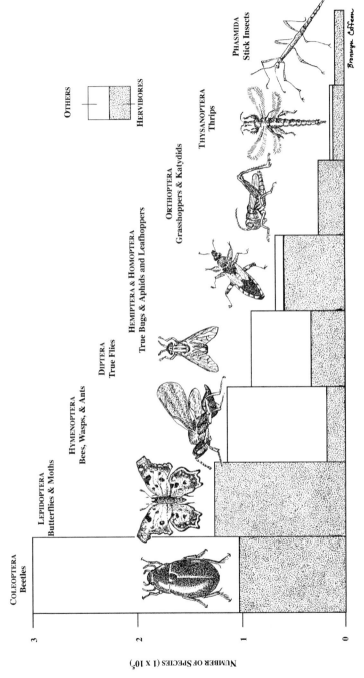

Figure 18-1   Phytophagous insect orders (from Romoser and Stoffolano 1998). Illustration by Bronwyn Coffeen and used with permission.

1983). Frass then is the third component in the equation for forest survival, a product of herbivory that subsequently feeds the decomposers on the forest floor below that, in turn, nourish the forest. These ancient canopy invertebrates represent a uniting thread in the nutrient cycling through forests via the influence of their waste stream on soil processes and organisms as well their impact on phytochemistry, parasites, and predators.

The Role of Herbivory in Canopy Processes

Herbivory, the feeding on living plant parts by animals, is a key ecological process (Schowalter 2000) that affects all canopy components either directly (primary consumption) or indirectly (secondary consumption). Leaves senesce and then decompose via bacteria, fungi, and microarthropods on the forest floor. Herbivory and senescence comprise pathways that link herbivory to nutrient cycling in the forest ecosystem. Foliage that is partially grazed by herbivores is called herbivory, whereas foliage that is grazed in its entirety (or grazed extensively, leading to senescence) is classified as defoliation. It is important to recognize that herbivory is the direct effect of grazing whereas defoliation results in mortality that may be partially a consequence of the grazing mechanism.

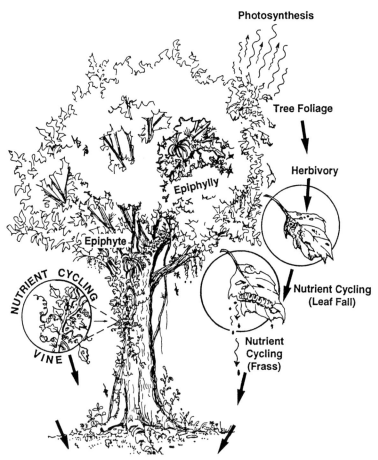

Figure 18-2 Canopy components and processes that are affected by herbivory in a forest stand (from Lowman 1995b; used with permission).

MEASURING FOREST HERBIVORY LEVELS USING CANOPY CRANES

Kristina A. Ernest

Comparing levels of herbivory among forest types poses many challenges for canopy scientists. One obvious difficulty is access into the canopy. Another is the time-consuming nature of measuring herbivory at many different spatial scales, from the leaf to the forest plot, to scale up estimates to the forest stand (Lowman 1985). To complicate matters, researchers have used a variety of approaches to measuring herbivory rather than using a single standardized protocol (Lowman 1995; Lowman 2001). To help solve these problems, a team of canopy researchers with diverse expertise (David C. Shaw, H. Bruce Rinker, Margaret D. Lowman, and myself) joined forces. We developed a novel technique for estimating the levels of herbivory in forest stands by sampling random locations within the three-dimensional canopy space of a forest (see Figure 1). This three-dimensional randomization avoids the complicated sub-sampling imposed by the traditional method of scaling up. For each randomly chosen XYZ coordinate with accessible foliage of any vascular plant species, we measured percent of leaf area consumed by herbivores on 10 randomly chosen leaves (or 50 conifer needles) within a 25 × 25 × 25 cm subplot (see Figure 2). Canopy cranes provided us rapid access to these sample locations. We tested this new method in two structurally and functionally dissimilar forests (tropical rainforest at Cape Tribulation, Queensland, Australia, and temperate conifer forest at Wind River, WA, USA).

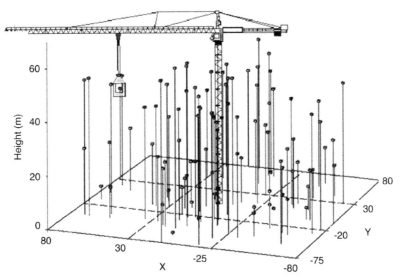

Figure 1   Spatial distribution of 104 actual sample locations at Wind River, WA.

We recorded an average herbivory level of 8.6 percent for 93 sample locations in tropical rainforest at the Australian Canopy Crane site, compared with 1.6 percent for 104 sample locations in temperate conifer forest at the Wind River Canopy Crane Research Facility. Although this method has limitations (e.g., not all random locations are accessible from the gondola of the crane), the advantage of fairly rapid access to all vertical

Continued

MEASURING FOREST HERBIVORY LEVELS USING CANOPY CRANES–*cont'd*

Figure 2   Subplot (25 × 25 × 25 cm) to randomly sample 10 leaves (or 50 needles) at each sample location. Photograph by D. Shaw.

levels within the forest allows researchers at any canopy crane site to compare forest stands around the globe in a standardized way.

We believe this method has the potential to make important advances in forest and herbivory research. First, we hope it will become a standard protocol for rapidly assessing the extent of herbivory in forest stands, and we plan to apply it at all canopy crane sites. Second, data collected using this protocol will help set baseline numbers for one of the key processes regulating ecosystem function. Primary productivity may be regulated by herbivory, yet we sorely lack measurements of herbivory at the stand level due to the difficulty in logistics and the extensive sampling required. Additionally, global climate change and increasing atmospheric CO_2 are likely to affect rates of herbivory (Coley 1998; McNaughton 2001). Standardized, quantitative data on rates of herbivory can be used in predictive models about how forest productivity will change under future climate scenarios.

References

Coley, P.D. (1998). Possible effects of climate change on plant/herbivore interactions in moist tropical forests. *Climatic Change* **39**, 455–472.

Lowman, M.D. (1985). Temporal and spatial variability in insect grazing of the canopies of five Australian rainforest tree species. *Australian Journal of Ecology* **10**, 7–24.

Lowman, M.D. (1995). Herbivory as a canopy process in rain forest trees. *In* "Forest Canopies" (M.D. Lowman and N.M. Nadkarni, Eds.), pp. 431–455. Academic Press, San Diego.

Lowman, M.D. (2001). Plants in the forest canopy: some reflections on current research and future direction. *Plant Ecology* **153**, 39–50.

McNaughton, S.J. (2001). Herbivory and trophic interactions. *In* "Terrestrial Global Productivity" (J. Roy, B. Saugier, and H.A. Mooney, Eds.), pp. 101–122. Academic Press, San Diego.

The study of herbivory as an integrated process throughout a forest stand requires information on many aspects of forest biology including plant phenology, demography of insect populations, leaf growth dynamics, tree architecture, foliage quality and density, physical environment, nutrient cycling, and plant succession. Interest in insect-plant relationships has emphasized interactions among a few species rather than within an entire ecosystem. It has also centered on studies of shrubs and herbaceous plants. Few studies exist on insect-tree interactions, and even fewer involving forest stands. Yet forest canopies comprise an ecological arena where some of the most complex insect-plant interactions occur in terms of spatial, taxonomic, and structural factors.

The consumption of plant materials by herbivores is a subject of great economic and ecological importance (Barbosa and Schultz 1987; Price et al. 1991; Schowalter 2000; Ribeiro 2003). Because forest canopies contain the bulk of terrestrial photosynthetic material involved in the maintenance of global biogeochemical cycles (e.g., carbon dioxide sequestering), the processes affecting canopy foliage have direct consequences on the health of our planet. Leaf predation is an example of herbivory. The loss of foliage by insect predators can occur by direct consumption or by less obvious impacts such as leaf mining, sap-sucking, and leaf tying. Herbivory affects foliage during all stages in the life of a leaf; over time, plants respond accordingly with the evolution of defenses against predation. Levels of herbivory range from negligible grazing to mortality of leaves, branches, whole crowns, and entire forest stands.

Herbivore populations fluctuate in the canopy and, in turn, affect the populations of other invertebrates and of vertebrates such as birds and mammals that feed on the herbivorous organisms (e.g., Woinarski and Cullen 1984). This fluctuation results from a legion of system variables including weather (Readshaw 1965), disturbances (Smith 1982), historical processes (Southwood 1961), topography (Ward 1979), tree density (Morrow and Fox 1980), plant structure (Lawton and Schröder 1977), plant secondary compounds (Macauley and Fox 1980), or even random processes (Clark 1962). Stand growth and dynamics may ultimately be affected by herbivory and by the susceptibility of a species to grazing (reviewed in Schowalter et al. 1986). The impact of leaf consumption on herbs, seedlings, and shrubs has been quantified in terms of mortality, succession, and compensatory growth (Lowman 1982; Coley 1983; Marquis 1991). Such factors are more difficult to measure for tall trees and across forest stands. Examples of herbivory that have been integrated with other aspects of forest dynamics include studies on the spatial distribution of canopy insect populations in the Australian rainforest tree *Argyrodendron actinophyllum* (Basset 1991), nutrient cycling via frassfall or litterfall pathways (Lowman 1988), pest outbreaks and stand mortality (Lowman and Heatwole 1992), and herbivory in relation to stand phenology (Schultz and Baldwin 1982). Underlying the variability between insect taxa and locations are discernible patterns to phytophagous insect distribution (Woinarski and Cullen 1984).

History of Herbivory Studies in Forest Canopies

Forests are not vast expanses of homogeneous green tissue. As we walk through woodlands, we usually focus our observations on a narrow band of foliage from ground level to 2 m in height. This represents, at most, 10 percent of the plant life in mature forests with the majority high above our heads and normally beyond our immediate observations. Because herbivore–plant interactions occur in the foliage, herbivory as a forest process remained relatively unknown until we developed safe and efficient methods of canopy access. Today we may best view forests, not as uniform expanses of green, but as mosaics of holes in leaves (Lowman 1995b).

Historically, most herbivory studies involved the measurement of levels of defoliation in forests at one point in time. Foliage was sampled typically near the ground level in temperate deciduous forests where annual losses of 3 to 10 percent leaf surface area were reported (Bray 1964;

Bray and Gorham 1964; Landsberg and Ohmart 1989). Most studies were extrapolated, however, to evergreen rainforests for three reasons: temperate deciduous forests have a comparatively simple phenology with an annual turnover of leaves; measurements were sometimes made from senescent leaves retrieved from the forest floor; and replicated stratified sampling was rarely attempted. In short, defoliation was treated as a discreet snapshot event (Diamond 1986), accounting for neither temporal nor spatial variability.

More recent studies expanded in scope to include temporal and spatial factors to explain the heterogeneity of herbivory throughout the canopy. There are five noteworthy discoveries in the history of herbivory research:

1. An important attribute affecting levels of foliage consumption is age of leaf tissue with soft, young leaves preferred over old, tough leaves (Coley 1983; Lowman 1985).
2. The most abundant herbivores in forests are insects in terms of both numbers and estimated impacts (reviewed in Schowalter et al. 1986; Lowman and Moffett 1993). In some ecosystems, however, mammals are also important (e.g., monkeys, koalas, and tree kangaroos, as reviewed in Montgomery 1978).
3. Canopy grazing levels are heterogeneous *between* forests, ranging from negligible losses to total foliage losses, and *within* forests, varying with plant and herbivore species, height, light regions, phenology, age of leaves, and individual crown (Lowman 1992).
4. The assumptions common in the 1960s (i.e., that herbivory averaged 5 to 10 percent annual leaf area loss and was homogeneous throughout forests) were oversimplified and underestimated, particularly for evergreen forests (e.g., Fox and Morrow 1983).
5. Foliage feeders are featured in the ecological literature as the most common type of herbivore. Sapsuckers may also be important, however, although they have not been as well studied. Reputedly, foliage consumption is easier to measure than sap consumption yet, even for measurement of folivory, standard protocols are not well established (see Lowman 1984).

These discoveries were facilitated by the development of efficient and safe access methods that expanded the scope of foliage sampling into the canopy. Subsequently, studies of herbivory in evergreen tropical forests increased and revealed the heterogeneous nature of plant/animal interactions.

A Comparison of Forest Herbivory

The texture, age class, and nutritional value of leaves vary considerably within and between individual trees. Phenological events, along with species composition and rarity along a latitudinal

Table 18-1   Stand-Level Herbivory (%) in Forests around the World

Forest Type	% Herbivory
Australian Dry Forest	15 to 300% (Lowman and Heatwole 1992)
Australian Wet Forest	8.6% (Shaw, Ernest, Rinker, and Lowman, pers. comm.)
Temperate Deciduous Forest	15% (Lowman 1999)
Pacific Northwest	1.6% (Shaw, Ernest, Rinker, and Lowman, pers. comm.)
Cloud Forest	26% (Lowman 1992)
Subtropical Forest	16% (Lowman 1987)
Warm Temperate Forest	21% (Lowman 1987)
Tropical Forest	12 to 30% (Lowman 1995a)

gradient, likewise fluctuate. All these variables result in patchy susceptibilities of trees to insect herbivores. Consequently, forest canopies support complex assemblages of phytophagous insects, usually inconspicuous except when population outbreaks produce widespread defoliation (Schowalter et al. 1986; Reynolds et al. 2000; Linsenmair et al. 2001). Herbivory studies that measure defoliation at a single point in time, historically the most common type of research (Lowman 1995b), provide only a limited window of understanding for plant–animal interactions in forests. The complex temporal and spatial patterns of leaves in forest canopies then necessitate long-term approaches to studies of insect herbivory.

Herbivory in canopies ranges from 1 to 5 percent of total leaf area production in temperate forests (Schowalter et al. 1981) to more than 300 percent in Australian eucalypt forests where trees re-foliated several successive times after leaf loss (Lowman 1992; Lowman and Heatwole 1992). Several studies (e.g., Seastedt and Crossley 1984; Lowman et al. 1998) and literature reviews (e.g., Schowalter et al. 1986; Lowman 1995b; Schowalter 2000) give details on herbivory rates at specific sites in tropical areas. Table 18-1 provides a summary of two researchers' fieldwork.

Arthropods are the most abundant herbivores in many terrestrial ecosystems (Erwin 1983; Seastedt and Crossley 1984; Lowman and Morrow 1998). Of the nine herbivorous insect orders, the three most important leaf-consuming taxonomic groups are Coleoptera (beetles), Lepidoptera (butterflies and moths), and Hymenoptera (bees, wasps, and ants) (Reichle et al. 1973; Lowman and Morrow 1998). The capacity of their populations to cause rapid, often dramatic change in forest dynamics such as tree growth and survival, evapotranspiration, and nutrient cycling speaks unequivocally about the significance of their connections to overall ecological processes (Schowalter 2000). Surprisingly, those links are poorly understood, especially those that connect forest canopies to forest soils.

Herbivory levels vary significantly between species and between forest types, as illustrated by long-term studies in Australian rainforests (see Figure 18-3). Lowman (1982) originally hypothesized that evergreen forests with lower diversity would have higher herbivory levels than neighboring evergreen forests with higher diversity. The cool temperate rainforest, where *Nothofagus moorei* dominated over 75 percent of the canopy, averaged an annual 26 percent leaf area loss to grazing insects (Selman and Lowman 1983). The majority of this was due to a host-specific chrysomelid beetle that fed exclusively on the young leaves of *N. moorei* during the spring flush. In contrast, the subtropical rainforests, where no species occupied more than 5 percent of the canopy within a stand (Lowman 1985), averaged only 15 percent annual leaf area loss to insect grazers. Herbivory also varied significantly among species. *Toona australis* (Meliaceae), which is relatively rare and is annually deciduous, average less than 5 percent loss, whereas neighboring canopies of *Dendrocnide excelsa* (Urticaceae) that colonizes gaps averaged over 40 percent annual leaf loss (Lowman 1992).

In addition to variation in grazing levels between species and stands, herbivory is variable within individual crowns. The heterogeneity of defoliation is a consequence of a leaf's environment and phenology with different leaf cohorts exhibiting different susceptibilities to grazing (*sensu* Whittam 1981). From long-term studies, Lowman successfully pinpointed "hot spots" in the canopy where grazing was high (see Figure 18-4). These hot spots included areas of predictable susceptibility to herbivores such as new leaf flushes, soft tissues for colonization, low shade regions of the canopy where insects aggregate to feed in the absence of predators, and canopy regions with flowers, epiphytes, and vines (Lowman 1992; Lowman et al. 1993). Long-term measurements of *Nothofagus moorei* (Nothofagaceae) showed crowns that exhibited different grazing levels with leaf age, different stands, and time (Selman and Lowman 1983). *Nothofagus moorei* had approximately eight cohorts of leaves present in the canopy at one point in time (see Figure 18-5), each with varying levels of susceptibility to insect attack. Young leaves that emerged during the spring

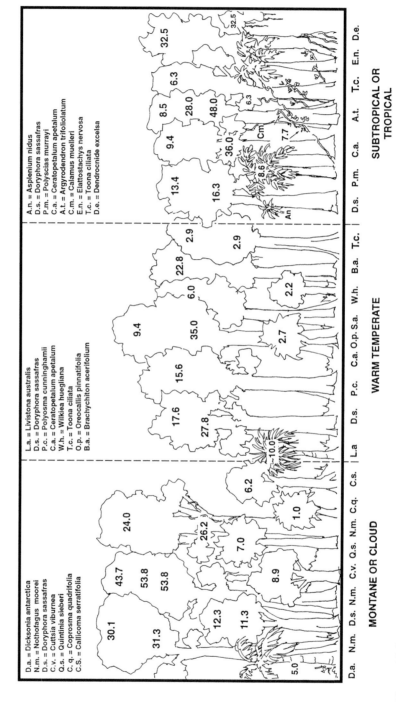

Figure 18-3 Variation in herbivory between species, heights, and sites along an elevational gradient in Australia (from Lowman 1995b; used with permission).

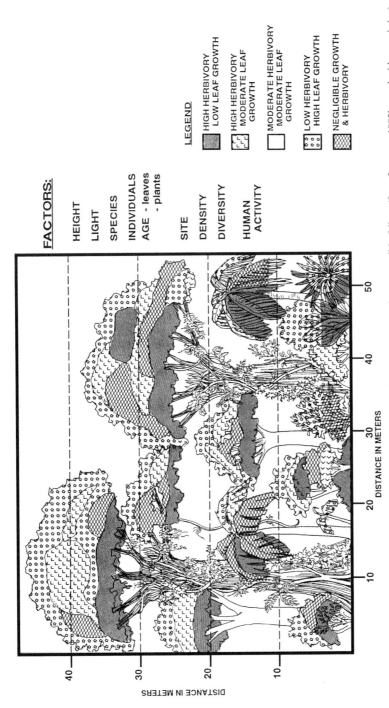

Figure 18-4 Schematic representation of canopy "hot spots" where herbivores are attracted to susceptible foliage (from Lowman 1995b; used with permission).

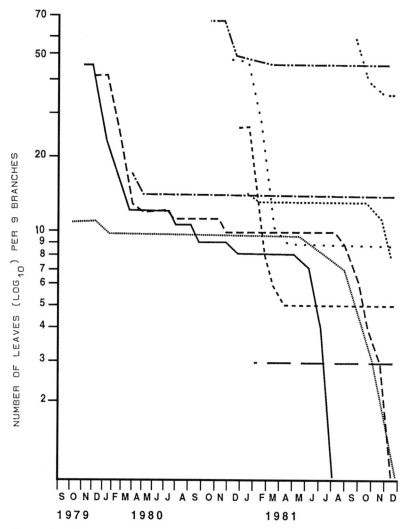

Figure 18-5 Leaf survivorship curves for leaves of *Nothofagus moorei* (Fagaceae) in Australia's cool temperate rainforests during a two-year period (September 1979 to September 1981), expressed on a $\log_{10}$ scale (from Lowman 1995b; used with permission).

(October and November) were the most preferred by the beetle larvae that emerged synchronously with flushing, whereas old leaves (greater than one year) from summer flushes and from the previous year were highly resistant to grazing. In addition, herbivory varied significantly between branches and individual crowns but not with light regimen or height (Selman and Lowman 1983).

More comparisons between forests are needed to understand the impact of herbivory on these ecosystems, though some progress was made during the last decade or so. For example, the annual levels of defoliation in Australian trees ranged from as low as 8 to 15 percent in subtropical rainforests to as high as 300 percent in dry sclerophyll (*Eucalyptus* spp.) stands (Lowman 1992; Lowman and Heatwole 1992). In this case, different mechanisms are clearly regulating insect defoliators and subsequent foliage responses in two forests. Gentry (1990) cited the need for comparisons between Central and South American forests as the stimulus for a symposium sponsored by the Association for Tropical Biology and his subsequent production of a volume on four neotropical forests. Similarly, a session for the American Institute of Biological Sciences was held to stimulate comparisons between neotropical and paleotropical forests (Lowman 1993). The prospect of increased ecological comparisons between forests, especially from different floristic zones, is an incentive to develop better protocols for field-sampling of events such as insect grazing. What species are appropriate to sample? Is there greater variation within or between forests? And how do we tackle these questions with statistical and biological rigor?

Herbivores as Mediators of Forest Processes

Herbivores have a variety of direct and indirect effects on plant communities (Seastedt and Crossley 1984; Ritchie et al. 1998). By consuming plant material, herbivores influence decomposition and nutrient cycling in communities (Pitelka 1964; Schultz 1964; Kitchell et al. 1979; Swank et al. 1981; Pastor and Cohen 1997). Experimental studies have confirmed the role of vertebrate herbivores as mediators of decomposition processes in terrestrial systems (McInnes et al. 1992; Molvar et al. 1993; Pastor et al. 1993; Ritchie et al. 1998). Both empirical (Ruess and Seagle 1994; Lovett and Ruesink 1995) and theoretical studies (Loreau 1995; De Mazancourt et al. 1998; De Mazancourt and Loreau 2000) suggest that herbivory, by both vertebrates and invertebrates, can have significant effects on the rates of decomposition and nutrient availability in soils.

Canopy processes (e.g., herbivory and defoliation) are coupled to forest floor processes (e.g., decomposition) through inputs of leaf and twig litter, canopy throughfall, and frass, the excretory products of insect digestion (Schowalter and Sabin 1991; Schowalter et al. 1991; Lovett and Ruesink 1995). Defoliation by insects in forests may impact primary productivity and nutrient cycling (Mattson and Addy 1975; Kitchell et al. 1979). Phytophagous insects in forest canopies drop materials into the soil community through two major pathways. First, defoliators introduce frass, greenfall (fragmented leaf tissue dropped during herbivory), and leaves abscised prematurely to the forest floor (e.g., Risley and Crossley 1993). Second, throughfall (rainwater modified by its passage through the forest canopy) is altered by the combination of dissolved frass and modified leachates from damaged leaves (e.g., Stadler and Michalzik 2000; Reynolds and Hunter 2001). These pathways combine to introduce increased amounts of carbon (C), nitrogen (N), and phosphorus (P) into the soil community (Reynolds et al. 2000; Reynolds and Hunter 2001). Increased activity by mites and springtails, comminuting these herbivore-derived inputs in the soil, will result in increased levels of decomposition.

Wilson (1987a) argued forcibly that insects in forests are vital components of the ecosystem with effects more pronounced than those of vertebrates. Although the impact of insects on forest systems is controversial (Terborgh 1988), it remains largely untested. A considerable proportion of forest canopies can be turned over annually by insect-herbivores (Lowman 1992), yet the consequences of herbivory for decomposers such as soil mesofauna are largely unexplored. Schowalter and Sabin (1991) reported increases in litter arthropod diversity and abundance following defoliation of saplings in a 10-year-old Douglas-fir forest, but the effects of defoliator inputs were not distinguished from defoliator-induced changes in microclimate.

As mentioned previously, evidence exists that herbivory can influence soil processes. Insect grazing can actually enhance nitrogen export from forest ecosystems (Swank et al. 1981; Reynolds et al. 2000). Low to moderate defoliation levels by forest insects can have significant effects on nutrient cycling (Schowalter et al. 1991). Some evidence suggests that nitrogen is immobilized in frass by fungal decomposers (Lovett and Ruesink 1995), although a field test of nitrogen dynamics following defoliation is sorely needed (Lerdau 1996). Likewise, the evidence is strong that defoliation influences the chemistry of throughfall (Seastedt and Crossley 1984;

THE LEIPZIG CANOPY CRANE PROJECT: BIODIVERSITY, ECOLOGY, AND FUNCTION IN A TEMPERATE DECIDUOUS FOREST

Wilfried Morawetz and Peter J. Horchler

The last decades' run on the biological investigation of forest canopies rarely included long-term studies on biodiversity in temperate forests. One exception is the Leipzig Canopy Crane Project, situated in a species-rich, near-natural deciduous forest in central-eastern Germany (Morawetz and Horchler 2002). The tower crane (Liebherr 71 EC), established in 2001, runs on rail tracks 120 m long (see Figure 1).

With its 45-m long jib, it covers an area of approximately 1.6 ha. With a gondola, scientists can reach almost any site in the three-dimensional forest space up to a height of 32 m. The project aims to identify spatio-temporal patterns of biodiversity and to link them functionally to environmental factors such as structure and microclimate. Since it is still in the pilot phase, only preliminary results are reported here.

Figure 1   Aerial photo of crane and rail track. Courtesy UFZ Leipzig-Halle AG.

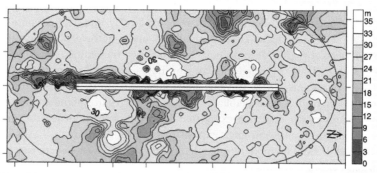

Figure 2   Map of the canopy topography in the crane plot. Note: The ellipsoid represents the area reached by the gondola (210 m × 90 m). The central depression (dark green and grey bar) reflects the rail track position, the other ones natural canopy gaps.

Stand Structure

The forest stand shows a variety of different stem classes and heights, and thus resembles a natural forest with a well-structured outer canopy, including natural tree fall gaps. The canopy topography has been measured in detail (see Figure 2), and some first vertical light (PAR) measurements allowed us to identify three distinct zones in the canopy similar to those found in an old Douglas-fir/western hemlock stand by Parker (1997). A very dark zone at the forest floor is followed by a zone of highly varying light conditions and leads to an upper zone of bright light.

Phenology

A detailed study of the trioecious tree species *Fraxinus excelsior L.* revealed a striking spatial and temporal variability in generative and vegetative phenology in the distribution of its three flower types as well as in its leaf and branching pattern and morphology. The investigation of this focal species will be continued and will include more detailed studies on reproductive ecology and population genetics.

Herbivory

Two preliminary studies aimed to assess the degree of stand-level herbivory revealed very low values (< 1 percent) in the range of coniferous forests. The data showed inter-specific differences (trends) in herbivory with high intra-specific variation. Some species also showed differences (trends) in herbivory at different heights (e.g., upper canopy vs. lower canopy).

Arthropods

Arthropods were collected by flight interception traps, barber traps, emergence traps, and trunk and branch traps. Preliminary results indicate a different fauna at the ground vs. the canopy, especially in the groups Carabidae, Lepidoptera, and Araneae. Arthropods mostly found in the canopy are often considered to be rare and threatened species in Germany. The majority of the collected animals still await identification.

Cryptograms

Lichens, which are very species-poor in Leipzig due to its dry climate and formerly bad air quality, are now colonizing the canopy, including some very rare species. Fungi are quite diverse (> 76 spp.), and some species show tendencies to settle in distinct canopy zones. Also, 15 species of slime molds (Myxomycota) were encountered in the canopy in a preliminary survey.

Bats

Preliminary studies of the flight activities of bats in the canopy revealed very complex adaptations to the habitat structure, with some tendency to use the upper canopy zone more frequently. Some completely new discoveries were also made simply due to the fact that formerly, almost no one reached this canopy habitat. The most notable finding has been the presence of the greenback frog (*Hyla arborea* L.) in the upper canopy (Schmidt et al. 2003).

Continued

THE LEIPZIG CANOPY CRANE PROJECT: BIODIVERSITY, ECOLOGY, AND FUNCTION IN A TEMPERATE DECIDUOUS FOREST–*cont'd*

References

Morawetz, W. and Horchler, P. (2002). The Leipzig Canopy Crane Project. *In* "The Global Canopy Handbook. Techniques of Access and Study in the Forest Roof" (A.W. Mitchell, K. Secoy, and T. Jackson, Eds.), pp. 54–57. Global Canopy Programme, Oxford.

Parker, G.G. (1997). Canopy structure and light environment of an old-growth Douglas-fir/Western Hemlock forest. *Northwest Science* **71**, 261–270.

Schmidt, C., Unterseher, M., and Grosse, W. R. (2003). Hoch hinaus—Sitzwarten beim Laubfrosch (*Hyla arborea* L.) in Baumkronen des Leipziger Auwalds. *Elaphe* **11**, Heft 2. S. 43–45.

Schowalter et al. 1991) and subsequent nutrient export (Swank et al. 1981). Interactions among canopy herbivores, soil mesofauna, and the processes of decomposition, however, remain to be quantified (Schowalter et al. 1986; Risley and Crossley 1993; Reynolds and Hunter 2001; Reynolds et al. 2003).

Thus, herbivory takes two major routes in influencing decomposition on the forest floor. First, solid materials drop to the floor during or following herbivory. Specifically, insect frass, green-fall, and prematurely abscised leaves represent major inputs to the soil community resulting from herbivory (Lovett and Ruesink 1995; Schowalter and Sabin 1991; Schowalter et al. 1991). Second, rainfall collects some products of herbivory and introduces those products to the soil in liquid form. This canopy throughfall represents the combined effects of dissolved insect frass and modified leachates from damaged foliage. Both pathways result in the input of carbon, nitrogen, and phosphorus to the decomposer community.

Forests as Regulators of Herbivory

Not only do herbivores affect plant communities, but the forests themselves influence herbivores through complex feedback mechanisms. Leaf emergence can have a "bottom-up" trophic effect on an ecosystem. Leaf abscission and leaf fall, too, can affect insect herbivores. Soil biologists have long recognized the influence of plant communities on ecosystem processes, but only recently have forest ecologists attempted to quantify those linkages (Wardle 2002).

Generalities about differences in leaf phenological patterns between temperate and tropical forests must first acknowledge the broad variation in a single latitude (Lowman 1995a). Nonetheless, most broad-leaved trees in temperate forests produce new foliage in spring and drop old foliage in fall. Foliage is largely absent for the rest of the year, as are insect folivores; and the availability of young expanding leaves shows a marked peak in the spring.

Trees exhibit significant variation in both budburst date and the timing of leaf abscission (Hunter 1992). Variation in the phenology of leaf flush and leaf loss occurs at several spatial scales with differences among forest types, among tree species, among individual trees within the same population, and even among canopy layers within individual plants (Lowman 1992, 1995a; Heatwole et al. 1997). The origins of such phenological variation (genetic, ontogenetic, and/or environmental) and the ecological consequences for the trees and the communities within which they live remain matters for debate (Phillipson and Thompson 1983; Hunter 1992). Phenological variation in leaf flush and leaf fall, however, clearly provides the kind of spatial and temporal heterogeneity in resource availability that determines the form of interactions among organisms in natural communities (reviewed in Dajoz 2000). For example, several authors have suggested that differences among individual trees in their budburst dates are related to insect herbivore performance and population densities and suffer greater levels of defoliation than those that burst bud late (Hunter 1990, 1992). Spatial variation in budburst phenology determines herbivore load and, consequently, influences canopy-wide defoliation levels and the distribution of frassfall and greenfall to the forest floor.

Though general agreement among aquatic ecologists exists for the widespread significance of trophic cascades (the effects of predators on the biomass of organisms at least two links away), the issue of whether cascades are similarly important in terrestrial communities is debatable (Pace et al. 1999; Holt 2000; Polis et al. 2000; Power 2000; Wardle 2002; Preisser 2003). In at least one system, variation in the timing of leaf expansion for forest trees is thought to have a cascading effect through the trophic system from plants through insect herbivores to avian predators (Hunter and Price 1992). In this case, the budburst phenology of *Quercus robur* varies spatially and temporally and drives the population dynamics of two spring lepidopterans. These herbivorous insects usually maintain populations below competitive levels; however, when competition does occur, the sensitivity of one species to budburst phenology reverses its competitive advantage it would have otherwise over the second species. Avian predators (e.g., *Parus* spp.) track the yearly changes in defoliator populations, exhibiting large clutches in years of high lepidopteran density.

Leaf abscission is another critical event in the dynamics of a forest. Spatial and temporal variation in the timing of leaf fall influences herbivore densities and may determine patterns of foraging by herbivores within and among trees. Leaf fall also represents one major pulse of resource input into the decomposer community on the forest floor. Thus, spatial and temporal variations in leaf abscission are reflected in variation in the activities of decomposers.

In European temperate forests, dominated by oaks, the temporal distributions of herbivores closely track leaf expansion. Feeny's (1970) work on the English oak, *Quercus robur*, demonstrated a spring-skewed species richness of oak herbivores. A second smaller peak in late summer/fall includes species that respond to a second flush of oak foliage resulting from spring defoliation. Yet more than 95 percent of the total defoliation on *Q. robur* occurs between budburst in April and the beginning of June, presumably because of seasonal declines in foliage quality (Feeny 1970; McNeill and Southwood 1978). As leaves age, they generally become lower in total nitrogen and water (Mattson 1980) and are often higher in fiber, lignin, and polyphenols than are younger leaves (Cates 1980). Inevitably, examples occur of insect-herbivores that prefer mature foliage (Cates 1980), but defoliation events are most usually associated with young leaves (Lowman 1992; Schowalter 2000). In rainforests, young leaves are more extensively grazed by insect-herbivores than are old leaves (Coley 1983; Lowman 1984, 1992; Coley and Aide 1991). In most deciduous forests in the eastern United States, folivory also is skewed toward newly emerged leaves (Reichle et al. 1973).

The tropics are not aseasonal. Most plants in the tropics produce new leaves periodically rather than continuously, and some synchrony occurs among different plant species, suggesting adaptive responses to biotic or abiotic variables (Van Schaik et al. 1993). Nonetheless, leaf emergence in tropical wet forests does not exhibit the same pronounced seasonal peak as in temperate forests. Much more inter-specific variability in leaf flush and leaf fall exists in tropical wet forests than in temperate forests. For example, in a French Guianan forest, each deciduous tree species appears to exhibit its own endogenous periodicity for shedding leaves (Loubry 1994). A 12-year study of flowering for 173 tree species at La Selva Biological Station in Costa Rica concluded that tree phenology was highly diverse, irregular, and complex (Newstrom et al. 1994). The authors also concluded that many tropical tree species show greater variation in phenology than do temperate species. Data on fruit fall from the Luquillo Experimental Forest in Puerto Rico suggest that, despite strong seasonal pulses in fruit production, a stand in some part of the forest is always in peak fruit production (Lugo and Frangi 1993). Tree species in Australian rainforest show a diversity of leaf flush phenologies from seasonal to continuous (Lowman 1992). Tropical dry forests show more seasonal patterns of leaf flush than do tropical wet forests, and the production of new foliage appears to peak toward the end of the dry season, perhaps avoiding the peak emergence of insect folivores that begins with the rains (Aide 1992). Even in tropical dry forest, however, more inter-specific variability in phenology is apparent than in temperate forests.

The responses of insect herbivores to leaf emergence are arguably even more pronounced in tropical wet forests than they are in temperate forests (Dajoz 2000). Most herbivory on tropical forest leaves occurs very early during expansion and may be even more skewed toward young leaves than defoliation in temperate forests (Reichle et al. 1973). The accumulation of herbivory throughout the forest should still occur more evenly through the year in wet tropical forests than in temperate forests (see Schowalter 2000). Despite clear pulses of leaf expansion in tropical wet forests, new leaves are still produced over extended periods. Data presented by Coley and Aide (1991) suggest, for example, that during eight months in Oak Ridge, Tennessee, no new leaves were produced on trees compared with only two months without new leaf production at La Selva, Costa Rica, and four months in Semego, Sarawak. Although both tropical wet forests and temperate forests exhibit seasonal peaks in leaf expansion, those peaks are broader in the tropics.

Methodologies to Assess Herbivory: Are They Accurate?

The turnover of photosynthetic tissues has direct impact on the growth and maintenance of the trees, vines, epiphytes, and herbaceous layers of forests (see Figure 18-6). The production of green tissue is also indirectly responsible for the maintenance of all animal life in the canopy. The ability of biologists to measure foliage and to predict photosynthetic activity has become an important topic for forest conservation. Obviously, it is important to make accurate measurements of

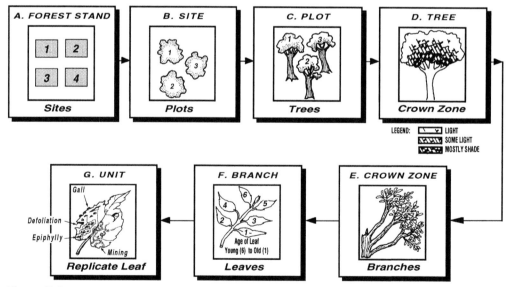

Figure 18-6 Experimental design for canopy foliage studies, illustrating the replication recommended at the spatial scales of forest stand, site, plot, tree, crown zone, branch, and leaf unit (from Lowman 1995b; used with permission).

both the production of foliage in a forest and of the removal of leaf material by herbivores to assess forest productivity.

The methods and spatial scales used to determine herbivory have direct consequences on our understanding of herbivore dynamics (see Figure 18-7). It should be noted forthwith that the source of variability in defoliation levels within and between forest sites might be a residual effect of different methodologies employed by researchers. Southwood (1961) examined insect species in tree canopies in Great Britain. For several species of temperate deciduous trees, he found that more diversity was associated with trees that had been established over a long period of time as compared to species introduced more recently. His insecticidal knockdown procedure has since been altered to include misting (Kitching et al. 1993), fogging (Erwin 1982), and restricted canopy fogging (Basset 1992) as more biologists become curious about the variety and numbers of insects in tree canopies. Similar to the artifacts of sampling for foliage consumption measurements, the

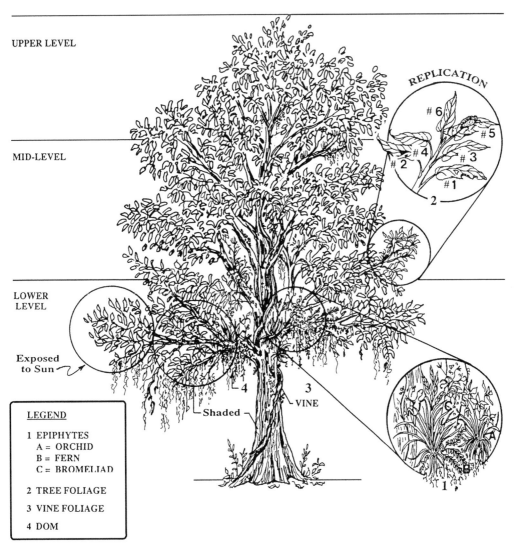

Figure 18-7 Schematic diagram of the components of a forest canopy to sample, including epiphytes (e.g., orchids, bromeliads, ferns), tree foliage, vines, and DOM (dead organic matter) (from Lowman 1995b; used with permission).

variability in methodologies to assess the diversity of insects in trees also leads to discrepancies in our estimates of herbivorous insects.

As biologists became interested in defoliation historically, a plethora of literature on herbivory emerged, much of which utilized different, but not entirely compatible, methodologies. For example, the techniques used to measure foliage losses included visual estimates (Wint 1983), graph paper tracings (Lowman 1984), templates in the field (Coley 1983), and leaf area meters (Lowman 1984). Similarly, the sampling designs varied dramatically, including collected leaves in litter traps (Odum and Ruiz-Reyes 1970), undefined leaf selection (Bray 1964), marked leaves in the understory (Coley 1983) and upperstory (Lowman 1985), marked leaves along a vertical transect (Lowman 1992), and collected frass (Ohmart et al. 1983). Obviously, such sampling designs may be adequate for a particular hypothesis, but they are not conducive to inter-site comparisons (Lowman 1987). In some cases, literature reviews have misquoted herbivory levels, perhaps because it is so difficult to interpret the various methodologies employed in different studies (Landsberg and Ohmart 1989).

Comparison of discrete versus long-term measurement techniques in rainforest canopies revealed discrepancies up to fivefold, with long-term studies producing significantly higher measurements than short-term ones. For example, estimates of herbivory levels in neotropical saplings were three times higher than in previous studies that used discrete harvested leaves (21 percent in Coley 1983 versus 7 percent in Odum and Ruiz-Reyes 1970). Similarly, long-term measurements of coral cay shrubs produced estimates of 21 percent missing area (Heatwole et al. 1981) compared to levels of 2 to 3 percent measured by discrete sampling (Lowman 1984). Grazing in some plant communities may be higher than previously recorded from discrete measurements of missing leaf area, resulting in an under-estimation of the impact of herbivory. Temporal variability in levels of herbivory further complicates our ability to monitor this canopy process.

In addition to the potential areas from methods that do not account for heterogeneity of foliage throughout the canopy, other methods must be used with caution. Daily rates of defoliation are useful but can be misleading if measured only over short durations, as they will not include seasonal differences. Ranking and other grazing categories are useful for rapid assessment, but such information is not transformable into statistical analyses. Further, the assumption that the absence of leaf petioles indicates 100 percent defoliation may be misleading because physical factors are also responsible for loss of leaf blades. The extent to which methods may alter results remains a critical issue in the literature on herbivory.

Conclusion: The Little Things vs. the Big Things that Sustain Forest Health

Aldo Leopold, an eloquent American pioneer for conservation, wrote in his now-famous *A Sand County Almanac* (1949):

> Land, then, is not merely soil; it is a fountain of energy flowing through a circuit of soils, plants, and animals. Food chains are the living channels that conduct energy upward; death and decay return it to the soil. The circuit is not closed; some energy is dissipated in decay, some is added by absorption from the air, some is stored in soils, peats, and long-lived forests; but it is a sustained circuit, like a slowly augmented revolving fund of life.

That sustained circuit is the life-blood of forest systems around the world.

In 1988, one year after E.O. Wilson inaugurated the invertebrate exhibit at the National Zoo, an article by John Terborgh appeared in *Conservation Biology* that seemingly disputed the previous argument by Wilson. Terborgh, then professor in Princeton University's biology department, reasoned that hunting out top carnivores and herbivores, while fragmenting the landscape into patches too small to maintain the whole interlocking ecological system, would erode biodiversity at all levels.

"In the end, this would work to the detriment of many of Professor Wilson's 'little things.' The essential point is that the big things are important, too; what is worrisome in these changing times is that they are so much more vulnerable" (Terborgh 1988). The "little things" may run the world, but the "big things" provide a stabilizing function in regional ecosystems via their functional longevity (Redford 1992). Yet, according to Terborgh, the "big things" are more susceptible to environmental perturbations and human exploitation than are fast-breeding, but short-lived, arthropods and worms. Thus, both the "little things" and the "big things" are indispensable to sustain the health and integrity of forest systems.

Terborgh's assertion seems almost disingenuous, or at least unconvincing, in the light of the sheer diversity, distribution, and abundance of arthropods. If there are indeed 30 million species of insects and just about 4,600 species of mammals, then beetles, bugs, ants, and flies far outweigh peccaries, pacas, and agoutis in both their biomass *and* their ecological significance to global food webs. If the "big things" were to disappear overnight, inevitably there would be short-term disruption of global ecosystems; however, after a geological flash, another kind of stability would probably settle across the planet. One need only remember the readjustment of woodland ecosystems after the rapid, poignant demise of the passenger pigeon, arguably a one-time keystone species for the forests of eastern North America. (e.g., see Wilcove 1989; Shaw 1995). On the other hand, if the "little things" were to vanish, one suspects that the ecological link between forest canopies and soils might be disrupted and, consequently, that both the health and integrity of the energy circuit be diminished. This chapter argues that insect herbivores in forest canopies and soil mesofauna are some of Leopold's little "cogs and wheels" that sustain this complex system.

Acknowledgments

The authors wish to thank the National Science Foundation (DEB-9815133), the Global Canopy Programme, and the Triad Foundation (Ithaca, New York) for their generous support for field research apropos to this chapter.

Appendix: Numbers of Estimated Host-Specific Beetle Species per Trophic Group on *Luehea semannii* (from Erwin 1983)

Trophic Group	Number of Species	Percent Host-Specific	Number Host-Specific
Herbivores	682	20%	136
Predators	296	5%	15
Fungivores	69	10%	7
Scavengers	96	5%	5
Total	1200+		163

Erwin's Estimation of 30 Million Species:

1. *Luehea* = 163/1200 − 13.6 percent host-specific species
2. Average Tropical Forest = 70 trees/ha
3. 163 × 70 = 11,410 host-specific beetle species/ha
4. 1200 total beetle species = 163 host-specific + 1037 transient species for one *Luehea*
5. 11,410 host-specific/ha + 1037 transients = 12,447 beetle species/ha
6. Beetles = 40 percent all insect species, 31,118 insect species/ha in canopy of tropical forest (12447/0.40)

7. Canopy species twice as rich as species on forest floor, add one-third more species to canopy figure: $31,118 + 10,373 = 41,491$ insect species/ha of tropical forest
8. $41,491$ insect species/70 trees/ha $= 593$ insect species/tree
9. 593 insect species/tree $\times$ 50,000 species of tropical trees $=$ approx. 30 million species

Areas of Uncertainty for Erwin's Calculations (see André et al. 1994):

1. What fraction of the beetle fauna on a given tree species is effectively specialized to it?
2. Do beetle species constitute the same fraction of insect fauna in the tropics that they do in better-studied temperate regions?
3. For each insect species in the canopy, how many other species are found elsewhere in or around a tree?
4. How do we scale up from the number of insect species associated with a given tree in a given place to a global total?
5. What is the relationship between insect fauna and tree phenology?

References

Aide, T.M. (1992). Dry season leaf production: an escape from herbivory. *Biotropica* **24**, 532–537.

André, H.M., Noti, M-I., and Lebrun, P. (1994). The soil fauna: the other last biotic frontier. *Biodiversity and Conservation* **3**, 45–56.

Barbosa, P. and Schultz, J.C. (1987). "Insect Outbreaks." Academic Press, Orlando, Florida.

Basset, Y. (1991). The spatial distribution of herbivory, mines and galls within an Australian rain forest tree. *Biotropica* **23**, 271–281.

Basset, Y. (1992). Host specificity of arboreal and free-living insect herbivores in rain forests. *Biol. J. Linn. Soc.* **47**, 115–133.

Basset, Y., Aberlenc, P.H., and Delvare, G. (1991). Abondance, diversite et stratification verticale de l'entomofaune d'une foret tropicale humide Africaine. *In* "Biologie d'Une Canopée de Forêt Équatoriale II" (F. Hallé and O. Pascal, Eds.), pp. 45–52. Opération Canopée, Lyons, France.

Basset, Y., Novotny, V., Miller, S.E., and Kitching, R.L. (2003). "Arthropods of Tropical Forests: Spatio-Temporal Dynamics and Resource Use in the Canopy." Cambridge University Press, Cambridge, UK.

Bray, J.R. (1964). Primary consumption in three forest canopies. *Ecology* **45**(1), 165–167.

Bray, J.R. and E. Gorham. (1964). Litter production in forests of the world. *Adv. Ecol. Res.* **2**, 101–157.

Cates, R.G. (1980). Feeding patterns of monophagous, oligophagous, and polyphagous insect herbivores: The effect of resource abundance and plant chemistry. *Oecologia* **46**, 22-31.

Clark, L. R. (1962). The general biology of *Cardiaspina albitextura* (Psyllidae) and its abundance in relation to weather and parasitism. *Australian Journal of Zoology* **10**, 537–586.

Coley, P.D. (1983). Herbivory and defensive characteristics of tree species in a lowland tropical forest. *Ecological Monographs* **53**, 209–233.

Coley, P.D. and T.M. Aide (1991). Comparison of herbivory and plant defenses in temperate and tropical broad-leaved forests. *In* "Plant-Animal Interactions: Evolutionary Ecology in Tropical and Temperate Regions" (P.W. Price, T.M. Lewinsohn, G.W. Fernandes, and W.W. Benson, Eds.), pp. 25–49. John Wiley and Sons, Inc., New York.

Colinvaux, Paul. (1997). Ancient sediments show rain forest covered Amazon Basin in last ice age. *Smithsonian Institution Research Report* **87**.

Dajoz, Roger. (2000). "Insects and Forests: The Role and Diversity of Insects in the Forest Environment." Lavoisier, Paris.

De Mazancourt, C., Loreau, M., and Abbadie, L. (1998). Grazing optimatization and nutrient cycling: When do herbivores enhance plant production? *Ecology* **79**, 2242–2252.

De Mazancourt, C.L. and Loreau, M. (2000). Effect of herbivory and plant species replacement on primary production. *The American Naturalist* **155**(6), 735–754.

DeVries, P.J. (1987). "The Butterflies of Costa Rica and Their Natural History: *Papilionidae, Pieridae, Nymphalidae*." Princeton University Press, Princeton, New Jersey.

Diamond, J. (1986). Overview: Laboratory experiments, field experiments and natural experiments. *In* "Community Ecology" (J. Diamond and T.J. Case, Eds.), pp. 3–22. Harper and Row, New York.

Erwin, T. (1982). Tropical forests: Their richness in Coleoptera and other arthropod species. *The Coleopterists Bulletin* **36**(1), 74–75.

Erwin, T. (1983). Tropical forest canopies: The last biotic frontier. *Bulletin of the Entomological Society of America* **29**, 14–19.

Feeny, P. (1970). Seasonal changes in oak leaf tannins and nutrients as a cause of spring feeding by winter moth caterpillars. *Ecology* **51**(4), 565–581.

Fox, L.R. and Morrow, P.A. (1983). Estimates of damage by insect grazing on *Eucalyptus* trees. *Australian Journal of Ecology* **8**, 139–147.

Gentry, A.H. (Ed.) (1990). "Four Neotropical Forests." Yale University Press, New Haven, CT.

Heatwole, H., Done, T., and Cameron, E. (1981). "Community Ecology of a Coral Cay, a Study of One Tree Island, Great Barrier Reef." Dr. W. Junk Publications, The Hague, The Netherlands.

Heatwole, H., Lowman, M.D., Donovan, C., and McCoy, M. (1997). Phenology of leaf-flushing and macroarthropod abundances in canopies of *Eucalyptus* saplings. *Selbyana* **18**(2), 200–214.

Hölldobler, B. and Wilson, E.O. (1990). "The Ants." Belknap Press of Harvard University Press, Cambridge, Massachusetts.

Holt, R.D. (2000). Trophic cascades in terrestrial ecosystems. Reflections on Polis *et al. Trends in Ecology and Evolution* **15**(11), 444–445.

Huffaker and Gutierrez. (1999). "Ecological Entomology." John Wiley and Sons, New York.

Hunter, M.D. (1990). Differential susceptibility to variable plant phenology and its roles in competition between two insect herbivores on oak. *Ecological Entomology* **15**, 401–408.

Hunter, M.D. (1992). A variable insect-plant interaction: The relationship between tree budburst phenology and population levels of insect herbivores among trees. *Ecological Entomology* **17**, 91–95.

Hunter, M.D. and Price, P.W. (1992). Playing chutes and ladders: Heterogeneity and the relative roles of bottom-up and top-down forces in natural communities. *Ecology* **73**, 724–732.

Janzen, D.H. (1983). "Costa Rican Natural History." University of Chicago Press, Chicago.

Kitchell, J.F., O'Neill, R.V., Webb, D., Gallep, G.A., Bartell, S.M., Koonce, J.F., and Ausmus, B.S. (1979). Consumer regulation of nutrient cycling. *Bioscience* **29**, 28–34.

Kitching, R.L., Bergelsohn, J.M., Lowman, M.D., MacIntyre, S., and Carruthers, G. (1993). The biodiversity of arthropods from Australian rain forest canopies: General introduction, methods, sites and ordinal results. *Australian Journal of Ecology* **18**, 181–191.

Koike, F. and Nagamitsu, Teruyoshi. (2003). Canopy foliage structure and flight density of butterfies and birds in Sarawak. *In* "Arthropods of Tropical Forests: Spation-Temporal Dynamics and Resource Use in the Canopy" (Y. Basset, N. Novotny, S. E. Miller, and R. L. Kitching, Eds.), pp. 86–91. Cambridge University Press, Cambridge, United Kingdom.

Kricher, J. (1997). "A Neotropical Companion: An Introduction to the Animals, Plants, and Ecosystems of the New World Tropics." Princeton University Press, Princeton, New Jersey.

Landsberg, J. and Ohmart, C.P. (1989). Levels of defoliation in forests: Patterns and concepts. *Trends in Ecology and Evolution* **4**, 96–100.

Lawton, J. H. and Schröder, D. (1997). Effects of plant type, size of geographic range, and taxonomic isolation on number of insect species associated with British plants. *Nature* **265**, 137–140.

Leopold, A. (1949/1966). "A Sand County Almanac." Oxford University Press, New York.

Lerdau, M. (1996). Insects and ecosystem function. *Trends in Ecology and Evolution* **11**, 151.

Linsenmair, K.E., Davis, A.J., Fiala, B., and Speight, M.R. (2001). "Tropical Forest Canopies: Ecology and Management." Kluwer Academic Publishers, Dordrecht, The Netherlands.

Loreau, M. (1995). Consumers as maximizers of matter and energy flow in ecosystems. *The American Naturalist* **145**, 22–42.

Loubry, D. (1994). Phenology of deciduous trees in a French Guianan forest (5 degrees latitude north): Case of a determinism with endogenous and exogenous component. *Canadian Journal of Botany* **72**, 1843–1857.

Lovett, G.M. and Ruesink, A.E. (1995). Carbon and nitrogen mineralization from decomposing Gypsy moth frass. *Oecologia* **104**, 133–138.

Lowman, M.D. (1982). "Leaf Growth Dynamics and Herbivory in Australian Rain Forest Canopies." Ph.D. Thesis, University of Sydney, Sydney, Australia.

Lowman, M.D. (1984). An assessment of techniques for measuring herbivory: Is rain forest defoliation more intense than we thought? *Biotropica* **16**(4), 264–268.

Lowman, M.D. (1985). Temporal and spatial variability in insect grazing of the canopies of five Australian rain forest tree species. *Australian Journal of Ecology* **10**, 7–24.

Lowman, M.D. (1987). Insect herbivory in Australian rain forests: is it higher than in the neotropics? *Proceedings of the Ecological Society of Australia* **14**, 109–119.

Lowman, M.D. (1988). Litterfall and leaf decay in three Australian rain forest formations. *Journal of Ecology* **76**, 451–465.

Lowman, M.D. (1992). Leaf growth dynamics and herbivory in five species of Australian rain forest canopy trees. *Journal of Ecology* **80**, 433–447.

Lowman, M.D. (1993). Forest canopy research: Old World, New World comparisons. *Selbyana* **14**, 1–2.

Lowman, M.D. (1995a). Herbivory in Australian forests: a comparison of dry sclerophyll and rain forest canopies. *Proceedings of the Linnaean Society* **115**, 77–87.

Lowman, M.D. (1995b). Herbivory as a canopy process in rain forest trees. In "Forest Canopies" (M.D. Lowman and N.M. Nadkarni, Eds.), pp. 431–455. Academic Press, San Diego.

Lowman, M.D. and Heatwole, H. (1992). Spatial and temporal variability in defoliation of Australian eucalypts. *Ecology* **73**, 129–142.

Lowman, M.D. and Moffett, M. (1993). The ecology of tropical rain forest canopies. *Trends in Ecology and Evolution* **8**(3), 104–107.

Lowman, M.D., Moffett, M., and Rinker, H.B. (1993). A new technique for taxonomic and ecological sampling in rain forest canopies. *Selbyana* **14**, 75–79.

Lowman, M.D. and Morrow, P. (1998). Insects and their environment: plants. In "The Science of Entomology" (W.S. Romoser and J.G. Stoffolano, Eds.), pp. 290–316. WCB McGraw-Hill, Boston, Massachusetts.

Lowman, M.D. and Nadkarni, N.M. (1995). "Forest Canopies." Academic Press, San Diego.

Lowman, M.D., Foster, R., Wittman, P., and Rinker, H.B. (1998). Herbivory and insect loads on epiphytes, vines, and host trees in the rain forest canopy of French Guinana. In "Biologie d'Une Canopée de Forêt Équatoriale III" (Francis Hallé, Ed.), pp. 116–128. Pro-Natura International and Opération Canopée, Paris, France.

Lowman, M.D. (1999). "Life in the Tree tops." Yale University Press, New Haven, Conn.

Lugo, A.E. and Frangi, J.L. (1993). Fruit fall in the Luquillo experimental forest, Puerto Rico. *Biotropica* **25**, 73–84.

MacArthur, R.H. (1969). Patterns of communities in the tropics. *Biological Journal of the Linnaean Society* **1**, 19–30.

MacArthur, R.H. (1972). "Geographic Ecology: Patterns in the Distribution of Species." Princeton University Press, Princeton, New Jersey.

Marquis, R. (1991). Herbivore fauna of *Piper* (Piperaceae) in a Costa Rican wet forest: Diversity, specificity and impact. In "Plant-Animal Interactions, Evolutionary Ecology in Tropical and Temperate Regions" (P. Price, T.M. Lewinsohn, G.W. Fernandes, and W.W. Benson, Eds.), pp. 179–208. Wiley, New York.

Mattson, W.J. (1980). Herbivory in relation to plant nitrogen content. *Annual Review of Ecology and Systematics* **11**, 119–161.

Mattson, W.J. and Addy, N.D. (1975). Phytophagous insects as regulators of forest primary production. *Science* **190**, 515–522.

Mawdsley, N.A. and Stork, N.E. (1997). Host-specificity and the effective specialization of tropical canopy beetles. In "Canopy Arthropods" (N.E. Stork, J. Adis, and R.K. Didham, Eds.), pp. 104–130. Chapman and Hall, London, UK.

McInnes, P.F., Naiman, R.J., Pastor, J., and Cohen, Y. (1992). Effects of moose browsing on vegetation and litter of the boreal forest, Isle Royale, Michigan, USA. *Ecology* **73**, 2059–2075.

McNeill, S. and Southwood, T.R.E. (1978). The role of nitrogen in the development of insect/plant relationships. In "Biochemical Aspects of Plant and Animal Coevolution: Proceedings of the Phytochemical Society Symposium, Reading, April 1977" (J.B. Harbourne et al., Ed.), pp. 77–98. Academic Press, New York.

Molvar, E.M., Boyer, R.T., and Vanballengerghe, V. (1993). Moose herbivory, browse quality, and nutrient cycling in an Alaskan treeline community. *Oecologia* **94**, 472–479.

Montgomery, G.G. (1978). "The Ecology of Arboreal Folivores." Smithsonian Institution, Washington, DC.

Morrow, P.A. and Fox, L.R. (1980). Effects of variation in *Eucalyptus* essential oil on insect growth and grazing damage. *Oecologia* **45**, 209–219.

Newstrom, L.E., Frankie, G.W., and Baker, H.G. (1994). A new classification for plant phenology based on flowering patterns in lowland tropical rain forest trees at La Selva, Costa Rica. *Biotropica* **26**(2), 141–159.

Novotny, V., Basset Y., and Kitching, R.L. (2003). Herbivore assemblages and their food resources. In "Arthropods of Tropical Forests: Spatio-Temporal Dynamics and Resource Use in the Canopy" (Y. Basset, V. Novotny, S.E. Miller, and R.L. Kitching, Eds.), pp. 40-53. Cambridge University Press, Cambridge, United Kingdom.

Odum, H.T. and Ruiz-Reyes, J. (1970). Holes in leaves and the grazing control mechanism. In "A Tropical Rain Forest" (H.T. Odum and R.F. Pigeon, Eds.), pp. I-69–I-80. U.S. Atomic Energy Commission, Oak Ridge, Tennessee.

Ohmart, C.P, Stewart, L.G., and Thomas, J.L. (1983). Phytophagous insect communities in the canopies of three *Eucalyptus* forest types in south-eastern Australia. *Australian Journal of Ecology* **8**, 395–403.

Pace, M.L., Cole, J.J., Carpenter, S.R., and Kitchell, J.F. (2000). Trophic cascades revealed in diverse ecosystems. *Trends in Ecology and Evolution* **14**(12), 483–488.

Pastor, J. and Y. Cohen. (1997). Herbivores, the functional diversity of plant species, and the cycling of nutrients in boreal ecosystems. *Theoretical Population Biology* **51**, 165–179.

Pastor, J., Dewey, B., Naiman, R.J., McInnes, P.F., and Cohen, Y. (1993). Moose browsing and soil fertility in the boreal forests of Isle Royale National Park. *Ecology* **74**, 467–480.

Phillipson, J. and Thompson, D.J. (1983). Phenology and intensity of phytophage attack on *Fagus sylvatica* in Wytham Woods, Oxford. *Ecological Entomology* **8**, 315–330.

Pitelka, F.A. (1964). The nutrient-recovery hypothesis for arctic microtine cycles. I. Introduction. *In* "Grazing in Terrestrial and Marine Environments" (D.J. Crisp, Ed.), pp. 55–56. Blackwell, Oxford, UK.

Polis, G.A., Sears, A.L.W., Huxel, G.R., Strong, D.R., and Maron, J. (2000). When is a trophic cascade a trophic cascade? *Trends in Ecology and Evolution* **15**(11), 473–475.

Power, M.E. (2000). What enables trophic cascades? Commentary on Polis *et al*. *Trends in Ecology and Evolution* **15**(11), 443–444.

Preisser, E.L. (2003). Field evidence for a rapidly cascading underground food web. *Ecology* **84**(4), 869–874.

Price, P.W., Lewinsohn, T.M., Fernandes, G.W., and Benson, W.W. (1991). "Plant-Animal Interactions, Evolutionary Ecology in Tropical and Temperate Regions." Wiley, New York.

Raven, P.H. and Johnson, G.B. (1996). "Biology." Wm. C. Brown, Dubuque, Iowa.

Readshaw, J.L. (1965). A theory of phasmatid outbreak realease. *Australian Journal of Zoology* **13**, 475–490.

Redford, K.H. (1992). The empty forest. *BioScience* **42**(6), 412–422.

Reichle, D.E., Goldstein, R.A., Van Hook, R.I., and Dodson, G.J. (1973). Analysis of insect consumption in a forest canopy. *Ecology* **54**, 1076–1084.

Reynolds, B.C., Crossley, D.A., and Hunter, M.D. (2003). Response of soil invertebrates to forest canopy inputs along a productivity gradient. *Pedobiologia* **47**, 127–139.

Reynolds, B.C. and Hunter, M.D. (2001). Responses of soil respiration, soil nutrients, and litter decomposition to inputs from canopy herbivores. *Soil Biology and Biochemistry* **33**(12, 13), 1641–1652.

Reynolds, B.C., Hunter, M.D., and Crossley, D.A. (2000). Effects of canopy herbivory on nutrient cycling in a northern hardwood forest in western North Carolina. *Selbyana* **21**(1,2), 74–78.

Ribeiro, S.P. (2003). Insect herbivores in the canopies of savannas and rainforests. *In* "Arthropods of Tropical Forests" (Y. Basset, V. Novotny, S.E. Miller, and R.L. Kitching, Eds.), pp. 348–359. Cambridge University Press, Cambridge, UK.

Rinker, H.B. (In preparation). The effects of canopy herbivory on soil microarthropods in a tropical rainforest.

Rinker, H.B., Lowman, M.D., Hunter, M.D., Schowalter, T.D., and Fonte, S.J. (2001). Literature review: canopy herbivory and soil ecology, the top-down impact of forest processes. *Selbyana* **22**(2), 225–231.

Risley, R.S. and Crossley, D.A. (1993). Contribution of herbivore-caused greenfall to litterfall nitrogen flux in several southern Appalachian forested watersheds. *American Midland Naturalist* **129**, 67–74.

Ritchie, M.E., Tilman, D., and Knops, J.M.H. (1998). Herbivore effects on plant and nitrogen dynamics in oak savanna. *Ecology* **79**, 165–177.

Romoser, W.S. and Stoffolano, J.G. (1998). "The Science of Entomology." WCB McGraw-Hill, Boston, Massachusetts.

Ruess, R.W. and Seagle, S.W. (1994). Landscape patterns in soil microbial processes in the Serengeti National Park, Tanzania. *Ecology* **75**, 892–904.

Schowalter, T.D. (2000). "Insect Ecology: An Ecosystem Approach." Academic Press, San Diego.

Schowalter, T.D., Hargrove, W.W., and Crossley, D.A. (1986). Herbivory in forested ecosystems. *Annual Review of Entomology* **31**, 177–196.

Schowalter, T.D. and Sabin, T.E. (1991). Litter microarthropod responses to canopy herbivory, season and decomposition in litterbags in a regenerating conifer ecosystem in western Oregon. *Biology and Fertility of Soils* **11**(93–96).

Schowalter, T.D., Sabin, T.E., Stafford, S.G., and Sexton, J.M. (1991). Phytophage effects on primary production, nutrient turnover, and litter decomposition of young Douglas-fir in western Oregon. *Forest Ecology and Management* **42**, 229–243.

Schowalter, T.D., Webb, J.W., and Crossley, D.A. (1981). Community structure and nutrient content of canopy arthropods in clearcut and uncut forest ecosystems. *Ecology* **62**, 1010–1019.

Schultz, A.M. (1964). The nutrient-recovery hypothesis for arctic microtine cycles. II. Ecosystem variables in relation to arctic microtine cycles. *In* "Grazing in Terrestrial and Marine Environments" (D.J. Crisp, Ed.), pp. 57–68. Blackwell, Oxford, England.

Schultz, J. and Baldwin, I. (1982). Oak leaf quality declines in response to defoliation by gypsy moth larvae. *Science* **217**, 149–151.

Seastedt, T.R. and Crossley, D.A. (1984). The influence of arthropods on ecosystems. *BioScience* **34**, 157–161.

Selman, B. and Lowman, M.D. (1983). The biology and herbivory rates of *Novacastria nothofagi* Selman (Coleoptera: Chrysomelidae), a new genus and species on *Nothofagus moorei* in Australian temperate rain forests. *Australian Journal of Ecology* **31**, 179–191.

Shaw, H.E. (1995). The return of the passenger pigeon. *Birding* **27**(3), 223–227.

Smith, A.P. (1982). Diet and feeding strategies of the marsupial sugar glider in temperate Australia. *Journal of Animal Ecology* **51**, 149–166.

Southwood, T.R.E. (1961). The number of species of insect associated with various trees. *J. Anim. Ecol.* **30**, 1–8.

Stadler, B. and Michalzik, B. (2000). Effects of phytophagous insects on micro-organisms and throughfall chemistry in forested ecosystems: herbivores as switches for the nutrient dynamics in the canopy. *Basic and Applied Ecology* **1**, 109–116.

Stork, N.E., Adis, J., and Didham, R.K. (Eds.) (1997). "Canopy Arthropods." Chapman and Hall, London.

Swank, W.T., Waide, J.B., Crossley Jr., D.A., and Todd, R.L. (1981). Insect defoliation enhances nitrate export from forest ecosystems. *Oecologia* **51**, 297–299.

Terborgh, J. (1992). "Diversity and the Tropical Rain Forest." Scientific American Library, New York.

Terborgh, John. (1988). The big things that run the world: a sequel to E.O. Wilson. *Conservation Biology* **2**, 402–403.

Van Schaik, C.P., Terborgh, J.W., and Wright, S.J. (1993). The phenology of tropical forests: Adaptive significance and consequences for primary consumers. *Annual Review of Ecology and Systematics* **24**, 353–377.

Ward, D.B. (1979), *Rare and Endangered Biota of Florida*, Volume Five: Plants, University Press of Florida, Gainesville.

Wardle, D.A. (2002). "Communities and Ecosystems: Linking the Aboveground and Belowground Components." Princeton University Press, Princeton.

Whittman, T.G. (1981). Individual trees at heterogeneous environments: Adaptation to herbivory or epigenetic noise? *In* "Species and Life History Patterns: Geographic and Habitat Variations" (R.F. Denno and H. Dingle, Eds.), pp. 9–27. Springer, New York.

Wilcove, D. (1989). In memory of Martha and her kind. *Audubon* **September**, 52–55.

Wilson, E.O. (1987a). The little things that run the world: The importance and conservation of invertebrates. *Conservation Biology* **1**, 344–346.

Wilson, E.O. (1987b). The earliest known ants: An analysis of the Cretaceous species and an inference concerning their social organization. *Paleobiology* **13**(1), 44–53.

Wilson, E.O. (1992). "The Diversity of Life." Belknap Press of Harvard University Press, Cambridge, MA.

Wint, G.R.W. (1983). Leaf damage in tropical rain forest canopies. *In* "Tropical Rain Forest: Ecology and Management" (S.L. Sutton, T.C. Whitmore, and A.C. Chadwick, Eds.), pp. 229–240. Blackwell, Oxford, UK.

Woinarski, J.C.Z. and Cullen, J.M. (1984). Distribution of invertebrates on foliage in forests of southeastern Australia. *Australian Journal of Ecology* **9**, 207–233.

CHAPTER 19

Nutrient Cycling

Barbara C. Reynolds and Mark D. Hunter

If the biota, in the course of aeons, has built something we like but do not understand,
then who but a fool would discard seemingly useless parts?
To keep every cog and wheel is the first precaution of intelligent tinkering.
—Aldo Leopold, A Sand County Almanac with Essays on
Conservation from Round River, *1966*

Introduction

Most chemical elements tend to circulate between biotic and abiotic components of the biosphere. Because these paths are usually circular, and because the role of organisms is so critical in the process, the paths are known as biogeochemical cycles. In particular, the movement of those elements and compounds essential to life is termed nutrient cycling (Odum 1971). Although the term "canopy" is frequently used in reference to all above-ground vegetation in a plant community (Moffett 2000; Rinker et al. 2001), for the purposes of this chapter, we will be referring in large part to the tree crown (the upper part of the tree), all its associated organisms, and the ecological processes such as nutrient cycling and herbivory (Lowman 1995) occurring among these components. Therefore, in this chapter on nutrient cycling in the canopy, we will address nutrient inputs to the canopy, nutrient cycling within the canopy, and, to broaden our perspective, nutrient inputs from the canopy to the forest floor in tropical and temperate forests (see Figure 19-1).

Nutrient Inputs to the Canopy

Abiotic Inputs

Although nutrient inputs to the canopy from the soil/root interface are substantial (Kimmins 1997), this chapter deals with direct inputs to the canopy. These inputs occur through wet deposition, dry deposition, and cloud deposition (Lovett 1994) (see Figure 19-1). Elements deposited by these mechanisms can include nitrogen, sulfur, sodium, chloride, and potassium. Wet deposition is the deposition of nutrients in rainfall or snowfall. It includes gases and particles that are incorporated into cloud droplets and later fall as rain or snow. Dry deposition refers to the direct deposition of atmospheric particles and gases to the canopy; cloud or fog deposition—sometimes referred to as occult precipitation (Weathers 1999)—is the deposition of non-precipitating droplets of clouds and fogs to surfaces (Lovett 1994).

Dry deposition, at least in the United States, is usually a significant source of deposition from the atmosphere and may sometimes dominate the total atmospheric deposition (Lovett 1994).

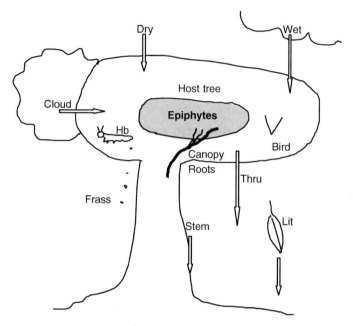

Figure 19-1 Canopy nutrient cycling model. Nutrients enter the canopy through wet deposition (Wet), dry deposition (Dry), cloud deposition (Cloud), or roosting/nesting of birds and other animals. In some systems, nutrients in the epiphytes may be transferred to the host tree via canopy roots. Nutrients may leave the canopy via stemflow (Stem), throughfall (Thru), litterfall (Lit), or droppings (Frass) of canopy herbivores (Hb). (After Coxson and Nadkarni 1995.)

Cloud moisture, intercepted as horizontal precipitation by tropical montane cloud forests, can supply a significant portion of water during periods of low rainfall. For example, in some tropical montane cloud forests with large cloud moisture inputs, more water reached the forest floor from deposition of cloud moisture than was measured in incident rainfall (Weaver 1972).

The physical structure of the canopy can have a strong influence on the interception of nutrients. For example, conifers have a higher canopy filtering efficiency than do hardwoods (Neirynck et al. 2002), and the complex layers of the rainforest canopy, including the many epiphytes and epiphyllous organisms (e.g., bacteria and lichens that live non-parasitically on the leaves), take up moisture and nutrients from precipitation (Coxson and Nadkarni 1995; Cavelier et al. 1997; Kimmins 1997).

Lovett (1994) reviewed a number of studies that indicate forest edges have higher rates of deposition than do nearby forest interiors. In fact, forest edges, which increase with forest fragmentation, act as traps and concentrators for airborne nutrients and pollutants. Weathers et al. (2001) found that unthinned forest edges increased atmospheric sulfur deposition compared to forest interiors and an open field.

Proximity to sources of atmospheric nutrients also affects canopy inputs. Ions such as sodium, chloride, and potassium may be more prevalent in aerosols from marine environments (Coxson and Nadkarni 1995; Cavelier et al. 1997) than from terrestrial systems. In coastal areas near upwelling zones, the ocean may be a major source of organic nitrogen (Weathers et al. 2000). Anthropogenic sources such as cities and power plants can increase deposition rates of pollutants like nitrate and sulfur hundreds of kilometers downwind (Lovett 1994). Canopies in high-elevation sites near sources of pollution, such as Mount Mitchell in the eastern United States, may suffer a "double-whammy," as mountains experience greater wind speeds, humidity, and cloud cover, leading to increased deposition rates (Lovett 1994). In addition (a triple-whammy?), the

Mount Mitchell canopy and other high-elevation forests in eastern North America are comprised of efficient-scavenging conifers. The needle-like leaves of these conifers act like a comb to extract more gases and particulates out of the atmosphere than broad-leaved trees can extract. Cloudwater pH at Mount Mitchell has been measured as low as 2.29 (Li and Aneja 1992).

Biotic Inputs

In addition to abiotic inputs of nutrients into the canopy, some canopy epiphytes are capable of taking nitrogen gas from the atmosphere and incorporating it into their tissues through the process of nitrogen fixation. Cyanobacteria associated with lichens and living on leaves (epiphyllous) have been found in forest canopies, both tropical and temperate. Some evidence indicates that a significant portion of the fixed nitrogen can be transferred from the epiphyll to the host leaf. In fact, as much as 25 percent of the total nitrogen in canopy leaves of a rainforest may be derived from epiphyllic microorganisms (Bentley and Carpenter 1984). However, the nitrogenase enzyme responsible for this nitrogen fixation is dependent on appropriate moisture levels and, thus, humidity is probably a major factor controlling nitrogen fixation. Another route by which nitrogen may be fixed is through nodules in above-ground adventitious roots found in red alders (Coxson and Nadkarni 1995).

What role might animals have in bringing nutrients into the canopy? It would seem that birds (see Figure 19-2), large bats (e.g., the flying fox, *Pteropus conspicillatus*), and insects, perching, roosting, or feeding in the canopy must account for some nutrient inputs. Evidence to support biotic inputs is presented by Weir (1969), who reported that rooks feeding in agricultural areas during the day brought in 6.1, 9.5, and 89.2 kg ha-1 of sodium, potassium, and calcium, respectively, to their woodland roosts over a two-month period. McCune et al. (2000) hypothesized that the presence of distinctive communities of epiphytes in the tallest trees in their Pacific Northwest study area was due to the delivery of lichen propagules and nutrients by birds. Immigrating insects that ultimately die and decompose within the canopy of trees are likely to represent significant nutrient inputs to forest canopies. However, explicit estimation of these inputs has yet to be made. With new techniques that increase the ease of canopy access for scientific study, the degree to which fauna act to import nutrients to the crowns of trees should be a subject for future research.

Nutrient Cycling within the Canopy

Many of the nutrients received by the canopy are taken up and cycled within canopy systems. These cycles occur along a continuum from open to closed, in terms of exposure to and leakage of nutrients. Open cycles are those in which a significant proportion of nutrients entrapped by the canopy are lost via stemflow (precipitation that drips down the boles of trees), throughfall (precipitation that falls through the canopy), or litterfall to the forest floor (see Figure 19-1). Open cycles are exemplified by those canopy epiphytes that trap and use nutrients from the atmosphere but also may enhance the deposition of water and nutrients to the forest floor (Knops et al. 1996). Examples of closed cycles are the microcosm ecology of the tank bromeliads (Richardson, this volume) and adventitious roots growing from tree stems into epiphytes and their associated organic matter (Nadkarni 1994). In these systems, nutrient cycling is tightly coupled and only a small proportion of nutrients reaches the forest floor. An excellent review of the roles of epiphytes in nutrient cycles was provided in Coxson and Nadkarni (1995).

Canopy epiphytes are plants and other photosynthetic organisms that live on the stems and branches of trees, receiving much of their nutritional support from the atmosphere and their physical support from the trees (phorophytes). Epiphytes may be vascular plants such as bromeliads, orchids, and ferns (see Figure 19-3) or nonvascular (cryptogamic) epiphytes such as lichens

Figure 19-2 Kookaburra *(Dacelo novaeguinae)* in Australia.

and mosses. Epiphytic material (EM, after Coxson and Nadkarni 1995) is comprised of living and dead epiphytes plus their associated organic matter, microbes, fungi, and invertebrates. As in forest floor soils, canopy invertebrates are of major importance in the decomposition of the organic matter (Paoletti et al. 1991; Fagan and Neville 1999). The concentration of EM organic matter in cloud forests of Costa Rica has recently been shown to be significantly higher than in terrestrial soil and, thus, could be of great utility in nutrient conservation for epiphytes, trees, and indeed the entire ecosystem (Nadkarni et al. 2002).

Although epiphytes have been used as examples of commensalism (benefiting from their partnership without affecting their partner) (Raven and Berg 2001), Benzing (1995) reasoned that epiphytes in relatively dry forests, using the CAM photosynthetic pathway, may intercept some nutrients from trees and reduce productivity of the whole system. Conversely, in humid forests, where epiphytes and their associated organic matter accumulate, adventitious roots from the phorophytes may exploit the nutrients in these epiphytic systems (Nadkarni 1981, 1994) (see Figure 19-1). Thus, some epiphytes may act as indirect parasites whereas others may be mutualists.

Open Cycle

As an example of an open nutrient cycle in the canopy, epiphytic lichens in a blue oak savanna woodland in California have been shown to increase the input of nitrogen and phosphorus from

Figure 19-3 Basket fern *(Drynaria rigidula),* an epiphyte in Australia.

the atmosphere and increase the deposition of nitrogen and other ions under the canopy (Knops et al. 1996). By comparing oak trees that had most of their epiphytic lichens experimentally removed with oak trees having an intact canopy, Knops et al. (1996) demonstrated that oaks with their canopy lichens intact intercepted approximately twice as much precipitation as the lichen-stripped oaks. In addition, throughfall collected under the intact canopy contained significantly greater concentrations of total nitrogen, organic nitrogen, calcium, magnesium, sodium, and chloride. Nutrients for which there was no significant difference measured included nitrate, ammonium, total phosphorus, and potassium. Litterfall from the lichen canopy also contributed to forest floor nutrients; 15 percent of the total nitrogen deposited by litterfall was derived from lichens. Dry deposition at this site was estimated to be ten times higher than wet deposition. Taking into account the long dry periods at this site, the authors estimate that enhanced deposition caused by the lichens in the canopy could contribute 2.85 kg N/ha/yr.

Closed Cycle

Phytotelmata (plant pools) may be considered to be examples of relatively closed nutrient cycles within the canopy in that nutrients entering into the plant base are cycled within a whole community of aquatic animals (Paoletti et al. 1991). These communities are essentially microcosms where nutrient inputs from precipitation and canopy-derived organic matter can be measured accurately, as can the resident aquatic organisms (Richardson et al. 2000a). Tank bromeliads in dwarf forests of Puerto Rico accumulate around 25 percent of potassium and phosphorus inputs into the forest and may comprise almost 13 percent of the forest net primary productivity. Within these bromeliads, invertebrates break down the canopy litter inputs into fine particulate organic matter that washes into the leaf bases as a fine organic soil (Richardson et al. 2000b).

Nutrient outputs from these phytotelmata are primarily litterfall, including whole plants. Few studies have been done separating epiphytic material from total litterfall. In those studies, Coxson and Nadkarni (1995) found that EM litterfall was distributed in extremely patchy fashion both in time and space. Living EM also reaches the forest floor, dislodged by wind or animals, or riding on a fallen branch or tree (Coxson and Nadkarni 1995). Tank bromeliads, having poorly developed root systems, tend to fall as individuals. Other epiphytes more often fall as a clump of intertwined plants and organic matter. The decomposition of these clumps, and subsequent release of nutrients, has been measured as greater than one year in the Monteverde Cloud Forest Reserve, Costa Rica (Matelson et al. 1993). Thus, there is a potential lag effect in nutrient release from fallen live epiphytes.

The role of canopy invertebrates in nutrient cycling needs further examination. Richardson et al. (2000a) have established the importance of invertebrates in the relatively closed systems of phytotelmata. Another example of invertebrates participating in nutrient cycling within the canopy is that of ant gardens, a mutualistic system that develops between an arboreal ant nest constructed around the roots of one or more canopy epiphytes, common in the canopy of Amazonian forests. Ants help to recycle nutrients by bringing in organic material such as dead insects, nectar, and seeds to feed their larvae in the nests. Roots of the epiphytes extract nutrients from some of this organic matter (Cedeno et al. 1999). Although recent studies of canopy arthropods in less specialized habitats (Nadkarni and Longino 1990; Rodgers and Kitching 1998; Halaj et al. 2000; and Basset 2001) have determined the presence of taxa well-known for their important functions in decomposition and nutrient cycling in the soil, little is known about their activities in the canopy. In fact, Basset (2001) pointed out that even the abundance of several significant groups such as Acari and Collembola has probably been seriously underestimated in several prominent tropical habitats. We do know that mites, scavenging on epiphyllous fungi (Walter and O'Dowd, 1995), mobilize nutrients sequestered in leaf fungi, algae, and other epiphylls as they graze and defecate.

Nutrient Inputs from the Canopy to the Soil

The literature on canopy nutrient inputs to the soil is extensive and may be divided into studies that deal with processes not known to be mediated by biota and studies that focus on herbivore-mediated inputs. Three major categories of inputs from the canopy, not mediated primarily by biota, include litterfall, throughfall, and stemflow (Parker 1983; Potter et al. 1991; Kimmins 1997) (see Figure 19-1). Litterfall, comprised of leaves, twigs, and EM, is one major source of nutrients. Other inputs, contained in water intercepted by the canopy, are primarily in the form of throughfall and stemflow and are also important components in nutrient cycling within the forest ecosystem. Elements such as potassium, sodium, and sulfur enter the forest soil primarily via throughfall and stemflow with little contribution from litterfall. However, some of these nutrients originate from leaching of canopy foliage, especially potassium and carbon compounds.

Throughfall typically contains significantly greater concentrations of some nutrients than the incident precipitation, especially cations such as potassium, sodium, and magnesium (Parker 1983). In a regenerating canopy in the southern Appalachians, throughfall contained greater amounts of sulfate, phosphate, potassium, and calcium than did the incident precipitation. However, nitrogen, in the forms of nitrate and ammonium, was removed by the canopy since throughfall concentrations of these ions were less than those in precipitation (Potter et al. 1991). In a tropical montane forest, ammonium, but not nitrate, was enriched in throughfall compared to precipitation (McDowell 1998).

Stemflow importance varies from 3 to 46 percent of total incident precipitation, depending on branch and stand morphology and on bark roughness. In forests where stemflow is considerable, precipitation is concentrated near the base of the trees (Kimmins 1997). Stemflow contains an average of 12 percent of waterborne nutrients and contributes large amounts of calcium, sulfate, sodium, and manganese in forests where stemflow is a significant portion of total precipitation (Parker 1983).

Unlike waterborne nutrients in throughfall and stemflow, nutrients in litterfall are not immediately available for plant uptake since the litter must first decompose. In temperate forests, most litterfall occurs during a few weeks out of the year. Litterfall typically contributes the majority of nitrogen, calcium, and magnesium that is lost from vegetation. Whereas stemflow and throughfall contribute from 1 to 11 kg ha-1 yr-1 nitrogen to the soil system, litterfall may contain between 11 and 200 kg ha-1 yr-1 (Kimmins 1997).

Role of Fauna

Insects and other herbivores feeding in the crowns of trees contribute to nutrient input from the canopy and probably increase the rate of nutrient cycling in many forest systems (see Hunter 2001 and Rinker et al. 2001 for literature reviews). Solid materials, such as insect frass (feces), greenfall (leaf fragments dropped by herbivores), and prematurely abscised leaves represent major inputs to the forest soil resulting from herbivory (Schowalter et al. 1991; Lovett and Ruesink 1995; Reynolds et al. 2000). Rainfall carries some products of herbivory as throughfall, combining the effects of dissolved insect frass and modified leachates from damaged foliage (Rinker et al. 2001). During outbreaks of insect herbivores, significant increases in nutrients such as nitrogen have been measured at the ecosystem level (Swank et al. 1981; Eshleman et al. 1998; Reynolds et al. 2000).

Frass

Insect frass and cadavers represent a potentially important source of nutrients entering the soil system during the growing season in temperate forests. In the southern Appalachians, nitrogen inputs in frass were estimated at between 0.3 and 1.1 kg ha-1 yr-1, approximately 2 to 4 percent of that in annual litterfall for the same forest under endemic (non-outbreak) conditions (Hunter et al. 2003). During a short outbreak of sawflies, *Periclista* sp. (Hymenoptera: Symphyta), nitrogen inputs in frass were as large as 30 kg ha-1 yr-1 (B.C. Reynolds and M.D. Hunter, unpublished data). It is not surprising, then, that nitrogen returning to the forest floor as frass can sometimes exceed that in litterfall (Grace 1986).

Nutrients from frass may stimulate populations of soil invertebrates. For example, experimental additions of frass to forest floor plots in the southern Appalachians resulted in increased numbers of Collembola, fungal-feeding nematodes, bacterial-feeding nematodes, and prostigmatid mites (Reynolds et al. 2003). These soil invertebrates are probably responding to fungal and bacterial blooms resulting from the additions of nutrient-rich frass (Coleman and Crossley 1996).

Throughfall

Insect frass may contribute to nutrient concentrations in throughfall (Hunter 2001; Rinker et al. 2001). Recent research in the southern Appalachians has shown that nitrate in throughfall was strongly correlated with nitrogen in frass at low elevations (Hunter et al. 2003). In contrast, under endemic levels of herbivory, Stadler et al. (2001) found concentrations of inorganic nitrogen in throughfall were reduced underneath trees fed upon by aphids and caterpillars compared to uninfested trees. These results are attributed to immobilization of nutrients by microbes living on the leaf surfaces.

General Conclusions and Future Studies

Nutrient cycling in the canopy is comprised of many complex processes, and several key questions remain to be answered in most canopy systems. First, we need to understand the relative magnitude of abiotic (nutrient deposition) and biotic (fixation, translocation) inputs of nutrients into canopies. While studies of pollutants have provided reliable estimates of abiotic deposition in some forest systems (Lovett 1994), we know relatively little about the contributions of flora and fauna to canopy inputs. Mobile organisms such as birds, bats, lizards, and insects seem to have considerable potential to translocate nutrients from outside canopy systems and contribute to nutrient dynamics. Second, we need better estimates of how "leaky" forest canopies are to nutrient losses. While we might consider phytotelmata as "closed" systems and folivore activities as "open" systems, it is likely that forest canopies vary considerably in the degree to which they exchange materials with the atmosphere and forest floor. It is not enough to measure import and export of materials. Rather, we must compare their relative importance to the cycling that occurs within forest canopies. Finally, as in forest floors, nutrient cycling in forest canopies cannot proceed without the participation of microbial flora and their associated food webs. While there has been much emphasis on the role of soil food webs in decomposition and nitrogen cycling in the forest floor (Coleman and Crossley 1996; Neutel et al. 2002), decomposer food webs in forest canopies are virtually unexplored. We see this as a critical focus for future work.

References

Basset, Y. (2001). Invertebrates in the canopy of tropical rain forests: how much do we really know? *Plant Ecology* **153**(1–2), 87–107.

Bentley, B.L. and Carpenter, E.J. (1984). Direct transfer of newly-fixed nitrogen from free-living epiphyllous microorganisms to their host plant. *Oecologia* **63**, 52–56.

Benzing, D.H. (1995). Vascular Epiphytes. *In* "Forest Canopies" (M.D. Lowman and N.M. Nadkarni, Eds.), pp. 225–254. Academic Press, San Diego.

Cavelier, J., Jaramillo, M., Solis, D., and deLeon, D. (1997). Water balance and nutrient inputs in bulk precipitation in tropical montane cloud forest in Panama. *Journal of Hydrology* **193**, 83–96.

Cedeno, A., Merida, T., and Zegarra, J. (1999). Ant gardens of Surumoni, Venezuela. *Selbyana* **20**(1), 125–132.

Coleman, D.C. and Crossley, Jr., D.A. (1996). "Fundamentals of Soil Ecology." Academic Press, San Diego.

Coxson, D.S. and Nadkarni, N.M. (1995). Ecological roles of epiphytes in nutrient cycles of forest ecosystems. *In* "Forest Canopies" (M.D. Lowman and N.M. Nadkarni, Eds.), pp. 495–543. Academic Press, San Diego.

Eshleman, K.N., Morgan II, R.P., Webb, J.R., Deviney, F.A., and Galloway, J.N. (1998). Temporal patterns of nitrogen leakage from mid-Appalachian forested watersheds: role of insect defoliation. *Water Resources Research* **34**(8), 2005–2116.

Fagan, L.L.W. and Neville, N. (1999). Arboreal arthropods: diversity and rates of colonization in a temperate montane forest. *Selbyana* **20**(1), 171–178.

Grace, J.R. (1986). The influence of gypsy moth on the composition and nutrient content of litter fall in a Pennsylvania oak forest. *Forest Sci.* **32**(4), 855–870.

Halaj, J., Ross, D.W., and Moldenke, A.R. (2000). Importance of habitat structure to the arthropod food-web in Douglas-fir canopies. *Oikos* **90**,139–152.

Hunter, M.D. (2001). Insect population dynamics meets ecosystem ecology: effects of herbivory on soil nutrient dynamics. *Agricultural and Forest Entomology* **3**, 77–84.

Hunter, M.D., Linnen, C.R., and Reynolds, B.C. (2003). Effects of endemic densities of canopy herbivores on nutrient dynamics along a gradient in elevation in the Southern Appalachians. *Pedobiologia* **47, XXXX**.

Kimmins, J.P. (1997). "Forest Ecology." Prentice-Hall, Upper Saddle River, New Jersey.

Knops, J.M.H., Nash III, T.H., and Schlesinger, W.H. (1996). The influence of epiphytic lichens on the nutrient cycling of an oak woodland. *Ecological Monographs* **66**(2), 159–180.

Li, Z. and Aneja, V.P. (1992). Regional analysis of cloud chemistry at high elevations in the eastern United States. *Atmospheric Environment Part A General Topics* **26**(11), 2001–2017.

Lovett, G.M. (1994). Atmospheric deposition of nutrients and pollutants in North America: an ecological perspective. *Ecological Applications* **4**(4), 629–650.

Lovett, G.M. and Ruesink, A.E. (1995). Carbon and nitrogen mineralization from decomposing gypsy moth frass. *Oecologia* **104**, 133–138.

Lowman, M.D. (1995). Herbivory as a canopy process in rainforest trees. *In* "Forest Canopies" (M.D. Lowman and N.M. Nadkarni, Eds.), pp. 431–455. Academic Press, San Diego.

Matelson. T.J., Nadkarni, N.M., and Longino, J.T. (1993). Longevity of fallen epiphytes in a neotropical montane forest. *Ecology* **74**(1), 265–269.

McCune, B., Rosentreter, R., Ponzetti, J., and Shaw, D.C. (2000). Epiphyte habitats in an old conifer forest in Western Washington, U.S.A. *Bryologist* **103**(3), 417–427.

McDowell, W.H. (1998). Internal nutrient fluxes in a Puerto Rican rain forest. *Journal of Tropical Ecology* **14**(4), 521–536.

Moffett, M.W. (2000). What's "Up"? A critical look at the basic terms of canopy biology. *Biotropica* **32**(4a), 569–596.

Nadkarni, N.M. (1981). Canopy roots: convergent evolution in rainforest nutrient cycles. *Science* **214**, 1023–1024.

Nadkarni, N.M. (1994). Factors affecting the initiation and growth of aboveground adventitious roots in a tropical cloud forest tree: an experimental approach. *Oecologia* **100**, 94–97.

Nadkarni, N.M. and Longino, J.T. (1990). Invertebrates in canopy and ground organic matter in a neotropical montane forest, Costa Rica. *Biotropica* **22**(3), 286–289.

Nadkarni, N.M., Schaefer, D., Matelson, T.J., and Solano, R. (2002). Comparison of arboreal and terrestrial soil characteristics in a lower montane forest, Monteverde, Costa Rica. *Pedobiologia* **46**(1), 24–33.

Neirynck, J., Van Ranst, E., Roskams, P., and Lust, N. (2002). Impact of decreasing throughfall depositions on soil solution chemistry at coniferous monitoring sites in northern Belgium. *Forest Ecology and Management* **160**, 127–142.

Neutel, A-M., Heesterbeek, J.A.P., and de Ruiter, P.C. (2002). Stability in real food webs: weak links in long loops. *Science* **296**, 1120–1123.

Odum, E.P. (1971). "Fundamentals of Ecology." W.B. Saunders Company, Philadelphia.

Paoletti, M.G., Taylor, R.A.J., Stinner, B.R., Stinner, D.H., and Benzing, D.H. (1991). Diversity of soil fauna in the canopy and forest floor of a Venezuelan cloud forest. *Journal of Tropical Ecology* **7**, 373–383.

Parker, G.G. (1983). Throughfall and stemflow in the forest nutrient cycle. *In* "Advances in Ecological Research" (A. Macfadyen and E.D. Ford, Eds.), pp. 58–120. Academic Press New York.

Potter, C.S., Ragsdale, H., and Swank, W. (1991). Atmospheric deposition and foliar leaching in a regenerating southern Appalachian forest canopy. *Journal of Ecology* **79**, 97–115.

Raven, P.H. and Berg, L.R. (2001). "Environment." Harcourt College Publishers, Fort Worth, Texas.

Reynolds, B.C., Hunter, M.D., and Crossley, Jr., D.A. (2000). Effects of canopy herbivory on nutrient cycling in a northern hardwood forest in Western North Carolina. *Selbyana* **21**(1–2), 74–78.

Reynolds, B.C., Crossley, D.A., and Hunter, M.D. (2003). Responses of soil invertebrates to forest canopy inputs along a productivity gradient. *Pedobiologia* **47**, 127–139.

Richardson, B.A., Richardson, M.J., Scatena, F.N., and McDowell, W.H. (2000a). Effects of nutrient availability and other elevational changes on bromeliad populations and their invertebrate communities in a humid tropical forest in Puerto Rico. *Journal of Tropical Ecology* **16**(2), 167–188.

Richardson, B.A., Rogers, C., and Richardson, M.J. (2000b). Nutrients, diversity, and community structure of two phytotelm systems in a lower montane forest, Puerto Rico. *Ecological Entomology* **25**, 348–356.

Rinker, H.B., Lowman, M.D., Hunter, M.D., Showalter, T.D., and Fonte, S.J. (2001). Literature Review: canopy herbivory and soil ecology. The top-down impact of forest processes. *Selbyana* **22**(2), 225–231.

Rodgers, D.J. and Kitching, R.L. (1998). Vertical stratification of rainforest collembolan (Collembola:Insecta) assemblages: description of ecological patterns and hypotheses concerning their generation. *Ecography* **21**(4), 392–400.

Schowalter, T.D., Sabin, T.E., Stafford, S.G., and Sexton, J.M. (1991). Phytophage effects on primary production, nutrient turnover, and litter decomposition of young Douglas-fir in western Oregon. *Forest Ecology and Management* **42**, 229–243.

Stadler, B., Solinger, S., and Michalzik, B. (2001). Insect herbivores and the nutrient flow from the canopy to the soil in coniferous and deciduous forests. *Oecologia* **126**, 104–113.

Swank, W.T., Waide, J.B., Crossley, Jr., D.A., and Todd, R.L. (1981). Insect defoliation enhances nitrate export from forest ecosystems. *Oecologia* **51**, 297–299.

Walter, D.E. and D.J. O'Dowd. (1995) Life on the forest phylloplane: Hairs, little houses, and myriad mites. *In* "Forest Canopies" (M.D. Lowman and N.M. Nadkarni, Eds.), pp. 325-351. Academic Press, San Diego.

Weathers, K.C. (1999). The importance of cloud and fog in the maintenance of ecosystems. *Trends in Ecology and Evolution* **14**(6), 214–215.

Weathers, K.C., Lovett, G.M., Likens, G.E., and Caraco, N.F.M. (2000) Cloudwater inputs of nitrogen to forest ecosystems in southern Chile: forms, fluxes, and sources. *Ecosystems* **3**, 590–595.

Weathers, K.C., Cadenasso, M.L., and Pickett, S.T.A. (2001). Forest edges as nutrient and pollutant concentrators: potential synergisms between fragmentation, forest canopies, and the atmosphere. *Conservation Biology* **15**(6), 1506–1514.

Weaver, P.L. (1972). Cloud moisture interception in the Luquillo Mountains of Puerto Rico. *Caribbean Journal of Science* **12**(3–4), 129–144.

Weir, J. (1969). Importation of nutrients into woodlands by rooks. *Nature* **221**, 487–488.

CHAPTER 20

Reproductive Biology and Genetics of Tropical Trees from a Canopy Perspective

Michelle L. Zjhra and Beth A. Kaplin

*The diversity of the contrivances adapted to favour
the intercrossing of flowers seems to be exhaustless.*
—*Charles Darwin,* The Various Contrivances by
which Orchids are Fertilised by Insects, *1877*

Reproductive Biology

Flowers are the sexual reproductive organ systems of angiosperms. Sexual reproduction is the fusion of male gametes (sperm carried in pollen) and female gametes (egg within the plant ovary) to form a zygote (carried and nourished in the seed). Male gametes are first dispersed as pollen to a maternal parent to form the embryo and then male gametes move with the seed. Female gametes move only once, i.e., with the seed.

Pollination is the transfer of pollen from the anther (male reproductive part) to the stigma (female reproductive part). Most tropical trees are hermaphroditic (flowers containing both male and female reproductive parts) (Ashton 1969; Bawa 1974, 1979) and self-incompatible (self-pollination does not result in seeds) (Bawa 1974; Chan 1981; Bawa et al. 1985a; Ha et al. 1988). Tropical trees are usually obligate outcrossers (Ashton 1969; Bawa 1974; Bawa and Opler 1975; Bawa and Opler 1977; Opler and Bawa 1978). Additional tropical forest reproductive ecology reviews include Bawa (1990), Bawa and Hadley (1990), and Loveless (1992).

Sexual reproduction is expensive, requiring investment in attracting and rewarding pollinators (animals that carry pollen between flowers). Sexual reproduction is also advantageous, resulting in new combinations of genetic traits that allow plants to adapt to their environment and colonize new areas.

Most tropical trees are pollinated by insects, birds, or mammals (Frankie 1975; Janzen 1975; Opler et al. 1980). Bawa et al. (1985b) estimate that 98 to 99 percent of lowland tropical forest trees are pollinated by animals. Medium- to large-size bees are especially important pollinators of canopy trees, although up to 31 percent of lowland tropical forest trees may have generalist pollination by an array of small insects (Bawa et al. 1985b). Important understory pollinators include hummingbirds (Stiles 1975, 1978; Feinsinger 1978), beetles, and moths (Bawa 1990). Other tropical forest pollinators include bats, butterflies, wasps, large flies, and thrips. The greatest array of pollinators is in the understory, reflecting the higher diversity of plants (Bawa 1990). Different sizes of bees tend to be found at different strata, with small-sized bees in the understory and medium- to large-sized bees prevalent in the canopy (Bawa 1990). The medium- to large-sized bees comprise a varied group representing five different families of bees (Roubik 1989) with a diversity of social and foraging behaviors (and subsequently a diversity of pollen movement patterns and distances).

How specialized are tropical flowers in their interactions with pollinators? Flower specialization is important both in its potential consequence to speciation and community function (discussed later). Flowers may receive a range of visitors, yet members of only one or two groups (e.g., bees and wasps) are effective pollen movers (Schemske and Horvitz 1984; Bawa et al. 1985b). Even generalist pollination systems—flowers visited by a wide range of small insects (including bees, butterflies, beetles, flies, and wasps)—may be more specialized than initial observations suggest (Bawa 1990). Schemske and Horvitz (1984) observed 10 species of bees and butterflies visiting *Calanthea ovandensis*, but only one species of bee represented 66 percent and a second species 14 percent of all fruit set. In other words, despite many visitors, only one or two (or three) were effective pollinators.

How far does pollen move? Tropical trees are often spatially isolated, existing at low densities. As most tropical trees are self-incompatible, there may be selection for long-distance pollen movement (Bawa 1990). Long-distance pollen movement via pollinators provides gene flow that maintains the genetic diversity in a population. Little is known about the extent to which pollinators move among trees within a forest due to the height of trees and abundance of plants and pollinators. We do know that bats forage over long distances (Heithaus et al. 1975; Start and Marshall 1976), bees up to a kilometer and a half (Apsit et al. 2001), hawkmoths 350 m (Chase et al. 1996), and hummingbirds 225 m. See also Loveless et al. (1998; *Tachigali*) and Nason (2002; *Ficus*) for additional estimates of pollen movement by insects. How frequent is long-distance pollen movement? Hamrick and Murawski (1990) found that 25 percent of pollination by bees of *Platypolium elegans* involved pollen movement of a distance greater than 750 m. Pollen movement—how far, how much, and how directly—is determined by the behavior of the pollinator. A species of tree that exists at very low densities in the forest (such as fig trees) may require a very specialized pollinator (such as the fig wasp) that moves the long distance expeditiously between individuals of the same species (see Nason 2002).

Canopy Case Brazil: Density and Distance Matter

In the Mogi Guaçu Ecological Reserve in Brazil, the shrub *Helicteres brevispira* is a colonizing species pollinated by hummingbirds. This species is also self-compatible. Franceschinelli and Bawa (2000) examined the various ecological factors that affect the mating system of this shrub by determining outcrossing rate, the effect of the number of flowers per plant on outcrossing rate, and the effect of plant density and number of flowers per plant on pollinator behavior. They found that plants growing in higher densities had higher outcrossing rates. The behavior of the hummingbird pollinator (that either visits more flowers at the same plant vs. flies to the next plant) was a consequence of plant density and the number of flowers per plant. At low densities, hummingbirds visited high numbers of flowers per plant, resulting in increased selfing (see diagram below). At high densities, hummingbirds visited several plants and moved onto a different area.

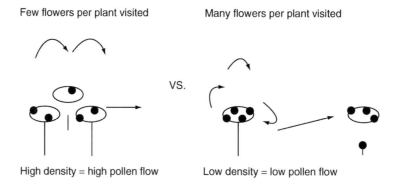

Few flowers per plant visited   Many flowers per plant visited

VS.

High density = high pollen flow   Low density = low pollen flow

Seed dispersal is the movement of seeds from the parent tree. The seeds of tropical trees are dispersed by a number of mechanisms, including via animals and less commonly explosively and via water or wind. In neotropical forests, 51 to 93 percent of the canopy trees and 77 to 98 percent of the subcanopy trees and shrubs bear fleshy fruits adapted for animal consumption (Howe and Smallwood 1982). Plants that rely on animals to disperse their seeds produce fruits that attract animals, with rewards including juicy or nutritious pulp, or high protein arils (edible outgrowths of the seed coat). While it is in the direct interest of plants to have their seeds dispersed, effectively dispersing seeds is irrelevant to the fruit-eating animal, and a consequence of this difference is that adaptations for attracting fruit-eating animals will occur in plants, but only fruit processing adaptations will occur in the animals (Howe 1986). While many animals are not effective seed dispersers, the general means by which plants attract frugivores is believed to be indicative of evolutionary adjustments to mutualists (Howe 1986).

How specialized are tropical fruits in their interactions with seed dispersers? Plants exhibit both morphological and ecological adaptations for seed dispersal (Howe 1986). Morphological attributes of suites of fruits associated with particular groups of dispersal agents are called *dispersal syndromes* (van der Pijl 1972; Howe 1986). For example, small, scentless dark fruits (blue, black, or red) are generally thought to be consumed by birds and are distinguished from aromatic green, yellow, or white 'mammal' fruits. However, dispersal syndromes have been found to be only marginally predictive, and broad overlap in fruit consumption has been found in many empirical studies (e.g., Gautier-Hion et al. 1985). There has been a general lack of tight evolutionary interactions between rainforest frugivores and fruiting plants (Herrera 1986; Jordano 1995).

Ecological adaptations that may influence which animals visit a plant to eat fruits include fruit presentation, nutritional content, seediness, phenology, toxin content, and taste (Howe 1986). Howe's (1993) paradigm, a synthesis of these ideas, predicts that trees with highly nutritious, large fruits with few seeds attract a limited assemblage of specialized dispersers, while highly fecund species with small, less nutritious fruits and many small seeds attract a broad suite of generalised fruit-eaters (Snow 1971; McKey 1975; Howe 1986). In general, most field studies to date have found diverse, non-specialized assemblages of frugivores visiting fruit crops.

How far do fruits move? Seeds taken by animals in the process of fruit consumption may be dropped directly under the parent canopy where mortality is usually extremely high, or moved long or short distances before being dropped, regurgitated, or defecated. *Seed rain* refers to the spatial distribution of seeds on the landscape. This pattern is affected by the disperser's behavior, which in turn is influenced by aspects of forest canopy structure. Recent work on wild olive fruits (*Olea eauropaea*) in southern Spain, for example, has shown that more complex seed rain patterns were found under dense, abundant foliage, while open spaces between shrubs collected few seeds (Alcántara et al. 2000). Seed dispersal can reduce seed and seedling predation either by removing propagules from the region of extremely high mortality near the parent, or by carrying seeds to an area with reduced mortality (Schupp 1988). Even moving seeds just 5 m from the parent crown reduced seed mortality rates in the neotropical subcanopy tree *Faramea occidentalis* (Schupp 1988).

Seed dispersers are evaluated for their effectiveness, or contribution to plant fitness. Effectiveness is comprised of the quantity of seeds dispersed and the quality of that dispersal (quality of seed treatment and quality of seed deposition) (Schupp 1993). While primary seed dispersers move seeds for the first time, other animals may move seed again, either to eat the seed or cache it, a process called *post-dispersal seed removal*. Post-dispersal seed removal, which plays an important role in determining composition and density of tree recruitment, is influenced by forest structure, including degree of canopy closure. Peña-Claros and De Boo (2002) found that seed removal rates decreased with an increase in age of forest in the Bolivian Amazon.

Secondary dispersers, including ants, dung beetles, and rodents, are often attracted to dung piles deposited by primary dispersers, where they remove and cache or bury seeds. Secondary dispersers can swamp any selective forces exerted by primary dispersers on plant traits. For example, ants that bury seeds in a Costa Rican rainforest are highly unpredictable because ant species composition changes dramatically between adjacent areas. The evolutionary implication of this is the effective masking of any selective pressures by the primary dispersers of those seeds (Byrne and Levey 1993).

Another example of the confounding of selective forces exerted by frugivores is demonstrated by a study of the seed diversity in bird droppings. Birds affect plant fitness not only by their behavior, but also by the combinations and densities of seeds in droppings and the nature of the resulting interspecific seedling competition (Loiselle 1990).

How animal seed dispersal correlates to seedling establishment and spatial patterns of trees is less well-studied (Howe and Smallwood 1982; Howe 1986; Jordano 1992; Wilson 1992). An exemplary study of seed dispersal by red howler monkeys indicated that their dispersal behavior significantly increased the seedling abundance of five shade tolerant, tropical forest species in French Guiana (Julliot 1997). Julliot (1997) suggested that the seed dispersal behavior of the howler monkeys contributes to the aggregated spatial pattern of these tree species, because seedlings of these plants were concentrated under howler monkey sleeping sites.

Canopy Case South Africa and Madagascar: Pollinator Assemblage and Behavior Matter

Bohning-Gaese and Bleher (1998) compared primary seed dispersal by birds, secondary seed dispersal by ants, seedling establishment, and spatial patterns of seedlings for South African and Malagasy species of *Commiphora*. They found that the South African species was visited by 13 species of frugivorous birds, whereas the Malagasy species was visited by only four. The diversity of South African birds mirrors the higher diversity of birds in South Africa and the relative dearth of birds in Madagascar. Furthermore, South African birds dispersed seeds by swallowing them and consequently tended to move seeds away from the maternal tree. The South African *Commiphora* did not appear to include secondary seed dispersal. In contrast, the Malagasy frugivorous birds did not swallow seeds and subsequently did not move seeds away from the maternal tree. The seeds of Malagasy *Commiphora*, however, did have secondary seed dispersal by ants. Ants carried the seed into their colonies, removed the aril, and discarded the seed at the edge of the colony (resulting in a high seed density around ant colonies). Nonetheless, the Malagasy species had low rates of seed establishment (determined indirectly by measuring the distance between seedlings in the population to the nearest tree of the same species). Those that did establish did so close to the maternal tree. This study demonstrates that not only is frugivore behavior important, but so is the frugivore community (e.g., different species of birds handle the seeds differently and therefore impact the distance seeds are moved). Furthermore, secondary dispersal plays an important role in moving seeds, although it appeared to have a negligible effect on the Malagasy species of *Commiphora*.

Phenology describes the timing of recurring biological events (Sakai et al. 1999), including bud formation, flowering, fruiting, seed germination, and leafing. Reviews on phenology include Bawa (1983), Borchert (1983), and Rathcke and Lacey (1985). Flowering and fruiting phenology impacts animal populations by causing temporal changes in resource availability (Sakai et al. 1999). From the plant perspective, flowering and fruiting phenology includes the spatial and temporal aspects of pollination and dispersal that allow for variability within a population.

Community-wide patterns of phenology shape animal foraging behavior. A study of the vertical stratification of fruit in a tropical lowland forest in southern Venezuela found that small-sized fruits were distributed in the understory and canopy but were nearly absent from the mid-story (Schaefer and Schmidt 2002). Correspondingly, small frugivorous species limited their foraging to the canopy strata, while large avian frugivores concentrated foraging in the mid-story (Schaefer and Schmidt 2002). Furthermore, the caloric content of the fruit crop was highest in both the lowest and top layers of the forest, and less in the middle layer (Schaefer and Schmidt 2002). Temporal displacements in leafing and flowering have been shown in tropical cloud forest, with the understory starting earlier than the canopy (Williams-Linera 2003).

Canopy Case Madagascar: Temporal and Spatial Patterns Matter

Zjhra (1998) examined an assemblage of 13 sympatric species of locally endemic Malagasy Bignoniaceae. How can this high diversity of related species co-exist? These tree species either utilize different groups of pollinators (bees, beetles, birds, or lemurs) or use the same pollinator group differently. Pollinators were spatially divided both between canopy and understory as well as within the understory (with different species of trees presenting flowers cauliflorously at different heights along the trunk). Species that shared a pollinator differed temporally: they flowered at different times and used different phenological strategies. For instance, more than one species shared small-sized bees by flowering at different times and presenting the flowers cauliflorously at different heights along the trunk of the tree. The ecological and evolutionary impact of sharing pollinators by differing in temporal (phenological) and spatial presentation of flowers is that a high diversity of closely related species can coexist.

How Does Reproductive Biology Relate to Genetics?

An organism's interactions with its environment result in the differential fitness of individuals with different phenotypes (Lande and Arnold 1983; Endler 1986; Schluter 1988). Individuals that are better adapted survive and produce more offspring, passing those traits on to offspring (natural selection). Phenotypes (e.g., flower shape, fruit color, seed size) that interact with the environment (e.g., pollinators, dispersers) are the result of genes, and thus genes are important to the evolution of traits associated with an organism's ability to interact with its environment (e.g., attract pollinators and dispersers). For example, a study of seed size and recruitment in *Cryptocarya alba*, a shade-tolerant, dominant canopy tree of central Chile, found that large seeds have a greater probability of recruitment and their seedlings obtained greater biomass than small seeds (Chacón and Bustamante 2001).

Seed and pollen are the two vectors for gene flow in plants. Genetic variation distributed among populations depends on gene flow via pollen and seeds (Levin and Kerster 1974; Loveless and Hamrick 1984). Plants can affect the behavior of pollinators (via flowering time, quantity and quality of floral rewards, and flower color, size, and odor), thereby affecting gene flow among plants, and ultimately plant fitness (see review in Zimmerman 1988). Similarly, plants affect frugivore and seed disperser behavior via timing of fruiting, quantity and quality of fruit rewards (e.g., pulp quality), fruit color, and fruit and seed size.

As long-lived organisms, trees must deal with pathogens, predators, and competitors as well as with abiotic agents. To successfully survive the attack of pathogens, predators, and competitors requires an arsenal of genetic variation on which natural selection can act (Janzen 1970; Levin 1975; Bawa 1990). Interacting with these agents may also select for gene flow over greater distances (larger pollen and seed shadows) (Bawa 1990), thereby increasing the amount of genetic variation in a population.

A number of factors affect the genetic diversity of a population, including reproduction, dispersal, and survival of seeds and seedlings over time. The genetic make-up of tropical tree populations has been investigated in numerous recent studies (reviewed in Loveless 1992; Hall et al.

1994; Boshier et al. 1995a, 1995b; Gibson and Wheelwright 1995). Gene flow (the exchange of gametes or genes among dispersed trees) determines the geographic scale of populations. Without gene flow, distinct populations may develop. Populations differ due to the degree of gene flow, as well as due to selection to local environments and random factors such as genetic drift. Natural selection can play a strong role in causing differentiation among populations, especially when selection pressure is high, gene flow is low or episodic, or there are small, isolated populations. However, gene flow in tropical trees is usually sufficient even between rare species to create large breeding units (i.e., sufficient gene flow) (Hamrick and Murawski 1990; Chase et al. 1996; Nason et al. 1996; Stacy et al. 1996).

How plants move in their environment is a combination of male gamete (pollen) and seed movement (Richards 1986). Therefore, patterns of plant movement genetically in the environment result from distribution and reproductive biology: pollen and seed dispersal (Hamrick and Loveless 1989; Hamrick and Murawski 1990). Most studies have focused on gene flow signified by pollen movement, as mating systems and geographic ranges of species are important factors affecting genetic variation among populations (Hamrick 1983). Results from pollen movement studies suggest that most of the genetic diversity of tropical tree populations is found within populations, although a small amount of variation is found among populations (Hamrick and Loveless 1989).

In examining gene flow as indicated by pollen movement, a number of factors may affect the genetic diversity of a population, including out-crossing rates (the frequency of pollination between individuals of a species), distances between trees (spatial variation), and timing of flowering (temporal variation). Nason (2002) observed an inverse relationship between tree density and breeding unit area. High-density trees are pollinated by insects that move short distances, whereas low-density trees are pollinated by insects that move greater distances. Stacy et al. (1996) observed that clumped individuals are more likely to mate with the nearest neighbor.

Canopy Case Panama: Temporal and Spatial Patterns Matter

Patterns of genetic differentiation using allozymes of the legume tree *Tachigali versicolor* were examined by Loveless et al. (1998) in Panama. This tree is common in the forest yet has low levels of genetic variation comparable with rare tropical tree species. Despite low levels of variation, populations shared little genetic differentiation, suggesting high levels of gene flow throughout the population. In fact, nearly one quarter of pollen arriving at a cluster of trees by pollinators arrived from a minimum of 500 m away, demonstrating long-distance pollen movement. Despite complete outcrossing (no instances of inbreeding), the trees exhibited little heterozygosity, a phenomenon usually associated with inbreeding. It appears that the entire larger population is genetically differentiated spatially (trees grow near and far) and temporally (different trees flower different years) (see diagram below). In fact, these trees are monocarpic: different adult trees

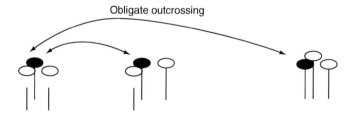

Long-distance pollen movement

Obligate outcrossing

Asymmetrical flowering within a spatial cohort

flower in different years. Individuals that flower in a particular year are not from a common seedling cohort (they are not siblings), thereby resulting both in genetic diversity (pollen is likely to arrive via long distance because the neighbor tree is not flowering) and a temporal genetic bottleneck (only a few trees and their subset of the gene pool exchange genetic material in a particular year). Because only a small fraction of the population's individuals produce flowers in a given year, the subsets of flowering individuals each year are likely to experience small population phenomenon such as genetic drift (and subsequent loss of genetic variation) as well as evolutionary events including temporal bottlenecks or speciation.

This study is important because it demonstrates that genetic structure may arise due to asynchrony of flowering (temporal effects), long-distance pollen movement (spatial effects), and outcrossing—even when neighboring adults frequently exchange pollen. This adds additional dimensions to the classic isolation by distance view of genetic structure (Levin and Kerster 1974). Seed dispersal can determine the genetic composition and structure of plant populations. However, little genetic data are available on seed dispersal. The abundance of animal-mediated seed dispersal may indicate that dispersal has a high variance (is not uniform), including both long-distance and very short seed movement from the parent tree (Hamrick and Loveless 1986; Hamrick et al. 1993), and further influenced by secondary dispersers and seed predators. It appears that there is more differentiation within populations of tropical trees than between populations (e.g., Loveless et al. 1998; see Canopy Case Panama, below), which may reflect limited distance of seed dispersal and clumping of seedlings around the maternal tree (Foster 1977; Kitajima and Augspurger 1989; Hamrick et al. 1993). Species with limited seed dispersal distance have genetic heterozygosity between patches, whereas species with extensive seed dispersal distance have less spatial genetic structure (Fleming and Heithaus 1981; Howe 1989, 1990).

Seed dispersal in *Tachigali versicolor* is local (Loveless et al. 1998), with most of the fruit falling within 150 m of the adult tree (Kitajima and Augspurger 1989). Growth rates among seedlings vary, even in those with similar genetic origins (due to chance and environmental events), resulting in a population with successful seedlings occurring relatively near the parent tree that flowered in a particular year. Over a number of years, the maternal contribution of seeds (from nearby or the same tree) and paternal contribution from pollen (of different distances and different temporal/flowering year cohorts) will create genetic diversity within this spatial cohort or population (see diagram below). In fact, being members of the same flowering cohort may be important. This demonstrates that flower phenology, flowering tree density, and patterns of pollinator movement are all important factors contributing to population size and genetic structure.

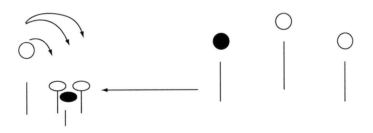

Maternal   Genetic differentiated cohort   Paternal trees

For dispersal to be successful, the seed must arrive in a suitable microhabitat and survive seed predators and herbivory at the seedling stage to successfully recruit into the adult population. Dispersal may increase a species' distribution if the seed dispersal event results in recruitment of adult trees into a new region that is accessible to the plant's pollinators. If the new location is not accessible to the plant's pollinators, a new population may be established or even a new species, via isolation or extinction.

Frugivores determine the pattern of seed distribution and density (Godoy and Jordano 2001). Seed dispersal by animal frugivores creates spatial patterns of where seeds land in relationship to maternal plants. Dispersal is important to understanding the pattern (density and distribution) of trees in a forest. Dispersal is one dimension of why some trees are common or rare. While most dispersal is short distance, long-distance dispersal results from infrequent, nonrandom foraging by a limited set of disperser species (Godoy and Jordano 2001). Unlike pollination, dispersal is not usually specialized. In other words, a number of frugivores may move seeds various distances and to many types of habitats, thereby increasing the chance of a new plant growing and surviving.

Recent research on seed dispersal distance has led to exciting new approaches (see Canopy Case Spain, below). Levels of gene flow traditionally have been assessed via indirect methods that look at differentiation between (sub)populations (e.g., Sork et al. 1999). Direct analysis of gene flow has used hypervariable genetic markers: pollen-mediated gene flow (e.g., Devlin and Ellstrand 1990) and seed-mediated gene flow (e.g., Godoy and Jordano 2001).

Canopy Case Spain: Frugivore Behavior Matters

Godoy and Jordano (2001) have pioneered methods to track unambiguously the distance seeds move from maternal trees. They combined the use of molecular tools (simple sequence repeats or SSR, microsatellites that determine and match the genetic fingerprinting of the tree and seed), ecological data, and geo-referencing to determine seed dispersal distances. Their research on Mediterranean *Prunus mahaleb* demonstrated that approximately 18 percent of the seeds in a habitat result from long-distance dispersal (3 to 5 km), suggesting that long-distance dispersal occurs but is infrequent (distance limitation). Most seeds, however, are dispersed beneath the canopy of the maternal trees. One to five maternal trees contributed seeds to a particular microhabitat site (clumping or aggregate seed dispersal pattern), creating a landscape mosaic with patches made up of different combination of seeds from different assortments of parents. These mosaics are created as a result of the frugivorous birds and mammals that visit the trees, as the frugivores select particular microhabitats (e.g., for shelter or protection). In other words, the dispersal pattern depends on the disperser's movement and behavior, which in turn may be determined by nearby microhabitats.

Phenology influences the genetic structure of a population. Phenology describes the temporal aspects of flowering and fruiting; these temporal changes influence genetic structure in a variable manner over different time scales. Phenological processes determine the spatial makeup of a population via pollinator interactions and seed disperser behaviors. Phenological processes determine or constrain what flowers and fruits are available to pollinators and dispersers and, therefore, can determine or constrain the mating pool and the genetic makeup of seeds and whether those seeds will be moved near or far from the parent tree.

Canopy Case Malaysia: Community Patterns Matter

Sakai et al. (1999) used tree towers and aerial walkways to record phenological data of canopy trees in the lowland mixed-dipterocarp forest of Sarawak. They examined plants as individuals, as populations, and as a community as well as considered correlations between pollination systems. They found that most individuals follow the same flowering pattern as those at the population level and that reproduction follows an aggregated pattern among species. In other words, numerous species flower during the same time interval. It is unclear whether this synchronized flowering among individuals, populations, and species results from environmental variables (e.g., hours of sunshine, El Niño occurrence, drop in temperature) or from the same flowering frequency. Additionally, canopy and subcanopy trees flowered annually or subannually, whereas

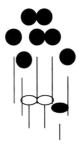

Major flowering
(most trees flower)

Non-flowering times
(some understory trees flower)

half the understory trees in their study did not flower during four years (see diagram above). These differences in the phenology of plants of different forest strata may be related to mortality (Momose et al. 1998b). Sakai et al. (1999) found that the community-wide increase in flowering resulted in an increase of pollinators due to migration of pollinators to the forest (especially by bees) and shifts in feeding niches (especially beetles shifting from feeding on leaves to flower rewards). This study and that of Momose et al. (1998a) on the pollination biology of species in the same forest are important in that they give a community view of pollination biology and phenology patterns and, thereby, allow for a better understanding of species interactions (plant-plant, plant-animal) in the forest.

How Do Breeding Systems Relate to Tropical Forest Patterns and Processes (Ecology and Evolution)?

Plant-pollinator interactions relate to gene flow and speciation, two topics that impact species richness and maintenance of diversity in the tropics (Bawa 1990). How might plant-pollinator interactions contribute to species richness? Limited pollen flow may result in fragmentation of populations into subpopulations. Due to ensuing small population size, the effects of inbreeding and genetic drift may contribute to the origin and maintenance of closely related species in the tropics (Fedorov 1966). Additionally, speciation may result from interactions with pollinators, as pollinator isolation may be a common form of reproductive isolation in plants (Grant 1949; Straw 1956; Grant 1981). Plant-animal interactions may promote speciation (Bawa 1990) due to genetic drift accelerating speciation and/or founder populations becoming reproductively isolated from parent populations (e.g., changes in pollinator may result in reproductive isolation). Flower phenotype changes that may lead to attraction or adaptation to a new pollinator may occur rapidly, as one or a few genes control differences in flower shape, color, and nectar reward (Prazmo 1965; Bradshaw et al. 1995). Hence speciation has the potential to occur rapidly (Coyne and Orr 1998).

How Does Reproductive Biology Relate to Conservation?

Tropical forests are generally characterized as comprised of high diversity of species at low densities. Species at low densities often suffer the fate of rare species. Rare species tend to exist in small populations that are geographically isolated and are more prone to extinction (Ellstrand 1992) due to inbreeding depression (a decrease in reproductive success) (Franklin

DNA SEQUENCES AND ORCHID CLASSIFICATION

Wesley E. Higgins

The ability to sequence DNA has led to a refinement of orchid classification (and other taxonomic groups). This has occurred in the same timeframe as the shift in cladistic theories from phenetic to parsimony algorithms for phylogenetic analyses. Phenetic analysis groups organisms based on overall similarity, whereas parsimony analysis forms groups based on shared evolved characteristics. While changes in classifications get much attention in journals, DNA analyses often confirm traditional classification schemes.

Traditional classification schemes were phenetic in nature with more weight given to sexual characters (Linnaeus 1751). This approach resulted in some classification schemes that inferred false evolutionary relationships. Current taxonomic thinking favors phylogenetic-based classifications. However, a phylogeny is not a classification and the application of rank levels—e.g., family, tribe, and genus—to monophyletic groups is still subjective. Since a classification scheme is an end-user tool for plant identification, morphology-based application of ranks is preferred.

The following three examples of changes in orchid classification prompted by DNA analyses demonstrate the vulnerability of morphology only–based classification to convergent evolution. These examples involve homoplasy in floral morphology.

Laelia was a genus with disjunct populations in Mexico and Brazil. These populations are clearly members of the same subtribe and were placed in the same genus on the basis of pollinia number (8). An analysis of nrDNA sequences found that there are many other genera in the subtribe related more closely to the Brazilian laelias than to the Mexican laelias (van den Berg et al. 2000). Therefore, the Brazilian laelias have been placed in the new genus *Hadrolaelia* (Chiron and Castro Neto 2002). This was no surprise to horticulturists, who already knew that the two populations responded differently to culture.

In the genus *Oncidium*, the pollinator is attracted by deception. The orchid flowers mimic Malpigiaceae flowers that grow sympatric. An analysis of five DNA regions in the nuclear and plastid genomes show this genus as currently delimited is not monophyletic (a natural group) (Williams et al. 2001). The plastic floral form is pollinator-driven convergent evolution. Reclassification of this large genus is ongoing based on a new understanding of mimicry and plasticity of flower morphology.

Species in the genus *Vanilla* have flowers that resemble the Epidendroid floral form but have soft mealy pollen instead of pollinia. The placement of this genus with the higher Epidendroids was accepted until DNA provided new information. An analysis of rbcL sequences places *Vanilla* in a clade with other more primitive orchids that also have soft mealy pollen (Cameron and Chase 2000). This discovery has led to the verification of a new subfamily, Vanilloideae.

These examples show that a taxonomist does not intuitively know which morphological characters are homoplasious. The use of molecular data helps to remove some of the subjectivity of character state assignment. Holomorphology is a new approach that uses total evidence by combining characters from morphology, anatomy, chemistry, and DNA as the best estimate of evolution.

References

Cameron, K.M. and Chase, M.W. (2000). Nuclear 18S rDNA sequences of Orchidaceae confirm the subfamilial status and circumscription of Vanilloideae. *In* "Monocots: Systematics and Evolution, Volume 1 of Proceedings of the Second International Conference on the Comparative Biology of the Monocots, Sydney" (K.L. Wilson and D.A. Morrison, Eds.), pp. 457–464. CSIRO Publishing, Melbourne.

Chiron, G.R. and Castro Neto, V.P. (2002). Révision des espéces brèsiliennes du genre Laelia Lindley. *Richardiana* **2,** 4–28.

Linnaeus, C. (1751). "Philosophia Botanica." Kiesewetter, Stockholm.

van den Berg, C., Higgins, W.E., Dressler, R.L., Whitten, W.M., Arenas, M.A.S., Culham, A., and Chase, M.W. (2000). A phylogenetic analysis of Laeliinae (Orchidaceae) based on sequence data from internal transcribed spacers (ITS) of nuclear ribosomal DNA. *Lindleyana* 15. 96–114.

Williams, N.H., Chase, M.W., Fulcher T., and Whitten, W.M. (2001). Molecular systematics of the Oncidiinae based on evidence from four DNA sequence regions: expanded circumscriptions of Cyrtochilum, Erycina, Otoglossum, and Trichocentrum and a new genus (Orchidaceae). *Lindleyana* 16, 113–139.

1980; Frankel and Soule 1981; Beardmore 1983; Simberloff 1988) and genetic drift. Genetic drift may result in fixation of harmful alleles (Franklin 1980) and/or reduction of genetic variation (Wright 1931; Karron 1987). Inbreeding depression is a reduction in reproductive performance due to breeding with only a few individuals, resulting in re-mixing the same limited gene pool. Both inbreeding and genetic drift result in less genetic variation to respond (via natural selection) to changes in the environment, and the population becomes more prone to extinction (Franklin 1980; Beardmore 1983; Lande and Barrowclough 1987; Simberloff 1988). These phenomena occur due to a decrease in gene flow. Low gene flow results in little genetic variability in a population. Conversely, gene flow into a small population from a population adapted to a different environment may disrupt local adaptations (Antonovies 1976; Simberloff 1988).

Reproduction can be episodic in small patches, thereby generating genetic and demographic mosaics. These mosaic effects disappear with wide-scale forest clearing. Species with restricted (or low) pollen or seed movement form mosaics with large between-population variations of long duration due to selection on gene frequencies within and among the populations. Wide-scale forest clearing can eliminate genetic and demographic mosaics within large stands.

Forest fragmentation also disrupts reproductive processes. The ability of plant populations to persist and expand in fragmented landscapes depends on pollination and seed dispersal processes (Tewksbury et al. 2002). Fragmentation of populations of animal-dispersed trees may create severe genetic bottlenecks due to isolation of patches of genetically similar trees (Nason and Hamrick 1997; Schnabel et al. 1998). Habitat fragmentation may decrease the number of pollinators (Klein 1989) or influence the composition of pollinators, thereby altering patterns of gene flow (Bawa 1990). Extinction of frugivore dispersers can seriously alter spatial patterns of seed distributions (Godoy and Jordano 2001). Cordeiro and Howe (2001) found that recruitment of endemic, animal-dispersed tree species was 40 times lower in small fragments than in continuous forest or large fragments. Fragmentation can affect dispersers in different ways: understory fruit bat species were more sensitive to fragmentation than canopy species, and subsequent changes in the frugivorous bat community may have indirect consequences on the demographic and genetic structure of plant populations in fragments (Cosson et al. 1999).

Reproductive biology studies also shed light on forest restoration efforts. Several authors have identified the importance of tree structure to seed rain in abandoned fields and pastures (e.g., Guevara and Laborde 1993; McClanahan and Wolfe 1993). A study of the successional development of abandoned farmland in Australia, formerly occupied by subtropical rainforest, showed that most tree species are recruited from seed dispersed into the site by birds or bats (Toh et al. 1999). The authors found that seed rain into the abandoned fields increased when trees taller than 6 m were present to serve as perches; the identity of the trees was not as important as their structure and suitability as a perch.

Limitations and Future Directions

Canopy trees are difficult to study because of their height. Hundreds of plants exist with thousands of species of pollinators and seed dispersers, creating complex relationships (Bawa 1990). However, plant-pollinator and seed disperser interactions allow for the study of gene flow (Ashton 1969; Bawa et al. 1985b; Janzen 1971), coevolution (Feinsinger 1983), and community function (Howe 1983). Better access to the canopy is facilitating breeding system studies that were previously not possible due to the logistics of reaching and observing flowers and fruits in tall trees. Canopy walkways and ropes, however, still make access to flowers and fruits in numerous trees difficult. Reaching flowers is necessary for nectar and odor sampling as well as for breeding system experiments. Observation of pollinator and disperser behavior, on the other hand, has been greatly facilitated by canopy walkways and rope access. Nonetheless, catching elusive pollinators for identification is still challenging 25 m from the ground.

The challenge of breeding system studies is to integrate the fields of breeding system genetics, pollination/seed dispersal ecology, and evolutionary patterns and processes. In other words, we need to understand pattern and process in pollinator and seed disperser visitations to flowers and fruits, the ecological and genetic consequences of pollen and seed movement, and the role of these events in the patterns and processes of speciation and distribution. Finally, we need to collaborate and standardize in order to describe broadly and understand what is occurring in forest canopies throughout the tropics.

References

Alcántara, J.M., Rey, P.J., Valera, F., and Sánchez-Lafuente A.M. (2000). Factors shaping the seedfall pattern of a bird-dispersed plant. *Ecology* **81**, 1937–1950.

Antonovies, J. (1976). The nature of limits to natural selection. *Ann Mo Bot Gard* **63**, 224–247.

Apsit, V.J., Hamrick, J.L., and Nason J.D. (2001). Breeding population size of a fragmented population of a Costa Rican dry forest tree species. *J Hered* **92**, 415–420.

Ashton, P.S. (1969). Speciation among tropical forest trees: some deductions in the light of recent evidence. *Biol. J. Linn. Soc.* 1: 155–196.

Bawa, K.S. (1974). Breeding systems of tree species of a lowland tropical community. *Evolution* **28**, 85–92.

Bawa, K.S. (1979). Breeding systems of tree in a tropical wet forest. *New Zeal. J. Bot.* **17**, 521–524.

Bawa, K.S. (1983). Patterns of flowering in tropical plants. In Handbook of Experimental Pollination Biology, ed. C.E. Jones, and R.J. Little, pp. 394–410. *New York: Van Nostrand, Reinhold.*

Bawa, K.S. (1990). Plant-pollinator interactions in tropical rain forests. *Ann Rev Ecol Syst* **21**, 399–422.

Bawa, K.S. and Hadley, M. eds. (1990). Reproductive Ecology of Tropical Forest Plants. Carnforth, England: Partrhenon.

Bawa, K.S. and Opler. P.A. (1977). Spatial relationships between staminate and pistillate plants of dioecious tropical forest trees. *Evolution* **31**, 64–68.

Bawa, K.S., Perry, D.R., and Beach. J.II. (1985a). Reproductive biology of tropical lowland rain forest trees. I. Sexual systems and self-incompatibility mechanisms. *Am J Bot* **72**, 331–45.

Bawa, K.S., Perry, D.R., Bullock, S.H., Coville, R.E., and Grayum M.H. (1985b). Reproductive biology of tropical lowland rain forest trees. II. Pollination mechanisms. *Am J Bot* **72**, 346–56.

Beardmore, J.A. (1983). Extinction, survival, and genetic isolation. In: C.M. Schonewald-Cox, S.M. Chambers, B. Bryde, and W.I. Thomas (eds). Genetics and conservation: a reference for managing wild animals and populations. Benjamin/Cummings. Menlo Park. Pp. 151.

Bohning-Gaese, K. and Bleher B. (1998). Seed dispersal in the genus *Commiphora* in Madagascar and South Africa. In Results of worldwide ecological studies. S.-W. Breckle, B. Schweizer, and U. Arndt (eds). Pp. 285–298. AFW Schimper Foundation Symposium.

Borchert, R. (1983). Phenology and control of flowering in tropical trees. *Biotropica* **15**, 81–89.

Boshier, D.H., Chase, M.R., and Bawa. K.S. (1995a). Population genetics of *Cordia alliodora* (Boraginaceae), a neotropical tree. 2. Mating system. *Am J Bot* **82**, 476–483.

Boshier, D.H., Chase, M.R., and Bawa. K.S. (1995b). Population genetics of *Cordia alliodora* (Boraginaceae), a neotropical tree. 3. Gene flow, neighborhood, and population substructure. *Am J Bot* **8**, 484–490.

Bradshaw, H.D., Wilbert, S.M., Otto, K.G., and Schemske D.W. (1995). Genetic mapping of floral traits associated with reproductive isolation in monkeyflowers (*Mimulus*). *Nature* **376**, 762–765.

Byrne, M.M., and Levey. D.J. (1993). Removal of seeds from frugivore defecations by ants in a Costa Rican rain forest. *Vegetatio* **107/108**, 363–374.

Chacón, P. and Bustamante R.O. (2001). The effects of seed size and pericarp on seedling recruitment and biomass in *Cryptocarya alba* (Lauraceae) under two contrasting moisture regimes. *Plant Ecology* **152**, 137–144.

Chan, H.T. (1981). Reproductive biology of some Malaysian Dipterocarps. III. Breeding systems. *Malaysian For* **44**, 28–34.

Chase, M.R., Moller, C., Kesseli, R., and Bawa. K.S. (1996). Distant gene flow in tropical trees. *Nature* **383**, 398–399.

Cordeiro, N.J. and Howe. H.F. (2001). Low recruitment of trees dispersed by animals in African forest fragments. *Conservation Biology* 1733–1741.

Cosson, J.-F., Pons, J.-M., and Masson. D. (1999). Effects of forest fragmentation on frugivorous and nectivorous bats in French Guiana. *Journal of Tropical Ecology* **15**, 515–534.

Coyne, J.A. and Orr. H.A. (1998). The evolutionary genetics of speciation. *Phil Trans R Soc London B* **353**, 287–305.

Devlin, B. and Ellstrand. N.C. (1990). The development and application of a refined method for estimating gene flow from angiosperm paternity analysis. *Evolution* **44**, 248–259.

Ellstrand, N.C. (1992). Gene flow by pollen: implications for plant conservation genetics. *Oikos* **63**, 77–86.

Endler, J.A. (1986). Natural selection in the wild. Princeton University Press, Princeton, N.J.

Fedorov, A.A. (1966). The structure of tropical rain forest and speciation in the humid tropics. *J. Ecol.* **54**, 1–11.

Feinsinger, P. (1978). Ecological interactions between plants and hummingbirds in a successional tropical community. *Ecol Monogr* **48**, 269–87.

Feinsinger, P. (1983). Coevolution and pollination. In Coevolution, ed. D.J. Futuyma, and M. Slatkin, pp. 282–310. Sunderland, Mass: Sinauer.

Fleming, T.H. and Heithaus. E.R. (1981). Fruigvorous bats, seed shadows, and the structure of tropical forests. *Biotropica* **13**, 45–53.

Foster, R.B. (1977). *Tachigalia versicolor* is a suicidal neotropical tree. *Nature* **268**, 624–626.

Franceschinelli, E.V. and Bawa. K.S. (2000). The effect of ecological factors on the mating system of a South American shrub species (*Helicteres brevispira*). *Heredity* **84**, 116–123.

Frankel, O.H. and Soule. M.E. (1981). Conservation and evolution. Cambridge University Press, Cambridge.

Frankie, G.W. (1975). Tropical forest phenology and pollinator plant coevolution. In Gilbert, L.E. and P.H. Raven (eds.). Coevolution of animals and plants. University of Texas Press, Austin.

Franklin, I.R. (1980). Evolutionary change in small populations. In: M.E. Soule and B.A. Wilcox (eds), Conservation biology: an evolutionary, ecological perspective. Sinauer, Sunderland, pp. 135–150.

Gautier-Hion, A., Duplantier, J.-M., Quris, R., Feer, F., Sourd, C., Decoux, J.-P., Dubost, G., Emmons, L., Erard, C., Hechestweiler, P., Moungazi, A., Roussilhon, C., and Thiollay. J.M. (1985). Fruit characters as a basis of fruit choice and seed dispersal in a tropical forest vertebrate community. *Oecologia* **65**, 324–337.

Gibson, J.P. and Wheelwright. N.T. (1995). Genetic structure in a population of a tropical tree *Ocotea tenera* (Lauraceae): influence of avian seed dispersal. *Oecologia* **103**, 49–54.

Gilbert, G.S., Harms, K.E., Hamill, D.N., and Hubbell. S.P. (2001). Effects of seedling size, E1 Niño drought, seedling density, and distance to nearest conspecific adult on 6-year survival of Ocotea whitei seedlings in Panamá. *Oecologia* **127**, 509–516.

Godoy, J.A. and Jordano. P. (2001). Seed dispersal by animals: exact identification of source trees with endocarp DNA microsatellites. *Molecular Ecology* **10**, 2275–2283.

Grant, V. (1949). Pollination systems as isolating mechanisms in angiosperms. *Evolution* **3**, 82–97.

Grant, V. (1981). Plant Speciation, 2nd ed. New York: Columbia University Press.

Guevara, S. and Laborde. J. (1993. Monitoring seed dispersal at isolated standing trees in tropical pastures: consequences for local species availability. *Vegetatio* **107/108**, 319–338.

Ha, C.O., Sands, V.E., Soepadmo, E., and Jong. K. (1988). Reproductive patterns of selected understorey trees in the Malaysian rain forest: the sexual species. *Bot J Linn Soc* **97**, 295–316.

Hall, P., Orrell, L.C., Bawa. K.S. (1994). Genetic diversity and mating system in a tropical tree, *Carapa guianensis* (Meliaceae). *Am J Bot* **81**, 1104–1111.

Hamrick, J.L. (1983). The distribution of genetic variation within and among natural plant populations. Pp. 335.348. In: C.M. Schonewald-Cox, S.M. Chambers, B. MacBryde, and L. Thomas (eds), Genetics and Conservation. Benjamin Cummings, Menlo Park, CA.

Hamrick, J.L. and Loveless. M.D. (1986). The influence of seed dispersal mechanisms on the genetic structure of plant populations. In: A. Estrada and T.H. Fleming (eds) Frugivores and Seed Dispersal, pp. 211–223. W. Junk Publ., The Hague.

Hamrick, J.L. and Loveless. M.D. (1989). The genetic structure of tropical tree populations: associations with reproductive biology. In: Bock, J.H. and Y.B. Linhart (eds), The Evolutionary Ecology of Plants, pp. 129–146. Westview Press, Boulder, CO.

Hamrick, J.L. and Murawski. D.A. (1990). The breeding structure of tropical tree populations. *Pl Sp Biol* **5**, 157–165.

Hamrick, J.L., Murawski D. A., and Nason. J.D. (1993). The influence of seed dispersal mechanisms on the genetic structure of tropical tree populations. *Vegetatio*. **107/108**, 281–297.

Heithaus, E.R., Fleming, T.H., and Opler. P.A. (1975). Foraging patterns and resource utilization in seven species of bats in a seasonal tropical forest. *Ecology* **56**, 841–54.

Herrera, C.M. (1986). Vertebrate-dispersed plants: why they don't behave the way they should. Pp. 5-20 *in* A. Estrada and T.H. Fleming (eds.), Frugivory and seed dispersal, Junk, The Hague.

Howe, H.F. (1983). Constraints on the evolution of mutualisms. *Am Nat* **123**, 764–77.

Howe, H.F. (1986). Seed dispersal by fruit-eating birds and mammals. In: D.R. Murray (ed): Seed dispersal. Academic Press, Sydney, Australia: 123–189.

Howe, H.F. (1989). Scatter- and clump-dispersal and seedling demography: hypothesis and implications. *Oecologia* **79**, 417–426.

Howe, H.F. (1990). Seed dispersal by birds and mammals Implications for seedling demography. In *Reproductive Ecology of Tropical Forest Plants*. Man and the Biosphere Series (K.S. Bawa and M. Hadley, eds.), Vol. 7, UNESCO, Paris and Parthenon Publishing, Carnforth, pp. 191–218.

Howe, H.F. (1993). Specialized and generalized dispersal systems: where does the 'paradigm' stand? *Vegetatio* **107/108**, 3–13.

Howe, H.F. and Smallwood. J. (1982). Ecology of seed dispersal. *Annual Review of Ecology and Systematics* **13**, 201–228.

Janzen, D.H. (1970). Herbivores and the number of tree species in tropical forests. *Am Nat* **104**, 501–28.

Janzen, D.H. (1971). Euglossine bees as long distance pollinators of tropical plants. *Science* **171**, 203–5.

Janzen, D.H. (1975). Ecology of plants in the tropics. Inst. Biol. Stud Biol. No 58 Edward Arnold, London.

Jordano, P. (1992). Fruits and frugivory. In M. Fenner (ed) Seeds. The ecology of regeneration in plant communities. CAB International, Wallingford, UK pp 105–156.

Jordano, P. (1995). Angiosperm fleshy fruits and seed dispersers: a comparative analysis of adaptation and constraints on plant-animal interactions. *American Naturalist* **145**, 163–191.

Julliot, C. (1997). Impact of seed dispersal by red howler monkeys *Alouatta seniculus* on the seedling populations in the understorey of tropical rain forest. *Journal of Ecology* **85**, 431–440.

Karron, J.D. (1987). A comparison of levels of genetic polymorphism and selfcompatibility in geographically restricted and widespread plant congeners. *Evol Ecol* **1**, 47–58.

Kitajima, A. and Augspurger. C.K. (1989). Seed and seedling ecology of a monocarpic tropical tree, *Tachigalia versicolor*. *Ecology* **70**, 1102–1114.

Klein, B.C. (1989). Effects of forest fragmentation on dung and carrion beetle communities in central Amazonia. *Ecology* **70**, 1715–25.

Lande, R. and Arnold. S.J. (1983). The measurement of selection on correlated characters. *Evolution* **37**, 1210–1226.

Lande, R. and Barrowclough. G.F. (1987). Effective population size, genetic variation, and their use in population management. In: M.E. Soule (ed.), Viable populations for conservation. Cambridge University Press, Cambridge. Pp. 87–123.

Levin, D.A. (1975). Pest pressure and recombination systems in plants. *Am Nat* 109: 437–51.

Levin, D.A. and Kerster. H.W. (1974). Gene flow in seed plants. *Evol Biol* 7: 139–220.

Loiselle, B.A. (1990). Seeds in droppings of tropical fruit-eating birds: importance of considering seed composition. *Oecologia* **82**, 494–500.

Loveless, M.D. (1992). Isozyme variation in tropical trees: patterns of genetic organization. New Forests 6: 67–94.

Loveless, M.D. and Hamrick. J.L. (1984). Ecological determinants of genetic structure in plant populations. *AnnRev Ecol Syst* **15**, 65–95.

Loveless, M.D., Hamrick, J.L., and Foster. R.B. (1998). Population structure and mating system in *Tachigali versicolor*, a monocarpic neotropical tree. *Heredity* **81**, 134–143.

McKey, D. (1975). The ecology of coevolved seed dispersal systems. Pp. 159–191 *in* L.E. Gilbert and P.H. Raven (eds.), Coevolution of animals and plants. Univ. of Texas Press, Austin, Texas.

McClanahan, T.R. and Wolfe. R.W. (1993). Accelerating forest succession in a fragmented landscape: the role of birds and perches. *Conservation Biology* **7**,279–287.

Momose, K., Yumoto, T., Nagamitsu, T., Kato, M., Nagamasu, H., Sakai, S., Harrison, R.D., Itioka, T., Hamid, A.A. and Inoue. T. (1998a). Pollination biology in a lowland dipterocarp forest in Sarawak, Malaysia. I. Characteristics of the plant-pollinator community in a lowland dipterocarp forest. *Am J Bot* **85**, 1477–1501.

Momose, K., Ishil, R., Sakai, S., and Inoue. T. (1998b). Reproductive intervals and pollinators of tropical plants. *Proceedings of the Royal Society of London* **265**, 2333–2339.

Nason, J.D. (2002). Genetic structure of tropical tree populations. Pp. 299–327 *in* M.R. Guariguata and G.H. Kattan (eds.), Ecology of Neotropical Rainforests. IICA Press, San Jose, Costa Rica.

Nason, J.D. and Hamrick. J.L. (1997). Reproductive and genetic consequences of forest fragmentation: two case studies of neotropical canopy species. *Journal of Heredity* **88**, 264–276.

Nason, J.D., Herre, E.A., and Hamrick. J.L. (1996). Paternity analysis of the breeding structure of strangler fig populations: evidence for substantial long-distance wasp dispersal. *J Biogeog* **23**, 501–512.

Opler, P.A., Baker, H.G., and Frankie. G.W. (1980). Plant reproductive characteristics during secondary succession in neotropical lowland forest ecosystems. *Biotropica* **12**, 40–46.

Opler, P.A. and Bawa. K.S. (1978). Sex ratios in tropical forest trees. *Evolution* **32**, 812–821.

Peña-Claros, M. and De Boo. H. (2002). The effect of forest successional stage on seed removal of tropical rain forest tree species. *Journal of Tropical Ecology* **18**, 261–274.

Prazmo, W. (1965). Cytogenetic studies on the genus *Aquilegia*. III. Inheritance of the traits distinguishing different complexes in the genus *Aquilegia*. *Acta Soc. Bot. Poloniae* **34**, 403–437.

Rathcke, B. and Lacey. E. L. (1985). Phenological patterns of terrestrial plants. *Annual Review of Ecology and Systematics* **16**, 179–214

Richards, A. J. (1986). *Plant breeding systems*. Unwin Hyman, London.

Roubik, D.W. (1989). Ecology and Natural History of Tropical Bees. Cambridge: Cambridge Univ. Press.

Sakai, S., Momose, K., Yumoto, T., Nagamitsu, T., Nagamasu, H., Hamid, A.A., Nakashizuka, T., and Inoue. T. (1999). Plant reproductive phenology over four years including an episode of general flowering in a lowland dipterocarp forest, Sarawak, Malaysia. *Am J Bot* **86**, 1414–1436.

Schaefer, H.M. and Schmidt. V. (2002). Vertical stratification and caloric content of the standing fruit crop in a tropical lowland forest. *Biotropica* **34**, 244–253.

Schemske, D.W. and Horvitz. C.C. (1984). Variation among floral visitors in pollination ability: a precondition for mutualism specialization. *Science* **225**, 519–21.

Schluter, D. (1988). Estimating the form of natural selection on a quantitative trait. *Evolution* **42**, 849–861.

Schnabel, A., Nason, J.D., and Hamrick. J.L. (1998). Understanding the population genetic structure of *Gleditsia triacanthos* L. seed dispersal and variation in female reproductive success. *Molecular Ecology* **7**, 819–832.

Schupp, E.W. (1988). Seed and early seedling predation in the forest understory and in treefall gaps. *Oikos* **51**, 71–78.

Schupp, E.W. (1993). Quantity, quality and the effectiveness of seed dispersal by animals. *Vegetatio* **107/108**, 15–29.

Simberloff, D. (1988). The contribution of population and community biology to conservation science. *Ann Rev Ecol Syst* **19**, 473–511.

Snow, D.W. (1971). Evolutionary aspects of fruit-eating birds. *Ibis* **113**, 194–202.

Sork, V.L., Nason, J., Campbell, D.R., Fernandez. J.F. (1999). Landscape approaches to historical and contemporary gene flow in plants. *Trends in Ecology and Evolution* **14**, 224.

Stacy, E.A., Hamrick, J.L., Nason, J.D., Hubbell, S.P., Foster, R.B. and Condit. R. (1996). Pollen dispersal in low density populations of three neotropical tree species. *Am Nat* **148**, 275– 298.

Start, A.N. and Marshall. A.G. (1976). Nectarivorous bats as pollinators of trees in west Malaysia. In Tropical Trees: Variation, Breeding and Conservation, ed. J. Burley, B.T. Styles. Pp. 141–50. New York: Academic.

Stiles, F.G. (1975). Ecology, flowering phenology and hummingbird pollinaton of some Costa Rican *Heliconia* species. *Ecology* **56**, 285–310.

Stiles, F.G. (1978). Temporal organization of flowering among the hummingbird food plants of a tropical wet forest. *Biotropica* **10**, 194–210.

Straw, R.M. (1956). Floral isolation in *Penstimon*. *Am Nat* **90**, 47–63.

Tewksbury, J.J., Levey, D.J., Haddad, N.M., Sargent, S., Orrock, J.L., Weldon, A., Danielson, B.J., Brinkerhoff, J., Damschen, E.I., and Townsend. P. (2002). Corridors affect plants, animals and their interactions in fragmented landscapes. *PNAS* **99**, 12923–12926.

Toh, I., Gillespie, M., and Lamb. D. (1999). The role of isolated trees in facilitating tree seedling recruitment at a degraded sub-tropical rainforest site. *Restoration Ecology* **7**, 288–297.

Van der Pijl, L. (1972). Principles of dispersal in higher plants. Berlin: Springer-Verlag.

Williams-Linera, G. 2003. Temporal and spatial phenological variation of understory shrubs in a tropical montane cloud forest. *Biotropica* **35**, 28–36.

Wilson, M.F. (1992). The ecology of seed dispersal. In M. Fenner (ed.) Seeds. The ecology of regeneration in plant communities. CAB International, Wallingford, UK pp. 61–85.

Wright, S. (1931). Evolution in Mendelian populations. *Genetics* **16**, 97–159.

Zimmerman, M. (1988). Nectar production, flowering phenology, and strategies for pollination. In: Lovett Doust, J. and L. Lovett Doust, eds. Plant Reproductive Ecology. Patterns and Strategies. New York: Oxford University Press, pp. 157–178.

Zjhra, M.L. (1998). Pollination biology, phenology, and biogeography of endemic Malagasy Coleeae (Bignoniaceae). Ph.D. dissertation. University of Wisconsin.

CHAPTER 21

Decomposition in Forest Canopies

Steven J. Fonte and Timothy D. Schowalter

*Der gut Herr Gott said, "Let there be rot," and hence bacteria and fungi sprang into existence
to dissolve to the knot of carbohydrates photosynthesis
achieves in plants, in living plants.*
—John Updike, Facing Nature, *1985*

Introduction

Decomposition is a fundamental ecological process, assuring turnover of organic matter and nutrients from dead organisms, and is a key to understanding ecosystem function and nutrient cycling in forest ecosystems. As the major source of available nutrients in most terrestrial ecosystems, this process is essential for plant growth and primary production in forests (Swift et al. 1979; Schlesinger 1997). By cleansing the forest continually of debris, decomposition also provides growing room and substrate for emerging trees and other plants. The breakdown of forest material produces essential habitat and nutritional resources for a succession of plant and animal species at every stage of decay and is an important factor in soil development (Harmon et al. 1986). Decomposers and associated organisms represent a wide array of taxonomic and functional groups and can constitute a substantial portion of total ecosystem biodiversity (Deyrup 1981). The decay process in forests is also important in determining how much carbon is stored in ecosystems or transferred to the atmosphere, thereby influencing global climatic trends (Harmon et al. 1986; Schlesinger 1997).

Most studies of decomposition have addressed the decay of litterfall components on the forest floor. However, the degradation of aboveground plant material generally begins in the canopy. Decay in the canopy may involve only minor, transitory alterations to senescing material (Osono 2002) or long-term degradation (Swift et al. 1976) and soil formation processes (Nadkarni 1981; Paoletti et al. 1991). Regardless of duration in the canopy, any alterations to organic matter in the canopy are important to consider. Some forms of organic matter degradation begin well before senescence and are relevant to all litterfall reaching the forest floor. Even after senescence, dead wood and leaves can remain attached for significant periods of time and are often colonized by a variety of decay organisms before reaching the ground (Swift et al. 1976; Stone 1987; Chapela and Boddy 1988a; Osono 2002). These early stages of colonization and decay influence subsequent stages of decay and may have long-term effects on the decomposer community and ecosystem nutrient dynamics. Therefore, an understanding of decomposition in forest ecosystems requires consideration of the contributions of decay processes in the canopy.

The Decomposition Process

Decomposition involves several degradative processes. Photooxidation is the abiotic catabolism resulting from exposure to solar radiation. Leaching is the loss of soluble materials as a result of percolation of water through decomposing material. Comminution is the fragmentation of organic material, largely as the result of feeding (detritivory) by animals. Mineralization involves the catabolism of organic molecules by saprophytic organisms and the subsequent release of nutrients in a mineral form that is available to plants and other organisms (Vossbrink et al. 1979; Seastedt 1984; Schowalter 2000).

The importance of each process varies with habitat conditions and stage of decomposition. For example, leaching is minimal in arid environments and maximum in wet environments; photooxidation is most important where litter is fully exposed to solar radiation. These two processes may be more important in forest canopies where precipitation is initially intercepted and solar radiation more intense than on the forest floor (e.g., Parker 1995). Conversely, compared to the forest floor, the larger temperature fluctuations and drier conditions of the canopy, resulting from solar exposure, evaporation, and wind, may reduce the roles of detritivory and mineralization. As decomposition proceeds and older litter is buried under newer litter, photooxidation becomes less important. Leaching initially removes soluble nutrients (e.g., K and P) and sugars. In wet environments, up to 90 percent of these nutrients can be lost from litter within a few weeks (e.g., Schowalter and Sabin 1991). Leaching also removes phenolics and other chemicals that defend plant tissues against herbivores (as evidenced by the tea-colored leachate collected under fresh litter), making these materials more available for detritivores. At more advanced stages of decomposition, leaching facilitates the transfer of mineralized nutrients into the soil, where they become available for plant uptake (Setälä and Huhta 1991).

Invertebrate detritivores and bacterial and fungal saprophytes are the major drivers of decomposition in most ecosystems (Seastedt and Crossley 1984). Vossbrink et al. (1979) found that litter lost only 5 percent mass over a nine-month period, due entirely to photooxidation and leaching during the first month, when arthropods and microbes were excluded from litter in an arid grassland ecosystem. Mineralization becomes more important as feeding by detritivores exposes more litter surface area and inoculates fresh surfaces with saprophytic bacteria and fungi.

Although fungi and bacteria are ultimately responsible for the decay of organic matter in forests, detritivorous insects and mites are key contributors to the decomposition process, both in the canopy and on the forest floor (Swift 1977; Seastedt and Crossley 1984; Coleman and Crossley 1996). In addition to comminution, forest animals also play an important role in the transport and inoculation of microorganisms in decaying materials (Schowalter et al. 1992; Müller et al. 2002). For example, ambrosia beetles intentionally inoculate decaying wood with a mutualistic fungus upon which they later feed (Batra 1966; French and Roeper 1972). Other animals tend to spread microbial life in a more passive manner by simply moving from substrate to substrate and making contact with exposed material. Given that decaying components in the canopy are generally isolated from the rest of the decomposer system, this role of transport and inoculation may be all the more vital in the canopy.

Wood-boring insects provide access for a multitude of other organisms as well as for plant roots and air. This serves to modify the microclimate, moisture status, and concentrations of gases inside of a tree bole. It also allows for aerobic decay (Müller et al. 2002) that proceeds at much higher rates than anaerobic decay. Decomposer animals are also important in the redistribution of nutrients and the direct modification of substrate quality via consumption, transport, and defecation (Ausmus 1977). The entry of smaller animals into wood may facilitate decay indirectly by attracting larger, predatory animals that can substantially fragment woody debris in search of their prey (for example, the destructive foraging by woodpeckers). Hence, animals can have important effects on the decay of canopy materials, and their indirect effects are perhaps more critical than their direct impacts (i.e., consumption) on organic matter degradation.

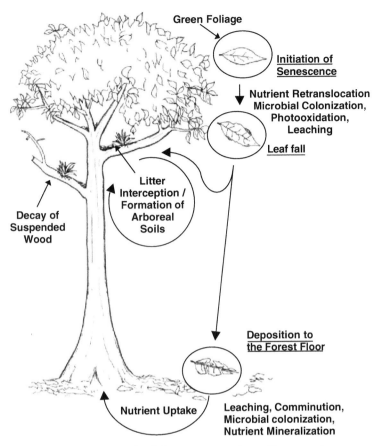

Figure 21-1 Stages of leaf decay and associated decay processes, along with canopy decay components (short-term alteration to foliage, wood decay, and litter interception/formation of arboreal soils).

Although herbivores are not traditionally considered decomposer organisms, they initiate the degradation of organic material and affect decomposition processes. Herbivores fragment, partially assimilate, and chemically alter various plant parts upon consumption. A number of studies have demonstrated that insect frass affects rates of decomposition and nutrient movement through soil (Lovett and Ruesink 1995; Fonte 2003; Hunter et al. 2003). Herbivores also transmit diseases to plants or cause mortality directly, thus playing a role in litter accumulation.

Inevitably, a fraction of litter organic material forms highly complex polyphenolic compounds that are extremely resistant to decay and persist indefinitely in soils. This material, humus, determines many of the physical and chemical properties of soils and is instrumental in water and nutrient retention in the soil. The structure and turnover dynamics of humus in soils remains a poorly understood but essential facet of soil science.

Factors Controlling Decomposition

Several factors regulate the decomposition of organic material. The primary factors are climate, litter quality, the organisms involved, and the nutrient status of the surrounding area. Although the relative influence of these factors may differ between the forest floor and canopy environments, the same governing principles dictate rates of decay in both locations.

On a global scale, climate is considered to have the greatest influence on decomposition (Meentemeyer 1978; Coûteaux et al. 1995; Aerts 1997). Temperature and moisture are two major components of climate that strongly determine the potential for microbial growth (Harmon et al. 1986; Paul and Clark 1996) and the activity of larger organisms (Coleman and Crossley 1996). Extreme temperature or moisture can severely inhibit decay. At smaller scales, minor differences in microclimate can influence decay processes significantly as well. Although decomposition rate generally increases with increasing litter moisture (Meentemeyer 1978), decomposition can be inhibited under saturated conditions. Such conditions are more likely on the ground than in the canopy, where greater evaporation and drainage minimizes saturation.

Litter quality varies greatly at multiple spatial scales and is perhaps the most important determinant of decay within a particular ecosystem (Gallardo and Merino 1993; Aerts 1997). Litter quality reflects nutrient content as well as structural characteristics (e.g., toughness) and defensive chemistry. Although no single measurement can accurately predict the relative decay rates for differing organic materials, a number of litter quality parameters have been proposed. These include the carbon to nitrogen ratio, phosphorus content, lignin content, and various indices of polyphenolics (Fogel and Cromack 1977; Schlesinger and Hasey 1981; Melillo et al. 1982; Loranger et al. 2002). Wood generally decays at a much slower rate than leaf material due to its complex cellular structure, the presence of recalcitrant molecules (e.g., lignin and cellulose), and its low nutrient content. Foliage quality varies greatly among species, and decay rates for different species differ by orders of magnitude (Enriquez et al. 1993).

The decomposer community plays a significant role in regulating decomposition as well (Harmon et al. 1986; Killham 1994). The presence or absence of organisms in various size classes or functional groups may greatly alter decay processes (Heneghan et al. 1999; González and Seastedt 2001). Large detritivores are instrumental in fragmenting larger litter particles into smaller particles, which can be further degraded by small detritivores and microbes, and in the mixing of the soil and litter. Nitrogen fixation and lignin degradation are two examples of ways in which particular microorganism functional groups might influence decomposition.

Finally, decay can be influenced by the nutrient status of the surrounding environment (Lavelle et al. 1993). If litter is of poor quality and decay is limited by certain elements (most often nitrogen or phosphorus), decomposition can be enhanced by nutrients obtained from the surrounding soil-litter matrix via diverse mechanisms, including the deposition of feces by soil animals or the importation of nutrients through fungal transport (Boddy and Watkinson 1995).

Decay in the Canopy

Although the processes of and controls on decomposition are important for both canopy and terrestrial environments, some aspects of decay are unique to the canopy environment. Decay can begin before death or senescence and often depends on these processes. The following discussion considers the senescence of canopy components and the aspects of decay that pertain specifically to the canopy environment. Decomposition does not proceed in the same way or at the same rate for all organic material. Because dead plant and animal tissues vary widely in quality and abundance and decay at different rates (Fogel and Cromack 1977; Fonte 2003), they are discussed separately.

Foliage

Leaves are relatively short-lived, transient units in the canopy and, although they may only comprise a small portion of the canopy biomass, their high nutrient content and turnover rates contribute disproportionately to canopy nutrient dynamics. Numerous physical and biological factors control the decay of foliage in the canopy. As a leaf ages, initial stages of degradation begin early. Both tissue damage by photooxidation and the microbial biomass in the phyllosphere increase with age and contribute to increases in leaching losses, even while leaves are still green (Rodrigues 1994; Carroll 1995; Stone et al. 1996; Beattie and Lindow 1999). Nonpathogenic fungi and bacteria living in leaves also contribute to decay since their utilization of intracellular leaf components (Carroll and Petrini 1983; Stone 1987; Beattie and Lindow 1999) represents a loss of leaf material. These interactions are, however, complicated, and the overall effect of these organisms on leaf survival is not always clear (Carroll 1988, 1995). In some cases, microorganisms may provide a benefit to the leaf such as protection from herbivory (Clay 1988).

With the onset of senescence, plants remobilize essential nutrients from leaves to the stem, resulting in the discoloration of leaves. During this period, microbial colonization of the phyllosphere increases rapidly (Osono 2002). Fungi and bacteria already present in the phyllosphere have early access to readily available energy sources in senescing leaves and thus possess an advantage over saprophytic decomposers on the ground (Stone et al. 1996; Osono 2002). These microorganisms can persist for many weeks after leaf abscission (Stone 1987) and are important contributors to initial stages of decomposition on the forest floor (Osono 2002). The importance of these organisms varies with species and the retention time of senescing leaves before abscission. For example, fungi in a marsh ecosystem have been shown to reduce leaf mass by almost 30 percent while the blade was still attached to a stem (Gessner 2001). Alterations to leaf structure and nutrient content following such prolonged canopy retention are likely to affect later stages of decay.

Woody Components

Although wood has relatively low nutrient concentrations and slow decay rate, woody components constitute most of the aboveground biomass in forests and can remain in the canopy for extended periods of time following death. Woody tissues vary in nutritional quality and decay rate (Harmon et al. 1986; Schowalter et al. 1998). Snags (standing dead trees) have been shown to persist for decades in some systems (Harmon et al. 1986; Lee 1998), and senescent branches can remain attached to a living bole for years before falling (Swift et al. 1976). The amount of dead woody material in the canopy can be extensive, especially in very old forests. Mature, live trees contain large portions of dead material in the form of heartwood and dead branches. In fact, living cells may comprise as little as 10 percent of the total tree biomass in some conifers (Franklin et al. 1987). Much of this material begins to rot well before reaching the forest floor. In some cases, heartrot alone can make up 20 percent of the total stand volume (Parks and Shaw 1996). Thus, we see that the decay of woody material in the canopy is much more extensive than it may seem and is essential in evaluating ecosystem C flux and course woody debris dynamics.

Processes of colonization and degradation of woody material in the canopy are complex and depend on a diverse range of factors, including the cause of death, the condition of the surrounding forest, the height of the decaying wood, and the seasonality of death. Traditionally, wood decay was associated primarily with wound-related infections and heartrot (Shigo 1972; Rayner and Boddy 1986). Damage from both physical and biotic agents increases the susceptibility of trees to disease and may initiate the preliminary phases of decay. However, decay also occurs in apparently undamaged trees (Chapela and Boddy 1988a). The weakening of a tree by drought or competition can trigger the growth of latent fungi in the wood, and activation of these fungi may be signaled by simple changes in moisture content of the wood (Boddy and Rayner 1983b; Chapela and Boddy 1988b).

Upon death, wood decay is carried out by a succession of organisms (Swift 1977) and is mediated by multiple changing factors such as moisture content (Boddy and Rayner 1983b) and substrate quality (Shigo 1972; Swift et al. 1976). Organisms already present at the time of death carry out the earlier stages of decay. These organisms may alter the substrate in a way that allows colonization by later successional groups (Boddy and Rayner 1983b). Many functional groups of microbes may be active at the same time with minimal competition since they often specialize on different resources. In fact, the presence of one may promote the activity of others (Blanchette and Shaw 1978). Although bacteria may have an important role in the early stages of wood decomposition (Blanchette and Shaw 1978) or in N-fixation (Larsen et al. 1978), fungi are by far the dominant microbial agents of wood decay (Boddy and Rayner 1983a; Harmon et al. 1986).

Several studies have addressed the decay of woody material suspended in the canopy and found it to proceed slower than decay on the forest floor (Swift et al. 1976; Rice et al. 1997). Other studies comparing the decay of snags with fallen tree boles have reported similar results (Harmon et al. 1986; Onega and Eickmeier 1991). Although these results are not conclusive, a number of key differences between the forest floor and canopy environment may be responsible for lower decomposition rates in the canopy. Fungal diversity has been shown to decrease with increasing canopy height (Lodge and Cantrell 1995), and a reduced diversity of decomposer organisms might retard decay. This trend may result from larger fluctuations in temperature and moisture experienced by the canopy environment or decreased accessibility for the decomposer community. Decaying material on the ground is in intimate contact with decay organisms, whereas wood in the canopy may require organisms with relatively long-range dispersal mechanisms for colonization. Under some circumstances, the delayed fall of woody materials in the canopy may actually speed up decay in the long term due to increased fragmentation upon impact (Swift et al. 1976). In wetter forest systems, woody materials in the canopy may decay more rapidly due to better drainage and the avoidance of anaerobic conditions that inhibit decay (Progar et al. 2000). Despite the importance of canopy wood decay, most of our knowledge comes from studies of coarse woody debris on the ground, and generalizations about wood decay may not be uniformly applicable to both environments.

Canopy Epiphytes and Arboreal Soils

Although most canopy material eventually reaches the forest floor, leaves and other fine litter components are frequently intercepted by branches, epiphytes, and subdominant plants (Rickson and Rickson 1986; Nadkarni and Matelson 1991; Alverez-Sanchez and Guevara 1999). Canopy architecture and epiphyte density are probably the most important determinants of fine litter retention in the canopy. Trees with upward sloping branches, more intricate dendrology, or dense foliage are likely to retain more fine litter. Epiphyte density depends largely on moisture and hence is highest in humid rainforest ecosystems (Richards 1996). The deposition of litter to stable canopy positions is generally low compared to total litterfall (Nadkarni and Matelson 1991; Richardson et al. 2000). However, in some systems, litter retention in the canopy can exceed deposition to the forest floor (Alverez-Sanchez and Guevara 1999). Litter-trapping fungi can also play an important role in canopy decomposition in tropical forests (Lodge and Cantrell 1995; Lodge 1996). These fungi not only form rhizomorph nets to catch fine litterfall, but anchor the material to the canopy and contribute to its decomposition (Hedger et al. 1993).

If litterfall comes to rest in stable canopy locations such as in large epiphytes or on branch unions, it may persist indefinitely and undergo substantial decay. This litter along with senescing epiphytic components can form soil in the canopy. In tropical forests in Borneo, bird's nest ferns (*Asplenium nidus*) are abundant in the canopy and can capture large amounts of litter. They form soil systems that can weigh more than 200 kg and house a diverse array of invertebrate fauna (Ellwood et al. 2002). Paoletti et al. (1991) estimated bromeliads in the canopy of a Venezuelan cloud forest to contain as much as 100 kg of canopy soil per hectare and found these soils to have higher nutrient concentrations than their terrestrial counterparts. They also found that the faunal assemblages and litter decomposition rates associated with these habitats were very similar to those found for terrestrial soils (Paoletti et al. 1991). The importance of these arboreal soils is further emphasized by the formation of adventitious canopy roots into the epiphyte masses (Nadkarni 1981). If we consider that most tropical soils are shallow, highly weathered, and relatively nutrient poor (Paoletti et al. 1991; Richards 1996), rich canopy soils may play an important role in contributing to the nutrient requirements of their host trees. Other studies that have compared the decay of litter between the canopy and forest floor have shown decay to be slower in the canopy, but these studies used litterbags attached to a branch and not in bromeliad tanks or in similar canopy soil systems (Nadkarni and Matelson 1991; Clark et al. 1998; Fagan and Winchester 1999). The slower rates of decay observed in these canopy locations may result from fluctuations in both moisture and temperature (Bohlman et al. 1995) as well as the lower diversity of decomposer organisms found in the canopy, compared to the forest floor (Nadkarni and Longino 1990; Lodge and Cantrell 1995; Fagan and Winchester 1999).

Conclusion

Decomposition is the primary mechanism for ensuring the turnover of carbon and nutrients from dead plant and animal tissues. Although decomposition has been most widely studied on the ground, much litter is retained in forest canopies for varying lengths of time, thereby affecting nutrient dynamics and formation of suspended soils in the canopy and the quality of litter reaching the forest floor. Decomposition processes may differ between the canopy and the ground as a result of differences in solar exposure, wind, amount of water intercepted, litter quality, and habitat conditions for detritivores and microorganisms responsible for fragmentation and mineralization. While decay rates generally may be higher on the ground, forest canopies may be more favorable sites for decomposition where very wet conditions inhibit decay on the ground. Therefore, decomposition in forest canopies may contribute greatly to turnover of organic matter and nutrients in forest ecosystems.

References

Aerts, R. (1997). Climate, leaf litter chemistry and the decomposition in terrestrial ecosystems: a triangular relationship. *Oikos* **79**, 439–449.

Alverez-Sanchez, J. and Guevara, S. (1999). Litter interception on *Astrocaryum mexicana* Liebm. (Palmae) in a tropical rain forest. *Biotropica* **31**, 82–92.

Ausmus, B.S. (1977). Regulation of wood decomposition by arthropod and annelid populations. *Eco. Bull. Stockholm* **25**, 180–192.

Batra, L.R. (1966). Ambrosia fungi: extent of specificity to ambrosia beetles. *Science* **153**, 193–195.

Beattie, G.A. and Lindow, S.E. (1999). Bacterial colonization of leaves: a spectrum of strategies. *Phytopathology* **89**, 353–359.

Blanchette, R.A. and Shaw, C.G. (1978). Associations among bacteria, yeast, and basidiomycetes during wood decay. *Phytopathology* **68**, 631–637.

Boddy, L. and Rayner, A.D.M. (1983a). Ecological roles of basidiomycetes forming decay communities in attached oak branches. *New Phytol.* **93**, 77–88.

Boddy, L. and Rayner, A.D.M. (1983b). Origins of decay in living deciduous trees: the role of moisture content and re-appraisal of the expanded concept of tree decay. *New Phytol.* **94**, 623–641.

Boddy, L. and Watkinson, S.C. (1995). Wood decomposition, higher fungi, and their role in nutrient redistribution. *Can. J. Bot.* **73**, S1377–S1383.

Bohlman, S.A., Matelson, T.J., and Nadkarni, N.M. (1995). Moisture and temperature patterns of canopy humus and forest floor soil of a montane cloud forest, Costa Rica. *Biotropica* **27**, 13–19.

Carroll, G. (1988). Fungal endophytes in stems and leaves: from latent pathogen to mutualistic symbiont. *Ecology* **69**, 2–9.

Carroll, G. (1995). Forest endophytes: pattern and process. *Can. J. Bot.* **73**, S1316–S1324.

Carroll, G. and Petrini, O. (1983). Patterns of substrate utilization by some fungal endophytes from coniferous foliage. *Mycologia* **75**, 53–63.

Chapela, I.H. and Boddy, L. (1988a). Fungal colonization of attached beech branches. I. early stages of development of fungal communities. *New Phytol.* **110**, 39–45.

Chapela, I.H. and Boddy, L. (1988b). Fungal colonization of attached beech branches. II. spatial and temporal organization of communities arising from latent invaders in bark and functional sapwood, under different moisture regimes. *New Phytol.* **110**, 47–57.

Clark, K.L., Nadkarni, N.M., and Gholz, H.L. (1998). Growth, net production, litter decomposition, and net nitrogen accumulation by epiphytic bryophytes in a tropical montane forest. *Biotropica* **30**, 12–23.

Clay, K. (1988). Fungal endophytes of grasses: a defensive mutualism between plants and fungi. *Ecology* **69**, 10–16.

Coleman, D.C. and Crossley, D.A. (1996). "Fundamentals of Soil Ecology." Academic Press, San Diego.

Coûteaux, M.M., Bottner, P., and Berg, B. (1995). Litter decomposition, climate and litter quality. *Trends Ecol. Evol.* **10**, 63–66.

Deyrup, M. (1981). Deadwood decomposers. *Nat. Hist.* **90**, 84–91.

Ellwood, M.D.F., Jones, D.T., and Foster, W.A. (2002). Canopy ferns in lowland dipterocarp forest support a prolific abundance of ants, termites, and other invertebrates. *Biotropica* **34**, 575–583.

Enriquez, S., Duarte, C.M., and Sand-Jensen, K. (1993). Patterns in decomposition rates among photosynthetic organisms: the importance of detritus C:N:P content. *Oecologia* **94**, 457–471.

Fagan, L.L. and Winchester, N.N. (1999). Arboreal arthropods: diversity and rates of colonization in a temperate montane forest. *Selbyana* **20**, 171–178.

Fogel, R. and Cromack, K. (1977). Effect of habitat and substrate quality on Douglas-fir litter decomposition in western Oregon. *Can. J. Bot.* **55**, 1632–1640.

Fonte, S.J. 2003. The influence of herbivore generated inputs on nutrient cycling and soil processes in a lower montane tropical rainforest of Puerto Rico. M.S. Thesis. Oregon State University, Corvallis.

Franklin, J.F., Shugart, H.H., and Harmon, M.E. (1987). Tree death as an ecological process. *Bioscience* **37**, 550–556.

French, J.R.J. and Roeper, R.A. (1972). Interactions of the ambrosia beetle, *Xyleborus dispar*, with its symbiotic fungus *Ambrosiella hartigii* (Fungi Imperfecti). *Can. Entomol.* **104**, 1635–1641.

Gallardo, A. and Merino, J. (1993). Leaf decomposition in two Mediterranean ecosystems of southwest Spain: influence of substrate quality. *Ecology* **74**, 152–161.

Gessner, M.O. (2001). Mass loss, fungal colonisation and nutrient dynamics of *Phragmites australis* leaves during senescence and early aerial decay. *Aqua. Bot.* **69**, 325–339.

González, G. and Seastedt, T.R. (2001). Soil fauna and plant litter decomposition in tropical and subalpine forests. *Ecology* **82**, 955–964.

Harmon, M.E., Franklin, J.F., Swanson, F.J., Sollins, P., Gregory, S.V., Lattin, J.D., Anderson, N.H., Cline, S.P., Aumen, N.G., Sedell, J.R., Lienkaemper, G.W., Cromack, K., and Cummins, K.W. (1986). Ecology of coarse woody debris in temperate ecosystems. *Adv. Ecol. Res.* **15**, 133–302.

Hedger, J., Lewis, P., and Gitay, H. (1993). Litter-trapping by fungi in moist tropical forest. *In* "Aspects of Tropical Mycology" (S. Isaac, J.C. Frankland, R. Watling, and A.J.S. Whalley, Eds.), pp. 15–35. Cambridge University Press, Cambridge.

Heneghan, L., Coleman, D.C., Zou, X., Crossley, D.A.J., and Haines, B.L. (1999). Soil microarthropod contributions to decomposition dynamics: tropical-temperate comparisons of a single substrate. *Ecology* **80**, 1873–1882.

Hunter, M.D., Linnen, C.R., and Reynolds, B.C. (2003). Effects of endemic densities of canopy herbivores on nutrient dynamics along a gradient in elevation in the southern Appalachians. *Pedobiologia* **47**, 231-244.

Killham, K. (1994). "Soil Ecology." Cambridge University Press, Melbourne.

Larsen, M.J., Jurgensen, M.F., and Harvey, A.E. (1978). N_2 fixation associated with wood decayed by some common fungi in western Montana. *Can. J. For. Res.* **8**, 341–345.

Lavelle, P., Blanchart, E., Martin, A., Martin, S., Spain, A., Toutain, F., Barois, I., and Schaefer, R. (1993). A hierarchical model for the decomposition in terrestrial ecosystems: application to soils of the humid tropics. *Biotropica* **25**, 130–150.

Lee, P. (1998). Dynamics of snags in aspen-dominated midboreal forests. *For. Ecol. Manage.* **105**, 263–272.

Lodge, D.J. (1996). Microorganisms. *In* "The Food Web of a Tropical Rainforest" (D.P. Reagan and J.B. Waide, Eds.), pp. 53–108. University of Chicago Press, Chicago.

Lodge, D.J. and Cantrell, S. (1995). Fungal communities in wet tropical forests: variation in time and space. *Can. J. Bot.* **73**, S1391–S1398.

Loranger, G., Ponge, J.F., Imbert, D., and Lavelle, P. (2002). Leaf decomposition in two semi-evergreen tropical forests: influence of litter quality. *Biol. Fert. Soils* **35**, 247–252.

Lovett, G.M. and Ruesink, A.E. (1995). Carbon and nitrogen mineralization from decomposing gypsy moth frass. *Oecologia* **104**, 133–138.

Meentemeyer, V. (1978). Macroclimate and lignin control of litter decomposition rates. *Ecology* **59**, 465–472.

Melillo, J.M., Aber, J.D., and Muratore, J.F. (1982). Nitrogen and lignin control of hardwood leaf litter decomposition dynamics. *Ecology* **63**, 621–626.

Müller, M.M., Varama, M., Heinonen, J., and Hallaksela, A.-M. (2002). Influence of insects on the diversity of fungi in decaying spruce wood in managed and natural forests. *For. Ecol. Manage.* **166**, 165–181.

Nadkarni, N.M. (1981). Canopy roots: convergent evolution in rainforest nutrient cycles. *Science* **214**, 1023–1024.

Nadkarni, N.M. and Longino, J.T. (1990). Invertebrates in canopy and ground organic matter in a neotropical montane forest, Costa Rica. *Biotropica* **22**, 286–289.

Nadkarni, N.M. and Matelson, T.J. (1991). Fine litter dynamics within the tree canopy of a tropical cloud forest. *Ecology* **72**, 2071–2082.

Onega, T.L. and Eickmeier, W.G. (1991). Woody detritus inputs and decomposition kinetics in a southern temperate deciduous forest. *Bull. Torrey Bot. Club* **118**, 52–57.

Osono, T. (2002). Phyllosphere fungi on leaf litter of *Fagus crenata*: occurrence, colonization, and succession. *Can. J. Bot.* **80**, 460–469.

Paoletti, M.G., Taylor, R.A., Stinner, B.R., Stinner, D.H., and Benzing, D.H. (1991). Diversity of soil fauna in the canopy and forest floor of a Venezuela cloud forest. *J. Trop. Ecol.* **7**, 373–383.

Parker, G.G. (1995). Structure and microclimate of forest canopies. *In* "Forest Canopies" (M.D. Lowman and N.M. Nadkarni, Eds.), pp. 73–106. Academic Press, San Diego.

Parks, C.G. and Shaw, D.C. (1996). Death and decay: a vital part of canopies. *Northwest Sci.* **70**, 46–53.

Paul, E.A. and Clark, F.E. (1996). "Soil Microbiology and Biochemistry." Academic Press, San Diego.

Progar, R.A., Schowalter, T.D., Freitag, C.M., and Morrell, J.J. (2000). Respiration from coarse woody debris as affected by moisture and saprotroph functional diversity in Western Oregon. *Oecologia* **124**, 426–431.

Rayner, A.D.M. and Boddy, L. (1986). Population structure and the infection biology of wood-decay fungi in living trees. *In* "Advances in Plant Pathology" (D.S. Ingham and P.H. Williams, Eds.), pp. 119–160. Academic Press, San Diego.

Rice, M.D., Lockaby, B.G., Stanturf, J. A., and Keeland, B.D. (1997). Woody debris decomposition in the Atchafalaya River Basin of Louisiana following hurricane disturbance. *Soil Sci. Soc. Am. J.* **61**, 1264–1274.

Richards, P.W. (1996). "The Tropical Rain Forest." Cambridge University Press, Cambridge.

Richardson, B.A., Richardson, M.J., Scatena, F.N., and McDowell, W.H. (2000). Effects of nutrient availability and other elevational changes on bromeliad populations and their invertebrate communities in a humid tropical forest in Puerto Rico. *J. Trop. Ecol.* **16**, 167–188.

Rickson, F.R. and Rickson, M.M. (1986). Nutrient acquisition facilitated by litter collection and ant colonies on two Malaysian palms. *Biotropica* **18**, 337–343.

Rodrigues, K.F. (1994). The foliar fungal endophytes of the Amazonian palm *Euterpe oleracea*. *Mycologia* **86**, 376–385.

Schlesinger, W.H. (1997). "Biogeochemistry: an analysis of global change." Academic Press, San Diego.

Schlesinger, W.H. and Hasey, M.M. (1981). Decomposition of chaparral shrub foliage: losses of organic and inorganic constituents from deciduous and evergreen leaves. *Ecology* **62**, 762–774.

Schowalter, T. D. (2000). "Insect Ecology: An Ecosystem Approach." Academic Press, San Diego.

Schowalter, T.D., Caldwell, B.A., Carpenter, S.E., Griffiths, R.P., Harmon, M.E., Ingham, E.R., Kelsey, R.G., Lattin, J.D., and Moldenke, A.R. (1992). Decomposition of fallen trees: effects of initial conditions and heterotroph colonization rates. *In* "Tropical Ecosystems: Ecology and Management" (K.P. Singh and J.S. Singh, Eds.), pp. 373–383. Wiley Eastern Unlimited, New Delhi.

Schowalter, T.D. and Sabin, T.E. (1991). Litter microarthropod responses to canopy herbivory, season and decomposition in litterbags in a regenerating conifer ecosystem in western Oregon. *Biol. Fert. Soils* **11**, 93–96.

Schowalter, T.D., Zhang, Y.L., and Sabin, T.E. (1998). Decomposition and nutrient dynamics of oak *Quercus* spp. logs after five years of decomposition. *Ecography* **21**, 3–10.

Seastedt, T.R. (1984). The role of microarthropods in decomposition and mineralization processes. *Annu. Rev. Entom* **29**, 25–46.

Seastedt, T.R. and Crossley, D.A. (1984). The influence of arthropods on ecosystems. *Bioscience* **34**, 157–161.

Setälä, H. and Huhta, V. (1991). Soil fauna increase Betula pendula growth: laboratory experiments with coniferous forest floor. *Ecology* **72**, 665–671.

Shigo, A.L. (1972). Successions of microorganisms and patterns of discoloration and decay after wounding in red oak and white oak. *Phytopathology* **62**, 256–259.

Stone, J.K. (1987). Initiation and development of latent infections by *Rhabdocline parkeri* on Douglas-fir. *Can. J.Bot.* **65**, 2614–2621.

Stone, J.K., Sherwood, M.A., and Carroll, G.C. (1996). Canopy microfungi: function and diversity. *Northwest Sci.* **70**, 37–45.

Swift, M J. (1977). The ecology of wood decomposition. *Sci. Prog.* **64**, 175–199.

Swift, M.J., Heal, O.W., and Anderson, J.M. (1979). "Decomposition in terrestrial systems." University of California Press, Berkeley.

Swift, M.J., Healey, I.N., Hibberd, J.K., Sykes, J.M., Bampoe, V., and Nesbitt, M.E. (1976). The decomposition of branch-wood in the canopy and floor of a mixed deciduous woodland. *Oecologia* **26**, 139–149.

Vossbrink, C.R., Coleman, D.C., and Woolley, T.A. (1979). Abiotic and biotic factors in litter decomposition in a semiarid grassland. *Ecology* **60**, 265–271.

CHAPTER 22

Survival Strategies: A Matter of Life and Death

Stephen R. Madigosky

Oh, what a tangled web we weave. When first we practice to deceive.
— *Sir Walter Scott,* Marmion, *1808*

For millions of years, the interplay of competitive forces between species and their environment has helped to hone the extraordinary physical and behavioral adaptations displayed within each organism. In doing so, organisms have advanced their survivability. Since life began on earth some 3.5 to 4 billion years ago, climatic cycles have played a critical role in the refinement of both form and function within organisms. Coupled with innate genetic elasticity within each species, this has contributed greatly in establishing the many wonderful and peculiar forms that have passed through geologic time. Nowhere is this true more than in the tropics, the epicenter of biological diversity. This is not to say that similar forces are not at work in other regions of the globe, but they appear more numerous in tropical regions owing to the extent of the diversity. Change directed through natural selection has also helped shape organisms in ways that we may never fully understand. Perhaps this alone provides impetus for examining strategies used by organisms faced with limited resources (Table 22-1).

Historically, scientists have focused on the descriptive nature of plants and animals rather than on the functional role they play within the environment. This certainly is a rational way of approaching the subject, given the tremendous diversity on Earth, especially in tropical regions. Even at that, our best efforts over the past 250 years have yielded the taxonomic description of some 1.75 million species (Morell 1999; Purvis and Hector 2000) of an estimated 3 to 100 million species or more (May 1992). It is little wonder that our knowledge of ecological interactions within and between organisms is limited. We did not consciously approach biological exploration with this intent but rather were forced to deal with an environment that even today remains biologically unexplored. This is certainly true of tropical forests.

A recent approach has been to concentrate on biological and ecological processes to better assess how species function, and current technology is helping to chart this new course. The use of sophisticated genetic techniques and biochemical analyses have helped forge new ideas and relationships within the context of biological and ecological science. We can now decipher greater relation within and between organisms and, on occasion, glimpse into the mechanisms of evolution. One unforeseen outcome has been to assign greater importance to species and ecological diversity. Ultimately, this may well determine our own destiny as a species and the type of environment we seek to promote (Madigosky and Grant 1996). The topic of how organisms survive is an intriguing area of study. In essence, it attempts to grapple with the difficult questions of how and why organisms behave as they do. This chapter will concentrate on a few of these strategies relating to biological and ecological interaction. It represents a short story of life in the long struggle for survival.

Table 22-1   A Select List of Organisms Known to Display Distinct Survival Strategies.
The Majority of Species Listed are Indigenous to the Tropics.

Classification	Survival Strategy	Citation
Chemical Protection		
Butterflies		
Danaus plexippus	larva sequester toxic cardiac glycosides	Owen 1980
Battus philenor	larvae sequester toxic aristolchic acids	Owen 1980
Eurytides marcellus	larvae sequester toxic acetogenins	Martin et al. 1999
Caterpillars		
Cysteodemus wislizeni	body fluids contain cantharidin	Borror and DeLong 1964
Lytta sp.	body fluids contain cantharidin	Borror and DeLong 1964
Beetles		
Thermonectus marmoratus	ejects volitile steroids: mirasorvone and cybisterone when disturbed	Meinwald et al. 1998
Brachinus spp.	ejects hot hydrogen peroxide and hydroquinine	Behe 1996; Penny and Arias 1982
Phasmid		
Oreophoetes peruana	glandular discharge of quinoline when disturbed	Eisner et al. 1997
Ants		
Solenopsis (Diplorhoptrum) sp.	alkaloids: pyrrolizidines	Jones et al. 1999
Monomorium pharaonis poison frogs (skin alkaloids)	alkaloids: indolizidines, pyrrolidines	Jones et al. 1999
Mantella betsileo	alkaloids: quinolizidine	Jones et al. 1999
Epipedobates tricolor	alkaloids: epibatidine	Badio et al. 1994
Dendrobates histrionicus	alkaloids: histrionicotoxin	Garaffo et al. 2001
Dendrobates lehmanni	alkaloids: pyrrolo[1,2-a]azepane	Garaffo et al. 2001
Dendrobates pumilio	alkaloids: pumiliotoxin	Daly et al. 2002
Pseudophryne sp.	pseudophrynamines/putriliotoxins	Smith et al. 2001
Epipedobates trivittatus	alkaloids: histrionicotoxins	Daly 1998
Salamander		
European fire salamander	alkaloid: samandarine	Daly et al. 2002
Toxic birds		
Pitohui spp.	alkaloids: batrachotoxins	Dumbacher et al. 2000
Ifrita kowaldi	alkaloids: batrachotoxins	Dumbacher et al. 2000
Chemical Communication		
Ants		
Pheidole cephalica	alarm recruitment	Wilson 1986
Solenopsis invicta	recruitment pheromone (alpha-farnesene)	Hölldobler and Wilson 1990
Camouflage/Eucrypsis		
Caterpillars		
Colias edusa	countershading by orientation	Cott 1940
Papilio machaon	disruptive pattern	Jarvi et al. 1981

Table 22-1—*Continued*

Classification	Survival Strategy	Citation
Arctiid		
Clemensia albata	disruptive coloration	McCabe 1981

Aposomatic Coloration

Neotropical rainbow katydid *Vestra* sp.	boldly colored (yellow/blue abdomen)	personal observation
Neotropical poison frogs		
Dendrobates ventrimaculatus	boldly colored (red/yellow stripes on body)	personal observation
Dendrobates reticulatus	boldly colored (orange/red/black body)	personal observation

Transparent Body Parts

Neotropical butterflies		
Cithaerias aurorina	clearwings	personal observation
Oleria sp.	clearwings	personal observation

Mimicry

Neotropical orthopterans		
Omura congrua	grass mimic	personal observation
Dysonia sp.	lichen mimic	Nickle and Castner 1995
Championica sp.	lichen mimic	Nickle and Castner 1995
Rhinischia sp.	bark mimic	Nickle and Castner 1995
Dasyscelidius sp.	twig mimic	Nickle and Castner 1995
Aganacris pseudosphex	wasp mimic	Castner 2000
Scaphura sp.	wasp mimic	Castner 2000
Pterochroza ocellata	dead leaf mimic	Castner 1995
Typophyllum bolivari	green/brown leaf mimic (highly iregular leaf margins)	Castner and Nickle 1995b
Fish		
Monocirrhus polyacanthus	leaf mimic	Owen 1980
Bird (neotropical Potoo)		
Nyctibius spp.	branch or snag mimic	Hilty and Brown 1986

Sex Pheromone Mimicry

Bolas spider		
Mastophora hutchinsoni	(Z)-9Tetradecenyl acetate and (Z,E)-9,12-tetradecadienyl acetate (sex pheromone of prey moth)	Haynes et al. 2001 Gemeno et al. 2000
Spider orchid		
Ophrys sphegodes	pollination by sexual deception (hydrocarbon production)	Schiestl et al. 2000
Rust fungi		
Puccinia monoica complex	(infection causes a pseudoflower) aromatics and fatty acid derivatives	Raguso and Roy 1998

Continued

Table 22-1—*Continued*

Classification	Survival Strategy	Citation
Lure		
Alligator snapping turtle		
Macroclemys temminckii	tounge wiggles to lure fish	Ernst et al. 1994
Defensive Display/Behavior		
Neotropical katydids		
Typophyllum bolivari	striation on wings revealed if disturbed	Castner and Nickle 1995b
Pterochroza ocellata	eye spots revealed when disturbed	Castner and Nickle 1995b
Panoploscelis specularis	loud vocalization/churp when handled	personal obsercation
Smoky jungle frog		
Leptodactylus pentadactylus	loud vocalization/scream when handled	personal observation
False Eye Spots		
Neotropical butterflies		
Caligo spp.	large eye spots on wings	personal observation
Morpho spp.	eye spots positioned on wing periphery	personal observation
Click beetles		
Alaus oculatus	large eye spots on thorax	Arnett and Jacques 1981
Alaus melanops		Arnett and Jacques 1981
Birds (pygmy owls)		
Glaucidium jardinii	black eye spots on back of head	De Schauensee and Phelps 1978
Glaucidium brasilianum	black eye spots on back of head	De Schauensee and Phelps 1978
Abstract Mimicry of Snakes		
Sphingid moth		
Hemeroplanes triptolemus	caterpilar inflates & reveals eyespots	Pough 1988
Pseudospinx tetrio	presumed coral snake mimic	Pough 1988
Nymphalid butterfly		
Dynastor darius	snake-like appearance	Pough 1988

The Nature of Mimicry

In 1862, Henry Walter Bates, while working in the Amazon Basin, described a phenomenon he referred to as "mimetic analogy," whereby members of distinct families of butterflies resemble each other in color, shape, and patterns. He surmised that palatable species of Amazonian butterflies were garnering protection by closely resembling noxious unpalatable butterfly species. The avoidance of both edible and inedible species by predators, essentially one at the expense of the other, is now referred to as *mimicry* or, more precisely, *Batesian mimicry*. While this accurately describes the affiliation between unrelated species of similar appearance (only one possessing noxious properties), Bates failed to discern reasons for mutual similarities among distinct unpalatable species of butterflies. Nearly 17 years later, the German zoologist Fritz Müller helped solve the mystery by proposing that both species had more to gain by sharing properties that rendered them unpalatable. He, too, was able to contribute additional information pertinent to what Bates had described after observing butterfly populations in Amazonia. According to his explanation, fewer individuals would be required to condition predators if they both possessed properties that would render them unpalatable or protected. This would help preserve even greater numbers of individuals within the gene pool of both populations. Quite appropriately, the phenomenon became known as *Müllerian mimicry*. Although butterflies initially provided the model that led to the discovery of these relationships, many organisms have since been implicated in similar associations. These, too, are not without controversy or question, but they have enabled us to better understand the complex nature of biological interactions that surround the phenomenon called mimicry.

Sometimes the relationship between two distinct species is blurred. In Africa, larvae of the monarch *Danaus chrysippus* (family Danaidae) are known to sequester glycosides alongside the acraeid butterfly *Acraea encededon* (family Acraeidae) that accumulates hydrogen cyanide within its tissues. Each bears a striking resemblance to the other, using combinations of black, orange, and white wing colors to announce their presence. In fact, it is sometimes difficult to discern model from mimic. Since the models are from each of two families and a great deal of variability exists among individuals relative to their toxicity, members have been implicated in both Batesian and Müllerian mimicry. Dozens of species from several other butterfly families (Nymphalidae, Lycaenidae, Papilionidae, and Satyridae) have also become a part of this mimicry ring, but they are Batesian mimics of the unpalatable Danaidae and Acraeidae butterflies.

In order for mimicry to work effectively, certain conditions need to be maintained. For one, mimics must be credible representations of their models. The divergence of a mimetic or aposomatic (boldly colored) hybrid too far from the norm may invite a greater degree of selection pressure. Second, mimics must be less common than their models. A failure to meet this criterion may augment the possibility of predators becoming educated about color and pattern associations with palatability versus unpalatability issues. In a Müllerian sense, models may emerge earlier in the season, be at least slightly more conspicuous with distinct color and patterns, maintain a gregarious existence, and occupy a more extended geographic range (Mallet 2001).

Automimicry

On occasion, instances occur where only some members of a particular species (especially butterfly larvae) may contain or sequester toxins. In many cases, this is due to the selection of the food plant on which they have been reared. An example of this is seen in *Danaus plexippus*, the orange-winged, black veined monarch butterfly (family Danaidae). The larvae of this species feed upon the foliage of milkweeds (Asclepiadaceae) that are known to contain toxic glycosides. In doing so, they accumulate toxins within their tissue, rendering them inedible or distasteful to birds. Because larvae sometimes utilize several milkweed species possessing different levels of glycosides, a fair amount of variation exists within larvae relative to glycoside content. At one extreme, they can be extremely unpalatable and, at the other, quite edible. Since a single species is involved whereby individual members differ in their palatability, the phenomenon is referred to as *automimicry*. For this to work well, more butterflies need to carry significant loads of glycosides to assure that predators are adequately educated upon eating them. If other palatable butterfly species sharing the same range mimic monarchs, they are considered Batesian mimics.

An example of this involves the North American viceroy butterfly (*Limenitis archippus*). The viceroy, a member in the family Nymphalidae, closely resembles the monarch both in wing coloration and in general markings. The viceroy, however, does not feed upon milkweed and therefore lacks the residual glycosides. As a result, it is considered a palatable species. Since it shares a similar range with the monarch, it is protected by virtue of its association, a classic case of Batesian mimicry. This story has some unusual twists that may deviate from typical Batesian mimicry. The queen (*Danaus gilippus*) also looks quite similar to the monarch and shares the same geographic range throughout southern North America. It, too, feeds upon the blunt-leaved milkweed (*Asclepias amplexicaulis*) and rambling milkweed (*Sarcostemma hirtellum*) and is able to sequester toxic glycosides. Here, considerable variation in glycoside accumulation exists within the tissue of the queen. Queens that are toxic are Müllerian mimics to unpalatable monarchs and palatable queens are Batesian mimics to the monarch. Both species may exhibit automimicry.

Abstract Mimicry

As noted, many instances occur in nature where a model and mimic bear an extreme likeness to each other. As a consequence, the mimic may be categorized according to the exact species it is modeling. More often than not, the association between two or more forms is blurred. This is referred to as *abstract mimicry* and has been documented as a strategy used by both plants and animals. Mimicry of this nature tends to be inaccurate, imprecise, and generalized. The mimic possesses the broad generalized features of the model without being precise enough to identify the association to species. This is widely noted among larvae and pupae in families of butterflies and moths found throughout the tropics. Immature forms of lepidopterans within the families Papilionidae, Geometridae, Noctuidae, Oxytenidae, Notodontidae, and Sphingidae are all known to resemble snakes (Pough 1988) both in gross morphology and even in behavior. In such cases, mimics resemble the general characteristics of venomous or semi-venomous snakes (see Figure 22-1). Some Nymphalid pupae have even been known to thrash back and forth when disturbed to produce a rhythmic hissing vocalization that is broadcast to startle predators. In other instances, leaf hoppers (Homoptera: Cicadellidae) possessing modified wings quickly take flight and encircle a leaf within a two-meter radius, occasionally landing and pulsating their wings in perfect butterfly fashion (Madigosky, personal observation). Even the small markings on the wings resemble eyespots that are typical of so many butterflies and moths. It is obvious that the leafhopper is mimicking a butterfly, but the exact purpose is difficult to assess. Certainly, the wings coupled with the extraordinary behavior can take it farther and faster than other members of its family.

Warning Coloration and Mimicry

The use of warning coloration among animals is a common theme displayed in almost every taxonomic group. Brightly colored or patterned animals are protected from attack as a result of openly broadcasting such endowments (see Figure 22-2). But the underlying theme that makes the color and/or the pattern effective is that many brightly colored organisms contain toxins, distasteful chemicals, noxious irritants, secretions, foul odors, or some other entity that predators have learned to avoid. In the majority of cases, organisms must learn to avoid potential prey items via trial and error and may even succumb to the indulgence of such food items. Those fortunate enough to live through such episodes will avoid these prey and will even shy away from organisms that remotely resemble the prey. In essence, predators appear to generalize the features of the noxious prey item and use this information to avoid future encounters with questionable prey items.

Figure 22-1 Pictured is the final larval stage of the hawk moth *Hemeroplanes* sp. prior to (top) and after (bottom) exhibiting its snake-like warning display.

Figure 22-2 Pictured is a gaudily colored short-horned grasshopper (Orthoptera: Acrididae) positioned on a solanaceous plant.

Figure 22-3 The stunningly patterned and colored venomous coral snake (*Micruroides euryxanthus*) climbing a rock encrusted with lichens.

The suggestion that non-venomous or slightly venomous snakes resemble the patterns and colors of the venomous coral snake originated well over 100 years ago. Over the years, this example has been portrayed as a classic case of vertebrate mimicry. In North America, the Arizona coral snake (*Micruroides euryxanthus*) is gaudily ringed and strikingly colored (see Figure 22-3). This description also corresponds to dozens of coral snake species found throughout Central and South America. In North America, the non-venomous banded king snake and shovel-nosed snake have patterns and/or colors that closely resemble the Arizona coral snake and are considered Batesian mimics.

Coral snakes are difficult to miss owing to their bright colors and patterns—or are they? Much controversy surrounds this topic as suggestions have been made as to whether they are cryptic or aposematic. At first glance it would seem that coral snakes are impossible to miss. They are boldly colored and in many cases banded with the presentation of red, orange, yellow, black, and white in a variety of combinations. These are all of the essential attributes required for aposematic function, especially when viewed in close proximity. Conversely, their color and banding are distracting and difficult to follow when they move through the congested underbrush. This is especially true during the twilight hours; they seem to vanish in an instant. Pough (1988) suggested that coral snakes are most likely cryptic and aposematic depending on the circumstances, and he provided sound evidence for both. Whatever the case, they serve as exceptional models for a number of similarly patterned colubrid snakes that share their range.

Mimicry in Heliconius Butterflies

Few organisms exist that attract the attention, admiration, and intrigue of scientists, naturalists, and laymen alike as do butterflies. They are conspicuous in nearly every biome. Undoubtedly, this contributed to the attention they received by Bates, Müller, and a host of other scientists. Yet their existence is obscured by the lack of information regarding their ecological significance, especially throughout the tropics. Perhaps this reflects their largely solitary existence or their minor role in pollinating food crops. Or it simply may reflect their preference toward a nomadic existence, thereby rendering them difficult to study. Whatever the case, little is known about the role they play in their environment. In a few instances, mutualistic relationships have been implicated as a result of close study. Such is the case with members of the genus *Heliconius*. Both Bates and Müller used members of this group as their models, and it is not difficult to see why. They are abundant throughout the neotropics, are conspicuous owing to their orange and black coloration/patterns, and are slow flyers as compared to most other species. Perhaps this is why they have been reasonably dealt with on more than a taxonomic level.

To the casual observer and expert alike, considerable similarity exists in coloration and patterns within *Heliconius* butterflies. In a classic example of Müllerian mimicry, *Heliconius melpomene* and *Heliconius erato* share nearly identical wing markings and coloration throughout a wide geographic area. In fact, they so closely resemble one another that there is disagreement as to which form acts as the model and which is the mimic. Apparently, this is complicated by their numerical dominance in different parts of their range. However, Mallet (2001) suggested that *H. erato* serves as the model because it is twice as abundant as *H. melpomene* throughout most of its range. Whatever the case, both species are unpalatable and, therefore, confirm Müllerian mimicry. Of greater interest is the degree of phenotypic variation noted throughout their range. Nearly a dozen different races are recognized throughout Central and South America and in almost every case, *H. melpomene* and *H. erato* possess a counterpart for each other (Turner 1975, 1981; Sheppard et al. 1985; Turner and Mallet 1996; Mallet 2001). These forms are so similar that some scientists speculate that they may have coevolved or codifferentiated from a common ancestral pattern. However, cladograms of races based on mitochondrial DNA sequences of *H. erato* and *H. melpomene* are not concordant and indicate that they do not share a common biogeographical history (Joron and Mallet 1998).

Even more interesting are the similarities noted among distinct families of butterflies resembling Heliconids. Members of the Ithomiidae, Danaidae, and Pieridae, and even in a few tailless species of swallowtail butterflies (Papillionidae) possess mimics of Heliconids. One of the more stunning examples occurs in the family Pieridae. Members of this family typically exercise sporadic flight patterns and possess solid yellow, white, or orange wings with few markings. They are fast flyers and are not shy about puddling on riverbanks or in open areas. But *Dismorphia amphione*, an unusual member of the Pieridae, is phenotypically aligned to the well-patterned orange, black, and yellow, slender bodied, elongate winged *Heliconius* butterfly, and is a presumed Batesian mimic. The disposition of this butterfly also favors the slow-moving heliconians that gently waft about the neotropical canopy.

Often, large groups of butterflies may simultaneously occupy a mimicry ring, a group of sympatric species sharing a common warning pattern. The extent of this phenomenon is worldwide. One well-known ring complex exists in the Ecuadorian Amazon. Here, ithomiine butterflies belong to some eight distinct mimicry rings harboring more than 120 distantly related species of butterflies, and in some instances day-flying moths (Joron and Mallet 1998). These relationships are by no means static, as species may move in and out of specific rings. There is considerable experimental evidence that butterflies and moths switch rings through mutations that control key color patterns and distribution features (Mallet and Gilbert 1995). These are usually a result of single gene mutations that result in phenotypic variation; this is thought to have driven the substantial racial variation among *H. erato* and *H. melpomene* (Mallet 1986; Turner 1988) throughout their extended range.

On top of being protected via mimicry, *Heliconius* butterflies possess unusual behavioral characteristics. Interestingly, they are the only butterflies known to eat pollen. They are long lived (with longevity extending for more than a year), often forage along the same routes day after day, and roost communally. Most intriguing are their efforts to avoid areas where they have been netted or harassed even days after the incident (personal observation). In studies assessing their ability to associate flower colors with a reward, it has been shown that they learn colors as quickly as many bees (Oliveira 1998). It is obvious that these butterflies are equipped to deal with a host of pressures not observed in other butterfly families.

Even the larval stages of *Heliconius* butterflies are protected. A primary strategy lies in the choice of foliage they ingest from the passion vine (Passifloraceae). This is a large plant family in the neotropics and is, therefore, easily accessible to larvae of *Heliconius* butterflies. A well-known property of this group is that the foliage synthesizes cyanogenic glycosides and cyanohydrins that are considered toxic. Since *Heliconius* eggs are deposited by the adult butterfly on the underside of the passion vine leaves, larvae hatch with the luxury of not having to seek out a food source. In the process of ingesting the foliage, larvae requisition toxins that render them unpalatable. Actually, the situation is a bit more twisted than described. The story really centers on the ability of the larvae to detoxify cyanogenic glycosides. Apparently, adult butterflies are able to distinguish between the many different species of passion vine (*Passiflora*) and choose to lay eggs only on those species that produce glycosides that their larvae can detoxify (Spencer 1984). This would seem not only to help protect their offspring but also to reduce competition between like family members that share interest in the same food source. Additionally, by laying eggs on specific vines, larvae may acquire permutations of the same basic toxin that may further help to confuse predators. In this respect, it is the adult that assures larvae will contain a distinct type of unpalatable property. Larvae feeding upon passion vine have the necessary enzymes to detoxify only certain species within this plant group. Therefore, this also tends to eliminate competition by adult female selection. In addition, the production of cyanogens among different species of passion vine may revolve around the degree and extent of attack between butterfly and plant over extended periods. Correspondingly, the ease at which butterflies detoxify components from these plants has undoubtedly helped to refine the relationship.

Many passion vine species have adapted to being parasitized by *Heliconius* butterflies. It is interesting to observe the variation in leaf morphology encountered within one passion vine. More than a dozen different-looking leaves within the same vine is not uncommon. This seems to be one way that the plant has adapted to tricking the butterfly to look elsewhere. However, if chemical cues direct homing to passion vines, a change in morphology would seem not to be of any significance.

Of related importance, some passionflower vines release attractants from tiny extrafloral nectaries found on the underside of the leaves. These structures produce important amino acids that are sought after and harvested by ants, bees, and wasps (Horn et al. 1984). The presence of these aggressive insects serves to discourage herbivory from invading caterpillars and other small insects. In addition, the nectaries look strikingly similar to butterfly eggs and may serve to trick gravid heliconid butterflies to look for other leaves without eggs (Gilbert 1971). This behavior appears to be important in reducing both sibling and intraspecies competition among larvae and is practiced by many insects. Gravid butterflies of many species often examine the underside of leaves thoroughly before depositing their eggs.

Several genera of *Heliconius* butterflies have larvae that are spiny. On top of being unpalatable, the spines further help to discourage predators from partaking of an easy meal. The voracious appetite of the larvae, especially in a collective sense, can inflict considerable damage to the host plant. In periods when larvae are not feeding, they sometimes congregate or huddle into a vertical mass that resembles a spiny brush. Individuals cling to a central twig or stripped petiole and bend acutely. This arrangement gives the collective appearance of a botanical structure, possibly that of a wilted or dried spiny inflorescence (see Figure 22-4). This is most likely an example of abstract mimicry (broad generalized features of a plant model in this instance).

Tactile and Wasmannian Mimicry

Without question, ants are one of the most often studied and interesting groups of insects known. They are well represented in nearly every biome, possess complex social organization, and utilize

Figure 22-4 *Heliconius* caterpillars at rest on a tendril. Collectively, they appear as a dried spiny inflorescence.

sensory modalities that employ chemical, visual, vibrational, and tactile cues. They attract a host of organisms that through symbiosis enter into complex interactions such as mutualism, commensalism, parasitism, and mimicry. They are a dynamic component in many systems and especially tropical forests, where they have historically exerted considerable selection pressure on their environment. Because of subterranean habits, much of the behavior of this group goes unnoticed. Given that they prefer an existence beneath the ground, much of the communication between members of their society is of a chemical and/or tactile nature. Fortunately, we are now beginning to unravel some of the intricate relationships that exist within ant communities in addition to their extraordinary associations with other animals and plants. A few examples illustrate this point.

To begin, ants communicate via trophallaxis (an exchange of digestive fluids), allogrooming, and body contact. They make contact intentionally and continuously with one another in the close confines of the nest. This provides them an opportunity to distribute their individual scents among siblings, thereby giving them collective recognition of self and/or colony. In this respect, they are able to recognize members of their colony from invaders. Moreover, individuals within colonies have been shown to possess separate individual odors based on their genetic predisposition. Many of these entities are volatile hydrocarbons that are maintained within the cuticle of the exoskeleton. Within some species, hydrocarbon profiles are uniform, although the relative proportions of these entities can fluctuate (Provost et al. 1993). Assuming this model, if for some reason colony members become isolated, they may have only a limited time to reunite with the mother colony before risking outcast status. This appears to be a function of the degree of change exhibited in hydrocarbon profiles over time. The observation of *Camponotus fellah* reintroduced into their former colonies after a short absence indicates frequent trophallactic solicitations by their nestmates; however, longer isolation periods (20 to 40 days) tend to make reconciliation more problematic (Boulay et al. 2000). Undoubtedly, there is a fair bit of flexibility to this relationship contingent upon the individual species involved. In other instances, it is likely that this relationship is driven by other mechanisms. As we delve further into the chemical nature of colony odor, we may come to a more complete understanding of how odors direct aspects of behavior.

Because ants tend to carry a somewhat consistent colonial odor, several organisms have been able to exploit their chemical wizardry. This is especially true of beetles (Coleoptera). In an unusual display of symbiosis, paussine carabid beetles maintain modified antennal segments impregnated with gland cells that secrete a fragrance that gains them acceptance into select host ant colonies. Beetles that participate in this relationship are termed *myrmecophiles* and their deceitful entrance into a colony can be quite profitable. They are able to feed freely upon ant larvae, pupae, and even the dismantled portions of adult ants. In similar fashion, ptiliid beetles are known myrmecophiles, some associating with neotropical army ants in the genus *Eciton*. Here, beetles feed without hindrance upon the surface oils of ant larvae, pupae, and eggs for at least a portion of their nourishment (Evans and Bellamy 1996). Beetles are not the only insects known to maintain myrmecophilic relations. Other examples include mutualistic associations between temperate zone *Myrmica* ants and several species of Palaerctic lycaenid butterflies in the genus *Maculinea*. In this instance, late 4th instar *Maculinea* larvae are taken to the ant nest for adoption, where they spend 10 to 23 months eating ant brood and being fed via trophallaxis (Elmes et al. 2002). This group also reports that unless butterflies are adopted by the proper species of *Myrmica* ant, they will die because *Maculinea* are host-specific to only one species of *Myrmica* ant.

In yet another example of mimicry, arthropods have come to resemble ants. This deception may occur in two basic forms. An organism (usually a beetle and, more uncommonly, a spider) may anatomically and behaviorally appear ant-like in nature but does not associate itself directly with a colony. This relationship has been observed among neotropical long-horned, wood-boring Cerambycid beetles and within other beetle families (Madigosky and Morgan 2001, personal observations). The similarity they bear to highly venomous ponerine bullet ants is striking. Their diurnal foraging habits also seem to support this premise. In the second instance, a mimic, again usually a beetle, actually lives among the ants in which it models. In this case, beetles possess the key surface features of their subterranean hosts such that ants are tricked into accepting these foreigners into their colony. This phenomenon, coined *Wasmannian mimicry*, most likely represents a unique form of Batesian mimicry and serves to protect and provide nourishment for the foraging beetle. The association is best observed among beetles in the family Staphylinidae.

A host of other organisms have come to exploit ants by harnessing their pheromones, hydrocarbons, and feeding behavior, as well as enticing ants into distributing their eggs, all without confrontation. Scarab beetles have a somewhat unusual way of encouraging ants to befriend them. Instead of actually producing chemical entities that mimic ant pheromones or hydrocarbons, scarabs are able to pirate and maintain these substances from their hosts. The chemicals are stored within the confines of their exoskeleton. In this fashion, they are recognized as part of the ant colony and are thus able to utilize ant food stores, feed upon newly developing brood, and even entice a delectable exchange of fluids from the ants via trophyallaxis. Since their presence within the colony is not contingent upon beetles internally producing the pheromone, they are free to move on and deceive other colonies of genetic similarity or difference.

Anyone who has ever observed a swarm of foraging army ants ravage the tropical underbrush understands well the commotion that they cause. Arthropods, annelids, reptiles, amphibians, and a host of other unsuspecting animals easily may perish in this event. In an attempt to escape the clutches of such activity, organisms attempt to flee as quickly as possible, often dragging dozens of biting ants along with them. The ants expeditiously dismantle the animals that do not escape.

The resulting tumultuous activity often draws the attention of antbirds looking to capitalize on the event. Birds will follow the ants for hours to pick off fleeing insects as they attempt to escape. This is a much-appreciated event for the birds since they easily locate insects that usually are well hidden. In the process of following the ant swarms, birds deposit their droppings, and this, too, may not go unnoticed. At least three species of ithomiine butterflies (*Mechanitis polymnia isthmia*, *Mechanitis lysimnia doryssus*, and *Melinaea lilis imitata*) have been implicated in following army ant swarms of *E. burchelli* to feed upon the droppings left behind by foraging antbirds (Ray and Andrews 1980). How butterflies are initially oriented to the areas of the swarm is not understood, but odors released by the ants may be responsible for the butterfly homing behavior.

In many instances, especially in those where ants maintain an established home base, considerable material is brought into the colony for processing. Depending on the individual species, ants gather plant material, whole and dismantled insects, carrion from fish, reptiles, amphibians, mammals, and even the tiny eggs of other arthropods. Their wandering lifestyle has opened up yet another possibility of exploiting their industrious nature, and this is exactly what some stick insects (Phasmids) have done. Stick insects are relatively sedentary organisms, maintaining a restricted range. When gravid females crawl onto a branch and ready themselves to lay eggs, they usually drop them directly below their roosting site. However, some phasmids are able to toss their eggs with the help of their abdomen; and this helps broadcast eggs a considerable distance from the point of release. This is the extent of maternal care. Without a doubt, many eggs never hatch and many fall victim to insect predators. Those that are found by ants are at an advantage. The eggs of the stick insect are tiny and look similar to seeds. They possess an outer shell rich in calcium that is attached to a small round packet (a capitulum) filled with lipids. This packet releases chemical compounds that tend to attract foraging ants (Agosta 2001). Upon locating the eggs, the ants transport them back to the colony, where the capitulum is severed and eaten by developing ant larvae. The removal of this structure does not affect the viability of the egg, however. Although data is sketchy, phasmid eggs take a considerable time to hatch. Experiments conducted by Creamer (1995) and Morgan and Schmidt (1997) indicate a gestation period of two to nine months, depending on the species. By this time it is likely that the eggs are forgotten, are maintained in the nest that subsequently has been abandoned, or are purposely placed outside the nest in a disposal arena. Whatever the outcome, one thing is certain: eggs are dispersed far from their point of origin, and this serves to lessen the competition among their siblings upon hatching.

Camouflage

The ability of organisms seemingly to disappear into their surroundings or behave in a particular manner that aids concealment is an effective strategy to avoid predators (see Figures 22-5 and 22-6). In this context, organisms resemble rocks, sticks, tree limbs or snags, leaves, flower parts, lichens, bird droppings, and a host of other objects. Here the organism models another organism or an inanimate object and in this manner is camouflaged, a subtle form of mimicry. This technique is the strategy of choice among many palatable animals.

There have been some extraordinary examples of animals that utilize this strategy to their advantage. The neotropical pyrgomorph grasshopper *Omura congrua* resembles a tiny blade of crab grass (see Figure 22-7). It is the identical shade of green as the crab grass upon which it resides. The head, thorax, and abdomen are dorsal/ventrally flattened, appear fused, and are collectively bowed and tapered, giving the impression of a blade of grass. The head is modified and the mouth parts are positioned ventrally when at rest so that the grasshopper can feed without exposing its head movement to potential predators. Yet the eyes are mounted on the side of the head

Figure 22-5 *Bufo typhonius* as seen on a light colored sandy matrix. Note the expanded cranial crests and stiff appearance that help camouflage the toad.

in a manner that allows a full range of view. *Omura* is very difficult to detect when among grass, owing to its cryptic coloration and sedentary disposition. If threatened, the grasshopper repositions itself, but only after considerable hesitation. It seems to sense its vulnerability when moving and is, therefore, reluctant to forfeit its concealment. If after repositioning it finds itself on a suitable matrix, it again will remain without movement. This is true even if repositioning is within inches of the original resting site. If the organism finds itself on a substrate that is in opposition to its color or pattern, it will retreat to another matrix until it is well matched. The attempt to readjust itself (jump) to a new position seems awkward and even stressful. This indicates an important link between morphology and behavior in delivering an overall appearance that is reliable.

Figure 22-6 The leptodactylid frog, *Ceratophrys cornuta*. The superb modeling throughout the body allows the frog to nicely match the surrounding substrate.

Figure 22-7   The grass mimic orthopteran *(Omura congrua)* on a preferred substrate.

In another example, a caterpillar of unknown affinity is positioned on a branch of a small tree (see Figure 22-8). It looks identical to the branch upon which it is positioned and has many of the characteristics exhibited by the tree. Drab coloration, false lenticels, and even an apparent break in the body of the caterpillar are used as a means of deceptive concealment. The caterpillar remains without movement throughout the day even when birds and ants forage about.

Members of the order Orthoptera (grasshoppers, crickets, mantids, cockroaches, katydids, stick insects) are some of the most diversely shaped and patterned insects known. The katydids are among the most intriguing of all. And, although this group has been taxonomically studied for well over a century (Brunner von Wattenwyl 1878), little information exists relative to its ecological significance, especially in tropical forests. Only recently has the examination of tropical Tettigoniidae orthpterans helped to reveal their ecological, adaptive, and behavioral significance (Belwood 1990; Castner 1995; Castner and Nickle 1995a, 1995b, 1995c; Nickle and Castner 1995). An interesting array of survival tactics has emerged. First, few insects compare with the degree of refined defense adaptations as those exhibited by katydids. They avoid predators through concealment (see Figure 22-9), shading, movement or the lack thereof, aposomatic coloration (see Figure 22-10), startle display (see Figure 22-11), auditory stridulation-acoustical alarm displays, and a host of other disguises, including extraordinary camouflage and mimicry. The twig mimics conceal themselves by hunkering down into depressions that they modified by chewing into the undersides of twigs (Nickle and Castner 1995). Coupled with their superb modeling and cryptic coloration, this makes them nearly impossible to discern. At the other extreme are katydids such as *Vestria* sp. that seem to advertize themselves via aposomatic coloration. *Vestria*, in addition to being brightly colored (green wings, yellow and blue banding on the abdomen, and red tipped tarsi), has a chemical composition that discourages attack for predators. In its defensive posture it raises its wings and arches its abdomen, which helps to reveal its stunning patterns of yellow and blue. If this does not discourage predators, *Vestria* may also emit an unpleasant array of chemicals that repel birds.

Most katydids are not strong flyers and are relatively slow moving. While this may initially seem like a disadvantage, their incredible disguises and mimicry serve as compensation. Many masquerade as bark mimics, lichen mimics, or even wasp mimics. In the case of the two previously

Figure 22-8   A cryptically colored caterpillar positioned on a branch of nearly the same diameter. Note the false break in the body to give the appearance of a broken twig along with markings that resemble plant lenticels.

Figure 22-9   A well-disguised katydid (Tettigoniidae) positioned closely against a tree that is similar in color and pattern.

Figure 22-10   The aposomatic rainbow katydid (*Vestria* sp.) in defensive posture as seen in profile.

Figure 22-11   A front view of the leaf-mimicking katydid *(Pterochroza ocellata)* in a defensive posture. The flashy spots situated high upon the wings are an effective strategy to startle prey. At rest, *Pterochroza* is highly cryptic.

mentioned katydids, they maintain themselves for long periods without movement. They forage under the cloak of darkness. If disturbed in lighted conditions, they quickly resort to matching themselves within the context of their immediate surroundings. This often means positioning or manipulating themselves vertically or laterally in relation to key substrate features on which they reside. They are quick to revert or submit to a posture resembling the external matrix.

In contrast, katydids that resemble wasps maintain a more active existence quite similar to their models. Information about this group is scant as there are only a few species known throughout the tropics. Anatomically and behaviorally, these are similar to tropical sphecid and pompilid wasps. Their antennae, much shorter than most katydids, resemble the form and movement of wasps and those of other diurnal hymenoptera. When foraging, they bob

Figure 22-12   *Panacanthus cuspidatus*, the spiny devil katydid, positioned in a defensive posture.

their antennae in a wasp-like fashion. Belwood (1990) suggests the pompilid wasp (*Hemipepsis mexicana*) as the probable model of *Aganacris insectivora* in Panama while Nickle and Castner (1995) infer several of the *Pepsis* sp. pompilid wasps as models for the Peruvian katydids *Aganacris nitida* and *P. mexicana*. Although the exact relationships have yet to be ascertained, the strategies employed by these katydids are more wasp-like than orthopteran.

Many organisms make superb use of their acoustical abilities to communicate and attract one another, to startle predators, and to lure prey. Insects are well equipped to employ this technique, and katydids are no exception. Using tegminal stridulation, several katydids are capable of broadcasting loud alarm signals. The robust spiny lobster katydid (*Panoploscelis specularis*), indigenous to the Amazon Basin, is a predatory insect that forages by night. By day it remains concealed in partially rotted logs, curled leaves, and the thick underbrush near the canopy floor. If disturbed from its refuge, it can loudly broadcast sharp brisk chirps that can startle a predator. If a predator attempts to remove the katydid from its perch, it tenaciously adheres to the branch or supporting substrate and retracts its body into a position whereby the legs cover most of the exposed surface. Spines covering the legs make it very difficult for predators to dislodge the katydid. Since it also packs a pair of robust mandibles, it is capable of biting with enough force to draw blood from a human finger. Other species such as the spiny devil katydid (*Panacanthus cuspidatus*) also behave in a similar manner. The devil not only has spiny legs but also has the periphery of the head region lined with spines, making it even more difficult for predators to subdue (see Figure 22-12).

Chemicals for Protection

Part of the prerequisites of being a biological entity means that there is an innate ability to synthesize substances that govern growth, development, and reproduction. But occasionally there are instances where organisms manufacture substances to protect themselves. These substances come in a variety of chemical forms and have even greater functional use. They serve to catalyze, incapacitate, digest, paralyze, deceive, enslave, and kill. When used by a skilled predator, they can be deadly.

Rather than using the conventional web-building technique that spiders are most noted for, the bolas spider selects a slightly more creative approach to obtain prey. The adult female spider constructs a sticky mass that it attaches to a strand of silk. Unlike most spiders, this predator is quite adept at hurling this mass with its forelegs at a moth or fly that comes within range. If the spider hits its prey, it can then quickly retract the silk to draw in its victim for a meal. It would initially seem that not having a web would be a disadvantage. After all, a strategically placed web can often fill with more prey items than a spider can eat. But a closer look at the bolas spider reveals another interesting facet of its behavior that may compensate for its offbeat tactics. The female bolas spider (*Mastophora hutchinsoni*) releases a carefully concocted blend of chemicals that mimic the long-range mate attractant of a moth she seeks (Gemeno et al. 2000). Since this substance mimics a chemical pheromone produced by female moths, only males are lured to the trap (Haynes et al. 2001). The lure is species specific so the spider's ability to attract different prey items is limited by its own ability to synthesize workable attractant pheromones. If the chemical pheromone is similar to those produced by other closely related moths, they too may be pulled into the scheme. When moths enter the spider's throwing radius, the spider readies itself by retracting and quickly flicking the bolas. Haynes et al. (2001) suggest that spiders are stimulated to produce a bolas as a result of detecting wing vibrations emitted by fluttering moths, and this further stimulates the spider to release mimetic pheromone. Without a bolas. the spider is helpless to capture its prey.

In another twist of chemical deception, the spider orchid (*Ophrys sphegodes*) releases hydrocarbons that attract its pollinator, a solitary bee (*Andrena nigroaenea*). Orchids within this genus are generally pollinated by one or a few very closely related species; in this case, *A. nigroaenea* serves as the main pollinator of the spider orchid. A comparison of the hydrocarbons found in both *A. nigroaenea* cuticle extracts and *O. sphegodes* leaf extracts revealed the presence of n-alkanes, n-alkenes, aldehydes, esters, l trans-farnesol, and all trans-farnesyl hexanoate in similar patterns (Schiestl et al. 2000). The similarity of these profiles helps explain why male bees have such a strong affinity for this orchid. The similarity in pattern may also serve to restrict the number of specific pollinators visiting this orchid.

The extent to which scents and fragrances are able to attract pollinators goes well beyond those produced by flowers. This may also be illustrated in fungi. Just as plants need to have their pollen distributed from one flower to the next, fungi must also have a way broadcasting their spores. There are several ways that this may be accomplished. The use of wind and water are widely used, but in many instances, animals mediate the process. Fungi are noted for emitting scents that mimic rotting flesh, decaying vegetation, and, to a lesser extent, floral aromas. Those that smell similar to rotting organic material often seek the assistance of flies and beetles to disperse their spores. The reward for their effort is the actual fungal tissue. Yet other fungi attract bees in a manner that is similar to flowers and in some instances may even alter the morphology of vegetation. The rust fungi in the *Puccinia monoica* complex (Pucinniniaceae, Uredinales, Basidiomycetes) are known to infect the leaves of nearly a dozen species of angiosperms. Once the fungus becomes established, it is able to modify the morphology of the leaves in such a way that it resembles a flower. Not only do these pseudoflowers look like actual flowers, but they also release a sweet pungent fragrance that is rich in fructose consumed by a host of insects including ants, bees, nymphalid butterflies, along with anthomyid, muscid, and sarcophagid flies (Raguso and Roy 1998). The attraction of such a broad group of insects assures ample distribution of spores throughout a multitude of habitats. In this way, the fungi assumes wide distribution.

A New Model with a Toxic Twist

The North American Zebra Swallowtail (*Eurytides marcellus*), named quite appropriately for the black and white markings on both sets of wings, has several means of protecting itself. If disturbed, the larvae expel butyric acids from specialized osmaterial glands to discourage the pursuit of small predators. This alone, however, is not effective enough to discourage large predators such as birds. Additional protection comes from the diet that the larvae ingest. Immature stages of the butterfly feed exclusively upon foliage of the North American paw paw tree (*Asimina triloba*). The foliage of this tree is unique in that it contains a group of biologically active wax-like fatty acid derivatives called acetogenins. These are toxic compounds that contain potent pesticidal and anti-neoplastic agents having emetic activity in vertebrates (Martin et al. 1999). Laboratory investigations indicate these compounds are cytotoxic with a profile approaching one billion times greater than the standard reference adriamycin (Lewis et al. 1993; Zhao et al. 1994; Zhao et al. 1996). The mechanism of acetogenins centers on the inhibition of a key enzyme (NADH-ubiquinone oxidoreductase) that plays an important role in diminishing energy production at the cellular level. Zebra swallowtail butterfly larvae feeding upon the foliage retain acetogenins throughout their bodies but are unaffected by these potent toxins. As the larvae begin to mature, they ingest greater amounts of vegetation containing acetogenins. Consequently, they accumulate significant amounts of the toxins, thereby rendering them protected against would-be predators. Additionally, this protection is passed along to the adult butterfly, even after the larvae pupate and metamorphose into adult butterfly (Martin et al. 1999). Of note, no reports exist of birds preying upon larvae of *Erytides marcellus* in the field, although the butterflies have been studied for extended periods (Damman 1986). Yet unconditioned captive starlings and catbirds restricted to diets of adult swallowtails usually succumbed within 24 to 36 hours after ingesting several butterflies (Martin et al. 1999). The subsequent analysis of select tissue post mortem revealed the presence of acetogenins.

A theme often employed by insects, especially those in the tropics, is the use of noxious chemicals as a means of protection against predators. Members of nearly every group of animals have the ability either to produce internally or to obtain from dietary means substances that are noxious, poisonous, and/or toxic to predators. Because the majority of insects are maintained on plant material, they have developed a multitude of ways to exploit the noxious properties to their benefit. The degree of chemical wizardry used among this group of animals is unmatched. One case involves *Oreophoetes peruana*, the Peruvian fire-stick (order Phasmatodea). As is typical of most phasmids, this organism has a slender sticklike body with fine wire-like legs that are barely noticeable, hence the name "stick insect." Many of the 2,500 species within this order are known to disguise themselves simply by remaining still among the dense understory. They are cryptic in both their appearance and their behavior. Along with their camouflage, many move slowly in an oscillating manner as to not call attention to themselves. In addition, they are predominately more active at night. There are, however, those such as the Peruvian fire stick that deviate from this norm. The fire stick is brightly colored and is highly conspicuous, possessing bright red bands on a predominantly black body and exhibiting diurnal tendencies. When placed on a background of intense green, as is commonly encountered in the wild, the organism appears as an advertisement. Secretively protecting the conspicuous presence of this insect is a pair of glands tucked directly behind the head that can instantly discharge a noxious irritant (quininoline) that is effective against ants, spiders, cockroaches, and frogs (Eisner et al. 1997). What is more amazing is that the glands responsible for producing quininoline do not lose their ability to defend the insect even during ecdysis when insects are typically vulnerable to attack (Eisner et al. 1997).

During the past decade, much interest and effort has been dedicated to understanding the nature of chemical defense in animals (Badio et al. 1994; Daly 1998; Meinwald et al. 1998; Jones et al. 1999; Garraffo et al. 2001; Daly et al. 2002; Smith et al. 2002; Daly 2003). One of the more intriguing examples of this came with the discovery of toxic birds, specifically the hooded pitohui (*Pitohui dichrous*) from the tropical forests of New Guinea. After handling birds captured in mist nets, researchers inadvertently introduced a substance from the birds on to their buccal and nasal tissues that elicited a burning and numbing sensation (Dumbacher et al. 1992). Upon analytical examination, they were able to isolate, assay, and characterize the compound responsible for causing these sensations. The compound implicated is a steroidal alkaloid called homobatrachotoxin. This was a bit surprising, given that the compound had only been known from a few species of poison-dart frogs in the genus *Phyllobates* (family Dendrobatidae). Even more intriguing is the extent of genetic and geographic separation between these species. Phyllobatid frogs are exclusively New World species, and *Pitohui* birds are Old World. The presence of identical alkaloids in such diverse species certainly provides some indication of its biological value. Bright colors accompany the presence of homobatrachotoxin in the phyllobatid frogs, but the color of the birds is drab orange and black. The association of bold colors with the presence of toxins is believed to be an effective combination in protecting many organisms, but the exact nature of the coloration in pitohui remains unresolved. Recently, a second toxic bird from New Guinea (*Ifrita kowaldi*) was discovered to possess steroidal batrachotoxin alkaloids (Dumbacher et al. 2000). The extent of toxic alkaloids found in frogs and birds over such geographic separation is most likely a result of independent evolutionary acquisition (Dumbacher et al. 1992).

Mutualism and Pollination

Plants face a host of problems that differ from those of their animal counterparts. One of the main obstacles centers on how to overcome immobility. Without movement, few options are available for plants to disperse and exchange genetic material, an element essential for species preservation. Consequently, plants have evolved ingenious reproductive strategies that, in many cases, have arisen with the assistance of animals, especially birds, bees, and flies. Nowhere else is this more pronounced than in flowering plants, the angiosperms, which are thought to have arisen some 124.6 million years ago from the Upper Jurassic/Lower Cretaceous (Sun et al. 2002). From this point, they radiated throughout every continent and now dominate most terrestrial landscapes worldwide. Some 235,000 angiosperm species are recognized in the phylum Anthophyta, and these share at least one characteristic in common: a reproductive structure called a flower. The definitive structure of the flower is the carpel, a structure containing the ovules that develop into seeds after fertilization. Insects are especially important because they help to transfer genetic material from flower to flower and from plant to plant. The sum total of transactions between plants and animals over millions of years has helped to drive floral evolution to what we see today. How and why pollinators seek flowers in a dense forest is yet another question.

From a humanistic perspective, flowers are attractive; they are colorful, wonderfully shaped, and often emit fragrances that are pleasing to us. These same attributes attract insects to flowers, but in ways that are sometimes not overtly apparent. For one, flowers found throughout the tropical canopy are sparse and are often difficult to find. Many are small, inconspicuous, and possess colors that seem to conceal their whereabouts. Still, they seem to attract bees, flies, beetles, and a host of other pollinators. We now know that insects perceive color well beyond our limited range of detection. Many forest flowers reflect electromagnetic radiation (wavelengths) that humans simply cannot detect. These same wavelengths are what insects find easy to detect, even within the dense growth of a primary tropical rainforest. Petal shape and pattern also help direct insects to flowers. A strategically positioned marking or pattern on a flower petal can serve to orient an insect to an exact location to encounter a reward. Ultraviolet patterns in flowers have been shown to provide orientation cues to foraging bees (Jones and Buckmann 1974; Houge 1993) and those possessing distinct UV patterns such as in the orchid genera *Arethusa*, *Calopogon*, and *Pogonia* may indicate floral convergence to similar types of pollinators (Thien and Marcks 1972). Finally, our ability to detect fragrances, especially those emitted by flowers, is not nearly as keen as insects. A few molecules of scent released from a canopy flower can be tracked with ease by receptive insects, even over considerable distances. But beyond color, shape, and fragrance as an initial attractant, flowers must reward their pollinators if they are to encourage repeat customers.

Some bees find pollen reason enough to visit flowers. Pollen is high in protein, can be readily transported and manipulated, is easily stored, and is a nutritious food source for adult bees and their larvae. This accounts for most visitations to flowers by bees. Since the plant produces an overabundance of pollen, it matters little that many insects use this as a source of food. An adequate supply of pollen remains for transport to assure pollination.

Other insects orient to floral shape or a combination of shape and aroma in an attempt to mate with the flower. Such is the case with the Australian Glossy-leafed Hammer Orchid (*Drakaea elastica*). Here the thynnid wasp is attracted to the orchid flower because of a release of fragrance that mimics the female wasp's sex pheromone. The flower is structurally similar to the female wasp and even has similar tactile properties to a wasp. The male wasp attempts to mate with the flower (a process referred to as *pseudocopulation*) and, in doing so, triggers a hinge mechanism that repeatedly manipulates the wasp between the petal and the column of the flower. In the process, the wasp picks up pollen that it may bring to the next flower with which it attempts to mate.

In a similar manner, drones of the South American eusocial bee *Plebeia droryana* have recently been observed attempting to copulate with sepals and petals of the orchid *Trigonidium obtusum* (Singer 2002). In this instance, bees cling to the flower until they eventually slip on the waxy perianth. As a result, they become trapped in a funnel-like floral tube. To escape, they are pressed between the column and the lip of the flower, and the orchid implants its pollinaria onto their exoskeleton. A repeat performance of this event in a similar flower is needed to assure successful pollination. In some species, the floral deceit is further augmented with the release of chemicals that mimic bee pheromones. This entices unsuspecting bees and wasps to the trap.

The large white flowers of the South American giant water lily (*Victoria amazonica*) (see Figure 22-13) emit a sweet fruity fragrance that attracts *Cyclocephala* scarab beetles (Lovejoy 1978). Flowers initially open at night and allow beetles the opportunity to explore the many intricate spaces within the complex floral structure. By daybreak the flowers close, thereby trapping the beetles. They begin to feed upon the anthers of the lily and, in the process, position pollen from previous floral encounters on the receptive stigma of the flower. This assures cross-pollination. Soon thereafter, the giant flower opens; however, before leaving, beetles move among the ripened anthers to pickup additional pollen that they will then transport to the next lily.

Other flowers lure insects by presenting them with potable water. The orchid *Coryanthus speciosa* has a small reservoir housed between petals that entices a bee toward a slippery slope and subsequently a unidirectional ride to the bottom. The only escape is through a tight space that assures the bee exits through a passage laden with pollen. Although elaborate, this is an efficient way to assure the development of successive generations. Orchids that force their pollinators in a unidirectional escape are sometimes referred to as *trap flowers*. Many modifications are seen throughout nature. Some use water to entice their pollinators, others employ the use of nectar; but, more often than not, fragrances are the most widely used lures.

Several species of bees visit flowers simply to gather fragrance. The best known of these are the orchid or euglossine bees, found exclusively throughout the neotropics. This group of insects has intrigued naturalists for well over a century. In fact, Darwin gave considerable attention to this group in his work addressing orchid pollination (Darwin 1890). Since then, considerable research has been directed toward this group of organisms. The Euglossini (Hymenoptera: Apidae) encompass some 175 brightly colored species (Agosta 2001), with many appearing glossy and almost metallic. Most prefer a solitary existence, although some species are known to be quasi-social or communal. They are seldom observed in nature unless enticed by aromatic attractants and baits (see Figure 22-14). Consequently, what little is known of their natural history has been obtained serendipitously or through observations of bees harvesting fragrances from lures

Figure 22-13   The giant Amazon water lily (*Victoria amazonica*) with flower.

Figure 22-14   The orchid bee *(Euglossa mixta)* homing to a locality upon which synthetic lure (methyl salyicylate) had been placed.

and/or orchids. Since male euglossines are credited with doing most of the fragrance collecting, they have been observed most frequently gathering from orchids and from a limited number of other unrelated plant families (Williams 1982).

To capture a fragrance, bees secrete lipids from a gland located in the head. This fluid acts to absorb fragrances emitted by orchids and from other sources. Using modified structures on the front tarsi, euglossines brush and manipulate their secretions to mop up molecules of fragrance released by the orchid. This fluid is then passed to the rear legs, where it is absorbed by a slit located on the inside of a large grossly modified triangular tibia. Hairs within the cavity accentuate the capture of fragrance for storage. In a mechanism not fully understood, fluids are recycled from this area and are returned to the area of origin. In this manner bees repeatedly visit orchids to acquire a smorgasbord of scents.

It is unclear why euglossines collect these substances, although we do know that without them their lives are apparently shorter (Dodson et al. 1969). Many scientists believe that the fragrances collected by male bees are precursor substances that are then modified by the bee to create pheromone attractants. And since it is the male that usually collects the scent, there is speculation that these may be used to entice females to mate—or possibly other males to engage in behavior that ultimately attracts females. Whatever the case, orchid bees have a propensity to collect scents and expend considerable energy in doing so. They aggressively compete for natural and synthetic fragrances placed in traps containing combinations of eugenol, methyl salicylate, cineole, skatole, vanillin, methyl cinnamate, and several other lures. On top of collecting these fragrances, each species has a unique hydrocarbon profile imbedded within the confines of the exoskeleton. A fair amount of speculation surrounds the specific function of these markers, but some observers suspect that they play a role in recognition and/or reproductive function. Although each species maintains a unique hydrocarbon profile, slight variation does exist among members of the same species within a restricted geographic location (see Figure 22-15). Conversely, evidence also exists that profiles within some species remain fairly consistent for distances of over 150 km (Madigosky and Betz, unpublished data).

The mating rituals of some euglossines may help shed light on the behavior of male-dominated perfume collecting. In several species, male euglossines form leks to attract receptive females. A lek consists of a group of male bees that congregate for the purpose of enticing a female to mate. With increased competition, it might be that the fragrances collected and stored are used to secure mating rights. However, female orchid bees are not known to home to orchid perfumes or combinations of scents gathered from orchids (Dressler 1982), so it seems unlikely that an unmodified perfume increases a female's receptivity to mate. Or it might be that bees gather fragrances because of their inability to synthesize a critical metabolite (Dodson et al. 1969). No matter how unusual or complex the relationship between bee and plant appears or becomes, the ultimate test is whether the plant gets pollinated. In the case of euglossine pollination, the relationship works well.

Conclusions

The relationships examined within this chapter represent a few of the more unusual that persist in nature. They serve to provide a vision, albeit a narrow one, of how wonderfully entangled natural systems can become. To understand the significance of each is to open a door that reveals unexpected insights into the natural world and the staggering variety of behavior exhibited by life-forms. Similarly, each one represents a pulse or snapshot of evolutionary history, an opportunity to assess complex relationships that may have taken millions of years to refine.

There exits many challenges relative to unraveling the nature of strategies used by organisms to assure their survivability. Most often, all but the most profound or extraordinary examples go

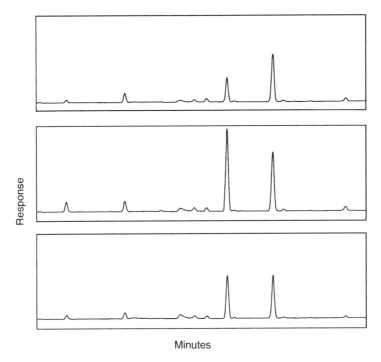

Response

Minutes

Figure 22-15 Cuticular hydrocarbon gas chromatographic profiles from three separate orchid bees (*Euglossa mixta*) are shown. Bees were collected from a forest locality in Iquitos, Peru. The pattern profiles are nearly identical in peak retention (Madigosky and Betz, unpublished data).

unnoticed. Yet there is great value in ascertaining the sum total of all interactions forged by organisms because they provide us with considerable insight into the dynamics of the evolutionary process. Ultimately, deciphering the nature of survival strategies even within the simplest of biological systems can only provide us with clues into our own past and our future.

Acknowledgments

The author wishes to extend appreciation to Widener University for financial support received through grants and departmental funds for well over a decade. I especially thank Randy Morgan (Cincinnati Zoo and Botanical Garden), who over the past decade has accompanied me often in exploration of unusual plants and animals throughout the upper Amazon Basin. Gratitude is also extended to Linda Betz and Martin Schultz for their expertise in determining hydrocarbon profiles of several orchid bees.

References

Agosta, W.C. (2001). "Thieves, Deceivers, and Killers: Tales of Chemistry in Nature." Princeton University Press, Princeton.

Arnett, R.H. and Jacques, Jr,, R.L. (1981). "Insects." Simon and Schuster, New York.

Badio, B., Garraffo, H.M., Spande, T.F., and Daly, J.W. (1994). Epibatidine: discovery and definition as a potent analgesic and nicotinic agonist. *Med. Chem. Res.* **4**, 440–448.

Behe, M.J. (1996). "Darwin's Black Box, The Biochemical Challenge to Evolution." Touchstone Book, Simon & Schuster, New York.

Belwood, J.J. (1990). Anti-predator defenses and ecology of neotropical forest katydids, especially the Pseudophyllinae. *In* "The Tettigoniidae: Biology, Systematics, and Evolution" (W. Bailey and D.C.F. Rentz, Eds.), pp. 8–26. Springer-Verlag. Heidelberg.

Borror, D.J. and DeLong, D.M. (1964). "An Introduction to the Study of Insects." Holt, Rinehart and Winston, New York.

Boulay, R., Hefetz, A., Soroker, V., and Lenoir, A. (2000). *Camponotus fellah* colony integration: worker individuality necessitates frequent hydrocarbon exchanges. *Animal Behaviour* **59**, 1127–1133.

Brunner von Wattenwyl, C. (1878). "Monographie der Phaneropteriden." Brockhaus, Vienna.

Castner, J.L. (1995). Defensive behavior and display of the leaf-mimicking katydid *Pterochroza ocellata* (L.) (Orthoptera: Tettigoniidae: Pseudophyllinae: Pterochrozini). *J. Orth. Res.* **4**, 89–92.

Castner, J.L. (2000). "Amazon Insects: A Photo Guide." Feline Press, Gainesville, Florida.

Castner, J.L. and Nickle, D.A. (1995a). Intraspecific color polymorphism in leaf-mimicking katydids (Orthoptera: Tettigoniidae: Pseudophyllinae: Pterochrozini). *J. Orth. Res.* **4**, 99–103.

Castner, J.L. and Nickle, D.A. (1995b). Note on the biology and ecology of the leaf-mimicking katydid *Typophyllum bolvari* Vignon (Orthoptera: Tettigoniidae: Pseudophyllinae: Pterochrozini). *J. Orth. Res.* **4**, 105–109.

Castner, J.L. and Nickle, D.A. (1995c). Observations on the behavior and biology of leaf-mimicking katydids (Orthoptera: Tettigoniidae: Pseudophyllinae: Pterochrozini) *J. Orth. Res.* **4**, 93–97.

Cott, H.B. (1940). "Adaptive Coloration in Animals." Methuen, London.

Creamer, K.D. (1995). Fantastic Phasmids: Walking stick biology, rearing and display techniques. 1995 American Associate of Zoos and Aquariums. Great Lakes Regional Conference Proceedings.

Daly, J.W. (1998). Thirty years of discovering arthropod alkaloids in amphibian skin. *Journal of Natural Products* **61**, 162–172.

Daly, J.W. (2003). Ernest Guenther Award in chemistry of natural products. Amphibian skin: a remarkable source of biologically active arthropod alkaloids. *Journal of Natural Products* **46**(4), 45–452.

Daly, J.W., Kaneko, T., Wilham, J., Garraffo, H.M., Spande, T.F., Espinosa, A., Donnelly, M.A. (2002). Bioactive alkaloids of frog skin: combinatorial bioprospecting reveals that pumiliotoxins have an arthropod source. *PNAS* **99**(22), 13996–14001.

Damman, H. (1986). The osmeterial glands of the swallowtail butterfly *Eurytides marcellus* as a defense against natureal enemies. *Ecol. Entomol.* **11**, 261–265.

Darwin, C. (1890). "The Various Contrivances by Which Orchids Are Fertilized by Insects, Second Edition." John Murray, London

De Schauensee, R.M. and Phelps, W.H., Jr. (1978). "Birds of Venezuela." Princeton University Press, Princeton, New Jersey.

Dodson, C.H., Dressler, R.L., Hills, H.G., Adams, R.M., and Williams, N.H. (1969). Biologically active compounds in orchid fragrances. *Science* **164**, 1243–1249.

Dumbacher, J.P., Beehler, B.M., Spande, T.F., Garraffo, H.M., and Daly, J.W. (1992). Homobatrachotoxin in the genus *Pitohui*: chemical defense in birds? *Science* **258**, 799–801.

Dumbacher, J.P., Spande, T.F., and Daly, J.W. (2000). Batrachotoxin alkaloids from passerine birds: a second toxic bird genus (*Ifrita kowaldi*) from New Guinea. *Proceedings of the National Academy of Sciences of the United States of America.* **97**(24), 12970–12975.

Dressler, R.L. (1982). Biology of the orchid bees (Euglossini). *Ann. Rev. Ecol. Syst.* **13**, 373–394.

Eisner, T., Morgan, R.C., Attygalle, A.B., Smedley, S.R., Hearath, K.B., and Meinwald, J. (1997). Defensive production of quinoline by a phasmid insect (*Oreophoetes peruana*). *The Journal of Experimental Biology* **200**, 2493–2500.

Elmes, G.W., Akino, T., Thomas, J.A., and Clarke, R.T. (2002). Interspecific differences in cuticular hydrocarbon profiles of *Myrmica* ants are sufficiently consistent to explain host specificity by *Maculinea* (large blue) butterflies. *Oecologia* **130**, 525–535.

Ernst, C.H., Lovich, J.E., and Barbour, R.W. (1994). "Turtles of the United States and Canada." Smithsonian Institution Press, Washington.

Evans, A.V. and Bellamy, C.L. (1996). "An Inordinate Fondness for Beetles." Henry Holt and Company, Inc., New York.

Garraffo, H.M., Jain, P., Spande, T.F., Daly, J.W., Jones, T.H., Smith, L.J., and Zottig, V.E. (2001). Structure of alkaloid 275A, a novel 1-azabicyclo[5.3.0]decane from a dendrobatid frog, *Dendrobates lehmanni*: synthesis of the tetrahydroiastereomers. *Journal of Natural Products* **64**, 421–427.

Gemeno, C., Yeargan, K.V., and Haynes, K.F. (2000). Aggressive chemical mimicry by the bolas spider *Mastophora hutchinsoni*: identification and quantification of a major prey's sex pheromone components in the spider's volatile emission. *J. Chem. Ecol.* **26**, 1235–1243.

Gilbert, L.E. (1971). Butterfly-plant coevolution: has *Passiflora adenopoda* won the selectional race with Heliconiine butterflies? *Science* **172**, 585–586.

Haynes, K.F., Yeargan, K.V., and Gemeno, C. (2001). Detection of prey by a spider that aggressively mimics pheromone blends. *J. Insect Behavior* **14**, 535–544.

Hilty, S.L. and Brown, W.L. (1986). "A Guide to the Birds of Columbia." Princeton University Press, Princeton, New Jersey.

Hölldobler, B. and Wilson, E.O. (1990). "The Ants." Harvard University Press, Cambridge, Massachusetts.

Horn, J.M., Spencer, K.C., and Smiley, J.T. (1984). The chemistry of extrafloral nectar of *Passiflora* and related species. *Bull. Ecol. Soc. Amer.* **65**, 265.

Houge, C.L. (1993). "Latin American Insects and Entomology." University of California Press, Berkeley.

Järvi, T., Silén-Tullberg, B., and Wiclund, C. (1981). The cost of being aposematic: an experimental study of predation on larvae of *Papilio machaon* by the great tit *Parus major*. *Oikos* **36**, 267–272.

Jones, C.E. and Buckmann, S.L. (1974). Ultraviolet floral patterns as functional orientation cues in hymenopterous pollination systems. *Animal Behavior* **22**, 481–485.

Jones, T.H., Gorman, J.S.T., Snelling, R.R., Delabie, J.H.C., Blum, M.S., Garraffo, H.M., Jain, P., Daly, J.W., and Spande, T.F. (1999). Further alkaloids common to ants and frogs: decahydroquinolines and a quinolizidine. *Journal of Chemical Ecology* **25**(5), 1179–1193.

Joron, M. and Mallet, J.L.B. (1998). Diversity in mimicry: paradox or paradigm? *Trends Ecol. Evol.* **13**, 461–466.

Lewis, M.A., Arnason, J.T., Philogene, B.J.R., Rupprecht, J.K., and McLaughlin, J.L. (1993). Inhibition of respiration at site I by asimicin, an insecticidal acetogenin of the paw paw, *Asimina triloba* (Annonaceae). *Pestic. Biochem. Physiol.* **45**, 15–23.

Lovejoy, T.E. (1978). Royal water lilies: truly Amazonian. *Smithsonian* **7**, 77–82.

Madigosky, S.R. and Grant, B.W. (1996). Biodiversity loss: a problem or symptom of our time? *In* "Forests—A Global Perspective" (S.K. Majumdar, E.W. Miller, and F.J. Brenner Eds.), pp. 226–241. The Pennsylvania Academy of Sciences, Easton.

Madigosky, S.R. and Morgan, R.C. (2001). Noted on several occasions while conducting research in the area of Iquitos, Peru.

Mallet, J. (1986). Hybrid zones of *Heliconius* butterflies in Panama and the stability and movement of warning colour clines. *Heredity* **56**, 191–202.

Mallet, J. (2001). Causes and consequences of a lack of coevolution in Müllerian mimicry. *Evolutionary Ecology* **13**, 777–806.

Mallet, J. and Gilbert, L.E. Jr. (1995). Why are there so many mimicry rings? Correlations between habitat, behaviour and mimicry in Heliconius butterflies. *Biological Society of the Linnean Society* **55**, 159–180.

Martin, J.M., Madigosky, S.R., Gu, Z., Wu, J., and McLaughlon, J.L. (1999). Chemical defense in the zebra swallowtail buttery, *Erytides marcellus*, involving annonaceous acetogenins. *Journal of Natural Products* **62**, 2–4.

May, R.M. (1992). How many species inhabit the Earth? *Scientific American* **267**, 42–48.

McCabe, T.L. (1981). *Clemensia albata*, an algal feeding arctiid. *J. Lepidopt. Soc.* **35**, 34–40.

Meinwald, J., Huang, Q., Vrkoc, J., Herath, K.B., Yang, Z., Schröder, F., Attygalle, A.B., Iyengar, V.K., Morgan, R.C., and Eisner, T. (1998). Mirasorvone: a masked 20-ketopregnane from the defensive secretion of a diving beetle (*Thermonectus marmoratus*). *Proc. Natl. Acad. Sci.* **95**, 2733–2737.

Morell, V. (1999). Biodiversity: The fragile web. *National Geographic* **195**(2), 12–23.

Morgan, R. and Schmidt, K. (1997). Biology and husbandry of an aposomatic phasmid: the Peruvian firestick *Oreophoetes peruana*. *In* "Invertebrates in Captivity Conference Proceedings," pp. 128–135. Sonoran Arthropod Studies Institute, Tucson.

Nickle, D.A. and Castner, J.L. (1995). Strategies utilized by katydids (Orthoptera: Tettigoniidae) against diurnal predators in rainforests of northeastern Peru. *J. Orth. Res.* **4**, 75–88.

Oliveira, E.G. (1998). Migratory and foraging movements in diurnal neotropical Lepidoptera: experimental studies on orientation and learning. Ph.D. Dissertation, University of Texas, Austin.

Owen, D. (1980). "Camouflage and Mimicry." Oxford University Press, Oxford.

Penny, N.D. and Arias, J.R. (1982). "Insects of an Amazon Forest." Columbia University Press, New York.

Pough, F.H. (1988). Mimicry and related phenomena. *In* "Biology of the Reptilia: Defense and Life History" (C. Gans and R.B. Huey, Eds.), pp. 154–234. Alan R. Liss, Inc., New York.

Provost, E., Riviere, G., Roux, M., Morgan, E.D., and Bagneres, A.G. (1993). Change in the chemical signature of the ant *Leptothorax lichtensteini* Bondroit with time. *Insect Biochemistry and Molecular Biology* **23**, 945–957.

Purvis, A. and Hector, A. (2000). Getting the measure of biodiversity. *Nature* **405**, 212–219.

Raguso, R.A. and Roy, B.A. (1998). "Floral" scent production by *Puccinia* rust fungi that mimic flowers. *Molecular Ecology* **7**, 1127–1136.

Ray, T.S. and Andrews, C.C. (1980). Antbutterflies: butterflies that follow army ants to feed on antbird droppings. *Science* **210**, 1147–1148.

Schiestl, F.P., Ayasse, M., Paulus, H.F., Lofstedt, C., Hansson, B.S., Ibarra, F., and Francke, W. (2000). Sex pheromone mimicry in the early spider orchid (*Ophrys sphegodes*): patterns of hydrocarbons as the key mechanism for pollination by sexual deception. *J. Comp. Physiol. A* **186**, 567–574.

Smith, B.P., Tyler, M.J., Kaneko, H.M., Garraffo, T.F., Spande, T.F., and Daly, J.W. (2002). Evidence for biosynthesis of pseudophrynamine alkaloids by an Australian myobatrachid frog (*Pseodophryne*) and for sequestration of dietary pumiliotoxins. *Journal of Natural Products* **65**, 439–447.

Sheppard, P.M., Turner, J.R.G., Brown, K.S., Benson, W.W., and Singer, M.C. (1985). Genetics and the evolution of müllerian mimicry in *Heliconius* butterflies. *Philos. Trans. Royal Soc. Lond. B* **308**, 433–613.

Singer, R.B. (2002). The pollination mechanism in *Trigonidium obtusum* Lindl (Orchidaceae: Maxillariinae): sexual mimicry and trap flowers. *Annals of Botany.* **89**, 157–163.

Spencer, K.C. (1984). Chemical correlates of coevolution: The *Passiflora/Helionius* interaction. *Bull. Ecol. Soc. Amer.* **65**, 231.

Sun, G., Ji, Q., Dilcher, D.L., Zheng, S., Nixon, K.C., and Wang, X. (2002). Archaefructaceae, a new basal angiosperm family. *Science* **296**, 899–904.

Thein, L.B. and Marcks, B.G. (1972). The floral biology of *Arethusa bulbosa*, *Calopogon tuberosus*, and *Pogonia ophioglossoides* (Orchidaceae). *Can J. of Bot.* **50**, 2319–2325.

Turner, J.R.G. (1975). A tale of two butterflies. *Nat. Hist.* **84**(2), 29–37.

Turner, J.R.G. (1981). Adaptation and evolution in *Heliconius*: A defense of neo-Darwinism. *Ann. Rev. Ecol. Syst.* **12**, 99–121.

Turner, J.R.B. (1988). The evolution of mimicry: a solution to the problem of punctuated equilibrium. *In* "Mimicry and the Evolutionary Process" (L. P. Brower, Ed.), pp. 42–65. University of Chicago Press, Chicago.

Turner, J.R.B. and Mallet, J. L. B. (1996). Did forest islands drive the diversity of warning coloured butterflies? Biotic drift and the shifting balance. *Philos. Trans. Royal Soc. Lond. B* **351**, 835–845.

Williams, N.H. (1982). The biology of orchids and euglossine bees. *In* "Orchid Biology, Reviews and Perspectives" (J. Arditti, Ed.), pp. 119–171. Cornell University Press, Ithica.

Wilson, E.O. (1986). The organization of food evacuation in the ant genus Pheidole (Hymenoptera: Formicidae). *Insectes Sociaux* **33**(4), 458–469.

Zhao, G.X., Chao, J.F., Zeng, L., Rieser, M.J., and McLaughlin, J.L. (1996). The absolute configuration of adjacent bis-THF acetogenins and asiminocin, a novel highly potent asimicin isomer form *Asimina triloba*. *Bioorg. Med. Chem.* **4**, 25–32.

Zhao, G.X., Miesbauer, L.R., Smith, D.L., and McLaughlin, J.L. (1994). Asimin, asiminacin, and asiminecin: novel highly cytotoxic asimicin isomers from *Asminina triloba*. *J. Med. Chem.* **37**(13), 1971–1976.

IV _____

Conservation and Forest Canopies

From local to global scales, we are losing biological diversity and ecosystem structure at unparalleled rates of decline. Proximate causes include forest fragmentation, water and air pollution, soil depletion, introduction of exotic species, and resource exploitation; however, the ultimate or root cause seems to be, in large part, a burgeoning human population. Superimposed on graphs for human population growth are often found graphs for annual species extinctions that show uncanny, near-perfect correspondence. Of course, correlation does not necessarily imply causation. Human population growth and species extinction rates are plainly related; however, the reasons for that linkage have varied greatly over the centuries. The number of humans on the planet has increased exponentially, along with our use of technology and natural resources. Given that 99 percent of all species that have ever lived on the Earth are now extinct, the demise of species is part and parcel to the history of life itself. Yet extinction has accelerated quickly and universally since the advent of humankind, especially in the 1970s as humans and technology marched into previously inaccessible tropical rainforest canopies and soils.

Conservation can be defined as ecology combined with an ethical bearing in response to the current extinction crisis. As a matter of course, it exemplifies the last phase of biological studies, but the most significant one in terms of the global responsibility of scientists who directly observe anthropogenic impact on natural resources over time. The conservation of forest canopies comes with an added moral weight relative to research in numerous other habitats: without a doubt, a disproportionate level of the biological diversity on Earth evolved and resides in this lofty region. The loss of forest canopies represents an obvious disappearance of numerous and important species and, less obviously, ecological processes such as nutrient cycling, carbon sequestering, and seed dispersal regimes that tie together the species components. From a human perspective, too, much of the biological richness in tropical forest canopies may contain goods and services essential to the welfare of culture and science.

Section IV represents a synthesis of many scientists working around the world on forest canopy structure, population biology, and processes. It also represents the ethics of this body of researchers, all of whom hope fervently that their collective efforts will lead to the conservation of the very habitat in which they work. This is not easy in our world of escalating technology, where a chainsaw can decimate a favorite research tree in moments or an acre of unspoiled wilderness in days. New approaches such as our moral and historical imperative to ascend into the treetops (Meg Lowman in Chapter 23), the economic values of forests (David Pearce in Chapter 24), eco-tourism (Meg Lowman in Chapter 25), emerging ethnobotanical products from tree canopies (a sidebar by Edward Burgess in Chapter 24), and a close philosophic scrutiny of canopy science (Bruce Rinker and David Jarzen in Chapter 26) are slowly changing the perspectives of many agencies and institutions that have viewed forests traditionally in terms of timber production. But will these new ideas come in time?

The current avalanche of extinctions and habitat destruction can be discouraging, but they can also be challenging for scientists and laypeople anxious to move toward positive action. The chapters and sidebars in Section IV offer some guidelines to meet this challenge. This final part of *Forest Canopies* is the conclusion of a textbook designed to help scientists, managers, conservationists, politicians, lawyers, naturalists, citizens' action groups, educators, ethicists, and others to collaborate on governing policies, cutting regimens, wilderness designations, and other broad issues related to the sensitivities of the ecological circuits operating in forest systems. Each author is a voice representing hundreds of professionals and laypeople attempting day-to-day to protect a vanishing resource.

CHAPTER 23

Tarzan or Jane? A Short History of Canopy Biology

Margaret D. Lowman

Growing up in the midwestern United States I knew trees well. I looped from one bare branch to the next in the backyard red maple with, I believed, the speed and grace of a monkey making its rounds. Like Kipling's Mowgli, I had the position and strength of each branch memorized. I learned how to rest my body comfortably among the orderly boughs in order to have a clear view of my mother, small as an ant, tending the garden below. The branches I favored became burnished from repeated scuffings. In time I identified with the monkey's world. I grew up to be a zoologist.
—*Mark Moffett*, The High Frontier, *1993*

Why Study the Treetops?

E.O. Wilson called it "the last frontier" of biological research on the planet (Wilson 1992). Andrew Mitchell referred to its invisible inhabitants as "a world I could only dream of" (Mitchell 2001). Tom Lovejoy confessed that "the canopy rendered me the biologist's equivalent of Tantalus from the very outside" (Lovejoy 1995). And Steve Sutton compared it to "Alice grows up" as canopy science moves from a sense of wonder to a reality of hypotheses (Sutton 2001). Nalini Nadkarni exclaimed about "tree climbing for grown-ups" (Nadkarni 2001) and I simply noted, "My career is not conventional. I climb trees" (Lowman 1999). In 1985, these six individuals may have represented almost half of the canopy scientists worldwide. Today, only two decades later, there are several hundred explorers of Wilson's last frontier.

The forest canopy is defined as "the top layer of a forest or wooded ecosystem consisting of overlapping leaves and branches of trees, shrubs, or both" (Art 1993). Studies of plant canopies typically include four organizational levels of approach: individual organs (leaves, stems, or branches), the whole plant, the entire stand, or the plant community (Ross 1981). Canopy biology is a relatively new discipline of forest science that incorporates the study of mobile and sessile forest organisms and the processes that link them as an ecological system.

Forest canopies have long eluded scientists because of the logistical difficulties of reaching tree crowns and the subsequent challenges of sampling once one gets up there. Only in the last decade have field biologists begun extensive exploration of this unknown world of plants, insects, birds, mammals, and their interactions. These logistic strides are attributed to the development of several innovative and creative techniques that facilitate ascent into tree crowns.

Biologists in the 19th and 20th centuries traditionally based their ideas about forests on observations made at ground level. These ground-based perceptions are summarized in a comment by Alfred R. Wallace in 1878:

Overhead, at a height, perhaps, of a hundred feet, is an almost unbroken canopy of foliage formed by the meeting together of these great trees and their interlacing branches; and this canopy is usually so dense that but an indistinct glimmer of the sky is to be seen, and even the intense tropical sunlight only penetrates to the ground subdued and broken up into scattered fragments . . . it is a world in which man seems an intruder, and where he feels overwhelmed.

Ideas about forest canopies had changed very little for a hundred years until the 1970s, when biologists first adapted technical mountain-climbing hardware for ascending tall trees. Termed SRT (single rope techniques), this versatile method enables scientists to reach the mid-canopy with ease, and hang suspended on a rope to make observations of pollinators, epiphytes, herbivores, birds, monkeys, and other biological phenomena.

There are a number of exciting reasons for the escalating priority in canopy research during the past two decades. First, as rainforests continue to decline due to human activities, the urgency of surveying the biodiversity of tree crowns challenges some researchers. There are reputedly many orchids, as well as other plants and countless invertebrates, that inhabit the treetops, and perhaps have escaped detection due to their aerial location. Many of these organisms are important not just as keystone species to the health of the rainforest ecosystem, but also as sources of medicines, foods, and materials. Second, canopy processes are essential to life on our planet—canopy organisms are integral to the maintenance of rainforest ecosystems, and the canopy is a major site of productivity in terms of photosynthesis, nutrient cycling, and exchange of carbon dioxide. As the economics of our planet become better understood, the rainforest has emerged as a critical region where ecosystem services abound. The rainforest contributes to our global economy by providing productivity (as a center for photosynthesis), medicines, materials, and foods; by housing a genetic library; by being a part of nutrient cycling, carbon storage, and other important sinks; by acting as a climate stabilizer; by participating in the conservation of water runoff; and by offering countless biodiversity and a vast cultural heritage (reviewed in Hawken et al. 1999). And third, many researchers confess to a simple curiosity to explore this previously inaccessible region of our planet. There are very few unknown frontiers left in the 21st century, but the treetops (like the ocean floor and the soil ecosystem) remain as yet little understood.

Chronology of the Development of Canopy Access Tools

Binoculars and telescopes were probably the first tools for canopy exploration. Charles Darwin, in the 19th century, looked into the tropical rainforest foliage, exclaiming:

Delight itself . . . is a weak term to express the feelings of a naturalist who, for the first time, has wandered by himself in a Brazilian forest. The elegance of the grasses, the novelty of the parasitical plants, the beauty of the flowers, the glossy green of the foliage, but above all the general luxuriance of the vegetation, filled me with admiration. A most paradoxical mixture of sound and silence pervades the shady parts of the wood. The noise from the insects is so loud, that it may be heard even in a vessel anchored several hundred yards from the shore; yet within the recesses of the forests a universal silence appears to reign. To a person fond of natural history, such a day as this brings with it a deeper pleasure than he can ever hope to experience again" (Darwin 1883).

Allee (1926) made the first published, quantified measurements of the canopy environment in Panama. Only three years later, Hingston and colleagues erected an observation platform in

British Guiana where they baited traps for canopy organisms. Sadly, no data were published; but the chronology of canopy access begins in these ideas of the 1920s.

Thirty years later in the 1950s, a steel tower was constructed in Mpanga Forest Reserve in Uganda to study gradients from the forest floor to the canopy. Towers provided access to monitor insect vectors of human diseases, which remain the first (and landmark) applied biological studies conducted in the forest canopy (Haddow et al. 1961). In the 1960s, engineers pioneered in canopy construction, including Ilar Muul, who built the first canopy walkway in Malaysia anchored in tree crowns (Muul and Liat 1970). Ladders were also used for studies of canopy vertebrates, such as McClure's studies of tree phenology and animal visitors (McClure 1966). Canopy walkways were resurrected and burgeoned later in the 1980s with the modular construction of Lowman and Bouricius (1995). Since then, canopy walkways and ladders used in conjunction with climbing ropes and other tools have become popular as permanent canopy field sites (Lowman and Wittman 1996).

The 1970s represented the era of SRT (single rope techniques). This portable, relatively inexpensive technique for canopy access allowed graduate students and others with a modest budget to survey life at the top. Perry (1986) first used SRT at LaSelva in Costa Rica, examining the ecology of a *Ceiba* tree. Perry went on to develop the canopy web, the aerial tram, and other methods that were creative extensions of a rope system (Perry 1995). Ropes were not effective, however, to reach the leafy perimeters of tree crowns, since the ropes had to be looped over sturdy branches usually close to the tree trunk. To access the leafy outer foliage of canopy trees, Appanah and Ashton invented the canopy boom, a horizontal bar with a bosun's chair at one end, which could be swung around into the leafy canopy away from the woody trunks. In Pasoh, Malaysia, booms solved the mystery of the pollination of dipterocarp flowers (Appanah and Chan 1981). In recent years, the "magic missile" was used in conjunction with conventional SRT to expand access throughout the canopy (Dial et al., this volume).

In the 1980s, biologists utilized walkways in combination with SRT and perhaps canopy booms, ladders, cherry pickers, or other creative means. Terry Erwin (1982) revolutionized our estimates of biodiversity by introducing fogging apparatus into canopy research. By misting the treetop of a *Luehea seemanii* in Panama, he collected the rain of insects and counted the diversity of species, especially beetles. Erwin's extrapolations raised our estimates of global biodiversity from almost 10 million to over 30 million. Fogging continues to be utilized extensively by rainforest biologists who need to estimate the diversity of life in the treetops. Canopy science moved from a "sense of wonder" and exploration to a more rigorous science where hypotheses were tested and vast databases were collected.

The last chapter in the development of tools for canopy research involves the ability for integrated, collaborative research projects. In general, SRT, booms, cherry pickers, scaffolding, ladders, and to some extent canopy walkways are more limited in scope, favoring solo work or small studies rather than large comprehensive studies (although some of the most recently constructed walkways are of sizeable dimension and less limited in carrying capacity). Two scientists (Francis Hallé and Alan Smith) expanded the scope of canopy research with their creative genius on two separate continents.

Hallé designed a colorful hot-air balloon, called *Radeau des Cimes* (or raft on the rooftop of the world). His inflatable raft is 27 m in diameter and forms a platform on top of the forest canopy that is utilized as a base camp for research on the uppermost canopy. A dirigible, or hot air balloon, is used in conjunction with the raft, serving to move the raft to new positions throughout the forests and also to move researchers within the above-canopy atmosphere for studies of the canopy-air interface. In 1991, the *Radeau des Cimes* expedition team pioneered a new canopy technique called the sled, or skimmer. This small (5 m) equilateral, triangular mini-raft was towed across the canopy by the dirigible, similar to a boat with a trawling apparatus in the water column

CANOPY WALKWAYS: HIGHWAYS IN THE SKY

James Burgess

Research in forest canopies has been limited by the challenges of access. During the 1980s, several relatively inexpensive but solo techniques were developed: single ropes, ladders, and towers (Lowman and Bouricius 1995; Moffett and Lowman 1995). In the early 1990s, collaborative canopy access techniques were developed, including canopy cranes, the raft and dirigible apparatus, and canopy walkways.

Walkways create permanent sites at moderate costs for long-term observations and data collection, allow collaborative research by a group of researchers within one region, and are the most effective compromise between inexpensive but inefficient access methods and those that are costly but productive. I recently completed a study of a herbivorous beetle in the bromeliads of the upper canopy of Amazonian Peru, where walkways provided canopy access through night and day, rain and sun, wind and calm, and over many seasons at the same site (Burgess et al., 2003).

Structurally, walkways have a simple, modular design incorporating bridges and platforms that interconnect to form a network in the treetops. Networks can have as many platforms and bridges as needed. Although the basic minimum module consists of one platform or bridge, walkways are usually supported by stainless steel cable with aluminum or rot-resistant wood treads. Their simple, but safe design can also accommodate researchers with varying skills and ages (see Figure 1).

Selection of sites depends on both engineering and biological factors. For engineering purposes, the site must contain healthy, mature canopy trees with upper branching conducive to platform support. The trees must also be in close proximity to one another and away from edges and tree falls, which could create dangerous wind patterns. Biologically, the site should include a species diversity that is representative of the forest type. Platforms should be placed to maximize observations of the crown area, but create minimal disturbance to the tree.

Figure 1   An enthusiastic representative of the next generation of canopy explorers, high off the forest floor in Williamstown, MA. Photograph by M.D. Lowman.

Walkways have been constructed in many different forest types, including temperate deciduous forests, tropical forests, and subtropical forests. In short, any area with a canopy is a potential candidate for walkway construction. Canopy Construction Associates (www.canopyaccess.com), based in Amherst, Massachusetts, played a large (but not exclusive) role in the development of a worldwide network of canopy walkways. Its sites include Upper Momon River, Peru (1990); Williams College, Massachusetts (1991); Hampshire College, Massachusetts (1992); Coweeta Hydrological Lab, North Carolina (1993); Blue Creek, Belize (1994); Mountain Equestrian Trails Lodge, Belize (1994); Marie Selby Botanical Gardens, Florida (1994); Millbrook School, New York (1995); Inhutani, Indonesia (1998); Tiputini Biodiversity Station, Ecuador (1998); Grandfather Mountain, North Carolina (1999); Myakka River State Park, Florida (2000); University of the South, Tennessee (2001); and Burgundy Center for Wildlife Studies, West Virginia (2001).

This global walkway network facilitates comparative canopy research among all major forest types, utilizing a standardized access technique.

References

Burgess, J., Burgess, E., and Lowman, M.D. (2003). Observation of Coleopteran herbivory on a bromeliad in the Peruvian Amazon. *J. Bromeliad Soc.* **53**, 221-224.

Lowman, M.D. and Bouricius, B. (1995). The construction of platforms and bridges for forest canopy access. *Selbyana* **16**(2), 179–184.

Moffett, M.W. and Lowman, M.D. (1995). Canopy access techniques. *In* "Forest Canopies" (M.D. Lowman and N.M. Nadkarni, Eds.), pp. 3–26. Academic Press, San Diego.

of the ocean. It facilitated the rapid collection of canopy leaves, flowers, vines, and epiphytes as well as their pollinators and herbivores (see Lowman et al. 1993; Rinker et al. 1995). In Madagascar in 2001, Hallé's team launched a new device that was essentially an individual cell within the crown of one tree whereby researchers could be dropped off by the balloon for temporary residence inside the metal frame of the canopy camp.

Alan Smith, during his research career with Smithsonian Tropical Research Institute in Panama, first considered the notion of using a construction crane for treetop exploration. A 40-m long crane was erected in a Panamanian dry forest, and since then, seven other crane operations have commenced. Cranes are quite expensive to install and operate (usually ranging from $1 to $5 million), but they offer unparalleled access to the uppermost canopy as well as to any section of the understory that is within reach of the crane arm (Parker et al. 1992).

In the next decade, Andrew Mitchell, director of the Global Canopy Programme, aspires to create the most ambitious canopy tool ever, Biotopia. This concept integrates several field methods, including cranes, walkways, canopy rafts, towers, and ropes, and it will essentially comprise a field station dedicated to canopy research.

Advances in Studies of the Canopy

The development of canopy research has been affected by several spatial and temporal constraints of this habitat:

1. Differential use of this geometric space by canopy organisms
2. Variation in microclimate of the canopy–atmosphere interface

3. Differences in ages within the canopy (e.g., soil/plant communities growing in crotches, leaves of different ages between sun and shade environments)
4. Heterogeneity of surfaces (e.g., bark, air, foliage, debris, water)
5. High diversity of organisms (many unknown to science)
6. Development of new techniques to study processes that exist in an aerial three-dimensional environment (see Figure 23-1).

Recognizing these constraints, biologists must now design useful sampling techniques within the canopy and address testable hypotheses. Canopy studies range from measuring sessile organisms (orchids, sedentary insects, trees) and mobile organisms (flying insects, birds, mammals) to canopy processes (studies of the interactions of organisms up in the treetops). All of these studies require sampling designs that are effective at heights, that can function in an air substrate, and that can be carried out while dangling from a rope or some otherwise awkward position (awkward for the researcher, that is). How are organisms detected and sampled in such a heterogeneous environment where humans are rendered less agile? In a scenario similar to the expansion of coral reef fish ecology in the 1970s with the advent of SCUBA, canopy biologists are developing sampling protocols to measure the spatial, temporal, and substrate heterogeneity of their environment.

Studies of sessile organisms in forest canopies pose fewer logistical challenges to measurement than other canopy components because they can usually be counted. Sessile organisms include the obvious groups such as trees and vines, as well as less apparent groups (mosses, lichens, scale insects). Trees are the largest sessile organisms in forest canopies, and they comprise the major substrate of the canopy ecosystem. Tree species—with their varying architecture, limb strength, surface chemistry, and texture—play fundamental roles in shaping the canopy community. Epiphytes (or air plants) live on the surfaces of trees, as do many epiphylls (tiny air plants such as lichens that live on leaf surfaces). In addition, trees serve as both shelter and food for many

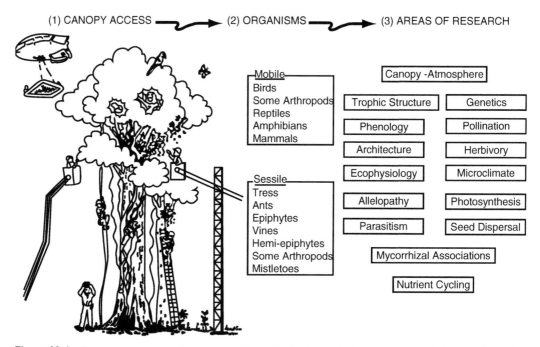

Figure 23-1   Forest canopy research has progressed from the development of canopy access techniques to descriptive studies of different types of organisms. Scientists are now able to employ a more rigorous experimental approach to study canopy interactions.

INTERNATIONAL CANOPY CRANE NETWORK

David C. Shaw

The International Canopy Crane Network (ICCN) is an affiliation of canopy crane sites around the planet (Basset et al. 2003). The purpose of the ICCN is to facilitate research collaboration and to synchronize long-term monitoring of forest canopies at many sites. The need for an international network of canopy cranes was conceived by the late Alan Smith who, with colleagues at the Smithsonian Institution, established the first canopy crane at Parque Natural Metropolitano in Panama in 1990. Since then, 10 other canopy cranes have been erected in various temperate and tropical forests of the world. The Global Canopy Programme, based in Oxford, UK, is actively collaborating with the ICCN to install more canopy cranes in important forest regions of the world (Mitchell et al. 2002).

Canopy cranes (Figure 1) are safe. Trees are often rotten in the center, and the tops and outer foliage in the upper crown are difficult or impossible to access safely. Canopy cranes provide the ability to come down from above, thereby permitting researchers access to dead

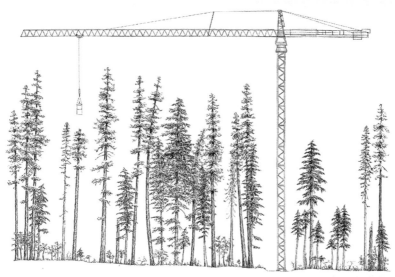

Figure 1   Line drawing of a canopy crane in a temperate coniferous forest. The gondola is attached to cables controlled from a jib that swings over the tallest trees in a forest stand, allowing access from above the canopy. In this illustration, the trees are 60 m tall and the crane is 75 m tall at the jib. The access portion of the jib is 85 m long, allowing 2.3 ha of forest to be studied from above. The mast of the crane tops at 87 m and can be used to place atmospheric and other sampling equipment. Illustration by R. Van Pelt and K. Bible, from Wind River Canopy Crane Research Facility image archives (http://depts.washington.edu/wrccrf).

wood, tops of dead trees, foliage in the outer/upper crown, and other trees unsafe to climb (see Figure 2). Stinging wasps, biting ants, snakes, and other nasties can also affect tree climbers, but canopy cranes usually avoid these aspects of canopy life. People in trees have an impact, which includes dislodging epiphytes off branches, breaking branches, disturbing vertebrates, and eliciting defense in social insects. Canopy cranes reduce this impact by allowing researchers non-invasive access without putting their weight on branches and trees.

Canopy cranes offer a unique research perspective that is not possible with other access systems. Long-term repeated access to the canopy is provided to all canopy positions,

Continued

INTERNATIONAL CANOPY CRANE NETWORK–*cont'd*

Figure 2   Sampling epiphytes at the top of a 50-m-tall dead tree (snag). This unique ability to access regions of the forest that are unsafe for climbers makes canopy cranes an important tool for whole forest studies. From Wind River Canopy Crane Research Facility image archives (http://depts.washington.edu/wrccrf).

including space between trees, individual leaves and branches, and randomized trapping and sampling locations. Large amounts of heavy sampling equipment can be hauled up in the gondola of the crane. Pollination exclusion nets can be installed and managed easily in large numbers; leaf physiology can be followed on hourly, daily, monthly, and annual time scales; and observations on avian foraging in the upper canopy are possible. Research at canopy crane sites has emphasized such topics as tree ecophysiology, CO_2 dynamics and climate change, biodiversity, invertebrate ecology, herbivory, forest structure and dynamics of canopy components, vertebrates (especially birds), pollination ecology, phenology, and ecology of forest diseases. Cross-site collaboration has occurred in tree ecophysiology, herbivory, and insect biodiversity.

Canopy cranes have limitations in that only one forest stand is sampled (N = 1).

However, the vision of the ICCN is to have each crane associated with a referenced, mapped forest stand that is linked to the surrounding forest landscape in a way that strategically places the canopy crane forest in an ecological perspective. Hypotheses generated from canopy crane sites can form the framework for new understandings in how forests function, which are then tested at larger spatial scales.

We know that forest canopies play an important role in maintenance of biodiversity and ecosystem services at a global scale (Ozanne et al. 2003), and the ICCN connects unique research facilities from around the globe that are at the forefront of scientific research in these areas. The installation of a canopy crane in a forest ecosystem provides for intensive interdisciplinary scientific focus, and by facilitating collaboration among these sites, the ICCN cuts across funding agencies, political boundaries, and scientific disciplines to provide a new view of forest ecosystems.

References

Basset, Y., Horlyck, V., and Wright, S.J. (2003). "Studying Forest Canopies from Above: The International Canopy Crane Network." Smithsonian Tropical Research Institute and UNEP, Panama.

Mitchell, A.W., Secoy, K., and Jackson, T. (2002). "The Global Canopy Handbook." Global Canopy Programme, Oxford, UK.

Ozanne, C.M.P., Anhuf, D., Boulter, S.L., Keller, M., Kitching, R.L., Körner, C., Meinzer, F.C., Mitchell, A.W., Nakashizuka, T., Silva Dias, P.L., Stork, N.E., Wright, S.J., and Yoshimura, M. (2003). Biodiversity meets the atmosphere: a global view of forest canopies. *Science* **301**, 183–186.

mobile organisms; and most canopy processes are directly dependent upon trees. The biggest obstacle to quantified studies of trees is the access to the growing tips (buds, flowers, root hairs) that often occur in the uppermost branches where the light is greatest. Some methods, such as the raft and the crane, provide access to this fragile region, whereas others, such as ropes and walkways, are limited to mid-canopy regions since they depend on nearness to sturdy branches for safety. Major achievements in studies of sessile organisms include the extensive research on canopy architecture (e.g., Hallé et al. 1978), the distribution and biology of epiphytes (Nadkarni 1984; Benzing 1990), mapping canopy surfaces from construction cranes (Parker et al. 1992), and the growth of vines (Putz and Mooney 1991).

Mobile organisms in the canopy pose great challenges for scientists because mobility in an air medium is very challenging. Most studies of vertebrates have been made from ground level, such as the extensive work on the biology and distribution of neotropical mammals by Emmons (1995). In contrast, Malcolm (1995) sampled small rodents in the tree canopies of Brazil and discovered that many have unexpected arboreal proclivities. He found that species exhibit distinct height preferences, and more mammals were arboreal than terrestrial. Similarly, Taylor censused small mammals in temperate deciduous oak-beech canopies and found vertical stratification of mammals there (Taylor and Lowman 1996). One species, a flying squirrel, had not previously been censused adequately due to its arboreal tendencies, and it was found to be an important predator on the gypsy moth. This is but one example where the neglect of the canopy environment may have led to misinformation about the distribution and abundance of forest organisms.

Ornithologists also face the challenge of trying to observe and/or capture birds in tree crowns. In New Guinea, Beehler hoisted nets up and down tall poles to quantify birds of paradise in the canopy (Beehler 1991). Lovejoy and Bierregaard (1990) found that birds will increase the size of their territories vertically to compensate for forest fragmentation. In Peru, Munn and Loiselle (1995) used a large slingshot to position aerial mist nets in the canopy; and Nadkarni and Matelson (1991) used SRT to observe 193 species of birds using epiphytes in the cloud forests of Costa Rica.

Reptiles and amphibians in tree canopies have been studied in Puerto Rico. Reagan (1995) monitored *Anolis* lizard populations in tree canopies; and Dial (1992) performed some of the first experimental canopy studies by excluding lizards from tree crowns and monitoring changes in insect populations. Invertebrate studies in forest canopies have perhaps received more attention and subsequently endured more controversy than any other component of aerial ecology. Erwin's most recent surveys of invertebrates in tropical tree crowns have led him to speculate that 32.5 million species exist on Earth, 2.5 million higher than his previous calculations (this volume). But the enormous spatial and temporal variability of insects in tree crowns, as well as artifacts of sampling, make studies of canopy arthropods difficult, with subsequent enormous volumes of data to analyze (Lowman et al. 1993; Lowman et al. 1995).

Processes in forest canopies are the most difficult to study because they require information about both sessile and mobile organisms as well as interactions between the two groups.

Reproductive biology is predominantly a canopy phenomenon in forests, although the pollinator and dynamics of flowering and fruiting in tropical trees is relatively unexplored (reviewed in Murawski 1995). Herbivory and insect–plant interactions have been quantified in several forests using a combination of access methods including SRT, hot-air balloons, walkways, cherry pickers, and rafts (Lowman 1995). Herbivores consume significantly less foliage in the upper crowns (sun leaves) as compared to the lower crowns (shade leaves), but young leaves (especially in the shade) are often completely consumed (Coley 1983; Lowman 1984). Differences in herbivory levels can arise from artifacts of sampling, although canopy access has increased the accuracy of results since the sun leaves can be measured as well as the understory (reviewed in Lowman and Wittman 1996).

Access to tree crowns has stimulated studies of canopy nutrient cycling, particularly with regard to epiphytes (Benzing 2000). Wind-blown litter provides nutrition for epiphytes that in turn influences other organisms in the canopy food webs (Nadkarni and Matelson 1991). As tropical forests are cleared, the epiphytes continue to decline, as does subsequently the entire community of biodiversity that they support (Benzing 2000). Other processes, such as photosynthesis, have not been measured for epiphytes or other canopy foliage, but construction cranes have begun to provide permanent sampling platforms for such measurements (Mulkey et al. 1996). The interaction of canopy processes is absolutely unknown. How does the seasonality of nutrient cycling affect pollination? How does herbivory alter decomposition on the forest floor? Extrapolation of small-scale data collection to later-scale ecosystem applications is untested for most aspects of ecological processes.

Future Directions

Now that scientists have solved the mysteries of canopy access tools, the more demanding challenges lie ahead. Canopy organisms—both mobile and sessile—must be mapped and measured. Canopy processes must be sampled with respect to the complex differences in light, height, tree species, and seasonality. Establishing rigorous sampling techniques and conducting long-term research while dangling from a rope are ambitious objectives. The next 10 years will be critical as scientists attempt to classify the biodiversity and ecology of forest canopies before habitat fragmentation takes an irreversible toll. Exciting new directions lie ahead—the extrapolation from leaf to canopy, from organisms to populations, from flower to entire crown, from seedling mortality to recruitment patterns throughout forest patches, from light levels in small gaps to photosynthesis to entire stands. Canopy access techniques will provide invaluable access to this exciting region of our planet and, hopefully, yield results that will facilitate the implementation of sound conservation practices.

References

Allee, W.C. (1926). Measurement of environmental factors in the tropical rain-forest of Panama. *Ecology* **7**, 273–302.

Art, H.W. (1993). "The Dictionary of Ecology and Environmental Science." Holt, New York.

Appanah, S. and Chan, H.T. (1981). Thrips: pollinators of some dipterocarps. *Malay. For.* **44**, 234–252.

Beehler. B. (1991). "A Naturalist in New Guinea." Texas Univ. Press, Austin, Texas.

Benzing, D.H. (1990). "Vascular Epiphytes." Cambridge University Press, Cambridge, Massachusetts.

Benzing, D.H. (2000). "Vascular Epiphytes." Cambridge University Press, Cambridge, Massachusetts.

Coley, P.D. (1983). Herbivory and defensive characteristics of tree species in a lowland tropical *Forst. Ecol. Monogr.* **53**, 209–233.

Darwin, C.R. (1883). "Journal of Researches into the Natural History and Geology of the Countries Visited during the Voyage of the H.M.S. Beagle Round the World." D. Appleton and Company, New York.

Dial, R. (1992). A food web for a tropical rain forest. The canopy view from Anolis. Ph.D. Dissertation, Stanford University, Stanford CA.

Dial R., Sillett, S., and Spickler, J. (2004). *In* "Forest Canopies" (M.D. Lowman and H.B. Rinker, Eds.). pp. XXX–XXX. Academic Press, Boston.

Emmons, L. (1995). Mammals of rain forest canopies. *In* "Forest Canopies" (M.D. Lowman and N.M. Nadkarni, Eds.), pp. 199–224. Academic Press, San Diego.

Erwin, T.L. (1982). Tropical forests: their richness in Coleoptera and other arthropod species. *Coleopterists Bull.* **36**, 74–75.

Haddow, A.J., Corbet, P.S., and Gilett, J.D. (1961). Studies from a high tower in Mpanga Forest, Uganda. *Trans. R. Ent. Soc. London* **113**, 249–368.

Hallé, F., Oldeman, R.A.A., and Tomlinson, P.B. (1978). "Tropical Trees and Forests: An Architectural Analysis." Springer-Verlag, Boston, MA.

Hawken, P., Lovins, A., and Lovins, L.H. (1999). "Natural Capitalism." Little, Brown and Co., Boston, MA.

Lovejoy, T.E. (1995). Foreword *In* "Forest Canopies" (M.D. Lowman and N.M. Nadkarni, Eds.), pp. xv–xvi. Academic Press, San Diego.

Lovejoy, T.E. and Bierregaard, R.O. (1990). Central Amazonian forests and the minimum critical size of ecosystems project. *In* "Four Neotropical Rainforests" (A.H. Gentry, Ed.), pp. 60-71. Yale Univ. Press, New Haven, CT.

Lowman, M.D. (1984). An assessment of techniques for measuring herbivory: Is rain forest defoliation more intense than we thought? *Biotropica* **16**(4), 264–268.

Lowman, M.D. (1995). Herbivory as a canopy process in rain forest trees. *In* "Forest Canopies" (M.D. Lowman and N.M. Nadkarni, Eds.), pp. 431–456. Academic Press, San Diego.

Lowman, M.D. (1999.) "Life in the Treetops." Yale University Press, New Haven, CT.

Lowman, M.D. and Bouricius, B. (1995). The construction of platforms and bridges for forest canopy access. *Selbyana* **16**, 179–84.

Lowman, M.D., Moffett, M. and Rinker, H.B. (1993). A new technique for taxonomic and ecological sampling in rain forest canopies. *Selbyana* **14**, 75–79.

Lowman, M.D. and Wittman, P.K. (1996). Forest canopies: methods, hypotheses, and future directions. Ann. Rev. Ecol. Syst. **27**, 55–81.

Lowman, M.D., Kitching, R.L., and Carruthers, G. (1996). Arthropod sampling in Australian subtropical rain forest: how accurate are some of the more common techniques? *Selbyana* **17**, 36–42.

Malcolm, J.R. (1995). Forest structure and the abundance and diversity of neotropical small mammals. *In* "Forest Canopies" (M.D. Lowman and N.M. Nadkarni, Eds.), pp. 179–198. Academic Press, San Diego.

McClure, H.E. (1966). Flowering, fruiting and animals in the canopy of a tropical rain forest. *Malay For.* **29**, 182–203.

Mitchell, A. (2001). Canopy science: time to shape up. *In* "Tropical Forest Canopies: Ecology and Management" (K.E. Linsenmair, A.J. David, B. Fiala, and M.R. Speight, Eds.), pp. 5–13. Kluwer Academic Publishers, The Netherlands.

Mulkey, S.S., Chazdon, R.L., and Smith, A.P. (Eds.) (1996). "Tropical Forest Plant Ecophysiology." Chapman & Hall, New York.

Muul, I. and Liat, L.B. (1970). Vertical zonation in a tropical forest in Malaysia: methods of study. *Science* **196**, 788–789.

Munn, C.A. and Loiselle, B.A. (1995). Canopy access techniques and their importance for the study of tropical forest canopy birds. *In* "Forest Canopies" (M.D. Lowman and N.M. Nadkarni, Eds.), pp. 165–178. Academic Press, San Diego.

Murawski, D.A. (1995). Reproductive biology and genetics of tropical trees from a canopy perspective. *In* "Forest Canopies" (M.D. Lowman and N.M. Nadkarni, Eds.), pp. 457–494. Academic Press, San Diego.

Nadkarni, N. (1984). Epiphyte biomass and nutrient capital of a neotropical elfin forest. *Biotropica* **16**, 249–56.

Nadkarni, N. (2001). Tropical rainforest ecology from a canopy perspective. *In* "Tropical Ecosystems" (K.N. Ganeshaiah, R. Uma Shaanker, and K.S. Bawa, Eds.), p. 433. Oxford and IBH Publishing, New Delhi, India.

Nadkarni, N. and Matelson, T. (1991). Fine litter dynamics within the tree canopy of a tropical cloud forest. *Ecology* **72**, 2071–2082.

Parker, G.G., Smith, A.P., and Hogan, K.P. (1992). Access to the upper forest canopy with a large tower crane. *Bioscience* **42**, 664–670.

Perry, D.R. (1986). "Life Above the Jungle Floor." Simon & Schuster, New York.

Perry, D. (1995). Tourism, economics, and the canopy: the perspective of one canopy biologist. *In* "Forest Canopies" (M.D. Lowman and N.M. Nadkarni, Eds.), pp. 605–608. Academic Press, San Diego.

Putz, F.E. and Mooney, H.A. (Eds.). (1991). "The Biology of Vines." Cambridge University Press, Cambridge, UK.

Reagan, D.P. (1995). Lizard ecology in the canopy of an island rain forest *In* "Forest Canopies" (M.D. Lowman and N.M. Nadkarni, Eds.), pp. 149–164. Academic Press, San Diego.

Rinker, H.B., Lowman, M.D., and Moffett, M.W. (1995). Africa from the treetops. *American Biology Teacher* **57**(7), 393–401.

Ross, J. (1981). "The Radiation Regime and Architecture of Plant Stands." W. Junk, The Hague.

Sutton, S.L. (2001). Alice grows up: canopy science in transition from wonderland to reality. *In* "Tropical Forest Canopies: Ecology and Management," pp. 13–23. Kluwer Academic Publishers, The Netherlands.

Taylor, P. and Lowman, M.D. (1996). Vertical stratification of small mammals in a northern hardwood forest. *Selbyana* **17**, 15–21.

Wallace, A.R. (1878). "Tropical Nature and Other Essays." MacMillan and Co., London.

Wilson, E.O. (1992). "The Diversity of Life." Harvard University Press, Cambridge, MA.

CHAPTER 24

Economics and the Forest Canopy

David W. Pearce

Conservation and development are inextricably linked. At the national and international levels, conservation is not affordable without economic development, whereas economic development cannot be sustained without conservation.
—R. Kramer and C. P. van Schaik,
Last Stand: Protected Areas and the Defense of Tropical Biodiversity, *1997*

Global rates of deforestation are disputed. The Food and Agriculture Organization (FAO, 2001) estimated global net rates of deforestation of around 9 million hectares in the 1990s, or 0.23 percent of total forest area. The World Resources Institute (Matthews 2001) disputes the figure, noting that FAO data include biodiversity-poor plantations as afforestation, offsetting natural forest loss. Net of plantation growth, annual losses are closer to 16 million hectares per annum, or 0.4 percent per annum of forest cover, nearly double the FAO figure. Whatever the correct figure, rates of loss are positive, and there appears to be little evidence that they have slowed over recent decades, despite a large and vocal forest conservation movement. Much of the loss is irreversible in the sense that any re-growth cannot replicate the stock of information and services in the original forest and which have been produced on an evolutionary time scale. Forests, and especially forest canopies, are also under-researched. What information forests have yet to yield is an unknown. Certainly, canopy research alone promises better understanding of the earth–atmosphere interface and knowledge of newly identified species. The issue, then, is how to make decisions in this complex context of uncertainty and irreversibility. Economic analysis has a good deal to say about such contexts.

Why Do We Need Forest Economics?

Many conservationists are ill at ease with economic approaches to forest loss. An economic approach argues that, if the benefits of deforestation exceed the costs, then deforestation has a social justification. Society is better off with forest loss than without it. Moreover, costs and benefits are defined in economics according to human preferences. If I prefer A to B, then A has a bigger benefit to me than B. The argument appears to "legitimize" deforestation and to do so on the flimsy basis of human preference. Anyone believing that forests have intrinsic value—i.e., value embedded in the forest asset rather than in the eyes of the beholder—will want to reject the economic approach. But there are powerful reasons for heeding economics.

First, regardless of who activates forest loss, whether it is agriculturists, loggers, or road builders, their motivation is primarily financial. They do it because the financial gains exceed the financial costs. Economics is best placed to analyze these financial gains and losses. Second, economics does not necessarily sanction an activity simply because it makes a profit. It sanctions activities that

ETHNOBOTANY IN FOREST CANOPIES

Edward A. Burgess

A large number of the trees forming these forests are still unknown to science, and yet Indians, these practical botanists and zoologists, are well acquainted, not only with their external appearance, but also with their various properties. It would greatly contribute to the progress of science if a systematic record were made of all information thus scattered throughout the land; an encyclopedia of the woods, as it were, taken down from the tribes which inhabit them.

—Professor and Mrs. Louis Agassiz (1868), quoted in Schultes 1990

Ethnobotany deals promisingly with the study of human interactions with local plants, especially their uses for medicinal purposes (Martin 1995). Rather than direct scientific investigation, ethnobotanists attempt to gain knowledge of forest plants through the long-term exploitation by indigenous peoples. This approach can yield excellent results because tribes have often lived in forested regions for hundreds of years and, thus, possess an immense knowledge of local plants and their uses (Balick 1990). Although interactions with plants take place more commonly on the forest floor, the numerous plants in the forest canopy also promise many medicinal prospects. Relatively little exploration of canopy plants for ethnobotanical applications has been conducted to date. This is now expected to change, however, with the advent of advanced canopy access techniques.

Bennett (1995) provided a review of the ethnobotany and economic botany of lianas, epiphytes, and many other host-dependent plants found in the canopy. He concluded that these plants are largely overlooked by ethnobotanists despite their great biological diversity and abundance. He observed that host-dependent plants are significant resources to many indigenous peoples, as they contribute to all the major uses classes such as fibers, foods, medicines, and fuel sources. Furthermore, Bennett noted that most of the ethnobotanical data on host-dependent plants are from Amazonia due to abundant epiphytes and to extensive ethnobotanical studies performed historically in this vast region.

Castner et al. (1998) reviewed some of the medicinal plants throughout the Amazon. The authors did not specifically address canopy plants, as most of their focus was on the species of the forest floor or understory. However, these researchers provided descriptions, uses, and locations of numerous species, some of which are connected to the treetops of Amazonia.

Schultes and Raffauf (1990) presented an extensive history of medicinal plants of the Amazon, yet they acknowledged that their efforts merely scratched the surface of the pharmaceutical "gold mine" in the Amazon. They also advocated the importance of ethnobotanical conservation because of the rapid depletion of forests as well as the indigenous tribes who bear a vast cultural knowledge of the plants found therein. Schultes and Raffauf's comprehensive study included many vines found in the canopy, but they did not focus on canopy plants.

Plotkin (2000) cited several case studies of important ethnobotanicals. He discussed some important plants of the shamans from which ayahuasca and curare are derived.

In summary, most current ethnobotanical studies have neglected plant species obtained from the forest canopy. However, it is clear that plants in forest canopies possess great chemical diversity and potential benefit to the field of ethnobotany. With growing interest in the pharmaceutical possibilities of plants, as well as new techniques for ethnobotanical analysis, the future of ethnobotany in the forest canopy shows great promise for modern medicine.

References

Balick, M.J. (1990). Ethnobotany and the identification of therapeutic agents from the rainforest. *In* "CIBA Foundation Symposium No. 154." Chichester: J. Wiley and Sons.

Bennett, B.C. (1995). Ethnobotany and economic botany of epiphytes, lianas, and other host-dependent plants: an overview. *In* "Forest Canopies" (M.D. Lowman and N.M. Nadkarni, Eds.), pp. 547–548. Academic Press, San Diego.

Castner, J.L., Timme, S.L., and Duke, J.A. (1998). "A Field Guide to Medicinal and Useful Plants of the Upper Amazon." Feline Press, Gainesville, Florida.

Martin, G.J. (1995). "Ethnobotany." Chapman and Hall, London.

Schultes, R.E., and Raffauf, R.F. (1990). "The Healing Forest." Dioscorides Press, Portland, Oregon.

Plotkin, M.J. (2000). "Medicine Quest." Viking, New York.

generate net *social* benefits, not net *financial* benefits. The difference is crucial. Forest loss that is financially profitable could easily be economically unsound if the loss imposes costs on others in the form of extra siltation of rivers, reduced flood protection, lost biological diversity, increased carbon dioxide emissions from released carbon, and so on. So what is financially good could easily be socially bad. Third, since forest converters have few or no incentives to take these third-party impacts, so-called "externalities," into account when they make decisions, what actually happens diverges from what is justified from an economic point of view. We can safely say that there is too much deforestation, but that the economically desirable level (the "optimal" level) is probably not zero. Fourth, economics is very much about incentive design, about restructuring markets so that anyone deciding to convert forest land faces the full social costs of doing so. That social cost acts like a price or a tax: it becomes part of the cost of conversion and hence reduces, perhaps dramatically, the profitability of forest loss. A final reason for taking economics seriously is that other approaches have not worked in the past, whereas the economic approach is showing signs that it may work. There is no rationality in rejecting an approach that has not yet been fully tested.

Economics also has something to say about the uncertainty–irreversibility context. It is not just a matter of adding up all the good things that forests do, then putting a money value on them and comparing them to the costs of conservation. That is an important part of the exercise, of course. Uncertainty and irreversibility provide an additional argument to delay deforestation. No delay means information is lost irretrievably. Perhaps those losses are not very large now because so much forest remains. The information lost perhaps resides in that part of the forest area that remains. But if forests are not homogeneous, which obviously they are not, and if rates of loss are rapid, which they are, then information losses could be serious. Delay permits more research and understanding of what forests do. Delay enables the generation of information, and this information has an economic value. Hence there are now at least two arguments for slowing deforestation rates:

1. The existence of externalities, or third-party losses, means that forest loss will often not pass a cost–benefit test;
2. The existence of uncertainty and irreversibility provides a rationale for delaying forest loss.

The first argument is a conventional cost benefit argument. Costs and benefits reflect human preferences, and preferences are most easily measured by looking at how people "vote" in marketplaces through their willingness to pay for goods and services. One obvious problem is that many forest services have no markets. Indeed, this is precisely *why* deforestation takes place. The apparent price of carbon dioxide, measured by what people are willing to pay to avoid global warming, is zero. Its true price is positive because we know carbon dioxide emissions result in global warming, which has detrimental consequences for human well-being. Hence the challenge is to find these true prices (in economics, they are often known as *shadow prices*) through a variety of techniques that elicit individuals' willingness to pay for those services.

Sometimes other markets reveal the sums in question: the value of subsistence fuelwood can be approximated by marketed fuelwood, or the equivalent energy value of kerosene. Sometimes the damage can be estimated directly, as with the loss of agricultural output or the rise in healthcare costs due to global warming. In other cases, it is necessary to "construct" a market by adopting a market research approach and asking people about their willingness to pay. This is the "stated preference"' approach to valuing environmental assets in economic terms. While still controversial, stated preference approaches have been used extensively.

The second argument is more recent and rests on what environmental economists call "quasi option value" and what investment theorists call "option value" (Fisher 2000). These concepts describe the money value of learning more about the future, a value that is lost by making an irreversible decision today. Interestingly, option value gives a rationale for the "precautionary principle," which is increasingly becoming embedded in environmental legislation, usually without too much thought about what it means.

Cost–benefit and option value are not the only arguments that an economist can use to reflect considerable caution about current rates of deforestation. There are powerful equity or environmental justice arguments. Poor people depend heavily on forests and their products. Table 24-1 shows that forest products can often make up substantial fractions of effective household income. Forests play a significant role in poverty alleviation.

A further reason for arguing that current rates of deforestation are excessive is that much timber harvesting and forest burning is subsidized. While subsidies can have justifiable social motives, in practice, most have a political basis and few actually benefit the poor (van Beers and de Moor 2001). Moreover, such subsidies are unproductive in that they do not produce economically justified output. The subsidies simply transfer wealth from one group in society (taxpayers) to another (farmers, loggers). Subsidies, therefore, make forest conversion more profitable but socially less desirable, a further instance of how financial and economic perspectives diverge.

Forest Economic Values

A substantial literature exists on the economic values of forest functions and services. This literature is reviewed in Pearce (2001) and more extensively in Pearce and Pearce (2001). Table 24-2

Table 24-1   Non-Timber Forest Products as Percentages of Total Household Income

Study	Site	NTFPs as % Household Income
Lynam et al. 1994	Zimbabwe:	
	Chivi	40–160%
	Mangwende	12–47%
Houghton and Mendelsohn 1996	Middle Hills, Nepal	Fodder, fuel and timber can yield as much net revenue as agriculture
Kramer et al. 1995	Mantandia, Madagascar	47% (lost forest products as % of household output)
Bahuguna 2000	Madhya Pradesh, Orissa, and Gujarat, India	49% (fuelwood and fodder = 31%, 10% employment, 6% other NTFPs, 2% timber and bamboo)
Cavendish 1999	Zimbabwe	35% (across many different environmental goods)

From Pearce and Pearce 2001.

Table 24-2 Summary Economic Values of Forests (US$ ha/p.a., unless otherwise stated)

Forest Good or Service	Tropical Forests	Temperate Forests
Timber		
conventional logging	200–4400 (NPV)*	
sustainable logging	300–2660 (NPV)*	−4000 to +700 (NPV)
conventional logging	20–440	
sustainable logging	30–266	
Fuelwood	40	—
Non-timber products	0–100	small
Genetic information	0–3000	—
Recreation	2–470 (general)	80
	750 (forests near towns)	
	1000 (unique forests)	
Watershed benefits	15–850	−10 to +50
Climate benefits	360–2200 (GPV)†	90–400 (afforestation)
Biodiversity (other than genetics)	?	?
Amenity	—	small
Nonuse values		
Option values	n.a.	70?
Existence values	2–12	12–45
	4400 (unique areas)	

From Pearce and Pearce 2001.

*NPV is net present value, the discounted flow of revenues minus costs over time.

†GPV is gross present value, i.e., the discounted sum of revenues over time.

provides a brief summary of the findings. Timber values are seen to be high. For a tropical forest, typical values might be US $20 to $440 per hectare. Note that sustainable forest revenues tend to be less than conventional logging revenues, reflecting the role that time plays in timber economics.

Sustainable rotations promise revenues in the distant future, whereas conventional logging tends to ignore replanting and conservation practices, increasing the returns now against yields in the future. Timber revenues can be thought of as one benchmark against which conservation has to compete. The large range for genetic information reflects the current debate about the value of forests as sources of information for pharmaceutical and agricultural research. While the popular view is that such information must have an extremely high value because it might lead to cures for major diseases, economic analysis provides a different perspective. The low values derive from studies that treat the information as the result of random searches in forest areas for biomass samples. The fact that there is a lot of forest, and the fact that forests are not the only source of the information, means that there are many substitutes for any given hectare of forest, lowering the economic value. In other economic models, however, high values are generated.

Nonuse values reflect individuals' willingness to pay for forest conservation independently of any use made of the forest. Someone not visiting forests and never intending to may still have an "existence" value, perhaps the closest that economics gets to the notion of individuals valuing something on behalf of the asset in question, i.e. a sort of intrinsic value. That existence values are positive is easily demonstrated by looking at donations to conservation societies, since few who donate actually make use of the assets in question. *Option value* (unfortunately and confusingly, the same name as is given to the information value from delay in making decisions) refers to the value placed on an asset by those who do not make use of it today but who may choose to do so in the future.

Two other issues stand out in Table 24-2. First, climate benefits appear to be huge, i.e., carbon "stored" in biomass and soil has a very large damage value if that carbon is released as a global warming gas. This confirms what many would argue anyway—namely that forest conservation plays a crucial role in any climate change policy. Second, the value of biodiversity is not known. The lack of information on the economic value of diversity simply reflects the slowness of economists in studying this issue. There are numerous studies of what people are willing to pay for conserving single species, especially exotic species such as the giant panda and the African elephant. What is missing is an understanding of how people's preferences behave in respect of diversity itself.

Some use values of forest ecosystem functions are derived from component parts of the forest, say a specific species or service. Other use values are very much dependent on the whole ecosystem. Nonuse values may similarly be heavily influenced by a single component (again, a single species, for example) or by an appreciation of the interconnected whole asset. Surprisingly, *what* it is that people value is seriously under-researched. In contrast to the "mixed" part–whole determinants of use and nonuse values, the value of ecosystem resilience derives from the diversity of the components. Thus, whereas the economic value derived from ecological functions and nonuse motives could be the value placed on *biological resources*, the economic value of ecosystem resilience to shocks and stresses is very much a value of *biological diversity*. Resources in a low-diversity system may still attract high use and nonuse values—it is a matter of perception. But low diversity systems are unlikely to contain high resilience values, although the precise links between diversity and resilience are debated and occupy a substantial ecological literature.

Views differ as to whether the value of resilience is local or global, or, as seems more likely, both. Perrings et al. (1995) are of the view that the main consequences of diversity show up in local losses rather than global losses. Resilience shifts the focus to the way ecosystems may change in the presence of stresses and shocks. These processes of change may not be "linear." For example, a modest change may result in some dramatic effect rather than a modest one. The process of change is marked by discontinuities and potential irreversibilities. Equally, some major changes may have little effect on the system. Resilience measures the degree of shock or stress that the system can absorb before moving from one state to another very different one. Diversity, it is argued, stimulates resilience perhaps because individual species threatened or affected by change can have their roles taken over by other species in the same system. The smaller the array of species, the less chance there is of this substitution process taking place.

From an economic standpoint, the issue is one of identifying and measuring this insurance value. Unfortunately, neither is easy. Identifying how close a system might be to collapse of some or all functions is itself extremely difficult, yet one would expect willingness to pay to avoid that collapse to be related in some way to the chances that the collapse will occur. If the chances are known, the value sought is then the premium that would be paid to conserve resilience. Suggestions include the entire cost of managing non-resilient systems since these costs can be avoided if more diverse and therefore more resilient systems are adopted. In the agricultural context, for example, this would make the premium equal to the entire costs of ensuring that intensive agriculture is maintained including fertilizer and pesticide costs. Inverting the process, it could be argued that the premium is approximated by the cost of all the losses incurred by maintaining a resilient system. If, as is widely suggested, diverse/resilient systems are lower productivity systems, then the loss of productivity from maintaining a resilient system might be thought of as the economic value of resilience—i.e., as the resources that have to be sacrificed to maintain diversity.

The economic benefits of forests therefore comprise three general components: contribution to ecosystem functions, nonuse values, and contribution to ecosystem resilience. Currently, least is known about the last category of value, resilience.

THE VALUE OF HERBARIA FOR PLANT CONSERVATION

Bruce K. Holst

An often repeated statement is that "you must know what you have before you can conserve it." But how do you know what you have in the first place? Field biologists have been working on this question for hundreds of years by conducting inventories, from general natural history inventories to small, highly detailed plot studies of flora and fauna. These inventories result in preserved specimens, which in the case of plants are placed into herbaria along with their associated original locality information. Herbarium specimens then serve as documented proof of a plant's occurrence at a specific time and place. Herbaria have an advantage over living collections; if well cared for, specimens can last for hundreds of years, if not longer. Although living collections are valuable for many purposes, they are expensive and difficult to maintain in cultivation for long periods of time. It is the vast number of preserved plants and their longevity in storage that make them ideally suited for the purpose of archiving information on plants and their habitats.

Worldwide, approximately 3,000 herbaria collectively hold 300 million specimens (extrapolated from Holmgren et al. 1990), nearly all with labels that include collector and collection locality information. The labels may also include a variety of information, such as autecology, flowering time, ethnobotany, geology, pollination, fragrances, and dispersal. Herbarium specimens can also provide insights into which plants grew where over time and when a particular invasive plant invaded a particular region. Herbarium label information is most often summarized and published in floras and monographs, though its potential for biodiversity mapping is being realized through the accumulation of its data on computers in combination with geographical information systems (Figure 1). One of the most valuable outcomes of this type of work is the ability to determine which areas are the most species-diverse or rich in endemic species, allowing them to be given higher conservation priorities. A great challenge for herbaria, and for biological sciences in general, is capturing plant label information electronically to make it widely available for conservation purposes.

Perhaps of greatest importance, herbaria are potential repositories of genetic information for every species of plant known to science. Techniques are continually refined to extract DNA and other chemical information from plants preserved decades or even centuries ago. Newer techniques of DNA extraction, particularly amplification, allow for genetic analysis from smaller and smaller pieces of herbarium material, thus preserving the herbarium specimens for other types of studies. Recognizing that some traditional preservation techniques may hinder the future extraction of DNA, herbaria and botanical gardens are adapting and placing more emphasis on the collection of materials in silica gel and the cryopreservation of seeds and tissues.

As a fundamental tool for plant taxonomy, herbaria play other roles in plant conservation efforts. The simple act of correctly identifying a plant species can assist for conservation, providing the language (Latin names) for biologists of far-flung regions and ethnicities to be able to communicate accurately. Herbaria facilitate taxonomic studies by enabling collections from diverse habitats and localities to be studied in one place. Plant species vary in size and shape across their geographical range, and this variation can be observed and studied easily in a herbarium. Herbaria also house voucher specimens from scientific studies where the correct identification of a plant may serve as a tool to teach the next generation of field biologists. Although considered as relics of the past by some biologists,

Continued

THE VALUE OF HERBARIA FOR PLANT CONSERVATION–*cont'd*

Figure 1   Researchers employ herbaria to document plant diversity distribution and facilitate their project needs; often herbaria data reveal ecological associations in forest canopies. Photograph by H. Bruce Rinker.

herbaria help us understand and conserve life on earth when they complement *in situ* management practices.

References

Holmgren, P.K., Holmgren N.H., and Barnett, L.C. (1990). "Index Herbariorum, Part I: The Herbaria of the World." 8th Edition. The New York Botanical Garden, New York.

Creating Forest Service Markets

Finding out what economic values are is only the first stage of an economic approach to forest conservation. Hardly anyone is going to be persuaded to change his or her decision about the use of forest land unless conservation is associated with some form of cash or real benefit flow. Market creation is the process of converting the "paper" values in a cost–benefit analysis into real benefit flows. There are various ways of classifying market creation schemes. The essential feature of all market creation activities is that payment is made directly for a service or "function" of the environmental asset in question. Assets may range from entire ecosystems to specific wildlife species. Payments are made by beneficiaries or agents for beneficiaries, with beneficiaries being defined as those who would lose if the service is degraded or not provided. Payments are made to owners of assets or to anyone involved in a conservation effort. Table 24-3 shows a taxonomy. This taxonomy excludes a second form of market creation that arises when those doing the damage are taxed or subjected to some form of regulation. For example, a tax on forest conversion would also be a form of market creation, but this is not developed further here.

Some of the most innovative examples of market creation can be found in Costa Rica. Costa Rica has given official recognition to the role that forests play in (a) biodiversity conservation, (b) carbon storage, (c) watershed protection, and (d) scenic beauty for ecotourism. In 1996, Costa Rica adopted a new Forest Law whereby forest land owners can be compensated for the provision of the services (a)–(d). Financing for payments comes from a gasoline tax and is directed via the national Forestry Fund. Landholders can secure payment for reforestation, sustainable management practices, forest regeneration, and forest conservation. The payment schedule is outlined in Table 24-4.

The agreement is for five years and the assumption is that at the end of the period, such prices could be renegotiated or the rights could be sold to others. Forest conversion is forbidden so that the options facing a landowner of an existing forest are some form of unsustainable forestry (but which does not result in wholesale conversion) and sustainable forestry. Essentially, then, the $200 must compensate for the higher profits of unsustainable forestry. The reforestation payments appear to be more generous and probably would act as a strong incentive to reforest. The regeneration incentive is thought to be about equal to the rental price for pasture (i.e., giving an incentive to regenerate rather than lease the land for cattle). As of 1998, there was excess demand for the program, the capacity being determined by the availability of finance. There are also monitoring, verification, and compliance problems. In particular, noncompliance effectively requires slow and

Table 24-3   A Taxonomy of Market Creation through Beneficiary Payments

Beneficiary	Direct Use Values	Indirect Use Values	Nonuse Values
Local			
	National environmental funds	Beneficiary payments, e.g., water conservation fees payable to upland forest owners Differential interest rates on "green" credit	
International	Certification Ecotourism Bio-prospecting Information payments	Certification	Certification Voluntary funds
Global		Global Environment Facility	Debt for Nature Swaps Global Environment Facility

Table 24-4   Schedule of Payments for Forest Services in Costa Rica

Activity	Payment per Hectare over 5 years	Annual Schedule, % Payments per Year
Reforestation	$480	50, 20, 15, 10, 5
Natural forest management	$320	50, 20, 10.10, 10
Forest regeneration	$200	20, 20, 20, 20, 20
Forest protection	$200	20, 20, 20, 20, 20

From Chomitz et al. 1998

difficult court processes for breach of contract. Some NGOs have emerged as brokers in an effort to reduce the otherwise significant transaction costs involved in setting up the contracts.

Conclusions

Economic analysis suggests several ways to justify the slowing of the rate of deforestation. A fairly conventional case can be made in terms of identifiable and measurable costs and benefits, especially the value of forests as carbon stores. The more challenging, and perhaps more persuasive, approach is to focus on the under-researched issues of the economic value of biological diversity itself, as opposed to biological resources, and on the value of delaying forest loss (the quasi-option value argument). The latter is especially relevant to canopy research because of the greater uncertainty about the information value of canopies. Economics has some powerful arguments in favor of both forest conservation and canopy research.

References

Bahuguna, V. (2000). Forests in the economy of the rural poor: an estimation of the dependency level. *Ambio* **29**, 126–129.

Cavendish, W. (1999). Empirical Regularities in the Poverty-Environment Relationship of African Rural Households. Working paper 99–21, Centre for the Study of African Economies, Oxford University.

Chomitz, K., Brenbes, E., and Constantino, L. (1998). "Financing Environmental Services: The Costa Rican Experience and its Implications." World Bank. Washington DC.

Fisher, A.C. (2000). Investment under uncertainty and option value in environmental economics. *Resource and Energy Economics* **22**, 197–204.

Food and Agriculture Organization. (2001). "Forest Resources Assessment 2000." FAO, Rome.

Houghton, K. and Mendelsohn, R. (1996). An economic analysis of multiple use forestry in Nepal. *Ambio* **25**, 156–159.

Kramer, R., Sharma, N., and Munasinghe, M. 1995. "Valuing Tropical Forests: Methodology and Case Study of Madagascar." Environment Paper No.13, Washington DC: World Bank.

Kramer, R., and van Schaik, C.P. (1997). Preservation paradigms and tropical rain forests. *In* (Kramer R., van Schaik, C.P and Johnson, J., Eds.) (1997). "Last Stand: Protected Areas and the Defense of Tropical Biodiversity." New York: Oxford University Press, 3-14.

Lynam, T., Campbell, B., and Vermeulen, S. (1994). "Contingent Valuation of Multipurpose Tree Resources in the Smallholder Farming Sector in Zimbabwe." Paper 1994:8, Gothenburg University: Department of Economics.

Matthews, E. (2001). "Understanding the Forest Resources Assessment 2000." World Resources Institute, Washington DC.

Pearce, D.W. (2001). The economic value of forest ecosystems. *Ecosystem Health* **7**, 1–11.

Pearce, D.W and Pearce, C. (2001). "The Economic Value of Forest Ecosystems." Convention on Biological Diversity, Ottawa. http://www.biodiv.org/doc/publications.

Perrings, C., Mäler, K-G., Folke, K., Holling, C., and Jansson, B-O. (1995). Biodiversity conservation and economic development: the policy problem. In "Biodiversity Conservation" (C. Perrings, K-G. Mäler, K. Folke, C. Holling, and B-O Jansson, Eds.), Dordrecht. Kluwer. 3–22.

van Beers, C and de Moor, A. (2001). "Public Subsidies and Policy Failures: How Subsidies Distort the Natural Environment, Equity and Trade, and How to Reform Them". Edward Elgar, Cheltenham.

CHAPTER 25

Ecotourism and the Treetops

Margaret D. Lowman

*Around the world, ecotourism has been hailed as a panacea: a way to fund conservation
and scientific research, protect fragile and pristine ecosystems, benefit rural communities,
promote development in poor countries, enhance ecological and cultural sensitivity, instill
environmental awareness and a social conscience in the travel industry, satisfy and educate
the discriminating tourist, and some claim, build world peace.*
—*M. Honey*, Ecotourism and Sustainable Development, *1999*

Principles of Ecotourism

A new century of environmental consciousness is dawning. Under the pressures of explosive
human population growth, our planet's natural communities are shriveling rapidly. They are
shrinking on all sides due to the expansion of agriculture, urbanization, damming, logging, road
building, and even more indirect human impacts such as the invasion by exotic species and the
distribution of genetic crops.

As a tropical biologist for over 30 years, I have witnessed the impact of tourism on many relatively pristine tropical rainforests. In contrast, I have also witnessed the salvation of exploited
tropical regions by the interests of conservation and the economy of ecotourism working collectively. In the not-too-distant future, our wilderness areas will be small islands of biodiversity amid
seas of domesticated landscape. Ecotourism creates both an impact on natural ecosystems and a
salvation for the conservation of these regions. As the planet's natural ecosystems become increasingly rare, more people aspire to see what isolated populations of wildlife remain. In Nepal, ecotourists flock to hike one of the remaining wilderness regions on the planet; but hikers have
stripped the landscape bare of sticks and twigs for fuel and have left trash that spoils the experience for future visitors. In the Galápagos, burgeoning numbers of visitors strain these sensitive and
fragile islands. Disease, fire, and theft have altered the natural balance of the island ecosystems.

Ecotourism is loosely defined as nature-based tourist experiences where visitors travel to regions
for the sole purpose of appreciating natural beauty. As early as 1965, responsible tourism was
defined to respect local culture, maximize benefits to local people, minimize environmental impacts,
and maximize visitor satisfaction (Hetzer 1965). The first formal definition, coining the term "ecotourism," was published in 1987: "traveling to relatively undisturbed or uncontaminated natural
areas with the specific objective of studying, admiring, and enjoying the scenery and its wild plants
and animals, as well as any existing cultural manifestations (both past and present) found in these
areas" (Ceballos-Lascurain 1987; see also Weaver 2001). Subsequently, many other definitions have
arisen (e.g., Valentine 1990, 1992; Figgis 1993; Orams 1995; Perry 1995; Higgins 1996; Lowman
1998; Weaver 1998). Ecotourism probably had its foundations in the ethics of conservation, but its

recent surge has certainly been economic, as developing countries have begun to recognize that nature-based tourism offers a means of earning money with relatively little exploitation of resources. It is this economic incentive, perhaps more than the consciousness of human ethics, that has given rise to the global expansion of environmentally responsible tourism activities.

The objectives of ecotourism are to provide a nature-based, environmental education experience for visitors and to manage this in a sustainable fashion. These requirements have made it increasingly difficult to provide a true ecotourism experience—as forests become logged, as streams become polluted, and as other signs of human activity become ubiquitous. To compensate for the "invasion" of human disturbance, ecotourism has promoted the educational aspects of the experience, including such opportunities as working with scientists to collect field data in a remote wilderness (e.g., Earthwatch) or traveling with a naturalist to learn the secrets of a tropical rainforest (e.g., Smithsonian Institution travel trips). Environmental education serves to provide information about the natural history and culture of a site; it also promotes a conservation ethic that may infuse tourists with aspirations of pro-environmental attitudes. The question of sustainability remains untested since many sites have relatively new initiatives of nature-based tourism and the long-term impacts are not yet measurable. The challenges of removing trash from remote wilderness lodges, of constructing low-impact electric wires across a beautiful valley, or of minimizing the introduction of exotic bacteria to Antarctica require the test of time to determine their success.

Types of Ecotourism

In most cases, ecotourism follows two important principles of sustainability: to support conservation of the natural ecosystems and to support local economies (Blamey 2001). These underlying principles are the pillars that will provide a lasting basis for ecotourism and will also create sound economic support for the conservation of natural resources. They provide ultimately competitive reasons for the expansion of ecotourism above and beyond other types of leisure activities. A challenge arises, however, when ecotourism becomes successful in the sense that too many tourists will destroy the reason for success. In the case of ecotourism, as different from many other marketed products in our western economy, economic success is a matter of limiting supply no matter how much the demand.

In just 20 years, this type of recreation has burgeoned to include many different intensities and levels of experiences (reviewed in Orams 2001). There is soft versus hard ecotourism, alluding to the physical rigor of the conditions experienced by the visitors. Trekking on the Inca Trail is much more rigorous than visiting Machu Picchu and staying in the lodge. There are arguments over the natural versus unnatural versions of ecotourism; in other words, proponents of ecotourism believe that humans are part of nature and their impact is part of the natural process, whereas critics of ecotourism uphold that people simply should not visit natural areas since they invariably degrade them. Ecotourism can be passive (viewing the Grand Canyon), active (rafting down the Colorado River), or exploitive (staying in the lodge on the rim of the canyon). And ecotourism can be mass tourism (where maximization of income is the most important factor, and expanding programs are measures of success) versus alternative tourism where environmental sustainability (therefore limiting the number of tourists) is the most important measure of success. In most cases, a continuum of both economic and ecological incentives results in many varying levels of ecotourism.

Global Impact of Ecotourism

Since Thomas Cook began the world's first travel agency in 1841 (Gartner 1996), the number of people who enjoy organized travel has continued to increase. Today, an estimated 1.6 billion people from all cultures and all walks of life participate in different avenues of tourism, spending over US $2 trillion (Hawkins and Lamoureux 2001). On a global scale, ecotourism is growing because of its international appeal. Tourists recognize that if they travel with sensitivity to the environment, they will not only contribute to conservation but also become educated about a new habitat, country, or culture.

Eco-labels or certification for authentic nature-based tourism have been established in some countries, such as *Green Globe* (international) or *Committed to Green* (Europe). This enhances the credibility of an experience for tourists who come to know specific reputable names. In contrast, some negativity in certain countries will affect tourism negatively, such as the 2002 bombing in Bali or the reputation of drug-trafficking in Colombia. Exotic countries with stable governments such as Belize are the true beneficiaries of ecotourism in contrast to some of their neighbors who cannot offer the same reliability of ecotourist experience. As politics continue to affect our ability (or lack thereof) to travel, regional tourism will be significantly affected by local government and policies related to stability.

Case Studies of Canopy Ecotourism

The Aerial Tram in Costa Rica

Many see Costa Rica as a shining example of conservation, but it is neither better nor worse than many other countries. Banana cultivators, lumbermen, and farmers are stripping those lowlands of rainforest trees. At this time, Costa Rica has proportionally less rainforest left than many other tropical countries. In fact, only a few islands of relatively untouched lowland Caribbean rainforest remain. Less than 10 percent of Costa Rica's forests are national parks. Moreover, both the national park service and FUNDACORE, a nonprofit organization devoted to the conservation of national parks, acknowledge there is not enough money to protect the national parks and their boundaries from encroachment and destruction (Perry, pers. comm.).

While Costa Rica is busy cutting the remaining 10 percent of its forest outside the parks, there is concern that Costa Rica will then begin to harvest timber inside the parks. At the current rate, non-park forests will only last five to 10 years in this country that supposedly represents the bastion of conservation in Latin America (Perry 1995).

Travel to Costa Rica to admire a country with tropical rainforests is popular. As tourism grows, the determination of Costa Ricans to protect their natural resources will also grow. The Aerial Tram had its origin in Don Perry's efforts to devise tree-climbing techniques for scientific investigation. After experiencing the limitations of single-rope techniques, he realized that to study the canopy effectively, researchers needed a vehicle for access. In 1982, the development of this vehicle became his primary objective. In 1983, Perry teamed up with the engineering expertise of John Williams, and together they created the Automated Web for Canopy Exploration (AWCE) at Rara Avis in Costa Rica. This device is composed of a power and winch station, support and control cables, and a radio-controlled steel platform that holds up to three people. The support cable spans a forested canyon and is about 300 m in length. The platform is suspended from the support cable and can move along its length. It can also carry scientists from ground level to above the treetops through approximately 22,000 cubic meters of forest. This was constructed at a cost of approximately $1.82 per cubic meter of access. Because the support cable is stationary, the AWCE is a linear system. It is also a prototype for a canopy vehicle for researchers to investigate the treetops.

Subsequent to the AWCE was the construction of the Aerial Tram, closer to San Jose and more suitable for ecotourists. The tram occupies a 1.3-km route through pristine lowland rainforest, and 24 cars each hold up to six people, including one guide. The cars are attached to a cable that rotates around two end stations—in actuality, the system is a converted ski lift.

Approximately 70 people per hour are carried through the canopy with an estimated 40,000 visitors per year (Perry 1995). Free rides are donated to students, and Costa Ricans now own and operate the aerial tram ecotourist operation. Displays at the site educate visitors about the tropical rainforest canopy and its inhabitants. This site is the first of its type worldwide, but more aerial trams and canopy walkways are now in operation in other countries such as Australia, Peru, Panama, and Florida, USA.

Climbing of Age in Samoa: Canopy Walkways for Conservation

Paul Cox, internationally recognized ethnobotanist, has worked for many decades on the islands of the South Pacific. In particular, Cox spent many years in the village of Falealupo on the island of Savai'i. He befriended the village healer, a wonderful botanist named Pele, who taught him how Samoans use their local plants as an apothecary. In the early 1990s, Cox partnered with Canopy Construction Associates on an ecotourism project that now serves as a model for the application of canopy research to conservation (Lowman et al. in press).

A monsoon demolished the school and many other buildings in Samoa during the early 1990s. As a consequence, the Samoan government mandated that all villages must build (or rebuild) schools for their children, and they must be solid structures made from cement. The cost of construction for the school at Falealupo was estimated at $65,000. The village had no cash economy since the local people fished for food and carefully harvested plants for medicines, clothing, and shelter. The per capita economy was less than US $100 per year, so the notion of paying for a school was beyond the villagers' comprehension.

The village was offered a large sum of money in exchange for logging rights to their forest. As an island, the forest composition was unique and the likelihood of restoration of the original forest composition was nil. Island ecosystems are chance events since the combination of species collects via drift, wind, and bird dispersal. In addition to the unusual diversity of an island ecosystem, the Samoan forests were sacred to the people. The villagers depended on the forest for everything—food, clothing, medicines, and homes. The forests nurtured even their ancestors, whose spirits were embodied in the flying foxes (*Pteropus samoensis*) that lived and bred within the forest. Spiritual, economic, biological—the forest provided all the needs of the village and had done so for many generations. The chiefs were not happy about the proposal to log their forest, but they needed to pay for their school.

Paul Cox had a novel suggestion for the chiefs. What about developing ecotourism to bring in a cash economy yet manage the forest sustainably? He suggested a canopy walkway to attract tourists, who would pay for the privilege of walking in the treetops of Samoa. A loan was obtained and some generous seed funding was offered from Seacology Foundation in the United States.

Three staff from Canopy Construction Associates (Meg Lowman, Bart Bouricius, and Phil Wittman) joined Cox for a reconnaissance trip to determine the feasibility of an ecotourism walkway in Falealupo. The walkway represented a relatively radical idea for the village. Not only was the school debt large and unprecedented, but also the notion of encouraging tourists to visit was a new and perhaps frightening notion. The walkway team was led into the center of the meetinghouse and met with fifteen village chiefs.

They honored the team with a kava ceremony. In the center of their circle was a large wooden bowl with a muddy liquid. The ceremonial drink, kava, was made from the roots of a tropical shrub (*Piper myristicum*), an important medicinal plant in the South Pacific. (I called it the consensus plant because after drinking several cups, all 15 chiefs voted "yes" to the walkway.)

Canopy Construction Associates spent many hours with the chiefs, with the schoolteachers, and with others in discussion and field reconnaissance for the canopy walkway. At long last, we agreed that an enormous emergent fig would represent the center of all construction from which a bridge could be spanned to adjacent trees.

A CLIMB FOR CONSERVATION

Stephen R. Madigosky

Approximately 160 km down the Amazon River and up the Napo River from Iquitos, Peru, the Amazon Conservatory for Tropical Studies (ACTS Field Station) is managed by the Peruvian non-profit organization CONAPAC (Conservacion de la Naturaleza Amazonica del Peru, A.C.).

Constructed in one of the most biologically diverse forests in the world, the ACTS serves as an open laboratory for tropical research, educational initiatives, workshops, and sustainable development projects that promote sound conservation practice throughout the region. The large thatched buildings at the conservatory provide both laboratory and rooming facilities for scientists as well as overnight lodging for tourists. A donation to maintain and enlarge the surrounding reserve and canopy walkway is included in the tariff paid by tourists for each night spent at the ACTS. Station facilities include a kitchen, dining area, the Alwyn H. Gentry Research Laboratory, a conference room, and 20 guestrooms each with two twin beds covered by mosquito netting. Solar batteries and a gasoline generator provide for the station's electrical needs. A short hike from the field station is the ACTS aerial canopy walkway network, more than 500 meters long and 36 meters high (see Figure 1). The walkway consists of a series of single and double platforms and bridges that connect more than a dozen trees, including emergents, in the Upper Amazon of Peru. It was constructed over a three-year period and officially opened in 1993. In addition to the canopy walkway, extensive trails in over 250,000 hectares of primary rainforest are available for exploration by visitors and scientists alike. The facility maintains a 5-ha medicinal plant garden that contains over 240 species of native plants valued for their pharmaceutical use and potential. Requests for research guidelines and applications for the ACTS Field Station may be made to Director of Research, ACTS Scientific Advisory Board, Widener University, One University Place, Chester, PA 19013.

Figure 1   ACTS aerial canopy walkway network. Photograph by Donna J. Krabill.

As predicted, the planning phases in the U.S. involved hundreds of hours of telephone time among the building team and a good deal of speculation about issues at the site that could not be double-checked. Some of the builders had grave concerns about liability: What if a storm hit the new walkway? Who would inspect the site each year? What if some equipment was damaged during shipment? In the case of an accident, how would a builder be evacuated from this remote island? What was the timetable for completion, and how would weather affect it? The logistics of the builders became more difficult than dealing with the chiefs, who spoke a different language altogether. In the midnight hour, Canopy Construction Associates handed the job over to another smaller construction team, Kevin Jordan and Stephanie Hughes, who were willing to take on the risks of a weather-controlled timetable, no liability, and innovation if some equipment were lost or damaged. The end result is a wonderful walkway (see Figure 25-1) that is contributing to the economy of the village of Falealupo in a sustainable fashion and insuring the conservation of their precious forest for future generations.

Biosphere 2 Canopy Access System

Most canopy construction in natural forests has involved estimating the load-carrying capacity of trees based on visual observation, judgment, and professional experience. A number of such systems have been designed and utilized for research and ecotourism over the past 15 years

Figure 25-1 Canopy Walkway in the village of Falealupo on the Samoan island Savai'i. Photograph by Stephanie Hughes.

FLORIDA FROM THE TREETOPS

Paula J. Benshoff

The Myakka Canopy Walkway spans 26 m through a subtropical hammock in a popular state park outside Sarasota, Florida. Though the aerial suspension bridge is just 7.6 m above ground level, the height is optimum to explore this typically low hardwood hammock canopy of the south Florida peninsula. A 23-m tower at one end provides access to all levels of the canopy and a view of the Myakka River, Upper and Lower Myakka Lakes, and the hammock/prairie interface (see Figure 1).

The project (touted as a vehicle for research, education, and ecotourism) is the result of an alliance among Marie Selby Botanical Gardens, TREE Foundation, Friends of Myakka River, and Myakka River State Park. It was financed primarily with funds from local foundations and benefited from contributions by community and service clubs, local businesses, park visitors, and school groups. Funding was also paired with another project to send

Figure 1   Views of the Myakka Canopy Walkway near Sarasota, FL, USA.

Continued

FLORIDA FROM THE TREETOPS–*cont'd*

24 disadvantaged students to visit the Amazon Conservatory for Tropical Studies (ACTS), a canopy walkway downriver from Iquitos, Peru. Each student raised $100 for the trip and another $100 for a donation toward the construction of the Florida walkway.

The structure was designed and primarily constructed by Canopy Construction Associates (Amherst, Massachusetts), a group that specializes in building walkways and towers without the use of cranes or heavy equipment. All tools and materials were transported to the site along a winding nature trail via a small pickup truck. Park volunteers contributed 47 percent of the 2,571 hours logged on the project, which kept construction costs within budget. Total expenditures were $98,860. A $15,000 endowment fund was also established for future maintenance and yearly inspections.

Acquired in 1934, Myakka River State Park is one of Florida's largest and oldest state parks. Fourteen miles of the "Florida Wild and Scenic" Myakka River flow through the park; and two shallow lakes attract large numbers of wading birds, migrating waterfowl, and shorebirds. Approximately 45 percent of the 14,973 hectares under park management is globally imperiled Florida dry prairie with a history of decades of fire exclusion. Park research efforts usually prioritize fire ecology, uplands restoration, and hydrological monitoring. Little attention has been devoted to studying the oak/palm hammocks, inventorying their inhabitants, or analyzing system functions. The canopy walkway focuses awareness on this subtropical forest system.

The walkway traverses a canopy of live oak *(Quercus virginiana)*, laurel oak *(Quercus laurifolia)*, and sabal palm *(Sabal palmetto)*. Other trees in the park's hardwood hammocks include elm *(Ulmus americana)*, sugarberry *(Celtis laevigata)*, willow *(Salix caroliniana)*, Carolina ash *(Fraxinus caroliniana)*, snowbells *(Styrax americana)*, water locust *(Gleditsia aquatica)*, and red maple *(Acer rubrum)*. Canopy epiphytes include six species of *Tillandsia*, two orchids, and four ferns. The hammock lies within the Myakka River floodplain and floods at least annually.

Over 250,000 people a year visit the park, and it is highly utilized by primary schools and universities as an outdoor laboratory for studying the state's unique natural communities and their inherent natural processes. Since completion in June 2000, the walkway and tower have become very popular for both ecotourism and education. A 33 percent increase in park visitation over the first summer and fall was attributed to the walkway. The walkway's popularity has not waned, as evidenced by the nature trail's continuously overflowing parking lot. It is also a primary destination of school buses and tour buses entering the park.

The walkway proved its practical value with a shocking discovery a few months after it opened (Benshoff 2002). An exotic weevil from Central America, accidentally released in Ft. Lauderdale about 1990, arrived at Myakka. *Metamasius callizona* lays its eggs in tank bromeliads, and the larvae consume the heart of the plant, killing the epiphyte before it can flower and reproduce. Ten species of native bromeliads are now considered endangered or threatened by the State of Florida because of the weevil. The weevil's discovery at this popular park and ensuing associated monitoring projects gained media attention. The publicity brought this cryptic insect into public view, making it easier for the University of Florida to obtain funding for developing a biological control. If the project is successful, the park could be the first release site for a biological control to save keystone species of Florida hardwood forests.

Another benefit of the canopy walkway project has been the ongoing partnership between the park, New College, and the TREE Foundation. A research station is planned that will provide opportunities for independent researchers to study subtropical canopy ecosystems. It also serves as an outdoor laboratory for local students. Meanwhile staff members at the park and the TREE Foundation in the development of interpretive brochures and signage to deliver their shared conservation message to the world.

Reference

Benshoff, P.J. (2002). "Myakka." Pineapple Press, Sarasota, Florida.

(Lowman and Bouricius 1995; Lowman and Wittman 1996). In most forest situations, canopy access structures utilize tall canopy tree trunks for structural support. In Biosphere 2, however, such mature trees did not exist in the rainforest biome, so the canopy access system needed to rely on support from the structural members of the glass pyramid itself. Access to the canopy also required a simple model with minimal light reduction, little or no impact on the vegetation, and maximum accessibility for researchers with safety compliance under U.S. Occupational Safety and Health Administration (OSHA) standards (Leigh et al. 1999).

The Biosphere 2 canopy access system is the first construction of its kind that utilized OSHA Standards for Fall Protection in its design (Marino and Odum 1999). Details of the canopy access system are described elsewhere (Grushka et al. 1999). In summary, the OSHA standards were achieved by the installation of a fall arrest system utilizing the strength of the supporting space-frame. A double cable system was erected that allowed a transect for the researcher and offered the safety element of the OSHA standards. Both horizontal and vertical access to all major tree canopies within the rainforest biome were possible from the cables, and the lateral load was tested to 5000 lb on the frame (if a climber fell), as per OSHA requirements. With this guaranteed level of safety, Biosphere 2 represented a unique contribution to the field of forest canopy research. All key components of equipment for the fall arrest system met with the standards of the American National Standards Institute, and were tested under dynamic conditions where possible. Specific equipment is described in Grushka et al. (1999) and can also be viewed by ecotourist visitors who are now allowed to tour the Biosphere 2 site near Oracle, Arizona. The staff employ a rigorous schedule of maintenance training for users of the canopy access system, and a strict requirement that all climbers have a group support person for extra safety. To date, gas exchange and other detailed data collection has been safety employed from the Biosphere 2 canopy access system.

Wheelchairs in the Canopy: An ADA-Sponsored Walkway

It was a lifelong aspiration for me to create canopy access for the public without fees or restrictions. At the Marie Selby Botanical Gardens, a handicapped-accessible canopy walkway was constructed in 1998, and in Myakka River State Park, a public canopy walkway and tower opened in June 2000. Both of these projects came from public funding and awareness of the importance of bringing people into forests.

The Selby Gardens walkway was the inspiration of a local builder, Michael Walker Associates. Walker donated a children's exhibit for a Gardens' theme-based activity whereby children's books were brought to life. Walker modeled his construction after the children's book, *The Most Beautiful Roof in the World,* by Kathryn Lasky. Walker carefully built a child-friendly structure to ADA-regulations, so that anyone—in a wheelchair, with a stroller, or with a walker—can enjoy the treetops. The bridge climbs into a fig tree on the Gardens' campus and was the site of the nation ADA 10th anniversary celebrations in 2000. A group of delegates in wheelchairs cut the ribbon, officially opening the walkway. Many tears of joy were shed by these wheelchair-bound adults who had always dreamed of "climbing" a tree. The Myakka River State Park walkway was built to educate the public about the importance of forest conservation and is open to the public (Lowman et al. in press; see Sidebar). It has received national acclaim and has been featured on several educational television programs.

Ecotourism, in partnership with research, has the potential to affect forest conservation significantly and in many positive ways. The influx to local economies, as well as their education potential, make canopy walkways a unique solution to potential deforestation trends in many regions.

References

Blamey, R.K. (2001). Principles of ecotourism. *In* "The Encyclopedia of Ecotourism" (D.B. Weaver, Ed.), pp. 5–22. CABI Publishing, New York.

Ceballos-Lascurain, H. (1987). The future of ecotourism. *Mexico Journal* January, 13–14.

Figgis, P. (1993). Ecotourism: special interest or major direction? *Habitat Australia.* February, pp. 8–11.

Gartner, W.C. (1996). "Tourism Development: Principles, Processes and Policies." Van Nostrand Reinhold, New York.

Grushka, M.M., Adams, J., Lowman, M., Lin, G., and Marino, B.D.V. (1999). The Biosphere 2 canopy access system. *In* "Biosphere 2—Research Past and Present" (B.D.V. Marine and H.T. Odum, Eds.), pp. 313–321. Elsevier Publishers, Ireland.

Hawkins, D.E. and Lamoureux, K. (2001). Global growth and magnitude of ecotourism. *In* "The Encyclopedia of Ecotourism" (D.B. Weaver, Ed.), pp. 63–72. CABI Publishing, New York.

Hetzer, W. (1965). Environment, tourism, culture. *Links*, July, 1–3.

Higgins, B.R. (1996). The global structure of the nature tourism industry: ecotourists, tour operators, and local businesses. *J. Travel Research* 35(2), 11–18.

Honey, M. (1999). "Ecotourism and Sustainable Development: Who Owns Paradise?" Island Press, Washington, DC.

Lasky, K. (1997). The most beautiful roof in the world: exploring the rainforest canopy. GulliverGreen/Harcourt Brace & Company, New York.

Leigh, S.L., Burgess, T., Wei, Y.D., and Marino, B.D.V. (1999). Tropical rainforest biome of Biosphere 2: structure, composition and results of the first 2 years of operation. *Ecol. Eng.* 13, 65–93.

Lowman, M.D. (1998). "Life in the Treetops." Yale University Press, Princeton, New Jersey.

Lowman, M.D. and Bouricius, B. (1995). The construction of platforms and bridges for forest canopy access. *Selbyana* 16(2), 179–184.

Lowman, M.D. and Wittman, P. K. (1996). Forest canopies: methods, hypotheses, and future directions. *Ann. Rev. Ecol. Syst.* 27, 55–81.

Lowman, M.D., Burgess, J., and Burgess, E. (In press). "Out on a Limb—More Life in the Treetops." Yale University Press, New Haven, CT.

Marino, B.V.D. and Odum, H.T. (Eds.) (1999). Biosphere 2—Research Past and Present. Elsevier Publishers, Ireland.

Orams, M.B. (1995). Toward a more desirable form of ecotourism. *Tourism Management* 16(1), 3–8.

Orams, M.B. (2001). Types of ecotourism. *In* "The Encyclopedia of Tourism." (D.B. Weaver, Ed.), pp. 23–36. CABI Publishing, New York.

Perry, D. (1995). Tourism, economics, and the canopy: the perspective of one canopy biologist. *In* "Forest Canopies" M.D. Lowman and N.M. Nadkarni, Eds.), pp. 605–609. Academic Press, San Diego.

Valentine, P.S. (1990). Nature-based tourism: a review of prospects and problems. *In* "Proceedings of the 1990 Congress on Coastal and Marine Tourism. Volume 2." (M.L. Miller and J. Auyong, Eds.), pp. 475–485. National Coastal Resources Research and Development Institute, Coravallis, Oregon.

Valentine, P.S. (1992). Review: nature-based tourism. *In* "Special Interest Tourism" (B. Weiler and C.M. Hall, Eds.), pp. 105–128. Belhaven Press, London.

Weaver, D.B. (1998). "Ecotourism in the Less Developed World." CABI Publishing, Wallingford, UK.

Weaver, D.B. (Ed.) (2001). "The Encyclopedia of Ecotourism." CABI Publishing, New York.

CHAPTER 26

Reintegration of Wonder into the Emerging Science of Canopy Ecology

H. Bruce Rinker and David M. Jarzen

No illumination can sweep all mystery out of the world. After the departed darkness the shadows remain.
—*Joseph Conrad,* The Arrow of Gold, *1919*

Introduction: A View from the Treetops

The word "canopy" is historically a botanical term and, for this chapter, refers to all aboveground vegetation in a plant community (Nadkarni 1995; Parker 1995; Moffett 2000). According to scientists in the field, each plant community has a canopy (Seastedt and Crossley 1984). Individual trees can have canopies (see Lowman 1995b; Reynolds and Crossley 1997; Sillett and van Pelt 2000) even though, in normal usage, trees have crowns and forests have canopies. A temperate forest and a tropical forest each have canopies. Technically speaking, even an orchard, a lawn, a golf course, and a kelp forest have canopies. This systems-wide term includes plants and all their aboveground associations. Contrast this with the word "crown" which, in the parlance of professional forestry, refers exclusively to the upper part of a tree, not to its attending flora – including epiphytes and lianas – and fauna (Winters 1977) and to its ecological processes. The term "canopy" denotes community architecture as well as species composition, nutrient cycling, energy transfer, plant-animal interactions, and conservation issues from the ground to the community-atmosphere interface for all plant assemblages. For the purposes of this chapter, canopy refers specifically to forest ecosystems.

Why study the canopy ecosystem? It is an unexplored frontier for scientific research and education. It is also a living laboratory. Most of the world's estimated 30 million species live in the treetops because the canopy is also that layer of forest containing most of its productive tissue. Like gigantic stands of lollipops, temperate and tropical forests are sugar factories that have the bulk of photosynthetically active foliage and biomass high off the ground. Sugar in the treetops means the presence of sugar-eaters, too, along with their predators, parasites, decomposers, and most of the forest biota. To study the canopy ecosystem is to enter a leafy frontier for discovery and opportunity.

Researching the treetops is also vital for forest conservation. "Understanding the canopy as part of whole-ecosystem processes is an obvious priority if we are to responsibly manage and conserve forests in the future" (Nadkarni 1995). A view from the treetops provides us with a much more integrated perception of forest ecology than a ground-based view. In fact, the study of life in the canopy is a key to the whole of forest ecology (Nadkarni 1995). A treetops vista is a stunning reordering of spatial and temporal perspectives on the workings of the forest community. The late British ecologist J.B.S. Haldane once wrote, "Our only hope of understanding the universe is to look at it from as many different points of view as possible" (Haldane 1928). Whether by rope, walkway, or crane, a canopy scientist—rising vertically through the vegetation—sees the forest from the perspectives of bird and biped, as Alexander Skutch (1992) encouraged us more than a quarter-century ago: "To know the forest, we must study it in all aspects, as birds soaring above its roof, as earth-bound bipeds creeping slowly over its roots." Such an effort sweeps away some of the mystery, but not the wonder, of our temperate and tropical woodlands to bring them fully within the devoted scrutiny of modern science.

Until recently, research on forest ecology was primarily ground-based and uninformed about canopy processes. Only during the past 25 years has our understanding of treetop ecology expanded substantially beyond this bipedal bias, due in large part to the dauntless efforts of a handful of tropical biologists working from ropes, walkways, airships, cranes, canopy webs, and towers sometimes 30 or 40 m off the forest floor (Rinker et al. 1995). These access systems now permit researchers to study forests from top to bottom, seeing them as integrated living systems, not as an illustrated series of discrete vegetative strata too often simplified in general biology textbooks.

The metaphors in the early scientific literature for the forest canopy were resplendent, atypically poetic, and capture some of our childlike intrigue with this borderland of science (Rinker 2000, 2001): tropical air castles, canopy oceans, hanging gardens, green mansions, aerial continents, highways in the trees, the eighth continent. Up there we were explorers in the roof of the world's forests and—until we could observe, quantify, replicate, and collect in this leafy realm—we waxed romantic, unsure about the outcomes of our enquiry but convinced of its merits. Often quoted in the early canopy literature was an observation from William Beebe's *Tropical Wild Life in British Guiana* (Beebe et al. 1917): "Yet another continent of life remains to be discovered, not upon the earth, but one or two hundred feet above it, extending over thousands of square miles. . . . There awaits a rich harvest for the naturalist who overcomes the obstacles—gravitation, ants, thorns, rotten trunks—and mounts to the summits of the jungle trees." A rallying cry for canopy scientists, Beebe's prose compressed all the reasons for canopy research into two poignant words: *discovery* and *opportunity*. These words registered on the heartstrings of every biologist, teacher, and student fortunate enough to enter the forest canopy. The canopy was a New World for us—or, more accurately, an Old World that our remote ancestors left behind millions of years ago. The treetops seem a siren song for our genetic memory. Now we return to the forest canopy, this time as researchers with probes, microscopes, and notebooks, but not so much an invasive return to our homeland as it is a search for our ancestral roots in the trees. This aerial eighth continent has much to teach about our primeval selves.

The 1990s represented the defining decade for canopy ecology, though, arguably, the science is little more than a quarter-century old. In its early days, the literature was filled with poetic metaphors in order to capture some of our intrigue with this borderland of science. We also used this language of wonder to attract the attention and imagination of readers around the world. Since then, and especially since the 1990s, we have removed much of the poetry and replaced it with language typical of established biological fields. In this chapter, we argue that wonder can be reintegrated into canopy ecology without jeopardizing its legitimacy or rigor in order to promote discovery and opportunity in the treetops.

A Sense of Wonder in the Treetops

Wonder is a rational response to the sublime. To borrow from the medievals, it is *semen scientiae* or seed of knowledge (see Heschel 1955). It is, however, more than just a precursor to knowledge; it is a form of thinking, an unceasing attitude toward the immediately unknowable. It is

THE INTERNATIONAL CANOPY NETWORK

Nalini M. Nadkarni

The International Canopy Network (ICAN) was established as a non-profit in 1994 to enhance communication among individuals and institutions concerned with research, education, and conservation of organisms and interactions in forest canopies. ICAN was formed in response to a perceived need to bring together the diverse group of people who were dedicating themselves to the emerging field of studies and conservation about the then little-known part of forest ecosystems (Nadkarni et al. 1995). A small cadre of international scientists and students created a tax-exempt corporation with 501(c)3 status and the networking capacity of electronic media. Although the initial focus was on bringing together canopy researchers in academia, its interests have included outreach, education, and conservation.

One of the first core activities of ICAN was to facilitate rapid communication by establishing and maintaining an electronic mail bulletin board. The logistics have been handled by the Network Office of the Long-Term Ecological Research Program, sponsored by the National Science Foundation. In 2003, its subscribership was over 750 people, living in 62 countries. The bulletin board is used to circulate information about meetings, job openings, new research findings, and recent publications of concern to canopy researchers. It has also been used for information exchange (e.g., on equipment and canopy access methods) and for lively debates (e.g., on the definitions of canopy-related terminology).

The ICAN also publishes a quarterly newsletter (titled "What's Up?") to members. Prior to publication of each issue, the editor calls for articles and items of interest. Occasionally, the newsletter publishes sets of research abstracts from recent projects. Conservation and education programs are also featured. One of the most critical features of each issue is a list of recent scientific citations from ecological and environmental journals. Because canopy research is scattered throughout the literature, this provides a focal ground for researchers to easily keep in touch with current work. ICAN also organizes and co-sponsors scientific symposia, usually in conjunction with larger ecological meetings, using the email bulletin board and newsletter as a springboard for efficient communication.

Another major activity of the ICAN office is the maintenance of a website. A key feature is the bibliographic database of scientific and popular aspects of canopy science. In 2003, over 4,500 keyworded citations were listed, searchable by 18 categories. The website also contains a growing images library; electronic images can be downloaded directly. Other parts of the website include a canopy researcher directory, information on safe methods of canopy access and canopy sampling methods, and interpretive materials for school children.

Because of the pressing need to conserve forest canopy biota beyond the current generation, the ICAN has instituted a number of children's programs. Staff members write articles about canopy plants and animals for children's magazines and give talks to school groups, with a focus on 8- to 10-year-olds. In 1998, the ICAN initiated the "Ask Dr. Canopy!" program, in which children are invited to write or email questions about the canopy or forests in general to Dr. Canopy. They get responses to their queries from the "collective persona" of six volunteer canopy researchers.

Other outreach efforts to the general public have involved writing numerous articles for popular magazines (e.g., *Natural History, National Geographic, Audubon, Glamour*) and newspapers. An extremely important function of ICAN is consulting to the media and press. The topic of rainforest canopies has been a highly appealing one for television documentaries, and the ICAN has made strong efforts to provide scientifically sound information and review of material that goes into such programs (e.g., *Heroes of the High Frontier*, National Geographic Society).

Although the mission of the ICAN as described in the bylaws does not include political activities, members have carried out programs to heighten awareness among politicians and other non-scientists on the importance of canopy biota and processes. For example, in 1997, the ICAN spearheaded an effort to get the signature of the Governor of Washington State on a Proclamation to declare July 17–24 as "Washington Forest Canopy Week." Similar symbolic efforts to link the canopy with applied issues to the general public have also been carried out, such as the "Legislators Aloft Program" in 2002, in which ICAN members taught 12 Washington State members of congress to climb into canopy-level platforms and discussed forest management and conservation.

The Board of Directors of ICAN consists of eight members who represent the constituent fields of research, education, conservation, and arboriculture. The Directors meet annually to provide oversight on short- and long-term directions. The Scientific Advisory Council, which consists of 15 members, takes part in decisions on an ad hoc basis.

ICAN is a self-supporting organization, funded by subscriber dues, donations, and grants. For more information on the ICAN, contact Nalini Nadkarni at nadkarnN@evergreen.edu.

Reference

Nadkarni, N.M., Parker, G.G., Ford, E.D., Cushing, J.B., and Stallman, C. (1996). The International Canopy Network: a pathway for interdisciplinary exchange of scientific information on forest canopies. *Northwest Science* **70**, 104–108.

what makes us "Stand still and consider" (Job 37:14). The opposite of wonder is the indifference that emerges from habituation to the laws governing one's world picture: "The reason why the adult no longer wonders is not because he has solved the riddle of life, but because he has grown accustomed to the laws governing his world picture. But the problem of why these particular laws and no others hold, remains for him just as amazing and inexplicable as for the child" (Max Planck in Heschel 1955). Solar eclipses and blood circulation may no longer be mysteries to 21st century society, but for most scientists—theists, agnostics, and atheists alike—they are still wonder-filled phenomena because of their sublime beauty and their connection to still other mysteries. Indifference to these phenomena is stifling, irrational, and shallow.

Many contemporary authors have decried the attenuation of wonder that often occurs in the aging of both individuals and societies. For example, Rachel Carson (1907–1964), a prophet for the modern environmental movement, noted this sad change as young people grow up: "A child's world is fresh and new and beautiful, full of wonder and excitement. It is our misfortune that for most of us that clear-eyed vision, that true instinct for what is beautiful and awe-inspiring, is dimmed and even lost before we reach adulthood" (Brooks 1972). Abraham Joshua Heschel, internationally recognized Jewish scholar and theologian, observed the same diminution in society at large: "As civilization advances, the sense of wonder declines. Such decline is an alarming symptom of our state of mind. Mankind will not perish for want of information; but only for want of appreciation. The beginning of our happiness lies in the understanding that life without wonder is not worth living. What we lack is not a will to believe but a will to wonder" (Heschel 1955). Rather than a necessary precursor to *maturity*, the diminution of wonder, whether in individuals or society at large, may actually be a vestige of our *immaturity* as we advance.

So, too, has this diminution occurred time after time in the sciences. Thomas Aquinas, the great Dominican apologist for the medieval Church, wrote in his *Summa contra gentiles*, that "[T]he astronomer does not wonder when he sees an eclipse of the sun, for he knows its cause, but the person who is ignorant of this science must wonder, for he ignores the cause. And so, a certain event is wondrous [*mirum*] to one person, but not so to another" (in Daston and Park 1998). What Aquinas seems to have overlooked in his erudition is that, for the sciences, a mystery solved is a wonder gained. In other words, the explanation of natural phenomena leads endlessly to more questions and more uncertainties about the universe. Like circus mirrors, this "intricate minuet of wonder and curiosity" (Daston and Park 1998) channels scientists into deeper and deeper mystery. Astronomers may understand eclipses, but then what do they know about solar flares and black holes and, after these, the expansion of the cosmos? Science is an indefatigable engine for wonder.

Forest biologists have always been curious about processes in the treetops. Yet, until the 1970s, they seemed to be content with *ex situ* investigations: collecting a beetle or a bird here, examining a fallen branch there that happened to be loaded with lofty epiphytes. *In situ* canopy studies began a quarter-century ago, especially with Donald Perry's canopy web in Costa Rica for which he later earned a Rolex Award and front-page coverage in *Scientific American* (Perry 1984). Because early canopy biology focused on access techniques and descriptive documentation of poorly known arboreal flora and fauna, many academics viewed the field as a "Tarzan and Jane" frill or a throwback to 19th-century descriptive biology (Nadkarni 1995), similar to the deprecating accusations by Rutherford (1871–1937) about biologists and postage-stamp collecting (see Mayr 1988). The 1990s, however, represented the defining decade for treetops ecology (Nadkarni 1995; Rinker et al. 1995). Scientists discovered new species and new processes. A plethora of access techniques, the gathering of comparative datasets via widely accepted protocols, the generation of rigorously tested hypotheses[1], and the launch of a global communication network (including two international canopy symposia[2] and a newsletter) promptly helped to define and organize the science. Unfortunately, with the movement from qualitative to quantitative science, canopy ecology forfeited some of the wonder so effortlessly expressed in its early literature.

Capturing the pre-access speculation about the wonders of treetop ecology, William Beebe's words quoted earlier in this chapter helped to rally the attention—and courage—of intrepid investigators. Decades later, Donald Perry forged an intellectual path with his tropical canopy web and associated writings. His book, *Life Above the Jungle Floor,* became a tour de force that popularized the field, while the scientific community scrutinized his technical works. One excerpt from the text highlights Perry's arboreal poetry:

> Twenty million years from now, long after the planet has crawled with billions of the hopelessly starving, long after man has driven nearly all other species to extinction, and beyond when we see to the depths of our despondent spirits and intellect, and past the stage when we judge ourselves incompetent to live and rule by using poisonous and morbid tools of destruction, then and only then will tropical forests again raise their crowns in luxury, to feast in the warm sun. In that canopy there will be arboreal beasts, and I do not doubt that some will descend from the trees to cross our clever course. Perhaps one will stoop to wonder over fossilized remains and discover that another bipedal species had preceded it from the trees. And just maybe it will be a marvelous creature that takes a higher step in the mental plane and treats the planet and its inhabitants in a manner about which we have only talked (Perry 1984).

---

[1] On 1 September 2000, the National Science Foundation and Marie Selby Botanical Gardens funded two of these hypotheses for four years, both related to the effects of forest canopy herbivory on soil processes in temperate and tropical forests.

[2] A third international symposium was held in Cairns, Australia, from 23 to 28 June 2002 for 250 delegates.

Two other first-generation canopy ecologists, Andrew Mitchell and Mark Moffett, also expressed their poetic zeal about the high frontier. For Mitchell, the canopy is a frontier of hope: "The discovery that nature's last frontier is richer than was ever before dreamed is a message of hope, a gift for us to enjoy, and use wisely" (Mitchell 1986). For Moffett, the treetops are a triumph of biodiversity where "[T]here is more feasting, more famine, more courtship and sex, more tender care of the young and of home, more combat and more cooperation in this arboreal realm than anyplace else on the globe. The tropical rainforest tapestry has only begun to capture our imagination" (Moffett 1993). Like Beebe, these scientists seemed convinced about the incontrovertible lessons of conservation and interconnection to be learned in the treetops.

These excerpts from pivotal texts represented the underpinnings of wonder for the defining decade of canopy ecology. Then the inevitable occurred for nearly every emerging science—a concerted movement from qualitative to quantitative study. Another volume, published in the mid-1990s, symbolized this movement in the tiny community of treetop scientists: Lowman and Nadkarni's *Forest Canopies* (First Edition). As a watershed text, it represented a fundamental shift in the scientific community toward analysis and synthesis. Thomas E. Lovejoy, author of the book's foreword, predicted its central nature: "There is no better evidence than canopy biology that the age of exploration is not over. We can anticipate a diverse panoply of discoveries emanating from this field. Some will be of serious practical benefit to ourselves as living organisms. Others will illuminate aspects of biology never before dreamed of. Yet others will astound with their beauty or be intrinsically fascinating. In all cases it will be clear that canopy biology, as a recognized field of intellectual endeavor, began with this book." (Lowman and Nadkarni 1995). Certainly, there were numerous articles in the scientific and popular presses, along with symposia and documentaries, throughout the 1980s and 1990s and into the 21st century that furthered the message of canopy ecology. But textbooks, as secondary or even tertiary summations of the state of affairs for science, can have far-reaching and long-lasting impact for an emerging field. Mitchell and Moffett's books made their way to coffee tables, while *Forest Canopies* quickly found a place in university libraries, research facilities, science classrooms, and offices of field biologists around the globe. First-generation canopy ecologists who contributed to the text went beyond the metaphorical, beyond the descriptive, beyond the "intricate minuet of wonder and curiosity" toward the much needed, but bittersweet, quantification of the High Frontier. Even the founders of the science quickly emptied their latest writings of their early testimonials to radical amazement and then crammed the *Forest Canopies* pages with exacting prose common to other biological sciences. Tables, equations, and graphs proliferated in the literature, but—for some—the childlike expression of wonder became almost gauche and unscientific.

The Battleground of Terminology

The lexicon of an emerging science is largely experimental. Old words can find new applications. New words can materialize as epithets for new concepts. But the lexicon for a new scientific field can also suppress a sense of wonder among its supporters if it sets too quickly. As a frontier science develops into an established discipline, vocabulary can become a battleground. Societies close ranks, individuals stake out personal territories, and practitioners attempt to standardize the profession. Inevitably, something surfaces from the mêlée that acts as an overseer, much like the French Academy in Paris or the Royal Academy of La Lingua in Madrid. Praised as "high courts of letters and rallying points for educated opinion" (Le Bars 1999), these learned academies have also been accused of hampering, even crushing, originality in their attempts to purify the language. As a clear warning to her colleagues against such insularity, Margulis noted that "our minds are incarcerated by our words" (Margulis 1990). Ill-defined words muddle an author's meaning. On the other hand, words too narrowly defined can throttle creative endeavor. Vocabulary for an emerging science needs to be a middle-ground affair—not too vague, not too specific—to ensure its lasting contribution to society, growing and changing as the application of the science suggests.

Words are also artificial constructs that have little or nothing to do with the natural world. They are human devices to help us organize our perceptions and experience. Allen and Hoekstra wrote evocatively that a definition is a formal description of a discontinuity that makes it easy to assign subsequent experience to the definition. After experiencing the world, an observer decides whether or not the experiences fit the definition.

> For all we know, *nature is continuous* [italics added], but to describe change, we must use definitions to slice the world into sectors. The world either fits into our definitions or not. Either way, all definitions are human devices, not parts of nature independent of human activity (Allen and Hoekstra 1992).

Nature is continuous. We have trouble with continuities, however, so we pigeonhole and divide for intellectual comfort and practical worldview. Words are verbal expressions of an ancient survival technique: scan the immediate environment, focus on a few facts, and make operational generalities about the world at large. It is a primitive and limiting (incarcerating, to borrow from Margulis) practice; but, historically, it has worked well for our clannish survival.

As one of the first-generation canopy ecologists, Mark Moffett (2000) attempted to unify the otherwise muddled lexicon of this frontier science. He proposed a unique précis on the basic terms of the field, attempting to address inconsistency and confusion with a standard set of terminology and definitions. In his introductory passage, Moffett bemoaned exactly some of the necessary qualities for an emerging science: inconsistency, misuse, inadequacy, and confusion. Rather than lamentable traits, these are early signs of illumination and right thought that can foster a sense of wonder. So long as individual authors clearly express the meanings of their technical vocabulary, discrepancies—even disagreements—with other authors can serve to advance the science, especially in its infancy. Discrepancies can also retain wonder in the foreground for its practitioners by highlighting ambiguities and calling attention to areas that require more scrutiny.

Epiphyte: A Divisive Definition

There is one seemingly impervious term among botanists, foresters, horticulturists, and canopy ecologists whose definition is incontrovertible: epiphyte. Or is it? Even a casual glance at the word's etiology reveals substantial disagreement among biologists about its meaning, yet some inaccurately connote a watertight definition. In his recent critique of canopy terminology, Moffett defined epiphyte as a "plant, fungus, or microbe sustained entirely by nutrients and water received nonparasitically from within the canopy in which it resides" (Moffett 2000). In the same article, however, he pointed out that "every definition of this term I have seen contains serious discrepancies with actual usage." Later Moffett admitted that "epiphyte is a term laden with semantic issues." Continuing the emphasis on the nonparasitic nature of this group of organisms, Nadkarni et al. wrote in the new *Encyclopedia of Biodiversity*: an epiphyte is a "nonparasitic plant that uses another plant as mechanical support but does not derive nutrients or water from its host" (Nadkarni et al. 2001). Historically, in the spectrum of ecological interrelationships called symbiosis, epiphytes were viewed by many biologists as mutualistic or commensal with their host plants but never parasitic.

David Benzing, one of the world's scientific authorities on vascular epiphytes, shunned such either/or distinctions and underscored the obvious ambiguities of epiphytism. "Plants with common ways of living usually share key qualities that set them apart from other vegetation. Occurrence on similar substrata under similar climate fosters evolutionary convergence . . . . Yet approximately 25,000 epiphytic species (mistletoes included) exhibit little obvious unifying basis; no growth form, seed type, kind of pollen vector, water-carbon balance regimen, nutrient source, or resource procurement mechanism is shared by all" (Benzing 1990). Later in his text, Benzing presented a classification of epiphytes based on their relationships to host plants. Epiphytes can be autotrophic, i.e., supported by woody vegetation without extracting nutrients from host vasculature. This group includes accidental, facultative, and hemi-epiphytic plants such as stranglers. Epiphytes can also be heterotrophic, i.e., subsisting on xylem contents and sometimes receiving a substantial part of their carbon supply from a host. This type of epiphyte includes parasites such as mistletoes. In addition to classification schema, Benzing provided a photographic series of epiphytes in his book, among them a native Florida orchid (*Encyclia tampensis*), a South Florida strangler fig (*Ficus aurea*), and a Kentucky mistletoe (*Phoradendron flavescens*), all captioned as "selected epiphyte types." Interestingly, both Moffett and Nadkarni et al. referenced Benzing's text in their essays. Yet they clearly promulgated the viewpoint that epiphytes are not parasites, whereas Benzing acknowledged that epiphytes are found in a muddled spectrum of relationships with the host plant: mutualism, commensalisms, *and* parasitism. Benzing's position, of course, agrees with Allen and Hoekstra's stance that nature is continuous. The dissimilarity proffered by Moffett, Nadkarni, and others is a common but unnecessary, and unconvincing, distinction. Living things manifest a fuzzy and idiosyncratic nature that repeatedly defies our nomenclature. The term "epiphyte" is laden with semantic issues because life itself, particularly among these so-called air plants, is ambiguous—often continuous, rarely dualistic. Perhaps we should avoid altogether terms such as mutualistic, commensal, and parasitic (see Margulis 1990) and use only the moniker "symbiotic" to describe the ecological relationship between two species. Further, we should recognize that this relation can be positive, neutral, or negative for the partners in surprising and wonder-filled ways (Figure 26-1).

Epiphytes include thousands of orchid and bromeliad species. Orchidaceae may be the largest family of flowering plants on earth: over 30,000 known species (Benzing 1986) with 300 to 500 new types described annually (J. Beckner and W. Higgins, pers. comm.). Bromeliaceae is a smaller family by an order of magnitude: 3,000 known species with approximately 40 new species annually. Though both families have terrestrial species, most known orchids and bromeliads are epiphytes (Benzing 1986) inhabiting tropical rainforests and montane cloud forests. As canopy plants, they are "uncoupled from soil" (Benzing 1995) and have evolved fascinating methods to obtain their nutrient and water needs in the treetops.

The trichomes and tank architecture of bromeliads act as efficient traps for non-soil nutrients, intercepting canopy litter and leachates from foliage (see Richardson et al. 2000a, 2000b). In spite of their close association with host vasculature, bromeliads have developed nonparasitic, alternative morphological features to tap the organic resources of their arboreal habitats. There is anecdotal

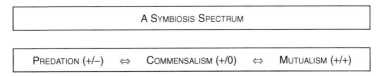

Figure 26-1 An epiphyte defined through a spectrum of symbiosis: is its ecological role static or dynamic, immutable or evolving? Predation includes a symbiosis subset called parasitism. "+" identifies a positive impact for one of the species involved in the symbiotic relationship; "−" associates a negative impact; and "0" signals no apparent impact, positive or negative.

evidence for the suppression of host vigor from heavy *Tillandsia* infestations (see Benzing 1978). They have neither direct nor indirect connection to their hosts' living tissue; however, they are not parasites, although they can be considered nutritional pirates (Benzing 1978, 1979), not just competitors, for their sustenance.

Orchids, on the other hand, are much more diverse in numbers of species and in their ecological strategies for acquiring water and nutrients. With such diversity in an ancient group of plants (a family between 100 and 110 million years old, according to Chase 2001), orchids show a "nature is continuous" fidelity that should not surprise us. There are numerous achlorophyllous orchids where heterotrophism, utilizing a kind of indirect parasitism, is clearly essential for their survival (e.g., see Campbell 1963; Furman and Trappe 1971; Leake 1994; Richardson and Curah 1995; Cullings et al. 1996; Szendrak and Read 1997; Yoder et al. 2000; International Orchid Conservation Congress 2001). In fact, over millions of years of evolution, orchids have become extraordinarily adept at solving disparate environmental challenges with unusually malleable morphology. The environmental heterogeneity among and within orchid habitats, especially forest canopies, doubtless explains the remarkable diversity of this large family of angiosperms. "Orchidaceae probably has more of the appropriate character combinations for life in tree crowns than does any other family . . . " (Benzing 1986). With so many species adrift in a lofty ocean of vegetation, far removed from ground-level nutrients and near to their hosts' tissues, can we really say with certainty that orchids cannot ever be parasitic? The theme "nature is continuous" is fully embodied in this diverse group of flowering plants. Orchids are icons of change for the natural world.

GLOBAL CANOPY PROGRAMME: A WORLDWIDE ALLIANCE FOR FOREST STUDIES

Andrew Mitchell and Katherine Secoy

The Global Canopy Programme (GCP) is a global alliance that links studies of forest canopies worldwide into a collaborative program of research, education, and conservation to address biodiversity, climate change, and poverty alleviation. Its mission is to integrate forest studies across the world into a 10-year program focused on understanding the critical role of forest canopies in biodiversity and climate change. It also aims to identify cultural benefits from forest canopies and to transmit information to key stakeholders. This initiative evolved from a European Science Foundation/National Science Foundation–funded International Canopy Science Workshop in Oxford held in November 1999 in collaboration with the International Canopy Network (ICAN). At the workshop, 29 international experts from 10 countries produced a template for the GCP. They concluded that, by working together, canopy researchers can leverage more funding for a major collaborative natural science project to investigate one of nature's last biotic frontiers. They also called for significant new funding on the scale of large physical science projects to undertake this pioneering task.

Such a program is now urgently needed to plug major gaps in our knowledge (Mitchell 2001). The structure, function, and resilience of the world's forest canopy environment are unknown. Almost half of all terrestrial life-forms exist in forest canopies, but only a small fraction has been documented. The influence of forest canopies on climate change and their role in maintaining the earth's biological diversity are based on very limited data. These roles are directly connected between the canopy interface with the atmosphere. The

Kyoto Protocol focused attention on the role of forest canopies in sequestering carbon from the atmosphere. Whether forests act as a sink or a source of carbon remains uncertain. How the mechanism works at the canopy atmosphere interface remains unclear. We do not know the economic value of canopy biodiversity or ecological services, the value of its products for human health, or its ecotourism potential for local communities. The GCP seeks to create a significant new international effort in support of the Convention on Biological Diversity, calling for research on forest canopies. The canopy–atmosphere interface is a critical environment for interrogating future models of global change (Ozanne et al 2003).

Information will be made available on-line to the public, policy-makers, and scientists through the state-of-the-art Big Canopy Database currently being developed with funds from the U.S. National Science Foundation. A Canopy Training School will offer capacity building courses for scientists, forest managers, and conservationists in biodiversity-rich nations that need extra skills to inspire new leadership in canopy science and conservation. A CanopyLIFE biodiversity rapid assessment program is in development to assess the value of the myriad life forms in different forested environments. CanopyWorld, an interactive virtual rainforest website, is planned for schools. A number of pilot projects on biodiversity and LIDAR laser scanning of canopy structure are already underway to demonstrate the comparative and collaborative capabilities of the GCP and the benefits they could bring. The *Global Canopy Handbook* (Mitchell et al. 2002) was compiled by using the expertise and experience of over 40 leading canopy researchers from around the world. It covers the who, what, and where of all current techniques and projects available for investigation of forest canopies.

Figure 1   Installation of the Lambri Hills Canopy Crane, Sarawak, Malaysia. Photograph courtesy of Tohru Nakashizuka.

Continued

GLOBAL CANOPY PROGRAMME: A WORLDWIDE ALLIANCE FOR FOREST STUDIES–*cont'd*

References

Big Canopy Database website: http://www.canopy.evergreen.edu/bcd

CanopyLIFE website: http://www.globalcanopy.org/core/

Canopy Training School website: http://www.globalcanopy.org/core/

CanopyWorld: Global Canopy Programme website: http://www.globalcanopy.org

Mitchell, A.W. (2001). Canopy science: time to shape up. *Plant Ecology* **153**, 5–11.

Mitchell, A.W., Secoy, K., and Jackson, T. (2002). "The Global Canopy Handbook: Techniques of Access and Study in the Forest Roof." Global Canopy Programme, Oxford, UK.

Ozanne, C.M.P., Anhuf, D., Boulter, S.L., Keller, M., Kitching, R.L., Korner, C., Meinzer, F.C., Mitchel, A.W., Nakashizuka, T., Silva Dias, P.L., Stork, N.E., Wright. S.J., and Yoshimura, M. (2003). Biodiversity meets the atmosphere: a global view of forest canopies. *Science* **301**, 183–186.

When scientists espouse a hard-edged attitude toward definitions, they hound out a sense of wonder for the natural world. Insistence that epiphytes cannot simultaneously be parasites is a manifest disregard for nature's spectrum of diversity and also contravenes future discovery and opportunity by incarcerating our minds' creativity. If we are convinced that there are no parasitic epiphytes, then we will not look for, nor will we expect to find, parasitic epiphytes. The workings of nature are best seen in the exception rather than in the rule. We are more likely to foster our sense of wonder about the natural world and our place in it if we accept as a training motto the dictum of Pasteur (1822–1895), who noted: "In the field of observation, chance favors the prepared mind."

Finally, a Word about the "Final Frontier"

Long ago, Old Testament prophets counseled us: "Beware lest we say, we have found wisdom" (Job 32:13). Aldo Leopold (1887–1948), in his classic *Sand County Almanac*, echoed this theme for modernity when he warned that "We all strive for safety, prosperity, comfort, long life, and dullness" (Leopold 1966). He continued with an important reference to Henry David Thoreau.

> Perhaps this [peace in our time] is behind Thoreau's dictum: In wildness is the salvation of the world. Perhaps this is the hidden meaning in the howl of the wolf, long known among mountains, but seldom perceived among men (Leopold 1966).

Perhaps we can now amend Thoreau's proverb without offense to the Master Naturalist by stating plainly that the salvation of a human-filled world lies in our sense of wonder. From the Inca Trail in the Peruvian High Andes to the summit of Mount Sinai, from the oceanic trenches off the shores of the Galápagos Islands to the subarctic coastline of Labrador, from the dark forest floor of Cameroon to the sunny treetops of Amazonia, humanity's ecological footprint is irrefutable and all pervading. Yet we discern new worlds ceaselessly—whether on a molecular or microscopic echelon or on a macroscopic scale. Nature is continuous, and so are our observations and discoveries.

It is, therefore, baffling that a modern scientist should point to any technical field and label it as a "last" or a "final" frontier. The history of science is jam-packed with innovative exploration and discovery in the natural world, even in areas studied exhaustively. For example, who could have guessed the far-reaching benefits of T.H. Morgan's meticulous genetics experiments on humble fruit flies in the early 1900s? Or, more apropos to this discussion, who could have guessed the windows of opportunity that would open when Donald Perry first climbed into the rainforest canopy of Costa Rica in the 1970s? Nevertheless, space exploration, marine biology, electron microscopy, and now canopy and soil studies have all been labeled recently as "final frontiers." From *Star Trek*'s embellished overture ("Space . . . the final frontier") to Smithsonian researcher Carole Baldwin's cinematic hyperbole about deep-sea trenches as the final frontier for biologists, scientists who make these assertions confuse the public and fetter further discovery. History has taught time and again that such statements are, indubitably, false prophecies.

In recent literature, forest canopies and forest soils have been labeled final frontiers by their enthusiasts as if no discoveries await us beyond dotting the i's and crossing the t's of these still-developing fields of research. Terry Erwin, entomologist at the National Museum of Natural History, did not equivocate when he called tropical forest canopies "the last biotic frontier" (Erwin 1983). Andrew Mitchell, British naturalist and television producer, subtitled his 1986 book on rainforest canopies as a "journey of discovery to the last unexplored frontier, the roof of the world's rainforests." Further, in the final paragraph of the text, Mitchell wrote: "The discovery that nature's last frontier is richer than was ever before dreamed is a message of hope, a gift for us to enjoy, and use wisely" (Mitchell 1986). Though an otherwise lovely, wonder-filled passage, its assertion that canopy ecology is a last frontier of discovery easily distracts from Mitchell's message of conservation. Edward O. Wilson, Pulitzer-prize-winning scientist and professor emeritus at Harvard University, wrote in his foreword to Mark Moffett's 1993 book, *The High Frontier*: "And within the rainforests, the canopy is the remote outland, the final frontier." Though poetic and engaging, such hyperbole diminishes a sense of wonder in the field. If the canopy is considered as the entire biological column in forests, then all the biodiversity and ecological processes from the treetops to bedrock are linked together closely and offer endless discovery and opportunity for the scientific community.

Nadkarni (1995) appropriately admonished the scientific community in the conclusion of her historical overview of canopy ecology: "And so a long climb awaits. The last biotic frontier is dead. Long live the next one." As long as humankind remains inquisitive, there will always be frontiers in science. And, as long as we are able to enter the treetops of the world's forests, there will always be discoveries awaiting the prepared—and wonder-filled—minds. The forest canopy is *a* frontier for exploration and discovery; however, like all other frontiers, it is far from being *the* final frontier for biology or for any other aspect of human study. André et al. (1994), of course, subtitled their pivotal article about soils "the other last biotic frontier" as a playful rebuke for canopy ecologists working at the top end of the forest's bio-column. Yet soil ecologists have also employed such incautious embellishments in recent literature. For example, Coleman and Crossley (1996) called soil "one of the last great frontiers in biological and ecological research." From the tops of forests to their soils, scientists move like faces in carnival mirrors, deeper and deeper into discovery. Time and again, throughout the history of science, we learn that no intellectual cul-de-sacs occur along its frontier borders.

Summary: Reintegration of Wonder into the Emerging Science of Canopy Ecology

Collectively, the world's treetops are often referred to as a leafy eighth continent filled with undescribed species and ecological processes. Such phenomena are life's riddles, arranged like nested boxes to lead scientists deeper and deeper into living mystery. Since the pre-access days of canopy ecology, scientists have described many new kinds of plants and animals. We've found canopy roots (also called aboveground adventitious roots; see Coxson and Nadkarni 1995) and arboreal earthworms. We've better detailed nutrient cycling, reproductive strategies, and symbiotic associations among plant communities. We've also developed a systems view of tropical and temperate forests, seeing trees as part of a three-dimensional mosaic of vegetation stretching from the ground to the canopy/atmosphere interface. Several hundred researchers worldwide have established more than 100 permanent or semi-permanent access sites across the globe to study forest canopy ecology. International expeditions, symposia, conferences, and publications are now important components of a global network for this emerging discipline. On a daily basis, canopy ecologists solve aerial puzzles and communicate this knowledge quickly to colleagues around the world.

Presently, there is an urgency to quantify the world's treetops due to rapid species loss and habitat fragmentation. Foundations, universities, research centers, schools and museums, companies, scientists, and laypeople alike are investing millions of dollars into the field. As we move from qualitative to quantitative canopy studies, early signs suggest that we are losing a sense of wonder. It is especially noted in the field's battleground of terminology, the lexicon used to communicate recent discoveries and opportunities in the treetops, where prematurely entrenched viewpoints exclude or, at least, postpone, creativity and wonder.

In the early 20th century, British scientist J.B.S. Haldane penned his famous quote now brandished as an emblem of the scientific method: "Now, my own suspicion is that the universe is not only queerer than we suppose, but queerer than we *can* suppose" (Haldane 1928). Later in *Possible Worlds*, he echoed this theme by stating, "I suspect that there are more things in heaven and earth than are dreamed of, or can be dreamed of, in any philosophy." In a chapter on wonder, Heschel, too, embraced the incomprehensibility of human perception. "What fills us with radical amazement is not the relations in which everything is embedded but the fact that even the minimum of perception is a maximum of enigma. The most incomprehensible fact is the fact that we comprehend at all" (Heschel 1955). There are enough mysteries in the cosmos to stoke the fires of wonder indefinitely.

In the 1950s, the National Audubon Society published a wonderful but little-known book that speaks clearly about the hallmarks of leading naturalists. Entitled *Homo sapiens auduboniensis: A Tribute to Walter VanDyke Bingham*[3], it lists 11 characteristics fundamental to this rare subspecies of human (National Audubon Society 1953), including a strong preference for the outdoors, an absorbing curiosity for nature, excellence in observation and description, a contagious enthusiasm, and a sensitivity to the beauty and worth of different forms of life. But, above all these, the Society listed the capability of retaining, in the presence of nature, an attitude of wonder as that enduring feature to separate leading naturalists from middling ones.

All these characteristics are probably common among canopy ecologists worldwide. Like their brethren in other emerging sciences, they are marked by their absorbing curiosity, skills at observation and intuition, and knack for putting physical discomfort and challenges into their proper places. In just over two decades, canopy ecologists have integrated a number of sub-disciplines and perspectives into their devoted work. The field of canopy ecology, newly emerged, is now answering questions about forest organisms and processes heretofore unknown. Let us hope that these remarkable scientists will retain in their endeavors, above all other attributes fundamental to leading naturalists, an attitude of wonder in the presence of nature.

---

[3] Walter VanDyke Bingham (1880-1952) was a prominent cognitive psychologist who supported nature studies, particularly the conservation efforts of the National Audubon Society.

References

Allen, T.F.H. and Hoekstra T.W. (1992). "Toward a Unified Ecology." Columbia University Press, New York.

André, H.M., Noti, M.-I., and Lebrun, P. (1994). The soil mesofauna: The other last biotic frontier. *Biodiversity and Conservation* **3**, 45–56.

Beebe, W., Hartley, G.I., and Howes, P.G. (1917). "Tropical Wild Life in British Guiana: Zoological Contributions from the Tropical Research Station of the New York Zoological Society." New York Zoological Society, New York.

Benzing, D.H. (1979). Alternative interpretations for the evidence that certain orchids and bromeliads act as shoot parasites. *Selbyana* **5**(2), 135–144.

Benzing, D.H. (1986). The genesis of orchid diversity: Emphasis on floral biology leads to misconceptions. *Lindleyana* **1**(2), 73–89.

Benzing, D.H. (1990). "Vascular Epiphytes: General Biology and Related Biota." Cambridge University Press, Cambridge, UK.

Benzing, D.H. (1995). The physical mosaic and plant variety in forest canopies. *Selbyana* **16**(2), 159–168.

Benzing, D.H. and Seeman, J. (1978). Nutritional piracy and host decline: A new perspective on the epiphyte-host relationship. *Selbyana* **2**, 133–148.

Brooks, P. (1972). "The House of Life: Rachel Carson at Work." Houghton Mifflin Company, Boston.

Campbell, E.O. (1963). *Gastrodia minor* Petrie, an epiparasite of Manuka. *Transactions of the Royal Society of New Zealand* **2**(6), 73–81.

Chase, M. (2001). First International Orchid Conservation Congress, Perth, Australia. September 2001.

Coleman, D.C. and Crossley, D.A. (1996). "Fundamentals of Soil Ecology." Academic Press, San Diego.

Coxson, D. S. and Nadkarni, N.M. (1995). Ecological roles of epiphytes in nutrient cycles of forest ecosystems. *In* "Forest Canopies" (M. D. Lowman and N.M. Nadkarni, Eds.), pp. 495–543. Academic Press, San Diego.

Cullings, K.W., Szaro, T.M., and Burns, T.D. (1996). Evolution of extreme specialization within a lineage of ectomycorrhizal epiparasites. *Nature* **379**(6560): 63–66.

Daston, L. and Park, K. (1998). "Wonders and the Order of Nature," pp. 1150–1750. Zone Books, New York.

Erwin, T. (1983). Tropical forest canopies: The last biotic frontier. *Bulletin of the Entomological Society of America* **29**, 14–19.

Furman, T.E. and Trappe, J.M. (1971). Phylogeny and ecology of mycotrophic achlorophyllous angiosperms. *Quarterly Review of Biology* **46**(3): 219–225.

Haldane, J.B.S. (1928). "Possible Worlds and Other Papers." Harper and Brothers, New York.

Heschel, A.J. (1955). "God in Search of Man: A Philosophy of Judaism." Farrar, Straus, and Giroux, New York.

International Orchid Conservation Congress (2001). "The Orchids of Australia."

Le Bars, J. (1999). The French Academy, The Catholic Encyclopedia. 2001. www.newadvent.org/cathen/01089a.htm.

Leake, J.R. (1994). Tansley Review No. 69: The biology of myco-heterotrophic ('saprophytic') plants. *The New Phytologist* **127**, 171–216.

Leopold, A. (1966). "A Sand County Almanac with Other Essays on Conservation from Round River." Oxford University Press, New York.

Lowman, M.D. (1995). Herbivory as a canopy process in rain forest trees. *In* "Forest Canopies" (M.D. Lowman and N.M. Nadkarni, Eds.), pp. 431–455.

Lowman, M. D. and Nadkarni, N.M. (Eds.) (1995). "Forest Canopies." Academic Press, San Diego.

Margulis, L. (1990). Words as battle cries—symbiogenesis and the new field of endocytobiology. *BioScience* **40**(9), 673–677.

Margulis, L. (2001). What is canopy biology? A microbial perspective. *Selbyana* **22**(2), 232–235.

Mayr, E. (1988). "Toward a New Philosophy of Biology: Observations of an Evolutionist." Belknap Press of Harvard University Press, Cambridge, Massachusetts.

Mitchell, A.W. (1986). "The Enchanted Canopy: A Journey of Discovery to the Last Unexplored Frontier, the Roof of the World's Rainforests." Macmillan Publishing Company, New York.

Moffett, M.W. (1993). "The High Frontier: Exploring the Tropical Rainforest Canopy." Harvard University Press, Cambridge, Massachusetts.

Moffett, M.W. (2000). What's "up"? A critical look at the basic terms of canopy biology. *Biotropica* **32**(4a), 569–596.

Nadkarni, N. (1995). Good-bye, Tarzan. *The Sciences* **35**(1), 28–33.

Nadkarni, N.M., Merwin, M.C., and Nieder, J. (2001). Forest canopies, plant diversity. *In* "Encyclopedia of Biodiversity 3" (S.A. Levin, Ed.), pp. 27–40. Academic Press, San Diego.

Nadkarni, N.M. and Lowman, M.D. (1995). Canopy science: a summary of its role in research and education. *In* "Forest Canopies" (M.D. Lowman and N.M. Nadkarni, Eds.), pp. 609–613. Academic Press, San Diego.

National Audubon Society. (1953). "*Homo sapiens auduboniensis:* A Tribute to Walter VanDyke Bingham." National Audubon Society, New York.

Parker, G.G. (1995). Structure and microclimate of forest canopies. *In* "Forest Canopies" (M.D. Lowman and N.M. Nadkarni, Eds.), pp. 73–106. Academic Press, San Diego.

Perry, D.R. (1984). The canopy of the tropical rain forest. *Scientific American* **251**,138–147.

Reynolds, B.C. and Crossley, D.A. (1997). Spatial variation in herbivory by forest canopy arthropods along an elevation gradient. *Environmental Entomology* **26**(6), 1232–1239.

Richardson, B.A., Richardson, M.J., Scatena, F.N., and McDowell, W.H. (2000a). Effects of nutrient availability and other elevational changes on bromeliad populations and their invertebrate communities in a humid tropical forest in Puerto Rico. *Journal of Tropical Ecology* **16**, 167–188.

Richardson, B.A., Rogers, C., and Richardson, M.J. (2000b). Nutrients, diversity, and community structure of two phytotelm systems in a lower montane forest, Puerto Rico. *Ecological Entomology* **25**, 348–356.

Richardson, K.A. and Currah, R.S. (1995). The fungal community associated with the roots of some rainforest epiphytes of Costa Rica. *Selbyana* **16**(1), 49–73.

Rinker, H.B. (2000). Conservation from the treetops: environmental action in the emerging science of canopy ecology, BioScience Productions, Inc. www.actionbioscience.org/environment/rinker.html.

Rinker, H.B. (2001). Halfway between heaven and earth: Bird conservation in the treetops. *Bird Watcher's Digest* **23**(5), 60–64.

Rinker, H. B., Lowman, M.D., and Moffett, M.W. (1995). Africa from the treetops. *American Biology Teacher* **57**(7), 393–401.

Seastedt, T. R. and Crossley, D.A. (1984). The influence of arthropods on ecosystems. *BioScience* **34**(3), 157–161.

Sillett, S. C. and van Pelt, R. (2000). A redwood tree whose crown is a forest canopy. *Northwest Science* **74**(1), 34–43.

Skutch, A.F. (1992). "A Naturalist in Costa Rica." University Press of Florida, Gainesville, Florida.

Szendrak, E. and Read, P.E. (1997). Discovery of an albino form of *Orchis morio* L. (Orchidaceae). *American Journal of Botany* **6** (Supplement), 253.

Winters, R.K. (1977). "Terminology of Forest Science Technology Practice and Products." Society of American Foresters, Bethesda, Maryland.

Yoder, J.A., Zettler, L.W., and Stewart, S.L. (2000). Water requirements of terrestrial and epiphytic orchid seeds and seedlings, and evidence for water uptake by means of mycotrophy. *Plant Science* **156**, 145–150.

INDEX

Page numbers followed by *f* denote figures; page numbers followed by *t* denote tables

Printed and bound by CPI Group (UK) Ltd, Croydon, CR0 4YY

03/10/2024

01040321-0012